Vascular Biomechanics

T. Christian Gasser

Vascular Biomechanics

Concepts, Models, and Applications

 Springer

T. Christian Gasser
KTH Solid Mechanics, Department of
Engineering Mechanics
KTH Royal Institute of Technology
Stockholm, Sweden

The book includes additional materials at sn.pub/lecturer-material

ISBN 978-3-030-70968-6   ISBN 978-3-030-70966-2   (eBook)
https://doi.org/10.1007/978-3-030-70966-2

This Springer imprint is published by the registered company Springer Nature Switzerland AG
The registered company address is: Gewerbestrasse 11, 6330 Cham, Switzerland

To my family

Foreword

Cardiovascular disease remains a leading cause of disability and death worldwide. Given that the primary function of the cardiovascular system is biomechanical—to transport blood and substances therein via a pressurized system—it is not surprising that vascular biomechanics is fundamental to understanding cardiovascular health and disease. Indeed, all primary vascular cells—endothelial cells of the intima, smooth muscle cells of the media, and fibroblasts of the adventitia—are exquisitely sensitive to changes in their local biomechanical environment, often changing gene expression accordingly. Knowledge of the biomechanics is thus critical both for understanding the molecular and cellular basis of vascular biology and for understanding tissue-level behaviors such as vasoconstriction-vasodilatation, the development of atherosclerotic plaques or aneurysms, and even catastrophic ruptures.

Vascular biomechanics builds upon the foundations of nonlinear continuum mechanics, hence one must understand well this fundamental area of study. Problems within vascular biomechanics often involve complex geometries, material properties, and applied loads, hence numerical methods are often essential for understanding health and disease. The physiology and pathophysiology further require an understanding of both the vascular wall and the hemodynamics, thus necessitating an understanding of both the solid mechanics and fluid mechanics and the associated anatomy and histology, that is, the study of gross and fine structures of tissues.

This book, *Vascular Biomechanics: Concepts, Models and Applications*, by Professor T. Christian Gasser appropriately emphasizes the foundations of nonlinear continuum mechanics and methods of numerical solution, primarily via the finite element method. It similarly considers together both the solid and fluid mechanics, which provides a balanced approach to formulating and solving critical problems important to both basic science studies and clinical translation. Discussions of the mechanics are balanced by those of the physiology and histology.

The reader is encouraged to study well the foundations—biological, mechanical, and computational—for it is only via reliance on the foundations that important new problems can be formulated and solved. In conclusion, it is critical to remember that biomechanics must include the development and extension of mechanics, not just application, hence the full power and promise of vascular biomechanics will be achieved only by understanding the basic principles and approaches so well that they can be developed further. In the words of Professor Y.C. Fung, regarded by many as the Father of Modern Biomechanics, "enjoy the work."

New Haven, CT, USA   Jay D. Humphrey
February 15, 2021

Preface

The development of the vascular system was a critical step in the evolution of species. It allowed organisms to meet the growing metabolic demands and opened the door for the development of more complex forms of life. Vascular function depends on the delicate interaction between blood flow and vascular tissue—biomechanical factors are therefore common denominators of many vascular pathologies. Even considering epidemic threats from viruses, such as COVID 19, it is vascular diseases that are by far the most common cause of death in the industrialized world. The biomechanics of the vascular system is therefore a very important object to study.

Aren't there enough books available on vascular biomechanics? Indeed, the importance of the topic led to a large number of texts, many of which are truly masterpieces. Unfortunately, most texts are rather narrow and poorly cover the many interdisciplinary aspects of the vascular system. Disciplines such as mathematics, physics, imaging, medicine, and biochemistry provide the foundation for successful vascular biomechanics applications, see Fig. 1.1. A fundamental understanding of the vascular system therefore requires the tight integration of these disciplines, and I truly believe that broader vascular biomechanics texts are needed to better train students and to solve vascular challenges.

What does this book cover? This book aims at providing the reader with holistic information of biomechanics-related aspects of the vascular system. It addresses analytical and numerical solving strategies and covers topics such as vascular perfusion, vascular tissue mechanics, blood flow mechanics, and vasoreactivity as well as tissue growth and remodeling. Whilst top-down approaches are followed to introduce the topic, bottom-up approaches, often linked to fundamental governing principles, are favored later in the text.

To whom is this book addressed? This text is primarily addressed to readers who have a basic background in mathematics and mechanics. The main target groups are students and researchers in classical and computational vascular biomechanics. The book is self-consistent and may be used as course text at undergraduate as well as graduate levels. The book could also be of interest to developers of vascular devices and experts working with the regulatory approval of biomedical simulations. Selective parts also cover topics of interest to students of continuum mechanics, physics, scientific computing, and medicine.

How is this book designed? This book follows the principle "learning by doing" and provides many fully through, calculated examples for active learning, immediate recall, and self-examination. The introduced concepts are illuminated through example applications. It is when the spark jumps between the concepts and their application that *learning* happens. Besides re-enforcing the theory, examples are designed to actively extend the content of the individual chapters, to proof statements and to gain new insights. You have to work hard, and I fully admit that some of these examples are quite difficult to solve, especially when seen for the first time. Creative thinking is required to crack them.

Some additional thoughts and information. Whilst the solutions to all examples are provided at the end of the book, do not look it up directly. Use your imagination and try to solve the examples by your own first. Any attempt, even the unsuccessful ones, will help you to master new skills into that most prestigious thing—*understanding*. If solving an example does not work out directly, find some distraction and try again later—you might even sleep over it to unfold your brain's hidden capacities [72]. Many of the examples in this book can be carried out with "paper and pen", others require software. I used MATHEMATICA (Wolfram Research Inc.) and COMSOL (COMSOL AB) to solve some more complex examples, and these input files are available as additional material at the publisher's website.

Stockholm, Sweden   T. Christian Gasser
July 7, 2022

The original version of this book was revised. The correction to this book is available at https://doi.org/10.1007/978-3-030-70966-2_8

Acknowledgments

Having an idea and turning it into a book is as hard as it sounds. The experience is both internally challenging and rewarding. I especially want to thank the individuals that helped make this happen.

I have to start by thanking my family, especially *Carmen*—without her support, encouragement, and advise, this book would simply not exist. Our children, *Antonia*, *Amelie*, and *Leo*, whose energy sparked my motivation day by day. My parents who taught me staying power and endurance and so much more, which has helped me to focus on a goal until it is delivered.

To all the students I have had the opportunity to teach courses, such as vascular biomechanics, continuum mechanics, the finite element method, and constitutive modeling—thank you for being the inspiration of starting this project. Special thanks go to doctoral students *Marta Alloisio*, *Christopher Miller*, *Andrii Gytsan*, *Giampaolo Martufi*, *Caroline Forsell*, and *Jacopo Biasetti*. In addition to the numerous editorial remarks, their feedback led to a much more deductive presentation of the material.

To the inspirational working environment at KTH Solid Mechanics and the many discussions at the coffee table. To the never-ending inspirations from my colleagues at Karolinska Institute and University Hospital—*Jesper Swedenborg*, *Joy Roy*, *Rebecka Hultgren*, *Ulf Hedin*, *Magnus Bäck*, and *Anders Arner*, and the PhD students *Moritz Lindquist-Liljeqvist*, *Antti Siika*, *Marko Bogdanovic*, *Andrew Buckler*, *Emma Larsson*, and *Maggi Folkesson*—their work had a huge impact on this book.

Without the experiences and support from my peers, this book would not at all look as it does. Thanks for the many thoughts on the original manuscripts go to *Stéphane Avril*, *Salvatore Federico*, *Sae-Il Murtada*, *Mårten Olsson*, *Lisa Prahl Wittberg*, and *Daniela Valdez-Jasso*. It was their in-depth expert knowledge that helped shape the individual chapters of this book.

To the many co-authors with whom I had the privilege to work over the past years. The content of this book reflects a summary of their very different approaches to vascular biomechanics. A special thank goes to *Ray W. Ogden*, whose shear inexhaustible knowledge and skills deeply impressed me. I would also like to thank *Stanislav Polzer* for his creativity and inspirations. A very special thanks also to the never-ending encouragement from my friend *Prashanth Srinivasa*—his numerous editorial remarks and content advices are also very much appreciated.

To the speakers at engineering and medical conferences and authors of papers that passed my desk. Thanks for the many exemplifications and inspirations that resulted in a number of examples in the book and allowed to cover so many different topics—your contribution made this book a truly multidisciplinary text.

Last but not least, I would like to thank the publishing team who helped me so much. Special thanks to the ever-patient Senior Editor *Merry Stuber*, the Project Manager *Mary Diana Manickam*, and the Administrator *Shabib Shaikh*.

Contents

Acronyms

18F-FDG	18-Fluoro-Deoxy-Glucose
1D/2D/3D	One/Two/Three-Dimensional; One/Two/Three-Dimensions
AAA	Abdominal Aortic Aneurysm
AD	Advection-Diffusion
ADP	Adenosine DiphosPhate
AI	Artificial Intelligence
ANN	Artificial Neural Network
ANP	Atrial Natriuretic Peptide
ATP	Adenosine Triphosphate
AVR	Ascending Vasa Recta
BAB	Blood Aqueous Barrier
BBB	Blood Brain Barrier
BC	Boundary Condition
BN	Bayesian Network
BNB	Blood Nerve Barrier
BRRA	Biomechanical Rupture Risk Assessment
BVP	Boundary Value Problem
Ca	Calcium
CaM	CalModulin
CDF	Cumulative Density Function
CDM	Continuum Damage Mechanics
CFL	Courant-Friedrichs-Lewy Criterion
cGMP	Cytosolic Cyclic Guanylyl MonoPhosphate
CI	Confidence Interval
CKD	Chronic Kidney Disease
CNS	Central Nervous System
CO	Cardiac Output
COPD	Chronic Obstructive Pulmonary Disease
CoU	Context of Use
CT-A	Computed Tomography-Angiography
DAG	DiAcylGlycerol
DL	Deep Learning
DICOM	Digital Imaging and Communications in Medicine
DNS	Direct Numerical Simulation

EC	Endothelial Cell
ECAP	Endothelial Cell Activation Potential
ECG	ElectroCardioGram
ECM	Extra Cellular Matrix
EDHF	Endothelium-Driven Hyperpolarization Factors
EEL	External Elastica Lamina
eNOS	Endothelial NO Synthase
FB	FibroBlast
FBD	Free Body Diagram
FDG	Fluoro-Deoxy-Glucose
FDM	Finite Difference Method
FEM	Finite Element Method
Fr	Froude Number
FSI	Fluid Structure Interaction
GAG	GlycosAminoGlycan
GLS	Galerkin Least-Square
HDL	High-Density Lipoprotein
HI	Hemolysis Index
HPC	High Performance Computing
HU	Hounsfield Unit
iBVP	Initial Boundary Value Problem
IEL	Internal Elastica Lamina
ILT	Intra-Luminal Thrombus
IMA	Intended Model Application
IP3	Inositol TriPhosphate
IQR	InterQuartile Range
IVUS	IntraVascular UltraSound
K	Potassium
LDL	Low-Density Lipoprotein
LFM	Linear Fracture Mechanics
MAP	Mean Arterial Pressure
ML	Machine Learning
MLC	Myosin Light Chain
MLCK	Myosin Light Chain Kinase
MLCP	Myosin Light Chain Phosphatase
MLU	Medial Lamellar Unit
MMP	Matrix MetalloProteinase
MR-A	Magnetic Resonance-Angiography
mRNA	Messenger RiboNucleic Acid
Na	Natrium
NCX	Na^+-Ca^{2+} eXchanger
NF-κB	Nuclear Factor Kappa-Light-Chain-Enhancer of Activated B Cells
NO	Nitric Oxide
NURBS	Non-Uniform Rational Basis Spline
O	Oxygen

OSI	Oscillatory Shear Index
PECAM1	Platelet Endothelial Cell Adhesion Molecule-1
PDE	Partial Differential Equation
PDF	Probability Density Function
Pe	Péclet Number
PG	ProteoGlycan
pH	Potential of Hydrogen
Pi	Phosphate ion
PKC	Protein Kinase C
PKG	Protein Kinase G
PMCA	Plasma Membrane Ca^{2+} ATPase
PUFEM	Partition of Unity Finite Element Method
PVW	Principle of Virtual Work
PWI	Pulse Wave Imaging
PWRI	Peak Wall Rupture Index
PWS	Peak Wall Stress
RANS	Reynolds-Averaged Navier–Stokes equations
Re	Reynolds Number
RNA	RiboNucleic Acid
ROCC	Receptor-Operated Calcium Channels
ROS	Reactive Oxygen Species
RRED	Rupture Risk Equivalent Diameter
RVE	Representative Volume Element
SERCA	Sarco/Endoplasmic Reticulum Ca^{2+}-ATPase
SD	Standard Deviation
SMC	Smooth Muscle Cell
SPECT	Single-Photon Emission Computed Tomography
Sr	Strouhal Number
St	Stokes Number
STL	STereo Lithography
SU	Streamline Upwind
SUPG	Streamline Upwind Petrov-Galerkin
TAA	Thoracic Aortic Aneurysm
TSL	Traction Separation Law
UTS	Ultimate Tensile Strength
VE-cadherin	Vascular Endothelial Cadherin
vWF	von Willebrands Factor
VEGF	Vascular Endothelial Growth Factor
VGCC	Voltage-Gated Ca^{2+} Channel
VS	Vortical Structure
WK	WindKessel
Wo	Womersley Number
WSS	Wall Shear Stress
XFEM	eXtended Finite Element Method

Modeling in Bioengineering

1

This chapter discusses concepts and strategies toward the development and testing of bioengineering models. We explore the complexity of vascular bioengineering problems and then introduce the Intended Model Application (IMA), a target that determines the design of all development protocols. Following the specification of models and their development, approaches to test bioengineering models are discussed — model sensitivity, verification, and validation are explored. Given their importance in bioengineering, statistical approaches are analyzed, and we cover topics, such as study design, hypothesis testing, correlation testing, and mean difference testing. In addition to statistical modeling, the basics of Artificial Intelligence-based (AI-based) modeling approaches are also discussed. A case study is then used to exemplify the aforementioned key features of bioengineering modeling. The study explores the rupture risk assessment of Abdominal Aortic Aneurysms (AAA), a clinical exercise in the treatment of AAA patients. Conclusions concerning optimal model complexity and future perspectives summarize this chapter.

1.1   Introduction

Bioengineering modeling plays a prominent role in the study of biological systems and processes and can be of great help to explore the vasculature. Specifically, the power of computational modeling has fundamentally changed both industry and health care. Already today, competitiveness in these fields is tightly linked to modeling expertise. Whilst some properties of a biological system can be directly measured, the exploration of others, and especially *in vivo* properties, requires

---

The original version of this chapter was revised: ESM has been added. The correction to this chapter is available at https://doi.org/10.1007/978-3-030-70966-2_8

Supplementary Information The online version contains supplementary material available at https://doi.org/10.1007/978-3-030-70966-2_1

models. In addition, prospective events, and questions such as *What would be the outcome from a certain clinical intervention in an individual patient?*, can only thoroughly be studied through model-based simulations.

A modeler should always focus on the Intended Model Application (IMA), also sometimes called Context of Use (CoU). *What bioengineering question should be answered?* IMA guides *all* development steps, from specifying the model requirements all along to its verification and validation [15, 491]. In addition, biological data is always uncertain, and designing a bioengineering model as well as drawing conclusions from it requires statistical methods; some of which are discussed in this chapter. A case study illustrates the integration of the different modeling concepts and concluding remarks then summarize the chapter. Whilst this chapter describes the vascular often through black-box modeling, the subsequent parts of this book will focus on white-box modeling.

Vascular biomechanics is a specialization of bioengineering and uses highly multidisciplinary problem solving strategies that may be illustrated by the tree in Fig. 1.1. The tree's roots represent branches of knowledge that merge to the stem, which then allows to form branches representing vascular bioengineering applications. The interaction amongst disciplines, such as mathematics, physics, imaging, medicine, and biochemistry, is the basis to address applications in the fields of diagnostics, device design, treatment planning, forensic analysis, and functional imaging. Given very different ways of thinking and curricula, the individual disciplines use very different methods and strategies for problem solving—they need to be *synchronized* towards successful bioengineering work.

1.1.1   Bottom-Up Approach

A bottom-up approach is the *assembling* of pieces to give rise to more complex systems. Engineering problem solving often follows such an approach, and the pieces may be seen as knowledge blocks. Frequently such knowledge blocks represent basic physical principles, and a bottom-up approach is a *principle-dominated* solution strategy. Principles, such as Newton's[1] laws of mechanics, are used, and a problem description that violates such basic principles would be unacceptable. *Inductive reasoning* integrates knowledge blocks leading to top-level systems, as complex as an airplane, a vehicle production line, or a Finite Element Method (FEM) model.

Given the vasculature being a complex hierarchical structure, the bottom-up analysis of a practically relevant question requires the integration of a large number of unit blocks. Whilst such a problem description still allows to draw qualitative conclusions, quantitative predictions are challenging. Almost always it is required to introduce phenomenological "scaling" parameters to *calibrate* the problem description to match experimental observations.

---

[1] Sir Isaac Newton, English mathematician, physicist, astronomer, theologian, 1642–1726.

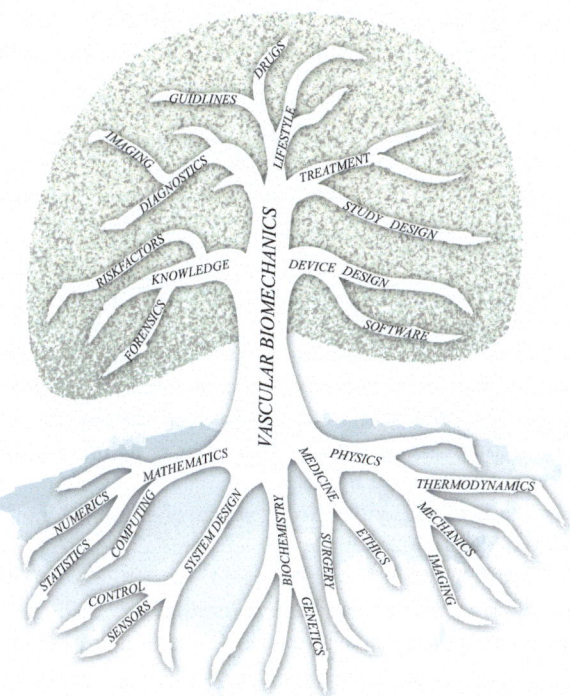

Fig. 1.1 The "vascular biomechanics tree" aims at illustrating the holistic approach of vascular biomechanical problem solving. Disciplines, such as mathematics, system design, biochemistry, medicine, and physics, form the roots, whilst the branches lead to new knowledge and applications in fields, such as diagnostics, device design, patient treatment, and drugs

1.1.2 Top-Down Approach

The patient is very complex, and modeling it through the assembly of fully reviewable knowledge blocks according to the bottom-up approach is never possible. Clinical science therefore often uses a top-down approach. It may be seen as the *breaking down* of a system into simpler sub-structures to gain better insight. This eventually allows for the exploration of bioengineering phenomena through *deductive reasoning*. Principles are less enforced, and a clinical method is regarded suitable as long as it is able to treat the patient, regardless of whether or not it violates any physical principles.

Clinical research questions are frequently posted retrospectively, such as *Which treatment did have a statistical effect on the outcome?* Given a treatment's benefit has been established through a retrospective study, its clinical implementation would still critically depend on the positive outcome from additional *prospective* validation studies.

Too much complexity hinders deductive reasoning, and the *pre-clinical approach* investigates a less complex system, such as an animal model or a wet laboratory experiment. Whilst such a system may be simple enough to draw conclusions, its relevance to a real clinical problem is in many cases greatly compromised.

1.1.3   Opportunities and Challenges

Biomechanical investigations may advance our understanding of vascular physiology and pathologies, interactions with medical devices, drug delivery pathways, the interplay between structure and function of vascular tissues, mechanotransduction, and many others. The success of such investigations is often related to the appreciation of the vasculature's multidisciplinary complexity. Whilst a *purely medical* perspective often misses physical principles, *purely engineering* approaches underestimate the clinical dimensions of the problem. In addition, medical investigations are often purely observational, whilst the engineering perspective is often too much focused on solving (complex) equations—the underlying equations may even lack experimental evidence.

In conclusion, medical and engineering approaches to vascular biomechanical problems are often isolated and not very much interacting. This naturally hinders their potential. Aside from the aforementioned *structural* challenges related to medical and engineering curricula, the exploration of vascular bioengineering problems faces many more *specific* challenges, some of them are listed below:

- *Engineering challenges.* Due to the complexity of the vasculature, traditional engineering methods are often not directly applicable. They require specific further developments and even the introduction of fundamentally new paradigms. The inherent property of vascular tissue to adapt to mechanical and biochemical environments remains one such challenging modeling task—it is far from solved. In addition, the investigation of an entire vascular organ and the comprehensive understanding of a biological process may require the coupling of bioengineering models amongst structural, fluid, chemical, and electrical fields.
- *Data acquisition challenges and uncertainty.* Laboratory testing of biological systems is challenging, especially with respect to human samples. Ethical aspects, sample harvesting constraints, and difficulties in providing an adequate *in vitro* testing environment are some of the challenges an investigator faces. Biological data is uncertain, and the intra-specimen and inter-specimen variability of input parameters, such as loading conditions and constitutive properties, weakens the predictability and benefit of bioengineering models.
- *Validation challenges.* The clinical acceptance of a bioengineering model requires sound prospective validation that demonstrates its clinical and economic benefits. Such validations are always time consuming and often challenged by ethical and other constraints. The successful clinical integration of a bioengineering model is often prevented by insufficient clinical validation.
- *Product development challenges.* An engineering solution might have a limited market or might be too expensive for its initial commercial deployment. It might neither be practical nor easy to be integrated into the clinical work flow. The proposed product might even threaten the market of operational medical

companies as well as the routine work of clinicians. Finally, a moderate clinical benefit might not justify the risks from the clinical integration of a bioengineering solution and then results in failure of medical device approval.

1.2 Model Design

A model represents the *real object* or *process* to some degree of completeness, and A. Einstein[2] states:

> everything should be made as simple as possible, but no simpler.[3] Any intelligent fool can make things bigger and more complex. It takes a touch of genius—and a lot of courage to move in the opposite direction.[4]

The *flat earth hypothesis* is an adequate model for many civil engineering applications. It is sufficient for the design of a house, and using a spherical gravitational field, and thus working with a spherical earth model, would clearly be an excess for such an exercise. *Newton's laws of mechanics* are also perfectly usable for many conditions, even known being only a special case of the theory of general relativity and to fail at velocities in the range of the light speed. Likewise, *continuum theories* are successful even though matter is discrete and inhomogeneous at small length-scales. However, the introduction of a representative length-scale, and thus a Representative Volume Element (RVE) over which properties are averaged, allows for the continuum-mechanical problem description.

1.2.1 Simplifications

A model introduces approximations to the real problem, and as with other modeling activities, the IMA guides making such simplifications. Figure 1.2 illustrates a typical vascular biomechanics model, and simplifications relate to the following aspects:

- *Vascular geometry.* Blood vessels of diameters ranging from micrometers to centimeters fill the entire human body. The modeler has to *separate the length-scales* and specify a length-scale above which the individual geometry of the vasculature is to be explicitly modeled. Below such length-scale the effect of "small vessels" is represented by a boundary model—or in some cases they may be simply neglected.
- *Material properties.* The vascular wall shows complex mechanical properties. They are influenced by histological, biological, and clinical factors. The modeler has to decide which mechanical properties are essential and which ones may be neglected.

---

[2] Albert Einstein, German-born theoretical physicist, 1879–1955.
[3] http://quoteinvestigator.com/2011/05/13/einstein-simple/#more-2363.
[4] http://www.alberteinsteinsite.com/quotes/einsteinquotes.html.

Fig. 1.2 Assumptions to be made on the input and output of vascular biomechanics simulations. Each assumption needs to be validated with respect to the Intended Model Application (IMA)

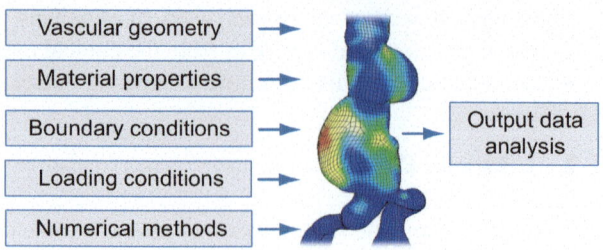

- *Boundary and initial conditions.* A simulation object will only cover a certain domain of the real problem. The modeler has to specify how the model interacts with its surrounding across such boundaries in time and space. Given multi-field models, boundary conditions also have to be specified across the different solution fields.
- *Loading conditions.* Blood vessels are exposed to complex time-dependent mechanical loadings, and the modeler has to select load cases that are of relevance to the IMA. Loading stems from factors, such as the blood pressure, body motion, and interactions with other organs and tissues.
- *Numerical methods.* A biomechanics model has to ensure governing principles and requires the solution of an initial Boundary Value Problem (iBVP) by an appropriate numerical method. It is up to the modeler to select an appropriate and stable numerical frame to solve the systems of governing equations.
- *Output data analysis.* A biomechanics model generates a large amount of data, such as stress and strain throughout the entire vessel wall, or blood flow velocity in the entire lumen. The modeler has to relate this information to the IMA, which again requires assumptions to be made.

In conclusion, a biomechanics model relies on a number of simplifications, all of which are made on the basis of *uncertain information*. A model therefore requires careful validation against the IMA.

1.2.2 Strategies

A model may always be seen as a *transformation matrix* that transforms the vector **x** of input variables to the vector **y** of output variables. Given a linear model, the transformation matrix is constant. However, for a non-linear model, the transformation matrix is not constant and may depend on the input **x**, the output **y**, or any other factor. Modelers follow very different strategies to develop the transformation matrix.

A model design may be ranked according to the feasibility to understand a model's *inner* working and how it reflects the *physics* of the bioengineering problem. A design that fully provides such information is called a *white-box* model, whilst a *black-box* model's structure does not correlate with the underlying physical processes of the bioengineering problem. Practically all models are *gray-box* models, and they are in between the limits of the white-box and black-box model, respectively.

1.2.2.1 White-Box Model

White-box models break down the response of the system into its underlying *physical mechanisms* and fully disclose the model's inner working. Such a model design provides full physical explanation of how model input is transformed into model output. White-box models implement physical processes, and they represent a *realistic*, but still *reduced*, description of the real problem. A model will always introduce approximations, and thus a fully accurate model *cannot* exist as it would be essentially a copy of reality.

White-box models allow for the *hierarchical* integration of physical processes and their interactions. They provide deep insights into a bioengineering problem and foster the acquisition of *qualitative* knowledge. A White-box model allows to describe a problem based on *limited* experimental data and is very flexible towards model revisions. However, the description of the problem's underlying physics may be complex, and the computation of model output requires often considerable computational resources.

1.2.2.2 Black-Box Model

Black-box models do not refer to any underlying physics—they use *empirical* descriptions or *sets of transfer parameters* to relate model input and output. Consequently, they do not allow for a physical explanation of how input transforms into output. Black-box models use purely *phenomenological* descriptions, and they provide quantitative data without deep insights into bioengineering processes. Back-box models require *sufficient* experimental data for robust model calibration, and their design is inflexible towards model revisions; the whole model needs a re-calibration after any minor change in model design. However, they *rapidly* transform model input into output, and the complexity of many bioengineering problems may require the black-box approach.

1.2.2.3 Gray-Box Model

A gray-box model implements a physical representation of the bioengineering problem, where phenomenological descriptions approximate *some* of the physics. A gray-box model may linearize a non-linear process or simplify physics by averaging over localized properties.

1.2.2.4 Surrogate Model

Model reduction techniques may be used to simplify a model towards *making it applicable* for a specific task, such as its integration in a clinical work flow. The model is then called a surrogate model. Given a number of applications, the time that is needed to transform model input into output is minimized at the loss of model accuracy and fidelity.

1.3   Model Development and Testing

Figure 1.3 illustrates the *model development cycle*. It starts with the description of the *real world problem* that is to be represented by the model. The task specifies the IMA, a list of model requirements, performance targets and risk assessment aspects. Given this information, the mathematical equations representing the real world problem are formulated. Often, they may not have a closed-form solution, and numerical methods are used to compute the model output. The model's predictions are then tested by *verification* and *validation*, exercises that conclude whether or not the model requires revision. Test protocols are determined by the IMA [491] and risk aspects. Given all tests are passed, the model is *deployed*, and the IMA specifies the domain within which it is allowed to operate.

1.3.1   Sensitivity Analysis

Biomechanical models are complex and depend on a large number of *uncertain* input parameters. Many biomechanical models are non-linear and require a careful investigation of how input uncertainties propagate towards the model output. A *parameter sensitivity analysis* is therefore an essential step in the development of a model. In addition to the specification of quality requirements for input data, it is also a critical exercise in the medical device approval process.

A single output variable of a model may always be represented by a function $y = f(\mathbf{x})$, where $\mathbf{x}$ is a vector of n uncertain model parameters $(x_1, x_2, \ldots, x_n)$ and y is a chosen univariate model output; several model outputs may be investigated consecutively. The sensitivity of y with respect to the model parameters $\mathbf{x}$ may be investigated either *locally* or *globally*. A local analysis draws conclusions at a

Fig. 1.3 Model development cycle. All development and testing activities are carried out in relation to the Intended Model Application (IMA)

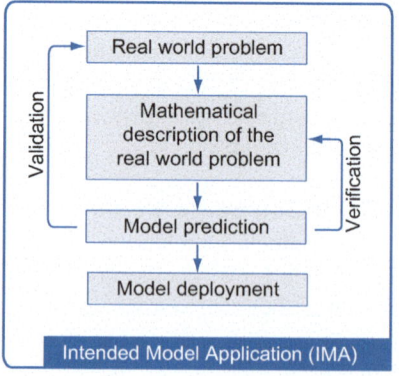

given position $\mathbf{x}_0$ in the parameter space. The variation of the output may then be expressed by

$$\Delta y = \mathbf{s}(\mathbf{x}_0) \cdot \Delta \mathbf{x}, \tag{1.1}$$

where $\mathbf{s}(\mathbf{x}_0) = \partial y/\partial \mathbf{x}|_{\mathbf{x}_0}$ denotes the sensitivity vector. The model is therefore linearized at $\mathbf{x}$ and the sensitivity explored at such linearization.

In contrary to the local analysis, a global sensitivity analysis explores the model's output for the entire parameter space, the domain covered by all (relevant) model parameters $\mathbf{x}$ [488].

Example 1.1 (Sensitivity of the Resistance of a Blood Vessel). The resistance R [Pa s m^{-3}] of a vessel describes the relation between the pressure p [Pa] and the flow q [m^{3}s^{-1}]. Given a cylindrical vessel of the length l and diameter d, the resistance is given by Hagen–Poiseuille's law

$$R = \frac{p_i - p_o}{q} = \frac{128 l \eta}{\pi d^4}, \tag{1.2}$$

where the term $p_i - p_o$ [Pa] denotes the pressure drop between inlet and outlet, and η [Pa s] describes the blood's dynamic viscosity, see Sect. 2.3.2.1.

(a) Compute the sensitivity vector $\mathbf{s} = \partial R/\partial \mathbf{x}$ to explore the local sensitivity of Hagen–Poiseuille's law to variations of $\mathbf{x} = [l\ \ d\ \ \eta]^{\mathsf{T}}$. Express the absolute resistance error ΔR and the relative resistance error $\Delta R/R$ in response to the variation of the input information.
(b) Consider the variation of a single parameter at the time and estimate the accuracy of the input to ensure R would not change by more than $\pm 10\%$.
(c) What is the largest possible relative error of R to be expected from the estimates in Task (b)?   ∎

Example 1.2 (Global Versus Local Sensitivity Measures). A biomechanical Finite Element Method (FEM) model is used to compute a risk index r. The FEM model uses several input parameters, out of which the wall thickness h and the tissue stiffness s are the most uncertain inputs. Through a sequence of computation, the data in Table 1.1 has been acquired. Given this information, the model's local and global sensitivity with respect to its two input parameters h and s should be investigated.

Table 1.1 Finite Element Method (FEM) model computations of the risk index r. Data represent 48 predictions by varying the model's wall thickness h and the tissue stiffness s, respectively

h [mm]	s [kPa]					
	100	400	700	1000	1300	1600
0.5	0.558955	0.964567	1.48654	2.08449	2.74171	3.44859
1.0	0.553621	0.966734	1.50520	2.12866	2.82038	3.57075
1.5	0.538621	0.964234	1.53021	2.19616	2.94538	3.76825
2.0	0.508955	0.952067	1.55654	2.28199	3.11171	4.03609
2.5	0.459621	0.925234	1.57921	2.38116	3.31438	4.36925
3.0	0.385621	0.878734	1.59321	2.48866	3.54838	4.76275
3.5	0.281955	0.807567	1.59354	2.59949	3.80871	5.21159
4.0	0.143621	0.706734	1.57521	2.70866	4.09038	5.71075

(a) Identify the coefficients a_i, b_j, c_k of the polynomial surrogate model

$$r = a_0 + a_1 h + a_2 h^2 + b_1 s + b_2 s^2 + c_1 hs + c_2 hs^2 + c_3 h^2 s + c_4 h^2 s^2 \tag{1.3}$$

from the data points listed in Table 1.1.

(b) Compute the local sensitivity at the point $h = 2.0$ mm and $s = 900.0$ kPa in the parameter space. How much would the wall thickness alteration of 0.6 mm and tissue stiffness alteration of 250 kPa influence the risk index?

(c) Consider the normal distributed wall thickness $h = 2.0(SD\ 0.6)$ mm and the tissue stiffness $s = 900.0(SD\ 250.0)$ kPa to investigate the global sensitivity of the model. Use Monte Carlo simulation and compute the distribution of the risk index r. The notation $a(SD\ b)$ denotes the mean a and the Standard Deviation (SD) of a sample, see Sect. 1.4.   ■

1.3.1.1 Sobol's Variance-Based Sensitivity Analysis

Sobol's variance-based sensitivity analysis [513] is a form of a *global* sensitivity analysis that allows to relate the variance of individual input parameters to the model's output variance. It considers the input parameter space of $0 < x_i < 1$, $i = 1, \ldots, n$ and relies on the *ANalysis Of VAriances (ANOVA)-representation* of the model response function

$$y = f(x_1, x_2, \ldots, x_n) = f_0 + \sum_i f_i(x_i) + \sum_{i<j} f_{ij}(x_i, x_j)$$

$$+ \cdots + f_{12\cdots n}(x_1, x_2, \ldots, x_n), \tag{1.4}$$

where all functions (other than f_0) satisfy the orthogonality condition

$$\int_0^1 f_{i\cdots j}(x_i, \ldots, x_j) dx_k = 0 \ \text{ for } \ k = i, \ldots, j \ .$$

With the definition of the variance

$$V = \text{Var}(y) = \int (f(\mathbf{x}) - f_0)^2 d\mathbf{x} = \int f^2(\mathbf{x}) d\mathbf{x} - f_0^2 \,,$$

the ANOVA-representation (1.4) yields the variance decomposition

$$V = \sum_i \underbrace{\int f_i^2(x_i) dx_i}_{V_i = \text{Var}(y|x_i)} + \sum_{i<j} \underbrace{\int f_{ij}^2(x_i, x_j) dx_i dx_j}_{V_{ij} = \text{Var}(y|x_i, x_j) - V_i - V_j} + \cdots, \tag{1.5}$$

where the notation $V_i = \text{Var}(y|x_i)$ denotes the variance of y under the variation of x_i, and $V_{ij} = \text{Var}(y|x_i, x_j) - V_i - V_j$ the variance of y that arises from mixed effects of varying x_i and x_j, respectively.

With the variance decomposition (1.5), *Sobol's sensitivity indices*

$$S_i = \frac{V_i}{V} \,, \quad S_{ij} = \frac{V_{ij}}{V} \,, \quad \cdots$$

can be introduced. The sum over all Sobol's sensitivity indices is one.

The *first-order sensitivity* (or main effect) index S_i specifies the relative effect of the input variable x_i on the model's output y. Similarly, S_{ij} denotes the *second-order effect* on y and it arises from the mixed interaction of the input variables x_i and x_j. Mixed effects are often neglected in a sensitivity analysis.

Example 1.3 (Sobol's Variance-Based Sensitivity Analysis). A biomechanical Finite Element Method (FEM) model computes a risk factor r, where the wall thickness H and tissue strength S are the most uncertain inputs. Sobol's variance-based sensitivity analysis should be used to explore the model's global sensitivity to these two input parameters. To this end, the surrogate model

$$r = \underbrace{0.555}_{f_0} \underbrace{-0.03(h - 1/2) - 0.4(h - 1/2)^3}_{f_h}$$

$$\underbrace{+0.25(s - 1/2)}_{f_s} \underbrace{+0.1(h - 1/2)(s - 1/2)}_{f_{hs}} \tag{1.6}$$

may be used, where the wall thickness H and the tissue strength S have been transformed, such that their dimensionless counterparts $0 < h < 1$ and $0 < s < 1$ cover the unit-square parameter space.

(a) Show that the model (1.6) is an ANalysis Of VAriances (ANOVA)-representation.
(b) Compute Sobol's sensitivity indices.
(c) How do the variances $\text{Var}(h) = 0.2$ and $\text{Var}(s) = 0.1$ translate into the variances $\text{Var}(r|h)$ and $\text{Var}(r|s)$ of the risk factor r? ∎

1.3.2 Verification

Verification aims at testing the correctness of the model's mathematical implementation, see Fig. 1.3. *Are the equations solved correctly?* is therefore asked by verification exercises. They test the model for analytical errors, software coding errors, parameter input failure, and ensure the appropriateness of the applied numerical technique to solve the mathematical equations. Verification is performed at each (sub) model level and uses black-box and white-box testing approaches. Black-box testing verifies that a given input results in the correct output, whilst white-box testing verifies the correct inner functioning of the model implementation. Verification exercises foresee approaches, such as testing model predictions against closed-form solutions, perform software code inspections and data plausibility checks.

1.3.3 Validation

Validation aims at testing that a model predicts the desired features of the real object or process, see Fig. 1.3. It ensures that the model represents the real object or process up to the level specified by the IMA. *Are the correct equations solved?* is asked by validation exercises. Whilst each modeling assumption needs to be validated separately, it is of utmost importance that the *global model output* is validated against experimental observations. A validation study should be well-designed and should not contain any *confounding factors*, see Sect. 1.3.3.1. In bioengineering, validation is often challenged by *ethical constraints* and cost aspects, factors that often form the bottleneck towards the successful clinical integration of bioengineering models.

1.3.3.1 Study Design

Generally, a *study* may be seen as a controlled approach to understand *cause and effect* of a *treatment*, such as a drug, a clinical intervention or a diagnostic parameter. The analyst tries then to control of how study units are assigned to groups and which treatment they receive. All studies consist of the following three parts.

- *Study units.* Recipients of treatment, such as people, animals, or plants.
- *Independent variables.* Factors that are controlled and manipulated by the analyst. Factors are causal or explanatory problem variables. The amount of a drug administered to the study units could be such a factor.
- *Dependent variables.* Response variables that represent the cause or effect of a treatment, such as the number of patients free from symptoms at the end of the study.

A *well-designed* study provides a *clean* test of causal connections between dependent and independent variables—it eliminates the influence of *extraneous* (or lurking) variables. Extraneous variables are variables that influence the study, often in an unknown way. They are neither independent variables nor dependent variables. Steps to reduce the effects of extraneous variables are important, and a well-designed study implements the following approaches:

- *Control.* Making a study as similar as possible for the study units in each treatment condition. *Double blinding* is such an approach, where neither study subjects nor analysts receive information about the treatment.
- *Randomization.* Using chance methods to assign study units to treatment groups, which in turn randomizes the influence of extraneous variables. Randomization also greatly enhances the appropriateness of statistical analysis methods, many of which assume random distribution of variables.
- *Replication.* Assigning each treatment to many study units. The more study units are in each treatment condition, the lower the variability of the dependent measures.

Given the influence of extraneous variables has not been eliminated, the study is *confounded.* A confounded study provides plausible alterative explanations for the observed relationship between independent and dependent variables.

1.4 Statistics-Based Modeling

Biological data is uncertain and *statistics-based modeling* is a common approach to describe bioengineering phenomena. A model is simply seen as a black-box and thus a mathematical function that correlates model input and output. This section discusses some methods of *defining* and *testing* such models, and the statistical preliminaries are listed in Appendix A.1.

1.4.1 Correlation Amongst Variables

Given observational data of a bioengineering phenomenon, a *correlation test* explores whether or not a correlation between variable exists—a correlation could then be expressed by a model. A correlation test provides a *correlation coefficient* $-1 \leq r \leq +1$. Positive and negative correlation coefficients denote positive and negative correlations, respectively. Given a positive correlation, the dependent variable increases with the increasing independent variable. The opposite holds for a negative correlation. The larger the absolute value $|r|$ of the correlation coefficient, the stronger is the correlation between the variables. Testing the dependence of a single variable y_i with respect to a single independent variable x_i is called *simple correlation testing*.

1.4.1.1   Pearson's Product-Moment Correlation Coefficient

Pearson's[5] *correlation coefficient* indicates the linear relation between two variables given in interval or ratio scales. With $x_i, y_i; i = 1, \ldots, n$, denoting a sample with the means $\overline{x} = (\sum_{i=1}^{n} x_i)/n$ and $\overline{y} = (\sum_{i=1}^{n} y_i)/n$, the Pearson's correlation coefficient reads

$$ r = \frac{\sum_{i=1}^{n} (x_i - \overline{x})(y_i - \overline{y})}{\sqrt{\sum_{i=1}^{n} (x_i - \overline{x})^2 \sum_{i=1}^{n} (y_i - \overline{y})^2}} . \tag{1.8} $$

Figure 1.6 shows scatter plots of distributed data points together with their Pearson's correlation coefficient r. The sets (a)–(c) show linearly correlated data, and the Pearson's correlation coefficients adequately describe such correlations. The sample in Fig. 1.6a shows the strongest possible positive correlation, the sample in Fig. 1.6b shows a strong negative correlation, and the sample in Fig. 1.6c shows uncorrelated data, respectively.

The sample shown in Fig. 1.6d illustrates the strongest possible, but *non-linear* correlation. Due to the non-linearity, any linear correlation measure, such as the Pearson's correlation coefficient, fails to address that. Given the particular case, the

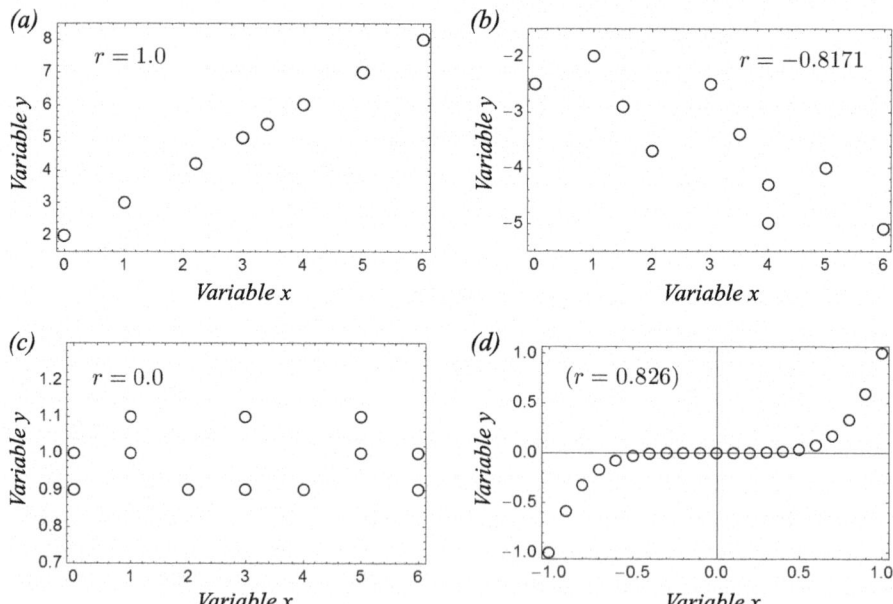

Fig. 1.6 Pearson's correlation coefficients r specifying linear correlations between the variables x_i and y_i. The linear correlation coefficient adequately specifies the data sets (**a**)–(**c**), but it fails to capture the perfect, but non-linear correlation shown in (**d**)

---

[5]Karl Pearson, English mathematician and biostatistician, 1857–1936.

Pearson's correlation coefficient yields $r = 0.826$, instead of the value of one for a perfect correlation.

1.4.1.2 Spearman's Rank Correlation Coefficient

Spearman's[6] correlation coefficient r_s indicates the *monotonic* (linear or non-linear) relation between two variables given in interval or ratio scales. The Spearman's correlation works by calculating the Pearson's correlation (1.8), but on the rank data. Consequently, it reads

$$r_s(x_i, y_i) = r(\text{rg}(x_i), \text{rg}(y_i)), \tag{1.9}$$

where $\text{rg}(x_i)$ and $\text{rg}(y_i)$ denote the rank variables of x_i and y_i, respectively. Ranking data is a transformation in which numerical values are replaced by their rank when the data are sorted. For example, the rank data of the set $x_i = \{2.7, 0.1, 1.9, 3.4\}$ reads $\text{rg}(x_i) = \{3, 1, 2, 4\}$.

Figure 1.7a,b illustrates Spearman's correlation coefficients for selective examples. Figure 1.7a shows the scatter plot of the raw data x_i, y_i, whilst Fig. 1.7b shows

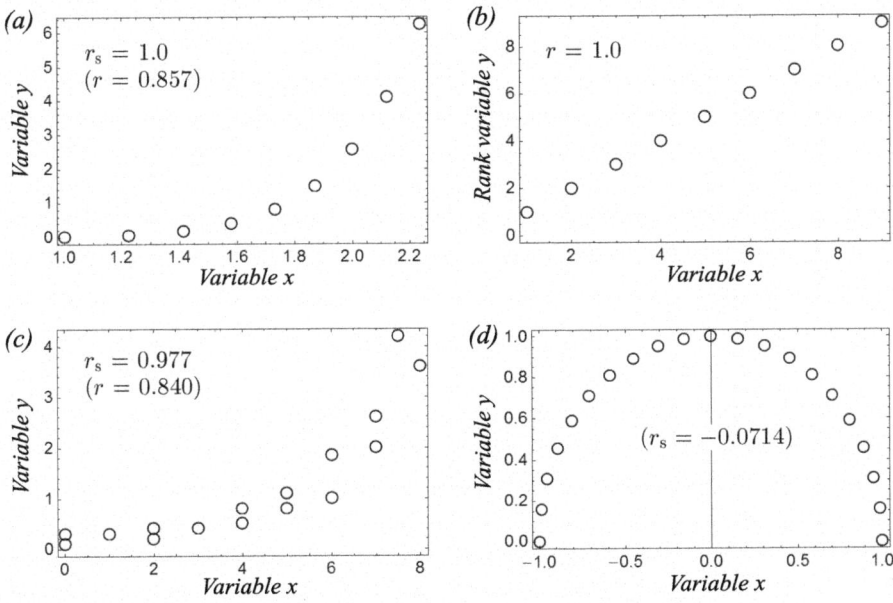

Fig. 1.7 Spearman's correlation coefficients r_s specifying the monotonic relations between the variables x_i and y_i. (**a**), (**b**) Perfect non-linear correlation, where (**a**) shows the raw data, and (**b**) the rank data, respectively. The non-linear correlation coefficient adequately determines the non-linear monotonic sample shown in (**c**), but it fails to describe the perfect, but non-monotonic correlation in (**d**)

---

[6]Charles Edward Spearman, English psychologist, 1863–1945.

the rank variables $rg(x_i)$, $rg(y_i)$. The rank data is always linearly related, which allows using the Pearson's correlation (1.8). Regardless a linear correlation analysis would be meaningless for the non-linearly correlated data in Fig. 1.7a, the Pearson's correlation coefficient is shown in brackets.

Figure 1.7c shows the Spearman's correlation coefficient (1.9) of dispersed data. Again, the (meaningless) Pearson's correlation coefficient is given in brackets. A perfect but *non-monotonic* correlation between x_i and y_i is shown in Fig. 1.7d. Given non-monotonic correlations, the Spearman's correlation coefficient is meaningless. Despite the perfect correlation, the Spearman's correlation coefficient would be $r_s = -0.0714$. A non-monotonic sample may be split into monotonic sub-samples, or more advanced correlation methods may be used to explore non-monotonic data samples.

1.4.2   Regression Modeling

Regression modeling is used to predict dependent variables based on the values of independent variables or a *cause* and *effect* relationship amongst them. Predicting one dependent variable from one independent variable is called *simple regression modeling*.

With $x_i, y_i; i = 1, \ldots, n$, denoting a sample with the means $\bar{x} = (\sum_{i=1}^{n} x_i)/n$ and $\bar{y} = (\sum_{i=1}^{n} y_i)/n$, a simple linear regression model reads

$$y_i = b_0 + b_1 x_i + e_i , \tag{1.10}$$

where e_i denotes the i-th *error* term or *residuum*. The coefficients b_0 and b_1 are identified through least-square optimization,

$$\sum_{i=1}^{n} e_i^2 = \sum_{i=1}^{n} (y_i - b_0 - b_1 x_i)^2 \rightarrow \text{MIN} ,$$

where the square of the error terms e_i is minimized. Figure 1.8 illustrates a scatter plot with the (linear) regression line on top of the sample data.

The *coefficient of determination* of a simple linear regression model reads

$$R^2 = \left[\frac{\frac{1}{n} \sum_{i=1}^{n} (x_i - \bar{x})(y_i - \bar{y})}{s_x s_y} \right]^2 \tag{1.11}$$

where

$$s_x = \sqrt{\frac{\sum_{i=1}^{n} (x_i - \bar{x})^2}{n - 1}} \quad ; \quad s_y = \sqrt{\frac{\sum_{i=1}^{n} (y_i - \bar{y})^2}{n - 1}}$$

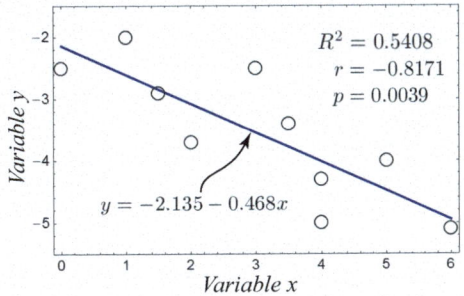

Fig. 1.8 Linear regression to predict the dependent variable y_i from the independent variable x_i. The coefficient of determination R^2 denotes the quality of the regression, and least-square optimization defines the regression line shown in blue. The Pearson's correlation coefficient is denoted by r, and p is the significance of the regression

denote the *Standard Deviation (SD)* of the sample. The coefficient of determination ranges from $0 \le R^2 \le 1$, where $R^2 = 0$ means that the data cannot be predicted, whilst for $R^2 = 1$, it can be predicted at no error, and thus at $e_i = 0$. The definitions (1.8) and (1.11) directly yield the relation

$$R^2 = [r(n-1)/n]^2 \qquad (1.12)$$

between the Pearson's product-moment correlation coefficient r and the coefficient of determination R^2.

1.4.2.1 Significance of a Regression

As for any probabilistic data analysis, the *significance* of the results should be investigated. Following *Hypothesis Testing* (see Sect. 1.4.3), the data may be tested against the Null Hypothesis $H_0 : b_1 = 0$, where b_1 denotes the parameter introduced by the regression model (1.10). Such a test investigates the probability that the slope of the regression line is zero, circumstances at which x_i and y_i are uncorrelated. Thus, the statistic reads

$$t = \frac{r\sqrt{n-2}}{\sqrt{1-r^2}}, \qquad (1.13)$$

and rejecting H_0 has the probability of $p = 2\int_{-\infty}^{t} \rho_t(x)dx$. Here, $\rho_t(x)$ denotes the *Probability Density Function (PDF)* of the *student*[7] *t-distribution* (A.3) of $n-2$ degrees of freedom. The determination of the two parameters b_0 and b_1 leads to

---

[7]Denoted after William Sealy Gosset, English statistician, 1876–1937, who used the pen name "student".

the deduction of two degrees of freedom from the data, and the test is a two-tailed significance test, which explains the factor 2 when computing the probability p.

Example 1.4 (Correlation of Vessel Wall Stiffness and Strength). An *in vitro* tissue characterization study used tensile testing to measure the vessel wall stiffness parameter x and the vessel wall strength y. The observed data comprises observations from vessel wall specimens taken from $n = 20$ animals, see Table 1.2.

Table 1.2 Stiffness parameter x and tensile strength y of vessel wall samples acquired from $n = 20$ animals

Specimen	Stiffness x [kPa]	Strength y [kPa]	Specimen	Stiffness x [kPa]	Strength y [kPa]
1	86.5	906.1	11	233.7	1907.2
2	191.7	1301.9	12	172.7	3060.1
3	193.3	1854.5	13	245.2	1871.4
4	228.2	627.5	14	175.4	1516.3
5	128.0	1299.1	15	300.5	2462.8
6	169.8	1751.3	16	266.7	1851.2
7	219.1	1538.5	17	190.0	2640.5
8	182.4	1815.9	18	350.7	1821.5
9	155.8	880.7	19	384.0	2272.3
10	194.3	1352.8	20	257.8	2491.5

A correlation analysis should be performed to draw conclusion from the acquired experimental data, addressing the following tasks:

(a) Investigate, whether the stiffness parameter x and the tensile strength y are linearly, or non-linearly correlated.
(b) Compute the correlation coefficient r and the coefficient of determination R^2.
(c) Consider the significance level of 5%, and test the statistical significance of a potential correlation between x and y.   ∎

1.4.2.2 Non-linear Regression
A simple linear regression model requires a linear correlation between x_i and y_i, which in turn yields a *normally distributed residuum* e_i of the regression model (1.10). Appendix A.1.3 outlines methods of testing the distribution of random

data. Given x_i and y_i would not be linearly correlated, one or both variables may be transformed, such that the transformed variables yield linear correlation. Such a transformation rule naturally depends on the individual data sample and the reader is referred to the statistics literature [382].

1.4.2.3 Multiple Regression

More than one independent variable may have an effect on the dependent variable. Predicting one dependent variable from a number of independent variables is called *multiple regression modeling*. Given two independent variables x_i^A and x_i^B, a linear multiple regression model reads

$$y_i = b_0 + b_1^A x_i^A + b_1^B x_i^B + e_i \, ,$$

where e_i denotes the i-th error term or residuum, and the coefficients b_0, b_1^A and b_1^B are identified through least-square optimization, and thus $\sum_{i=1}^{n} e_i^2 \rightarrow$ MIN. It may be difficult to identify a set of variables causing *independent effects* on the response variable—variables often interact in an unknown (and hidden) way.

Stepwise regression is one way to solve multiple regression problems. An automatic procedure adds or removes independent variables (explanatory variables) of the regression model, a selection process that is governed by optimizing the model's significance [289]. Techniques, collectively known as *Artificial Intelligence (AI)* or *Machine Learning (ML)*, may also be used to deal with interacting variables, see Sect. 1.5.

1.4.2.4 Multivariant Regression

A *multivariant* regression model uses a number of independent (explanatory) variables and predicts several outcome variables. Such models aim at understanding the relationships between variables and their relevance to the problem being studied.

1.4.3 Hypothesis Testing

The best way to test the significance of a model would be to examine the entire population, and thus all data eventually processed by the model. Since that is often impractical, conclusions have to be drawn from a random sample taken from the population, a method called *hypothesis testing*. It refers to the formal procedures to accept or reject statistical hypotheses. The test uses sample data taken from populations to answer questions related to observations in the populations. Given the sample data are not consistent with the statistical hypothesis, the hypothesis is *rejected*. There are two types of statistical hypotheses:

Null Hypothesis H_0. The Null Hypothesis is usually the hypothesis in which sample observations result purely from chance.

Alternative Hypothesis H_a. The Alternative Hypothesis is the hypothesis in which sample observations are influenced by some non-random cause.

The Null Hypothesis H_0 and the Alternative Hypothesis H_a are mutually exclusive.

Let us consider the problem of assessing whether or not a dice is fair, and thus each face $f = 1, 2, 3, 4, 5, 6$ has the same probability $p(f)$ of appearance. This could be tested with

$$H_0 : p(f) = 1/6 \; ; \quad H_a : p(f) \neq 1/6 \tag{1.17}$$

with H_0 and H_a being the Null Hypothesis and the Alternative Hypothesis, respectively. Rejecting H_0 would mean that the dice *is not* fair, and rejecting H_a would mean that the dice *is* fair.

The strength of evidence in support of a Null Hypothesis H_0 is measured by the probability and thus the *p-value*. If the *p*-value is less than the (predefined) significance level, we reject the hypothesis. A test of a statistical hypothesis, for which the region of rejection is on one or both sides of the sampling distribution, is called a *one-tailed* or a *two-tailed* test, respectively. Figure 1.10 illustrates that.

In summary, statistical hypothesis testing commonly follows the following three distinct steps:

- State the hypotheses to be tested and specify the statistics that represent the problem.
- Formulate an analysis plan, and calculate the probability p under the Null Hypothesis H_0.
- Given p is below the (selected) significance level, reject the Null Hypothesis, and draw conclusions.

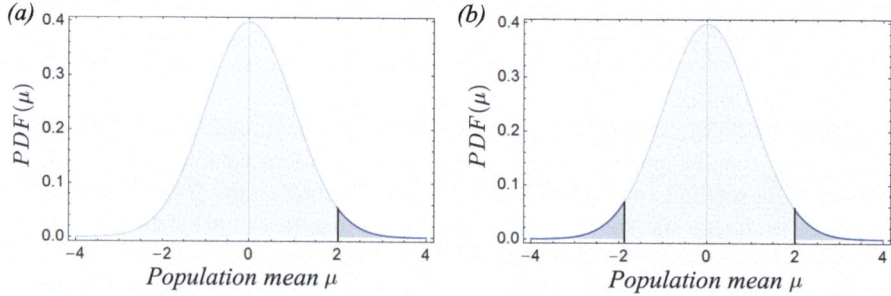

Fig. 1.10 Statistical hypothesis testing the population mean μ. (**a**) Rejecting the Null Hypothesis $H_0 : \mu \leq 2$ is illustrated by the highlighted domain and yields a one-tailed test. (**b**) Rejecting the Null Hypothesis $H_0 : |\mu| \leq 2$ is illustrated by the highlighted domain and yields a two-tailed test

1.4.3.1 Error and Power of a Hypothesis Test

A *Type I error* occurs when the Null Hypothesis H_0 is incorrectly rejected, and thus a *false positive* is predicted. The probability of committing a Type I error is called the significance level. For the aforementioned example with the dice, H_0 would be rejected although the dice would in fact be fair.

A *Type II error* occurs when the Null Hypothesis H_0 fails to reject a negative observation, and thus a *false negative* is predicted—H_0 would not be rejected although the dice would in fact be unfair. The probability of committing a Type II error is called the *power* of the test. The greater the sample size and the higher the significance level (i.e., the lower the probability of committing a Type I error), the greater is the power of the test. In addition, the greater the difference between the "true" value of a parameter and the value specified in the Null Hypothesis, greater is the power of the test.

Example 1.5 (Testing for Clairvoyance). A person is shown the reverse of $n = 25$ randomly chosen playing cards and asked which of the four suits it belongs to. The person should be tested for clairvoyance according to the following tasks:

(a) Formulate a suitable Null Hypothesis H_0 and an Alternative Hypothesis H_a.
(b) Draw a tree diagram and compute the number of false positives in such a test.
(c) Compute the significance level for at least $x = 9$ correct answers.
(d) Compute how many answers have to be correct to reach a significance level of less than 5%. ∎

1.4.4 Mean Difference Test

Whether or not two *samples* stem from *different populations* may be investigated by a mean difference test. Such a test is typically appropriate for normally distributed populations and a sample size of 15 or more. The sample should not contain outliers and samples of smaller sizes should be symmetric. Preliminaries of data distribution testing in statistics are summarized in Sect. A.1.3.

1.4.4.1 One-Sample t-test

The *one-sample t-test* tests the Null Hypothesis $H_0 : \mu = \mu_0$ that the population mean μ is equal to a specific value μ_0. Given a sample of the size n, the mean $\bar{x}$, and the SD s, the statistic reads

$$t = \frac{\bar{x} - \mu_0}{s/\sqrt{n}} . \tag{1.18}$$

Consequently, the two-tailed significance test of rejecting H_0 has the probability $p = 2 \int_{-\infty}^{t} \rho_t(x) \mathrm{d}x$, where $\rho_t(x)$ denotes the PDF of the student t-distribution (A.3).

The one-sample t-test may also be used to test the difference $x = x_2 - x_1$ of observations x_1 and x_2 taken from n *paired samples*.

1.4.4.2 Two-Sample t-test of Independent Samples

Given two independent samples x and y, the *two-sample t-test* specifies whether or not the two samples stem from *different populations*. The Null Hypothesis H_0 : $\mu_x = \mu_y$ is commonly used, where μ_x and μ_y denote the population means related to the samples x and y. Consequently, rejecting H_0 would suggest that the two samples stem from different populations. In general, the two samples would be of different sizes n_x, n_y and SD s_x, s_y, such that the statistic reads

$$t = \frac{\bar{x} - \bar{y}}{\bar{s}} \; ; \; \bar{s} = \sqrt{\frac{s_x^2}{n_x} + \frac{s_y^2}{n_y}}, \tag{1.19}$$

where $\bar{s}$ denotes a weighted SD. Such a test is called *Welch*[8] *t-test*, a generalization of the *student's t-test*.

In order to compute the probability of rejecting H_0, the nearest integer of the expression

$$(s_x^2/n_x + s_y^2/n_y)^2 / \left[\frac{(s_x^2/n_x)^2}{n_x - 1} + \frac{(s_y^2/n_y)^2}{n_y - 1} \right] \tag{1.20}$$

determines the degrees of freedom ν of the problem. Finally, rejecting H_0 has the probability of $p = 2 \int_{-\infty}^{t} \rho_t(x) dx$ of the two-tailed significance test, where $\rho_t(x)$ denotes the PDF of the student t-distribution (A.3).

Example 1.6 (Conclusions from Vessel Wall Stiffness Data). Let us consider an *in vivo* tissue characterization study to investigate the effect of hypertension on the stiffness of the rat aorta. The vessel's stiffness has been acquired at baseline, and the rats have then been medicated to increase their blood pressure. After six weeks the stiffness was measured again in the same animals. The data was acquired independently by two experimentalists (experimentalist A and experimentalist B) and in a number of different animals. It is reported in Table 1.3, and the significance level of $p = 0.05$ may be used for hypothesis testing.

Conclusions from the data in Table 1.3 should be drawn through the analysis of the following tasks:

---

[8]Bernard Lewis Welch, British statistician, 1911–1989.

Table 1.3 Aorta wall stiffness acquired through an *in vivo* tissue characterization study. The samples A1–A20 have been acquired by experimentalist A, and the samples B1 to B23 by experimentalist B

Sample	Baseline [kPa]	Medicated [kPa]	Change [kPa]	Sample	Baseline [kPa]	Medicated [kPa]	Change [kPa]
A1	1235.6	827.5	−408.1	B1	634.8	837.6	202.8
A2	595.6	705.1	109.5	B2	771.3	1121.8	350.5
A3	833.0	974.9	141.9	B3	1496.3	1118.4	−378.0
A4	498.5	955.4	456.9	B4	554.3	569.8	15.5
A5	709.7	1194.2	484.6	B5	1346.6	909.7	−436.9
A6	996.7	912.4	−84.3	B6	1217.5	1018.8	−198.7
A7	540.8	944.7	403.9	B7	535.6	1010.9	475.3
A8	612.3	867.5	255.2	B8	1246.3	715.9	−530.5
A9	592.2	964.4	372.2	B9	298.2	1075.4	777.1
A10	1193.7	1161.7	−32.0	B10	978.0	1110.7	132.7
A11	579.9	1010.7	430.8	B11	1203.3	872.6	−330.7
A12	707.8	753.5	45.6	B12	1181.9	1004.1	−177.9
A13	763.7	946.0	182.3	B13	512.0	1275.8	763.8
A14	1089.4	1121.6	32.2	B14	785.6	925.7	140.1
A15	883.6	954.9	71.3	B15	483.3	1115.5	632.2
A16	628.6	911.1	282.5	B16	590.1	1068.0	477.8
A17	614.9	590.4	−24.5	B17	1122.8	717.4	−405.4
A18	909.3	1108.7	199.5	B18	814.4	1008.6	194.2
A19	628.2	1243.5	615.3	B19	1709.0	979.5	−729.5
A20	953.0	1169.2	216.2	B20	466.2	964.4	498.2
				B21	45.2	1041.6	996.4
				B22	543.0	1056.0	513.0
				B23	776.5	797.9	21.5

(a) Is the experimental data normally distributed?
(b) Is the experimental data independent of the experimentalist who acquired it?
(c) Does the medication change the stiffness of the aorta in the rats? ■

1.5 Artificial Intelligence

Artificial Intelligence (AI), also called Machine Learning (ML), refers to the area of computer science that creates machines (software) to perform a specific task without using explicit instructions. It may be defined as *a system's ability to correctly interpret external data, to learn from such data, and to use those learnings to achieve specific goals and tasks through flexible adaptation* [297]. AI algorithms often mimic cognitive functions, such as *learning* and *problem solving* by using algorithms based on Artificial Neural Network (ANN), Bayesian Networks (BN), mathematical optimization, support vector networks, decision trees, and nearest neighbor. Such approaches are able to process large amounts of data and use it to make worthwhile predictions [377]. AI algorithms do not simply learn from data but enhance themselves and their predictive capability by learning new heuristics. The solution of practically relevant problems is often linked to the availability of High Performance Computing (HPC) resources and large data sets of experimental observations. This field of modeling is rapidly progressing with numerous mathematical algorithms being available [257].

1.5.1   Learning and Prediction

AI algorithms acquire their knowledge through *learning* or *training* using different approaches. *Supervised learning* acquires knowledge from data that contains both the inputs and the desired outputs. It is very similar to a regression analysis, see Sect. 1.4.2, where the independent variables are the input and the dependent variables are the output. The knowledge governed from learning is then represented by the coefficients of the regression model. However, AI approaches are superior in dealing with potential dependence among the independent variables as well as any non-linearity associated with the problem. In *unsupervised learning*, the AI algorithm acquires knowledge from a set of data that contains only inputs and *no* desired outputs.

Real world data is often polluted, and *anomaly detection*, also known as *outlier detection*, is used to clean the data from "suspicious" observations, prior to the commencement learning phase.

1.5.2   Artificial Neural Network

ANNs are models that are vaguely inspired by the structure of the brain. *Nodes* represent *neurons* and are connected by *edges*, which themselves represent the *synapses* in a biological brain, see Fig. 1.14a. Each edge can transmit a *signal* (information) from one node to another. A node processes the received information

Fig. 1.14 Two common model approaches used in Artificial Intelligence (AI). (**a**) An Artificial Neural Network (ANN) is an interconnected group of nodes that mimic a network of neurons in a biological brain. (**b**) A Bayesian Network (BN) models conditional dependence of variables in a network

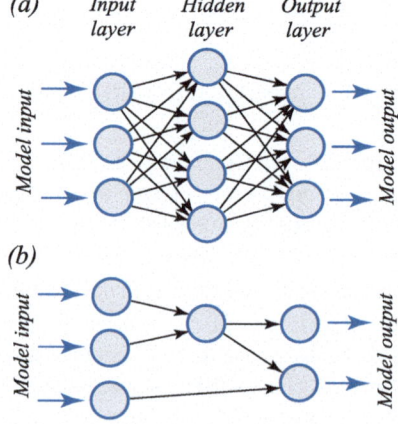

and then signals the nodes connected to it. In common ANN implementations, the signal that is transmitted through edges is a real number, and the output of each node is a non-linear function of the sum of its inputs. Edges have a *weight* that increases or decreases the strength of its signal. The weights are specified during the model's *learning* or *training* phase. A node may have a threshold, whereby a signal is only sent if it exceeds said threshold.

Typically, nodes are aggregated into layers. Different layers may perform different transformations upon their inputs. The signals travel from the *input layer*, through *hidden layers*, to the *output layer*. An ANN that uses multiple hidden layers uses a *deep learning* approach and may require advanced learning algorithms as well as HPC resources.

ANN training, and thus the identification of weights, can be seen as an optimization problem. The *backpropagation algorithm* follows this concept and allows for the efficient training of an ANN, following a gradient descent approach that exploits the chain rule [481].

Example 1.7 (Training an Artificial Neural Network). Figure 1.15 shows part of an Artificial Neural Network (ANN). The inputs x_1 and x_2 are weighted by w_1 and w_2, and the node N then determines the output y. The node N is represented by the logistic function

$$y = [1 + \exp(-z)]^{-1}, \tag{1.22}$$

where $z = \sum_{i=1}^{2} x_i w_i$ denotes the cumulative input.

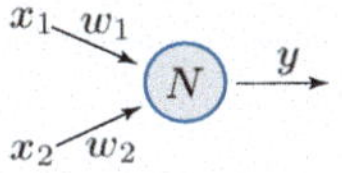

Fig. 1.15 Part of an Artificial Neural Network (ANN) with x_1, x_2, and y denoting the inputs and the output, respectively. The weights w_1 and w_2 are identified by training

Table 1.4 Set of training data. The input is denoted by x_1, x_2, whilst $\tilde{y}$ is the desired output of node N

x_1	x_2	$\tilde{y}$
−1.0	0.3	0.2
−0.9	0.1	0.3
2.0	2.0	0.9
3.0	1.0	1.0

(a) Express the "learning process" of the ANN through a least-square optimization problem.
(b) Identify the weights w_1 and w_2 through supervised learning, where Table 1.4 lists the set of training data. ∎

1.5.3 Bayesian Network

A BN, also known as a Bayes[9] network, or a decision network, is a probabilistic model that represents *a set of variables* and their *conditional* dependencies.

Let us consider a patient with a condition who may be either treated with Drug A or Drug B. Both drugs have the risk for dangerous interactions and must not be administered together. The BN shown in Fig. 1.16 represents this problem, where each variable (administration of Drug A, administration of Drug B, and Treatment Success) can either be True (T) or False (F). The administration of both drugs is conditionally dependent (they must not be administered together), and both drugs influence the probability of the Treatment Success.

A *Conditional Probability Table* expresses the conditional dependence amongst all variables, see Table 1.5. The left table reports that Drug A is administered to 20% of the patients, whilst 80% do not receive it. The middle table reports the administration of Drug B. Regardless of dangerous interactions amongst the two drugs, some patients received both drugs, see the bottom line in the table. The right table reports the probability of the Treatment Success in relation to the administration of the two drugs.

In conclusion, a BN is a *complete* model for its variables and their relationships that can be used to answer any probabilistic queries. Given AI applications, the information held by the Conditional Probability Tables is acquired through *learning* or *training*. Similar to ANN, optimization methods may be used to train a BN.

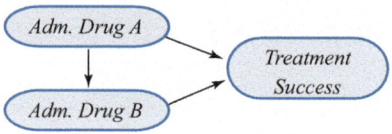

Fig. 1.16 Treating a condition with two different drugs. Variables and their conditional dependencies are represented by a Directed Acyclic Graph

Table 1.5 Conditional probability tables. The three binary variables administration of Drug A, administration of Drug B, and Treatment Success describe the problem

Adm. Drug A		Adm. Drug B			Treatment Success			
T	F	Adm. Drug A	T	F	Adm. Drug A	Adm. Drug B	T	F
0.2	0.8	F	0.4	0.6	F	F	0.0	1.0
		T	0.01	0.99	F	T	0.8	0.2
					T	F	0.9	0.1
					T	T	0.99	0.01

---

[9]Thomas Bayes, English statistician and philosopher, 1701–1761.

1.5.4 Decision Tree

A large number of observations may be used to "grow" a *decision tree* through such data, a process called decisions tree learning. Figure 1.17 illustrates the application of a decision tree towards the prediction of the vascular wall stiffness. The decision tree starts at its root: a cohort of patients with properties, such as patients' gender, their smoking status, and their age. In addition, each patient entry contains a vessel wall stiffness k_i [kPa], a variable that depends on patient attributes. It is this parameter that should be predicted by the decision tree model. For the present example, the mean wall stiffness in the root cohort is $\overline{k} = 98.3\,\text{kPa}$, see Fig. 1.17. The cohort is then split into two sub-cohorts, $\mathcal{A}$ and $\mathcal{B}$, towards the homogenization of either sub-cohort with respect to k_i. Commonly, this involves the minimization of a cost function

$$\sum_{i \in \mathcal{A}} (k_i - \overline{k}_\mathcal{A})^2 + \sum_{i \in \mathcal{B}} (k_i - \overline{k}_\mathcal{B})^2 \rightarrow \text{MIN}\,, \qquad (1.24)$$

where $\overline{k}_\mathcal{A}$ and $\overline{k}_\mathcal{B}$ denote the means of k_i in the sub-cohorts, respectively. This determines the optimal partitioning of the data. The process is *inherently recursive* and concludes with a number of sub-cohorts and their respective mean vascular wall stiffness.

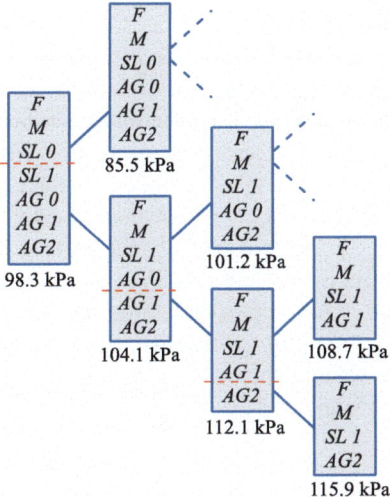

Fig. 1.17 Decision tree learning towards the prediction of the stiffness of the vascular wall. A box denotes a patient cohort with a number of attributes (*F*-female; *M*-male; *SL 0*-smoking level 0; *SL 1*-smoking level 1; *AG 0*-age group 0; *AG 1*-age group 1; *AG 2*-age group 2), and below each box the mean vessel wall stiffness within such cohort is reported. Branches develop through recursive splitting towards homogenizing the vessel wall stiffness in the (emerging) sub-cohorts. The red dashed lines indicate where the cohorts are split

Different criteria are used to *cease* the splitting and to *prune* the tree, operations that are necessary in order to avoid *overfitting* and to reduce the complexity of the tree. Given that the stiffness of the vascular wall is a continuous variable, the aforementioned example represents a *regression* decision tree. The prediction of a particular disease status would use a binary variable $p_i \in \{0, 1\}$, and a *classification* decision tree then describes the problem.

1.6   Case Study: Biomechanical Rupture Risk Assessment

This case study integrates some of the aforementioned concepts towards an Abdominal Aortic Aneurysm (AAA) rupture risk assessment model. AAA disease is a serious condition and causes many deaths, especially in men exceeding 65 years of age. Progressive treatment, and thus either surgical or endovascular AAA repair, cannot (and should not) be offered to all patients. AAA repair is recommended if AAA rupture risk is deemed to exceed the risk of intervention. Whilst the hospital-specific treatment risks are reasonably predictable, assessing the risk of *in vivo* AAA rupture in individual patients remains the bottleneck in clinical decision making.

1.6.1   Shortcomings of the Current AAA Risk Assessment

The current clinical practice for the assessment of AAA rupture risk relates to the aneurysm's largest transverse diameter, as well as its change over time. AAA repair is generally recommended if the largest diameter either exceeds 55 mm or grows faster than 10 mm per year [76, 225, 547]. The majority of clinicians follow this advice[10] and use both the aforementioned indication criteria for clinical decision making. This somewhat crude assessment of rupture risk is however the subject of much debate—AAAs with diameters smaller than 55 mm can and do rupture (even under surveillance), whilst many aneurysms larger than 55 mm remain quiescent [110, 393]. Given especially the *poor sensitivity* of the diameter-based criteria, the cost-effectiveness of AAA patient treatment is sub-optimal. The drawback of the present AAA repair indicator has also triggered considerable research in the field, and a more *individualized* AAA repair indication would be of great clinical benefit. The Biomechanical Rupture Risk Assessment (BRRA) [159, 193, 206, 353, 449, 509, 555, 567, 575] is one such concept and forms the basis of our cases study.

---

[10]See the survey amongst vascular clinicians at https://www.artecdiagnosis.com/files/ Vascops_survey2006.pdf

1.6.2 Intended Model Application

The BRRA model seeks to provide a biomechanics-based rupture risk index, which could be integrated directly into the clinical decision making process towards advising the clinician whether or not progressive aneurysm treatment is necessary. The diagnostic information should be derived from routinely taken Computed Tomography-Angiography (CT-A) images together with other patient information, such as age, gender, and medical history. The diagnostic information should also be derived directly by clinical operators, using standard computational facilities that are available in the hospital.

1.6.3 Failure Hypothesis

Raising the tension in the vessel wall to supra-physiological levels, leads to the formation of local *stress concentrations* that eventually damage the tissue at the micro-scale. Given the compromised biological integrity of aneurysm tissue [86], such *micro-defects* cannot heal, and the vessel wall continues to accumulate weak links. The micro-defects then merge together and lead to the formation of *macro-defects* that then eventually rupture the AAA wall.

A number of engineering concepts are available towards studying the initiation and propagation of failure (macro-defects) in materials [399]. The introduction of a risk index $\xi = \sigma_M / Y$ that relates the von Mises[11] stress σ_M in the wall to the wall strength Y is one such concept. The index ξ is calculated throughout the aneurysmatic aorta, with its maximum serving as a diagnostic risk index called Peak Wall Rupture Index (PWRI).

1.6.4 Work Flow and Diagnostic Information

A robust, fast, and operator-insensitive simulation pipeline is required to implement the BRRA in the clinical work flow, and Fig. 1.18a illustrates such an approach using the A4clinics software (VASCOPS GmbH, Graz, Austria). An *Image Segmentation* step acquires a 3D geometrical model, and a *Mesh Generation* step prepares the structure towards its numerical analysis [17]. The user sets patient-specific properties, and a *Structural Analysis* then computes the AAA wall stress that is required to carry the blood pressure for the individual aneurysm anatomy. A *Data Analysis* step extracts key geometrical and biomechanical parameters, information that is then *reported* together with other examination data in the *Analysis Report*. The whole analysis takes approximately 10 min using standard laptops or PCs.

The individual risk of rupture may be related to the risk of the mean population AAA patient through the *Rupture Risk Equivalent Diameter (RRED)*, see Fig. 1.18b.

---

[11]Richard von Mises, Austrian–Hungarian mathematician, 1883–1953.

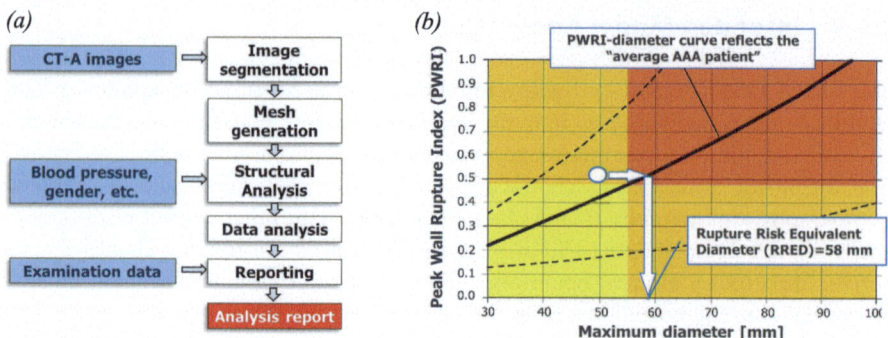

Fig. 1.18 (**a**) Workflow of the Biomechanical Rupture Risk Assessment (BRRA) of Abdominal Aortic Aneurysms (AAAs) using the A4clinics software (VASCOPS GmbH, Graz, Austria). (**b**) Definition of the Rupture Risk Equivalent Diameter (RRED) for an individual AAA patient. The RRED denotes the diameter of an average AAA that experiences the same Peak Wall Rupture Index (PWRI) as the individual case

The RRED reflects the size of the average aneurysm that experiences the same biomechanical rupture risk as the individual case [206]. It translates the individual biomechanical risk into a diameter risk, the currently applied risk stratification parameter in the clinics. The RRED connects the individual biomechanical assessment with the outcome of large diameter-based clinical trials, such as the UK small aneurysm trial [206, 547].

1.6.5   Key Modeling Assumptions

Any modeling assumption should be assessed against the IMA and the uncertainty of model input information. An efficient BRRA model should only include modeling details that improve the clinical outcome and *disregard* all the other knowledge of the biomechanical problem.

1.6.5.1 Organ-Level Model

A continuum approach at the macroscopic length-scale of centimeters, is used to model AAA biomechanics. The model encompasses the infrarenal aorta, and an accurate 3D geometrical representation has been reported to be the most critical parameter for reliable wall stress predictions [123, 449]. The model is fixed at the renal arteries and the aortic bifurcation, no contact with surrounding organs is considered, and the wall stress is computed at Mean Arterial Pressure (MAP). The discretized structural biomechanical problem model is then solved using the non-linear FEM [618].

1.6.5.2 Vascular Tissue Model

The histology of the vessel wall is very complex, giving rise to the highly non-linear, anisotropic, and time-dependent mechanical tissue properties, see Sect. 5.3. Not all of this complexity *needs* to be modeled.

AAA wall stress computations are not particularly sensitive to constitutive descriptions [123, 449], as long as the wall's low initial stiffness is followed by the stiffening observed at higher strains [436]. In addition, whilst Intra-Luminal Thrombus (ILT) tissue is highly porous [5, 196] and contains a lot of water, a single phase model that neglects fluid flow within the tissue predicts AAA wall stress at sufficient accuracy in the context of the BRRA [18, 433]. The isotropic two-parameter Yeoh model [608], calibrated to mean population data of the AAA wall [449] as well as the ILT [196], has therefore been used in the modeling of AAA tissue.

The strength of the AAA wall is highly dispersed and influenced by many factors [143, 171, 172, 352, 357, 458, 504]. Wall strength and wall thickness are strongly negatively correlated [203], an observation that favors the use of a uniform wall thickness in the BRRA model.

1.6.6   Clinical Validation

The BRRA reflects the real object up to a limited degree of completeness. Whether or not the model addresses the most important aspects of the real world problem needs to be *validated*—the model is to be tested against its IMA. As a minimum requirement, the BRRA diagnosis has to demonstrate an *improvement* over state-of-the-art clinical practice. The implementation into the regular clinical work flow would then be justified. Validation addresses always different aspects; some of such exercises are discussed below.

Operator Variability Intra-operator and inter-operator variability of the A4clinics rupture risk assessment system has been tested in clinical environments [279, 545]. Model assessments showed an intra-operator variability of 2.7% [497] for PWRI predictions and of 1.5% for maximum diameter measurements. This high precision could only be achieved with active (deformable) image segmentation models [17, 606].

Retrospective Comparison Between Ruptured and Non-ruptured AAAs The diagnostic value of the BRRA method has been studied for almost 20 years [159, 193, 206, 353, 435, 449, 555, 567, 575]. Given these studies, Peak Wall Stress (PWS) has been regularly shown to be higher in ruptured/symptomatic AAAs than in intact/non-symptomatic AAAs [305, 508]. The integration of wall strength, based on a statistical model that correlates mechanical *in vitro* tests with patient characteristics [570], then further improved the discrimination between ruptured and non-ruptured cases [193, 353]. Figure 1.19 summarizes these results and shows

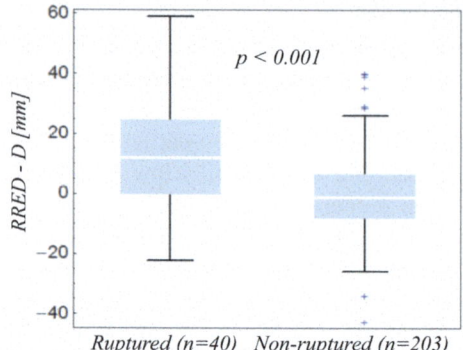

Ruptured (n=40)   Non-ruptured (n=203)

Fig. 1.19 Size-adjusted comparison between ruptured and non-ruptured Abdominal Aortic Aneurysms (AAAs). The comparison is based on the difference between Rupture Risk Equivalent Diameter (RRED) and the maximum transversal diameter (D). The number of patients and the one-tailed p-value are denoted by n and p, respectively. Image has been adjusted from [206]

that the RRED was, on average, 14.0 mm larger in patients who experienced AAA ruptured than in non-ruptured cases.

Quasi-prospective Comparison Between Ruptured and Non-ruptured AAAs CT-A acquisitions from patients before the rupture of their AAA provide ideal information to validate the BRRA. Given such data, the BRRA was able to discriminate between AAAs that ruptured as compared to a baseline-matched control group of stable cases [144]. In more than half of the cases, the rupture sites correlated even with the location, where PWRI has been predicted. The authors therefore concluded that asymptomatic AAA patients with high PWRI and RRED values have an increased rupture risk.

Female Versus Male AAA Rupture Risk Whilst AAA prevalence in females is several times lower than in males, female aneurysms rupture at *smaller* diameters [120, 247, 328]. *In vitro* AAA wall characterization also shows a lower strength of the female AAA wall [566, 570], a wall weakening effect that is considered by the BRRA—the computational biomechanical risk of a, on average, 53 mm large female AAA relates to a, on average, 66.2 mm male AAA [206].

Correlation of PWRI and FDG-Uptake 18-Fluoro-Deoxy-Glucose (18F-FDG), a tracer for Positron Emission Tomography (PET) imaging [485], allows for the indirect evaluation of the AAA wall's biological activity through its energy consumption. Vascular cells respond to the mechanical load [11], and although the aneurysmatic wall loses its biological vitality [86, 373], PET-scan images still showed a reasonable correlation between wall stress and 18F-FDG-uptake [352, 391].

Correlation of PWRI and Wall Histopathology The AAA's complex geometry and morphology leads to highly inhomogeneous wall stress. AAA wall segments that experienced high wall stress showed fewer Smooth Muscle Cells (SMCs), fewer elastic fibers, more soft and hard plaques, as well as a trend towards heightened fibrosis [143].

1.7 Summary and Conclusion

Biomechanical modeling is a key component in the exploration of life science problems. A model represents the real object or process up to the degree of complexity as it is determined by the IMA [491]. The IMA guides model development and testing, and a modeler should always try to keep a model as simple as possible. No model is complete, and one can criticize any model. *Is the model correct?* is a meaningless question, instead one should ask *Does the model fulfill its intended task?* Modeling does not seek to replicate reality, which would be impossible. The quote, *every model is wrong, but only a few are useful*[12], is a fitting statement to describe the nature of a model.

The model's *complexity* is usually increased in an effort to improve its accuracy, and thus towards decreasing the *systematic model error*, see Fig. 1.20. However, a complex model is not necessary more accurate than a simple one—frequently the opposite is true; a *(simple)* linear model will always propagate input uncertainties linearly to the output, whilst a *(complex)* non-linear model could amplify such input errors, potentially leading to vastly incorrect predictions.

The model's input information will always be uncertain, especially when data is required to be collected in a clinical environment. Uncertainties propagate, and thus more input information increases the *random model error*, see Fig. 1.20. The optimal model complexity is a trade-off between a model's systematic and random errors, and practically it is the quality of the input information that limits the level of model complexity. Measures to improve the quality of the input data, such as

Fig. 1.20 Model error as a function of the model's complexity. The systematic model error decreases with model complexity, whilst its uncertainty and thus random errors increase with model complexity. The optimal model complexity is a compromise between both errors

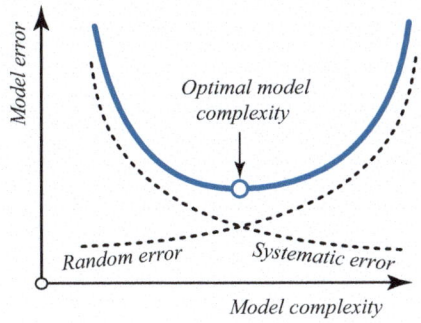

---

[12]George E. P. Box, British statistician, 1919–2013.

standardized data acquisition protocols, are often more successful, when compared to an increased model complexity.

Up till now, vascular biomechanical modeling has not gained much clinical acceptance, with very few vascular biomechanics models proposed in the literature having been at any assistance to clinicians. Even following decades of vascular biomechanical research a chasm between engineering and clinical approaches remains. To maneuver biomechanical simulation technology towards their clinical application, models require *rigorous validation* with respect to the IMA. Validation studies must be *well-designed* to avoid *confounding factors*. Without sound validation, the assistance a biomechanical model is able to provide is limited. Bioengineering work has to shift from model development towards model validation.

Many biomechanical models are *overloaded* with mechanical complexities and fail to solve or explain a bioengineering problem. In bioengineering, *a model can only be a tool to explore a certain biological phenomenon*[13], and bioengineering work should not be limited to the exercise of designing such models.

Given the complexity of the vascular system, black-box modeling approaches have gained significant clinical interest. Whilst such approaches are straightforward to implement, a *statistical correlation must not be confused with causation*, and black-box models can only provide limited biological insights. White-box modeling approaches account for the physics of the vascular system. They are often more challenging but allow for a much deeper understanding of physiological and pathological vascular mechanisms—the development of such models will cover the forthcoming chapters.

---

[13]Rik Huiskes, Professor of Biomedical Engineering, 1944–2010.

The Circulatory System

2

This chapter analyzes the macrocirculation, the microcirculation, and the lymphatic system of the vascular apparatus. We discuss the structure of the vessel wall, the main principles of hemodynamic control, and the mechanisms of vascular exchange. We look at the circulation from a system's perspective and introduce mechanical properties, such as pressure, capacity, flow, and vascular bed resistance. In addition, we explore the structure of the capillary wall toward the description of transcapillary transport mechanisms in microcirculation. The final part of this chapter introduces a number of lumped parameter models in the description of the macro- and microcirculation. In addition to WindKessel (WK) models, two and three-element models are derived toward the representation of vessel networks. Such models aim at capturing the steady-state, steady-periodic, and transient description of vasculature domains. With respect to microcirculation, hydrostatic and osmotic effects are examined, and vascular exchange is described by linear and nonlinear filtration models. Conclusions regarding the advantages and limitations of the discussed lumped-parameter models and future perspectives summarize this chapter.

2.1   Introduction

The evolution of species led to a more and more organized circulatory system. Simple diffusion of extracellular liquid evolved towards a highly *organized circulatory system* in mammals. This was made possible by the heart's pumping ability and regulated by peripheral resistances, which together generated the arterial blood

The original version of this chapter was revised: ESM has been added. The correction to this chapter is available at https://doi.org/10.1007/978-3-030-70966-2_8

Supplementary Information The online version contains supplementary material available at https://doi.org/10.1007/978-3-030-70966-2_2

pressure and *flow*. Hemodynamics is therefore a fundamental organizing principle selected for diversification and adaptation of life.

The circulatory system is set up by two separate systems: the *cardiovascular system*, which distributes blood, and the *lymphatic system*, which collects lymph and returns it into the cardiovascular system, see Fig. 2.1. The cardiovascular system supplies blood to the body's organs and quickly adjusts to sudden changes in demand for oxygen, nutrients, and other factors in response to the organism's activity. The lymphatic system is open and essentially recycles blood plasma after it has been filtered from the *interstitial fluid*, the fluid situated between cells. Both systems cooperate in immune response.

The physiology and pathophysiology of the cardiovascular system have been extensively studied and excellent texts are available [394, 548]. The present chapter aims at introducing the topic to bioengineers, and the reader should then be able to understand and model key properties of the vascular system.

2.1.1   Vascular System

The vascular (or cardiovascular) system has three main functions:

- *Supply.* Distribution and exchange of oxygen, nutrients, and other substances
- *Cleaning.* Removal of waste products
- *Immune response.* Delivery of leucocytes to organs in response to pathogens, anything that can produce disease

In vertebras the cardiovascular system is closed and formed by the *systemic* and the *pulmonary* circuits, see Fig. 2.1. The systemic circulation transports oxygenated blood away from the left ventricle through the aorta to the rest of the body and then returns oxygen-depleted (deoxygenated) blood back to the right ventricle. Note that oxygen-depleted blood still contains approximately 75% of oxygen of oxygenated blood. The pulmonary circulation transports this blood through the lungs, where it is oxygenated. It then returns into the left ventricle and enters again the systemic circuit. Absolute values for oxygen consumption depend on body size, and young healthy humans at rest consume somewhere between 0.15 and 0.4 l of oxygen per minute, a demand that can increase by 10 to 15 folds during exercise [288].

The essential components of the cardiovascular system are the heart, blood, and blood vessels. An average adult contains roughly 5.5 l of blood, accounting for approximately 7% of its total body weight. At rest, approximately 4 l min^{-1} or 80% of cardiac output is directed to the brain, heart, kidneys, and liver. Despite the cardiovascular system is closed, oxygen, nutrients, and macromolecules move across the wall of small blood vessels and enter the interstitial fluid on their way to the target cells. In return, carbon dioxide and wastes pass from the interstitial fluid directly back into small blood vessels, or through the lymphatic system back into the cardiovascular system. The transport of substances in and out of the vascular system is collectively called *exchange*.

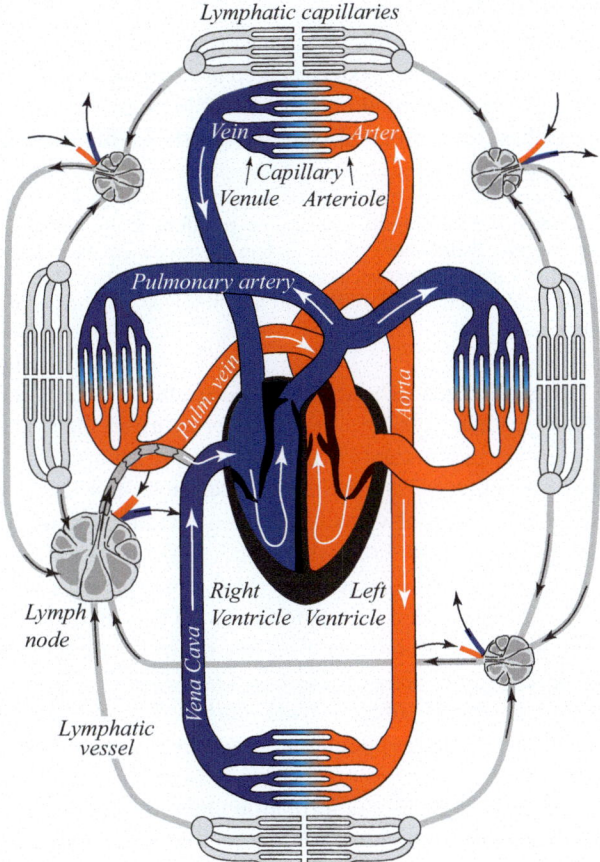

Fig. 2.1 The circulatory system comprises the cardiovascular system (systemic and pulmonary circuits) and the lymphatic system. *Systemic circuit:* The left ventricle pumps oxygenated blood through the aorta into all organs but the lungs. In the microcirculation blood is deoxygenated, and then the collected blood flows through the vena cava and returns into the right ventricle. *Pulmonary circuit:* The right ventricle pumps oxygen-depleted blood through the pulmonary artery into the lungs. In the microcirculation of the lungs, blood is oxygenated and returns through the pulmonary vein into the left ventricle. *Lymphatic system:* A one-way low-resistance network of drainage channels (lymphatic capillaries) returns lymph from the interstitial to lymph nodes and then, through the lymphatic venous anastomosis, into the venous system

The vascular system can also be seen to function in two parts: a *macrocirculation* and a *microcirculation*.

2.1.2 Key Concepts

Although the cardiovascular system shows large anatomical variability across species, key concepts are preserved. Some of these concepts are discussed in the following.

2.1.2.1 Form Follows Function

Given normal conditions, the cardiovascular system *continuously adapts* towards optimal system performance. The system is then at *homeostasis* and remains stable in time. Homeostasis is an essential component of biological evolution [552], and Cannon[1] [66] describes this biological system property as the following:

> The highly developed living being is an open system having many relations to its surroundings. Changes in the surroundings excite reactions in this system, or affect it directly, so that internal disturbances of the system are produced. Such disturbances are normally kept within narrow limits, because automatic adjustments within the system are brought into action and thereby wide oscillations are prevented and the internal conditions are held fairly constant.

Mechanical stress also excites vascular tissue reactions and explains a number of properties of the vascular system. The blood pressure in the systemic circuit is *much higher* than in the pulmonary circuit, and the left ventricle is therefore more muscular and has a thicker wall than the right ventricle. For the same reason the wall of arteries is thicker than of veins. The blood pressure determines the tension in the vessel wall, and the circumferential [91,363,598] and axial [215] *tensile force* in the vessel wall correlate with its thickness. Aside from pressure-related adaption, the diameter of blood vessels also adjusts to the blood flow in the vessel. The blood flows over the endothelium and induces *Wall Shear Stress (WSS)*, a quantity that is kept constant by adjusting the vessel's diameter [74,236].

The cardiovascular system uses a wide range of actions towards reaching homeostatic targets. We may group them into four classes of mechanisms:

- *Passive response.* Purely passive deformation under the action of forces
- *Vasoreactivity.* Vasoconstriction or vasodilation due to the action or relaxation of contractile cells in the vessel wall
- *Arteriogenesis.* Increase or decrease of the vessel's diameter and wall thickness in response to the turnover of tissue constituents
- *Angiogenesis.* Formation of new vessels sprouting out from pre-existing vessels

These adaptation mechanisms are linked to characteristic *time scales* and allow the system to adjust quickly (passive response) or very slowly (angiogenesis) towards meeting system needs. Whilst the aforementioned mechanisms are local, they have distinct system-level, and thus global implications. The homogenization of WSS would be one such example. It requires the total cross-sectional area of the vasculature to increase from the aortic cross-section of 3 to $5\,cm^2$ to the total cross-section of the capillary bed of 4500 to $6000\,cm^2$. Such a configuration of the vascular tree is energy-efficient and keeps the blood pressure relatively low in the systemic circuit.

---

[1] Walter Bradford Cannon, American physiologist, 1871–1945.

2.1.2.2 Blood Flows in Closed Loops

The vascular system is closed and local alterations have global implications—blood flowing through one organ affects the flow through another organ. Likewise, the pulmonary circulation influences the systemic circulation and *vice versa*, and the venous flow influences the arterial flow and *vice versa*.

The vascular system may also be divided into sub-loops with dedicated organ-supply function, which is most clearly observed in the kidney, the heart, and the brain. The pressure and flow within such sub-loops are separately controlled, a mechanism known as *autoregulation*, and allows for the (partly) independent operation of organs. The formation of sub-loops also explains the anatomical organization of the vascular tree with arteries and (deep) veins often running in parallel, and in close proximity to each other.

2.1.2.3 Vascular Network Is "Space Filling"

Only a very few tissues in mammals, such as the ligaments, the valve leaflets, and the cornea, are *avascular* and do not contain blood vessels. They are entirely perfused by *diffusion*. All other tissues contain vessels, with the vascular network being "space filling". Given limited resolutions, most of the vasculature remains invisible to *image modalities*, such as Computed Tomography (CT), Magnetic Resonance (MR), and ultrasound.

2.1.3   Cells in the Vascular System

The vascular system performs many very different tasks, such as transport and exchange, immune response, regulation of pressure and flow, Extra Cellular Matrix (ECM) maintenance, and the control of blood clotting and wound healing. These functions are carried out by *cells* together with their delicate interactions with ECM and blood plasma. Cells of the same type often perform multiple tasks to achieve proper system function.

2.1.3.1 Endothelium Cell

Endothelium Cells (ECs) are joined together to form a single-cell layer (monolayer) called *endothelium* that provides a clear separation between the blood and the vessel wall. ECs are flat and 0.2 to 2 μm thick. Given exposure to laminar blood flow, they align with the flow within 12 to 14 days [162] and adapt towards an elongated shape. The length of ECs ranges then from 1 to 20 μm. In humans, ECs make up approximately 1.0 kg and cover a surface of approximately 7000 m^2. ECs in arteries, capillaries, veins, and lymphatics are exposed to markedly different hemodynamic environments and must perform distant functions.

ECs sense WSS, in response to which they secrete vasoactive agents that control the tonus of adjacent contractile vascular cells, such as SMCs and pericytes. ECs also play a crucial role in response to inflammation. They express adhesion molecules towards capturing circulating leukocytes and promoting their transport into the tissue. ECs are also involved in immune response and tissue remodeling.

In their vicinity, ECs prevent blood from clotting by secreting vasoactive agents. In capillaries, ECs form a semipermeable membrane to allow oxygen, nutrients, and other factors to move into peripheral tissues whilst retaining blood cells and plasma in the circulation.

2.1.3.2 Smooth Muscle Cell

Smooth Muscle Cells (SMCs) can present either at the *contractile phenotype* or the *synthetic phenotype*. At the contractile phenotype, SMC serves as a contractile cell of arteries, arteriole, and veins. They appear at a spindle-shaped configuration, measuring approximately 2 to 5 μm in diameter and 100 to 500 μm in length. At the synthetic phenotype, the SMC synthesizes ECM proteins and has a more cobblestone-type shape. The cell appears then less elongated than at the contractile phenotype. In the vessel wall, SMCs are mainly aligned in circumferential vessel direction and communicate with each other though *tight junctions* and *gap junctions*.

2.1.3.3 Pericyte

Pericytes are the *contractile* cells of capillaries and venules. They regulate capillary blood flow and, together with ECs, the permeability of the vessel wall. Communication between pericytes and ECs is facilitated by *integrins*. Pericytes appear at an elongated shape of approximately 5 to 10 μm in length.

2.1.3.4 FibroBlast

FibroBlast (FB) *synthesizes ECM proteins*, out of which collagen is the most important one. FBs are 10 to 15 μm large and have a branched cytoplasm surrounding an elliptical nucleus. Active FBs have abundant rough endoplasmic reticulum, whereas inactive FBs, also denoted fibrocyte, appear more spindle-shaped. The active FB is attached to collagen fibers and puts them under tension—it pulls on collagen fibers. Given crowded FBs, they are often locally aligned in parallel clusters.

2.1.3.5 Erythrocyte

Erythrocytes (or red blood cells) are the most common type of blood cells, constituting almost half of the volume of blood. Their principal aim is to *deliver oxygen*. Erythrocytes have no nucleus, and they are highly deformable bi-concave-shaped discs, measuring approximately 6 to 8 μm in diameter and 2 to 4 μm in height.

2.1.3.6 Leukocyte

Leukocytes (or white blood cells) are cells of the *immune system* that are involved in protecting the body against infectious disease and foreign invaders. Leukocytes present in very different types, such as plasma cells, lymphocytes, and macrophages.

Plasma cells secrete large volumes of antibodies. They produce *antibody molecules* that bind to foreign substance (target antigen) and initiate its neutralization or destruction. They are 12 to 15 μm large, ovoid-shaped, and transported in blood plasma as well as in lymph. *Lymphocytes* are the main type of cell found in lymph and include natural killer cells, T cells, and B cells. They are approximately 6 to 30 μm large. *Macrophages* engulf and digest pathogen, participate in the initiation and resolution of inflammation, and in the maintenance of tissues. They are approximately 20 μm in diameter and show very different shapes, adapted to the functions to be carried out.

2.1.3.7 Platelets

Platelets (or thrombocytes) are tiny blood cells, which together with coagulation factors, stop bleeding by *forming a blood clot*. They have no nucleus, and non-activated platelets are approximately 2 μm large and of compact shape. On activation, platelets turn into octopus-like shapes, with multiple arms and legs. Minutes after activation, platelets start aggregating with each other and/or depositing on surfaces that are not covered by ECs.

2.1.3.8 Dendritic Cell

Dendritic cells process antigen material and present it on the cell surface to the T cells of the immune system. Dendritic cells are 10 to 15 μm large and have a very large surface-to-volume ratio.

2.1.4 Macrocirculation

The macrocirculation *transports* blood through the cardiovascular system. The *systemic circuit* carries it through all organs, but the lungs, and the *pulmonary circuit* through the lungs.

The first part of the systemic circulation is the *aorta*, a massive and thick-walled artery that origins at the aortic valve. It then arches and gives branches supplying the upper part of the body. After passing through the aortic opening of the diaphragm, it enters the abdomen and supplies branches to abdomen, pelvis, perineum, and the lower limbs. The renal circulation by itself is supplied with approximately 20% of the cardiac output.

Along the vascular tree, the arterial lumen continuously decreases until blood flows though arterioles and passes capillaries, where the *exchange* of oxygen, nutrients, and other substances takes place. Capillaries are often organized in a *capillary bed*, an interweaving network of capillaries supplying tissues and organs. The blood is then collected by venules, before veins return it to the heart, see Fig. 2.1. The properties of the different types of blood vessels are adapted to their function:

- An *artery* carries blood away from the heart. The luminal diameter ranges up to centimeters and the thick wall is designed to cope with high blood pressure.
- An *arteriole* connects arteries to capillaries. The lumen is approximately 10 to 100 μm in diameter, and vasoreactivity (vasoconstriction or vasodilation) allows it to control the blood flow into the capillaries.
- A *capillary* has a luminal diameter of approximately 5 to 8 μm, just wide enough to allow erythrocytes passing/squeezing through. The wall is permeable and

allows the capillary to supply tissues with factors, such as oxygen and nutrients, and to remove waste products in return.

- A *venule* connects capillaries to veins and has a luminal diameter of approximately 10 to 200 μm. The wall is thinner than of arterioles and equipped with a highly permeable endothelium layer.
- A *vein* carries blood to the heart. The wall shows a thin media and a thick adventitia. Veins have a diameter that ranges up to centimeters and they are often larger than arteries.

The blood flow velocity changes by *four* orders of magnitudes along the arterial tree. In large arteries the blood shows phases of *forward and backward* flow at velocities of tens of centimeters per second, whilst the blood flow in capillaries is *unidirectional* and of tens of micrometers per second. In veins the blood flow is generally more uniform than in arteries, a condition partly supported by *valves*. Given limb veins, valves counteract gravitation and prevent from back flow. Blood flow velocities in the systemic and pulmonary circuits are similar.

In addition to the transport function of the macrocirculation, the *elasticity* of the large blood vessel is of fundamental importance to the proper physiological function of the cardiovascular system. Especially the aorta contributes almost the entire *volume compliance* to the cardiovascular system. Veins are much more distensible than arteries, which allows them to serve as *venous compartment*. Approximately 60% of the blood is stored in the veins, a compartment controlled by the autonomous nervous system.

2.1.4.1 Blood Vessel Structure and Function

Blood vessels are *distensible*, a key feature to lower the pulse pressure and to support continuous flow into the distal tissue. The distensibility is a consequence of the wall's elasticity. It is determined by the ECM through the delicate interaction of structures, such as elastin, collagen, ProteoGlycans (PGs), fibronectin, and fibrillin. Whilst the ECM determines the vessel wall's structural integrity, cells maintain its vasoreactivity, metabolism, and immune response. Vascular cells are also able to alter the elasticity of the vessel wall. Contractile cells can augment vessel wall properties within seconds, whilst the effect from newly synthesized ECM appears at a delay of weeks.

The wall of arteries, arterioles, veins, and venules is built up by *three* distinct vessel wall layers: intima, media, and adventitia, see Fig. 2.2. In contrast to larger vessels, the glycocalyx, the endothelium, and a basal membrane form the single-layered wall of capillaries. The structure of veins and arteries is very similar. Given the low pressure in the venous system, veins have a thinner wall than arteries. They may also be equipped with passive valves to prevent the back flow of blood.

2.1.4.2 The Intima and the Endothelium

The interaction of the glycocalyx, the endothelium (a monolayer of EC), and a subendothelial layer forms the *intima*. The glycocalyx binds different anti-inflammatory and anti-coagulant factors, and its disruption results in thrombin

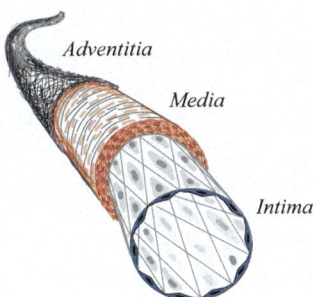

Fig. 2.2 The layered structure of large blood vessels. The adventitia is a collagen-rich fibrous layer that shields the inner layers from excessive mechanical forces and anchors the vessel to its surrounding. The media is a Smooth Muscle Cell (SMC)-rich layer that determines the physiological properties of large elastic vessels. The media is designed to withstand mechanical load acting along the circumference. The intima is dominated by a single layer of Endothelial Cells (ECs) that forms a lining between blood and tissue. The intima has important barrier functions

generation and platelet adhesion. The glycocalyx contributes also to EC mechanotransduction, and thus to the transduction of biomechanical forces into the biomolecular response of EC. Loss of glycocalyx apparently contributes to impaired sensing and transduction of WSS.

The endothelium provides an anti-thrombogenic and low-resistance lining between the blood and the vessel wall tissue. It responds to WSS and produces a host of chemical substances, such as Nitric Oxide (NO), endothelin, prostacyclin (PGI_2), and angiotensinogen, all are designed to maintain vascular homeostasis [394]. They modify the ability of platelets to adhere to the vascular wall and to aggregate with the formation of a blood clot. The endothelium is also a selective barrier for substances such as oxygen, nutrients, leukocytes, lipoproteins and influences factors, such as vessel wall permeability, tonus of contractile cells, inflammation, blood clotting, and tissue remodeling.

2.1.4.3 The Media

The *media* contains 30 to 60% vascular SMCs that are embedded in the ECM. The media's ECM itself contains 5 to 25% elastin, 15 to 40% collagen, and 15 to 25% other connective tissue. The media is formed by *Medial Lamellar Units (MLU)*, a structure that is clearly visible in elastic arteries but disappears towards muscular arteries. Elastic arteries are rich on elastin and found at the beginning of the vascular tree, whilst muscular arteries contain a large amount of vascular SMC and appear downwards the vascular tree. The media is the dominating layer in large elastic arteries and of utmost physiological relevance to the proper function of the cardiovascular system. The media is designed to cope with stress primarily along its circumferential direction.

SMC in the media are specialized in tonic contraction (contractile phenotype) but also in the production of ECM constituents (synthetic phenotype). SMC tonus is controlled by the autonomous nervous system and regulates blood flow through vasoreactivity in response to the body's activity. The cells in the inner media (mainly SMCs) are entirely fed through *transmural flow*, fluid flow that establishes from the pressure differences in the lumen and the interstitial space. SMCs also participate in inflammatory reactions by modulating vascular tone, but they have a limited capacity in direct immune response.

2.1.4.4 The Adventitia

The *adventitia* is an ECM-rich layer with collagen fibers covering approximately 60 to 80% of its volume. Another 10 to 25% is occupied by other connective tissue components. In addition to numerous macrophages providing immune response, FB is the primary cell type found in the adventitia. It maintains the ECM and covers approximately 10% of the adventitia's volume. Tiny blood vessels, the *vasa vasorum*, perfuse the adventitia together with the outer media and deliver cells, such as leucocytes for immune response. The adventitia anchors blood vessels to surrounding tissues, and its dense mesh of collagen protects the biologically vital medial and intimal layers from overextension. The adventitia is penetrated by nerves that control the SMCs in the media, and it is often thinner in veins than arteries.

2.1.4.5 Wall Shear Stress (WSS)

NO has an important signaling function in the vessel wall. It is produced from L-arginine by activity of endothelial nitric oxide synthase (eNOS), an enzyme that is continually released from healthy EC. After the diffusion of NO into the media, it relaxes SMCs and maintains vascular patency and distensibility. Stimuli for the release of NO from EC include WSS, exerted directly on the EC membrane or on the endothelial layer [398], see Fig. 2.3. The expression of NO is also influenced by the blood flow conditions, and thus the temporal occurrence of the WSS, see Fig. 2.3b. Periodic flow stimulates greatly NO expression, whilst turbulent flow shows similar expression to static conditions.

Aside from regulating the arterial diameter through the production of vasoactive mediators, WSS is also an important determinant of endothelial gene expression [394]. WSS regulates factors, such as transcription factors, growth factors, adhesion molecules, and enzymes. Endothelium function is optimal during youth and the absence of cardiovascular disease. With age endothelial function progressively deteriorates, which is then associated with the reduction of the bio-availability of NO and anatomical changes, such as thickening of the endothelia layer. The most obvious disfunction of the endothelium is seen with age-related diseases, such as atherosclerosis [394]. Endothelium function is defective, not only in patients with developed atherosclerosis, but already in persons with risk factors for atherosclerosis [557].

Fig. 2.3 Expression of Nitric Oxide (NO) in relation to the Wall Shear Stress (WSS) that is applied to the membrane of Endothelial Cells (ECs) (**a**) or at the endothelial layer (**b**). NO expression is illustrated through the formation of [^{3}H] L-citrulline, a by-product of NO expression

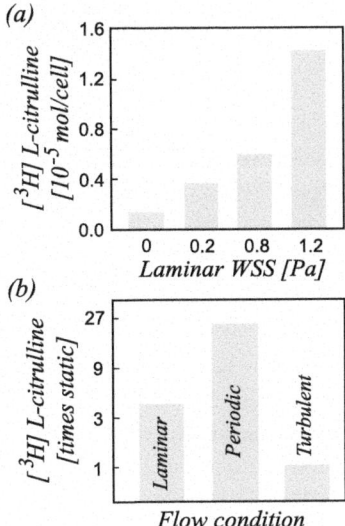

2.1.5 Lymphatic System

The lymphatic system constitutes a *one-way* low-resistance network of drainage channels that operates in conjunction with the cardiovascular system and returns lymph from the interstitial to the venous system, see Fig. 2.1. It plays a major role in helping the *immune system* to defend the body against diseases and serves as a "highway" for fast and efficient delivery of immune cells, as well as free antigens [63]. In a healthy human, lymph flow of approximately eight liters per day is expected, from which approximately half can be absorbed by lymph node microvessels, leaving four liters per day post-nodal lymph flow left.

The lymphatic system is composed of lymphatic vessels and lymphoid organs, such as the bone marrow, thymus, lymph nodes, spleen, Peyer's[2] patches, tonsils, and the appendix. Lymph flow is unidirectional and establishes through the rhythmic contraction of lymphatic contractile cells. Skeleton muscle contractions and arterial pulsations support the synchronized opening and closing of intra-luminal lymphatic valves—the *lymph propulsion*. In addition to nerves and chemicals, mechanical factors, such as the streamwise pressure gradient, transmural pressure, preload and afterload influence lymph propulsion.

Lymphatic vessels are *absorptive* vessels and found in almost all organs— recently they have also been identified in the brain [3]. Lymph nodes are located at the intersections of collecting lymphatics. Lymph nodes filter the interstitial flow and break down bacteria, viruses, and waste. Under normal conditions, interstitial fluid pressure is below atmospheric pressure whilst fluid in lymphatic

---

[2]Johann Conrad Peyer, Swiss anatomist, 1653–1712.

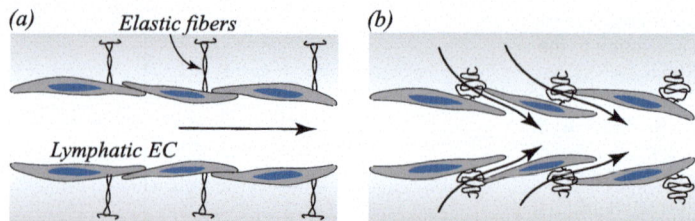

Fig. 2.4 Functioning of primary valves of lymphatics capillaries. Vessel configurations at low (**a**) and high (**b**) interstitial pressures. Arrows indicate fluid flow

capillaries is slightly above atmospheric pressure. Intra-lymphatic pressure slowly *increases* along the drainage route, and thus towards the larger collecting vessels and ultimately the thoracic duct or right lymphatic trunk, where lymph is returned to the blood circulation. Lymph propulsion establishes flow therefore against a positive streamwise pressure gradient.

The biomechanics of the lymphatic system are not yet very well explored, and further details, including its modeling, are reported elsewhere [285, 355, 454].

2.1.5.1 Lymphatic Vessels

Lymphatics *capillaries* are approximately 10 to 60 μm in diameter. A single layer of partially overlapping lymphatic EC forms their approximately 50 to 100 nm thick wall, see Fig. 2.4. Whilst it has neither a basal lamina nor contractile cells, it is equipped with active valves, so-called *primary valves*, to collect interstitial fluid. The lymphatic EC are oak-leaf-shaped and joined together by "button-like" junctions to form such primary valves. The lymphatic ECs are also anchored to the surrounding ECM through elastic fibers that function as mechanosensors. The fibers detect increased tissue pressure and open the primary valves to allow interstitial fluid to enter, see Fig. 2.4. Once the surrounding tissue swells, the primary valves open and fluid, macromolecules, and immune cells enter the lymphatic capillary.

The pre-collecting lymphatics connect the capillaries to the collecting lymphatics. Collecting lymphatics have diameters of 1 to 2 mm, contain intra-luminal valves, and their ECs are highly interconnected. Similar to blood vessels, the wall of collecting lymphatics is formed by three distinct wall layers: *intima, media*, and *adventitia*.

As with blood vessels, lymphatics adjust in response to mechanical and biochemical stimuli [189]. NO [190], histamine [396], and endothelin [484] are known to influence lymphatic contractility. Lymph flow is complex and corresponds to conditions of no flow, slow flow, and retrograde flow. The endothelium is therefore exposed to a wide range of WSS, and endothelial-derived NO is expected to have an important role in the orchestrated propulsion of the lymphatic system [127, 128].

2.1.6 Microcirculation

Organs are perfused by *feed arteries* that branch off a major conduit artery. Four to six branch orders are then counted before the terminal arterioles give rise to capillaries, where the *exchange* with the interstitial tissue appears.

Arterioles, capillaries, and venules regulate vascular *pressure* and *divert* blood flow to meet local metabolic needs. Their diameters are controlled by the tonus of pericytes, which in turn determine the pressure drop along these vessels. It is the arterioles that play a major role in the distribution of blood towards the most metabolically stressed areas. In contracting skeletal muscles, for example, marked dilation is seen in the smallest arterioles [562], whilst approximately 80% of capillaries are perfused at rest.

Delicate alterations in the capillary "forces" and vessel properties determine fluid exchange characteristics and allow for moment-to-moment regulation of *transcap-illary fluid flow*, a mechanism knows as *filtration*. Whilst diffusion determines the transport of the small molecules, filtration controls the advection of large solutes. The permeability of capillaries to water and solutes is often regarded constant, but this property is known to change at least in response to volume regulatory hormones and WSS that is sensed by ECs [307, 347].

Aside from exchange, the microvasculature restores blood pressure towards normal levels and serves as an autotransfusion compartment—at vascular volume overload fluid it automatically removed from the bloodstream and *vice versa*.

2.1.6.1 Exchange
Any factor in the blood that will be delivered from the capillary into the tissue has to *pass* the vascular wall, formed by the glycocalyx, endothelium, and basal membrane. This structure is adapted to support molecular exchange. Transport of liquid and solutes between the intravascular space and the interstitium, and thus across the semipermeable vascular endothelial barrier, is accomplished by:

* *Diffusion.* Transport of a substance down its concentration gradient. It is the primary mechanism for oxygen and lipids, and partial mechanism for proteins.
* *Advection.* Transport of a substance together with bulk motion of water and determined by gradients of hydrostatic and osmotic pressures. It is the primary mechanism for water and ions, and the partial transport mechanism for proteins.
* *Transcytosis.* Macromolecules, such as proteins, are captured in vesicles on one side of the EC, drawn across the cell, and ejected on the other side.

Under normal conditions, the balance of "forces" acting across the walls of exchange vessels favors the net flux of fluid *from the bloodstream to the interstitium*, a process commonly referred to as *capillary filtration*. Based on Starling's[3] equation (see Sect. 2.4.2), it has been estimated that in a healthy human, approximately

---

[3]Ernest Henry Starling, British physiologist, 1866–1927.

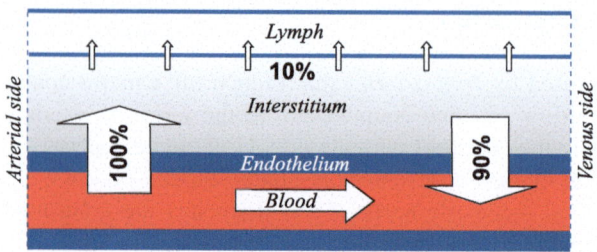

Fig. 2.5 Historical understanding of microcirculatory fluid fluxes between vascular (red), interstitial (gray), and lymphatic (white) spaces. More recent data suggests much less reabsorption back into the vascular space in some tissues, and much more fluid is then transferred into the lymphatic system

90% of the fluid that moves from the vascular capillaries into the interstitium will be *reabsorbed* back into the vascular system, whereas 10% will move into the lymphatic system, see Fig. 2.5. This quantification has been proposed by Starling in 1896 [524] and provided useful insights in vascular exchange. However, it also failed to explain some experimental observations [338], and additional experimental data [371] indicated much less reabsorbtion back into the vascular capillary. We may therefore conclude that the drainage of capillary filtrate by the lymphatic system is another dominating factor in interstitial volume homeostasis [338].

2.1.6.2 Colloid Osmotic Pressure and the Role of Albumin

Osmosis is the spontaneous flow of a solvent across a semipermeable membrane *towards* a more concentrated solution, see Appendix E.2. Osmotic pressure is the pressure that must be applied to the side of the more concentrated solution to stop such a flow. In the vasculature, the solvent is water, and solutes are typically macromolecules. The osmotic pressure usually tends to pull water *into* the vascular system, and as such opposes the hydrostatic pressure pushing water through the capillary wall *out* of the vascular system.

The *Colloid Osmotic Pressure (COP)*, or oncotic pressure, is the osmotic pressure exerted by proteins and largely determined by the concentration of *albumin*. The total COP of an average capillary is approximately 28 mmHg with albumin contributing approximately 22 mmHg. Albumin is produced in the liver. It is the most abundant blood plasma protein and constitutes approximately 50% of human plasma proteins. It is essential for maintaining COP, and as such responsible for proper distribution of body fluids between blood vessels and body tissues. With approximately 10 nm in diameter, albumin is smaller than most other proteins, which allows it to pass the capillary wall relatively easy. Therefore, approximately 50 to 60% of albumin content resides in the interstitium at an average concentration of approximately 15 g l^{-1}. In adipose tissue concentrations of 4.3 to 10.7 g l^{-1}, and in skeletal muscle 9.7 to 15.7 g l^{-1} have been reported [138]. In addition, glycosaminoglycans (GAGs) and collagen exclude albumin from up to 50% of interstitial space, such that local albumin concentration approaches 20 to 30 g l^{-1}

in the interstitium. This is approximately half of 35 to 50 g l^{-1}, the albumin concentration seen in the capillary lumen [40].

Besides controlling COP, albumin transports a wide variety of substances, such as fatty acids, calcium, phospholipids, bilirubin, enzymes, hormones, drugs, metabolites, and ions. Like other proteins in the interstitial spaces, albumin returns to the circulation via lymph.

2.1.6.3 Functional Adaptation of Capillaries

Regardless of capillaries being the smallest vessels, they have the *highest cumulative surface area* available for exchange. Exchange of oxygen occurs primarily from erythrocyte "packets" as they pass through the vessel, whilst CO, fluids, and molecules up to the size of the plasma proteins are exchanged directly between plasma and the interstitial space. Given the different exchange functions, capillaries may be classified as *continuous, fenestrated*, and *discontinuous*, see Fig. 2.6.

Continuous capillaries have a low hydraulic conductivity and feature strong barriers between blood and tissue. The tightest continuous capillaries form barriers known as the Blood–Brain Barrier (BBB), the Blood–Aqueous Barrier (BAB), the Blood–Nerve Barrier (BNB), and the blood–testes barrier. Such barriers involves the formation of specialized adherens and tight junctional structures between adjacent ECs. Given the BBB, it features transcytosis by specialized transporters to facilitate one-way and selective movement of the glucose and other small solutes. Pathological changes of the BBB are associated with stroke, Central Nervous System (CNS) inflammation, and neuropathologies including Alzheimer's disease, Parkinson's disease, epilepsy, multiple sclerosis, and brain tumors.

Fenestrated capillaries are equipped with fenestrae of the size of 20 to 100 nm that penetrate the endothelium and conduct fluid with considerable ease. Fenestrated capillaries are found in the kidney, area postrema, carotid body, endocrine and

Fig. 2.6 Types of capillaries. (**a**) Continuous capillaries have a continuous endothelium and a continuous basal membrane. Endothelial cells (ECs) are connected via tight junctions. (**b**) Fenestrated capillaries display endothelium with fenestrae on top of a continuous basal membrane. (**c**) Discontinuous capillaries are larger than the other capillaries and show large fenestrae and a fragmented basal membrane

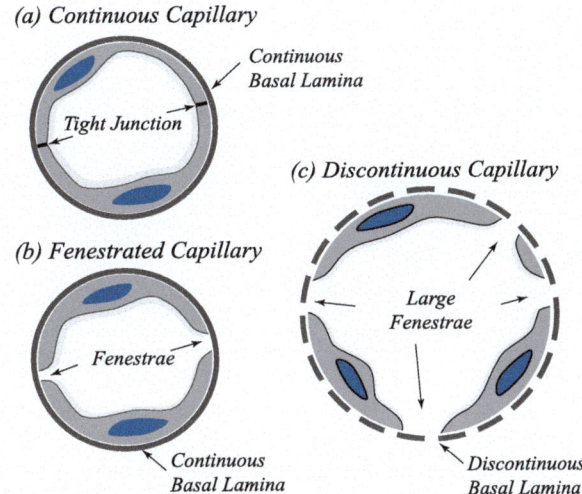

exocrine pancreas, thyroid, adrenal cortex, pituitary, choroid plexus, small intestinal villi, joint capsules, and epididymal adipose tissue.

Discontinuous capillaries show wide spacing between ECs, ranging up to micrometers, on top of a fenestrated basal membrane. Discontinuous capillaries have a very high hydraulic conductivity and are found in organs involved in the sequestration of formed vascular cells, such as spleen, bone marrow, or in the synthesis and degradation of fats and proteins, such as the liver.

2.1.6.4 The Glycocalyx

Fig. 2.7 illustrates the *glycocalyx*, a layer that plays a central role as physical barrier at the blood–endothelial interface. It is a negatively charged polysaccharide-rich surface layer that covers the luminal side of the endothelium. The glycocalyx is approximately one micrometer thick; in electron microscopy its thickness appears 50 to 300 nm and in confocal microscopy 2.5 to 4.5 μm. The glycocalyx layer is the first barrier that is permeable to water and solutes, such as electrolytes and small molecules. However, it prevents erythrocytes from contact with the EC surface and retains plasma proteins and inflammatory leukocytes in the vascular space, before any trans- or paracellular transfers appear. In continuous capillaries, the filtration of species is tightly controlled by the glycocalyx layer, and its interpolymer spaces function as a system of small pores with radii of approximately 5 nm. Given fenestrated and discontinuous capillaries, fenestrae provide an additional pathway for solvent and solutes, see Fig. 2.6.

2.1.6.5 Controlling Blood Pressure and the Role of Resistance Vessels

The small diameters of arterioles, capillaries, and venules poses considerable resistance to blood flow—they are therefore also called *resistance vessels*. Resistance vessels are highly vasoreactive, and the tonus of the pericytes in their walls controls their diameters, a mechanism to maintain an almost constant system pressure. During heavy exercise cardiac output is increasing four- to eightfold, whilst the Mean Arterial Pressure (MAP) rises by about 15 to 20 mmHg, and thus by less than 20% [366].

The resistance vessels are able to divert bloodstreams, an observation already reported in the late 1700s by Hunter:[4] *"blood goes to where it is needed"*. Given a local need of blood supply, the diameter of the local resistance vessels is controlled in response to the *metabolic tension* of the surrounding tissue. At heavy exercise up

Fig. 2.7 Schematic illustration of the glycocalyx layer, a barrier at the blood-endothelial interface

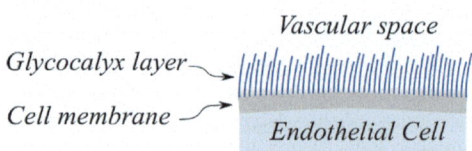

---

[4]William Hunter, Scottish anatomist and physician, 1718–1783.

to two liters of blood flow can be redirected to the skeletal muscles [288]. Given the onset of muscle contraction, vasodilation in the muscle microcirculation is delayed by only 5 to 20 s [359].

Whilst MAP increases only slightly during exercise, pulse pressure can increase dramatically. The peripheral vasodilation reduces the diastolic pressure, and the larger blood volume ejected by the left ventricle increases the systolic pressure [477], both of which contributes to the pulse pressure increase.

2.1.7 Hemodynamic Regulation

Hemodynamic regulation aims at maintaining the local equilibrium between delivery and consumption of blood-borne substances. Blood flow distribution through the vascular system can either be controlled by the CNS, or locally through neural impulses and hormonal cues. The endothelium, a semipermeable barrier sitting at the strategic position between blood and wall, plays a central role and responds to mechanical as well as chemical signals.

The tonus of contractile cells allows for the control of the vessel's diameter and thus the local delivery of blood to the tissue. At the physiological tonus, the vessel has its physiological diameter, which may be increased or decreased through the expression of *vasodilators* and *vasoconstrictors*, respectively. Whilst NO is a dominant vasodilator in large arteries, endothelium-driven hyperpolarization factors (EDHF), such as hydrogen peroxide, epoxyeicosatrienoic acids, prostacyclin, prostaglandin, and others contribute to the dilation of resistance vessels. The concentration of vasoconstrictors, such as catecholamines, Atrial Natriuretic Peptide (ANP), vasopressin, bradykinin, also affects the status of contractile vascular cells. Many of these factors may also influence EC's release of NO, and the net effect from vasodilators and vasoconstrictors determines the final vessel diameter.

2.1.7.1 Autoregulation Mechanisms

Hemodynamic regulation establishes at different levels, and individual vascular regions are regulated autonomously from other parts of the vascular system.

Given *myogenic regulation*, the vessel dilates or constricts in response to changing intravascular pressure [282, 498]. An elevated pressure causes paradoxically vasoconstriction and augments the arterial resistance. This mechanism enables matching blood supply to tissue demand over the pressure range of approximately 8 to 20 kPa. Myogenic regulation is mediated by contractile vascular cells and independent from ECs. A pressure increase in most resistance arteries involves stretch-induced activation of nonselective cation channels. This activation causes cell membrane depolarization, calcium influx, and cell contraction. The *Bayliss*[5] *effect* is a special example of myogenic regulation of arterioles, see Fig. 2.8.

---

[5]Sir William Maddock Bayliss, English physiologist, 1860–1924.

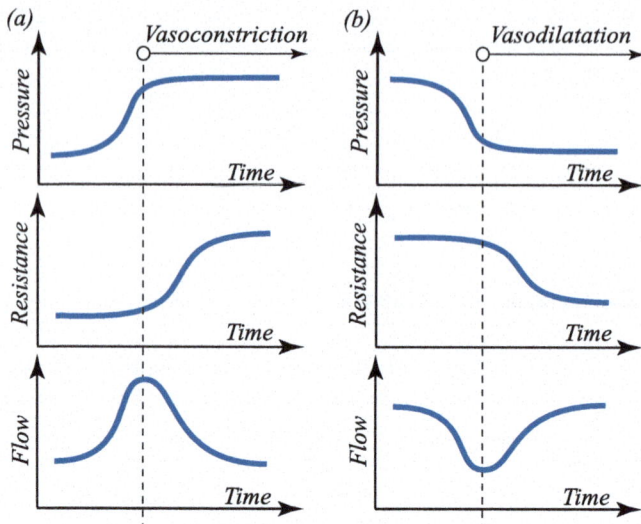

Fig. 2.8 The Bayliss effect is a particular myogenic regulation mechanism of arterioles. (**a**) A sudden increase of the intravascular pressure causes vasoconstriction and thus increases the vessel's resistance towards maintaining the flow. (**b**) A sudden pressure drop causes vasodilatation and a decreased resistance towards maintaining the flow

Blood flow-dependent regulation uses the ability of the vessel to sense WSS. The endothelium responds to WSS with the release of NO that in turn relaxes the contractile cells in the vessel wall.

The tonus of contractile vascular cells can also be controlled by other factors, such as upstream and downstream transmission of messengers along the vessel walls as well as the level of local metabolism. Regulatory messengers, manufactured and released from a site of metabolic activity, influence the activation of contractile vascular cells, especially in the wall of arterioles [548].

2.1.7.2 Short-Term Nervous Control of the Blood Pressure

Body motion and activity require the *independent* nervous control of the blood pressure for the different parts of the vascular system. Given an upright standing position for example, a gravitational shift towards the lower limbs needs to be compensated within seconds to avoid postural hypotension, also known as orthostatic hypotension. The vascular system is therefore equipped with pressure sensors that continuously record the hemodynamic regime, transduce signals, and feed the information to corresponding afferent neurons. Such receptors are called *baroreceptors* in the high-pressure circulation, and *voloreceptors* in the low-pressure circulation. Prominent baroreceptors are found in the carotid sinus and the aortic arch, whilst voloreceptors are in the pulmonary artery, the atria, the ventricles, and the vena cavae.

2.1.7.3 Long-Term Control of the Blood Pressure

The long-term control of blood pressure involves the indirect monitoring of *blood volume*. Hormonal-based control restores the blood volume and subsequently the blood pressure. The blood volume is controlled by fluid and electrolytes, such as sodium and potassium, excreted from the kidney. Other organs involved in blood volume control are the hypothalamus, sympathetic nerves, adrenal gland, and others.

2.2   Mechanical System Properties

The circulatory system relies on the pumping heart and the resistance in the vascular bed. These two effects together generated the arterial blood *pressure p* [Pa] and *flow q* [m³ s⁻¹]—the most fundamental mechanical properties of the circulatory system. Pressure and flow appear as waves and propagate along the vascular tree.

The heartbeat forces the vascular system to oscillate and it is normally in near-periodic steady-state oscillation. Given the heart suddenly stops beating, the vascular system stops oscillating and pressure and flow decrease smoothly to zero. This is characteristic for an *over-damped system*.

Although arteries have complex geometries, in this section we consider them as long, thin-walled tubes. This 1D approximation also ignores the variation of the velocity across the cross-section, necessarily abandoning the *no-slip* condition at the wall.

2.2.1   Waves in the Vascular System

Whilst the definition of a wave is commonly linked to the specific physical phenomena it represents, waves may generally be seen as disturbances that propagate in space and time. A wave has a waveform, or profile that changes along with its propagation. The profile may be decomposed into sub-waves. Such decomposition is not unique, and historically, the most common way to represent cardiovascular waveforms is the *Fourier*[6] *decomposition*, see Appendix A.3. It treats the waveform as the superposition of sinusoidal waves at the fundamental frequency and all of its harmonics. Since Fourier analysis is carried out in the *frequency domain*, it can be difficult to relate features of the Fourier representation to specific times in the cardiac cycle.

The *wave speed c* [m s⁻¹] in the cardiovascular system is determined by the area distensibility $D = (\mathrm{d}A/\mathrm{d}p)/A$ [Pa⁻¹] of the vessel, and given by the relation

$$c = \frac{1}{\sqrt{\rho D}}, \qquad (2.1)$$

---

[6]Jean-Baptiste Joseph Fourier, French mathematician and physicist, 1768–1830.

where ρ [kg m^{-3}] denotes the density of blood, and A [m^2] is the cross-section of the vessel. The derivation of (2.1) is given in Sect. 6.6.3. The waves are advected by the blood velocity v, so that the observed speed of propagation is $v + c$ in the downstream direction and $v - c$ in the upstream direction. Given normal arteries, the blood velocity v is always lower than the wave speed c.

Example 2.1 (Upstream Pressure Wave Propagation). Blood of density $\rho = 1060.0$ kg m^{-3} is under the pressure p and flows at the velocity v in an artery of the cross-section A, see Fig. 2.9. A pressure wave propagates at the speed $c - v$ in upstream direction and changes the velocity and pressure by Δv and Δp, respectively.

Fig. 2.9 Upstream propagating of a pressure wave in a blood vessel

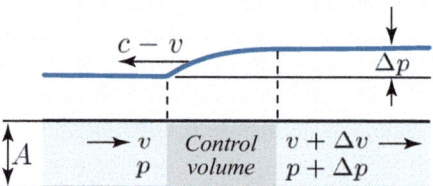

(a) Express the mass flow rate that passes through the control volume as shown in Fig. 2.9. The wave speed c in the vessel may be regarded much larger than the blood flow velocity v.
(b) Use Newton's second law of mechanics and apply it to the control volume towards the derivation of the relation between the wave speed c, the increments Δv, Δp and the blood density ρ, a relation called water hammer equation.
(c) Consider a vessel of distensibility $D = 0.0301$ kPa^{-1} and compute the wave speed c.
(d) Compute the change of velocity Δv that is caused by a pressure wave of $\Delta p = 0.23$ kPa.   ∎

Wave speed in the aorta has traditionally been determined by measuring the time it takes for the pulse wave to travel between two measurement sites—usually from the carotid to the femoral artery [394]. Although the peak of the pressure or the velocity is probably the easiest to measure, it is more accurate to measure the time of the foot of the wave. This measure alters less as the waveform changes with its propagation, and such methods are generally known as *foot-to-foot* measurements. The pressure–velocity loop provides an alternative method to measure wave speed, see Sect. 2.2.5.

The blood vessel's area distensibility D is not constant, but a function of the blood pressure, a factor that influences the wave speed (2.1). Let us consider the aorta with the aortic valve opening and closing at diastolic and late-systolic pressures, respectively. The opening and closing of the valve trigger waves that then travel at *different* speeds along the aorta, information that may be used to identify the non-linear stress–strain property of the vessel wall properties [301].

2.2.2 Vascular Pressure

Given a position x along the vascular path, the integration over the pressure waveform $p(x, t)$ [Pa] defines the *mean pressure*

$$p_{\mathrm{mean}}(x) = \frac{1}{T} \int_0^T p(x, t)\mathrm{d}t \approx \frac{1}{3} p_{\mathrm{syst}}(x) + \frac{2}{3} p_{\mathrm{diast}}(x), \qquad (2.3)$$

where T [s] is the duration of a cardiac cycle, and p_{syst} and p_{diast} denote systolic and diastolic blood pressures, respectively. For practical reasons one would integrate not only over one, but a number of cardiac cycles.

The mean pressure p_{mean} continuously decreases from the aorta towards the vena cava. However, the pressure gradient is not continuous all along the vascular path but appears almost exclusively in arterioles, capillaries, and venules—the vessels of the smallest diameters, see Fig. 2.10. The *vascular bed* houses these vessels and therefore determines the resistance of the vascular system. This key role of the vascular bed has already been noticed by Hales.[7]

Aside from the mean pressure, the *pulse pressure* $p_{\mathrm{p}} = p_{\mathrm{syst}} - p_{\mathrm{diast}}$ is another important hemodynamic property of the vascular system. Hales seems again to be the first to measure blood pressure and notice that pressure in the arterial system is not constant, but varies over the heartbeat. The pressure wave pulse, or waveform

Fig. 2.10 Change of pressure along the vascular path. Mean pressure p_{mean} (thick line) falls quickly at the level of the smallest vessels. The pulse pressure p_{p} (hatched area) increases towards distal arteries as a consequence of wave reflection, before it dissipates at the level of the smallest vessels. The capillaries and the entire venous system are free from pulsatility. The exchange of oxygen, nutrients, and other substances appears at the level of capillaries (gray area)

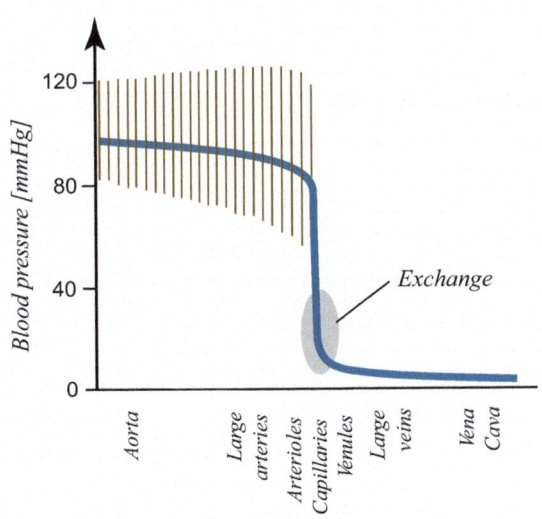

---

[7]Stephen Hales, English clergyman, 1677–1761.

is not only determined by the heartbeat, but it is also a direct manifestation of vascular properties. It is almost exclusively the elastic properties of the *thoracic aorta* together with the heartbeat that determines the pressure pulse.

The pressure wave travels over the arterial tree at the wave speed c, much faster than the blood flow velocity v. The vessel's distensibility and diameter determine the *wave speed* according to Eq. (2.1). The pulse wave travels two times faster in large arteries, and even four times faster in small arteries as compared to the aorta.

2.2.2.1 Pressure Waveform

A number of factors shape the pressure wave, resulting in its characteristic appearance. Fig. 2.11 illustrates the typical shape of the pressure wave in the aorta. It establishes from the superposition of the *forward* and *backward* traveling waves, see Fig. 2.11b. The forward wave originates at the ventricle, whilst the backward wave stems from wave reflections at downstream arterial branch points, where the aortic bifurcation is most dominant. The reflections explain that the pulse pressure p_p increases from the aorta towards the distal arteries and that the pressure wave looks very different in young and old subjects, see Fig. 2.11c. The aorta in old subjects is stiffer, and thus less distensible, and pressure waves travel therefore faster. The backward wave arrives then earlier and contributes more to the *systolic pressure augmentation*.

When reaching the level of arterioles, the pressure wave flattens out due to the high viscous dissipation of the flow in small vessels. Consequently, capillaries and the entire venous system are free of pulsatility, see Fig. 2.10.

2.2.3   Vascular Capacity

The *capacity* C [m^3 Pa^{-1}], also known as *volume compliance*, determines the vasculature's ability to increase the volume of blood it holds, and thus its reservoir/buffering function. Given its definition

$$C = \frac{\Delta V}{\Delta p}, \tag{2.4}$$

it relates the increase of blood volume ΔV [m^3] to the increase in blood pressure Δp [Pa]. The compliance, and thus the elasticity and size of the largest blood vessels determine the capacity of the vascular system.

Fig. 2.12 shows the blood volume in the arterial and venous systems as a function of the pressure [235]. The tangent to these curves is the respective capacity, and $C_{art} = 2$ ml mmHg^{-1} and $C_{ven} = 100$ ml mmHg^{-1} approximate the arterial and venous capacities of an adult human. The venous system stores approximately five times more blood than the arterial system and its capacity is approximately fifty times larger than that of the arterial system.

The aorta contributes almost the entire capacity to the vascular system, out of which the thoracic segment alone covers 85% [232]. The capacity of the

Fig. 2.11 The pressure pulse. (**a**) Typical pressure waveform. The systolic pressure is augmented by the backward wave. The diacrotic notch denotes the closure of the aortic valve. (**b**) The superposition of forward and backward traveling waves determines the pressure pulse. (**c**) Typical pressure waveforms in young and old subjects

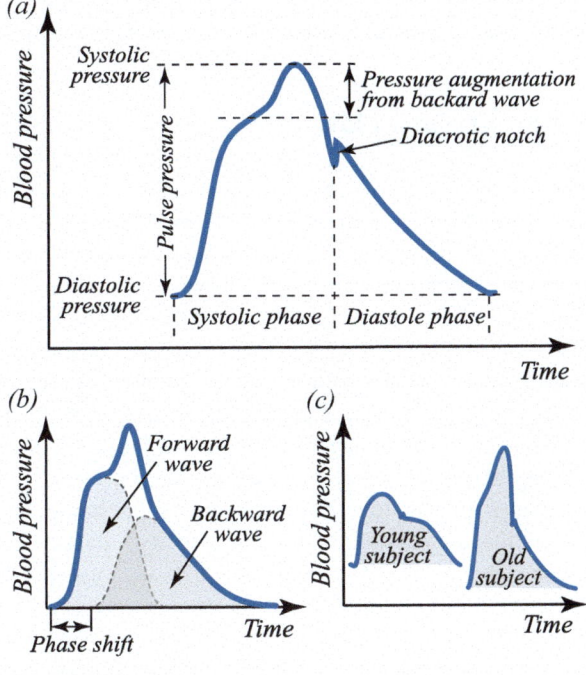

Fig. 2.12 Pressure–volume relationships of (**a**) the arterial and (**b**) the venous vascular system. Vasoreactivity influences the relationship as shown by the dashed curves

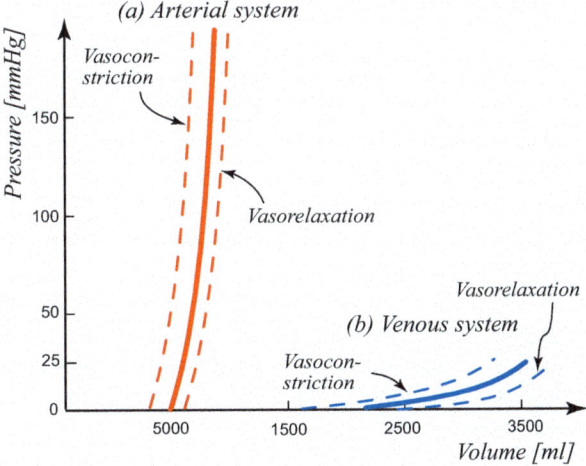

aorta is constant over a wide range of pressures and determines the circulation's *Windkessel (WK) properties*. The capacity of the aorta, and thus its elasticity, is of utmost importance to the entire cardiovascular system. A stiff aorta increases left ventricular load that may result in cardiac complications, such as cardiomyopathy. In addition to genetic or elastinopathies [94], the aorta also stiffens naturally with age. It is elastic lamellae that undergo fragmentation and thinning, leading to ectasia and a gradual transfer of mechanical load to collagen, which is 100 to 1000 times stiffer than elastin and then reduces the capacity of the vascular system [224].

2.2.4 Vascular Flow

The aorta is the first arterial segment of the systemic circuit, directly connected to the heart. A unidirectional flow of the blood ejected from the left ventricle into the aorta is maintained through the *aortic valve*. It passively opens and closes with each heartbeat. The discontinuous inflow of blood together with the aortic capacity defines the *pulsatile blood flow* in the aorta. Given peak systole, blood flows unidirectional and at velocities of approximately $60 \, \text{cm s}^{-1}$, whilst back flow establishes at the diastolic phase. Back flow in the aortic arch and the abdominal aorta reaches velocities of approximately $-20 \, \text{cm s}^{-1}$ and $-10 \, \text{cm s}^{-1}$, respectively.

Blood flow in the large arteries is similar to the flow in the aorta. For example in the iliac artery, the velocities over the cardiac cycle range from approximately -7.5 to $60 \, \text{cm s}^{-1}$. The flow in veins is much more homogeneous as compared to arteries. Given the saphenous vein, the velocity changes only between approximately 20 and $30 \, \text{cm s}^{-1}$. The distribution of the blood flow velocity over the vessel's cross-section is complex and influenced by factors, such as the form of the pressure wave, the vessel's diameter and centerline curvature, upstream and downstream flow properties, Vortical Structure (VS) dynamics, and others. Such effects are beyond a 1D flow description and will be discussed in Chap. 6.

Flow is inverse proportional to the vasculature's cross-sectional area. At the level of the capillaries the largest cross-sectional area appears, and blood flows at velocities as low as tens of micrometers per second. A Stokes[8] flow is then an adequate model of blood flow. The very low flow velocity is important to provide enough time for the exchange of oxygen, nutrients, and other substances in the capillaries. The blood flow is linked to the vessel's biochemical activity through WSS, and changes in response to factors, such as the oxygen tension of the surrounding tissue.

Given a 1D description, the flow q and the velocity v in a vessel are related through $q = Av \, [\text{m}^3 \, \text{s}^{-1}]$, where A denotes the luminal cross-section of the vessel. Similar to the pressure $p(x, t)$, the flow $q(x, t)$ also appears as a wave in the vascular system, where x and t denote the position along the vascular path and the time, respectively.

2.2.4.1 Venous Return

Given homeostasis, the time-averaged cardiac output equals the flow into the atrium, the *venous return*. Venous return and cardiac output are therefore interdependent, a relation known as the *Frank–Starling mechanism*. In addition to factors, such as rhythmical contraction of limb muscles during normal locomotory activity, vasoreactivity, respiration, and gravitation, the (partial) collapse of veins has an

---

[8] Sir George Gabriel Stokes, English/Irish physicist and mathematician, 1819–1903.

important influence on the venous return. It appears at negative ambient pressures and has been extensively studied [259, 422]. See also Fig. 2.13 that illustrates some factors that influence venous return. At *negative* atrial pressure, veins start to collapse and the linearity between pressure and flow is broken. Venous return can then no longer increase at increasing pressure gradient, and the pressure–flow curve flattens out.

2.2.5 The Pressure–Velocity Loop

Given the pressure $p(x, t)$ and the velocity $v(x, t)$ at the position x in a vessel, the pressure–velocity loop may be plotted, see Fig. 2.14. At the beginning of the loop when the pressure and velocity waves start, the tangent to the pressure–velocity

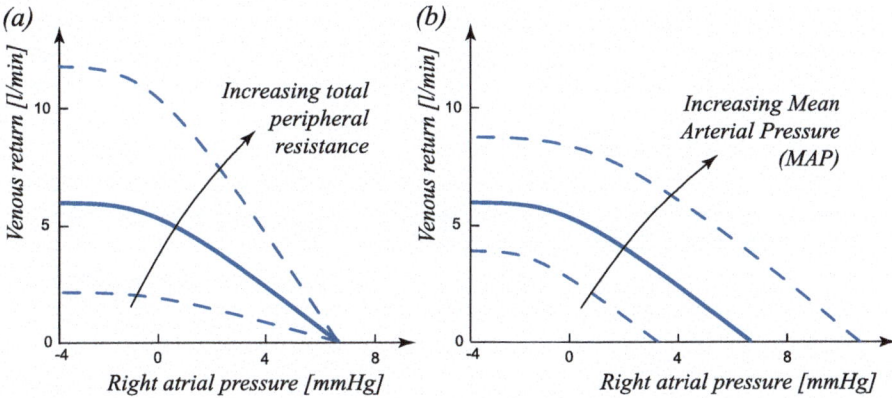

Fig. 2.13 The non-linear relation between the venous return and the atrial pressure. (**a**) Influence of the total peripheral resistance. (**b**) Influence of the Mean Arterial Pressure (MAP)

Fig. 2.14 The pressure–velocity loop illustrating the weighted wave speed ρc. The dot indicates the beginning of the pressure and velocity waves

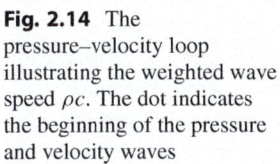
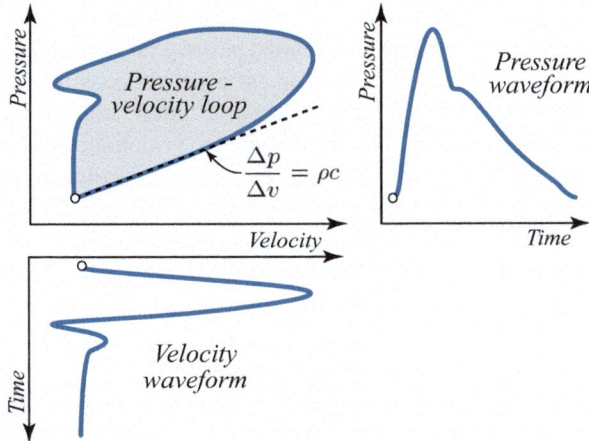

loop expresses the weighted wave speed ρc, where ρ denotes the density of blood and c is the wave speed. The *water hammer equation* (2.2) $\Delta p / \Delta v = \rho c$ explains this property of the pressure–velocity loop, where $c \gg v$ has been assumed in the derivation of (2.2). It relates the pressure and velocity increments Δp and Δv of wave propagation and goes back to the work of von Kries,[9] and pulse wave investigations in blood vessels [571]. Given $p(x, t)$ and $v(x, t)$ can be measured accurately and simultaneously, the pressure-volume loop represents a simple and accurate method of measuring the wave speed [304].

2.2.6 Vascular Resistance

The resistance $R = \Delta p / q$ [Pa s m^{-3}] of a vessel against the flow q is expressed by the pressure drop Δp between the inlet and outlet of the segment. The viscosity of the blood and the flow conditions in the vessel determine its resistance. Given laminar steady-state tube flow, the Hagen[10]–Poiseuille[11] law expresses the hydraulic resistance, see Sect. 2.3.2.1. It states that a tube of diameter d has a resistance that is proportional to $1/d^4$, and therefore only the smallest vessels, the resistance vessels, can provide noticeable resistance to blood flow [609].

As with an individual vessel, also a network of vessels provides resistance to flow. The dimensions of the vessels together with their organization within the network then determine the resistance. The arrangement of vessels in series increases the resistance, whilst their parallel arrangement reduces the resistance. Given an adult human, $R_{\mathrm{art}} = 1\,\mathrm{mmHg\ s\ ml}^{-1}$ and $R_{\mathrm{ven}} = 0.06\,\mathrm{mmHg\ s\ ml}^{-1}$ approximate the resistances of the arterial and venous systems, respectively.

2.2.7 Transcapillary Transport

The changes of hydrostatic and osmotic pressures across the capillary wall direct the *transcapillary fluid flux*, and thus the fluid exchange between the vascular and interstitial spaces. The fluid is in principle *water* that solves proteins and electrolytes, which then is called plasma. Factors such as solute size and its electrical charge determine whether or not they can pass the semipermeable capillary wall. The fluid flux is therefore always filtrated, and the transcapillary transport is also called *filtration*. Together with the lymphatic system, filtration determines the transcapillary solute concentrations and controls interstitial (volume) homeostasis.

Whilst the muscle tonus controls the hydrostatic pressure in the microvasculature, the transcapillary solute concentrations determine the osmotic pressure. Small solutes can easily pass the capillary wall, and in most vascular beds only the

---

[9]Johannes von Kries, German physiologist, 1853–1928.

[10]Gotthilf Heinrich Ludwig Hagen, German civil engineer, 1797–1884.

[11]Jean Léonard Marie Poiseuille, French physicist, and physiologist, 1797–1869.

macromolecular solutes appear at significant different concentrations across the vessel wall. The transcapillary osmotic pressure $\Delta\Pi$ can therefore be approximated by the sum of pressure differences that are exerted by such macromolecules. *Albumin* is the most important macromolecule in this context and accounts for approximately 80% of $\Delta\Pi$.

Aside from transcapillary pressure differences, the wall's leakiness, and thus its *hydraulic conductivity* L_p [m Pa^{-1}s^{-1}] determines how much fluid passes through it. The conductivity is defined by Darcy's[12] law through the relation

$$L_p = \frac{k}{\eta L},$$
(2.5)

where k [m^2] and L [m] denote the intrinsic permeability and thickness of the capillary wall, whilst η [Pa s] is the viscosity of water, see Appendix E.2. The hydraulic conductivity can also be directly measured by laboratory experiments [250].

2.3 Modeling the Macrocirculation

This section follows a top-down approach, where *lumped parameter models* describe parts of the vascular system. They represent vascular complexity by a low number of parameters yet capturing salient system features. A topology consisting of discrete entities, representing *resistance, capacity,* and *inductance*, describes the spatially distributed vascular system. Such models do not consider the anatomical organization of the vessels and cannot represent features, such as wave propagation. Lumped parameter modeling of the vascular system is well documented with excellent reviews [596] available in the literature.

2.3.1 WindKessel Models

Whilst Weber[13] seems to be the first who proposed the comparison of the *capacity (volume compliance)* of the large arteries with the *WindKessel (WK)* present in fire engines, it was Frank[14] [173] who quantitatively formulated and popularized the so-called two-element WK model.

2.3.1.1 Two-Element WindKessel Model

The *two-element WK model* represents the systemic vascular circuit by two lumped parameters—its total *capacity* C [m^3 Pa^{-1}], and its total peripheral *resistance* R [Pa m^{-3}], see Fig. 2.15. The capacity $C = \Delta V/\Delta p$ describes the intake of the

---

[12]Henry Philibert Gaspard Darcy, French engineer, 1803–1858.

[13]Ernst Heinrich Weber, German physician, 1795–1878.

[14]Otto Frank, German doctor and physiologist, 1865–1944.

blood volume ΔV [m^3] into the (elastic) arterial system in response to the pressure increase Δp [Pa]. In contrary, the resistance $R = \Delta p_{mean}/q_{CO}$ relates the drop of the mean pressure Δp_{mean} from the arterial side to venous side, to the cardiac output q_{CO} [m^3 s^{-1}]. Given the much higher pressure in the arterial than in the venous system, the simplification $R \approx p_{mean\,art}/q_{CO}$ may be made, where $p_{mean\,art}$ denotes the MAP. Whilst the elasticity of the aorta and the largest conduit arteries determines the system's capacity C, the resistance vessels govern the system's resistance R.

The total flow $q(t)$ through the system splits into the flow $q_R(t)$ through the resistor R and the flow $q_C(t)$ into the capacitor C, see the electrical representation of the two-element WK model in Fig. 2.15b. The flow balance then reads

$$q(t) = q_R(t) + q_C(t) = \frac{p(t)}{R} + C\frac{dp(t)}{dt}, \tag{2.6}$$

where $p(t)$ denotes the time-dependent arterial pressure, and the relations $q_R = p/R$ and $q_C = C(dp/dt)$ describe the resistor and the capacitor, see Appendix E.1. The governing equation of the two-element WK model (2.6) relates the pressure $p(t)$ and flow $q(t)$ of the systemic circuit, a system described by the properties C and R, respectively.

Given the pressure $p(t)$, relation (2.6) is an algebraic expression that directly yields $q(t)$. In contrary, given $q(t)$, it represents a first-order linear differential equation in $p(t)$ and may be solved (numerically) together with the initial condition $p(0) = p_0$. Fig. 2.16b illustrates such a (transient) solution for the pressure $p(t)$. Table 2.1 reports the systemic circuit parameters, whilst Fig. 2.16a shows

Fig. 2.15 (a) Hydraulic and (b) electric representations of the two-element WindKessel (WK) model. The flow $q(t)$ and pressure $p(t)$ describe the system state, and R and C denote vascular bed resistance and arterial capacity, respectively

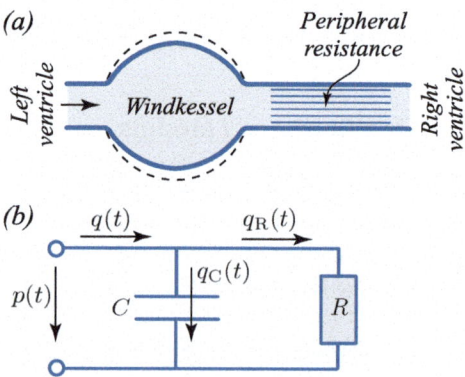

Table 2.1 System parameters used for WindKessel (WK) models

Vascular bed resistance R	1.1 mmHg s ml^{-1}
Arterial capacity C	0.7 ml mmHg^{-1}
Aortic impedance[a] Z	0.1 mmHg s ml^{-1}
Arterial inertance[b] L	0.02 mmHg s^2 ml^{-1}

[a] Only used by the three-element and four-element WK models.
[b] Only used by the four-element WK model.

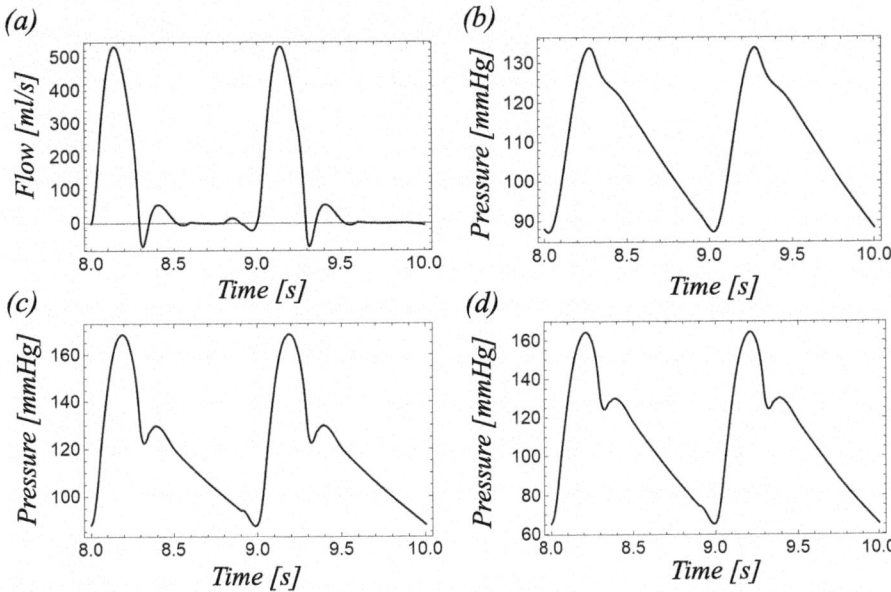

Fig. 2.16 WindKessel (WK) modeling of the systemic circuit. (**a**) Prescribed flow profile $q(t)$. (**b**) Pressure profile $p(t)$ according to the two-element WK model (2.6). (**c**) Pressure profile $p(t)$ according to the three-element WK model (2.20). (**d**) Pressure profile $p(t)$ according to the four-element WK model (2.30). Table 2.1 reports the parameters used for the WK models

the prescribed flow $q(t)$ through the system. The cardiac cycle time of $T = 1.0\,\mathrm{s}$ has been used, and the flow waveform was interpolated between a number of data points. The two-element WK model (2.6) has been solved at the initial condition $p_0 = 80\,\mathrm{mmHg}$, and Fig. 2.16b shows $p(t)$ for the time interval from 8.0 to 10.0 s. It is the ninth and tenth cardiac cycle, and the system has reached its steady-state periodic condition.

Example 2.2 (Two-Element Windkessel Model Predictions). A vascular system may be represented by a two-element Windkessel model, where $R = 50.1\,\mathrm{mmHgs}\,\mathrm{ml}^{-1}$ and $C = 0.018\,\mathrm{ml}\,\mathrm{mmHg}^{-1}$ describe the system's resistance and capacity, respectively. Given the cardiac cycle period $T = 1\,\mathrm{s}$, the flow

$$q(t) = \begin{cases} q_0 \sin(6\pi t) & 0 \le t \le 1/6\,\mathrm{s}, \\ 0 & 1/6 < t < 1\,\mathrm{s}, \end{cases} \tag{2.7}$$

where $q_0 = 18.2\,\mathrm{ml}\,\mathrm{s}^{-1}$ passes the system.

(a) Use the backward-Euler time discretization method and provide a discretized version of the two-element Windkessel model (2.6).

(b) Iteratively solve the discretized governing equation and predict the pressure $p(t)$ in the system at steady-state periodic conditions. Use different numbers k of time steps towards exploring the convergence with respect to this parameter. ■

Example 2.3 (Systemic Implication of EVAR Treatment). Given EndoVascular Aortic Repair (EVAR), a stent-graft of diameter $d_{sg} = 2.5\,\text{cm}$ is inserted in the $l_{aorta} = 35\,\text{cm}$ long thoracic aorta to cover an aneurysm, see Fig. 2.17. The stent-graft has the radial stiffness of $k_{sg} = \Delta d/\Delta p = 1.2 \cdot 10^{-3}\,\text{cm}\,\text{kPa}^{-1}$ and covers in total 70% of the thoracic aorta. The EVAR treatment changes the capacity C of the systemic circuit, and before treatment, the capacity $C_n = 9.7\,\text{cm}^3\,\text{kPa}^{-1}$ and the resistance $R = 0.18\,\text{kPa}\,\text{s}\,\text{cm}^{-3}$ determined the patient's systemic circulation.

Fig. 2.17 Schematic illustration of a thoracic aortic aneurysm that has been treated with EndoVascular Aortic Repair (EVAR). The stent-graft covers the $l_{sg} = \alpha l_{aorta}$ long aortic segment, where l_{aorta} and α denote the total length of the thoracic aorta and a dimensionless parameter, respectively

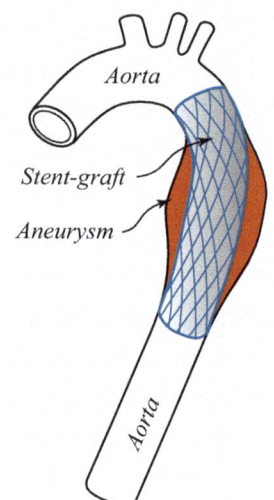

(a) Provide the relation of the capacity $C_{EVAR}(\alpha)$ as a function of the stent-graft coverage α for the EVAR-treated patient. The stent-graft coverage $\alpha = l_{sg}/l_{aorta}$ is the ratio between the stent-graft length and the length of the thoracic aorta.
(b) Consider the simplified cardiac output $q(t) = Q\sin(\pi t)^2$ with $Q = 150\,\text{cm}^3\,\text{s}^{-1}$ and use a two-element Windkessel (WK) model to study the systemic implication of EVAR treatment. Consider the initial pressure $p(0) = 13.3\,\text{kPa}$ and solve the WK governing equation for $\alpha = 0$ and $\alpha = 1$, respectively. Plot the aortic pressure over the time for said parameters. The result

$$I = \int \exp(x/a)\sin^2(\pi x)\,dx$$

$$= \frac{a\exp(x/a)\left[1 + 4a^2\pi^2 - \cos(2\pi x) - 2a\pi\,\sin(2\pi x)\right]}{2 + 8a^2\pi^2} + K$$

may be used to solve the linear first-order differential equation of the WK model, where K denotes an integration constant.   ∎

2.3.1.2 Homogeneous Solution

In most cases, only the *steady-state periodic, or homogenous* solution of the problem (2.6) is of interest. The analysis in the complex plane, or Argand's[15] diagram is then convenient, see Appendix A.2. The flow and the pressure are described by complex numbers, represented by the vectors $\mathbf{q}$ and $\mathbf{p}$ in the complex plane, respectively.

Let us first consider the case, where the pressure $p(t) = \mathrm{Re}(\mathbf{p})$ is known, whilst the flow $q(t) = \mathrm{Re}(\mathbf{q})$ through the system is unknown. At steady state, the pressure $p(t)$ (as the flow $q(t)$) is periodic and may be expressed by the Fourier series (see Appendix A.3)

$$p(t) = \mathrm{Re}(\mathbf{p}) = \mathrm{Re}\left(\sum_{n=-\infty}^{+\infty} \mathbf{P}_n \exp[i\omega t] \right), \qquad (2.9)$$

where $\mathbf{P}_n$ and $i = \sqrt{-1}$ denote the complex Fourier coefficients and the imaginary unit, respectively. Given the additive representation (2.9) of the pressure waveform through the superposition of its harmonics, it is sufficient to consider a single complex vector $\mathbf{p} = \mathbf{P}\exp(i\omega t) = P\exp(i\omega t)$ with $P = |\mathbf{P}|$ pointing in the real direction at the time $t = 0$. With the properties of the resistor and capacitor (E.1), the governing equation (2.6) then reads

$$\begin{aligned} \mathbf{q} &= R^{-1}P\exp(i\omega t) + iC\omega P\exp(i\omega t) \\ &= P(R^{-1} + iC\omega)\exp(i\omega t) \\ &= \mathbf{Q}\exp(i\omega t), \end{aligned} \qquad (2.10)$$

where $|\mathbf{Q}| = Q = P\sqrt{R^{-2} + C^2\omega^2}$ and $\phi = \arg(CR\omega) = \arctan(CR\omega)$ denote the *amplitude (norm)* and the *argument* of the complex flow vector $\mathbf{q}$, respectively. The time-dependent flow then reads $q(t) = \mathrm{Re}(\mathbf{q}) = Q\cos(\omega t + \phi)$, and Fig. 2.18a illustrates it in the complex plane.

In contrary to the aforementioned analysis, we consider now the system flow $q(t)$ to be given, whilst the pressure $p(t)$ is unknown. The flow $q(t) = \mathrm{Re}(\mathbf{q})$ may again be expressed as Fourier series $\mathbf{q} = \sum_{n=-\infty}^{+\infty} \mathbf{Q}_n \exp(i\omega t)$ with $\mathbf{Q}_n$ denoting the

---

[15]Jean-Robert Argand, French amateur mathematician, 1768–1822.

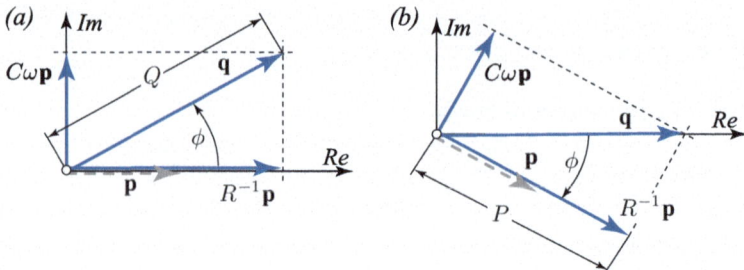

Fig. 2.18 Representation of flow $q(t) = \mathrm{Re}(\mathbf{q})$ and pressure $p(t) = \mathrm{Re}(\mathbf{p})$ in the complex plane. The relation amongst them is described by a two-element WindKessel (WK) model that describes the systemic circuit of resistance R and capacity C. (**a**) The pressure $\mathbf{p}$ is prescribed, and the WK model governs the flow $\mathbf{q}$. (**b**) The flow $\mathbf{q}$ is prescribed, and the WK model governs the pressure $\mathbf{p}$

Fourier coefficients. Again, it is sufficient to consider Eq. (2.6) for a single complex vector $q(t) = Q\exp(i\omega t)$ with $Q = |\mathbf{Q}|$ and $\mathbf{Q}$ pointing in the real direction at $t = 0$. It yields then the governing equation

$$Q\exp(i\omega t) = R^{-1}P\exp[i(\omega t + \phi)] + iC\omega P\exp[i(\omega t + \phi)]$$
$$= P\left\{R^{-1}\exp(i\phi) + C\omega\exp[i(\phi + \pi/2)]\right\}\exp(i\omega t), \qquad (2.11)$$

where the Ansatz $p(t) = \mathbf{P}\exp(i\omega t) = P\exp[i(\omega t + \phi)]$ has been used. Whilst $P = |\mathbf{P}|$ denotes the *pressure amplitude* (norm), ϕ is the *phase angle* between pressure and flow. Both parameters need to be identified from the complex equation (2.11). Equation (2.11)$_1$ already presents real and imaginary contributions, and $P = Q/\sqrt{R^{-2} + C^2\omega^2}$ denotes the norm of the flow. Towards the specification of the phase angle ϕ, we may consider the expression (2.11) at the time $t = 0$. The flow $\mathbf{q}$ points then in the real direction, and thus the imaginary part of (2.11) vanishes. The condition

$$0 = \mathrm{Im}\left[P\left\{R^{-1}\exp(i\phi) + C\omega\exp[i(\phi + \pi/2)]\right\}\exp(i\omega t)\right]_{t=0}$$
$$= R^{-1}\sin\phi + C\omega\underbrace{\sin(\phi + \pi/2)}_{\cos\phi}$$

then determines $\tan\phi = -RC\omega$ to be the phase angle. Fig. 2.18b illustrates flow and pressure in Argand's diagram at the time $t = 0$ when the flow $\mathbf{q}$ points into the real direction. The governing equations (2.10) and (2.11) describe both the two-element WK model, and the vector diagrams in Fig. 2.18a,b are rotated versions of each other.

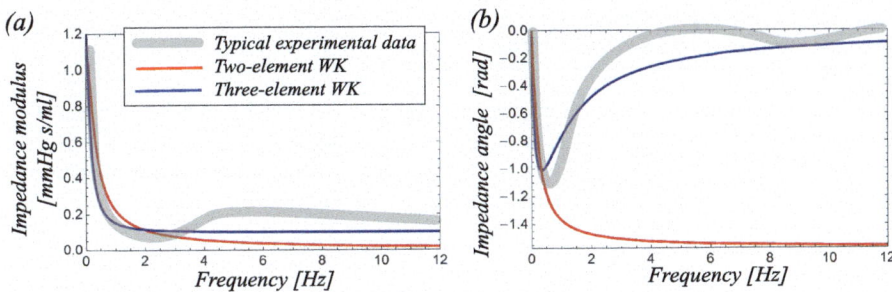

Fig. 2.19 (a) Impedance modulus Z and (b) impedance angle ϕ predicted by the two-element WindKessel (WK) (red) and the three-element WK (blue) models. The models use the parameters listed in Table 2.1, and the gray curves show typical experimental data

2.3.1.3 Impedance

The *impedance* **z** is a complex vector that relates a system's input and output. In the description of the vasculature, it compares the pressure **p** and the flow **q**. The *impedance modulus* $|\mathbf{z}| = Z = |\mathbf{p}|/|\mathbf{q}| = P/Q$ [Pa s m^{-3}] is the quotient of the pressure and flow amplitudes, whilst the *impedance angle* $\phi =$ arg**q** $-$ arg**p** [rad] is the phase difference between the two complex vectors **q** and **p**, respectively. Given the flow and pressure of the two-element WK model derived in Sect. 2.3.1.2,

$$Z = (R^{-2} + C^2\omega^2)^{-1/2} \quad \text{and} \quad \phi = \arctan(RC\omega) \tag{2.12}$$

express its impedance modulus Z and angle ϕ, respectively. Fig. 2.19 shows these quantities as a function of the system frequency $f = \omega/(2\pi)$ and based on the parameters listed in Table 2.1. At steady state $f = 0$, the entire flow runs over the resistor; the system's impedance is then equal to its resistance, $Z = R$.

Example 2.4 (Impedance of the Vascular System). Table 2.2 reports measurements of aortic pressure $p(t)$ and flow $q(t)$ in the ascending ferret aorta. Given these measurements, the vascular system's impedance **z** should be computed.

(a) Provide a Fourier series approximation of $p(t)$ and $q(t)$ up to $M = 10$ harmonics. Plot the Fourier series approximation on top of the original signal.

Table 2.2 Measured flow and pressure waves in the ascending aorta of an individual ferret. Data is extracted from plots reported elsewhere [62]

Time [s]	Flow [ml s^{-1}]	Pressure [mmHg]	Time [s]	Flow [ml s^{-1}]	Pressure [mmHg]
0.0	0.00	78	0.2	0.16	90
0.0125	0.00	78	0.2125	0.20	90
0.025	0.04	77	0.225	0.30	89
0.0375	0.23	77	0.2375	0.20	87
0.05	0.48	78	0.25	0.16	86
0.0625	3.00	95	0.2625	0.13	84
0.075	4.61	97	0.275	0.10	84
0.0875	4.50	97	0.2875	0.08	83
0.1	3.96	96	0.3	0.07	82
0.1125	3.20	93	0.3125	0.06	81
0.125	2.10	90	0.325	0.06	81
0.1375	0.00	81	0.3375	0.04	80
0.15	−0.32	83	0.35	0.02	80
0.1625	−0.08	85	0.3625	0.00	79
0.175	0.14	87	0.375	0.00	78
0.1875	0.15	89			

(b) Compute the system's impedance modulus Z and impedance angle ϕ, and plot them *versus* the signal frequency f. ∎

2.3.1.4 Parameter Identification

The vascular bed's resistance R determines the relation between the mean flow q_{mean} and the mean pressure p_{mean} over the cardiac cycle. Given the time T of the cardiac cycle, the expression

$$R = \frac{p_{\mathrm{mean}}}{q_{\mathrm{mean}}} = \frac{\int_0^T p(t)\mathrm{d}t}{\int_0^T q(t)\mathrm{d}t} \qquad (2.14)$$

allows us therefore to compute the resistance R from the flow $q(t)$ and pressure $p(t)$, respectively. The cardiovascular system is an over-damped system, and the heart directly determines its pulsatility. Therefore, $p(t)$, $q(t)$, and T may vary from cycle to cycle, and their averages over a number of cardiac cycles should be used to compute R through (2.14).

In addition to the resistance R, the capacity C of the two-element WK model needs to be identified. An approach known as *pressure decay method* considers the late diastolic phase, where the flow is approximately zero, see Fig. 2.16a. The governing equation (2.6) then reads $q(t) = p(t)/R + C\mathrm{d}p(t)/\mathrm{d}t = 0$, and the capacity

$$C = \frac{\Delta t}{R \ln\left(\frac{p_0}{p_1}\right)} \qquad (2.15)$$

is given by the pressure decay over the time period $\Delta t = t_1 - t_0$. The pressures $p_0 = p(t_0)$ and $p_1 = p(t_1)$ are commonly taken at time t_0 right after the diacrotic notch, as well as at the time t_1 at the end of the diastolic phase. Alternative methods to identify C by exploring the late diastolic phase have been discussed elsewhere [596].

Example 2.5 (Decay Method to Estimate the Vascular Resistance). Given the organ's arterial capacity $C = 0.012\,\text{ml mmHg}^{-1}$, the experimental set-up shown in Fig. 2.20 is used to measure the vascular bed resistance R. At the time $t = 0$ the valve closes and stops the inflow, such that $q_{\text{in}}(t) = 0$ holds for $t > 0$. At the times $t_0 = 0.1\,\text{s}$ and $t_1 = 0.9\,\text{s}$, the manometer measures the pressures $p_0 = p_{\text{in}}(t_0) = 112.0\,\text{mmHg}$ and $p_1 = p_{\text{in}}(t_1) = 75.0\,\text{mmHg}$, respectively.

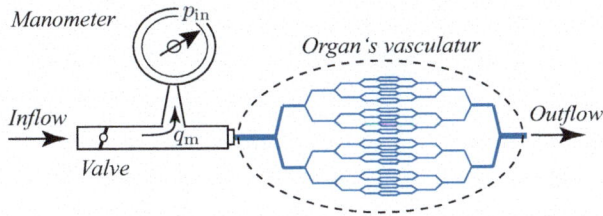

Fig. 2.20 Schematic illustration of an experimental set-up to estimate the vascular resistance R of an organ

(a) Design a lumped parameter model that represents the problem at negligible inflow $q_{\text{m}} = 0$ into the manometer. Derive the model's governing equation and estimate the resistance R from the two pressure measurements p_0 and p_1.
(b) Consider the flow $q_{\text{m}}(t) = \xi\, dp_{\text{in}}/dt$ to be proportional to the pressure change, where ξ denotes a manometer-dependent parameter. Provide the governing equation for this problem and estimate R from the two pressure measurements p_0 and p_1. Compute the relative error $e = 100(R - R_{\text{exact}})/R_{\text{exact}}$ [%] for $0 < \xi < 0.5C$, where R_{exact} denotes the resistance at $q_{\text{m}} = 0$.
(c) Consider the manometer to be an uptake tube with the inner diameter of $d_{\text{i}} = 1.0\,\text{mm}$ that is filled with water of the density $\rho = 1000.0\,\text{kg m}^{-3}$. Compute ξ for this device and estimate the resistance R of the vascular bed from the two pressure measurements p_0 and p_1. ∎

Least-square parameter identification is a popular approach in the identification of model parameters. Given n measurements at the times t_i of the pressure p_i and flow q_i, the minimization problem

$$\sum_{i=0}^{n} \left[\alpha(p(R, C; t_i) - p_i)^2 + (q(R, C; t_i) - q_i)^2 \right] \to \text{MIN} \qquad (2.17)$$

allows the identification of the least-square optimized parameters R and C of the two-element WK model. Here, α denotes a scaling/weighting parameter to account for differences concerning the values of pressure and flow and to ensure that both errors contribute to the objective function. Aside from directly using the pressure and flow measurements, the impedance modulus $Z = (R^{-2} + C^2\omega^2)^{-1/2}$ and the impedance angle $\phi = \arctan(RC\omega)$ may also be used to estimate R and C. The Fourier series of the pressure and flow waves provides the pairs (Z_i, ω_i) and (ϕ_i, ω_i) for $0 \leq i \leq M$, see Example 2.4. Here, M is the number of harmonics considered by the parameter identification, whilst ω denotes the angular velocity. The minimization problem

$$\sum_{i=0}^{M} \left[\alpha(Z(R, C; \omega_i) - Z_i)^2 + (\phi(R, C; \omega_i) - \phi_i)^2 \right] \to \text{MIN} \qquad (2.18)$$

allows then the identification of the least-square optimized parameters R and C, where α adjusts for data range difference.

Whilst the least-square-identified parameters yield the model that best agrees with the experimental data, the physical interpretation of R as vascular bed resistance, and C as volume compliance cannot be guaranteed.

2.3.1.5 Three-Element WindKessel Model

At higher frequencies, Fig. 2.19 demonstrates qualitative disagreement between experimental data and the predictions of the two-element WK model. Whilst the two-element WK model approaches the impedance modulus $Z = 0$ and the impedance angle $\phi = -\pi/2$ for the frequency $f \to \infty$, this is not supported by experimental data. The two-element WK model does not consider inertia effects of the blood, which is the main reason for this shortcoming.

The acceleration and deceleration of the large blood mass in the aorta influence the vascular system; it presents significant vascular resistance at higher frequencies. The *three-element WK model* introduced therefore an additional resistance $Z_a = v_{pw}\rho/A$, where v_{pw}, ρ, and A denote aortic pulse wave velocity, blood density, and aortic cross-section, respectively. The resistance Z_a is also called *aortic impedance*, and Fig. 2.21 shows the hydraulic and electric representations of the three-element WK model.

Given the flow $q(t)$ across the resistor Z_a, the pressure drops from $p(t)$ to $\overline{p}(t)$, and

$$\overline{p}(t) = p(t) - Z_a q(t) \qquad (2.19)$$

holds. In addition, the total flow splits into the part $q_R(t)$ through the resistor and the part $q_C(t)$ into the capacitor. It yields the relation

$$q(t) = q_R(t) + q_C(t) = \frac{\overline{p}(t)}{R} + C\frac{d\overline{p}(t)}{dt},$$

Fig. 2.21 (a) Hydraulic and (b) electric representations of the three-element WindKessel (WK) model. Flow $q(t)$ and pressure $p(t)$ describe the system state, whilst R, Z_a, and C denote the vascular bed resistance, aortic impedance, and arterial capacity, respectively

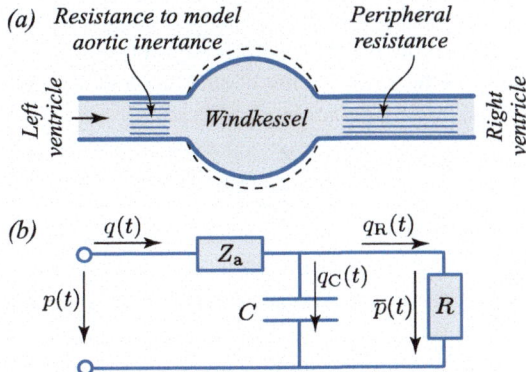

and the substitution of $\overline{p}(t)$ by (2.19) then leads to the governing equation

$$\frac{dp(t)}{dt} + \frac{p(t)}{RC} = Z_a \frac{dq(t)}{dt} + \frac{R + Z_a}{RC} q(t) \tag{2.20}$$

of the three-element WK model. Given either $p(t)$ or $q(t)$, the relation (2.20) yields a first-order linear differential equation. Fig. 2.16c illustrates the corresponding (transient) solution. It is based on the prescription of the flow shown in Fig. 2.16a and the parameters listed in Table 2.1. The solution starts at the initial condition $p_0 = 80\,\text{mmHg}$, and Fig. 2.16c shows $p(t)$ at steady-state periodic conditions.

The three-element WK model is probably the most widely used model to mimic the vascular system. It is often realized as a test rig to test vascular medical devices.

Example 2.6 (Two-Element Versus Three-Element WK Models). Table 2.2 reports measurements of the pressure $p(t)$ and flow $q(t)$ in the ascending ferret aorta, data that should be used to estimate model parameters of the two-element and three-element WK models.

(a) Estimate the peripheral resistance R of the ferret's vascular system.
(b) Estimate the total arterial capacity C of the ferret's vascular system.
(c) Estimate the ferret's aortic impedance Z_a. The aortic diameter of 6.0 mm, the aortic pulse wave velocity of 6.3 m s^{-1}, and the blood density of $\rho = 1060\,\text{kg m}^{-3}$ may be used.
(d) Prescribe the flow given in Table 2.2 and predict the pressure through the numerical solution of the governing equations of the two-element and three-element WK models. ∎

2.3.1.6 Homogeneous Solution

The steady-state periodic analysis of the problem (2.20) is conveniently investigated in the complex plane, where the pressure and flow waves assemble from the superposition of $\mathbf{p} = P \exp(i\omega t)$ and $\mathbf{q} = Q \exp[i(\omega t + \phi)]$, respectively. The complex vectors $\mathbf{p}$ and $\mathbf{q}$ rotate at the angular velocity ω in the complex plane and may be seen as sub-waves, the harmonics. The phase angle ϕ between both vectors denotes the system's impedance angle. Given this Ansatz for the pressure and flow, the governing equation (2.20) yields

$$i\omega Z + \frac{Z}{CR} = i\omega Z_\mathrm{a} \exp(i\phi) + \frac{R + Z_\mathrm{a}}{RC} \exp(i\phi), \tag{2.21}$$

where the definition of the system's impedance modulus $Z = P/Q$ has been used.

Without loss of generality, the expression (2.21) is then investigated at the time $t = 0$ towards the identification of the system unknowns, the impedance modulus Z, and the impedance phase ϕ, respectively. Euler's[16] formula $\exp(i\phi) = \cos\phi + i\sin\phi$ allows us to split (2.21) into imaginary and real parts

$$\left. \begin{aligned} \omega Z &= \omega Z_\mathrm{a} \cos\phi + \frac{R + Z_\mathrm{a}}{RC} \sin\phi, \\ \frac{Z}{RC} &= -\omega Z_\mathrm{a} \sin\phi + \frac{R + Z_\mathrm{a}}{RC} \cos\phi, \end{aligned} \right\} \tag{2.22}$$

which then results in a system with four solutions for Z and ϕ. The only physically admissible solution is

$$Z = \frac{\sqrt{\alpha\beta}}{1 + C^2 R^2 \omega^2} \; ; \; \phi = \begin{cases} -\arccos\left(\dfrac{R + Z_\mathrm{a} + C^2 R^2 \omega^2 Z_\mathrm{a}}{\sqrt{\alpha\beta}}\right) & \omega > 0 \\[3mm] \arccos\left(\dfrac{R + Z_\mathrm{a} + C^2 R^2 \omega^2 Z_\mathrm{a}}{\sqrt{\alpha\beta}}\right) & \omega < 0 \end{cases} \tag{2.23}$$

with $\alpha = 2RZ_\mathrm{a} + Z_\mathrm{a}^2 + R^2(1 + C^2 \omega^2 Z_\mathrm{a}^2)$ and $\beta = 1 + C^2 R^2 \omega^2$.

Given the parameters listed in Table 2.1, the impedance modulus and angle (2.23) are plotted against the system frequency $f = \omega/(2\pi)$ in Fig. 2.19. The three-element WK qualitatively captures the experimental observations, and at high frequencies the system's impedance approaches the aortic impedance $Z \to Z_\mathrm{a}$. At steady state $f = 0$, the entire flow passes the total resistance $R + Z_\mathrm{a}$, and the system's impedance then is $Z = R + Z_\mathrm{a}$.

Example 2.7 (Impedance-Based Estimation of WK Parameters). A ferret vascular system has a cardiac cycle of $T = 0.375$ s, and Table 2.4 shows its impedance $\mathbf{z}$ up to the frequency f of approximately $10\,\mathrm{Hz}$. Given this information, the parameters of the two-element and three-element WK models should be identified through least-square optimization.

(a) Define an objective function Φ and identify the least-square-optimized model parameters. Given the two-element WK model, the optimization problem $\Phi(R, C; \omega) \to \mathrm{MIN}$ determines the resistance R and the capacity C. Given

---

[16]Leonhard Euler, Swiss mathematician, physicist, astronomer, logician and engineer, 1707–1783.

Table 2.4 Impedance modulus Z and impedance angle ϕ of an individual ferret vascular system, see also Example 2.4

Frequency	Impedance modulus [mmHg s ml^{-1}]	Impedance angle [rad]
0.00000	108.415	0.00000
2.66667	4.12022	−0.96618
5.33333	3.34739	0.172411
8.00000	5.01376	0.132572
10.6667	4.60703	0.056390

Table 2.5 Fourier coefficients of a cyclic flow wave $q(t)$ with the period of $T = 0.375\,\text{s}$

Frequency [Hz]	Fourier coefficients [ml s^{-1}]
0.00000	0.786333
2.66667	0.0276938 − 0.646413 i
5.33333	−0.555816 − 0.112961 i
8.00000	−0.0896779 + 0.408525 i
10.6667	0.216635 + 0.0557698 i

the three-element WK model, the optimization problem $\Phi(R, C, Z_a; \omega) \to$ MIN determines the resistance R, the capacity C, and the characteristic aortic impedance Z_a.

(b) Plot the impedance modulus Z and impedance angle ϕ as predicted by the WK models on top of the data given in Table 2.4.

(c) Consider the flow wave $q(t)$ represented by the Fourier coefficients in Table 2.5 and compute the WK model-predicted pressure $p(t)$. Use a steady-state periodic analysis and compare the pressure with the predictions of a transient analysis. The numerical solution of the WK governing equations over a sufficiently large number of cardiac cycles, may be used. ∎

2.3.1.7 Four-Element WindKessel Model

Whilst the three-element WK model is suitable for many applications, the physical interpretation of the aortic impedance may be questioned [527]. Motivated by this shortcoming, the *four-element WK model* includes the *total arterial inertance* as an *inductor* element L in the circuit, see Fig. 2.22. It influences the system only at low frequencies.

The total flow $q(t)$ splits into the flow q_R through the vascular bed resistor R and the flow q_C into the capacitor C and determines the relation

$$q(t) = q_R + q_C = \frac{\overline{p}(t)}{R} + C\frac{d\overline{p}(t)}{dt}. \tag{2.26}$$

Fig. 2.22 (a) Hydraulic and (b) electric representations of the four-element WindKessel (WK) model. Flow $q(t)$ and pressure $p(t)$ describe the system state. The model parameters R, Z_a, C and L denote vascular bed resistance, aortic impedance, arterial capacity, and total arterial inertance, respectively

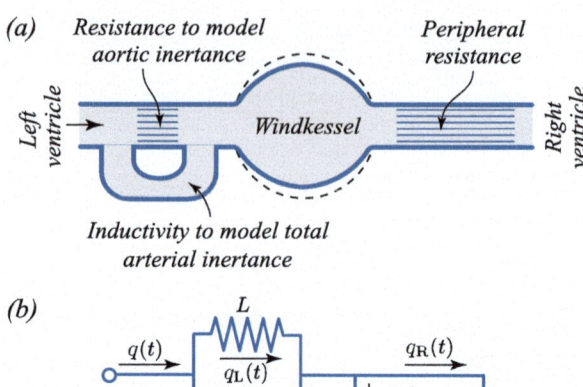

Given the flow $q_Z(t)$ over the aortic impedance Z_a, the total pressure $p(t)$ drops and

$$\overline{p}(t) = p(t) - Z_a q_Z(t) \tag{2.27}$$

holds. The pressure drop over Z_a is equal to the pressure drop over the inductor L,

$$L \frac{dq_L}{dt} = Z_a q_Z(t) , \tag{2.28}$$

where $p - \overline{p} = L(dq_L/dt)$ has been used to describe the inductor element, see Appendix E.1. The flow $q_L(t)$ and the flow $q_Z(t)$ together determine the total system flow

$$q(t) = q_L(t) + q_Z(t) , \tag{2.29}$$

which then closes the mathematical description of the four-element WK model.

The four equations (2.26)–(2.29) form the system

$$
\left.
\begin{aligned}
\frac{dp(t)}{dt} + \frac{p(t)}{CR} &= Z_a \frac{dq_Z(t)}{dt} + \frac{Z_a}{RC} q_Z(t) + \frac{q(t)}{C} , \\
\frac{dq_Z(t)}{dt} &= \frac{dq(t)}{dt} - \frac{Z_a}{L} q_Z(t) ,
\end{aligned}
\right\}
\tag{2.30}
$$

of linear differential equations that governs the four-element WK model. The expressions (2.26) and (2.27) lead to the first statement, whilst the second one follows from (2.28) and the time derivative of (2.29).

Given either $p(t)$ or $q(t)$, the system (2.30) may be solved numerically, and Fig. 2.16d illustrates such a (transient) solution. It considers the flow shown in Fig. 2.16a and uses the parameters listed in Table 2.1. The solution starts at the initial condition $p_0 = 80$ mmHg, and Fig. 2.16d shows $p(t)$ at steady-state periodic conditions.

2.3.2 Vessel Network Modeling

Lumped parameter models may also be used for the analysis of a *network of vessels*. Such a model facilitates the exploration of how changes in one part of the network influence the pressure $p(t)$ and flow $q(t)$ somewhere else in the network. They may therefore test the outcome of vascular interventions, for example. A lumped parameter model expresses the pressure and flow as functions of the time t and neglects their dependence on the vascular path coordinate x—a network model can therefore not simulate phenomena, such as wave propagation.

Given a network of n vessel segments, the individual segments are represented by their *capacity* C_i, *resistance* R_i, and *inertance* L_i; $i = 1, \ldots, n$, and then connected at m *network nodes*. Different designs of lumped parameter models have been proposed to describe the biomechanics of a vessel segment. The *three-element vessel model* illustrated in Fig. 2.23 is one possible design. It models a vessel segment of diameter d [m], length l [m], and wall thickness h [m], which is entirely filled by blood of the density ρ [kg m^{-3}] and the dynamic viscosity η [Pa s].

2.3.2.1 Vessel Segment Resistance
A laminar, steady-state and fully developed flow in a cylindrical vessel results in a *parabolic velocity profile* over the vessel's cross-section called a Poiseuille flow. The WSS that develops in response to the fluid flowing over the vessel wall presents resistance to the flow q. Given Poiseuille flow, the flow q and WSS τ_w are related through $q = -r^3 \pi \tau_w / 4$, where r denotes the vessel radius, see Chap. 6.

Fig. 2.23 (**a**) Schematic and (**b**) electric representations of the three-element vessel model. The flows $q_{in}(t)$, $q_{out}(t)$ and the pressures $p_{in}(t)$, $p_{out}(t)$ describe vessel inlet and outlet conditions. The vessel's biomechanical properties are expressed by its capacity C, resistance R, and inertance L, respectively

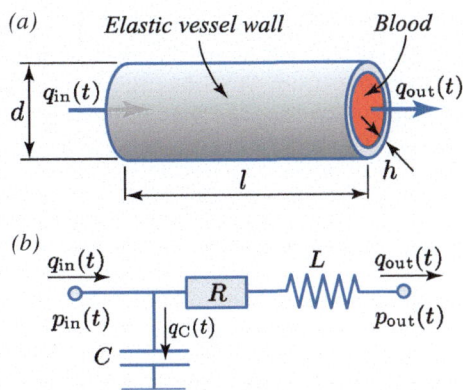

The flow q through the vessel depends on the pressure drop Δp between the vessel's inlet and outlet, and

$$R = \frac{\Delta p}{q} = \frac{128\eta l}{\pi d^4} \tag{2.31}$$

determines the *resistance* R against flow, a relation known as the *law of Hagen–Poiseuille*. Poiseuille flow provides a good description in smaller vessels and veins, whilst in large arteries the blood flow is highly influenced by inertia effects. A parabolic velocity profile is then only present during the systolic phase of the cardiac cycle, see Chap. 6.

The vessel's resistance R represents energy dissipation, and in addition to (2.31), dissipative sources may also relate to unsteady flow, flow separation, vessel curvature, and vessel bifurcations [376]. Finally, we note that laboratory experiments may also be used for the direct measurement of the resistance R of a vessel.

2.3.2.2 Vessel Segment Capacity

The vessel segment's *capacity* expresses the increase ΔV of blood volume that is inside the vessel in response to the increase Δp in pressure. This property can either be measured experimentally or predicted through the modeling of the elasticity of the vessel wall. Given linear elasticity, or Hooke's[17] law in the description of the vessel wall,

$$C = \frac{\Delta V}{\Delta p} = \frac{3d^3\pi l}{16hE} \tag{2.32}$$

expresses the vessel's capacity, where E [Pa] denotes the vessel wall's Young's[18] modulus, see Sect. 3.5.2. We may also derive the alternative expression (5.2) that considers any non-linear elastic description of the vessel wall.

2.3.2.3 Blood Inertance

Given the mean blood flow velocity $v_{\text{mean}} = q(t)/A$, the force equilibrium of the blood segment along the axial direction reads $\Delta p A = -\rho A l \dot{v}_{\text{mean}}$, where the contribution from WSS has been neglected. It allows us to express the pressure increment $\Delta p = \rho l \dot{q}/A$ as a function of the change of flow $\dot{q}$, where the cross-section $A = d^2\pi/4$ was assumed to be constant along the vessel. The *inertance*

$$L = \frac{\Delta p}{\dot{q}} = \frac{4\rho l}{d^2\pi} \tag{2.33}$$

then describes the inertia of the blood in the vessel.

---

[17]Robert Hooke, English natural philosopher, architect, and polymath, 1635–1703.
[18]Thomas Young, English polymath and physician, 1773–1829.

Example 2.8 (Renal Artery Adaptation to Partial Nephrectomy). Figure 2.24 schematically illustrates the surgical removal of a part of the kidney, an intervention called partial nephrectomy. It increases the kidney's vascular bed resistance R by the factor α. At baseline the renal artery has the radius r_0 and wall thickness h_0, properties that alter in response to the intervention towards r and h, respectively. Homeostasis drives the adaptation, and r and h change until the renal artery's Wall Shear Stress (WSS) τ_w as well as its circumferential wall stress σ_θ return to their homeostatic values. It may be assumed that the arterial pressure and blood properties are not influenced by the surgical intervention.

Fig. 2.24 Schematic illustration of partial nephrectomy, with the dark area indicating the removed section of the kidney

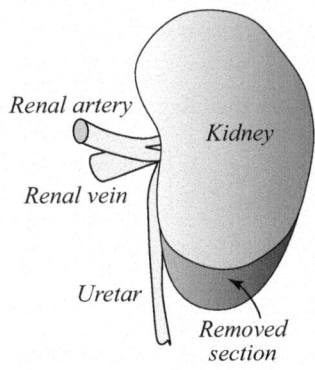

(a) Derive the relation between the factor α that augments the resistance and the mean flow q through the kidney.
(b) Assume Poiseuille flow and derive the relation between the WSS τ_w and the mean flow q through the renal artery.
(c) Express the circumferential wall stress σ_θ as a function of vessel's dimensions and the blood pressure. Consider the thickness of the vessel to be much smaller than its diameter.
(d) At homeostatic conditions of τ_w and σ_θ, r/r_0 and h/h_0 should be expressed as functions of α. ∎

2.3.2.4 Governing Equation

Each lumped parameter model of a vessel has its specific governing equation, and this section discusses the design shown in Fig. 2.23. The sum of the pressure drop Δp_R over the resistance as well as Δp_L over the inertance, determines the total pressure change

$$p_{in} - p_{out} = \Delta p_R + \Delta p_L = R q_{out} + L \dot{q}_{out}, \qquad (2.37)$$

where p_{in} and p_{out} denote the pressure at the vessel inlet and outlet, respectively. The flow continuity

$$q_{in} - q_{out} = q_C = C\dot{p}_{in} \tag{2.38}$$

closes the mathematical description, and in the derivation of (2.37) and (2.38) the properties of circuit elements have been used, see Appendix E.1. The model also assumes a negligible ambient (interstitial) pressure, and the capacity C is therefore directly exposed to the inlet pressure p_{in}.

Equations (2.37) and (2.38) may be rearranged towards

$$p_{out} = p_{in} - R(q_{in} - C\dot{p}_{in}) - L(\dot{q}_{in} - C\ddot{p}_{in}),$$

$$q_{out} = q_{in} - C\dot{p}_{in},$$

which then expresses the relation between input and output by the matrix equation

$$\begin{bmatrix} p_{out} \\ q_{out} \end{bmatrix} = \begin{bmatrix} 1 & -R \\ 0 & 1 \end{bmatrix} \begin{bmatrix} p_{in} \\ q_{in} \end{bmatrix} + \begin{bmatrix} RC & -L \\ -C & 0 \end{bmatrix} \begin{bmatrix} \dot{p}_{in} \\ \dot{q}_{in} \end{bmatrix} + \begin{bmatrix} LC & 0 \\ 0 & 0 \end{bmatrix} \begin{bmatrix} \ddot{p}_{in} \\ \ddot{q}_{in} \end{bmatrix}. \tag{2.39}$$

In symbolic notation it reads

$$\mathbf{d}_{out} = \mathbf{K}\mathbf{d}_{in} + \mathbf{D}\dot{\mathbf{d}}_{in} + \mathbf{M}\ddot{\mathbf{d}}_{in}. \tag{2.40}$$

The set $\{p, q, \dot{p}, \dot{q}, \ddot{q}\}$ of state variables describe the system, and a time-marching algorithm may be used to solve the governing equation (2.40) at prescribed boundary and initial conditions.

We may for example consider a *backward-Euler* discretization over the time step Δt, and the first and second time derivatives are then approximated by

$$\left. \begin{aligned} \dot{x}_{in} &= \frac{x_{in} - x_{in\,n}}{\Delta t} = \frac{x_{in}}{\Delta t} - \frac{x_{in\,n}}{\Delta t}, \\ \ddot{x}_{in} &= \frac{\dot{x}_{in} - \dot{x}_{in\,n}}{\Delta t} = \frac{x_{in}}{\Delta t^2} - \frac{x_{in\,n}}{\Delta t^2} - \frac{\dot{x}_{in\,n}}{\Delta t}, \end{aligned} \right\} \tag{2.41}$$

where $x = p, q$ and $(\bullet)_n$ denotes a quantity at the previous time step. Given such a discretization, the system (2.40) of differential equations leads to the algebraic set of equations

$$\begin{bmatrix} p_{out} \\ q_{out} \end{bmatrix} = \begin{bmatrix} 1 + \frac{RC}{\Delta t} + \frac{LC}{\Delta t^2} & -R - \frac{L}{\Delta t} \\ -\frac{C}{\Delta t} & 1 \end{bmatrix} \begin{bmatrix} p_{in} \\ q_{in} \end{bmatrix} + \mathbf{H}, \tag{2.42}$$

where the *history vector*

$$\mathbf{H} = \begin{bmatrix} -\frac{RC}{\Delta t} - \frac{LC}{\Delta t^2} & \frac{L}{\Delta t} \\ \frac{C}{\Delta t} & 0 \end{bmatrix} \begin{bmatrix} p_{in\,n} \\ q_{in\,n} \end{bmatrix} - \begin{bmatrix} \frac{LC}{\Delta t} & 0 \\ 0 & 0 \end{bmatrix} \begin{bmatrix} \dot{p}_{in\,n} \\ \dot{q}_{in\,n} \end{bmatrix}, \tag{2.43}$$

stores information from the previous time step. Consequently, the system (2.42) uniquely specifies the relation between input and output variables.

Example 2.9 (Two-Element Vessel Segment Model). An arterial vessel segment of the resistance R and the capacity C is modeled by the lumped parameter model shown in Fig. 2.25. The model is used for a steady-state periodic analysis of a vascular network.

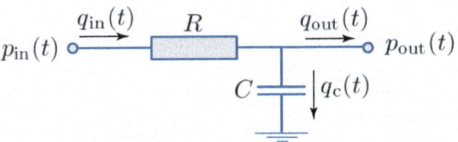

Vessel diameter d	5.2 mm
Vessel segment length l	12.7 cm
Vessel wall thickness h	0.5 mm
Vessel wall Young's modulus E	34.0 kPa
Dynamic blood viscosity η	4.0 mPa s

Fig. 2.25 Electrical representations of a two-element vessel model with the corresponding vessel segment parameters. The flows $q_{in}(t)$, $q_{out}(t)$ and the pressures $p_{in}(t)$, $p_{out}(t)$ describe the vessel inlet and outlet conditions. The vessel's biomechanical properties are expressed by the capacity C and resistance R, respectively

(a) Derive the governing equations of the lumped parameter model shown in Fig. 2.25 and rearrange them according to the system of equations

$$\begin{bmatrix} \mathbf{p}_{out} \\ \mathbf{q}_{out} \end{bmatrix} = \begin{bmatrix} K_{11} & K_{12} \\ K_{21} & K_{22} \end{bmatrix} \begin{bmatrix} \mathbf{p}_{in} \\ \mathbf{q}_{in} \end{bmatrix} + \begin{bmatrix} D_{11} & D_{12} \\ D_{21} & D_{22} \end{bmatrix} \begin{bmatrix} \dot{\mathbf{p}}_{in} \\ \dot{\mathbf{q}}_{in} \end{bmatrix} ,$$

where $\mathbf{p}_{in}$, $\mathbf{q}_{in}$ and $\mathbf{p}_{out}$, $\mathbf{q}_{out}$ are complex vectors that describe the inlet and outlet, respectively.
(b) Given the data in Fig. 2.25, compute the resistance R and the capacity C of the vessel segment.
(c) Compute the flow and pressure in the vessel in response to the cyclic boundary conditions $\mathbf{p}_{out} = |\mathbf{p}_{out}| \exp[i(\omega t + \pi/6)]$ and $\mathbf{q}_{in} = |\mathbf{q}_{in}| \exp[i\omega t]$, where $|\mathbf{p}_{out}| = 12.5$ Pa and $|\mathbf{q}_{in}| = 4.3$ ml s^{-1} are the pressure and flow amplitudes, whilst $\omega = 73\pi$ denotes the angular velocity of the imaginary vectors. Use Argand's diagram to draw pressure and flow in the complex plane. ∎

Example 2.10 (Three-Element Vessel Segment Model). The lumped parameter model in Fig. 2.26 uses the resistance R, the capacity C, and the inductance L to describe an arterial vessel segment.

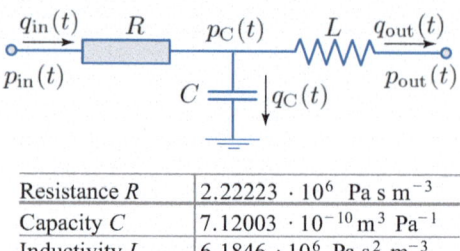

Resistance R	$2.22223 \cdot 10^6$ Pa s m^{-3}
Capacity C	$7.12003 \cdot 10^{-10}$ m^3 Pa^{-1}
Inductivity L	$6.1846 \cdot 10^6$ Pa s^2 m^{-3}

Fig. 2.26 Electrical representations of a three-element vessel model with vessel segment parameters. The flows $q_{in}(t)$, $q_{out}(t)$ and the pressures $p_{in}(t)$, $p_{out}(t)$ describe vessel inlet and outlet conditions. The vessel's biomechanical properties are expressed by the resistance R, capacity C, and inductance L, respectively

(a) Derive the governing equations of the lumped parameter model shown in Fig. 2.25 and rearrange them according to

$$\mathbf{d}_{out} = \mathbf{K}\mathbf{d}_{in} + \mathbf{D}\dot{\mathbf{d}}_{in} + \mathbf{M}\ddot{\mathbf{d}}_{in} \, ,$$

where $\mathbf{d}_{out} = [p_{out} \, q_{out}]^T$ denotes the outlet condition, whilst $\mathbf{d}_{in} = [p_{in} \, q_{in}]^T$, $\dot{\mathbf{d}}_{in} = [\dot{p}_{in} \, \dot{q}_{in}]^T$ and $\ddot{\mathbf{d}}_{in} = [\ddot{p}_{in} \, \ddot{q}_{in}]^T$ describe the inlet conditions.

(b) Compute the flow and pressure for the steady-state periodic inflow $\mathbf{q}_{in} = |\mathbf{q}_{in}| \exp[i\omega t]$ with $|\mathbf{q}_{in}| = 4.3$ ml s^{-1} and $\omega = 2\pi$, and against the constant outlet pressure $\mathbf{p}_{out} = |\mathbf{p}_{out}| = 1000.0$ Pa. Use Argand's diagrams to illustrate the magnitude and phase angle of each complex quantity. ∎

2.3.2.5 Assembly of Vessel Networks

Vessel segment models, such as the three-element model described by Eq. (2.40), may be connected at *nodes* to form a network of vessels. The *compatibility conditions* relate then the flow q and the pressure p (and their time derivatives) across network nodes. Given a single vessel connects to another single vessel, the compatibility condition at the node simply reads $q_1 = q_2$ and $p_1 = p_2$ with the index denoting the vessel number.

Given a single vessel that bifurcates into two vessels, the compatibility condition reads $q_1 = \xi_q q_2 + (1 - \xi_q)q_3$ and $p_1 = p_2 = p_3$, where the index denotes the vessel number. The system state variable $0 \le \xi_q \le 1$ describes how the flow splits in the bifurcation, and identical to the other state variables, ξ_q is identified by the solution of the system of equations that represents the entire network of vessels. It might also be required to consider a *pressure drop* at network nodes to capture *energy dissipation* from significant flow disturbance at the bifurcation. The pressure compatibility conditions then read $p_1 = \xi_p p_2$. Here, $0 \le \xi_p \le 1$ accounts for the pressure drop, and p_1 and p_2 denote the pressure upstream and downstream the bifurcation, respectively.

Example 2.11 (Connected Vessel Segments). Figure 2.27 illustrates the connection of three vessels and their properties. The inflow $q_{in} = q_0[1 + \sin(\omega t)]$ with $q_0 = 0.05 \, \text{ml s}^{-1}$ and $\omega = 2\pi$ is prescribed, whilst the constant pressures $p_2 = 11.5 \, \text{kPa}$ and $p_3 = 11.0 \, \text{kPa}$ are applied at the outlets. The vessels are filled with blood of the density $\rho = 1060 \, \text{kg m}^{-3}$ and the dynamic viscosity $\eta = 3.5 \, \text{mPa s}$.

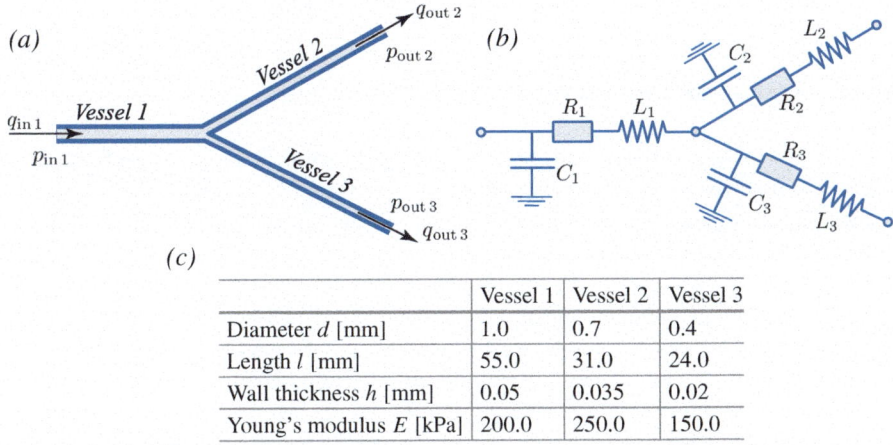

	Vessel 1	Vessel 2	Vessel 3
Diameter d [mm]	1.0	0.7	0.4
Length l [mm]	55.0	31.0	24.0
Wall thickness h [mm]	0.05	0.035	0.02
Young's modulus E [kPa]	200.0	250.0	150.0

Fig. 2.27 (a) Schematic and (b) electrical representation of a lumped parameter model that represents three connected vessel segments (c) Geometrical and mechanical properties of vessel segments

(a) Compute the resistance R_i, the capacity C_i, and the inductance L_i of the three vessels $i = 1, 2, 3$.
(b) Derive the governing equation of the i-th vessel according to the lumped parameter model shown in Fig. 2.27b.
(c) Provide incremental governing equations of the connected vessels by considering the time derivatives of flow and pressure according to the backward-Euler discretization (2.41).
(d) Propose an algorithm for the iterative solution of the incremental governing equations. At the bifurcation a dissipation-free flow may be considered, and the factor $0 \leq \xi \leq 1$ determines the flow split.
(e) Plot the pressure at the inlet $p_{in\,1}(t)$, the flows at the outlets $q_{out\,1}(t)$, $q_{out\,2}(t)$ as well as the flow split factor $\xi(t)$ over the time.   ∎

2.4   Modeling the Microcirculation

This section describes models towards the analysis of exchange aspects in the micro-circulation. They are limited to passive microfluidic transport mechanisms, and a bottom-up approach aims at modeling the physics of microcirculatory exchange. Exchange models are well documented with excellent reviews [40, 338, 493] available in the literature.

2.4.1   Transcapillary Concentration Difference

The vessel wall of capillaries may be seen as a rigid porous body that is perforated by a large number of micro-channels connecting the vessel lumen with the interstitial space. The transport of *substances or solutes* through such micro-channels deter-mines the transcapillary concentration difference of a solute. Towards the analysis of this property, we consider a micro-channel of length L filled with fluid that contains a solute at the concentration $c(x)$ [mol m^{-3}], where $0 \leq x \leq L$ denotes the Cartesian coordinate along the micro-channel, and thus across the capillary wall. Given microvascular exchange, the fluid is essentially *water*.

Diffusive and *advective* transport govern the *solute flux* J_s [mol s^{-1}m^{-2}] along the micro-channel. At steady state, J_s is constant all along the channel and governed by the first-order partial differential equation

$$J_s = \text{const} = \underbrace{-D \operatorname{grad} c(x)}_{\text{Diffusion}} + \underbrace{c(x)v}_{\text{Advection}} \quad , \tag{2.48}$$

where v [m s^{-1}] denotes the transport velocity of solute particles, whilst D [m^2 s^{-1}] is the diffusion constant for solute particles in water. Towards a dimensionless analysis, we may introduce $\xi = x/d$ and normalize the solute particle path length x with the solute particle diameter d. Equation (2.48) has then the solution $c(\xi) = J_s/v + H \exp(\xi Pe)$, where $Pe = vd/D$ and H denote the Péclet[19] number and an integration constant, respectively.

The Péclet number is the ratio between advective and diffusive transport. At large Péclet numbers the solvent moves together with the fluid flow, whilst at low numbers it moves independently from the motion of the fluid. The solute particles may not be spherical, and d then denotes the Stokes diameter—the diameter of the hydrodynamically similar, but spherical particle.

The aforementioned integration constant H can be identified from the solute concentration c_v in the vascular space, and thus at the inlet of the micro-channel. The solute concentration then reads $c(\xi) = J_s/v + \exp(\xi Pe)(c_v - J_s/v)$. We may also introduce the interstitial solute concentration $c_i = c(\xi = L/d)$, such that

---

[19]Jean Claude Eugène Péclet, French physicist, 1793–1857.

$$J_s = c_v v - \frac{v \Delta c}{\exp[Pe(L/d)] - 1} \tag{2.49}$$

finally expresses the solute flux across the capillary wall, where $\Delta c = c_i - c_v$ is the transcapillary concentration difference.

At steady state, the interstitial solute concentration c_i depends on the transcapillary filtration flux q_f [m s^{-1}] that determines the flow rate across the unit area of the capillary wall, see Sect. 2.4.2. Experimental data suggest that c_i is proportional to the solute flux J_s but inverse proportional to the filtration flux q_f [543]. It therefore justifies the relation $c_i = J_s/q_f$ and allows for the substitution of J_s in (2.49). The transcapillary concentration difference then reads [370]

$$\Delta c = c_v - c_i = \sigma c_v \frac{\exp[Pe(L/d)] - 1}{\exp[Pe(L/d)] - \sigma}, \tag{2.50}$$

where the solute particle velocity $v = (1 - \sigma)q_f$ has been related to the filtration flux q_f through Staverman's[20] osmotic reflection coefficient σ [526], see Appendix E.3.

Example 2.12 (LDL Transport Through a Micro-channel). Low-Density Lipoprotein (LDL) is a protein of about 25 nm in size and transports fat molecules around the body. Increased LDL concentration has been strongly associated with the development of atherosclerosis—LDL that invades the vessel wall is oxidized and then poses a risk for the development of atherosclerosis. Endothelial Cells (EC) junctions form micro-channels, and LDL eventually slowly "leaks" across the endothelial barrier into the vessel wall. The model system shown in Fig. 2.28 may be used to investigate LDL transport through EC junctions.

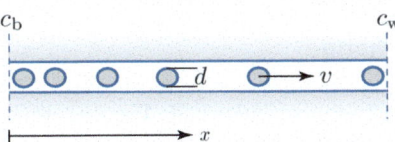

Fig. 2.28 Transport of Low-Density Lipoprotein (LDL) of diameter d and velocity v through a micro-channel. LDL concentrations in the bloodstream and the arterial wall are denoted by c_b and c_w, respectively

(a) Consider the velocity $v = q_f(1 - \sigma)$ of the LDL particles and express the LDL flux. Here, $q_f = 1.0$ μm s^{-1} and $\sigma = 0.78$ denote the fluid velocity and Staverman's osmotic reflection coefficient, respectively.
(b) Plot the LDL flux at the Péclet numbers $Pe = 0.01; 0.1; 1.0$ and the boundary concentrations of $c_b = 2.0$ mol m^{-3} and $c_w = 0.1$ mol m^{-3}. ∎

---

[20] Albert Jan Staverman, Dutch chemist, 1911–1993.

2.4.2   Filtration

Filtration is the transport of water through the capillary wall in response to *hydrostatic* and *osmotic* pressure differences. From the vascular space to the interstitial space, the hydrostatic pressure falls from p_v to p_i, whilst COP increases from Π_v to Π_i. The exchange of water between the two spaces is determined by the *filtration flux* q_f [m s^{-1}], and thus the flow rate that passes through the unit area of wall, see Appendix E.2. Given the flow through the vessel wall is aligned with the vessel's radial direction, the flux q_f is identical to the flow velocity.

2.4.2.1 Starling's Filtration Model

Filtration in the microvasculature may be described by Starling's filtration model [524]

$$q_f = L_p \left(\Delta p - \sigma \Delta \Pi \right) , \tag{2.51}$$

where $\Delta p = p_v - p_i$ is the transcapillary hydrostatic pressure, whilst $\Delta \Pi = \Pi_v - \Pi_i$ denotes the transcapillary COP. The term $\Delta p - \sigma \Delta \Pi$ is called *net filtration pressure* and positive for flow from the vascular system into the interstitium. In (2.51), L_p [m Pa^{-1}s^{-1}] and σ denotes the capillary wall's hydraulic conductivity and its Staverman's osmotic reflection coefficient [526], respectively. Whilst the hydraulic conductivity describes the leakiness of the wall to water, the reflection coefficient σ corrects the theoretical COP difference to match the effective one, see Appendices E.2 and E.3.

According to (2.51), the filtration is governed by the four "Starling forces" p_v, p_i, $\sigma \Pi_v$, and $\sigma \Pi_i$. The model is also "symmetric"—the increase in p_v (or Π_v) or the decrease of p_i (or Π_i) by the same amount affects the flux equally.

2.4.2.2 Predicted Exchange

The net effect from the inflow and outflow of water across the capillary walls determines the exchange in the vascular bed. The direct measurement of exchange is difficult, and a model, such as Starling's filtration law (2.51), helps to interpret (incomplete) experimental data.

The linear relation between q_f and Δp has been shown in a population of vessels [321], whilst much less experimental data confirmed the linearity between q_f and $\Delta \Pi$ [417]. The development of flow requires the hydrostatic pressure p_v to decrease along the vascular tree. Given human nailfold skin capillaries, pressures of $p_v = 35$ to 45 mmHg at the arterial side and $p_v = 12$ to 15 mmHg at the venous sides have been reported [322]. In addition, the interstitial hydrostatic pressure $p_i = -4$ to 0 mmHg is slightly below the atmospheric pressure in many tissues, and $\Pi_v = 25$ to 28 mmHg is believed to represent vascular COP in humans. Table 2.9 summarizes this information, and Starling's law (2.51) then predicts the fluid exchange that is shown in Fig. 2.29.

Table 2.9 Representative parameters for the description of a number of soft biological tissues

Hydrostatic pressure in the capillary p_v	Arterial/venous side: 35.0/12.0 mmHg
Hydrostatic pressure in the interstitium p_i	−2.0 mmHg
Colloid Osmotic Pressure (COP) in the capillary Π_v	Arterial/venous side: 28.0/25.0 mmHg
COP in the interstitium Π_i	1.0 mmHg
Hydraulic conductivity of the capillary wall L_p	$1.5 \cdot 10^{-9}$ m s^{-1}mmHg^{-1}
Reflection coefficient of the capillary wall σ	1.0

Fig. 2.29 Exchange of fluid along capillaries. (**a**) Transcapillary hydrostatic pressure Δp and transcapillary Colloid Osmotic Pressure (COP) $\Delta \Pi$. (**b**) Filtration flux q_f across the capillary wall. Exchange according to Starling's filtration model (2.51) and the data listed in Table 2.9. (**c**) Net outward flux at the arterial side and net inward flux at the venous side

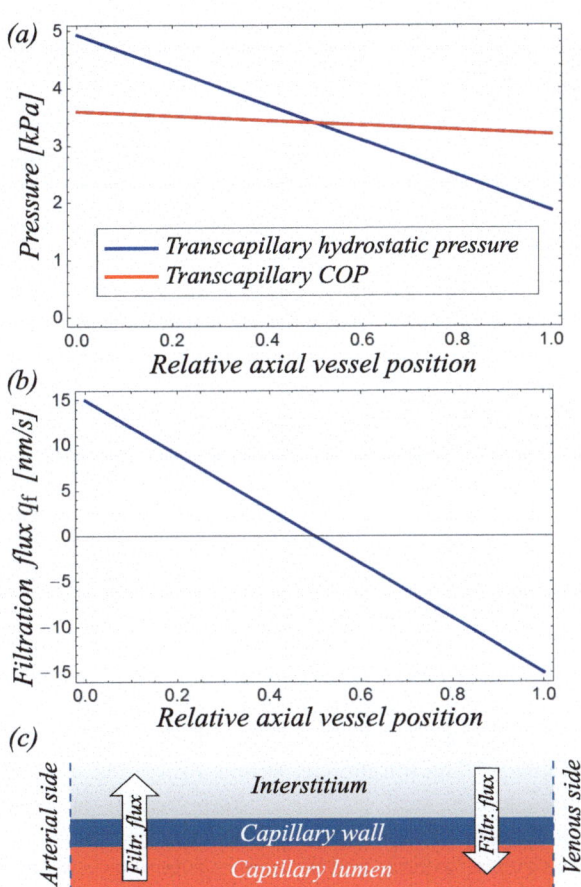

These findings suggest the exchange mechanism proposed by Starling in 1896 [524]: fluid transport into the interstitial over the arterial half of the capillary where $\Delta p > \sigma \Delta \Pi$, and reabsorbtion over the venous half where $\Delta p < \sigma \Delta \Pi$, see Fig. 2.29c. At transient conditions this mechanism has been confirmed by more advanced experiments [371], but counter-evidence emerged in a number of tissues at steady-state exchange conditions. A revised understanding of vascular exchange is presented in Sect. 2.4.2.3.

Example 2.13 (Transport Across the Ascending Vasa Recta Wall). The model system shown in Fig. 2.30 should be used to investigate the exchange of solutes between the Ascending Vasa Recta (AVR) of the renal medulla and its interstitial space. The AVR segment is $L = 1.2$ mm long, has a diameter of $D = 29\,\mu$m, and blood flows through it at the rate $q = 35$ nl min^{-1}.

Fig. 2.30 Model system to investigate the exchange between the Ascending Vasa Recta (AVR) and the interstitium

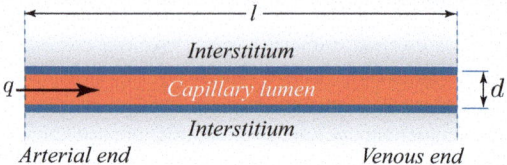

(a) Compute the distribution of the hydrostatic pressure p_v along the AVR segment. Blood may be regarded as a Newtonian fluid with the dynamic viscosity $\eta = 3.5$ mPa s, and the pressure $p_{v\,art} = 7.8$ mmHg applies at the AVR's arterial end.

(b) Compute the exchange along the AVR according to Starling's filtration law. Within the interstitium, the hydrostatic pressure $p_i = 6.0$ mmHg and the Colloid Osmotic Pressure (COP) $\Pi_i = 3.7$ mmHg are given. In addition, COP changes linearly from $\Pi_{v\,art} = 26.0$ mmHg at the arterial end to $\Pi_{v\,ven} = 16.7$ mmHg at the venous end. The Staverman's osmotic reflection coefficient $\sigma = 0.78$ and the hydraulic conductivity $L_p = 1.5 \cdot 10^{-9}$ m s^{-1}mmHg^{-1} may be assumed in the description of the vessel wall.   ∎

2.4.2.3 Current Understanding of Microvascular Exchange

Filtration measurements in response to a sudden change of the capillary hydrostatic pressure p_v [371] challenged the aforementioned concept of microvascular exchange. The filtration experiment recorded *transient results* right after the pressure step as well as *steady-state results* at least 2 min after it. Whilst the transient results confirmed earlier experiments [417], and once again validated the linearity between q_f and the transcapillary hydrostatic pressure Δp_v, the steady-state results did not show such a linearity and therefore contradicted Starling's filtration law (2.51). The observed slight imbalance in the transcapillary hydrostatic pressure at steady state favors a positive net filtration pressure. The fluid moves therefore into the interstitial space and is then *almost exclusively* drained via the lymphatics. Given the interstitial fluid volume remains constant, the filtration flux times capillary surface is equal to lymph flow. Whilst such steady-state exchange is observed in many tissues, the net filtration pressure is also negative in some tissues [338]. It then facilitates the function of organs, such as the kidney and intestinal mucosa. The lion part of the interstitial fluid is then absorbed back into the vascular compartment.

2.4.2.4 Colloid Osmotic Pressure

The macromolecule *albumin* is the main contributor to the COP in the vasculature. In the vascular space the albumin concentration remains approximately constant, and transcapillary COP is therefore primary controlled by the *interstitial* albumin concentration.

Given albumin's transcapillary concentration Δc_{alb}, Hoff's[21] law relates it to the transcapillary COP

$$\Delta \Pi_{alb} = \phi \overline{R} \theta \Delta c_{alb} , \tag{2.53}$$

where the universal gas constant $\overline{R} = 8.3145$ J K^{-1}mol^{-1}, the absolute temperature θ [K], and the osmotic coefficient ϕ have been introduced, see Appendix E.2. Albumin is a relatively small protein and passes the capillary wall together with water, and Δc_{alb} develops according to transport discussed in Sect. 2.4.1. Given (2.50) and (2.53), the transcapillary albumin COP may therefore be expressed by

$$\Delta \Pi_{alb} = \sigma \Pi_{v\,alb} \frac{\xi - 1}{\xi - \sigma} \; ; \; \xi = \exp[Pe(L/d_{alb})] , \tag{2.54}$$

where d_{alb} and L are the albumin's Stokes diameter and the capillary wall thickness, respectively. In addition, $\Pi_{v\,alb} = \phi \overline{R} \theta c_{v\,alb}$ is the albumin's COP in the vascular space, whilst $Pe = (1 - \sigma)q_f d_{alb}/D$ is the Péclet number with D [s^{-1}m^{-1}] denoting albumin's diffusion constant in water. The calculation of Pe assumes that albumin moves at the velocity $v = (1 - \sigma)q_f$ across the capillary wall, where σ and q_f denote Staverman's osmotic reflection coefficient and the filtration flux, respectively. Equation (2.54) may easily be adapted to proteins other than albumin.

2.4.2.5 The Non-linear Filtration Law

Albumin is the main contributor to osmosis in the microcirculation, and $\Delta \Pi$ in Starling's filtration model (2.51) may therefore be approximated by (2.54), which then yields the relation

$$q_f = L_p \left[\Delta p - \sigma^2 \Pi_{v\,alb} \frac{1 - \xi}{1 - \sigma \xi} \right] \; ; \; \xi = \exp[Pe(L/d_{alb})] . \tag{2.55}$$

Given the Péclet number $Pe = (1 - \sigma)q_f d_{alb}/D$, the expression (2.55) is implicit in the filtration flux q_f. Fig. 2.31 plots it against the transcapillary hydrostatic pressure Δp, where the properties in Table 2.10 have been used. The plot illustrates the *non-linearity* between q_f and Δp. It is caused by the second term at the right side of Eq. (2.55)$_1$ and determines the *diffusion-related* contribution to Δp. Diffusion is only significant at small Péclet numbers, whilst at high Péclet numbers, this term

---

[21] Jacobus Henricus van 't Hoff, Jr., Dutch physical chemist, 1852–1911.

Fig. 2.31 Filtration flux q_f in relation to transcapillary hydrostatic pressure Δp, as predicted by the non-linear filtration model (2.55). At high positive Δp, the vessel's hydraulic conductivity L_p determines exchange, whereas almost no flux appears at negative Δp

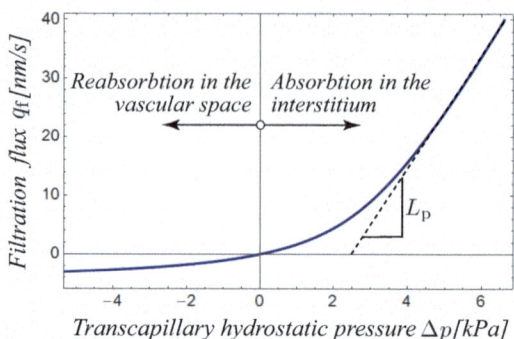

Table 2.10 Set of parameters used in the non-linear filtration model (2.55)

Diffusion coefficient D for albumin in water	$1.0\cdot10^{-14}\,\mathrm{m^{-2}s^{-1}}$
Thickness of the capillary wall L	$2.0\cdot10^{-6}\,\mathrm{m}$
Colloid Osmotic Pressure (COD) in the vascular space $\Pi_{v\,alb}$	$28.0\,\mathrm{mmHg}$
Hydraulic conductivity of the capillary wall L_p	$1.5\cdot10^{-9}\,\mathrm{m\,s^{-1}mmHg^{-1}}$
Staverman's osmotic reflection coefficient σ	0.95

vanishes, and the relation between q_f and Δp is fully determined by the hydraulic conductivity L_p of the wall. The curve in Fig. 2.31 is therefore linear at high transcapillary hydrostatic pressures.

The non-linear filtration model (2.55) holds for steady-state conditions—given enough time for the albumin concentration to settle down, steady-state conditions establish, and the assumptions made to derive (2.54) hold.

Fig. 2.31 shows minimal reabsorption back into the vascular space, conditions that holds for most, but not all tissues. In the kidney and the intestinal mucosa, interstitial fluid is continuously renewed by protein-free fluid, which in turn breaks the dependence of the filtration flux and the solute concentration. The assumptions made to derive equation (2.54) are then not valid, and fluid flux is best predicted by Starling's filtration model (2.51).

2.4.2.6 Two-Pore Models

The filtration flux (2.55) suggests minimal reabsorption at the venous side of capillaries [338], which is confirmed by experimental data in many tissues [371]. The lack of reabsorption backs into the vascular system, making it difficult to reconcile with low *in vivo* values for whole-body lymph flow rates. A class of models, known as *two-pore models*, aim at resolving this shortcoming [464]. Whilst albumin is not amongst the largest proteins, it still cannot pass the capillary wall through most pores. Most fluid flux q_f occurs through small pores that hinder albumin transport, whilst large pores that support the advection of albumin transport only a minute fraction of water through the wall.

The glycocalyx–cleft model illustrated in Fig. 2.32 is a specific case of a two-pore model. It assumes that the glycocalyx, *not* the entire capillary wall, dominates the filtration properties and represents the effective osmotic barrier. The glycocalyx almost entirely determines the reflection coefficient σ and the transcapillary COP. The COP underneath the glycocalyx layer is therefore very low, which in turn yields levels of net filtration fluxes that would compare to reported lymph drainage rates.

Example 2.14 (Glycocalyx–Cleft Model). Starling's filtration model (2.51) over-predicts the filtration flux q_f and has therefore been further developed towards the glycocalyx–cleft model shown in Fig. 2.32. The glycocalyx layer presents a system of ultra-fine pores in series to the larger pores formed by the endothelial clefts and thus the spaces between adjacent Endothelial Cells (ECs). Given the strong washout of albumin upon the filtration flux through the endothelial clefts, albumin concentration c_{gc} becomes very low underneath the glycocalyx layer. Along the filtration path three distinct albumin concentrations therefore establish: c_v in the vascular space, c_{gc} underneath the glycocalyx layer, and c_i in the interstitial space, see Fig. 2.32.

Given a $L = 3.5$ mm long capillary, which wall has the hydraulic conductivity of $L_p = 2.3 \cdot 10^{-9}$ s^{-1}mmHg^{-1}m and the Staverman's osmotic reflection coefficient $\sigma = 0.8$, the glycocalyx–cleft model should be analyzed. In addition, the hydrostatic pressure $p_i = -1.0$ mmHg and the Colloid Osmotic Pressure (COP) $\Pi_i = 3.7$ mmHg are approximately constant in the interstitium. Given the vascular space, the hydrostatic pressure changes linearly from $p_{v\,art} = 9.4$ mmHg to $p_{v\,ven} = 6.7$ mmHg, and the COP from $\Pi_{v\,art} = 15.8$ mmHg to $\Pi_{v\,ven} = 13.9$ mmHg between the capillary's arterial and venous ends, respectively.

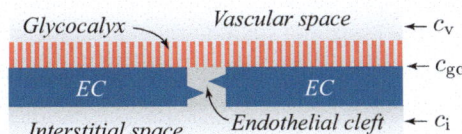

Fig. 2.32 Schematic illustration of the glycocalyx–cleft model. The glycocalyx layer provides a system of ultra-fine pores in series to the larger pores formed by the Endothelial Cell (EC) clefts

(a) Compute the fluid flux q_f across the capillary wall according to Starling's filtration model (2.51).
(b) Given $c_{gc} = (1 - \alpha)c_v + \alpha c_i$ describes the albumin concentration underneath the glycocalyx layer, compute the fluid flux q_f according to the glycocalyx–cleft model for $\alpha = 1.0; 0.9; 0.8$. ∎

2.5   Summary and Conclusion

Following the introduction of circulatory physiology, this chapter introduced a number of concepts and models towards the exploration of vascular mechanisms and system responses. The models have been either 1D in space or time, or the combination of both dimensions then led to 2D problem descriptions. Such approaches are simple and *very effective* in the investigation of how system parameters interact and influence vascular physiology and pathology. We assumed that key vasculature properties can be described by parameters, such as resistance, capacity, and hydraulic conductivity. Complex vascular mechanisms have therefore been lumped into key system parameters towards the description of *surrogate vascular functions*. The introduced system parameters can either be directly measured by tailored experiments or indirectly identified through parameter calibration methods.

The cardiovascular system is equipped with a large number of hemodynamic regulation mechanisms towards diverting blood flows and optimally perfusing the body [548]. It has also been hypothesized that in health homeostasis, a *functional crosstalk* between central and peripheral segments of the circulation is required for optimal operation [9]. This communication may be compromised, and a vicious cycle of minute alterations in central arterial stiffness and peripheral resistance starts, leading to the dramatic changes in arterial properties observed in response to diseases and aging—the mother of all diseases.

Whilst the modeling approaches discussed in this chapter fit a number of vascular applications, they may often fail to provide in-depth explanations of how local physical mechanisms determine and alter vascular parameters and function. In addition, the analysis of vascular biomechanical problems may require a multi-dimensional space-time description of individual vessel segments. This is clearly beyond the ability of the approaches discussed in this chapter, and the remaining parts of this book concerns tools and models for the analysis of individual vessel segments. It allows us then to explore localized vascular phenomena to further our understanding of vascular function. The up-scaling or homogenization of local vessel properties determines then the surrogate system parameters used in the present chapter. The generalized distributed lumped parameter framework recently reported [376] would be such an example. It allows one to compute the flow and pressure dynamics in blood vessels upon various sources of energy dissipation mechanisms.

Continuum Mechanics

3

This chapter concerns linear and non-linear Continuum Mechanics. We introduce the Representative Volume Element (RVE) and provide the kinematic description for the deformation of a body between its reference and spatial configuration, respectively. The presentation of strain measures is followed by the introduction of the Cauchy stress. An analysis of the individual stress/strain components, coordinate transformations and the derivation of the principal stresses/strains aims at acquainting the reader with the basic concepts of tensor analysis. We then use the Piola transformation to introduce the first and second Piola-Kirchhoff stresses, respectively. Motivated by the objective description of rate effects, material time derivatives of strains and stresses are developed. In addition, we introduce a number of constitutive models that cover linear, non-linear, finite strain and viscoelastic descriptions. Many of which concern the description of an incompressible material – the stress then contains a Lagrange pressure contribution. Two sections of this chapter address the general principles of Continuum Mechanics. The reader is first familiarized with Free Body Diagrams (FBD), mass conservation, momentum balance, the first and second laws of thermodynamic, Maxwell transport and localization, as well as the strong and weak forms of the Boundary Value Problem (BVP). Concepts towards the description of damage and failure are then discussed, and a summary concludes the chapter.

The original version of this chapter was revised: ESM has been added. The correction to this chapter is available at https://doi.org/10.1007/978-3-030-70966-2_8

Supplementary Information The online version contains supplementary material available at https://doi.org/10.1007/978-3-030-70966-2_3

T. C. Gasser, *Vascular Biomechanics*, https://doi.org/10.1007/978-3-030-70966-2_3

3.1   Introduction

In the previous chapter, we used lumped parameter models and other low-dimensional descriptions in the exploration of the vasculature system—a small number of parameters represented key system features, such as resistance, capacity, and hydraulic conductivity. A number of investigations, such as the study of many vascular diseases, however require a more *localized approach* and *continuum mechanics* represents then a more powerful analysis tool.

Continuum mechanics is a well developed discipline with plenty of excellent and comprehensive textbooks [53,185,271,354,401]. The present chapter covers *linear* and *non-linear* continuum approaches for *solid* mechanical and *fluid* mechanical applications. It is self-consistent and develops all knowledge to solve a wide range of vascular biomechanical problems. The in-depth understanding of continuum mechanics is not only fundamental to the analytical solution of vascular problems, but also for the effective use of advanced numerical approaches, such as the Finite Element (FE) method.

The back bone of continuum mechanics is the introduction of the *Representative Volume Element (RVE)* and thus the description of a material's continuum properties. The RVE should be large enough to allow a homogenized characterization of material properties, but at the same time, the RVE should be much smaller than the characteristic dimension of the vascular problem. A problem to be analyzed by continuum mechanics must therefore allow to *separate* at least two scales: (i) the global scale, at which the vascular problem is resolved by a discretization method such as the FE method, and (ii) the local scale, at which all heterogeneity is homogenized through the RVE. Vascular tissue, and to so some extend also blood, shows a hierarchical structure that may challenge such *separation of scales*.

3.2   Kinematics

We consider a continuous body that occupies the *reference configuration* Ω_0 within the Cartesian coordinate system $\{\mathbf{e}_1, \ldots, \mathbf{e}_{n_{\mathrm{dim}}}\}$, where n_{dim} denotes the problem's spatial dimension. The body moves in space, and given the time t, it occupies the spatial configuration Ω. The *motion* $\chi(\mathbf{X}, t)$ maps a material particle from its reference position $\mathbf{X}$ into its spatial (or current) position $\mathbf{x}(t) = \chi(\mathbf{X}, t)$, see Fig. 3.1. The material particle's position $\mathbf{x}(t)$ fully describes a Cauchy[1] or Boltzmann[2] continuum. Higher-order continua, such as the Cosserat[3] continuum, use in addition to the position $\mathbf{x}(t)$, also the rotation of the particle.

---

[1] Augustin-Louis Cauchy, French mathematician and physicist, 1789–1857.
[2] Ludwig Eduard Boltzmann, Austrian physicist and philosopher, 1844–1906.
[3] Eugène-Maurice-Pierre Cosserat, French mathematician and astronomer, 1866–1931.

Fig. 3.1 The motion $\chi(\mathbf{X}, t)$ maps the body from its reference configuration Ω_0 to its spatial configuration Ω

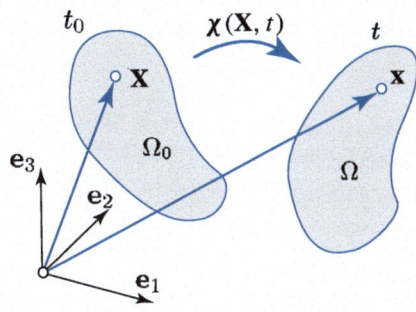

3.2.1 Deformation Gradient

At time t, the deformation in the vicinity of a particle, is defined by the *deformation gradient*

$$\mathbf{F}(\mathbf{X}, t) = \mathrm{Grad}\chi(\mathbf{X}, t) = \partial\chi(\mathbf{X}, t)/\partial\mathbf{X} \; ; \quad F_{iJ} = \partial\chi_i/\partial X_J \;. \tag{3.1}$$

It contains information of the *rigid body motion* as well as the *deformation* of the material particle and satisfies $\det\mathbf{F} > 0$ to avoid self-penetration of the continuum.

The deformation gradient's eigenvalue representation reads

$$\mathbf{F} = \lambda_i \widehat{\mathbf{n}}_i \otimes \widehat{\mathbf{N}}_i \; ; \quad i = 1, \dots, n_{\mathrm{dim}} \;,$$

where λ_i denotes the i-th *principal stretch*. The unit vectors $\widehat{\mathbf{n}}_i$ and $\widehat{\mathbf{N}}_i$ represent the i-th *principal stretch directions* in the spatial Ω and reference Ω_0 configuration, respectively. The deformation gradient $\mathbf{F}$ is therefore a *two-point tensor* that "connects" Ω with Ω_0. This is also indicated by its index notation F_{iJ}—the lower case index i relates to Ω, whilst the upper case index J relates to Ω_0.

3.2.2 Multiplicative Decomposition

Any motion $\chi(\mathbf{X})$ may be split into a number of incremental motions χ_i; $i = 1, \dots, n$, each of which successively applied, and $\chi(\mathbf{X}) = \chi_n(\chi_{n-1}(\cdots\chi_1(\mathbf{X})\cdots))$ then yields the total motion. The *multiplicative decomposition*

$$\mathbf{F}(\mathbf{X}) = \partial\chi(\mathbf{X})/\partial\mathbf{X} = \mathbf{F}_n\mathbf{F}_{n-1}\cdots\mathbf{F}_1(\mathbf{X}) \tag{3.2}$$

reflects then the motion, where the total deformation gradient is the product of the incremental deformation gradients $\mathbf{F}_i = \partial\chi_i(\mathbf{X})/\partial\chi_{i-1}(\mathbf{X})$; $i = 1, \dots, n$, respectively.

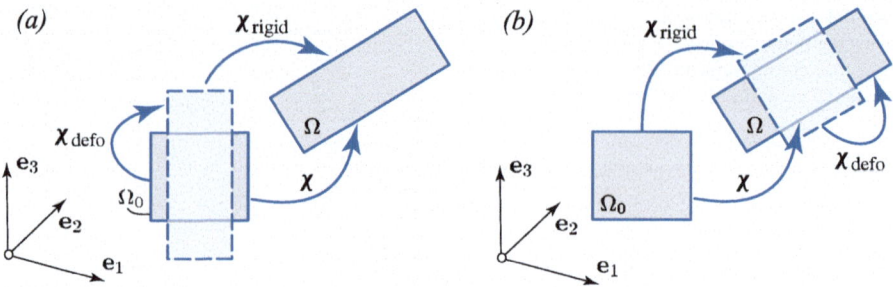

Fig. 3.2 Polar decomposition of the local motion $\chi(\mathbf{X})$. (**a**) Right polar decomposition $\chi(\mathbf{X}) = \chi_{\text{rigid}}(\chi_{\text{defo}}(\mathbf{X}))$. (**b**) Left polar decomposition $\chi(\mathbf{X}) = \chi_{\text{defo}}(\chi_{\text{rigid}}(\mathbf{X}))$. Intermediate configurations are dashed

3.2.3 Polar Decomposition

Any local motion $\chi(\mathbf{X})$ may be decomposed into χ_{rigid} and χ_{defo}, mappings that relate to *rigid body motion* and *deformation*, respectively. The *right polar decomposition* $\chi(\mathbf{X}) = \chi_{\text{rigid}}(\chi_{\text{defo}}(\mathbf{X}))$ or the *left polar decomposition* $\chi(\mathbf{X}) = \chi_{\text{defo}}(\chi_{\text{rigid}}(\mathbf{X}))$ expresses then the total motion of the material particle, see Fig. 3.2. With (3.1), we may therefore use the right polar decomposition $\mathbf{F} = \mathbf{R}\mathbf{U}$ or the left polar decomposition $\mathbf{F} = \bar{\mathbf{v}}\mathbf{Q}$ to express the deformation gradient, where $\mathbf{U}$ and $\bar{\mathbf{v}}$ denote the *right and left stretch tensors*, respectively. The rotation tensors $\mathbf{R}, \mathbf{Q}$ denote the rigid body rotations with the properties $\mathbf{R}^{-1} = \mathbf{R}^{\mathrm{T}}, \mathbf{Q}^{-1} = \mathbf{Q}^{\mathrm{T}}$ and $\det\mathbf{R} = \det\mathbf{Q} = +1$.

The eigenvalue representations of the right and left stretch tensors read

$$\mathbf{U} = \lambda_i \widehat{\mathbf{N}}_i \otimes \widehat{\mathbf{N}}_i \; ; \; \bar{\mathbf{v}} = \lambda_i \widehat{\mathbf{n}}_i \otimes \widehat{\mathbf{n}}_i \; ; \; i = 1, \ldots, n_{\text{dim}} \, ,$$

where λ_i denotes the principal stretches, whilst $\widehat{\mathbf{n}}_i$ and $\widehat{\mathbf{N}}_i$ are the principal stretch directions in Ω and Ω_0. The right and left stretch tensors are therefore one-point referential and spatial tensors, respectively. It is illustrated by their index notations U_{IJ} and $\bar{v}_{ij}$—the upper case indices relate to Ω_0, whilst the lower case indices relate to Ω.

3.2.4 Deformation of the Line Element

A *line element* $\mathrm{d}L$ at the position $\mathbf{X}$ in Ω_0 may be seen as a fiber segment of a fiber-reinforced continuum, see Fig. 3.3. The unit direction vector $\mathbf{a}_0(\mathbf{X})$ with $|\mathbf{a}_0(\mathbf{X})| = 1$ denotes its orientation, and the motion $\chi(\mathbf{X})$ maps it to its spatial position $\mathbf{x}$ in Ω. The linear transform $\mathbf{a} = \mathbf{F}\mathbf{a}_0$, with $\mathbf{F}$ denoting the deformation gradient, specifies the transformation of the line element, and

$$\lambda_{\mathbf{a}} = \frac{\mathrm{d}l}{\mathrm{d}L} = \frac{|\mathbf{F}\mathbf{a}_0|}{|\mathbf{a}_0|} = \sqrt{\mathbf{a}_0 \mathbf{F}^{\mathrm{T}} \mathbf{F} \mathbf{a}_0} = \sqrt{\mathbf{A} : \mathbf{C}} = \sqrt{I_4} = |\mathbf{a}| \tag{3.3}$$

Fig. 3.3 Deformation of the line element during the motion $\chi(\mathbf{X})$ between the reference and spatial configurations Ω_0 and Ω, respectively. The unit direction vector $\mathbf{a}_0$ describes the orientation of a fiber segment at the position $\mathbf{X}$ in Ω_0

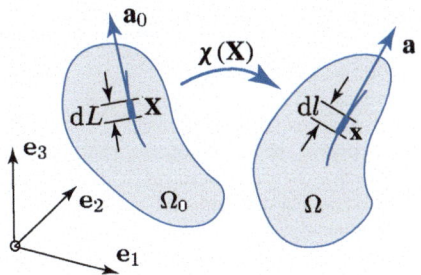

then determines the stretch of the fiber segment. Here, $\mathbf{C} = \mathbf{F}^{\mathsf{T}}\mathbf{F}$ denotes the right Cauchy–Green[4] strain (see Sect. 3.2.7.2), whilst $\mathbf{A} = \mathbf{a}_0 \otimes \mathbf{a}_0$ is a *structural tensor* and reflects the structure of the fiber-reinforced continuum at the position $\mathbf{X}$. In (3.3), $I_4 = \mathbf{C} : \mathbf{A}$ represents an invariant that is formed by the symmetric second-order tensors $\mathbf{C}$ and $\mathbf{A}$. It is commonly denoted as the *fourth invariant*, in addition to the standard three invariants of a second-order tensor in 3D, see Appendix A.7.

3.2.5 Deformation of the Volume Element

Let us consider the referential *volume element* $dV = dX_1 dX_2 dX_3$ that represents a 3D material particle at the position $\mathbf{X}$ in Ω_0, see Fig. 3.4. The motion $\chi(\mathbf{X})$ maps the particle to its spatial configuration $\mathbf{x}$, and it deforms from a cuboid in Ω_0 into a parallelepiped in Ω. The particle's spatial volume reads

$$d\upsilon = (d\mathbf{x}_1 \times d\mathbf{x}_2) \cdot d\mathbf{x}_3 = (\mathbf{F}d\mathbf{X}_1 \times \mathbf{F}d\mathbf{X}_2) \cdot \mathbf{F}d\mathbf{X}_3$$

$$= \left(\begin{bmatrix} F_{11} \\ F_{21} \\ F_{31} \end{bmatrix} \times \begin{bmatrix} F_{12} \\ F_{22} \\ F_{32} \end{bmatrix} \right) \cdot \begin{bmatrix} F_{13} \\ F_{23} \\ F_{33} \end{bmatrix} dX_1 dX_2 dX_3 = \det\mathbf{F}\, dV ,$$

where

$$d\upsilon = J dV \tag{3.4}$$

expresses the volume deformation, and $J = \det\mathbf{F}$ then denotes the *volume ratio*. It implies the condition $\det\mathbf{F} > 0$ of the deformation gradient (3.1) to avoid a negative volume element $d\upsilon$ from a self-penetrating continuum.

3.2.6 Deformation of the Area Element

Let us consider the referential *area vector* $d\mathbf{S} = \mathbf{N}dS$ that represents the area element dS at the position $\mathbf{X}$ in Ω_0, see Fig. 3.5. The motion $\chi(\mathbf{X})$ maps it to $d\mathbf{s} = \mathbf{n}ds$ in the spatial configuration.

---

[4]George Green, British mathematical physicist, 1793–1841.

Fig. 3.4 Deformation of the volume element during the motion $\chi(\mathbf{X})$ between the reference and spatial configurations Ω_0 and Ω, respectively. The cuboid of the volume dV in Ω_0 deforms towards the parallelepiped of the volume dv in Ω

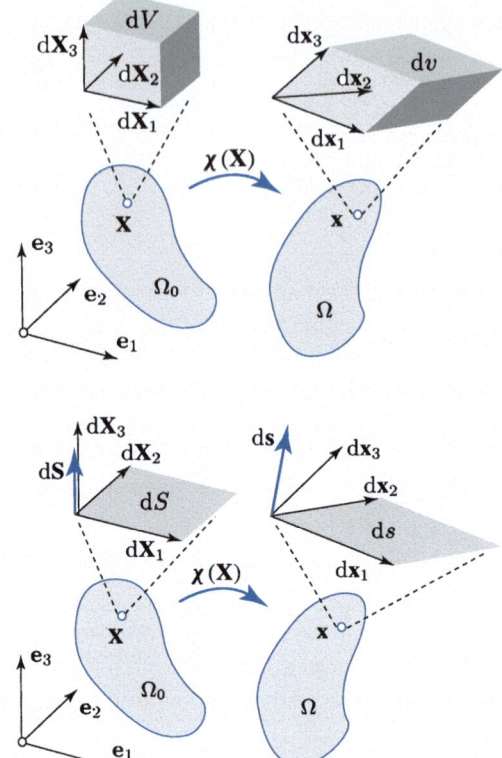

Fig. 3.5 Deformation of the area element during the motion $\chi(\mathbf{X})$ between the reference and spatial configurations Ω_0 and Ω, respectively. The square of area dS in Ω_0 deforms towards the parallelogram of area ds in Ω. The corresponding area vectors are denoted by $d\mathbf{S}$ and $d\mathbf{s}$, respectively

We may use the spatial volume element $dv = d\mathbf{x}_3 \cdot d\mathbf{s} = \mathbf{F}d\mathbf{X}_3 \cdot d\mathbf{s}$ towards the derivation of the relation between $d\mathbf{S}$ and $d\mathbf{s}$. Given the volume ratio (3.4), it reads $dv = JdV = Jd\mathbf{X}_3 \cdot d\mathbf{S}$, and

$$d\mathbf{s} = J\mathbf{F}^{-\mathrm{T}}d\mathbf{S} \;\; ; \;\; ds_i = JF_{Ai}^{-1}dS_A \tag{3.5}$$

then expresses the relation between the spatial and referential area elements. Whilst this expression is known as *Nanson's*[5] *formula*, it has been used earlier by others, such as Piola.[6]

Example 3.1 (Deformation of Line, Volume, and Area Elements). The homogenous deformation gradient

$$\mathbf{F} = \begin{bmatrix} 3 & 1 \\ 0 & 1.3 \end{bmatrix} \tag{3.6}$$

---

[5]Edward J. Nanson, British mathematician, 1850–1936.
[6]Gabrio Piola, Italian mathematician, 1794–1850.

specifies the mapping of a 2D body between its reference Ω_0 and spatial Ω configurations, respectively. Given a homogenous deformation, all particles of the body are equally deformed, and $\mathbf{F}$ is therefore constant throughout the body.

(a) The unit direction vector $\mathbf{a}_0 = [0.9553 \ \ 0.2955]^T$ determines the orientation of elastic fibers in Ω_0 that are embedded in the body. Compute the stretch within such fibers, where the assumption of affine deformation between the fibers and the continuum may be used.

(b) Given the body has the referential area $A = 2.0\,\mathrm{m}^2$, compute its spatial area a. Note that the area in 2D corresponds to the volume in 3D and therefore the transformation of the volume element is to be applied.

(c) Consider the normal vector $d\mathbf{S} = [0.7\,dx_1 \ \ 3.2\,dx_2]^T$ that is perpendicular to a line element $d\mathbf{x}$ in Ω_0 and compute the corresponding normal vector $d\mathbf{s}$ in Ω. Note that a line element in 2D corresponds to the area element in 3D and therefore the transformation of the area element is to be applied. ■

3.2.7 Concept of Strain

Strain is a dimensionless measure of the deformation of the material particle—it expresses the average deformation over the RVE. Normal strain reflects the change in length, whilst shear strain expresses the change in angle between pairs of lines initially perpendicular to each other, see Fig. 3.6. Normal strains and shear strains are collectively represented by a second-order *strain tensor*. In 3D, such a strain tensor has six independent components. With strains being normalized displacements, the use of different references results in different strain definitions.

As outlined in Sect. 3.2.3, the motion $\chi(\mathbf{X})$ can always be split into rigid body motion and deformation-related motion. Given both are small, a geometrically *linear* analysis may be carried out, whilst, if both are finite, a geometrically *exact* analysis is needed. Some problems may be determined by a small particle deformation on top of large rigid body motions; a *finite rotation* analysis may then be used.

Fig. 3.6 Shape changes of a material particle due to (**a**) normal and (**b**) shear strains. The dashed line indicates the material particle's undeformed configuration, whilst u_1 and u_2 denote the respective displacements

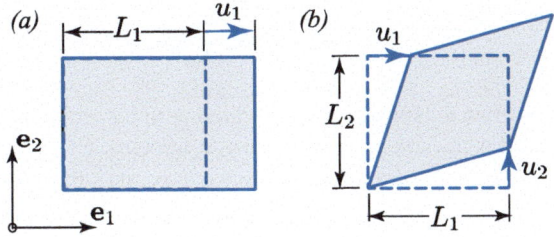

3.2.7.1 Engineering, or Linear Strain

At small deformations, the body's undeformed and deformed configurations are almost identical, $\Omega_0 \approx \Omega$, and the *engineering (or linear) strain* is an adequate strain measure.

Figure 3.6 illustrates the deformation of the material particle due to normal and shear strains, respectively. For illustration purposes the displacements u_1 and u_2 have been magnified. Their normalized counterparts determine the engineering *normal* strain $\varepsilon_{11} = u_1/L_1$ and the engineering *shear* strain $\varepsilon_{12} = (u_1/L_2 + u_2/L_1)/2$. Given an infinitesimal small material particle, and thus the kinematics at the limit $L_1, L_2 \to 0$, the engineering strain tensor reads

$$\boldsymbol{\varepsilon} = \frac{1}{2}\left[\text{grad}\mathbf{u} + (\text{grad}\mathbf{u})^{\mathrm{T}}\right] . \tag{3.9}$$

It represents a symmetric second-order tensor, which in Cartesian coordinates has the components $\varepsilon_{ij} = \left(\partial u_i/\partial x_j + \partial u_j/\partial x_i\right)/2; i, j = 1, \ldots, n_{\text{dim}}$, where n_{dim} denotes the spatial dimension. The eigenvalue representation $\boldsymbol{\varepsilon} = \varepsilon_i \widehat{\mathbf{N}}_i \otimes \widehat{\mathbf{N}}_i; i = 1, \ldots, n_{\text{dim}}$ expresses the engineering strain through the principal stains ε_i and principal strain directions $\widehat{\mathbf{N}}_i$.

The individual strain components cannot be independent from each other, and they have to satisfy the *strain compatibility* conditions. Given a 2D problem in Cartesian coordinates, the strain compatibility reads

$$\frac{\partial^2 \varepsilon_{11}}{\partial x_2^2} - 2\frac{\partial^2 \varepsilon_{12}}{\partial x_1 \partial x_2} + \frac{\partial^2 \varepsilon_{22}}{\partial x_1^2} = 0 ,$$

whilst in 3D *six* such compatibility conditions are to be satisfied [185].

3.2.7.2 Non-linear Strain Measures

At finite deformations, the body's reference configuration Ω_0 is different from its spatial configuration Ω, and *non-linear* strain measures are to be used to correctly express the strain of the material particle. Different non-linear strain measures may be defined and the most common ones are listed below.

The *right Cauchy–Green strain*

$$\mathbf{C} = \mathbf{F}^{\mathrm{T}}\mathbf{F} \; ; \quad C_{IJ} = F_{aI}F_{aJ} \tag{3.10}$$

is a strain tensor that is formulated with respect to the body's reference configuration Ω_0. It is the square of the right stretch tensor $\mathbf{U}$. Given a n_{dim}-dimensional problem, $\mathbf{C} = \lambda_i^2 \widehat{\mathbf{N}}_i \otimes \widehat{\mathbf{N}}_i; i = 1, \ldots, n_{\text{dim}}$ expresses therefore its eigenvalue representation, where λ_i and $\widehat{\mathbf{N}}_i$ denote the principal stretches and the principal stretch directions in Ω_0, respectively.

The *left Cauchy–Green strain*

$$\mathbf{b} = \mathbf{F}\mathbf{F}^{\mathsf{T}} \quad ; \quad b_{ij} = F_{iA}F_{jA} \tag{3.11}$$

is a strain tensor that is formulated with respect to the body's spatial configuration Ω. It is the square of the left stretch tensor $\bar{\mathbf{v}}$, and its eigenvalue representation therefore reads $\mathbf{b} = \lambda_i^2 \widehat{\mathbf{n}}_i \otimes \widehat{\mathbf{n}}_i; i = 1, \ldots, n_{\text{dim}}$ with $\widehat{\mathbf{n}}_i$ denoting the principal stretch directions in the spatial configuration Ω. The right and left Cauchy–Green strains are relative strain measures, and given a strain-free body, $\mathbf{C} = \mathbf{b} = \mathbf{I}$ holds.

The *Green–Lagrange*[7] *strain*

$$\mathbf{E} = \frac{1}{2}(\mathbf{C} - \mathbf{I}) = \frac{1}{2}(\mathbf{F}^{\mathsf{T}}\mathbf{F} - \mathbf{I}) \quad ; \quad E_{IJ} = \frac{1}{2}(F_{aI}F_{aJ} - \delta_{IJ}) \tag{3.12}$$

is a strain tensor that is formulated with respect to the body's reference configuration Ω_0, and δ_{IJ} denotes the Kronecker[8] delta. The Green–Lagrange strain's eigenvalue representation reads $\mathbf{E} = E_i \widehat{\mathbf{N}}_i \otimes \widehat{\mathbf{N}}_i; i = 1, \ldots, n_{\text{dim}}$, where $E_i = (\lambda_i^2 - 1)/2$ expresses the relation between the eigenvalues E_i and the principal stretches λ_i.

Another often used strain measure is the *Euler–Almansi*[9] *strain*

$$\mathbf{e} = \frac{1}{2}(\mathbf{I} - \mathbf{b}^{-1}) = \frac{1}{2}(\mathbf{I} - \mathbf{F}^{-\mathsf{T}}\mathbf{F}^{-1}) \quad ; \quad e_{ij} = \frac{1}{2}(\delta_{ij} - F_{Ai}^{-1}F_{Aj}^{-1}). \tag{3.13}$$

It is formulated with respect to the spatial configuration Ω, and its eigenvalue representation reads $\mathbf{e} = e_i \widehat{\mathbf{n}}_i \otimes \widehat{\mathbf{n}}_i; i = 1, \ldots, n_{\text{dim}}$, where the principal strains e_i and the principal stretches λ_i are related by $e_i = (1 - \lambda_i^{-2})/2$. The Green–Lagrange strain and the Euler–Almansi strain are absolute strain measures, and given a strain-free body, $\mathbf{E} = \mathbf{e} = \mathbf{0}$ holds. At the small-strain limit, both strain measures approach the engineering strain, and thus $\mathbf{E} \to \boldsymbol{\varepsilon}; \mathbf{e} \to \boldsymbol{\varepsilon}$ holds for $\Omega \approx \Omega_0$.

Example 3.2 (Non-linear Deformation and Strain Measures). The 2D motion $\mathbf{x} = \boldsymbol{\chi}(\mathbf{X}, t) = [(1.5X_1^2 + X_2)t \quad 1.3X_2t]^{\mathsf{T}}$ determines the mapping between the reference configuration Ω_0 and the spatial configuration Ω of a continuum body, where $\mathbf{X}$ and t denote the referential particle position and the time, respectively.

(a) Compute the deformation gradient $\mathbf{F}$, and discuss its physical validity in space $\mathbf{X}$ and time t.
(b) Compute the velocity gradient, a quantity defined by $\mathbf{l} = \dot{\mathbf{F}}\mathbf{F}^{-1}$.
(c) Compute the right Cauchy–Green strain $\mathbf{C}$, the left Cauchy–Green strain $\mathbf{b}$, the Green–Lagrange strain $\mathbf{E}$, and the Euler–Almansi strain $\mathbf{e}$.　■

---

[7]Joseph-Louis Lagrange, Franco-Italian mathematician and astronomer, 1736–1813.

[8]Leopold Kronecker, German mathematician, 1823–1891.

[9]Emilio Almansi, Italian mathematician, 1869–1948.

Example 3.3 (Linear Versus Non-linear Strain Measures). A 2D rigid body rotation **R** maps a material particle from its reference position **X** to its spatial position $\mathbf{x} = \mathbf{RX}$.

(a) Introduce the displacement vector $\mathbf{u} = \mathbf{x} - \mathbf{X}$ of the transformation and compute the engineering strain $\boldsymbol{\varepsilon}$. Discuss the validity of this strain measure.
(b) Compute the Green–Lagrange strain **E** and discuss the result. ∎

Example 3.4 (Independence of Strain Measures from Rigid Body Rotation). Use the polar decomposition theorem and show that the below listed strain measures are independent from the rigid body rotation that is superimposed on the spatial configuration Ω of a body.

(a) Right and left Cauchy–Green strains
(b) Green–Lagrange and Euler–Almansi strains ∎

Example 3.5 (Simple and Pure Shear Deformation Kinematics). A unit cube of incompressible material moves from its reference configuration Ω_0 to its current configuration Ω in response to the shear stress τ that acts at its faces, see Fig. 3.7a. The corresponding motion $\boldsymbol{\chi}(\mathbf{X})$ may be presented within the principal stretch coordinate system $\{\widehat{\mathbf{e}}_1, \widehat{\mathbf{e}}_2, \widehat{\mathbf{e}}_3\}$ as it is shown in Fig. 3.7b. Given small deformation, this deformation can also be expressed within the coordinate system $\{\mathbf{e}_1, \mathbf{e}_2, \mathbf{e}_3\}$ that is rotated at the angle $\pi/4$, see Fig. 3.7c. The deformation kinematics illustrated in Fig. 3.7b,c are commonly called pure shear and simple shear, respectively.

Fig. 3.7 Shear deformation kinematics. (**a**) The shear stress τ causes the motion $\boldsymbol{\chi}(\mathbf{X})$ that maps the unit cube from its reference Ω_0 to its spatial configuration Ω. (**b**) Pure shear kinematics. (**c**) Simple shear kinematics

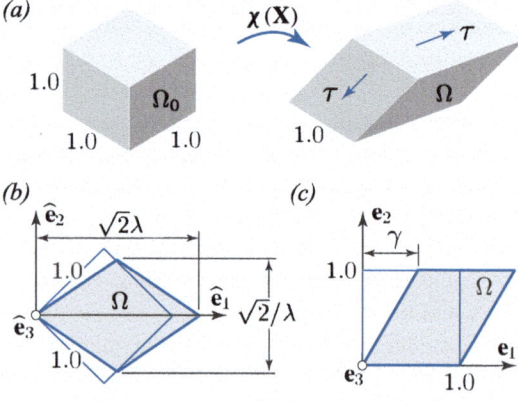

(a) Consider pure shear kinematics and describe the corresponding motion $\widehat{\boldsymbol{\chi}}_{\mathrm{ps}}$ and deformation gradient $\widehat{\mathbf{F}}_{\mathrm{ps}}$.
(b) Consider simple shear kinematics and describe the corresponding motion $\boldsymbol{\chi}_{\mathrm{ss}}$ and deformation gradient $\mathbf{F}_{\mathrm{ss}}$.

(c) Consider pure shear kinematics and describe the corresponding motion χ_{ps} and deformation gradient $\mathbf{F}_{\mathrm{ps}}$.
(d) Given small strains, show that pure shear and simple shear yield the same right Cauchy–Green strain.
(e) Show that pure shear, but not simple shear is free of rigid body rotation. ∎

3.2.7.3 Particular Strain States

Some applications may not support an unconstrained development of the deformation, which then results in distinct strain states. *Plain strain* is such a case and determined by the components $\varepsilon_{13} = \varepsilon_{23} = \varepsilon_{33} = 0$ of the engineering strain $\boldsymbol{\varepsilon}$. Given non-linear strain measures, equivalent conditions hold. *Rotational symmetry* defines another particular strain state, where no shear strain in the plane perpendicular to the axis of symmetry is able to develop.

3.3 Concept of Stress

Stress represents the load that acts at the material particle in average over the RVE. It expresses the force that passes through the material particle, divided by the area through which it passes. The force $\mathrm{d}P_{\mathrm{n}}$ that acts perpendicular to the area $\mathrm{d}s$ leads to the *normal stress* $\sigma_{11} = \mathrm{d}P_{\mathrm{n}}/\mathrm{d}s$ [Pa], whilst the force $\mathrm{d}P_{\mathrm{s}}$ that acts in parallel to the area $\mathrm{d}s$ leads to the *shear stress* $\sigma_{12} = \mathrm{d}P_{\mathrm{s}}/\mathrm{d}s$ [Pa], see Fig. 3.8a,b. Given a general load case, normal stress appears together with shear stress, and the state of stress may be represented by a *second-order tensor*—the Cauchy stress $\boldsymbol{\sigma}$. The balance of angular momentum

$$\boldsymbol{\sigma} = \boldsymbol{\sigma}^{\mathrm{T}} \tag{3.19}$$

requires the Cauchy stress tensor to be *symmetric*.

Given the 3D Cartesian coordinate system $\{\mathbf{e}_1, \mathbf{e}_2, \mathbf{e}_3\}$, three normal stress components $\sigma_{11}, \sigma_{22}, \sigma_{33}$ together with three shear stress components $\sigma_{12}, \sigma_{23}, \sigma_{13}$ define the stress state of the material particle, see Fig. 3.8c. They act at the faces that are perpendicular to the Cartesian base vectors $\mathbf{e}_i$. The notation σ_{ij} denotes the stress component that acts at the face perpendicular to $\mathbf{e}_i$ and that points into the

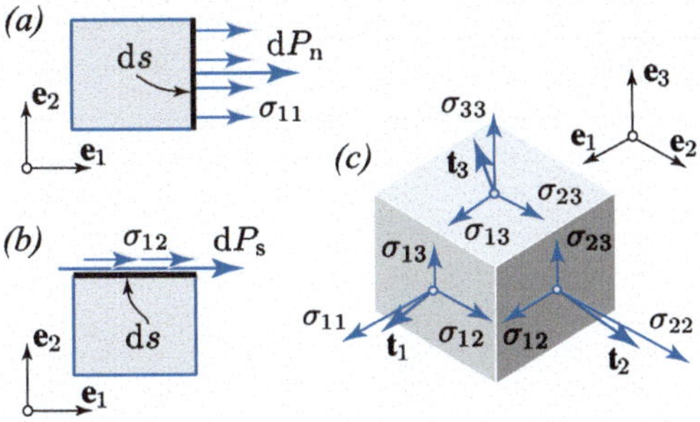

Fig. 3.8 Definition of stress. (a) Normal stress $\sigma_{11} = \mathrm{d}P_{\mathrm{n}}/\mathrm{d}s$, and (b) shear stress $\sigma_{12} = \mathrm{d}P_{\mathrm{s}}/\mathrm{d}s$. (c) The normal stress components $\sigma_{11}, \sigma_{22}, \sigma_{33}$ and the shear stress components $\sigma_{12}, \sigma_{23}, \sigma_{13}$ determine the traction vectors $\mathbf{t}_1, \mathbf{t}_2, \mathbf{t}_3$, respectively

Table 3.1 Basic 3D stress states

Simple tension	$\sigma_{11} = \sigma$ and $\sigma_{ij} = 0$ otherwise
Biaxial tension	$\sigma_{11} = \sigma_1, \sigma_{22} = \sigma_2$ and $\sigma_{ij} = 0$ otherwise
Plane stress	$\sigma_{13} = \sigma_{23} = \sigma_{33} = 0$ and $\sigma_{ij} \neq 0$ otherwise
Simple shear*	$\sigma_{12} = \tau$ and $\sigma_{ij} = 0$ otherwise
Pure bending	$\sigma_{11} = cx_2$ and $\sigma_{ij} = 0$ otherwise
Pure torsion**	$\sigma_{12} = g(r)$ and $\sigma_{ij} = 0$ otherwise

* Corresponds to the deformation kinematics "simple shear" at small deformations.
** The radius $r^2 = x_1^2 + x_2^2$ is the distance from the center of rotation, and $g(r)$ is a function of r.

direction $\mathbf{e}_j$, see Fig. 3.8c. The sum of the stress components that act at the i-th face of the material particle determines the traction vector $\mathbf{t}_i$ with $i = 1, \ldots, n_{\text{dim}}$ denoting the problem's spatial dimension. A stress state $\boldsymbol{\sigma}$ has to satisfy the local equilibrium conditions (see Sect. 3.6.2), and some basic stress states that are *a priori* at equilibrium are listed in Table 3.1. At finite deformations, the body's reference configuration Ω_0 differs from its spatial configuration Ω, and stress tensors other than the Cauchy stress may be defined, see Sects. 3.3.7 and 3.3.8.

Example 3.6 (Symmetry of the Cauchy Stress Tensor). Figure 3.9 shows the spatial configuration Ω of a material particle at homogeneous stress and with respect to the 2D Cartesian coordinate system $\{\mathbf{e}_1, \mathbf{e}_2\}$. The material particle is at rest and free of body forces.

Fig. 3.9 Material particle loaded by the normal stress σ_{11}, σ_{22} and the shear stress σ_{12}, σ_{21}

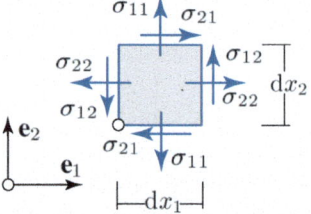

(a) Use the balance of linear momentum, also known as Euler's first principle, and show that the particle is at linear equilibrium.
(b) Use the balance of angular momentum, also known as Euler's second principle, and show that the particle is at angular equilibrium given the Cauchy stress tensor is symmetric $\boldsymbol{\sigma} = \boldsymbol{\sigma}^{\text{T}}$. ∎

Fig. 3.10 The traction
vector $\mathbf{t} = \boldsymbol{\sigma}\mathbf{n}$ acts at the area
element ds perpendicular to
the unit normal $\mathbf{n}$. It
maintains the stress $\boldsymbol{\sigma}$
underneath a hypothetical cut
through the spatial
configuration Ω of a
continuum body

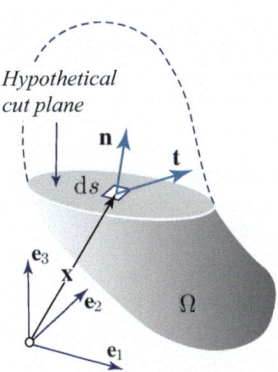

3.3.1   Cauchy Stress Theorem

We consider a body at its spatial configuration Ω, where the Cauchy stress $\boldsymbol{\sigma}(\mathbf{x})$
acts at a material particle that is located at the spatial position $\mathbf{x}$. A hypothetical
cut through this body would release the stress underneath the cut, and towards
reestablishing the stress state of before cutting, the traction $\mathbf{t}(\mathbf{x})$ is to be applied, see
Fig. 3.10. The traction $\mathbf{t}$ acts at the area element ds that is perpendicular to the unit
normal vector $\mathbf{n}(\mathbf{x})$, and according to *Cauchy's stress theorem*, the linear transform

$$\mathbf{t}(\mathbf{x}) = \boldsymbol{\sigma}(\mathbf{x})\mathbf{n}(\mathbf{x}) \; ; \quad t_i = \sigma_{ia}n_a \, , \tag{3.20}$$

provides the relation between the traction $\mathbf{t}$ and the Cauchy stress $\boldsymbol{\sigma}$. It determines
the traction $\mathbf{t}$ needed to counterbalance the stress underneath the hypothetical cut.

3.3.2   Principal Stresses

Whilst the stress state $\boldsymbol{\sigma}$ is *independent* from the coordinate system, its components
σ_{ij} change when changing the coordinate system. It is *always* possible to find
a coordinate system $\{\widehat{\mathbf{n}}_1, \ldots, \widehat{\mathbf{n}}_{n_{\dim}}\}$, within which the shear stress components
disappear, and $\sigma_{ij} = 0$ for $i \neq j$ holds. Such a stress state is called *principal
stress state*, and the traction vectors $\mathbf{t}_i; i = 1, \ldots, n_{\dim}$ are in parallel to the
corresponding base vectors $\widehat{\mathbf{n}}_i$ of the coordinate system, see Fig. 3.11. The condition
$\mathbf{t} = \lambda\widehat{\mathbf{n}}$ then holds, where $\lambda = |\mathbf{t}|$ is the magnitude of the traction vector and
represents the normal stress at the face of the material particle. Given Cauchy's
stress theorem (3.20), this condition then reads $\mathbf{t} = \boldsymbol{\sigma}\widehat{\mathbf{n}} = \lambda\widehat{\mathbf{n}}$ and may be rewritten
as a classical *eigenvalue problem*

$$(\boldsymbol{\sigma} - \lambda\mathbf{I})\widehat{\mathbf{n}} = \mathbf{0} \, . \tag{3.21}$$

Given the symmetry of the Cauchy stress tensor $\boldsymbol{\sigma} = \boldsymbol{\sigma}^{\mathrm{T}}$, the statement (3.21)
represents a symmetric eigenvalue problem with $n_{\dim}$ real solutions $(\lambda_i, \widehat{\mathbf{n}}_i); i =$

Fig. 3.11 Material particle
at principal stress within the
coordinate system
$\{\widehat{\mathbf{n}}_1, \widehat{\mathbf{n}}_2, \widehat{\mathbf{n}}_3\}$ of principal stress
directions

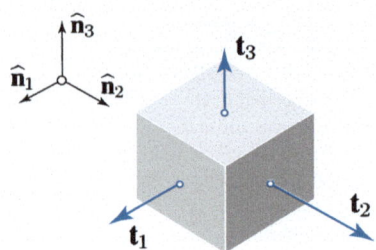

$1, \ldots, n_{\text{dim}}$. In addition, the eigenvectors are perpendicular to each other, $\widehat{\mathbf{n}}_i \cdot \widehat{\mathbf{n}}_j = \delta_{ij}$, where δ_{ij} denotes the Kronecker delta. The eigenvalues λ_i and the eigenvectors $\widehat{\mathbf{n}}_i$ are then the *principal stresses* and *principal stress directions* of the stress state $\boldsymbol{\sigma}$.

Given a 3D problem, the roots of the characteristic equation $\lambda^3 - I_1\lambda^2 + I_2\lambda - I_3 = 0$ are the solution of the eigenvalue problem (3.21), where the three *stress invariants*

$$\left. \begin{aligned} I_1 &= \text{tr}\boldsymbol{\sigma} = \sigma_{11} + \sigma_{22} + \sigma_{33} \ , \\ I_2 &= \left[(\text{tr}\boldsymbol{\sigma})^2 - \text{tr}\boldsymbol{\sigma}^2 \right]/2 \\ &= \sigma_{11}\sigma_{22} + \sigma_{11}\sigma_{33} + \sigma_{22}\sigma_{33} - \sigma_{12}^2 - \sigma_{13}^2 - \sigma_{23}^2 \ , \\ I_3 &= \det\boldsymbol{\sigma} \ , \end{aligned} \right\} \tag{3.22}$$

have been introduced.

Most commonly, the principal stresses are denoted by σ_i (instead of λ_i), and

$$\boldsymbol{\sigma} = \sigma_i \widehat{\mathbf{n}}_i \otimes \widehat{\mathbf{n}}_i \ ; \quad i = 1, \ldots, n_{\text{dim}} \tag{3.23}$$

denotes then the eigenvalue representation of the stress $\boldsymbol{\sigma}$.

3.3.3 Coordinate Rotation and Stress Components

We may consider the rod shown in Fig. 3.12 and explore the stress during coordinate system rotations. The rod may be seen as 2D continuum, and in response to the external load P, the internal stress components $\sigma_{11}, \sigma_{22}, \sigma_{12}$ develop. The linear momentum balance allows us to express the stress tensor by

$$\boldsymbol{\sigma} = \begin{bmatrix} \sigma_{11} = P/a & \sigma_{12} = 0 \\ \sigma_{12} = 0 & \sigma_{22} = 0 \end{bmatrix} ,$$

where the components $\sigma_{11}, \sigma_{22}, \sigma_{12}$ are taken with respect to the Cartesian coordinate system $\{\mathbf{e}_1, \mathbf{e}_2\}$, and a denotes the rod's cross-section. The only non-zero stress component is $\sigma_{11} = P/a$ [Pa], a stress that acts along the $\mathbf{e}_1$ direction. The inset in Fig. 3.12a shows the material particle and the corresponding stress components $\sigma_{11}, \sigma_{22}, \sigma_{12}$ at its faces. They are aligned with the coordinate base vectors $\mathbf{e}_1, \mathbf{e}_2$.

Fig. 3.12 2D continuum at simple tension. (**a**) Stress components acting at a material particle with respect to the Cartesian coordinate system $\{\mathbf{e}_1, \mathbf{e}_2\}$. (**b**) Stress components acting at the plane ϵ that is rotated by the angle α against the $\mathbf{e}_1$ direction

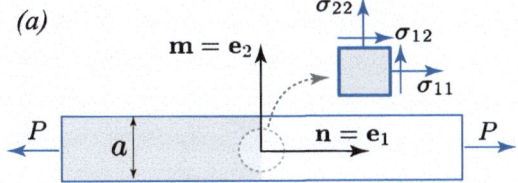

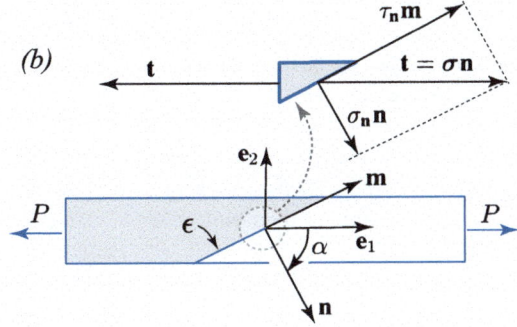

We can now explore the stress components at the face ϵ that is rotated by the angle α against the $\mathbf{e}_1$ base vector, see Fig. 3.12b.

The normal stress $\sigma_{\mathbf{n}}$ and the shear stress $\tau_{\mathbf{n}}$ act at ϵ, a plane determined by the unit normal vector $\mathbf{n} = [\cos\alpha \;\; -\sin\alpha]^{\mathrm{T}}$. The Cauchy stress theorem (3.20) allows us then to compute the traction vector

$$\mathbf{t} = \sigma\mathbf{n} = \big[\,(P/a)\cos\alpha \;\; 0\,\big]^{\mathrm{T}}$$

that acts at the face ϵ, and basic vector algebra yields the normal stress

$$\sigma_{\mathbf{n}} = \mathbf{t}\cdot\mathbf{n} = (\sigma\mathbf{n})\cdot\mathbf{n} = (P/a)\cos^2\alpha\,. \tag{3.24}$$

Given the tangential vector $\mathbf{m} = [\sin\alpha \;\; \cos\alpha]^{\mathrm{T}}$ at the face ϵ, the vector subtraction

$$\tau_{\mathbf{n}}\mathbf{m} = \mathbf{t} - \sigma\mathbf{n} = (P/a)\cos\alpha\,\big[\,1 - \cos^2\alpha \;\; \sin\alpha\cos\alpha\,\big]^{\mathrm{T}}$$

allows us to express the shear stress

$$\tau_{\mathbf{n}} = (\tau_{\mathbf{n}}\mathbf{m})\cdot\mathbf{m} = (P/a)\cos\alpha\sin\alpha \tag{3.25}$$

that acts at the face ϵ.

We may introduce a diagram with the x-axis and y-axis denoting the normal stress (3.24) and shear stress (3.25), respectively. Given such a stress space, the curve $(\sigma_{\mathbf{n}}(2\alpha),\, \tau_{\mathbf{n}}(2\alpha)) = (P/a\cos^2\alpha,\, P/a\cos\alpha\sin\alpha)$ determines a circle, known

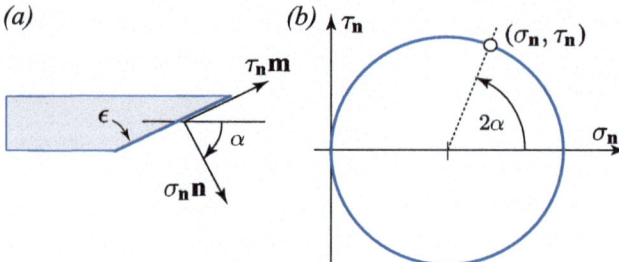

Fig. 3.13 Stress state at the rotated face ϵ in (**a**) the physical space and (**b**) Mohr's stress circle in the stress space

as Mohr's[10] stress circle, see Fig. 3.13. The *clockwise* rotation of α in the physical space corresponds to the *counter-clockwise* rotation of 2α in the stress space.

Example 3.7 (Cauchy Stress State in 2D). A material particle is loaded by the Cauchy stress

$$\boldsymbol{\sigma} = \begin{bmatrix} -5 & 3 \\ 3 & 10 \end{bmatrix} [\text{MPa}]$$

with respect to the Cartesian coordinate system $\{\mathbf{e}_1, \mathbf{e}_2\}$.

(a) Sketch the stress components that act at the faces of the material particle.
(b) Draw Mohr's stress circle and determine the principal stresses σ_1, σ_2 together with the principal stress directions $\widehat{\mathbf{n}}_1, \widehat{\mathbf{n}}_2$.
(c) Compute σ_1, σ_2 and $\widehat{\mathbf{n}}_1, \widehat{\mathbf{n}}_2$ through the solution of the corresponding eigenvalue problem.
(d) Compute the coordinate transformation $\mathbf{R}$ that rotates the Cartesian base vectors $\mathbf{e}_1, \mathbf{e}_2$ into the principal stress directions $\widehat{\mathbf{n}}_1, \widehat{\mathbf{n}}_2$.
(e) Draw a general conclusion regarding the stress state, and the components of the stress tensor, upon coordinate transformation. ∎

3.3.4 Isochoric and Volumetric Stress

For some applications, it is convenient to split the stress

$$\boldsymbol{\sigma} = \overline{\boldsymbol{\sigma}} + \boldsymbol{\sigma}_{\text{vol}} = \overline{\boldsymbol{\sigma}} + \underbrace{(\text{tr}\boldsymbol{\sigma}/n_{\text{dim}})}_{-p}\mathbf{I} \qquad (3.27)$$

---

[10]Christian Otto Mohr, German civil engineer, 1835–1918.

in *isochoric (or deviatoric)* $\overline{\sigma} = \text{dev}\sigma$ and *volumetric (or hydrostatic)* $\sigma_{\text{vol}} = -p\mathbf{I}$ contributions, where p denotes the negative hydrostatic pressure, and n_{dim} is the spatial dimension of the problem. The description of ductile materials would be one such example. Plastic yielding depends on the isochoric stress $\overline{\sigma}$ rather than on the total stress σ, and the invariants of $\overline{\sigma}$ are therefore of particular interest in the description of ductile materials. Given the definition of $\overline{\sigma}$, its first invariant $J_1 = \text{tr}\overline{\sigma}$ is identically zero, whilst the others may be expressed through

$$J_2 = 1/2[(\text{tr}\overline{\sigma})^2 - \text{tr}\overline{\sigma}^2] = -1/2\overline{\sigma}_{kl}\overline{\sigma}_{lk} = -I_1^2/3 + I_2 ;$$

$$J_3 = \det\overline{\sigma} = 2I_1^3/27 - I_1 I_2/3 + I_3 ,$$

where the stress invariants I_1, I_2, and I_3 according to the definitions (3.22) have been used.

3.3.5 Octahedral Stress and von Mises Stress

The *octahedral stress* is the stress that acts at the octahedral plane ϵ, a plane perpendicular to $\mathbf{n} = (1/\sqrt{3})[1\ 1\ 1]^\mathsf{T}$ in the *principal* stress coordinate system $\{\widehat{\mathbf{n}}_1, \widehat{\mathbf{n}}_2, \widehat{\mathbf{n}}_3\}$, see Fig. 3.16a. Given the Cauchy stress theorem (3.20), the normal stress σ_{oct} and the shear stress τ_{oct} at ϵ read

$$\sigma_{\text{oct}} = \mathbf{t} \cdot \mathbf{n} = (\sigma\mathbf{n}) \cdot \mathbf{n} = \frac{1}{3}(\sigma_1 + \sigma_2 + \sigma_3) = \frac{1}{3}I_1 = -p ;$$

$$\tau_{\text{oct}} = \sqrt{\mathbf{t} \cdot \mathbf{t} - \sigma_{\text{oct}}^2} = \sqrt{\frac{1}{3}(\sigma_1^2 + \sigma_2^2 + \sigma_3^2) - \frac{1}{9}(\sigma_1 + \sigma_2 + \sigma_3)^2}$$

$$= \frac{1}{3}\sqrt{(\sigma_1 - \sigma_2)^2 + (\sigma_2 - \sigma_3)^2 + (\sigma_1 - \sigma_3)^2} = \frac{1}{3}\sqrt{2I_1^2 - 6I_2} = \sqrt{-\frac{2}{3}J_2} ,$$

where σ_i; $i = 1, 2, 3$ denotes the principal stresses of σ.

Fig. 3.16 (a) Octahedral normal stress σ_{oct} and shear stress τ_{oct} in the principal stress space $\{\widehat{\mathbf{n}}_1, \widehat{\mathbf{n}}_2, \widehat{\mathbf{n}}_3\}$. (b) The cylinder of the radius $r = \sqrt{2/3}\,\sigma_M$ around the hydrostatic axis $\sigma_1 = \sigma_2 = \sigma_3$ describes a stress state of constant von Mises stress σ_M in $\{\widehat{\mathbf{n}}_1, \widehat{\mathbf{n}}_2, \widehat{\mathbf{n}}_3\}$

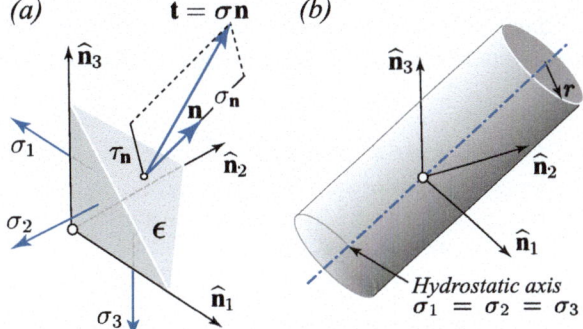

The octahedral *shear* stress τ_{oct} is a particularly important stress. It is often used towards assessing the risk of plastic failure (yielding) of a ductile material. Given simple tension, $\sigma_{11} = \sigma$ and $\sigma_{ij} = 0$ for all other stress components, the octahedral shear stress then yields $\tau_{oct} = \sqrt{2}/3\,\sigma$. The *von Mises*[11] *stress* σ_M scales the octahedral shear stress τ_{oct}, such that $\sigma_M = \sigma$ holds at simple tension. Given a general 3D stress state, the von Mises stress then reads

$$\sigma_M = \frac{3}{\sqrt{2}}\tau_{oct} = \sqrt{-3J_2}$$

$$= \frac{1}{\sqrt{2}}\sqrt{(\sigma_{11} - \sigma_{22})^2 + (\sigma_{22} - \sigma_{33})^2 + (\sigma_{11} - \sigma_{33})^2 + 6(\sigma_{12}^2 + \sigma_{23}^2 + \sigma_{13}^2)}$$

$$= 1/\sqrt{2}\sqrt{(\sigma_1 - \sigma_2)^2 + (\sigma_2 - \sigma_3)^2 + (\sigma_1 - \sigma_3)^2}\,, \tag{3.28}$$

where σ_{ij}; $i, j = 1, 2, 3$, are the components of $\boldsymbol{\sigma}$ with respect to the Cartesian coordinate system $\{\mathbf{e}_1, \mathbf{e}_2, \mathbf{e}_3\}$, whilst the principal stresses are denoted by σ_i. In the principal stress space $\{\widehat{\mathbf{n}}_1, \widehat{\mathbf{n}}_2, \widehat{\mathbf{n}}_3\}$, a constant von Mises stress σ_M prescribes a cylinder of radius $r = \sqrt{2/3}\,\sigma_M$ around the hydrostatic axis $\sigma_1 = \sigma_2 = \sigma_3$, see Fig. 3.16b.

Example 3.8 (Octahedral Stress and von Mises Stress). A material particle is loaded at the Cauchy stress

$$\boldsymbol{\sigma} = \begin{bmatrix} 10 & 5 & 0 \\ 5 & 25 & 5 \\ 0 & 5 & -14 \end{bmatrix} \text{ [MPa]}$$

with respect to the Cartesian coordinate system $\{\mathbf{e}_1, \mathbf{e}_2, \mathbf{e}_3\}$.

(a) Compute the second invariant J_2 of the deviatoric stress $\overline{\boldsymbol{\sigma}}$.
(b) Compute the octahedral normal σ_{oct} and shear τ_{oct} stresses.
(c) Compute the von Mises stress σ_M.   ∎

Example 3.9 (Octahedral and von Mises Stresses of Basic Stress States). Consider a continuum body at (i) simple tension, (ii) equi-biaxial tension, and (iii) simple shear.

(a) Use the matrix notation to express these stress states with respect to the Cartesian coordinate system $\{\mathbf{e}_1, \mathbf{e}_2, \mathbf{e}_3\}$.
(b) Compute the von Mises stress σ_M and the octahedral shear stress τ_{oct} of the aforementioned stress states.   ∎

---

[11]Richard von Mises, Austrian–Hungarian mathematician, 1883–1953.

3.3.6 Cauchy Stress in Rotated Coordinates

The *Cauchy stress* relates the force $d\mathbf{t} = \mathbf{t}ds$ to the area element $d\mathbf{s} = \mathbf{n}ds$, where the traction vector $\mathbf{t}$ and the normal $\mathbf{n}$ are defined in the spatial configuration Ω of the continuum body. At the time t, Ω is the physically existing configuration, and the Cauchy stress is therefore also called *"true stress"*. Given the rotation of the coordinate system, or alternatively the rotation of Ω, the components σ_{ij} of the Cauchy stress tensor $\boldsymbol{\sigma}$ change. Towards the exploration of this transformation we consider the rigid body rotation $\mathbf{R}$ that maps the body's spatial configuration Ω into a new spatial configuration $\widetilde{\Omega}$, see Fig. 3.17. The traction vector $\mathbf{t}$ and the unit normal vector $\mathbf{n}$ rotate into $\widetilde{\mathbf{t}} = \mathbf{R}\mathbf{t}$ and $\widetilde{\mathbf{n}} = \mathbf{R}\mathbf{n}$, respectively. The application of Cauchy's stress theorem (3.20) to the rotated configuration $\widetilde{\Omega}$ then reads $\widetilde{\mathbf{t}} = \widetilde{\boldsymbol{\sigma}}\widetilde{\mathbf{n}} = \mathbf{R}\mathbf{t} = \widetilde{\boldsymbol{\sigma}}\mathbf{R}\mathbf{n}$, and therefore $\mathbf{t} = \mathbf{R}^{\mathrm{T}}\widetilde{\boldsymbol{\sigma}}\mathbf{R}\mathbf{n}$ holds. The comparison of this result with (3.20) reveals that

$$\widetilde{\boldsymbol{\sigma}} = \mathbf{R}\boldsymbol{\sigma}\mathbf{R}^{\mathrm{T}} \quad ; \quad \widetilde{\sigma}_{ij} = R_{ia}\sigma_{ab}R_{jb} \tag{3.30}$$

determines the transformation of the Cauchy stress tensor with respect to the rigid body rotation $\mathbf{R}$. The Cauchy stress transforms therefore according to the transformation of a *spatial one-point* tensor.

3.3.7 First Piola–Kirchhoff Stress

The *first Piola–Kirchhoff*,[12] or *engineering stress* $\mathbf{P}$, relates the force $d\mathbf{t} = \mathbf{t}ds$ in the spatial configuration Ω to the area element $d\mathbf{S} = \mathbf{N}dS$ in the reference configuration Ω_0, see Fig. 3.18a. Cauchy's stress theorem (3.20) together with Nanson's formula (3.5) then yields $d\mathbf{t} = \boldsymbol{\sigma} J\mathbf{F}^{-\mathrm{T}}\mathbf{N}dS = \mathbf{P}\mathbf{N}dS$ and reveals the relation

Fig. 3.17 Change of the traction vector $\mathbf{t}$ and the normal vector $\mathbf{n}$ under rigid body rotation $\mathbf{R}$ of the body's spatial configuration from Ω to $\widetilde{\Omega}$

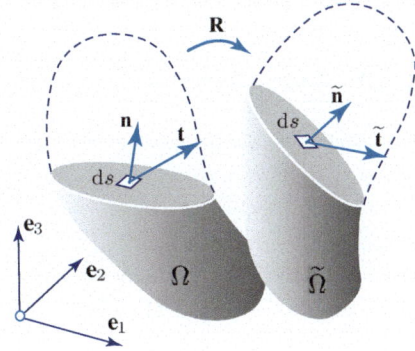

---

[12]Gustav Robert Kirchhoff, German physicist, 1824–1887.

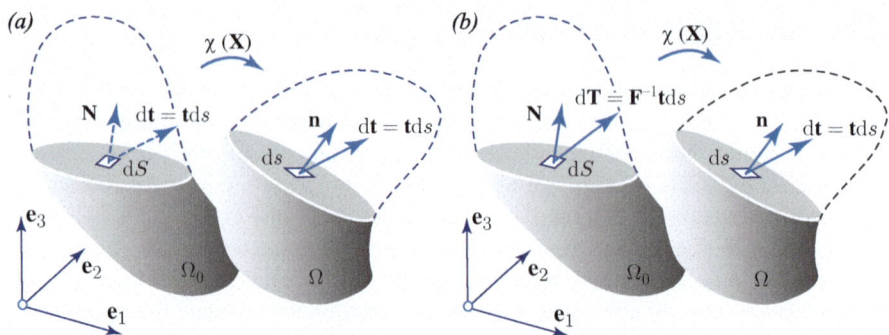

Fig. 3.18 Definition of (**a**) the first Piola–Kirchhoff stress **P** through the relation $d\mathbf{t} = \mathbf{P}\mathbf{N}dS$, and (**b**) the second Piola–Kirchhoff stress **S** through the relation $d\mathbf{T} = \mathbf{S}\mathbf{N}dS$

$$\mathbf{P} = J\boldsymbol{\sigma}\mathbf{F}^{-\mathrm{T}} \quad ; \quad P_{iI} = J\sigma_{ia}F_{Ia}^{-1} \ . \tag{3.31}$$

It is known as the *first backward Piola transform* of the Cauchy stress and relates the first Piola–Kirchhoff stress **P** to the Cauchy stress $\boldsymbol{\sigma}$.

The first Piola–Kirchhoff stress is a *two-point* tensor, and it relates to the spatial configuration Ω as well as to the reference configuration Ω_0. Given the principal first Piola–Kirchhoff stresses $P_i; i = 1, \ldots, n_{\mathrm{dim}}$, its eigenvalue representation reads $\mathbf{P} = P_i \widehat{\mathbf{n}}_i \otimes \widehat{\mathbf{N}}_i$, where $\widehat{\mathbf{n}}_i$ and $\widehat{\mathbf{N}}_i$ denote the eigenvectors in Ω and Ω_0, respectively. The first Piola–Kirchhoff stress is a two-point tensor, and the question regarding its symmetry does not make much sense—one cannot compare referential $\widehat{\mathbf{N}}_i$ to spatial $\widehat{\mathbf{n}}_i$ base vectors. Nevertheless, the coefficients of the first Piola–Kirchhoff stress may be represented by a matrix, and such a matrix would not be symmetric.

A rigid body rotation **R** of the body's spatial configuration Ω influences only one "leg" of the first Piola–Kirchhoff stress. It transforms only the traction vectors $\widetilde{\mathbf{t}} = \mathbf{R}\mathbf{t}$, but not the normal **N** of the referential area element. The rotated first Piola–Kirchhoff stress therefore reads $\widetilde{\mathbf{P}} = \mathbf{R}\mathbf{P}$.

3.3.8 Second Piola–Kirchhoff Stress

Towards the derivation of the *second Piola–Kirchhoff stress* **S**, the force $d\mathbf{t} = \mathbf{t}ds$ in the spatial configuration Ω is "pulled-back," which then defines the force $d\mathbf{T} = \mathbf{F}^{-1}\mathbf{t}ds$ in the reference configuration Ω_0, see Fig. 3.18b. The force $d\mathbf{T}$ is then related to the area element $d\mathbf{S} = \mathbf{N}dS$ in Ω_0 and defines the second Piola–Kirchhoff **S**. Given Cauchy's stress theorem (3.20) together with Nanson's formula (3.5), the expression $d\mathbf{T} = \mathbf{F}^{-1}\boldsymbol{\sigma}J\mathbf{F}^{-\mathrm{T}}\mathbf{N}dS = \mathbf{S}\mathbf{N}dS$ holds, and the relation

$$\mathbf{S} = J\mathbf{F}^{-1}\boldsymbol{\sigma}\mathbf{F}^{-\mathrm{T}} \quad ; \quad S_{IJ} = J F_{Ia}^{-1}\sigma_{ab}F_{Jb}^{-1} \tag{3.32}$$

may be deduced. It is known as the *second backward Piola transform* and relates the second Piola–Kirchhoff stress **S** to the Cauchy stress $\boldsymbol{\sigma}$. Given (3.31) and (3.32), the expression

$$\mathbf{S} = \mathbf{F}^{-1}\mathbf{P} \; ; \; S_{IJ} = F_{Ia}^{-1} P_{aJ} \tag{3.33}$$

follows and provides the relation between the first and second Piola–Kirchhoff stresses.

The second Piola–Kirchhoff stress is a *referential one-point* tensor that is entirely defined in the reference configuration Ω_0. Its eigenvalue representation reads $\mathbf{S} = S_i \widehat{\mathbf{N}}_i \otimes \widehat{\mathbf{N}}_i$, where S_i and $\widehat{\mathbf{N}}_i, i = 1, \ldots, n_{\mathrm{dim}}$ denote the principal second Piola–Kirchhoff stresses and eigenvectors in Ω_0, respectively. The definition (3.32) results in the *symmetry* of the second Piola–Kirchhoff stress, $\mathbf{S} = \mathbf{S}^{\mathrm{T}}$, and given its definition in Ω_0, it is not affected by a rigid body rotation **R** of the spatial configuration, $\widetilde{\mathbf{S}} = \mathbf{S}$.

Example 3.10 (Stress Measures at Finite Deformations). The 2D motion $\boldsymbol{\chi}(\mathbf{X}) = [3.0X_1 + X_2 \quad 1.3X_2]^{\mathrm{T}}$ determines the kinematics of a material particle, and the Cauchy stress

$$\boldsymbol{\sigma} = \begin{bmatrix} 1 & 5 \\ 5 & -10 \end{bmatrix} \, [\mathrm{MPa}]$$

specifies its loading in the spatial configuration Ω.

(a) Compute the deformation gradient **F**, the volume ratio J, and the inverse deformation gradient $\mathbf{F}^{-1}$.
(b) Express the loading of the material particle through the first Piola–Kirchhoff stress **P** and the second Piola–Kirchhoff stress **S**, respectively. ∎

3.3.9 Implication of Material Incompressibility on the Stress State

Given an *incompressible material*, the volumetric stress $\boldsymbol{\sigma}_{\mathrm{vol}} = -p\mathbf{I}$ with $p = -\mathrm{tr}\boldsymbol{\sigma}/n_{\mathrm{dim}}$ denoting the negative hydrostatic pressure, is no longer a function of the strain (or strain rate) but appears as a *Lagrange contribution* to enforce the incompressibility. This property of an incompressible material motivates the decoupling of the Cauchy stress $\boldsymbol{\sigma} = \overline{\boldsymbol{\sigma}} - p\mathbf{I}$ as introduced in Sect. 3.3.4.

Given the second Piola–Kirchhoff stress (3.32) and the incompressibility $J = 1$, the decoupled stress representation then reads $\mathbf{S} = \overline{\mathbf{S}} - p\mathbf{C}^{-1}$, where $\overline{\mathbf{S}} = \mathbf{F}^{-1}\overline{\boldsymbol{\sigma}}\mathbf{F}^{-\mathrm{T}}$ and $\mathbf{C} = \mathbf{F}^{\mathrm{T}}\mathbf{F}$ denote the isochoric second Piola–Kirchhoff stress and the right Cauchy–Green strain, respectively.

3.4   Material Time Derivatives

The time derivative of a property that is connected to a material particle is called
a *material time derivative*. We may also see the material time derivative as the
derivative that is "felt" by the material particle. Such derivatives appear in balance
laws, and they are also used to describe the constitutive properties of rate-dependent
materials. The material time derivative has to be objective and thus independent
from the choice of the underlying coordinate system. Given that the body's spatial
configuration Ω moves in space, the computation of the material time derivative of
a spatial quantity may be more challenging than of a quantity that is defined in the
body's *fixed* reference configurations Ω_0.

3.4.1   Kinematic Variables

3.4.1.1 Velocity Gradient
The *spatial* gradient of the particle velocity $\mathbf{v}(\mathbf{x})$ is called the *velocity gradient*

$$\mathbf{l} = \text{grad}\mathbf{v} = \frac{\partial \mathbf{v}(\mathbf{x}, t)}{\partial \mathbf{x}} = \frac{\partial \mathbf{v}(\mathbf{x}, t)}{\partial \mathbf{X}}\frac{\partial \mathbf{X}}{\partial \mathbf{x}} = \dot{\mathbf{F}}\mathbf{F}^{-1} \;\; ; \;\; l_{ij} = \dot{F}_{iA}F_{Aj}^{-1}, \quad (3.34)$$

where $\dot{\mathbf{F}} = \text{d}(\partial \boldsymbol{\chi}/\partial \mathbf{X})/\text{d}t = \partial(\text{d}\boldsymbol{\chi}/\text{d}t)/\partial \mathbf{X} = \text{Grad}\mathbf{v}$ denotes the material time
derivative of the deformation gradient. The velocity gradient specifies how fast two
neighboring points move relative to each other upon the motion $\boldsymbol{\chi}(\mathbf{X}, t)$. Through
$\mathbf{l} = \mathbf{w} + \mathbf{d}$, the velocity gradient can be decomposed into the spin tensor $\mathbf{w}$ and the
rate of deformation tensor $\mathbf{d}$, respectively.

3.4.1.2 Rate of Deformation
The Green–Lagrange strain $\mathbf{E}$ is defined in Ω_0 and its material time derivative reads
$\dot{\mathbf{E}} = (\dot{\mathbf{F}}^T\mathbf{F} + \mathbf{F}^T\dot{\mathbf{F}})/2$. The multiplication of $\dot{\mathbf{E}}$ from the left with $\mathbf{F}^{-T}$ and from the
right with $\mathbf{F}^{-1}$ may be seen as the "push forward" of $\dot{\mathbf{E}}$ to the spatial configuration
Ω. It defines the *rate of deformation* tensor

$$\mathbf{d} = \frac{1}{2}\left(\mathbf{F}^{-T}\dot{\mathbf{F}}^T + \dot{\mathbf{F}}\mathbf{F}^{-1}\right) = \frac{1}{2}\left(\mathbf{l}^T + \mathbf{l}\right) \;\; ; \;\; d_{ij} = (l_{ji} + l_{ij})/2 , \quad (3.35)$$

the symmetric part of the velocity gradient (3.34).

3.4.1.3 Spin Tensor
The skew-symmetric part of the velocity gradient (3.34) is the *spin tensor*

$$\mathbf{w} = \frac{1}{2}(\mathbf{l} - \mathbf{l}^T) = \frac{1}{2}\left(\dot{\mathbf{F}}\mathbf{F}^{-1} - \mathbf{F}^{-T}\dot{\mathbf{F}}^T\right) \;\; ; \;\; w_{ij} = (l_{ji} - l_{ij})/2 . \quad (3.36)$$

Fig. 3.19 Rotation of a
material particle at position $\mathbf{x}$.
The angular velocity $\boldsymbol{\omega}$
specifies the particle's
velocity $\mathbf{v} = \mathbf{r} \times \boldsymbol{\omega}$

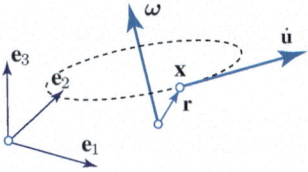

The right polar decomposition $\mathbf{F} = \mathbf{RU}$ allows us to express the spin tensor through

$$\mathbf{w} = \dot{\mathbf{R}}\mathbf{R}^{\mathsf{T}} + \frac{1}{2}\mathbf{R}\left(\dot{\mathbf{U}}\mathbf{U}^{-1} - \mathbf{U}^{-1}\dot{\mathbf{U}}\right)\mathbf{R}^{\mathsf{T}}, \tag{3.37}$$

where the symmetry $\mathbf{U}^{\mathsf{T}} = \mathbf{U}$ of the right stretch tensor and the orthogonality $\mathbf{RR}^{\mathsf{T}} = \mathbf{I}$ of the rotation tensor have been used. The derivation of (3.37) also used $\dot{\mathbf{R}}\mathbf{R}^{\mathsf{T}} = -\mathbf{R}\dot{\mathbf{R}}^{\mathsf{T}}$, a relation that directly follows from the time derivative of the orthogonality $\mathbf{RR}^{\mathsf{T}} = \mathbf{I}$. Given a rigid body motion, $\mathbf{U} = \mathbf{I}$, and the spin tensor then reads

$$\mathbf{w} = \dot{\mathbf{R}}\mathbf{R}^{\mathsf{T}} = \begin{bmatrix} 0 & -w_{12} & w_{13} \\ w_{12} & 0 & -w_{23} \\ -w_{13} & w_{23} & 0 \end{bmatrix} = \begin{bmatrix} 0 & -\omega_3 & \omega_2 \\ \omega_3 & 0 & -\omega_1 \\ -\omega_2 & \omega_1 & 0 \end{bmatrix},$$

where $\omega_i; i = 1, 2, 3$ denotes the components of the *angular velocity* $\boldsymbol{\omega}$ in 3D, see Fig. 3.19.

3.4.1.4 Rate of Volume Change

Given Jacobi's[13] formula $\partial \det\mathbf{F}/\partial\mathbf{F} = J\mathbf{F}^{-\mathsf{T}}$ and the definition $J = \det\mathbf{F}$ of the volume ratio, $\dot{J} = \partial J/\partial\mathbf{F} : \dot{\mathbf{F}} = J\mathbf{F}^{-\mathsf{T}} : \dot{\mathbf{F}} = J\mathbf{I} : \dot{\mathbf{F}}\mathbf{F}^{-1}$ expresses the *rate of volume change*. We may then use the velocity gradient $\mathrm{grad}\mathbf{v} = \mathbf{l} = \dot{\mathbf{F}}\mathbf{F}^{-1}$ and derive the expressions

$$\dot{J} = J\mathbf{I} : \dot{\mathbf{F}}\mathbf{F}^{-1} = J\mathbf{I} : \mathbf{l} = J\mathbf{I} : \mathrm{grad}\mathbf{v} = J\mathrm{div}\mathbf{v} = J\mathrm{tr}\mathbf{d}, \tag{3.38}$$

where $\mathbf{d} = (\mathbf{l} + \mathbf{l}^{\mathsf{T}})/2$ is the rate of deformation tensor, whilst $\mathrm{div}(\bullet) = \mathbf{I} : \mathrm{grad}(\bullet)$ denotes the divergence operator. With the property $\mathrm{tr}\mathbf{d} = \mathbf{C}^{-1} : \dot{\mathbf{E}}$ from Appendix C, $\dot{J} = J\mathbf{C}^{-1} : \dot{\mathbf{E}}$ yields another alterative expression of the rate of volume change.

Example 3.11 (The Physical Meaning of the Rate of Deformation Tensor). The homogeneous 2D motion $\boldsymbol{\chi}(t)$ has the components

$$\chi_1 = x_1 = X_1 + \gamma X_2 t \; ; \; \chi_2 = x_2 = X_2, \tag{3.39}$$

a mapping that determines simple shear kinematics, where the scalar γ denotes the amount of shear.

---

[13]Carl Gustav Jacob Jacobi, German mathematician, 1804–1851.

Fig. 3.20 Simple shear determines the mapping of convective vectors between their reference configurations in Ω_0 and their spatial configurations in Ω. (**a**) The vectors $\mathbf{m}_0$ and $\mathbf{n}_0$ are aligned and (**b**) rotated by $\pi/4$ with respect to the Cartesian base vectors $\mathbf{e}_1$ and $\mathbf{e}_2$

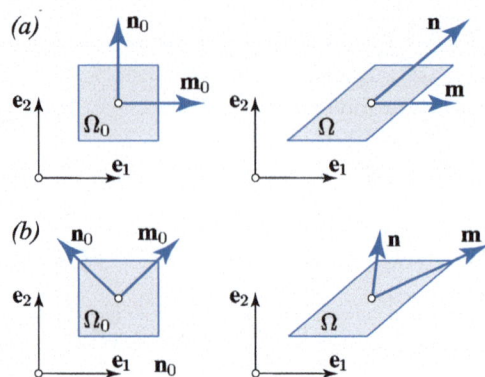

(a) Compute the rate of deformation tensor $\mathbf{d}$ and the spin tensor $\mathbf{w}$ that follow from the motion (3.39).

(b) Consider the convective vectors $\mathbf{m}$ and $\mathbf{n}$ that are connected to the body and follow its (homogeneous) motion $\chi(t)$. Compute the time derivatives of the inner products $\mathbf{m} \cdot \mathbf{n}$ and $\mathbf{m} \cdot \mathbf{m}$, and provide physical interpretations of these quantities.

(c) Study the deformation of the pairs $(\mathbf{m}_0 = [1\ 0]^\mathsf{T}, \mathbf{n}_0 = [0\ 1]^\mathsf{T})$ and $(\mathbf{m}_0 = [\sqrt{2}\ \sqrt{2}]^\mathsf{T}, \mathbf{n}_0 = [-\sqrt{2}\ \sqrt{2}]^\mathsf{T})$ of convective vectors, see Fig. 3.20. ∎

3.4.2 Stress Rates

Stress rates are used in the description of time-dependent physical processes—they have therefore to be *objective*. Towards the exploration of their objectivity, let us consider the spatial configuration $\widetilde{\Omega}$ that establishes through the rigid body rotation $\mathbf{R}$ applied to Ω. Given the Cauchy stress $\boldsymbol{\sigma}$ in Ω and the transformation (3.30),

$$\frac{\mathrm{d}}{\mathrm{d}t}\widetilde{\boldsymbol{\sigma}} = \overline{\dot{\mathbf{R}\boldsymbol{\sigma}\mathbf{R}^\mathsf{T}}} = \dot{\mathbf{R}}\boldsymbol{\sigma}\mathbf{R}^\mathsf{T} + \mathbf{R}\dot{\boldsymbol{\sigma}}\mathbf{R}^\mathsf{T} + \mathbf{R}\boldsymbol{\sigma}\dot{\mathbf{R}}^\mathsf{T} \tag{3.40}$$

expresses the time derivative of the Cauchy stress. Given the Cauchy stress is a spatial one-point tensor, we expect its time derivative also to transform like a spatial one-point tensor, and thus according to $\widetilde{\boldsymbol{\sigma}} = \mathbf{R}\dot{\boldsymbol{\sigma}}\mathbf{R}^\mathsf{T}$. The expression (3.40) shows that said property does not hold, and the time derivative of the Cauchy stress is therefore *not* objective and *must not* be used for the description of physical processes at finite deformations.

In contrary to the Cauchy stress $\boldsymbol{\sigma}$, the second Piola–Kirchhoff stress $\mathbf{S}$ is a referential one-point tensor and independent from a rigid body rotation $\mathbf{R}$ of the spatial configuration Ω. Its time derivative $\dot{\mathbf{S}}$ is therefore *a priori* an objective stress rate. Other objective stress rates are the *Truesdell*[14] *rate* of the Cauchy stress

$$\overset{\circ}{\sigma} = J^{-1}\mathbf{F}\dot{\mathbf{S}}\mathbf{F}^{\mathrm{T}} = J^{-1}\mathbf{F}\left[\mathrm{d}\left(J\mathbf{F}^{-1}\sigma\mathbf{F}^{-\mathrm{T}}\right)/\mathrm{d}t\right]\mathbf{F}^{\mathrm{T}} = \dot{\sigma} - \mathbf{l}\sigma - \sigma\mathbf{l}^{\mathrm{T}} + (\mathrm{trl})\sigma$$

or the *Jaumann*[15] *rate* of the Cauchy stress

$$\overset{\triangle}{\sigma} = \dot{\sigma} + \sigma\mathbf{w} - \mathbf{w}\sigma$$

with $\mathbf{l} = \mathrm{grad}\mathbf{v}$ and $\mathbf{w} = (\mathbf{l} - \mathbf{l}^{\mathrm{T}})/2$ denoting the velocity gradient and the spin tensor, respectively. The reasoning behind these definitions is given elsewhere [354].

3.4.3 Power-Conjugate Stress and Strain rates

Given different definitions of stress and strain in finite deformation theory, deformation power results from the multiplication of so-called *power-conjugate* pairs of stress and strain rate. Here, deformation power refers to the *Helmholtz*[16] *free energy* (or *strain energy*) per unit time and denotes the portion of the material's internal power that can be turned into *mechanical power*. Section 3.5.3 provides more information of the Helmholtz free energy.

The Coleman[17]–Noll[18] relation (3.129) $\mathbf{S} = 2\partial\Psi/\partial\mathbf{C}$ allows us to compute the second Piola–Kirchhoff stress $\mathbf{S}$, where Ψ [J m^{-3}] denotes the Helmholtz free energy per unit *undeformed* material volume, whilst $\mathbf{C}$ is the right Cauchy–Green strain. The deformation power per *unit undeformed material volume* therefore reads

$$\dot{\Psi} = \frac{\mathrm{d}\Psi}{\mathrm{d}t} = \frac{\partial\Psi}{\partial\mathbf{C}} : \frac{\mathrm{d}\mathbf{C}}{\mathrm{d}t} = \frac{1}{2}\mathbf{S} : \dot{\mathbf{C}}, \qquad (3.41)$$

and thus $\mathbf{S}$ and $\dot{\mathbf{C}}/2$ are power-conjugate quantities.

Given the definition (3.33) of the second Piola–Kirchhoff stress together with the strain definitions (3.10) and (3.12), the alternative expression $\dot{\Psi} = \mathbf{S} : \dot{\mathbf{E}} = \mathbf{P} : \dot{\mathbf{F}}$ follows from (3.41). The second Piola–Kirchhoff stress $\mathbf{S}$ is therefore also power-conjugate to the Green–Lagrange strain rate $\dot{\mathbf{E}}$, whilst the first Piola–Kirchhoff stress $\mathbf{P}$ is power-conjugate to rate of the deformation gradient $\dot{\mathbf{F}}$.

Towards the derivation of a power-conjugate pair that relates to the body's *spatial configuration* Ω, we consider the deformation power $\dot{\Gamma}$ [J s^{-1}] of the entire

---

[14]Clifford Ambrose Truesdell III, American mathematician, natural philosopher, and historian of science, 1919–2000.

[15]Gustav Jaumann, Austrian physicist, 1863–1924.

[16]Hermann Ludwig Ferdinand von Helmholtz, German physician and physicist, 1821–1894.

[17]Bernard D. Coleman, American physician, 1929-date.

[18]Walter Noll, American mathematician, 1925–2017.

continuum body. Given the specific deformation power (3.41), the definition (3.32) of the second Piola–Kirchhoff stress and the time derivative $\dot{\mathbf{C}} = \dot{\mathbf{F}}^{\mathrm{T}}\mathbf{F} + \mathbf{F}^{\mathrm{T}}\dot{\mathbf{F}}$ of the right Cauchy–Green strain, the expression

$$\dot{\Gamma} = \frac{1}{2} \int_{\Omega_0} \mathbf{S} : \dot{\mathbf{C}} \mathrm{d}V = \frac{1}{2} \int_{\Omega_0} J\mathbf{F}^{-1}\boldsymbol{\sigma}\mathbf{F}^{-\mathrm{T}} : (\dot{\mathbf{F}}^{\mathrm{T}}\mathbf{F} + \mathbf{F}^{\mathrm{T}}\dot{\mathbf{F}})\mathrm{d}V =$$

$$= \frac{1}{2} \int_{\Omega} [\mathrm{tr}(\mathbf{F}^{-1}\boldsymbol{\sigma}\mathbf{F}^{-\mathrm{T}}\dot{\mathbf{F}}^{\mathrm{T}}\mathbf{F}) + \mathrm{tr}(\mathbf{F}^{-1}\boldsymbol{\sigma}\mathbf{F}^{-\mathrm{T}}\mathbf{F}^{\mathrm{T}}\dot{\mathbf{F}})]\mathrm{d}v =$$

$$= \frac{1}{2} \int_{\Omega} [\mathrm{tr}(\boldsymbol{\sigma}\mathbf{F}^{-\mathrm{T}}\dot{\mathbf{F}}^{\mathrm{T}}) + \mathrm{tr}(\boldsymbol{\sigma}\dot{\mathbf{F}}\mathbf{F}^{-1})]\mathrm{d}v =$$

$$= \frac{1}{2} \int_{\Omega} \boldsymbol{\sigma} : (\mathbf{l}^{\mathrm{T}} + \mathbf{l})\mathrm{d}v = \int_{\Omega} \underbrace{\boldsymbol{\sigma} : \mathbf{d}}_{\dot{\psi}} \mathrm{d}v$$

determines the body's deformation power. The derivation used the volume ratio $J = \mathrm{d}v/\mathrm{d}V$, the rate of deformation tensor $\mathbf{d} = (\mathbf{l} + \mathbf{l}^{\mathrm{T}})/2$, as well as the properties $\mathbf{I} : \mathbf{A} = \mathrm{tr}\mathbf{A}$ and $\mathrm{tr}(\mathbf{ABC}) = \mathrm{tr}(\mathbf{BCA})$. The Cauchy stress $\boldsymbol{\sigma}$ and the rate of deformation $\mathbf{d}$ are therefore power-conjugate, and $\dot{\psi} = \boldsymbol{\sigma} : \mathbf{d}$ determines the power per *unit deformed material volume*.

3.5   Constitutive Modeling

A constitutive model is the *mathematical description* of the material's mechanical properties. A material may exhibit many, often complex, physical phenomena, and it is the Intended Model Application (IMA) that specifies which ones should be captured by a particular constitutive model. This section presents the basics of constitutive modeling including linear and non-linear descriptions—many more constitutive descriptions are discussed in the following chapters of this book.

3.5.1   Some Mechanical Properties of Materials

3.5.1.1 Incompressibility
An *incompressible* material preserves its volume during the deformation. The volumetric stress $\boldsymbol{\sigma}_{\mathrm{vol}} = -p\mathbf{I}$ is then independent from the deformation and may be represented by a *penalty* or *Lagrange* contribution. Given an incompressible material, its bulk modulus, and thus the resistance against volume change, is infinite.

3.5.1.2 Isotropy and Anisotropy
If the mechanical properties are independent of the spatial orientation, the material is *isotropic*, otherwise it is *anisotropic*. Most vascular tissues are anisotropic. The

characterization of an anisotropic material requires more experimental tests to be performed than of an isotropic material. Aside from its dependence on the strain, the stress in an anisotropic material depends also on *structural information*, such as the orientations along which the material is softest and stiffest, known as the *principal material axes*. Given an isotropic material, the principal stress directions coincide with the principal strain directions. This is in general not the case for an anisotropic material.

3.5.1.3 Strain Energy in Solids

The deformation of a body results in the transformation of external work W into *strain energy* Γ. Given simple tension of a *linear-elastic* bar of length l, the external work $W = Pu/2$ is "used" to elongate the bar by $u = l\varepsilon_{11}$, where $P = \sigma_{11}s$ denotes the force acting on the rod's cross-section s, see Fig. 3.21a. An ideal *reversible* process transforms the external work completely into strain energy, and $W = \Gamma = Pu/2 = \sigma_{11}sl\varepsilon_{11}/2$ holds. The strain energy per unit volume of a material, or its *specific strain energy*, then reads $\Psi = \Gamma/V = \sigma_{11}\varepsilon_{11}/2$.

Given simple shear of a *linear-elastic* block of the dimensions $l \times s$, the external work $W = Pu/2$ is "used" to distort the block by $u = 2l\varepsilon_{12}$, where $P = \sigma_{12}s$ denotes the shear force acting at the area s, see Fig. 3.21b. It is noted that the shear angle $\gamma_{12} = 2\varepsilon_{12}$ is twice the shear strain ε_{12}. An ideal *reversible* process acquires the strain energy $\Gamma = Pu/2 = \sigma_{12}sl2\varepsilon_{12}/2$, and the specific strain energy therefore reads $\Psi = \Gamma/V = \sigma_{12}\varepsilon_{12}$.

The generalization of the aforementioned results towards a multiaxial stress state of a *linear-elastic* material yields the specific strain energy $\Psi = \boldsymbol{\sigma} : \boldsymbol{\varepsilon}/2$, an expression that reads

$$\Psi = \sigma_{ab}\varepsilon_{ab}/2$$
$$= \sigma_{11}\varepsilon_{11}/2 + \sigma_{22}\varepsilon_{22}/2 + \sigma_{33}\varepsilon_{33}/2 + \sigma_{12}\varepsilon_{12} + \sigma_{23}\varepsilon_{23} + \sigma_{13}\varepsilon_{13}$$

in index notation of a 3D problem.

Given a *non-linear* material, the specific strain energy reads $\Psi = \int_t \boldsymbol{\sigma}(\boldsymbol{\varepsilon}) : d\boldsymbol{\varepsilon}(t)$, where the integral is taken over the deformation path of the material particle. The strain energy of an elastic material is *path-independent*, and Ψ is independent of how the final strain state $\boldsymbol{\varepsilon}(t)$ has been reached. At finite deformations the strain energy may be either defined with respect to the *undeformed* material volume and denoted by Ψ, or with respect to the *deformed* material volume and denoted by ψ.

Fig. 3.21 External work $W = Pu/2$ to deform a linear-elastic body according to (**a**) simple tension and (**b**) simple shear, respectively

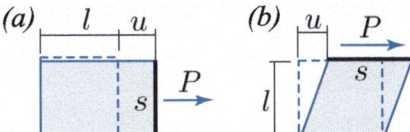

3.5.1.4 Dissipation in Isothermal Solids

At isothermal conditions, the Clausius[19]–Duhem[20] inequality (3.124) results in the dissipation

$$\mathcal{D} = -\dot{\psi} + \boldsymbol{\sigma} : \mathbf{d} \geq 0 \tag{3.42}$$

per unit spatial volume, and thus per unit volume of the deformed material. Here, the Cauchy stress $\boldsymbol{\sigma}$ multiplied with the rate of deformation $\mathbf{d}$ determines the *deformation power* $\boldsymbol{\sigma} : \mathbf{d}$, whilst $\dot{\psi}$ denotes the rate of the solid's Helmholtz free energy per unit spatial volume.

Given a viscoelastic material, the stress $\boldsymbol{\sigma} = \boldsymbol{\sigma}_e + \boldsymbol{\sigma}_v$ may be decomposed into the *elastic* stress $\boldsymbol{\sigma}_e$ and the *viscose*, or *over* stress $\boldsymbol{\sigma}_v$. The viscose stress refers to contributions from dashpot-like elements in the rheological description of viscoelastic solids, see Sect. 3.5.4. The viscose stress $\boldsymbol{\sigma}_v$ depends on the strain, the strain rate, and the *loading history* of the material. At fixed deformation and increasing time t, $\boldsymbol{\sigma}_v$ tends to zero, and at the limit $t \rightarrow \infty$, the elastic stress $\boldsymbol{\sigma}_e$ characterizes the material.

The elastic part of the deformation power is fully recoverable, it is therefore equal to the rate of the Helmholtz free energy, $\boldsymbol{\sigma}_e : \mathbf{d} = \dot{\psi}$, and the dissipation of a viscoelastic material then reads

$$\mathcal{D} = \boldsymbol{\sigma}_v : \mathbf{d} \geq 0 . \tag{3.43}$$

The dissipation has to be *non-negative* for all possible deformations to comply with the second law of thermodynamics, see Sect. 3.6.3.2.

3.5.1.5 Dissipation in Isothermal Fluids

In contrary to solids, (inelastic) fluids *cannot* store strain energy, and thus $\dot{\psi} = 0$. At isothermal conditions, the Clausius–Duhem inequality (3.124) then reads

$$\mathcal{D} = \boldsymbol{\sigma} : \mathbf{d} \geq 0 \tag{3.44}$$

and determines the dissipation $\mathcal{D}$ per unit spatial volume—the *entire* deformation power $\boldsymbol{\sigma} : \mathbf{d}$ is then dissipated.

We may consider an incompressible Newtonian fluid (see Sect. 3.5.4.1) to further elaborate on the dissipation inequality. Given such a rheological description, the stress $\boldsymbol{\sigma} = 2\eta\mathbf{d} - p\mathbf{I}$ develops in the fluid, where η and p denote the dynamic viscosity and the fluid pressure, respectively. The power $\mathcal{D} = \boldsymbol{\sigma} : \mathbf{d} = (2\eta\mathbf{d} - p\mathbf{I}) : \mathbf{d} = 2\eta\mathbf{d} : \mathbf{d}$ is then dissipated. In the derivation of this expression, the incompressibility (3.38) $\mathbf{I} : \mathbf{d} = 0$, also known as the continuity condition of an

---

[19]Rudolf Julius Emanuel Clausius, German physicist and mathematician, 1822–1888.

[20]Pierre Maurice Marie Duhem, French physicist, mathematician, historian, and philosopher of science, 1861–1916.

incompressible fluid, has been used. Note that the dissipation $\mathcal{D}$ of the Newtonian fluid is strictly positive for any rate of deformation **d**.

3.5.1.6 Stiffness

The *stiffness* describes how the stress changes in response to the strain. The internal structure of vascular tissue is continuously renewed, and its stiffness may therefore change over time in response to factors, such as aging and disease. The stiffness is described by the *fourth-order tensor* $\mathbb{c}$ that relates the stress increment $\Delta\boldsymbol{\sigma}$ to the strain increment $\Delta\boldsymbol{\varepsilon}$ according to

$$\Delta\boldsymbol{\sigma} = \mathbb{c} : \Delta\boldsymbol{\varepsilon}\,.$$

Given a 1D problem, the stiffness is the tangent to the stress–strain curve, see Fig. 3.22. The stiffness tensor of an *isotropic linear-elastic* material is constant and determined by two material parameters—the Young's modulus and the Poisson's[21] ratio, for example. See Sect. 3.5.2 for alternative sets of parameters. The stiffness of a non-linear material is not constant and depends on the strain. It is therefore *not* an intrinsic material property.

The stiffness tensor preserves always *minor symmetry* $c_{ijkl} = c_{jikl} = c_{ijlk}$, and for most materials also *major symmetry* $c_{ijkl} = c_{klij}$. The stiffness tensor of an elastic material is *positive definite*, and $\Delta\boldsymbol{\varepsilon} : \mathbb{c} : \Delta\boldsymbol{\varepsilon}$ remains positive for any possible strain increment $\Delta\boldsymbol{\varepsilon}$.

Given a geometrically non-linear analysis, it is convenient to use the stiffness $\mathbb{C}$ with respect to the *undeformed* material. It is a referential one-point tensor that relates to the body's reference configuration Ω_0. The referential stiffness expresses the increment of the second Piola–Kirchhoff stress by

$$\Delta\mathbf{S} = \mathbb{C} : \Delta\mathbf{E} = 2\mathbb{C} : \Delta\mathbf{C}\,,$$

where $\Delta\mathbf{E}$ and $\Delta\mathbf{C}$ denote the increments of the Green–Lagrange strain and the right Cauchy–Green strain, respectively. The push-forward operation $c_{ijkl} = J^{-1}F_{iI}F_{jJ}F_{kK}F_{lL}C_{IJKL}$ relates the referential stiffness tensor $\mathbb{C}$ to the spatial stiffness tensor $\mathbb{c}$, where **F** and J denote the deformation gradient and the volume ratio, respectively.

Given a hyperelastic material (see Sect. 3.5.3), the referential stiffness tensor

Fig. 3.22 Stress *versus* strain properties of a strain-stiffening material in 1D. The stiffness $k = \Delta\sigma/\Delta\varepsilon$ is not constant but increases with the strain

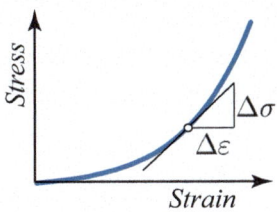

[21] Baron Siméon Denis Poisson, French mathematician, engineer, and physicist, 1781–1840.

$$\mathbb{C}(\mathbf{C}) = 4\frac{\partial^2 \Psi(\mathbf{C})}{\partial \mathbf{C} \partial \mathbf{C}} \qquad (3.45)$$

derives from the strain energy $\Psi(\mathbf{C})$ per unit reference volume. The expression (3.45) preserves major and minor symmetries and leads to *21 independent* components of the stiffness tensor. The decoupling of the stiffness tensor in the *isochoric (or deviatoric)* $\overline{\mathbb{C}}$ and *volumetric (or hydrostatic)* $\mathbb{C}_{\mathrm{vol}}$ contributions may be motivated for some applications.

Given an incompressible material, for example, $\mathbb{C}_{\mathrm{vol}}$ is *Lagrange* contribution, whilst it is a *penalty* contribution in the description of a quasi-incompressible material.

3.5.1.7 Strength
The *strength* denotes the *stress level* at which a material mechanically *fails*. As with the stiffness, the strength of vascular tissue may change over time.

A material may fail by a number of different mechanisms. A ductile material *yields* under shear stress, whilst a brittle material *fractures* under normal stress. A failure hypothesis often introduces an *equivalent (scalar) stress* that represents the material's vulnerability to failure and characterizes its strength under *multi-axial* stress conditions. The von Mises stress and the Tresca[22] stress are widely used equivalent stress measures for ductile materials, whilst the maximum principal stress, also known as the Rankine[23] criterion, may be used to describe brittle materials. A number of other failure mechanisms and theories have also been proposed, and some concepts are discussed in Sect. 3.8.

3.5.1.8 Fracture Toughness
The *fracture toughness* is a property that describes the ability of a material to *resist* fracture. The propagation of a fracture requires *energy* towards the creation of new fracture surface, and releases energy due to elastic unloading of the body in the vicinity of the crack tip. Given the energy needed to generate fracture surface is lower than the energy released through elastic unloading, a fracture will *spontaneously* propagate. The fracture toughness is tightly linked to the internal dissipation or *energy release rate* $\mathcal{D}_{\mathrm{int}}$ [J m^{-2}] and thus the energy that dissipates per unit of newly created fracture surface area.

3.5.2   Linear-Elastic Material

A *linear-elastic* or *Hooke*[24] material exhibits linear stress–strain properties. Given simple tension, it behaves very similar to a linear-elastic spring. The normal stress

---

[22]Henri Édouard Tresca, French mechanical engineer, 1814–1885.

[23]William John Macquorn Rankine, Scottish mechanical engineer, 1820–1872.

[24]Robert Hooke, English natural philosopher, architect, and polymath, 1635–1703.

σ_{11} and the normal strain ε_{11} are related by $\sigma_{11} = E\varepsilon_{11}$, where the proportionality factor E is called *Young's modulus*. The normal strain along one direction, however, influences the normal strain in the perpendicular directions, a phenomenon known as *Poisson's effect*. Given the aforementioned simple tension conditions, the material is therefore not only strained by ε_{11}, but also in the perpendicular directions, $\varepsilon_{22} = \varepsilon_{33} = -\nu\varepsilon_{11}$. The proportionality factor ν is called *Poisson's ratio*.

Given simple shear, $\sigma_{12} = 2G\varepsilon_{12}$ expresses the relation between the shear stress σ_{12} and the shear strain ε_{12}, where the proportionality factor G is called *shear modulus*. In contrary to the normal strains, the shear strain within one plane does *not* influence the shear strains perpendicular to it—the shear strains are uncoupled.

The Young's modulus E, the shear modulus G, and the Poisson's ratio ν are not independent from each other. Given an isotropic linear-elastic material, the kinematics relation

$$E = 2(1 + \nu)G \tag{3.46}$$

holds amongst them. The isotropic linear-elastic material is therefore fully described by *two* independent material parameters.

Example 3.12 (Coupling Between Material Parameters). The Young's modulus E, the shear modulus G, and the Poisson ratio ν describe the properties of an isotropic linear-elastic material. They are not independent, and the relation amongst these parameters should be explored at small deformations.

Fig. 3.23 Quadratic 2D material point at simple tension. (**a**) Reference configuration. (**b**) Deformed configuration with ε and γ denoting normal and shear strains. (**c**) Normal strain ε and normal stress σ. (**d**) Shear strain γ and shear stress τ.

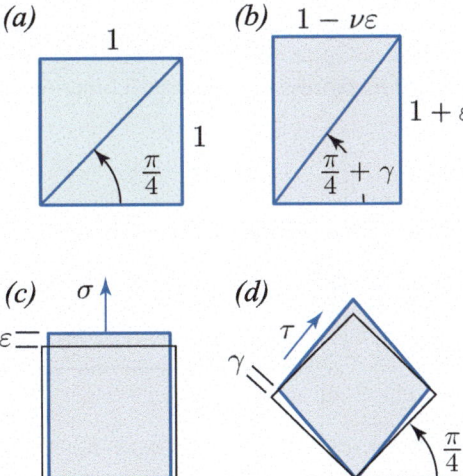

(a) Figure 3.23a,b illustrates the deformation of a 2D material particle. The particle is represented by a square of unit edge length and loaded at simple tension.

Given such kinematics and the assumption of small deformations, derive the relation between the normal strain ε, the shear strain γ, and Poisson's ratio v.

(b) Consider the deformed material particle once in principal stress directions, and twice at $\pi/4$ rotated to it, see Fig. 3.23c,d. Given this access, derive the relation (3.46) between E, G, and v of the linear-elastic isotropic material. ∎

3.5.2.1 Hooke's Law in Voigt's Notation

Towards the expression of a linear-elastic material, it is convenient to introduce *Voigt's notation*. Given a 3D problem, the stress and strain are then expressed by the six-dimensional vectors $\boldsymbol{\sigma} = [\sigma_{11}\ \sigma_{22}\ \sigma_{33}\ \sigma_{12}\ \sigma_{23}\ \sigma_{13}]^T$ and $\boldsymbol{\varepsilon} = [\varepsilon_{11}\ \varepsilon_{22}\ \varepsilon_{33}\ \varepsilon_{12}\ \varepsilon_{23}\ \varepsilon_{13}]^T$, respectively. Consequently, an isotropic linear-elastic material may be expressed by

$$
\begin{bmatrix} \varepsilon_{11} \\ \varepsilon_{22} \\ \varepsilon_{33} \\ \varepsilon_{12} \\ \varepsilon_{23} \\ \varepsilon_{13} \end{bmatrix} = \frac{1}{E} \underbrace{\begin{bmatrix} 1 & -v & -v & 0 & 0 & 0 \\ -v & 1 & -v & 0 & 0 & 0 \\ -v & -v & 1 & 0 & 0 & 0 \\ 0 & 0 & 0 & 1+v & 0 & 0 \\ 0 & 0 & 0 & 0 & 1+v & 0 \\ 0 & 0 & 0 & 0 & 0 & 1+v \end{bmatrix}}_{\mathbf{g}} \begin{bmatrix} \sigma_{11} \\ \sigma_{22} \\ \sigma_{33} \\ \sigma_{12} \\ \sigma_{23} \\ \sigma_{13} \end{bmatrix}, \quad (3.49)
$$

where $\mathbf{g}$ denotes the *compliance matrix*—it is constant and, here, expressed by the Young's modulus E and the Poisson's ratio v. The linear relation (3.49) may be inverted to express the stress as a function of the strain. It then reads

$$
\begin{bmatrix} \sigma_{11} \\ \sigma_{22} \\ \sigma_{33} \\ \sigma_{12} \\ \sigma_{23} \\ \sigma_{13} \end{bmatrix} = \widehat{E} \underbrace{\begin{bmatrix} (1-v) & v & v & 0 & 0 & 0 \\ v & 1-v & v & 0 & 0 & 0 \\ v & v & 1-v & 0 & 0 & 0 \\ 0 & 0 & 0 & 1-2v & 0 & 0 \\ 0 & 0 & 0 & 0 & 1-2v & 0 \\ 0 & 0 & 0 & 0 & 0 & 1-2v \end{bmatrix}}_{\mathbf{c}} \begin{bmatrix} \varepsilon_{11} \\ \varepsilon_{22} \\ \varepsilon_{33} \\ \varepsilon_{12} \\ \varepsilon_{23} \\ \varepsilon_{13} \end{bmatrix}, \quad (3.50)
$$

where $\mathbf{c}$ denotes the *stiffness matrix*, and $\widehat{E} = E/[(1-2v)(1+v)]$ is the effective Young's modulus.

3.5.2.2 Hooke's Law in Tensor Notation

The relation $\boldsymbol{\sigma} = \mathbb{C} : \boldsymbol{\varepsilon}$ expresses a linear-elastic material, where $\mathbb{C}$ denotes the *fourth-order elasticity tensor*. It reads

$$c_{ijkl} = \frac{E\nu}{(1+\nu)(1-2\nu)}\delta_{ij}\delta_{kl} + \frac{E}{2(1+\nu)}(\delta_{ik}\delta_{jl} + \delta_{il}\delta_{jk}) , \qquad (3.51)$$

where E and ν are the Young's modulus and the Poisson's ratio, respectively. The Kronecker delta is denoted by δ_{ij}, and the expression (3.51) allows for the easy verification that the elasticity tensor $\mathbb{C}$ satisfies major symmetry $c_{ijkl} = c_{klij}$ and minor symmetry $c_{ijkl} = c_{jikl} = c_{ijlk}$.

3.5.2.3 Hooke's Law with Decoupled Shear and Bulk Contributions

The linear-elastic material may also be expressed by

$$\boldsymbol{\sigma} = \mathbb{C} : \boldsymbol{\varepsilon} = \overline{\boldsymbol{\sigma}} + \boldsymbol{\sigma}_{\text{vol}} = (\overline{\mathbb{C}} + \mathbb{C}_{\text{vol}}) : \boldsymbol{\varepsilon} ,$$

where $\overline{\boldsymbol{\sigma}}$ and $\boldsymbol{\sigma}_{\text{vol}}$ denote the isochoric and volumetric stress contributions, respectively. The *decoupled* elasticity tensor $\mathbb{C}$ then reads

$$c_{ijkl} = \underbrace{G[\delta_{ik}\delta_{jl} + \delta_{il}\delta_{jk} - (2/3)\delta_{ij}\delta_{kl}]}_{\overline{c}_{ijkl}} + \underbrace{K\delta_{ij}\delta_{kl}}_{c_{\text{vol}\,ijkl}} ,$$

where δ_{ij} is the Kronecker delta, whilst $G = E/[2(1+\nu)]$ and $K = E/[3(1-2\nu)]$ denote the *shear* and *bulk* moduli, respectively. Whilst G specifies the material's resistance against shape changes, such as shearing, K represents its resistance against volume changes.

Example 3.13 (Hooke Material at Specific Load Cases). Consider Voigt's notation (3.49) of a linear-elastic material in 3D, and derive particular stress–strain expressions for the conditions of (a) simple tension, (b) simple shear, (c) plane stress, and (d) plane strain. ∎

3.5.3 Hyperelasticity

The *hyperelastic* description of a material assumes that the deformation energy, and thus the mechanical work to deform the solid, is *fully* recovered upon unloading and *no* energy dissipation then appears. Towards its description, it is convenient to introduce the *Helmholtz free energy*, or *strain energy* $\Psi = U - \theta S$ [J m^{-3}], where θ is the temperature, whilst U and S denote the *internal energy* and the *entropy*, respectively. The latter two quantities are related to the unit *undeformed* volume of the material. The linear-elastic material at small deformations is a special case of a hyperelastic material that is described by the strain energy $\Psi = \boldsymbol{\sigma} : \boldsymbol{\varepsilon}/2 = \boldsymbol{\varepsilon} : \mathbb{C} : \boldsymbol{\varepsilon}/2$.

The material's strain energy is either dominated by the energetic contribution U or by the entropic contribution $-\theta S$. A crystalline material is typically an energetic-elastic material, where the strain energy $\Psi \approx U$ is stored through the change of the distance between the atoms of the crystals. In contrary, rubber is an entropic-elastic material and the strain energy $\Psi \approx -\theta S$ is stored upon straightening out the macromolecular chains, rather than stretching them.

3.5.3.1 Coupled Formulation

The Helmholtz free energy Ψ is a function of the deformation, and it is convenient to use a *referential* strain measure as the argument of such a function. The energy is then *a priori* independent from the rigid body rotation $\mathbf{R}$ upon the body's spatial configuration Ω. The right Cauchy–Green strain $\mathbf{C}$ is such a strain measure, and the Helmholtz free energy then reads $\Psi = \Psi(\mathbf{C})$. With the three invariants $I_1 = \mathrm{tr}\mathbf{C}$, $I_2 = \left[(\mathrm{tr}\mathbf{C})^2 - \mathrm{tr}\mathbf{C}^2\right]/2$, and $I_3 = \det\mathbf{C}$ of a 3D problem, the free energy may be expressed by $\Psi = \Psi(I_1, I_2, I_3)$.

Given an *anisotropic* material, the Helmholtz free energy is also a function of the internal material structure. We may use n structural tensors $\mathbf{A}_i, i = 1, \ldots, n$ to describe such structure, and the Helmholtz free energy can again be expressed through the invariants $I_i, i = 1, \ldots, m$; $\Psi(\mathbf{C}, \mathbf{A}_i) = \Psi(I_1, \ldots, I_m)$. The invariants I_i are then formed by $\mathbf{C}$ and any possible combinations of $\mathbf{C}$ and $\mathbf{A}_i$ [519].

Coleman and Noll's procedure (3.129) allows us to compute the second Piola–Kirchhoff stress

$$S = 2\frac{\partial \Psi(I_1, I_2, I_3, \ldots, I_m)}{\partial \mathbf{C}} = 2\frac{\partial \Psi(I_1, I_2, I_3, \ldots, I_m)}{\partial I_i}\frac{\partial I_i}{\partial \mathbf{C}}, \qquad (3.54)$$

where the terms $\partial I_i/\partial \mathbf{C}$ may be derived from the deformation kinematics. Appendix C reports these expressions for $i = 1, 2, 3$ and more general expressions are reported elsewhere [256]. Given the second Piola–Kirchhoff stress, the Piola transform (3.32) allows to compute the Cauchy stress $\boldsymbol{\sigma} = J^{-1}\mathbf{F}\mathbf{S}\mathbf{F}^{\mathrm{T}}$.

For most applications, the Helmholtz free energy is a strongly convex potential with respect to the deformation gradient $\mathbf{F}$. It then leads to a *positive-definite* elasticity tensor, and the material obeys the physical requirement of increasing stress at increasing deformation. The design of such free energy functions is challenging. As an example, see the constitutive framework of *a priori* polyconvex transverse isotropic strain energy functions [497].

3.5.3.2 Volumetric–Isochoric Decoupled Formulation

It may be convenient to decompose the motion $\chi(\mathbf{X}) = \overline{\chi}(\chi_{\mathrm{vol}}(\mathbf{X}))$ of the material point $\mathbf{X}$ in *volumetric* (volume-changing) χ_{vol} and *isochoric* (shape-changing) $\overline{\chi}$ contributions, see Fig. 3.24. The total deformation gradient $\mathbf{F} = \mathrm{Grad}\chi(\mathbf{X}) = \overline{\mathbf{F}}\mathbf{F}_{\mathrm{vol}}$ is then expressed by the isochoric $\overline{\mathbf{F}} = \mathrm{Grad}\overline{\chi}(\chi_{\mathrm{vol}}(\mathbf{X})) = J^{-1/3}\mathbf{F}$ and respective volumetric $\mathbf{F}_{\mathrm{vol}} = \mathrm{Grad}\chi_{\mathrm{vol}}(\mathbf{X}) = J^{1/3}\mathbf{I}$ contribution, where J denotes the volume ratio. Such an approach links to the decomposition of the stress discussed in Sect. 3.3.4. It motivates the split of the Helmholtz free energy

Fig. 3.24 The total motion $\chi(\mathbf{X}) = \overline{\chi}(\chi_{\text{vol}}(\mathbf{X}))$ is decomposed in the volumetric motion χ_{vol} followed by the isochoric motion $\overline{\chi}$

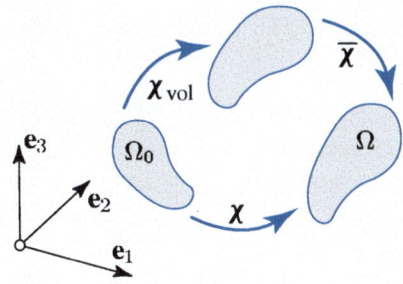

$$\Psi(J, \overline{\mathbf{C}}) = \Psi_{\text{iso}}(\overline{\mathbf{C}}) + \Psi_{\text{vol}}(J) \tag{3.55}$$

in isochoric Ψ_{iso} and volumetric Ψ_{vol} contributions, where $\overline{\mathbf{C}} = \overline{\mathbf{F}}^{\mathrm{T}}\overline{\mathbf{F}}$ denotes the *isochoric* right Cauchy–Green strain. Given the definition of $\overline{\mathbf{C}}$, its third invariant is zero, and an invariant-based description of (3.55) then reads $\Psi = \Psi_{\text{iso}}(\overline{I}_1, \overline{I}_2) + \Psi_{\text{vol}}(J)$ with $\overline{I}_1 = \operatorname{tr}\overline{\mathbf{C}}$ and $\overline{I}_2 = (\overline{I}_1{}^2 - \operatorname{tr}\overline{\mathbf{C}}^2)/2$ denote the first and second invariants of $\overline{\mathbf{C}}$, respectively.

Whilst the additive decomposition (3.55) is generally applicable to isotropic materials regardless whether they are compressible or incompressible, it can only describe anisotropic materials at their incompressible limit. Given incompressibility, Ψ_{vol} "degenerates" to a Lagrange contribution that enforces the incompressibility independent from the constitutive properties.

Given the Helmholtz free energy (3.55), Coleman and Noll's procedure (3.129) defines the second Piola–Kirchhoff stress

$$\mathbf{S} = 2\frac{\partial \Psi(J, \overline{\mathbf{C}})}{\partial \mathbf{C}} = \underbrace{2\frac{\partial \Psi_{\text{iso}}(\overline{\mathbf{C}})}{\partial \overline{\mathbf{C}}} : \frac{\partial \overline{\mathbf{C}}}{\partial \mathbf{C}}}_{\overline{\mathbf{S}}} + \underbrace{2\frac{\partial \Psi_{\text{vol}}(J)}{\partial J}\frac{\partial J}{\partial \mathbf{C}}}_{\mathbf{S}_{\text{vol}}}, \tag{3.56}$$

also decomposed into isochoric $\overline{\mathbf{S}}$ and volumetric $\mathbf{S}_{\text{vol}}$ contributions, respectively. The deformation kinematics determine the terms $\partial J/\partial \mathbf{C} = J\mathbf{C}^{-1}/2$ and $\partial \overline{\mathbf{C}}/\partial \mathbf{C} = J^{-2/3}[\mathbb{I} - (\mathbf{C} \otimes \mathbf{C}^{-1})/3]$ (see Appendix C), and the second Piola–Kirchhoff stress (3.56) then reads

$$\mathbf{S} = 2J^{-2/3}\operatorname{Dev}\left(\frac{\Psi_{\text{iso}}(\overline{\mathbf{C}})}{\partial \overline{\mathbf{C}}}\right) - Jp\mathbf{C}^{-1}, \tag{3.57}$$

where $p = -\partial \Psi_{\text{vol}}(J)/\partial J$ is the negative hydrostatic pressure, whilst $\operatorname{Dev}(\bullet) = (\bullet) - [\mathbf{C} : (\bullet)]\mathbf{C}^{-1}/3$ denotes the deviator operator in the *referential description*. Its "push-forward" yields the spatial deviator operator $\operatorname{dev}(\bullet) = (\bullet) - [\mathbf{I} : (\bullet)]\mathbf{I}/3$.

The Piola transform $\boldsymbol{\sigma} = J^{-1}\mathbf{FSF}^{\mathrm{T}}$ allows us then to compute the Cauchy stress

$$\boldsymbol{\sigma} = J^{-1}\mathbf{FSF}^{\mathrm{T}} = \underbrace{2\overline{\mathbf{F}}\left[\mathrm{Dev}\left(\frac{\Psi_{\mathrm{iso}}(\overline{\mathbf{C}})}{\partial\overline{\mathbf{C}}}\right)\right]\overline{\mathbf{F}}^{\mathrm{T}}}_{\overline{\boldsymbol{\sigma}}}\underbrace{-p\mathbf{I}}_{\boldsymbol{\sigma}_{\mathrm{vol}}}, \tag{3.58}$$

where $\overline{\boldsymbol{\sigma}}$ and $\boldsymbol{\sigma}_{\mathrm{vol}}$ are the respective isochoric and volumetric contributions, whilst $\overline{\mathbf{F}} = J^{-1/3}\mathbf{F}$ denotes the isochoric deformation gradient.

The stress expressions (3.56) and (3.58) may also be used in the description of quasi-incompressible materials. The volumetric strain energy $\Psi_{\mathrm{vol}}(J)$ serves then as a penalty term towards approximately enforcing the incompressibility $J = \det\mathbf{F} \approx 1$.

3.5.3.3 Incompressible Formulation

The description of incompressible solids may be derived from the afore-introduced volumetric–isochoric decoupled formulation. Given incompressibility, the volumetric stress contribution $\mathbf{S}_{\mathrm{vol}}$ "degenerates" towards $\mathbf{S}_{\mathrm{vol}} = -p\mathbf{C}^{-1}$ with the hydrostatic pressure p set to satisfy the incompressibility condition, $J = \det\mathbf{F} = 1$. In contrast to Sect. 3.5.3.2, p is now *not* derived from a strain energy function but a deformation-independent *Lagrange parameter* and determined by the particular Boundary Value Problem (BVP).

Coleman and Noll's procedure for incompressible materials (3.132) then yields the second Piola–Kirchhoff stress

$$\mathbf{S} = \underbrace{2\mathrm{Dev}\left(\frac{\partial\Psi_{\mathrm{iso}}(\overline{\mathbf{C}})}{\partial\overline{\mathbf{C}}}\right)}_{\overline{\mathbf{S}}}\underbrace{-p\mathbf{C}^{-1}}_{\mathbf{S}_{\mathrm{vol}}}. \tag{3.59}$$

Given incompressibility $J = 1$, the right Cauchy–Green strain is now identical to the isochoric right Cauchy–Green strain $\mathbf{C} = J^{-2/3}\mathbf{C} = \overline{\mathbf{C}}$, and therefore only *isochoric* strain energy appears, $\Psi = \Psi_{\mathrm{iso}}$.

The Piola transform (3.32), together with the incompressibility condition $J = 1$, allows us then to compute the Cauchy stress

$$\boldsymbol{\sigma} = \mathbf{FSF}^{\mathrm{T}} = \underbrace{2\mathrm{dev}\left(\mathbf{F}\frac{\partial\Psi_{\mathrm{iso}}(\overline{\mathbf{C}})}{\partial\overline{\mathbf{C}}}\mathbf{F}^{\mathrm{T}}\right)}_{\overline{\boldsymbol{\sigma}}}\underbrace{-p\mathbf{I}}_{\boldsymbol{\sigma}_{\mathrm{vol}}}, \tag{3.60}$$

where $\overline{\boldsymbol{\sigma}}$ and $\boldsymbol{\sigma}_{\mathrm{vol}}$ denote the isochoric and volumetric stress contributions, respectively.

3.5.4 Viscoelasticity

The *viscoelastic* description of a material assumes that the work done by external forces is not entirely transformed into elastic deformation energy, but a part *dissipates* into heat. The work that entered the mechanical system can therefore not been fully recovered upon mechanical unloading.

3.5.4.1 Newtonian Viscosity Model

A viscose fluid may be seen as a viscoelastic material that lacks the ability to store elastic energy. We consider it at the stress $\sigma = \bar{\sigma} - p\mathbf{I}$, where $\bar{\sigma}$ and p denote the *isochoric stress* and the negative *hydrostatic pressure*, respectively. Whilst the stress $\bar{\sigma}$ depends on the shear rate, the pressure p is independent from the deformation but determined by the BVP.

The constitutive model of a viscose fluid describes the isochoric stress $\bar{\sigma}$ as a function of the shear rate and thus in response to the velocity gradient gradv. The *Newtonian*[25] *fluid* model considers a linear relation between stress and strain rate, an assumption that holds for many homogeneous liquids. Given a 1D flow at the shear rate $\dot{\varepsilon}_{12}$, the model yields the shear stress $\sigma_{12} = 2\eta\dot{\varepsilon}_{12} = \eta(\partial v_1/\partial x_2 + \partial v_2/\partial x_1)$, where η [Pa s] denotes the *viscosity*, a constant for a Newtonian fluid. We may generalize this expression, and the system

$$
\sigma = \eta \underbrace{\begin{bmatrix} \dfrac{\partial v_1}{\partial x_1} & \dfrac{1}{2}\left(\dfrac{\partial v_1}{\partial x_2} + \dfrac{\partial v_2}{\partial x_1}\right) & \dfrac{1}{2}\left(\dfrac{\partial v_1}{\partial x_3} + \dfrac{\partial v_3}{\partial x_1}\right) \\[3mm] \dfrac{1}{2}\left(\dfrac{\partial v_1}{\partial x_2} + \dfrac{\partial v_2}{\partial x_1}\right) & \dfrac{\partial v_2}{\partial x_2} & \dfrac{1}{2}\left(\dfrac{\partial v_2}{\partial x_3} + \dfrac{\partial v_3}{\partial x_2}\right) \\[3mm] \dfrac{1}{2}\left(\dfrac{\partial v_1}{\partial x_3} + \dfrac{\partial v_3}{\partial x_1}\right) & \dfrac{1}{2}\left(\dfrac{\partial v_2}{\partial x_3} + \dfrac{\partial v_3}{\partial x_2}\right) & \dfrac{\partial v_3}{\partial x_3} \end{bmatrix}}_{\bar{\sigma}} - \begin{bmatrix} p & 0 & 0 \\ 0 & p & 0 \\ 0 & 0 & p \end{bmatrix}
$$

of equations then describes the Newtonian fluid in 3D Cartesian coordinates. In addition to shear stress components, the isochoric stress $\bar{\sigma}$ may also contain normal stress components. Given symbolic notation, the Newtonian fluid reads

$$
\sigma = 2\eta\mathbf{d} - p\mathbf{I}, \tag{3.61}
$$

where $\mathbf{d} = (\text{grad}\mathbf{v} + \text{grad}^{\text{T}}\mathbf{v})/2$ denotes the *rate of deformation* tensor. The continuity $\text{tr}\mathbf{d} = 0$ of an incompressible fluid verifies that p in (3.61) satisfies $p = -\text{tr}\sigma/n_{\text{dim}}$ and thus represents the negative hydrostatic pressure, where n_{dim} denotes the spatial dimension of the fluid problem.

---

[25] Sir Isaac Newton, English mathematician, astronomer, theologian, author, and physicist, 1642–1726.

3.5.4.2 Linear Viscoelasticity

We consider a 1D *linear* viscoelastic material that satisfies the (Boltzmann) superposition principle. The material is loaded by the stress $\sigma(t) = \Delta\sigma \mathcal{H}(t - t_0)$, where $\mathcal{H}(x)$ denotes the Heaviside[26] step function with the values 0 and 1 for $x < 0$ and $x \geq 0$, respectively. The stress step $\Delta\sigma$ at the time t_0 causes the strain $\varepsilon(t) = \mathcal{J}(t - t_0)\Delta\sigma$, where $\mathcal{J}(x)$ [Pa^{-1}] is the *creep compliance* or *creep function*, a function that determines the material's response to the *unit step* in stress.

Given the linearity of the problem, the application of the stress $\sigma(t) = \Delta\sigma\mathcal{H}(t - t_1)$, and thus the step $\Delta\sigma$ at the time t_1, then causes the strain $\varepsilon(t) = \mathcal{J}(t - t_1)\Delta\sigma$, and a sequence of n stress increments $\Delta\sigma$ at the times t_i yields the strain $\varepsilon(t) = \sum_{i=1}^{n} \varepsilon_i(t) = \sum_{i=1}^{n} \mathcal{J}(t - t_i)\Delta\sigma$. A sequence of infinitesimally small stress increments $d\sigma = \dot{\sigma}(t)dt$ defines the strain

$$\varepsilon(t) = \int_{\xi=-\infty}^{t} \mathcal{J}(t - \xi)\dot{\sigma}(\xi)d\xi , \qquad (3.62)$$

where $\dot{\sigma}(t)$ may be seen as the *history* of stress rates. The *convolution* or *hereditary* integral in (3.62), known as *Boltzmann's superposition integral*, relates stress and strain of a linear viscoelastic material—it represents the *constitutive information* of the material.

We may also load the linear viscoelastic body by the infinitesimal small-strain increments $d\varepsilon = \dot{\varepsilon}dt$, and arguments similar to aforementioned ones then lead to the stress

$$\sigma(t) = \int_{\xi=-\infty}^{t} \mathcal{G}(t - \xi)\dot{\varepsilon}(\xi)d\xi , \qquad (3.63)$$

where $\dot{\varepsilon}(t)$ denotes the *history* of strain rates, whilst the *relaxation function* $\mathcal{G}(x)$ [Pa] represents the material's response to the *unit step* in strain and thus contains the constitutive information.

The creep function $\mathcal{J}(t)$ and the relaxation function $\mathcal{G}(t)$ have to satisfy a number of physical constraints. Given ordinary materials, $\mathcal{J}(t)$ increases, whilst $\mathcal{G}(t)$ decreases with the time t. The functions also satisfy the conditions $\mathcal{J}(t \rightarrow \infty) \geq 0$ and $\mathcal{G}(t \rightarrow \infty) = 0$. Laplace[27] and Fourier transforms may be used to solve the convolution integrals (3.62) and (3.63), see Appendices A.4 and B.

---

[26]Oliver Heaviside, electrical engineer, mathematician, and physicist, 1850–1925.

[27]Pierre-Simon, marquis de Laplace, French mathematician, physicist, and astronomer, 1749–1827.

3.5.4.3 Maxwell Rheology Element

A *dashpot* and a *spring* that are arranged in series constitute a *Maxwell element*, see Fig. 3.25a. With the viscosity η [Pa s] of the dashpot and the stiffness E [Pa] of the spring, $\sigma_d = \eta(d\varepsilon_d/dt)$ and $\sigma_s = E\varepsilon_s$ expresses the stress in these devices, where t is the time, whilst ε_d and ε_s denote the strains in the dashpot and the spring, respectively. The time derivative of the kinematic compatibility $\varepsilon = \varepsilon_d + \varepsilon_s$ then reads

$$\frac{d\varepsilon}{dt} = \frac{1}{E}\frac{d\sigma}{dt} + \frac{\sigma}{\eta} \qquad (3.64)$$

and expresses the relation between the stress σ and the strain rate $\dot{\varepsilon}$ of the Maxwell element. The equilibrium $\sigma = \sigma_d = \sigma_s$ has also been used in the derivation of this expression.

The dissipation $\mathcal{D} = \sigma(d\varepsilon_d/dt) = \sigma^2/\eta > 0$ of the Maxwell rheology element appears entirely in the dashpot. At very high as well as at very low strain rates, the model does not dissipate any energy. At very high strain rates the dashpot "locks", whilst at very low strain rates no stress develops in the Maxwell model. Both of these conditions suppress the dissipation of energy.

Given a constant strain rate $\dot{\varepsilon}$, the governing equation (3.64) of the Maxwell element becomes an inhomogeneous linear ordinary differential equation. It may be multiplied with the integrating factor $g(t) = \exp[\int(1/\tau)dt] = \exp(t/\tau)$ and then reads

$$\frac{d\sigma}{dt}\exp(t/\tau) + \frac{\sigma}{\tau}\exp(t/\tau) = \frac{d}{dt}\left[\sigma\exp(t/\tau)\right] = E\frac{d\varepsilon}{dt}\exp(t/\tau),$$

where $\tau = \eta/E$ [s] denotes the relaxation time. The integration of this differential equation yields the stress

$$\sigma(t) = \eta\frac{d\varepsilon}{dt} + C\exp(-t/\tau), \qquad (3.65)$$

where the integration constant C is to be determined form the initial condition.

Let us consider the development of the stress in a *relaxation* experiment to further explore the properties of the Maxwell element. The strain increment $\Delta\varepsilon$ is applied

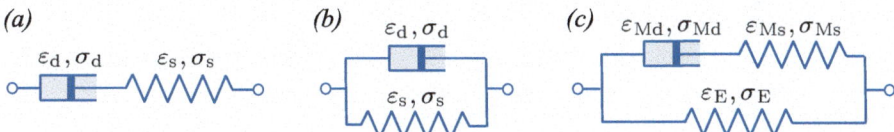

(a)   *(b)* ε_d, σ_d *(c)* $\varepsilon_{Md}, \sigma_{Md}$ $\varepsilon_{Ms}, \sigma_{Ms}$

ε_d, σ_d ε_s, σ_s

ε_s, σ_s ε_E, σ_E

Fig. 3.25 Basic rheological elements that use dashpot and spring devices towards the description of the viscoelastic properties of solids and fluids. (**a**) Maxwell element. (**b**) Kelvin–Voigt element. (**c**) Standard Solid element

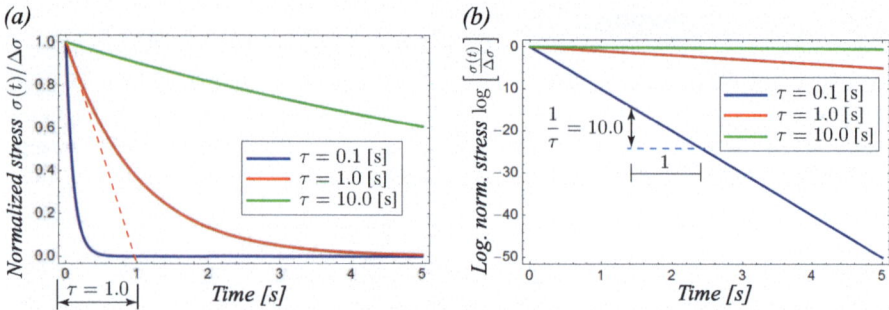

Fig. 3.26 Stress relaxation predicted by the Maxwell rheology element. (**a**) Normalized stress $\sigma(t)/\Delta\sigma$ and (**b**) logarithmic normalized stress $\log[\sigma(t)/\Delta\sigma]$ as a function of the time. The relaxation time is denoted by τ, and its physical meaning is illustrated in the diagrams

at the time $t = 0$ and then kept constant to allow the stress to relax over the time t. Given $\Delta\varepsilon$ is applied within the infinitely short time domain $0 < t < 0^+$, the dashpot "locks", and the spring accommodates the entire strain. The Maxwell element then instantaneously responds with the stress $\Delta\sigma = E\Delta\varepsilon$ and allows us to identify the integration constant in (3.65) from $\Delta\sigma = E\dot{\varepsilon} + C\exp(-t/\tau)$. Given the property $\dot{\varepsilon} = 0$ for $t \geq 0^+$ of the relaxation experiment, the integration constant reads $C = \Delta\sigma$, and

$$\sigma(t) = \Delta\sigma \exp(-t/\tau) \tag{3.66}$$

determines the stress of the Maxwell element. Figure 3.26 illustrates the relaxation response, where the normalized stress $\sigma(t)/\Delta\sigma$ as well as the logarithmic normalized stress $\log[\sigma(t)/\Delta\sigma]$ over time are shown. The figure also illustrates the physical meaning of the relaxation time τ.

The linearity of the Maxwell element allows the superposition of stress increments. The application of $\Delta\varepsilon$ at the times $t_i; i = 1, \ldots, n$ then yields the resulting stress

$$\sigma(t) = \begin{cases} 0 & ; \quad t < t_0, \\ \Delta\sigma_0 \exp(-(t - t_0)/\tau) & ; \quad t_0^+ \leq t < t_1, \\ \Delta\sigma_0 \exp(-(t - t_0)/\tau) + \Delta\sigma_1 \exp(-(t - t_1)/\tau) & ; \quad t_1^+ \leq t < t_2, \\ \ldots & \end{cases}$$

$$\tag{3.67}$$

Given infinitesimally small "stress steps" $d\sigma = E\dot{\varepsilon}dt$, the series (3.67) approaches the convolution integral

$$\sigma(t) = E \int_{\xi=-\infty}^{t} \exp(-(t - \xi)/\tau)\dot{\varepsilon}d\xi . \tag{3.68}$$

Fig. 3.27 Development of the normalized stress according to the Maxwell rheological element, where E and τ denote the elastic stiffness and the relaxation time, respectively. At the time $0 < t < 1\,\text{s}$, the constant strain rate k is prescribed, whilst the strain is constant at $t > 1\,\text{s}$

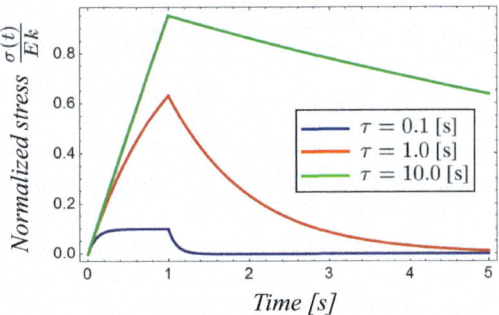

The comparison of Eqs. (3.68) and (3.63), reveals that $\mathcal{G}(x) = E \exp(-x/\tau)$ determines the *relaxation function* of the Maxwell element.

Towards the further exploration of the Maxwell element, we will now investigate the stress in response to the strain

$$\varepsilon(t) = \begin{cases} 0 & \text{for } t < 0, \\ kt & \text{for } 0 \leq t \leq 1\,\text{s}, \\ k & \text{for } t > 1\,\text{s}, \end{cases}$$

where $k\,[\text{s}^{-1}]$ is a constant proportionality factor. During the time interval $0 \leq t \leq 1\,\text{s}$, relation (3.68) yields the stress

$$\sigma(t) = Ek \int_{\xi=0}^{t} \exp[-(t - \xi)/\tau]\,\mathrm{d}\xi = Ek\tau \{\exp[-(t - \xi)/\tau]\}_{\xi=0}^{t}$$

$$= Ek\tau[1 - \exp(-t/\tau)] \quad \text{for} \quad 0 \leq t \leq 1\,\text{s}, \tag{3.69}$$

where $\dot{\varepsilon}(t) = k$ and the initial condition $\sigma(0) = 0$ have been used. At $t > 1$, the strain rate $\dot{\varepsilon}(t) = 0$ holds, and (3.68) then yields the stress

$$\sigma(t) = Ek \int_{\xi=0}^{1.0} \exp[-(t - \xi)/\tau]\,\mathrm{d}\xi = Ek\tau \{\exp[-(t - \xi)/\tau]\}_{\xi=0}^{1}$$

$$= Ek\tau \{\exp[-(t - 1)/\tau] - \exp(-t/\tau)\} \quad \text{for} \quad 1 < t < \infty.$$

Figure 3.27 plots the normalized stress response of the Maxwell rheological element for different relaxations times τ.

The analysis of the stress response (3.69) allows us to explore the physical properties of the Maxwell element. At its limits (3.69) reads

$$\sigma(t) = \begin{cases} Ek\tau = \eta k & \text{for } \tau \ll t, \\ Ekt = E\varepsilon & \text{for } \tau \gg t, \end{cases}$$

where the series representation $\exp(x) = 1 + x + x^2/2! + \cdots$ has been used in the derivation of these expressions. Given a very slow process $\tau \ll t$, the Maxwell element responds like a viscous fluid, whilst for a very fast process $\tau \gg t$ it has elastic spring-like properties.

3.5.4.4 Kelvin–Voigt Rheology Element

A *dashpot* and a *spring* that are arranged in parallel constitute a *Kelvin–Voigt element*, see Fig. 3.25b. As in Section 3.5.4.3, $\sigma_d = \eta(d\varepsilon_d/dt)$ and $\sigma_s = E\varepsilon_s$ are the stresses in the dashpot and the spring, respectively. The equilibrium

$$\sigma = E\varepsilon + \eta \frac{d\varepsilon}{dt}$$

then expresses the stress σ as a function of the strain ε and the strain rate $\dot{\varepsilon}$. The kinematic compatibility $\varepsilon = \varepsilon_d = \varepsilon_s$ has also been used in the derivation of this expression.

The dissipation $\mathcal{D} = \sigma_d(d\varepsilon_d/dt) = \sigma_d^2/\eta > 0$ appears entirely in the dashpot. At very low strain rates no stress develops in the dashpot, and no energy is therefore dissipated. The Kelvin–Voigt element yields infinite stress in response to an instantaneous strain change and acts then as a *rigid* body.

Example 3.14 (Linear Viscoelasticity—Kelvin–Voigt Element). Let us consider the Kelvin–Voigt rheology element of Sect. 3.5.4.4, whose viscoelastic properties are governed by the ordinary differential equation $\dot{\varepsilon} + \varepsilon/\tau = \sigma/\eta$. It has constant coefficients, and the ratio $\tau = \eta/E$ of the viscosity η and stiffness E denotes the model's retardation time.

(a) Given a constant stress σ, solve the governing equation of the Kelvin–Voigt element and express the strain $\varepsilon(t)$ up to an unknown integration constant C.
(b) Consider a creep experiment and apply the stress increment $\Delta\sigma$ at the time $t = 0$. The stress is then kept constant and the strain $\varepsilon(t)$ develops over the time t. Identify the integration constant C of the creep test, and plot the normalized strain $\varepsilon(t)/\Delta\varepsilon$ over the time t for different retardation times τ. Here, $\Delta\varepsilon = \Delta\sigma/E$ denotes the strain that is reached at the time $t \gg \tau$ and thus at the thermodynamic equilibrium of the Kelvin–Voigt element.
(c) Derive the strain $\varepsilon(t)$ in response to the application of n stress increments $\Delta\sigma$ at the times t_i; $i = 1, \ldots, n$. Generalize the discrete expression towards infinitesimally small stress increments $d\sigma = \dot{\sigma} dt/E$, where $\dot{\sigma}(t)$ denotes the stress rate.
(d) Compute the strain that the Kelvin–Voigt element predicts in response to the stress

$$\sigma(t) = \begin{cases} 0 & \text{for } t < 0, \\ kt & \text{for } 0 \le t \le 1\,\text{s}, \\ k & \text{for } t > 1\,\text{s}, \end{cases}$$

where k [Pa s^{-1}] is a constant proportionality factor. Plot the strain response within the time interval $0 < t < 5$ s as well as for the retardation time $\tau = 0.1$; 1.0; 10.0 s.

(e) Consider the loading $\sigma(t) = kt$ and describe the properties of the Kelvin–Voigt element at the limits $\tau \gg t$ and $\tau \ll t$, respectively. ∎

3.5.4.5 Standard Solid Rheology Element

A *mainspring* that is arranged in parallel to a *Maxwell element* (see Sect. 3.5.4.3) constitutes the *Standard Solid element*, see Fig. 3.25c. Let us consider a mainspring of the stiffness E_e [Pa], whilst the Maxwell element contains a dashpot of the viscosity η_M [Pa s] and a spring of the stiffness E_M [Pa]. At the dashpot strain ε_{Md} and the spring strain ε_{Ms}, the stress $\sigma_M = \eta(d\varepsilon_{Md}/dt) = E_M \varepsilon_{Ms}$ appears in the Maxwell element, where t denotes the time. In addition to the stress of the Maxwell element, the mainspring contributes the stress $\sigma_e = E_e \varepsilon$, where ε denotes the strain of the Standard Solid element.

The kinematic compatibility $\varepsilon = \varepsilon_{Md} + \varepsilon_{Ms}$, together with the equilibrium $\sigma = \sigma_M + \sigma_e$, governs the properties of the Standard Solid element. The time differentiation of the kinematic compatibility then reads

$$\frac{d\varepsilon}{dt} = \frac{d\varepsilon_{Md}}{dt} + \frac{d\varepsilon_{Ms}}{dt} = \frac{\sigma - E_e \varepsilon}{\eta} + \frac{1}{E_M}\left(\frac{d\sigma}{dt} - E_e \frac{d\varepsilon}{dt}\right),$$

where the constitutive descriptions of the dashpot and springs as well as the equilibrium of the Standard Solid element have been used. After some algebraic manipulation this relation yields the governing equation

$$\frac{d\varepsilon}{dt} = \frac{\frac{d\sigma}{dt} + \frac{E_M}{\eta}(\sigma - E_e \varepsilon)}{E_M + E_e} \tag{3.74}$$

of the Standard Solid element. Given the fixed strain ε, the *over-stress* $\sigma_M = E_M \varepsilon_{Ms}$ of the Maxwell element tends to zero at $t \to \infty$ and defines the *thermodynamic equilibrium* of the Standard Solid element.

The Standard Solid element has *solid-like* properties. Given constant stress σ, the strain approaches the stable limit of $\varepsilon = \sigma/E_e$ in time, whilst the model shows elastic properties in response to instantaneous strain changes. The dissipation $\mathcal{D} = \sigma_M(d\varepsilon_{Md}/dt) = \sigma_M^2/\eta > 0$ appears entirely in the dashpot of the Maxwell element, and, as with the Maxwell element, the Standard Solid element does not dissipate energy at very high as well as very low strain rates.

An alternative to the governing equation (3.74) of the Standard Solid element may be derived, which has advantages towards its numerical implementation. We may decouple the elastic and inelastic deformations and consider the over stress $\sigma_M = E_M \varepsilon_{Ms}$, the stress contributed by the Maxwell element. Its time derivative then yields the rate equation

$$\frac{d\sigma_M}{dt} + \frac{\sigma_M}{\tau} = \beta \frac{d\sigma_e}{dt} ,\qquad(3.75)$$

where the kinematics relation $\varepsilon = \varepsilon_{Md} + \varepsilon_{Ms}$ together with the constitutive expressions $\sigma_M = \eta(d\varepsilon_{Md}/dt)$ and $\sigma_e = E_e\varepsilon$ has been used. In (3.75), $\beta = E_M/E_e$ denotes a dimensionless *stiffness factor*, whilst $\tau = \eta/E_M$ [s] is the *relaxation time* of the Maxwell element. The rate Eq. (3.75) determines the development of the over stress σ_M, which is then to be added to the stress σ_e of the mainspring.

3.5.4.6 Generalized Models

The generalization of the aforementioned 1D rheology models leads to the description of n_{dim}-dimensional viscoelastic materials. Given *small* deformations, ε and σ may be simply replaced by the respective n_{dim}-dimensional tensorial quantities—the *engineering strain* $\boldsymbol{\varepsilon}$ and the *Cauchy stress* $\boldsymbol{\sigma}$, see Table 3.2.

Given *finite* deformations, the *objectivity* of the constitutive description is to be preserved, and we therefore use *a priori* objective strain and stress measures. The strain ε and the stress σ of the 1D descriptions are then replaced by the *Green–Lagrange strain* $\mathbf{E}$ and the *second Piola–Kirchhoff stress* $\mathbf{S}$, see Table 3.2. They are defined in the body's reference configuration Ω_0, which ensures the objectivity of the constitutive model. Given applications at small and finite deformations, the Young's modulus E is substituted by the *elasticity tensors* $\mathbb{c}$ and $\mathbb{C}$, respectively.

Table 3.3 lists governing equations of a number of viscoelastic models, which can be derived through the generalization of the Maxwell model, the Kelvin–Voigt model, and the Standard Solid model, respectively. Given the Standard Solid model, the continuum may be seen as the superposition of a *Maxwell body* and a purely *elastic body* [194]. These bodies correspond to the Maxwell element and the mainspring of the respective 1D Standard Solid Model, see Fig. 3.25c. The Maxwell body is described by the relaxation time τ, and the parameter β that relates its stiffness to the stiffness of the elastic body; $\mathbb{c}_M = \beta\mathbb{c}_e$ at small deformations and $\mathbb{C}_M = \beta\mathbb{C}_e$ at finite deformations. Such a generalized description avoids the explicit introduction of stiffness tensors (see last row in Table 3.3) and supports the description of non-linear materials.

Table 3.2 Stress, strain, and stiffness definitions of 1D as well as n_{dim}-dimensional rheology models at small and finite deformations

Parameter	1D small defo.	n_{dim} small defo.	n_{dim} finite defo.
Strain	ε	$\boldsymbol{\varepsilon} = \frac{1}{2}(\text{grad}\,\mathbf{u} + \text{grad}^T\mathbf{u})$	$\mathbf{E} = \frac{1}{2}(\mathbf{F}^T\mathbf{F} - \mathbf{I})$
Stress	σ	$\boldsymbol{\sigma}$	$\mathbf{S}$
Stiffness	E	$\mathbb{c} = \partial\boldsymbol{\sigma}/\partial\boldsymbol{\varepsilon}$	$\mathbb{C} = \partial\mathbf{S}/\partial\mathbf{E}$

E: Young's modulus; $\boldsymbol{\varepsilon}$: engineering strain; $\mathbf{E}$: Green–Lagrange strain; $\boldsymbol{\sigma}$: Cauchy stress; $\mathbf{S}$: second Piola–Kirchhoff stress; $\mathbb{c}$: spatial elasticity tensor; $\mathbb{C}$: referential elasticity tensor; $\mathbf{u}$: displacement; $\mathbf{F}$: deformation gradient

Table 3.3 Governing equations of 1D as well as n_{dim}-dimensional rheology models at small and finite deformations

Model	1D small defo.	n_{dim} small defo.	n_{dim} finite defo.
Maxwell	$\dot{\varepsilon} = \dot{\sigma}/E + \sigma/\eta$	$\dot{\boldsymbol{\varepsilon}} = \mathbb{C}^{-1} : \dot{\boldsymbol{\sigma}} + \boldsymbol{\sigma}/\eta$	$\dot{\mathbf{E}} = \mathbb{C}^{-1} : \dot{\mathbf{S}} + \mathbf{S}/\eta$
Kelvin–Voigt	$\sigma = E\varepsilon + \eta\dot{\varepsilon}$	$\boldsymbol{\sigma} = \mathbb{C} : \boldsymbol{\varepsilon} + \eta\dot{\boldsymbol{\varepsilon}}$	$\mathbf{S} = \mathbb{C} : \mathbf{E} + \eta\dot{\mathbf{E}}$
Standard Solid	$\sigma = \sigma_{\text{M}} + \sigma_{\text{e}}$;	$\boldsymbol{\sigma} = \boldsymbol{\sigma}_{\text{M}} + \boldsymbol{\sigma}_{\text{e}}$;	$\mathbf{S} = \mathbf{S}_{\text{M}} + \mathbf{S}_{\text{e}}$;
	$\dot{\sigma}_{\text{M}} + \sigma_{\text{M}}/\tau = \beta\dot{\sigma}_{\text{e}}$	$\dot{\boldsymbol{\sigma}}_{\text{M}} + \boldsymbol{\sigma}_{\text{M}}/\tau = \beta\dot{\boldsymbol{\sigma}}_{\text{e}}$	$\dot{\mathbf{S}}_{\text{M}} + \mathbf{S}_{\text{M}}/\tau = \beta\dot{\mathbf{S}}_{\text{e}}$

E: Young's modulus; η: viscosity; τ: relaxation or retardation time; ε: engineering strain; **E**: Green–Lagrange strain; σ: Cauchy stress; **S**: second Piola–Kirchhoff stress; $\mathbb{C}$: spatial elasticity tensor; $\mathbb{C}$: referential elasticity tensor. Indexes e and M denote elastic and viscose contributions to stress and strain, and superimposed dot denotes the material time derivative, respectively

3.5.4.7 Visco-hyperelasticity for Incompressible Materials

A hyperelastic description may be used to define the *elastic stress* and the *elasticity modulus* of finite strain viscoelastic models. The elastic properties are then represented by the Helmholtz free energy function $\Psi(\mathbf{C}, \boldsymbol{\Xi})$ per unit undeformed material, where $\mathbf{C}$ and $\boldsymbol{\Xi}$ denote the right Cauchy–Green strain and a strain-like *internal*, and therefore *hidden* variable.

We consider the generalized Standard Solid model of an incompressible material that is illustrated in Fig. 3.30. The free energy $\Psi_{\text{iso}}(\mathbf{C}, \mathbf{C}_{\text{M}}) = \Psi_{\text{iso E}}(\mathbf{C}) + \Psi_{\text{iso M}}(\mathbf{C}_{\text{M}})$ describes the material, where $\Psi_{\text{iso E}}(\mathbf{C})$ and $\Psi_{\text{iso M}}(\mathbf{C}_{\text{M}})$ denote the energies that are stored in the elastic body (mainspring) and the spring of the Maxwell body, respectively. The hidden internal variable is then the right Cauchy–Green strain of the spring of the Maxwell body, $\boldsymbol{\Xi} = \mathbf{C}_{\text{M}}$.

Coleman and Noll's procedure for incompressible materials (3.131) allows us to express the second Piola–Kirchhoff stress

$$\mathbf{S} = 2\frac{\partial \Psi_{\text{iso E}}(\mathbf{C})}{\partial \mathbf{C}} + 2\frac{\partial \Psi_{\text{iso M}}(\mathbf{C}_{\text{M}})}{\partial \mathbf{C}_{\text{M}}} : \frac{\partial \mathbf{C}_{\text{M}}}{\partial \mathbf{C}} - \kappa \mathbf{C}^{-1} , \tag{3.76}$$

where κ denotes the Lagrange pressure that is required to enforce incompressibility $J = 1$. The pressure κ is to be distinguished from the hydrostatic pressure p.

In the relation (3.76), the term $\partial \mathbf{C}_{\text{M}}/\partial \mathbf{C}$ describes the relation between the strain increment $\Delta \mathbf{C}_{\text{M}}$ of the spring of the Maxwell body and the total strain increment $\Delta \mathbf{C}$. It is usually an implicit relation and specified through a *rate equation*. Given its *a priori* objectivity, the Green–Lagrange strain **E** may be used in the formulation of the rate equation. It could read

$$\dot{\mathbf{E}}_{\text{M}} + \frac{1}{\tau}\mathbf{E}_{\text{M}} = \dot{\mathbf{E}} , \tag{3.77}$$

where τ denotes the relaxation time. The rate equation (3.77) satisfies the physical limits at very fast and at very slow deformations. At the limit $t \ll \tau$ the dashpot

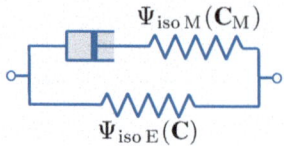

Fig. 3.30 Generalized Standard Solid rheology element with $\Psi_{\mathrm{iso\,E}}(\mathbf{C})$ and $\Psi_{\mathrm{iso\,M}}(\mathbf{C_M})$ representing the strain energy stored in the two (hyperelastic) springs/bodies

locks, and $\dot{\mathbf{E}}_M = \dot{\mathbf{E}}$, whilst at $t \gg \tau$ the deformation is too slow to activate the spring of the Maxwell body, and $\dot{\mathbf{E}}_M = \mathbf{0}$ follows from (3.77).

Towards closing the viscoelastic description, a Helmholtz free energy $\Psi_{\mathrm{iso}}(\mathbf{C}, \mathbf{C_M})$ that satisfies the Clausius–Duhem inequality (3.124) $-(\partial \Psi_{\mathrm{iso}}(\mathbf{C}, \mathbf{C_M})/ \partial \mathbf{C_M}) : \dot{\mathbf{C}}_M \geq 0$ of isothermal processes is selected. The model then obeys the second law of thermodynamics, see Sect. 3.6.3.2.

Example 3.15 (Strain-Based Viscoelastic Generalization of the Incompressible neo Hookean Material). We consider an incompressible viscoelastic material whose properties are represented by the rheological element shown in Fig. 3.30. The Helmholtz free energy $\Psi_{\mathrm{iso}}(\mathbf{C}, \mathbf{C_M})$ per unit (undeformed) material depends on the right Cauchy–Green strain $\mathbf{C}$, as well as the hidden strain variable $\mathbf{C_M}$, the right Cauchy–Green strain of the Maxwell body. The material is characterized by the Helmholtz free energy

$$\Psi_{\mathrm{iso}}(\mathbf{C}, \mathbf{C_M}) = \frac{G}{2} \left[(\mathrm{tr}\mathbf{C} - 3) + \beta(\mathrm{tr}\mathbf{C_M} - 3) \right] \tag{3.78}$$

and may be seen as the generalization of the classical neo Hookean material model, where G [Pa] denotes the material's elastic small-strain shear modulus. The dimensionless parameter β represents the ratio between stiffness of the elastic body (main spring) and the stiffness of the Maxwell body.

(a) Consider simple tension and compute the Cauchy stress at the material's thermodynamic limit.
(b) Assume the Green–Lagrange strain $\mathbf{E}_M = (\mathbf{C_M} - \mathbf{I})/2$ in the Maxwell body to be described by the rate equation (3.77), and consider simple tension through the description of the Green–Lagrange strain component

$$E_{11}(t) = \begin{cases} 0 & \text{for } t < 0, \\ kt & \text{for } 0 \leq t \leq 1\,\mathrm{s}, \\ k & \text{for } t > 1\,\mathrm{s} \end{cases} \tag{3.79}$$

in tensile direction, where k [s^{-1}] is a constant proportionality factor. Plot the development of the normalized strain component $E_{M\,11}/k$ of the Maxwell spring at $0 < t < 5$ s and $\tau = 0.1;\ 1.0;\ 10.0$ s.

(c) Given the parameters $k = 1.0$ s^{-1} and $\beta = 0.6$, compute the normalized Cauchy stress σ_{11}/G in response to simple tension determined by (3.79). Plot the normalized stress at $0 < t < 5$ s and $\tau = 0.1;\ 1.0;\ 10.0$ s. ■

3.5.4.8 Stress-Decomposed Visco-hyperelasticity for Incompressible Materials

Alternatively to the derivation directly from the Helmholtz free energy $\Psi_{\text{iso}}(\mathbf{C},\ \Xi)$ outlined in Sect. 3.5.4.7, visco-hyperelastic descriptions may also be based on the *additive decomposition* $\mathbf{S} = \overline{\mathbf{S}}_{\text{E}} + \overline{\mathbf{S}}_{\text{M}} - p\mathbf{C}^{-1}$ of the second Piola–Kirchhoff stress. The deviatoric stress contributions $\overline{\mathbf{S}}_{\text{E}}$ and $\overline{\mathbf{S}}_{\text{M}}$ represent constitutive information, and the term $p\mathbf{C}^{-1}$ enforces the incompressibility, where p and $\mathbf{C}$ denote the negative hydrostatic pressure and the right Cauchy–Green strain, respectively. Whilst $\overline{\mathbf{S}}_{\text{E}} = 2J^{-2/3}\text{Dev}[\partial\Psi_{\text{iso E}}(\mathbf{C})/\partial\mathbf{C}]$ follows from the free energy $\Psi_{\text{iso E}}$ with $\text{Dev}(\bullet) = (\bullet) - [\mathbf{C} : (\bullet)]\mathbf{C}^{-1}/3$ denoting the referential deviator operator in 3D (see Sect. 3.5.3.2), the over stress $\overline{\mathbf{S}}_{\text{M}}$ develops implicitly through a rate equation, such as

$$\dot{\overline{\mathbf{S}}}_{\text{M}} + \frac{1}{\tau}\overline{\mathbf{S}}_{\text{M}} = \beta\dot{\overline{\mathbf{S}}}_{\text{E}}, \qquad (3.89)$$

where τ [s] and β denote the relaxation time and a stiffness parameter, respectively. The over stress $\overline{\mathbf{S}}_{\text{M}}$ may be related to the Maxwell body of the Standard Solid rheology model shown in Fig. 3.30. The factor β in (3.89) therefore denotes the ratio between the stiffness of the Maxwell body and the stiffness of the elastic body. Given the linear governing equation (3.89), the convolution integral

$$\overline{\mathbf{S}}_{\text{M}}(t) = \beta \int_0^t \exp[-(t - \xi)/\tau]\dot{\overline{\mathbf{S}}}_{\text{E}}\,\text{d}\xi \qquad (3.90)$$

represents the closed-form solution of the over stress.

The stress-decomposed visco-hyperelastic approach leads to an extremely efficient numerical implementation [194, 222, 291, 506]. Whilst the description has not been derived from the free energy potential $\Psi_{\text{iso}}(\mathbf{C},\ \Xi)$, its algorithmic tangent $\mathbb{C} = 2\partial\mathbf{S}/\partial\mathbf{C}$ shows major symmetry and therefore proves the existence of the potential $\Psi_{\text{iso}}(\mathbf{C},\ \Xi)$.

Example 3.16 (Stress-Based Viscoelastic Generalization of the Incompressible neo Hookean Material). We consider an incompressible viscoelastic material whose elastic properties are determined by the neoHookean Helmholtz free energy

$$\Psi_{\mathrm{iso\,E}}(\mathbf{C}) = \frac{G}{2}(I_1 - 3),\tag{3.91}$$

where $I_1 = \mathrm{tr}\mathbf{C}$ and G denote the first invariant of the right Cauchy–Green strain $\mathbf{C}$ and the small-strain shear modulus, respectively.

(a) Consider simple tension kinematics and compute the Cauchy stress at the material's thermodynamic limit.

(b) The decomposed representation $\mathbf{S} = \overline{\mathbf{S}}_{\mathrm{E}} + \overline{\mathbf{S}}_{\mathrm{M}} - p\mathbf{C}^{-1}$ of the second Piola–Kirchhoff stress may be used, where the potential (3.91) determines the isochoric elastic stress $\overline{\mathbf{S}}_{\mathrm{E}}$, whilst the rate equation (3.90) governs the over stress $\overline{\mathbf{S}}_{\mathrm{M}}$. Compute the Cauchy stress at simple tension according to the stretch

$$\lambda(t) = \begin{cases} 1 & \text{for } t < 0, \\ 1 + kt & \text{for } 0 \le t \le 1\,\mathrm{s}, \\ 1 + k & \text{for } t > 1\,\mathrm{s}, \end{cases}\tag{3.92}$$

as a function of the time t, where k is a constant proportionality factor. Given the parameters $k = 1.0\ s^{-1}$ and $\beta = 0.6$, plot the normalized Cauchy stress component σ_{11}/G for the time $0 < t < 5\,\mathrm{s}$ and the relaxation times $\tau = 0.1;\ 1.0;\ 10.0\,\mathrm{s}$. ∎

3.5.5 Multiphasic Continuum Theories

A body may consist of *multiple material phases* that influence each other during deformation. Given tissue biomechanics, the case of a solid phase (skeleton) that is immersed in fluid is of substantial importance, and several continuum theories have been developed towards the description of such a body. The two most popular theories are the *mixture theory* and the *theory of poroelasticity*. Despite the mixture theory is more general and provides better conceptual mechanisms to integrate physical situations [102], poroelasticity can still handle complexities, such as blood flow through the beating myocardial tissue [77].

3.5.5.1 Mixture Theory
The *mixture theory* generalizes the principles of continuum mechanics upon a number of *interpenetrable* continua. It uses the basic assumption that, at any time, all phases *co-exist* at the material point and therefore occupy simultaneously the same position within the RVE [55]. Figure 3.33 aims at illustrating a two-phase mixture. The transport of phases relative to each other is then captured by diffusion models, such a Fick's law, see Appendix E.2.

As with other models, the mixture theory requires constitutive relations to close the system of balance equations. Given the right Cauchy–Green strain $\mathbf{C}$ determines the deformation of all phases that co-exist at a material point, the strain energy of the mixture of n phases then reads

Fig. 3.33 Representative Volume Element (RVE) of a two-phase (fluid and solid) material used in (**a**) poroelasticity and (**b**) mixture theory. In poroelasticity the phases determine separate domains within the RVE, whilst in mixture theory, they co-exist in space and occupy simultaneously the entire RVE

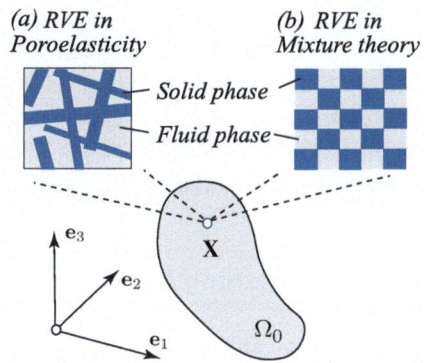

$$\Psi(\mathbf{C}) = \sum_{i=1}^{n} \xi_i \Psi_i(\mathbf{C}) \quad \text{with} \quad \sum_{i=1}^{n} \xi_i = 1, \tag{3.98}$$

where ξ_i and Ψ_i denote the *referential volume fraction* and the strain energy per unit reference volume of the i-th phase, respectively. Most mixture models of soft biological tissues assume *affine* deformation. The interfaces amongst the n constituents are then rigid and $\mathbf{C}$ directly determines the deformation of the individual phases.

3.5.5.2 Poroelasticity Theory

Poroelasticity regards an elastic solid phase (skeleton) immersed in a fluid phase—both phases occupy two *separate* domains of the RVE, see Fig. 3.33. Let us denote the volume fraction of fluid by ξ. The continuity of the fluid (see Sect. 3.6.1) at the material point then reads $\dot{\xi} = -\mathrm{div}\mathbf{q}$, where $\mathbf{q}$ [m s^{-1}] denotes the fluid flux—a property that is equivalent to the fluid filtration velocity discussed in Sect. 2.4.2. Given a Darcy-type filter law, we may use $\mathbf{q} = -L_\mathrm{p}\mathrm{grad}p - \mathbf{b}_\mathrm{ff}$ to relate the fluid flux $\mathbf{q}$ to the fluid (or pore) pressure p, see Appendix E.2. Here, L_p and $\mathbf{b}_\mathrm{ff}$ denote the hydraulic conductivity and the body force that acts at the fluid phase, respectively.

According to Biot[28] [48], we may decompose the stress $\boldsymbol{\sigma} = \boldsymbol{\sigma}_\mathrm{e} - \alpha p\mathbf{I}$, where $\boldsymbol{\sigma}_\mathrm{e}$ denotes the effective solid Cauchy stress, α is *Biot's effective stress coefficient*, and p is the negative hydrostatic pressure. At quasi-static conditions, the conservation of linear momentum (see Sect. 3.6.2)

$$\mathrm{div}\boldsymbol{\sigma} = \mathrm{div}\boldsymbol{\sigma}_\mathrm{e} - \alpha p\mathbf{I} = \mathbf{b}_\mathrm{f} \tag{3.99}$$

describes then the poroelastic material, where $\mathbf{b}_\mathrm{f}$ denotes the bulk body force per unit volume.

---

[28]Maurice Anthony Biot, Belgian–American applied physicist, 1905–1985.

As with other models, poroelasticity theory requires constitutive relations to express $\boldsymbol{\sigma}_\mathrm{e}$ and to close the system of balance equations. Despite Biot's theory being derived for small deformations, it may be generalized towards finite deformations. *Objective stress rates*, such as Jaumann's rate of the Cauchy stress $\overset{\triangle}{\boldsymbol{\sigma}} = \dot{\boldsymbol{\sigma}} + \boldsymbol{\sigma}\mathbf{w} - \mathbf{w}\boldsymbol{\sigma}$, are then used to implement (3.99), where $\mathbf{l} = \mathrm{grad}\mathbf{v}$ and $\mathbf{w} = (\mathbf{l} - \mathbf{l}^\mathrm{T})/2$ denote the velocity gradient and the spin tensor, respectively [433].

3.6   Governing Laws

A number of physical principals govern the motion $\boldsymbol{\chi}$ of a continuum body. It has to obey the balance of *mass* and *momentum*, as well as the *first* and *second laws of thermodynamics*. Such principles may be formulated within a *Lagrangian* description or an (*advective*) *Eulerian* description, see Fig. 3.34. Whilst both descriptions monitor the material particle at the spatial configuration Ω, they use different reference systems. Given the Lagrangian description, the *observer* is moving together with the continuum body, whilst in the Eulerian description, the observer does not move and is fixed to the coordinate system. The Lagrangian description is commonly used for solids, whilst fluids are most often described within an Eulerian setting.

In addition to the procedures discussed in this section, Maxwell transport and localization provides an alternative approach in the derivation of governing laws, see Sect. 3.7.1.

3.6.1   Mass Balance

The mass $\mathrm{d}m = \rho\mathrm{d}v$ [kg] of each material particle remains constant, and therefore

$$\frac{\mathrm{D}m}{\mathrm{D}t} = 0 \qquad\qquad (3.100)$$

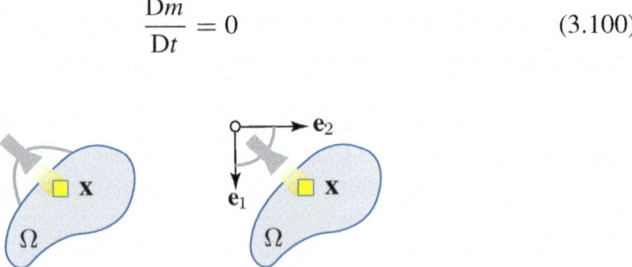

Lagrangian   *Eulerian*
description   *description*

Fig. 3.34 Observation of a material particle at its spatial position $\mathbf{x}$ using the Lagrangian description and the Eulerian description. A flash lamp denotes the observer, which is either fixed to the moving body Ω (Lagrangian description) or to the (stationary) coordinate system $\{\mathbf{e}_1, \mathbf{e}_2\}$ (Eulerian description)

holds, where $D(\bullet)/Dt$ denotes the *material time derivative*. It corresponds to the change that is "felt" by the material particle.

The conservation of mass describes vascular tissue at the time scale of the cardiac cycle. However, over the duration of hours and days, factors, such as turnover of tissue constituents and cell migration may result in mass change. The material point is then no longer described by a *closed system* but by an *open system*, as discussed in Chap. 7.

3.6.1.1 Lagrangian Description

In Lagrange description the material time derivative is equal to the partial derivative, $D(\bullet)/Dt = \partial(\bullet)/\partial t$. Given a material particle of the mass dm, the conservation of mass then reads

$$\frac{Dm}{Dt} = \frac{\partial(\rho d\upsilon)}{\partial t} = \frac{\partial\rho}{\partial t}d\upsilon + \rho\frac{\partial(d\upsilon)}{\partial t} = 0, \tag{3.101}$$

where ρ [kg m^{-3}] and $d\upsilon$ [m^3] denote the density and volume of the material particle at its spatial position $\mathbf{x}$ and thus properties of the deformed particle. The conservation of mass may also be expressed by

$$\frac{\partial\rho}{\partial t} + \rho\mathrm{div}\mathbf{v} = 0, \tag{3.102}$$

where $\mathbf{v}$ denotes the velocity of the material particle. The derivation of this relation considered the volume element $d\upsilon = JdV$ as well as the rate of volume change $(3.38)_4$, $\dot{J} = J\mathrm{div}\mathbf{v}$, where J denotes the volume ratio.

Given an incompressible material, the density ρ is a constant, and

$$\frac{\partial J}{\partial t} = 0$$

expresses the conservation of mass. Other representations of the conservation of mass directly follow from the alternative relations of the rate of volume change (3.38).

3.6.1.2 Eulerian Description

In Eulerian description, the conservation of mass reads

$$\frac{Dm}{Dt} = \frac{D(\rho d\upsilon)}{Dt} = \frac{D\rho}{Dt}d\upsilon + \rho\frac{D(d\upsilon)}{Dt} = 0, \tag{3.103}$$

which with the *material time derivative* $D(\bullet)/Dt = \partial(\bullet)/\partial t + (\bullet)\mathrm{div}\mathbf{v}$ then reads

$$\frac{Dm}{Dt} = \left(\frac{\partial\rho}{\partial t} + \rho\,\mathrm{div}\mathbf{v}\right)d\upsilon + \left(\frac{\partial(d\upsilon)}{\partial t} + d\upsilon\,\mathrm{div}\mathbf{v}\right)\rho = 0.$$

Given an *incompressible* material, the condition $D\rho/Dt = 0$ may be used, and the mass balance then reads $D(dv)/Dt = 0$. It may also be expressed by

$$\text{div}\,\mathbf{v} = 0\,, \tag{3.104}$$

where $\mathbf{v}$ denotes the velocity of the material particle, and $dv = J\,dV$ as well as the rate of volume change $(3.38)_4$ has been used. The conservation of mass of an incompressible material therefore defines a *divergence-free* velocity field. In fluid mechanics, the conservation of mass is commonly called *flow continuity*.

Example 3.17 (Continuity of the Incompressible Flow). An incompressible fluid flows at the velocity $\mathbf{v}$ through a control volume (domain) Ω. Given an Eulerian problem description, the domain Ω is fixed in space, and the linear expansion of the flow velocity $\mathbf{v}(\mathbf{x} + \Delta\mathbf{x}) = \mathbf{v} + \text{grad}\mathbf{v}(\mathbf{x})\Delta\mathbf{x}$ yields the velocity components shown in Fig. 3.35.

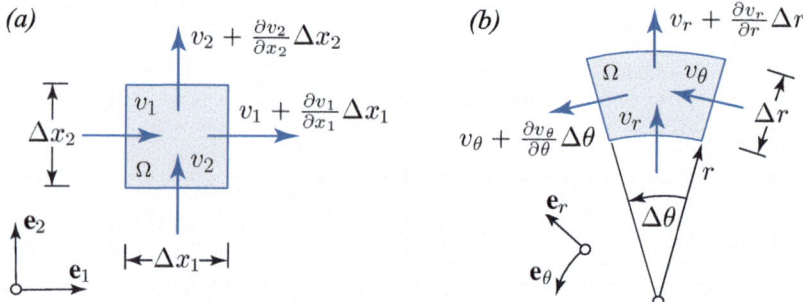

Fig. 3.35 Incompressible flow through a control volume Ω that is fixed in space. Problem description with respect to (**a**) the 2D Cartesian coordinate system $\{\mathbf{e}_1, \mathbf{e}_2\}$ and (**b**) the cylindrical coordinate system $\{\mathbf{e}_r, \mathbf{e}_\theta\}$

(a) Given the Cartesian coordinate system $\{\mathbf{e}_1, \mathbf{e}_2\}$ shown in Fig. 3.35a, derive the flow continuity and verify the expression (3.104).
(b) Given the cylindrical coordinate system $\{\mathbf{e}_r, \mathbf{e}_\theta\}$ shown in Fig. 3.35b, derive the flow continuity equation and verify the expression (3.104).   ∎

3.6.2 Balance of Linear Momentum

We may hypothetically free the material particle and use a Free Body Diagram (FBD; see Sect. 3.7.2) to relate the external forces that act at the material particle to the internal forces (stresses) that appear within the particle. Given the balance of momentum, these forces have to be in *equilibrium* at any time t. A material particle of the density ρ that moves at the velocity $\mathbf{v}$ therefore has to satisfy *Cauchy's linear momentum equation*

$$\rho \frac{D\mathbf{v}}{Dt} = \text{div}\boldsymbol{\sigma} + \mathbf{b}_f, \tag{3.107}$$

where $\boldsymbol{\sigma}$ denotes the Cauchy stress tensor, whilst $\mathbf{b}_f$ are body forces with respect to the unit *spatial* volume. Cauchy's momentum equation (3.107) represents the application of *Newton's second law of mechanics* [392] to a continuum body and holds for all its material particles.

In addition to the balance of linear momentum, the material particle has to be at *angular momentum* equilibrium. With the symmetry of the Cauchy stress tensor $\boldsymbol{\sigma} = \boldsymbol{\sigma}^T$, the balance of angular momentum is *a priori* satisfied.

3.6.2.1 Lagrangian Description
In Lagrangian description, $D\mathbf{v}/Dt = \partial\mathbf{v}/\partial t$ denotes the material particle's acceleration, and the system

$$\rho \begin{bmatrix} \dfrac{\partial v_1}{\partial t} \\[2mm] \dfrac{\partial v_2}{\partial t} \\[2mm] \dfrac{\partial v_3}{\partial t} \end{bmatrix} = \begin{bmatrix} \dfrac{\partial \sigma_{11}}{\partial x_1} + \dfrac{\partial \sigma_{12}}{\partial x_2} + \dfrac{\partial \sigma_{13}}{\partial x_3} \\[2mm] \dfrac{\partial \sigma_{21}}{\partial x_1} + \dfrac{\partial \sigma_{22}}{\partial x_2} + \dfrac{\partial \sigma_{23}}{\partial x_3} \\[2mm] \dfrac{\partial \sigma_{31}}{\partial x_1} + \dfrac{\partial \sigma_{32}}{\partial x_2} + \dfrac{\partial \sigma_{33}}{\partial x_3} \end{bmatrix} + \begin{bmatrix} b_{f1} \\[2mm] b_{f2} \\[2mm] b_{f3} \end{bmatrix} \tag{3.108}$$

expresses Cauchy's momentum equation (3.107) within 3D Cartesian coordinates. It determines the distribution of the stress within the body.

Example 3.18 (Equilibrium of the Material Particle in 2D). Let us consider a material particle of the density ρ [kg m^{-3}] that accelerates at $\dot{\mathbf{v}}$ [m s^{-2}]. The particle is part of a 2D continuum body at inhomogeneous plane stress conditions.

(a) Consider the Cartesian coordinate system $\{\mathbf{e}_1, \mathbf{e}_2\}$ and apply all stresses that act at the material particle.
(b) Provide the linear momentum equilibrium along the $\mathbf{e}_1$ and $\mathbf{e}_2$ directions and compare the result to (3.107).
(c) Consider the cylindrical coordinate system $\{\mathbf{e}_r, \mathbf{e}_\theta\}$, and apply all stresses that act at the material particle of a rotational symmetric problem.
(d) Provide the linear momentum equilibrium along the $\mathbf{e}_r$ and $\mathbf{e}_\theta$ directions and compare the result to (3.107). ∎

Example 3.19 (Inflated Thick-Walled Linear-Elastic Cylinder). Figure 3.37 shows a thick-walled circular tube that represents an artery, where $r_i = 2.0$ mm and $r_o = 3.0$ mm denote its inner and outer radii, respectively. The tube is inflated at the internal pressure p_i, and the linear-elastic material with the Young's modulus $E =$

625 kPa and the Poisson's ration $v = 0.49$ describes the vessel wall properties. Given small deformations, the radial σ_r and the circumferential σ_θ Cauchy stress components derive from the fundamental solution

$$\sigma_r(r) = \frac{a_0}{r^2} + b_0 + c_0(2\ln r + 1) \; ; \; \sigma_\theta(r) = \frac{-a_0}{r^2} + b_0 + c_0(2\ln r + 3)$$

$$(3.111)$$

of the rotational symmetric plane stress problem, where a_0, b_0, c_0 denote constants to be identified from the underlying Boundary Value Problem (BVP) [532].

Fig. 3.37 Inflated thick-walled cylinder representing an artery. The radius r points to the radial position of a material particle, whilst p_i denotes the inflation pressure

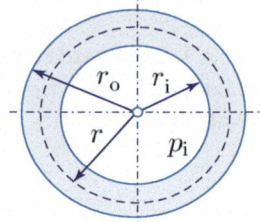

(a) Show that the fundamental solution (3.111) satisfies Cauchy's momentum equation (3.107).
(b) Identify the constants a_0, b_0, and c_0 from the boundary conditions, together with the strain compatibility at rotational symmetry

$$\frac{\partial(r\varepsilon_\theta)}{\partial r} - \varepsilon_r = 0 \; .$$

$$(3.112)$$

Plot the radial σ_r and the circumferential σ_θ stresses throughout the vessel wall at diastolic $p_{i\,d} = 75$ mmHg and systolic $p_{i\,s} = 120$ mmHg blood pressures, respectively.
(c) Compute the radial displacement pulsation Δr at the inside and at the outside of the vessel that appears in response to the pulse pressure. ■

3.6.2.2 Eulerian Description

Given an Eulerian description, the material particle's acceleration reads $Dv/Dt = \partial v/\partial t + \mathrm{grad}(\mathbf{v}) \cdot \mathbf{v}$, where $\mathbf{v}$ denotes its velocity. In addition to the *local* rate of velocity change $\partial \mathbf{v}/\partial t$, the *advective* (or *convective*) rate of velocity change $\mathrm{grad}(\mathbf{v})\cdot\mathbf{v}$ contributes to the acceleration. Figure 3.39 aims at illustrating the physical meaning of the advective rate of velocity change. It shows a steady-state, $\partial \mathbf{v}/\partial t = 0$, flow through a nozzle. According to the Eulerian-type problem description, the nozzle is fixed in space. The velocity $\mathbf{v}_i$ at the inlet *accelerates* towards the velocity $\mathbf{v}_o$ at the outlet, which, at steady state, can only be achieved through the advective term $\mathrm{grad}(\mathbf{v}) \cdot \mathbf{v}$.

Fig. 3.39 Steady-steady
flow through a nozzle that
illustrates advective
acceleration from the inflow
velocity $\mathbf{v}_i$ towards the
outflow velocity $\mathbf{v}_o$

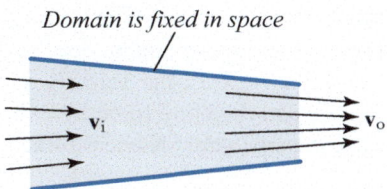

Domain is fixed in space

Given a Newtonian fluid, we may substitute the stress term in Cauchy's momentum equation (3.107) by the Newtonian viscosity model $\boldsymbol{\sigma} = 2\eta\mathbf{d} - p\mathbf{I}$, where $\mathbf{d} = (\mathrm{grad}\mathbf{v} + \mathrm{grad}^T\mathbf{v})/2$ denotes the rate of deformation tensor, whilst p is the fluid pressure, see Sect. 3.5.4.1. Cauchy's momentum equation (3.107) then leads to the *Navier*[29]–*Stokes*[30] equation. It reads $\rho(\partial\mathbf{v}/\partial t + \mathbf{v} \cdot \mathrm{grad}\mathbf{v}) = \mathrm{div}(2\eta\mathbf{d} - p\mathbf{I}) + \mathbf{b}_f$ and may also be expressed by

$$\rho\left(\frac{\partial\mathbf{v}}{\partial t} + \mathbf{v} \cdot \mathrm{grad}\mathbf{v}\right) = \eta\,\mathrm{div}(\mathrm{grad}\mathbf{v}) - \mathrm{grad}\,p + \mathbf{b}_f, \qquad (3.115)$$

where $\mathrm{div}(\mathrm{grad}\mathbf{v})$ denotes the vector Laplacian of $\mathbf{v}$. Given 3D Cartesian coordinates $\{\mathbf{e}_1, \mathbf{e}_2, \mathbf{e}_3\}$, the Navier–Stokes equations read

$$\rho\begin{bmatrix}\dfrac{Dv_1}{Dt}\\[2mm]\dfrac{Dv_2}{Dt}\\[2mm]\dfrac{Dv_3}{Dt}\end{bmatrix} = \eta\begin{bmatrix}\dfrac{\partial^2 v_1}{\partial x_1^2} + \dfrac{\partial^2 v_1}{\partial x_2^2} + \dfrac{\partial^2 v_1}{\partial x_3^2} + \dfrac{\partial^2 v_1}{\partial x_1\partial x_2} + \dfrac{\partial^2 v_1}{\partial x_1\partial x_3}\\[3mm]\dfrac{\partial^2 v_2}{\partial x_1^2} + \dfrac{\partial^2 v_2}{\partial x_2^2} + \dfrac{\partial^2 v_2}{\partial x_3^2} + \dfrac{\partial^2 v_2}{\partial x_1\partial x_3} + \dfrac{\partial^2 v_2}{\partial x_2\partial x_3}\\[3mm]\dfrac{\partial^2 v_3}{\partial x_1^2} + \dfrac{\partial^2 v_3}{\partial x_2^2} + \dfrac{\partial^2 v_3}{\partial x_3^2} + \dfrac{\partial^2 v_3}{\partial x_1\partial x_3} + \dfrac{\partial^2 v_3}{\partial x_2\partial x_3}\end{bmatrix} - \begin{bmatrix}\dfrac{\partial p}{\partial x_1}\\[2mm]\dfrac{\partial p}{\partial x_2}\\[2mm]\dfrac{\partial p}{\partial x_3}\end{bmatrix} + \begin{bmatrix}b_{f1}\\[1mm]b_{f2}\\[1mm]b_{f3}\end{bmatrix},$$

where the material time derivative

$$\begin{bmatrix}\dfrac{Dv_1}{Dt}\\[2mm]\dfrac{Dv_2}{Dt}\\[2mm]\dfrac{Dv_3}{Dt}\end{bmatrix} = \begin{bmatrix}\dfrac{\partial v_1}{\partial t}\\[2mm]\dfrac{\partial v_2}{\partial t}\\[2mm]\dfrac{\partial v_3}{\partial t}\end{bmatrix} + \begin{bmatrix}v_1\dfrac{\partial v_1}{\partial x_1} + v_1\dfrac{\partial v_1}{\partial x_2} + v_1\dfrac{\partial v_1}{\partial x_3}\\[3mm]v_2\dfrac{\partial v_2}{\partial x_1} + v_2\dfrac{\partial v_2}{\partial x_2} + v_2\dfrac{\partial v_2}{\partial x_3}\\[3mm]v_3\dfrac{\partial v_3}{\partial x_1} + v_3\dfrac{\partial v_3}{\partial x_2} + v_3\dfrac{\partial v_3}{\partial x_3}\end{bmatrix}$$

is to be used.

The different terms in the Navier–Stokes equation represent different *physical forces*, and the normalization of Eq. (3.115) allows the assessment of their individual

---

[29] Claude-Louis Navier, French engineer and physicist. 1785–1836.

[30] Sir George Gabriel Stokes, English–Irish physicist and mathematician, 1819–1903.

influence on a particular fluid flow problem. One possible normalization introduces the characteristic (problem-specific) length L, velocity V, and time T. The spatial position $\mathbf{x} = \mathbf{x}^* L$, the velocity $\mathbf{v} = \mathbf{v}^* V$, the time $t = t^* T$, the pressure $p = p^* \rho V^2$, and the body force $\mathbf{b}_f = \mathbf{b}_f^* \rho g$ may then be substituted in (3.115). It leads to the *dimensionless* Navier–Stokes equation

$$Sr \frac{\partial \mathbf{v}^*}{\partial t^*} + \mathbf{v}^* \cdot \text{grad}^* \mathbf{v}^* = \frac{\text{div}^*(\text{grad}^* \mathbf{v}^*)}{Re} = -\text{grad}^* p^* + \frac{\mathbf{b}_f^*}{Fr^2} , \qquad (3.116)$$

where the divergence $\text{div}^*(\bullet)$ and the gradient $\text{grad}^*(\bullet)$ have been defined with respect to the normalized dimension $\mathbf{x}^*$. The *Strouhal*[31] $Sr = L/(TV)$, the *Reynolds*[32] $Re = VL\rho/\eta$, and the *Froude*[33] $Fr = V/\sqrt{gL}$ numbers indicate the extent to which mass effects, viscous forces, and body forces influence the fluid mechanical problem, respectively.

3.6.3 Thermodynamic Laws

The motion χ of a body may be regarded as a *thermodynamic process*, where the body passes through a number of *thermodynamic states*. Such states are then governed by the laws of thermodynamics.

3.6.3.1 The First Law of Thermodynamics
The *first law of thermodynamics* states that the energy of a closed system is preserved. The change of *system energy* $\dot{e}$ equals then the *heat supply* h_{input} and *power input* p_{input}; $\dot{e} = h_{input} + p_{input}$. The system energy contains two parts, the *internal energy* u, such as elastic and chemical energies, and the *kinetic energy* $\rho |\mathbf{v}|^2/2$. They are taken with respect to the unit spatial volume, and the energy conservation of the subdomain Ω_s shown in Fig. 3.40 then reads

$$\underbrace{\frac{D}{Dt} \int_{\Omega_s} \left(u + \rho \frac{|\mathbf{v}|^2}{2} \right) dv}_{\text{Change of system energy}} = \underbrace{\int_{\Omega_s} h_{input} dv + \int_{\Omega_s} p_{input} dv}_{\text{Input of heat and power}} . \qquad (3.117)$$

The heat may be supplied by the *heat flux* $\mathbf{q}_h$ or the *heat source* r_h. The net heat supply to the subdomain Ω_s is therefore expressed by

$$\int_{\Omega_s} h_{input} dv = \int_{\Omega_s} r_h dv - \int_{\partial \Omega_s} \mathbf{q}_h \cdot \mathbf{n} ds = \int_{\Omega_s} (r_h - \text{div} \mathbf{q}_h) dv , \qquad (3.118)$$

---

[31] Vincenc Strouhal, Czech physicist, 1850–1922.
[32] Osborne Reynolds, Irish innovator, 1842–1912.
[33] William Froude, English engineer, 1810–1879.

Fig. 3.40 Subdomain Ω_s of
a continuum body that is
exposed to the heat flux **q**, the
heat source r, and the surface
traction **t**. The outward
normal vector to the system
boundary $\partial\Omega_s$ is denoted by **n**

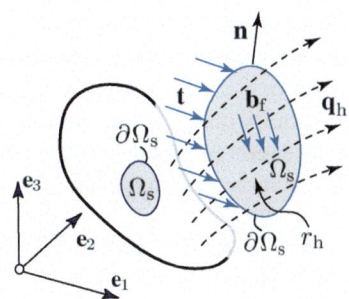

where **n** denotes the outward normal vector to the subdomain's surface $\partial\Omega_s$, and the divergence theorem (A.8.2) has been used in the derivation of this expression.

The *power input* p_{input} is associated with the *body force* $\mathbf{b}_f$ that acts within Ω_s, and the *traction* $\mathbf{t} = \boldsymbol{\sigma}\mathbf{n}$ at the boundary $\partial\Omega_s$. The last expression in (3.117) then reads

$$\int_{\partial\Omega_s} p_{input} dv = \int_{\Omega_s} \mathbf{v}\cdot\mathbf{b}_f dv + \int_{\partial\Omega_s} \mathbf{v}\cdot\boldsymbol{\sigma}\mathbf{n} ds$$

$$= \int_{\Omega_s} [\mathbf{v}\cdot\mathbf{b}_f + \mathrm{div}(\mathbf{v}\boldsymbol{\sigma})]\,dv$$

$$= \int_{\Omega_s} [\mathbf{v}\cdot(\mathbf{b}_f + \mathrm{div}\boldsymbol{\sigma}) + \boldsymbol{\sigma}:\mathbf{d}]\,dv = \int_{\Omega_s}\left(\frac{\rho}{2}\frac{\mathrm{D}|\mathbf{v}|^2}{\mathrm{D}t} + \boldsymbol{\sigma}:\mathbf{d}\right)dv\,,$$

$$(3.119)$$

where the rate of deformation $\mathbf{d} = (\mathrm{grad}\,\mathbf{v} + \mathrm{grad}^{\mathrm{T}}\mathbf{v})/2$ and the divergence theorem (A.8.2) have been used. In addition, Cauchy's momentum equation (3.107) allowed the substitution of $\mathbf{v}\cdot(\mathbf{b}_f + \mathrm{div}\boldsymbol{\sigma})$ through the term $(1/2)\mathrm{D}|\mathbf{v}|^2/\mathrm{D}t$.

Given the heat supply (3.118) and the power input (3.119), the energy conservation (3.117) reads

$$\int_{\Omega_s}(\dot{u} + \mathrm{div}\mathbf{q}_h - r_h - \boldsymbol{\sigma}:\mathbf{d})\,dv = 0\,. \qquad (3.120)$$

This integral condition holds for any spatial subdomain Ω_s of the body, and localization therefore yields the *strong condition*

$$\dot{u} + \mathrm{div}\mathbf{q}_h - r_h - \boldsymbol{\sigma}:\mathbf{d} = 0 \qquad (3.121)$$

of the energy balance of the material particle. Given a *Lagrange description*, the internal energy's material time derivative $\dot{u}$ corresponds to its partial time derivative $\partial u/\partial t$, whilst it reads $\partial u/\partial t + u\mathrm{div}\mathbf{v}$ in the *Eulerian description* of the thermodynamical process.

Example 3.20 (Conservation of Energy in Material Description). Similar to the spatial version of the conservation of energy (3.121), the first law of thermodynamics may also be expressed with respect to the body's reference configuration and thus in its material description. Let us consider a subdomain of a continuum body whose reference and spatial configurations are denoted by Ω_{s0}, and Ω_s, respectively. The body's motion is slow, and kinetic effects may be neglected.

(a) Given the internal energy U per unit reference volume, express the subdomain's change of system energy.
(b) Given the heat flux $\mathbf{Q}_h$ per unit reference area and the heat source R_h per unit reference volume, express the net accumulation of heat within the subdomain
(c) Pull-back equation $(3.119)_4$ to Ω_{s0}, and express the power input that enters the subdomain.
(d) Given the results of the previous tasks, derive the first law of thermodynamics in material description. ∎

3.6.3.2 The Second Law of Thermodynamics

We consider a body Ω at the absolute temperature θ that undergoes a thermodynamical process. The heat flux $\mathbf{q}_h$ and the heat source r_h express the influx $-\mathrm{div}(\mathbf{q}_h/\theta)$ and the respective supply r_h/θ of *entropy s* per unit spatial volume. The *second law of thermodynamics* states that the rate γ of production of entropy s is never negative, leading to the inequality

$$\gamma = \dot{s} - \left[\frac{r_h}{\theta} - \mathrm{div}\left(\frac{\mathbf{q}_h}{\theta}\right)\right] \geq 0. \tag{3.122}$$

The relation $\mathrm{div}(\mathbf{q}_h/\theta) = (\mathrm{div}\mathbf{q}_h)/\theta - (\mathbf{q}_h \cdot \mathrm{grad}\theta)/\theta^2$ together with the heat source $r_h = \dot{u} + \mathrm{div}\mathbf{q}_h - \boldsymbol{\sigma} : \mathbf{d}$, an expression that derives from the energy conservation (3.121), then gives

$$\gamma\theta = \theta\dot{s} - \dot{u} + \boldsymbol{\sigma} : \mathbf{d} - \frac{\mathbf{q}_h \cdot \mathrm{grad}\theta}{\theta} \geq 0, \tag{3.123}$$

where u denotes the internal energy, whilst $\boldsymbol{\sigma}$ and $\mathbf{d}$ are the Cauchy stress and the rate of deformation, respectively.

With the definition of the Helmholtz free energy $\psi = u - s\theta$ per unit spatial volume, the rate of internal energy reads $\dot{u} = \dot{\psi} + s\dot{\theta} + \theta\dot{s}$, and the second law of thermodynamics then reads

$$\gamma\theta = -\dot{\psi} - s\dot{\theta} + \boldsymbol{\sigma} : \mathbf{d} - \frac{\mathbf{q}_h \cdot \mathrm{grad}\theta}{\theta} \geq 0, \tag{3.124}$$

a relation known as *Clausius–Duhem inequality*.

The inequality (3.124) may also be expressed in its material description,

$$-\dot{\Psi} - S\dot{\theta} + \mathbf{P} : \dot{\mathbf{F}} - \frac{\mathbf{Q}_h \cdot \text{Grad}\theta}{\theta} \geq 0, \tag{3.125}$$

where S denotes the entropy per unit *reference* volume. Here, the Helmholtz free energy $\Psi = U - S\theta$ and the internal energy U are related to the unit reference volume. In addition, $\mathbf{Q}_h$ is the heat flux per unit reference area, whilst $\text{Grad}\,\theta$ and $\text{Div}\mathbf{Q}_h$ denote the material gradient of the temperature θ and the material divergence operator, respectively. Given the Cartesian coordinate system $\{\mathbf{e}_1, \mathbf{e}_2, \mathbf{e}_3\}$, $\text{Grad}\theta = \partial\theta/\partial X_I \, \mathbf{e}_I$ (no summation) expresses the gradient, and the divergence reads $\text{Div}\mathbf{Q}_h = \partial Q_{hI}/\partial X_I$. The material version of the Clausius–Duhem inequality (3.125) expresses the stress power $\mathbf{P} : \dot{\mathbf{F}}$ per unit reference volume, where $\mathbf{P}$ and $\mathbf{F}$ denote the first Piola–Kirchhoff stress and the deformation gradient, respectively.

Given a *Lagrange description*, the material time derivative $\text{D}(\bullet)/\text{D}t$ of the scalar $(\bullet)$ is equal to its partial time derivative $\partial(\bullet)/\partial t$, whilst in *Eulerian description*, it reads $\partial(\bullet)/\partial t + (\bullet)\text{div}\mathbf{v}$, where $\mathbf{v}$ denotes the velocity.

3.6.4 The Relation Between the Stress and the Helmholtz Free Energy

We consider a body that is described by a Helmholtz free energy Ψ per unit reference volume. The formulation is objective and of the form $\Psi = \Psi(\mathbf{F}, \mathbf{H}_1, \ldots, \mathbf{H}_n)$, where $\mathbf{F}$ and $\mathbf{H}_i$ denote the deformation gradient and n internal (hidden) variables, respectively. The body undergoes the motion $\chi(t)$, and like any other admissible thermodynamic process, it obeys the Clausius–Duhem inequality (3.125). Given most vascular biomechanical applications, such a process is *isothermal*, and $\dot{\theta} = 0$ as well as $Grad\theta = \mathbf{0}$ applies. It then leads to the isothermal version $-\dot{\Psi} + \mathbf{P} : \dot{\mathbf{F}} \geq 0$ of the Clausius–Duhem inequality, and with $\Psi(\mathbf{F}, \mathbf{H}_1, \ldots, \mathbf{H}_n)$, it yields the inequality

$$\left(\mathbf{P} - \frac{\partial\Psi}{\partial\mathbf{F}}\right) : \dot{\mathbf{F}} - \frac{\partial\Psi}{\partial\mathbf{H}_i} : \dot{\mathbf{H}}_i \geq 0 \tag{3.126}$$

of an admissible process. The inequality (3.126) holds for arbitrary deformations and then implies the constitutive relation

$$\mathbf{P} = \frac{\partial\Psi(\mathbf{F}, \mathbf{H}_1, \ldots, \mathbf{H}_n)}{\partial\mathbf{F}} \tag{3.127}$$

between the first Piola–Kirchhoff stress $\mathbf{P}$ and the free energy function Ψ.

We may also introduce n stress-like variables $\mathbf{L}_i = \partial\Psi/\partial\mathbf{H}_i$ that are work-conjugate to the hidden variables $\mathbf{H}_i$. Given (3.126) holds for any deformation, the inequality

$$-\mathbf{L}_i : \dot{\mathbf{H}}_i \geq 0 \tag{3.128}$$

ensures the body's motion obeys the *second law of thermodynamics*. The afore-mentioned approach to derive the constitutive relation (3.127) and the dissipation inequality (3.128) is known as *Coleman and Noll's procedure*.

The expressions (3.127) and (3.128) use two-point tensors, and equivalent expression with referential one-point tensors may be derived. We therefore introduce the *a priori* objective free energy $\Psi = \Psi(\mathbf{C}, \mathbf{G}_1, \ldots, \mathbf{G}_n)$ as a function of the right Cauchy–Green strain $\mathbf{C}$ and the strain-like internal variables $\mathbf{G}_i$. As with $\mathbf{C}$, the internal variables $\mathbf{G}_i$ are symmetric second-order tensors that are defined in the body's reference configuration Ω_0. Given the first Piola transform (3.31), the constitutive relation (3.127) expresses then the *second Piola–Kirchhoff stress* by

$$\mathbf{S} = \mathbf{F}^{-1}\mathbf{P} = 2\frac{\partial\Psi(\mathbf{C}, \mathbf{G}_1, \ldots, \mathbf{G}_n)}{\partial\mathbf{C}}, \tag{3.129}$$

where the chain rule $\mathbf{P} = (\partial\Psi/\partial\mathbf{C}) : (\partial\mathbf{C}/\partial\mathbf{F})$ and the property $(\partial\mathbf{C}/\partial\mathbf{F}) : \mathbf{F}^{-1} = 2\mathbf{I}$ have been used in the derivation of this expression. In equivalence to the dissipation inequality (3.128), the dissipation per unit reference volume then reads $-\mathbf{M}_i : \dot{\mathbf{G}}_i \geq 0$, where $\mathbf{M}_i = \partial\Psi/\partial\mathbf{G}_i$ are n stress-like variables that are work-conjugate to $\mathbf{G}_i$.

3.6.4.1 Stress of an Incompressible Material

For a number of applications, the deformation of vascular tissue can be considered to be incompressible. The motion χ is then *constrained*, and the constraint equation $J - 1 = 0$ is to be enforced in the material's constitutive description. Following the *Lagrange multiplier method*, we introduce the Lagrangian $\Psi = \Psi_{\mathrm{iso}}(\mathbf{C}, \mathbf{G}_1, \ldots, \mathbf{G}_n) - \kappa(J - 1)$, where the isochoric free energy Ψ_{iso} represents constitutive information, whilst the term $\kappa(J-1)$ serves as Lagrangian contribution. It enforces the incompressibility, and κ is the Lagrange multiplier. The Coleman and Noll procedure then yields

$$\mathbf{S} = 2\frac{\partial\Psi_{\mathrm{iso}}(\mathbf{C}, \mathbf{G}_1, \ldots, \mathbf{G}_n)}{\partial\mathbf{C}} + 2\kappa\frac{\partial(J - 1)}{\partial J}\frac{\partial J}{\partial\mathbf{C}}, \tag{3.130}$$

and, with $\partial J/\partial\mathbf{C} = J\mathbf{C}^{-1}/2$ and $J = 1$, the second Piola–Kirchhoff stress of an incompressible material reads

$$\mathbf{S} = 2\frac{\partial\Psi_{\mathrm{iso}}(\mathbf{C}, \mathbf{G}_1, \ldots, \mathbf{G}_n)}{\partial\mathbf{C}} - \kappa\mathbf{C}^{-1}. \tag{3.131}$$

We emphasize that the Lagrange parameter κ in the afore-derived expressions does *not* represent the hydrostatic pressure p, and therefore the two contributions at the right-hand side of (3.131) do *not* represent isochoric and hydrostatic stress contributions, respectively. The decoupled representation would instead read

$$\mathbf{S} = \underbrace{2\mathrm{Dev}\left[\frac{\partial\Psi_{\mathrm{iso}}(\mathbf{C},\mathbf{G}_1,\ldots,\mathbf{G}_n)}{\partial\mathbf{C}}\right]}_{\overline{\mathbf{S}}}\underbrace{-p\mathbf{C}^{-1}}_{\mathbf{S}_{\mathrm{vol}}},\tag{3.132}$$

where $\overline{\mathbf{S}}$ and $\mathbf{S}_{\mathrm{vol}}$ are the isochoric and respective volumetric second Piola–Kirchhoff stresses, whilst $\mathrm{Dev}(\bullet) = (\bullet) - [\mathbf{C}:(\bullet)]\mathbf{C}^{-1}/3$ denotes the deviator operator in the referential description.

Given (3.131) and (3.132), the application of the second Piola transform (3.32) for incompressible materials $J = 1$ then yields the two corresponding expressions

$$\boldsymbol{\sigma} = 2\mathbf{F}\frac{\partial\Psi_{\mathrm{iso}}(\mathbf{C},\mathbf{G}_1,\ldots,\mathbf{G}_n)}{\partial\mathbf{C}}\mathbf{F}^{\mathrm{T}} - \kappa\mathbf{I}\tag{3.133}$$

and

$$\boldsymbol{\sigma} = \underbrace{2\mathrm{dev}\left[\mathbf{F}\frac{\partial\Psi_{\mathrm{iso}}(\mathbf{C},\mathbf{G}_1,\ldots,\mathbf{G}_n)}{\partial\mathbf{C}}\mathbf{F}^{\mathrm{T}}\right]}_{\overline{\boldsymbol{\sigma}}}\underbrace{-p\mathbf{I}}_{\boldsymbol{\sigma}_{\mathrm{vol}}}\tag{3.134}$$

of the Cauchy stress, where $\overline{\boldsymbol{\sigma}}$ and $\boldsymbol{\sigma}_{\mathrm{vol}}$ denote its isochoric and volumetric contributions, respectively. In the derivation of these expressions the identity $\mathrm{dev}[\mathbf{F}(\bullet)\mathbf{F}^{\mathrm{T}}] = \mathbf{F}\{\mathrm{Dev}[(\bullet)]\}\mathbf{F}^{\mathrm{T}}$ has been used, see Appendix C.

The hydrostatic pressure p in (3.132) and (3.134) may be seen as a *Lagrange parameter* that enforces the incompressibility. As with κ, it is independent of the material's deformation and entirely defined by the individual BVP.

Example 3.21 (The Incompressible neoHookean Material). The neoHookean Helmholtz free energy $\Psi_{\mathrm{iso}}(\mathbf{C}) = G(I_1 - 3)/2$ per unit (reference) volume describes an incompressible body, where $I_1 = \mathrm{tr}\mathbf{C}$ denotes the first invariant of the right Cauchy–Green strain $\mathbf{C}$, whilst G is the shear modulus in the reference configuration. Compute the material's normalized Cauchy stress as a function of the deformation, given the following four motions $\boldsymbol{\chi}(\mathbf{X})$:

(a) Simple tension. The components

$$\chi_{st\,1} = \lambda X_1 \;\; ; \;\; \chi_{st\,2} = X_2/\sqrt{\lambda} \;\; ; \;\; \chi_{st\,3} = X_3/\sqrt{\lambda} \tag{3.135}$$

describe the motion, where λ denotes the stretch in tensile direction.

(b) Equi-biaxial tension. The components

$$\chi_{bt\,1} = \lambda X_1 \;\; ; \;\; \chi_{bt\,2} = \lambda X_2 \;\; ; \;\; \chi_{bt\,3} = X_2/(\lambda_1\lambda_2) \tag{3.136}$$

describe the motion, where λ denotes the stretch in the two tensile directions.

(c) Simple shear. The components

$$\chi_{ss\,1} = X_1 + \gamma X_2 \;\; ; \;\; \chi_{ss\,2} = X_2 \;\; ; \;\; \chi_{ss\,3} = X_3 \tag{3.137}$$

describe the motion, where γ denotes the amount of shear.

(d) Pure shear. The components

$$\chi_{ps\,1} = \frac{X_1 - X_2 + (X_1 + X_2)\lambda^2}{2\lambda} \;\; ; \;\; \chi_{ps\,2} = \frac{X_2 - X_1 + (X_1 + X_2)\lambda^2}{2\lambda} \;\; ;$$

$$\chi_{ps\,3} = X_3 \tag{3.138}$$

describe the motion, where λ denotes the principal stretch. Given small deformations, $\lambda = 1 - \gamma/2$ relates the stretch λ and the amount of shear γ. ∎

3.7 General Principles

As with other mechanics disciplines, continuum mechanics uses a number of principles towards the solution of specific applications, and some of them are discussed in this section.

3.7.1 Maxwell Transport and Localization

The balance equations in continuum mechanics discussed in the previous sections may also have been derived through a two-step approach, called *Maxwell transport* and *localization procedure*. Let us consider the derivation of the conservation of mass. Given the subdomain Ω_s shown in Fig. 3.42,

$$\frac{\partial}{\partial t} \int_{\Omega_s} \rho(\mathbf{x}, t) \mathrm{d}v = 0 \tag{3.140}$$

holds, where $\rho(\mathbf{x}, t)$ denotes the density per unit spatial volume. The subdomain Ω_s is not fixed in space, it evolves over time, and the time derivative of the integral cannot directly be computed. In a first step, we therefore pull-back the integral to the subdomain's fixed reference configuration $\Omega_{s\,0}$. Given the transformation of the volume element $dv = JdV$, the conservation of mass in the reference configuration then reads

$$\frac{\partial}{\partial t} \int_{\Omega_{s0}} \rho(\mathbf{x}(\mathbf{X}, t), t) J dV = 0 \, .$$

The integrand may now be differentiated, and (3.140) then reads

$$\int_{\Omega_{s0}} \frac{\partial}{\partial t}(J\rho) dV = \int_{\Omega_{s0}} \left(\frac{\partial \rho}{\partial t} + \rho \mathrm{divv} \right) J dV = 0 \, ,$$

where the rate of volume change (3.38)$_4$ has been used. In the second step, the integral is pushed forward to the spatial configuration Ω_s. The conservation of mass then reads

$$\int_{\Omega_s} \left(\frac{\partial \rho}{\partial t} + \rho \mathrm{divv} \right) dv = 0 \, ,$$

and the shrinkage of the domain Ω_s towards a point, a process called localization, yields

$$\frac{\partial \rho}{\partial t} + \rho \mathrm{divv} = 0 \, ,$$

in accordance with the expression (3.102).

Example 3.22 (Equilibrium in Material Description). We consider a body under the action of the body force $\mathbf{b}_f$ per unit spatial volume. The motion $\chi(\mathbf{X})$ maps a subdomain from its reference configuration Ω_{s0} to its spatial configuration Ω_s, see Fig. 3.42. The subdomain's surfaces and unit normal vectors are denoted by $\partial \Omega_{s0}$, $\partial \Omega_s$ and $\mathbf{N}$, $\mathbf{n}$ in the respective configurations.

Fig. 3.42 The motion $\chi(\mathbf{X})$ maps a subdomain from its referential configuring Ω_{s0} to its spatial configuration Ω_s

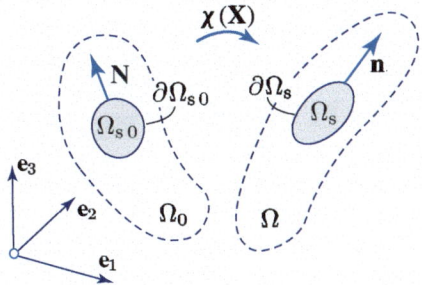

(a) Given the spatial configuration Ω_s, derive the linear momentum equation and show that the localization of this expression then yields Cauchy's (static) momentum equation (3.107).

(b) Show that the linear moment equilibrium with respect to the body's reference
     configuration may be expressed by

$$\text{Div}\mathbf{P} + \mathbf{B}_\text{f} = \mathbf{0}, \tag{3.141}$$

where $\text{Div}(\bullet)$ denotes the divergence with respect to the reference configuration,
whilst $\mathbf{P}$ and $\mathbf{B}_\text{f}$ are the first Piola–Kirchhoff stress and the body forces with
respect to unit referential volume, respectively.   ∎

3.7.2 Free Body Diagram

A Free Body Diagram (FBD) is the simplest abstraction of the forces that act at a
physical object and may be used to derive the relation between *external* and *internal*
forces of a continuum body. The internal forces are introduced by the hypothetical
sectioning of the continuum body and can then be related to stress components
through Euler's first and second principle, respectively. Simplifications, such as
plane stress, are often used in the application of FBD.

We may consider Fig. 3.43a towards the illustration of a FBD. It shows a
sphere of radius r [m] and a wall thickness h [m] at its spatial and thus deformed
configuration Ω. Given the symmetry of the problem, the inflation at the pressure
p_i [Pa] determines an equi-biaxial stress state in the wall, described by the Cauchy
stress σ [Pa]. The radius of the sphere is much larger than its wall thickness, $r \gg h$,
and therefore the condition of *plane stress* (or a *membrane stress* state) represents a
reasonable approximation to the stress in the wall.

As shown in Fig. 3.43b, the introduced sectioning "frees" the stress σ, which
in turn leads to the FBD shown in Fig. 3.43c. The inflation pressure p_i causes
the external force $P_\text{p} = r^2 \pi p_\text{i}$, whilst the stress σ leads to the internal force
$P_\sigma = 2r\pi h\sigma$. Given the symmetry of the problem, only the forces along the axial
direction are of relevance and appear in the FBD. The multiplication of the pressure

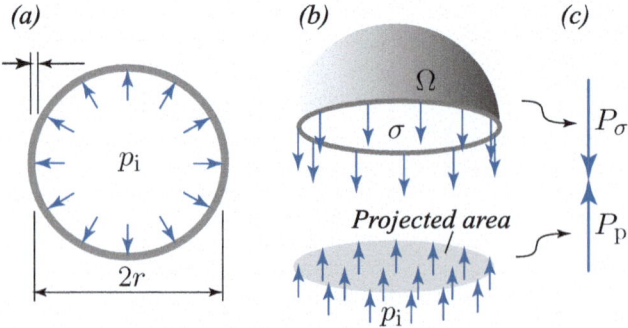

Fig. 3.43 Inflated thin-walled sphere. (**a**) Problem definition. (**b**) Hypothetically dissected structure to free internal forces. (**c**) Free Body Diagram (FBD)

p_i with the sphere's projected area $r^2\pi$ determines therefore the external force P_p, see Fig. 3.43b. Euler's first principle, and thus the equilibrium in the axial direction, then reads $P_p - P_\sigma = 0$. It allows us to express the stress through

$$\sigma = \frac{rp_i}{2h}, \tag{3.144}$$

a function of the inflation pressure p_i and the spatial properties r, h of the deformed sphere. This continuum mechanical problem is called *statically determined*, and the function (3.144) does not involve any material properties of the sphere.

Example 3.23 (The Inflated Thin-Walled and Linear-Elastic Circular Tube). We consider an artery that is represented by a thin-walled circular tube of the diameter $d = 5\,\mathrm{mm}$ and the wall thickness $h = 0.4\,\mathrm{mm}$. The tube is inflated at the pressure p_i, and the wall may be described by a linear-elastic material with the Young's modulus $E = 625\,\mathrm{kPa}$ and the Poisson's ratio $\nu = 0.49$. Given this problem, small deformation theory may be used.

(a) Express the wall stress as a function of the pressure p_i and the geometrical properties d and h.
(b) Given the diastolic $p_{id} = 75\,\mathrm{mmHg}$ and the systolic $p_{is} = 120\,\mathrm{mmHg}$ blood pressures, compute the corresponding change Δd of the vessel's diameter.
(c) Compute the capacity C [$\mathrm{mm^3\,mmHg^{-1}}$] of a vessel segment of the length $l = 8.0\,\mathrm{cm}$. It may be assumed that the vessel's length remains constant during pressure pulsations.   ■

3.7.3 Boundary Value Problem

The *Boundary Value Problem (BVP)* is the mathematical description of a problem in continuum mechanics. Let us consider a solid mechanics problem, where a body occupies the spatial domain Ω, and the displacement $\mathbf{u}$ denotes the *primary*, or *essential unknown*. Within Ω, Cauchy's momentum equation (3.107) $\rho\left(D^2\mathbf{u}/Dt^2\right) = \mathrm{div}\,\boldsymbol{\sigma} + \mathbf{b}_f$ applies, where $\mathbf{b}_f$ denotes the body force per unit spatial volume.

The body's surface $\partial\Omega$ is split into $\partial\Omega_u$ and $\partial\Omega_t$, such that $\partial\Omega_u \cup \partial\Omega_t = \partial\Omega$ and $\partial\Omega_u \cap \partial\Omega_t = \{0\}$ holds. At Ω_u a *Dirichlet*[34] or *essential* boundary condition prescribes the displacement $\mathbf{u} = \bar{\mathbf{u}}$, whilst at Ω_t a *Neumann*[35] or *natural* boundary condition prescribes the traction $\mathbf{t} = \boldsymbol{\sigma}\mathbf{n} = \bar{\mathbf{t}}$, where Cauchy's stress theorem (3.20) has been used and $\mathbf{n}$ denotes the normal vector to the boundary, see Fig. 3.45a. The specification of Dirichlet and Neumann boundary conditions relates

---

[34]Johann Peter Gustav Lejeune Dirichlet, German mathematician, 1805–1859.
[35]Carl Gottfried Neumann, German mathematician, 1832–1925.

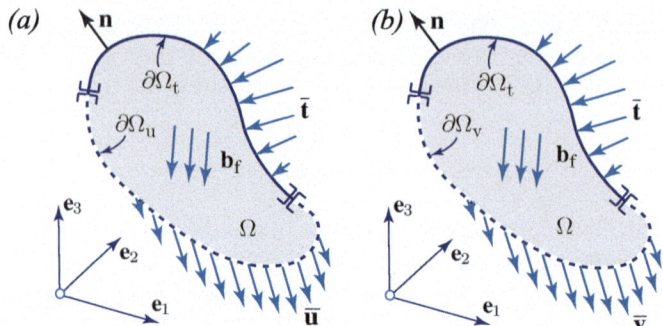

Fig. 3.45 Boundary Value Problem (BVP) in continuum mechanics. (**a**) BVP in solid mechanics using a Lagrange description, where Ω moves with the body. (**b**) BVP in fluid mechanics using an Eulerian description, where Ω is fixed in space

independently to the individual displacement components. For example, at a node the first displacement component may be specified by a Dirichlet condition, whilst a Neumann condition sets the second component of the nodal force vector.

Given a *transient problem*, the mathematical description is closed by the specification of the velocity and displacement over Ω at the time t_0, $\mathrm{D}\mathbf{u}/\mathrm{D}t = \mathbf{v}_0$ and $\mathbf{u} = \mathbf{u}_0$. The BVP is then called an initial BVP (iBVP) and described by the set

$$\left.\begin{aligned}
\operatorname{div}\boldsymbol{\sigma} + \mathbf{b}_{\mathrm{f}} &= \rho\frac{\mathrm{D}^2\mathbf{u}}{\mathrm{D}t^2} && \text{in } \Omega\,, \\[4pt]
\mathbf{u} &= \overline{\mathbf{u}} && \text{at } \partial\Omega_{\mathrm{u}}\,, \\[4pt]
\mathbf{t} = \boldsymbol{\sigma}\mathbf{n} &= \overline{\mathbf{t}} && \text{at } \partial\Omega_{\mathrm{t}}\,, \\[4pt]
\frac{\mathrm{D}\mathbf{u}}{\mathrm{D}t} = \mathbf{v}_0; \mathbf{u} &= \mathbf{u}_0 && \text{in } \Omega \text{ at } t = t_0\,.
\end{aligned}\right\} \qquad (3.146)$$

A Lagrangian description is commonly used for solid mechanics problems. The domain Ω follows then the motion of the body, and $\mathrm{D}\mathbf{u}/\mathrm{D}t = \partial\mathbf{u}/\partial t$ and $\mathrm{D}^2\mathbf{u}/\mathrm{D}t^2 = \partial^2\mathbf{u}/\partial t^2$ express the first and second material time derivatives of the displacement.

In the description of a fluid mechanical problem, the velocity $\mathbf{v}$ denotes the primary, or essential unknown, see Fig. 3.45b. The iBVP then reads

$$\left.\begin{aligned}
\operatorname{div}\boldsymbol{\sigma} + \mathbf{b}_{\mathrm{f}} &= \rho\frac{\mathrm{D}\mathbf{v}}{\mathrm{D}t} && \text{in } \Omega\,, \\[4pt]
\mathbf{v} &= \overline{\mathbf{v}} && \text{at } \partial\Omega_{\mathrm{v}}\,, \\[4pt]
\mathbf{t} = \boldsymbol{\sigma}\mathbf{n} &= \overline{\mathbf{t}} && \text{at } \partial\Omega_{\mathrm{t}}\,, \\[4pt]
\mathbf{v} &= \mathbf{v}_0 && \text{in } \Omega \text{ at } t = t_0\,,
\end{aligned}\right\} \qquad (3.147)$$

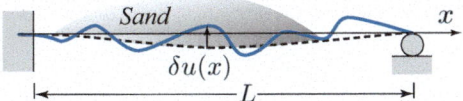

Fig. 3.46 Principle of Virtual Work (PVW). The beam's equilibrium configuration (dashed line) is perturbed by an admissible virtual displacement $\delta u(x)$. It is compatible with the Dirichlet boundary conditions at the left and right ends of the beam

and an Eulerian description is commonly used. The domain Ω is then fixed in space, and $D\mathbf{v}/Dt = \partial\mathbf{v}/\partial t + \mathrm{grad}\,\mathbf{v}\cdot\mathbf{v}$ expresses the material time derivative of the velocity.

The sets (3.146) and (3.147) represent the *strong* form of the iBVP, and the equilibrium equation is directly enforced at the material point level. Whilst such a description is used by a number of numerical methods, the Finite Element Method (FEM) uses the *weak* or *integral form* of the iBVP, see Chap. 4.

3.7.4 Principle of Virtual Work

The *Principle of Virtual Work (PVW)* is an energy principle that may be used to solve problems in continuum mechanics. Let us consider the beam in Fig. 3.46 towards the illustration of the PVW. The beam is clamped at the left end and simply supported at its right end. Given the weight of a pile of sand that rests on it, the displacement $u(x)$ deforms the beam into its equilibrium configuration, indicated by the dashed line in Fig. 3.46. We may now "freeze" the loading and introduce the admissible *virtual displacement* $\delta u(x)$ that (hypothetically) perturbs the beam around its equilibrium configuration. Aside from being continuous, $\delta u(x)$ has to be compatible with the *Dirichlet boundary conditions*. Given the problem shown in Fig. 3.46, an admissible virtual displacement therefore satisfies $\delta u(0) = \delta u'(0) = \delta u(L) = 0$.

The perturbation $\delta u(x)$ moves the external forces and deforms the beam, which then results in the *external* work δW_{ext} and the respective *internal work* δW_{int}. If and only if $\delta W_{\mathrm{ext}} = \delta W_{\mathrm{int}}$, the PVW concludes that the perturbation $\delta u(x)$ has been superimposed onto the beam's *equilibrium configuration*. The PVW is an alternative version of Newton's second law of mechanics and represents the weak form of the BVP. Calculus of Variations represents an alternative, and more general method, to derive a problem's description in the weak form, see Sect. 4.3.

Example 3.24 (Applications of the Principle of Virtual Work). The Principle of Virtual Work (PVW) $\delta W_{\mathrm{ext}} = \delta W_{\mathrm{int}}$, where W_{ext} and W_{int} denote the external and respective internal virtual works, should be used to derive the equilibrium conditions of the following mechanical systems:

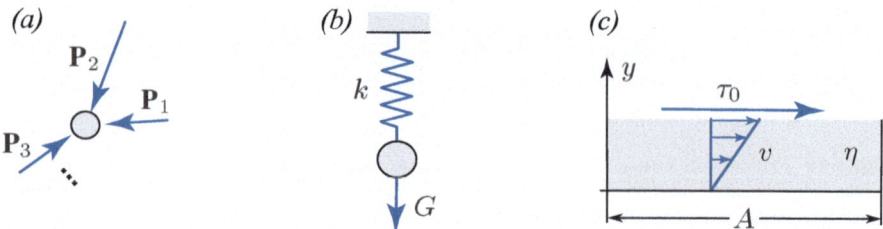

Fig. 3.47 Mechanical systems at equilibrium. (**a**) Rigid body loaded by the forces $\mathbf{P}_i$. (**b**) Spring of stiffness k loaded by the gravitational force G. (**c**) Fluid of viscosity η that is loaded by the shear stress τ_0 and moves at the velocity v

(a) A rigid body that is loaded by n constant forces $\mathbf{P}_i$, $i = 1, \ldots, n$, see Fig. 3.47a.
(b) A spring of stiffness k that is loaded by the constant gravitational force G, see Fig. 3.47b.
(c) A fluid layer of the thickness H and the base area A that is loaded by the shear stress τ_0, see Fig. 3.47c. The fluid within the layer moves at the velocity $v = Vy/H$, where V denotes the surface velocity. The layer is at steady state, and the fluid has the constant viscosity η.   ∎

3.7.4.1 Principle of Virtual Work for Small Deformation Problems

Towards the application of the PVW in solid mechanics, we consider the BVP in Fig. 3.45a at small deformations. The equilibrium displacement $\mathbf{u}$ is perturbed by the admissible virtual displacement $\delta\mathbf{u}$—given small deformation theory, $\delta\mathbf{u}$ is small. The perturbation then leads to the virtual work relation

$$\underbrace{\int_\Omega \boldsymbol{\sigma} : \delta\boldsymbol{\varepsilon}\,dv}_{\delta W_{\text{int}}} = \underbrace{\int_\Omega \mathbf{b}_f \cdot \delta\mathbf{u}\,dv + \int_{\partial\Omega} \mathbf{t} \cdot \delta\mathbf{u}\,ds}_{\delta W_{\text{ext}}}\,, \qquad (3.148)$$

where $\delta\boldsymbol{\varepsilon} = [\partial\delta\mathbf{u}/\partial\mathbf{X} + (\partial\delta\mathbf{u}/\partial\mathbf{X})^{\mathsf{T}}]/2$ denotes the *virtual engineering strain* in response to the virtual displacement $\delta\mathbf{u}$. The expression (3.148) reflects the *weak* or *integral form* of the BVP (3.146) at small deformations and static conditions.

3.7.4.2 Principle of Virtual Work for Finite Deformation Problems

Using work-conjugate stress and strain measures, the PVW may be directly applied to finite deformation problems. It then reads

$$\underbrace{\int_\Omega \boldsymbol{\sigma} : \text{grad}_s\delta\mathbf{u}\,dv}_{\delta W_{\text{int}}} = \underbrace{\int_\Omega \mathbf{b}_f \cdot \delta\mathbf{u}\,dv + \int_{\partial\Omega} \mathbf{t} \cdot \delta\mathbf{u}\,ds}_{\delta W_{\text{ext}}}\,, \qquad (3.149)$$

where $\text{grad}_s\delta\mathbf{u} = [\partial\delta\mathbf{u}/\partial\mathbf{x} + (\partial\delta\mathbf{u}/\partial\mathbf{x})^{\mathsf{T}}]/2$ denotes the symmetric gradient of the virtual displacement $\delta\mathbf{u}$. Expression (3.149) represents the PVW in its *spatial* formulation. The integrations are therefore taken over the spatial configuration Ω, and the body force $\mathbf{b}_f$ and the traction $\mathbf{t}$ relate to the unit spatial volume and the unit spatial surface, respectively.

The spatial domain Ω moves in time and complicates the application of (3.149), difficulties prevented with the *material* version of the PVW. Towards its derivation, we use the second Piola transform (3.32) $\boldsymbol{\sigma} = J^{-1}\mathbf{FSF}^{\mathrm{T}}$, the volume transform $\mathrm{d}v = J\mathrm{d}V$, Cauchy's stress theorem (3.20) $\mathbf{t} = \boldsymbol{\sigma}\mathbf{n}$, and Nanson's formula (3.5) $\mathbf{n}\mathrm{d}s = J\mathbf{F}^{-\mathrm{T}}\mathbf{N}\mathrm{d}S$. The PVW equation (3.149) may then be expressed by

$$\int_{\Omega_0} J^{-1}\mathbf{FSF}^{\mathrm{T}} : \mathrm{grad}_\mathrm{s}\delta\mathbf{u}J\mathrm{d}V = \int_{\Omega_0} \mathbf{b}_\mathrm{f} \cdot \delta\mathbf{u}J\mathrm{d}V + \int_{\partial\Omega_0} \boldsymbol{\sigma} J\mathbf{F}^{-\mathrm{T}}\mathbf{N} \cdot \delta\mathbf{u}\mathrm{d}S,$$

where $\mathbf{S}$ denotes the second Piola–Kirchhoff stress, whilst $\mathbf{N}$ is the normal vector to $\partial\Omega_0$. All integrations are now taken over the *referential* and thus fixed configuration. Given the body force per unit reference volume $\mathbf{B}_\mathrm{f} = J\mathbf{b}_\mathrm{f}$ and Cauchy stress theorem $\mathbf{T} = \mathbf{PN}$, the material version of the PVW then reads

$$\underbrace{\int_{\Omega_0} \mathbf{S} : \delta\mathbf{E}\mathrm{d}V}_{\delta W_{\mathrm{int}}} = \underbrace{\int_{\Omega_0} \mathbf{B}_\mathrm{f} \cdot \delta\mathbf{u}\mathrm{d}V + \int_{\partial\Omega_0} \mathbf{T} \cdot \delta\mathbf{u}\mathrm{d}S}_{\delta W_{\mathrm{ext}}}, \qquad (3.150)$$

where $\delta\mathbf{E} = \mathbf{F}^{\mathrm{T}}\mathrm{grad}_\mathrm{s}\delta\mathbf{u}\mathbf{F}$ denotes the virtual Green–Lagrange strain, see Appendix D. The expressions (3.149) and (3.150) reflect the weak or integral form of the BVP (3.146) at static conditions in the *spatial* and *material* description, respectively.

3.8   Damage and Failure

At loading beyond a body's elastic limit, the integrity of the material's microstructure is harmed and *damage* develops. Given vascular tissue, micro-defects such as breakage and/or pull-out of collagen fibrils start to appear. If healing is not able to repair such micro-defects, the tissue continues to accumulate weak links— its *stiffness* and *strength* diminish. Exceeding a certain threshold micro-defects coalesce towards the formation of macro-defects, and a single macro-defect then eventually propagates and fractures the material. A number of engineering concepts, such as Continuum Damage Mechanics (CDM), Linear Fracture Mechanics (LFM), and cohesive zone modeling allow the study of the formation and propagation of failure in materials.

CDM follows a *Kachanov-type* formulation [290], where a *damage parameter* describes the state of damage. Given isotropic damage, the damage parameter is a scalar and represents the density of micro-defects, whilst in anisotropic damage, a damage tensor describes the density together with the orientation of micro-defects. The damage parameter relates the stress $\boldsymbol{\sigma}$ of the continuum to the stress $\boldsymbol{\sigma}_{\mathrm{eff}}$ of a virtually undamaged material. For isotropic damage it reads

$$\sigma = (1 - D)\sigma_{\text{eff}}, \tag{3.151}$$

where D denotes the scalar damage parameter. In anisotropic damage similar relations are used, where a symmetric second-order (or higher-order) damage tensor determines the state of damage.

3.8.1 Physical Consequences of Damage

The accumulation of damage eventually leads to *strain softening*. The stress then decreases at increasing strain, and the material's stiffness tensor is no longer positive definite—the physics of the mechanical problem changes fundamentally.

We consider the rod at simple tension shown in Fig. 3.48 to demonstrate the consequences of strain softening. The material is described by the strain-dependent Young's modulus $E(\varepsilon)$, and the rod has the constant cross-section A. Equilibrium along the axial direction reads $-A\rho dx(\partial^2 u/\partial t^2) - A\sigma(x) + A\sigma(x + dx) = 0$ and determines the governing equation

$$\frac{\partial^2 u(x, t)}{\partial t^2} - c\frac{\partial^2 u(x, t)}{\partial x^2} = 0 \quad \text{with} \quad c = E(\varepsilon)/\rho, \tag{3.152}$$

where the linear expansion $\sigma(x + dx) = \sigma(x) + d\sigma = \sigma(x) + E(\varepsilon)(\partial^2 u/\partial x^2)dx$ has been used. Equation (3.152) represents a wave propagation problem and the parameter c determines the physics of the problem. For at $E(\varepsilon) > 0$ (and thus $c > 0$) the problem is *elliptic* and waves *can* propagate along the rod, whilst at $E(\varepsilon) < 0$ (and thus $c < 0$), the material exhibits strain softening, the problem is *hyperbolic*, and waves *cannot* propagate along the rod.

Given a multi-dimensional small-strain problem, the aforementioned 1D condition $E(\varepsilon) > 0$ relates to the *strong ellipticity condition*. Thus, $\Delta\varepsilon : \mathbb{C}(\varepsilon) : \Delta\varepsilon > 0$ for all possible strain increments $\Delta\varepsilon$, where $\mathbb{C}(\varepsilon)$ denotes the (non-constant) elasticity tensor [354]. A similar condition refers to finite deformation problems [46, 402].

Fig. 3.48 Forces acting at the cross-section of a rod at simple tension

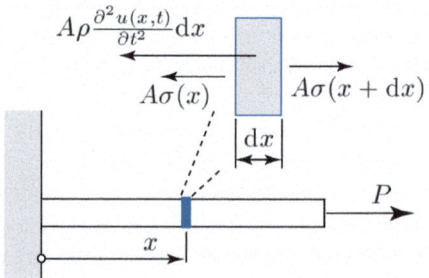

Fig. 3.49 Naive rupture test showing markedly different strain *versus* force F responses in the strain softening phase. The different responses are linked to three pairs of markers that are used to calculate the (average) strain in the tensile sample

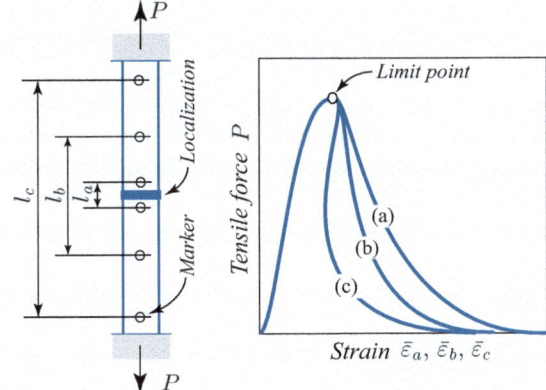

3.8.2 Strain Localization

Strain softening results in strain localization, and the deformation then localizes in a narrow domain within which all irreversible processes appear. The study of strain localizations requires appropriate computational and experimental tools. The classical (non-polar) continuum is *inadequate* to analyze strain localizations and requires regularization to avoid non-physical results. Given the non-polar continuum, strain localizes within an *infinitesimally small* volume and *no* energy is then dissipated by an inherently dissipative process, such as damage [31]. In contrary, the failure in a real material localizes always within a finite volume, which size is determined by the local stress state and the material's microstructure— parameters that determine the *internal length-scale* of the failure process. The non-polar continuum contains no internal length-scale and can therefore not be directly used to analyze material failure [287].

Strain localization challenges also the experimental characterization of materials, as schematically illustrated by the tensile test in Fig. 3.49. The set-up uses markers in the acquisition of the (averaged) strain $\bar{\varepsilon} = (l - L)/L$, where L and l denote the distance between corresponding markers in the respective configuration. The experimental measurements dependent on the *marker positions*. The strain calculated from markers that are close to the localization zone (case (a) in Fig. 3.49) results in a more ductile response as from markers further away from the localization zone (case (c) in Fig. 3.49).

Example 3.25 (Strain Localization in a Rod at Tension). The development of a strain localization in a rod at tension is a well discussed problem [618]. Figure 3.50 illustrates such a rod of the length $L = 10.0$ m and the cross-section $A = 1.0$ m². It is discretized by n equal sections of the length $l = L/n$, and Fig. 3.50b shows the bi-linear stress–strain properties of the material. The stress increases at the stiffness $E = 10.0$ MPa until it reaches the elastic limit $Y = 1.0$ MPa and then decreases at the softening stiffness $H = 1.0$ MPa.

We gradually increase the displacement u at the rod's right end. The load P then develops up to the limit $P = P_{\text{max}}$, a configuration indicated by State (II) in Fig. 3.50a, and all n sections have the same strain $\varepsilon = u/L$. Given the limit load is reached, a further increase of u leads to the reduction of P. The rod starts then to develop a strain localization and the solution bifurcates, see State (III) in Fig. 3.50. The section with the (numerically) smallest cross-section (in Fig. 3.50 this section is shaded) follows the strain softening path, whilst all the other sections elastically unload until the structure is completely stress-free at State (IV).

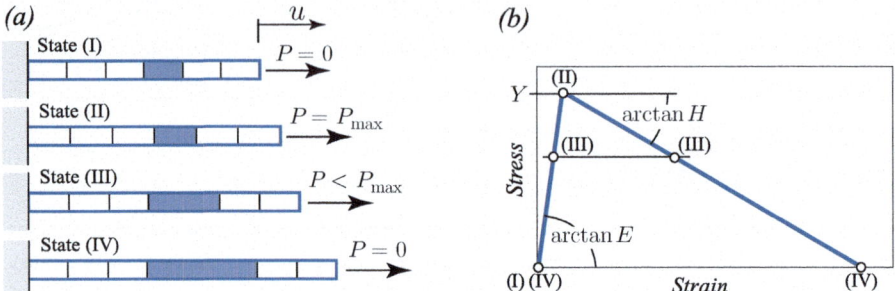

Fig. 3.50 Rod at simple tension that develops a strain localization. (**a**) The deformations at a number of loading states. (**b**) Bi-linear stress–strain properties with the corresponding loading states illustrated

(a) Given the rod in the strain softening phase, express the averaged strain $\bar{\varepsilon} = u/L$ as a function of the material properties E and H, as well as the number of sections n.
(b) Compute the energy $\mathcal{D}$ per unit volume that has been dissipated at State (IV).
(c) Regularize the problem by substituting H with the "section size-dependent" softening modulus $H_{\text{reg}} = H/n$. Compute then the regularized averaged strain $\bar{\varepsilon}_{\text{reg}}$ and the regularized dissipation $\mathcal{D}_{\text{reg}}$. ∎

A number of approaches have been reported to regularize the non-polar continuum towards preventing non-physical results in the description of strain softening materials. The gradient-enhanced damage model [425] would be one such concept. The evolution of the damage parameter D in a material point is then influenced by the deformation in the vicinity of the point, a non-local effect that may be incorporated by diffusion equations that are coupled to the other governing equations of the problem [415].

3.8.3 Linear Fracture Mechanics

In Linear Fracture Mechanics (LFM), the fracture toughness is expressed through the *stress intensity factor* K_i [Pa m$^{1/2}$], a theoretical construct to capture the stress

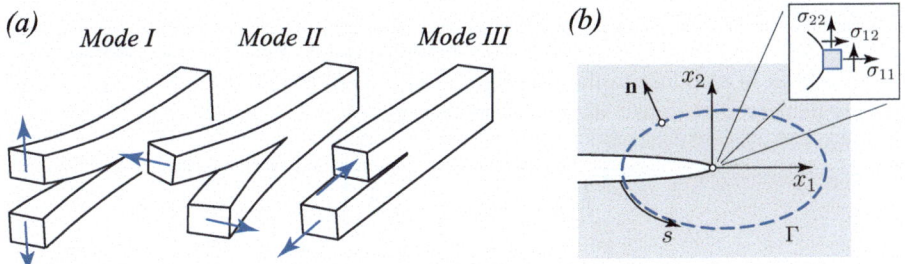

Fig. 3.52 Mechanical concepts used in the analysis of fracture problems. (**a**) Modes $i = I, II, III$ of the loading at the crack tip to define the stress intensity factors K_i. (**b**) Path Γ around a crack tip to compute the J-integral in 2D

near the crack tip. The subscript $i = I, II, III$ denotes the loading of the crack tip according to Fig. 3.52a. The corresponding stress intensity factors then read

$$K_I = \lim_{x_1 \to 0} \sqrt{2\pi x_1} \sigma_{22} \ ; \ \ K_{II} = \lim_{x_1 \to 0} \sqrt{2\pi x_1} \sigma_{12} \ ; \ \ K_{III} = \lim_{x_1 \to 0} \sqrt{2\pi x_1} \sigma_{23} \ ,$$

where x_1 and σ denote the distance from the crack tip and the stress at the crack tip, respectively. Figure 3.52b illustrates the stress at the crack tip in 2D towards the definition of these parameters.

Given a linear-elastic material (see Sect. 3.5.2) with the Young's modulus E and the Poisson's ratio v, the expression

$$\mathcal{D} = K_I^2 \left(\frac{1 - v^2}{E} \right) + K_{II}^2 \left(\frac{1 - v^2}{E} \right) + K_{III}^2 \left(\frac{1 + v}{E} \right)$$

relates the stress intensity factors K_I, K_{II}, K_{III} to the dissipation, or energy release rate $\mathcal{D}$ of the propagating fracture. The *energy release rate* $\mathcal{D}$ [J m^{-2}] represents the energy that is dissipated per unit of newly created fracture surface area. In addition to small strains, LFM requires the formation of a *sharp* crack tip, both of which limits the application of LFM in vascular biomechanics.

3.8.4 Non-linear Fracture Mechanics

The non-linear mechanical properties of vascular tissue and the formation of a blunted crack tip hinder the application of LFM. Despite failure mechanisms of vascular tissues are still poorly understood, a number of tools are available to analyze non-linear fracture problems—some of them are discussed below.

3.8.4.1 J-Integral

The *J-integral* represents a technique for the calculation of the energy release rate $\mathcal{D}$ of a material in response to the propagation of a fracture [81, 460]. Given the strain energy density $\psi = \int_0^{\varepsilon} \boldsymbol{\sigma} : \mathrm{d}\bar{\boldsymbol{\varepsilon}}$ of a solid, the J-integral in 2D reads

$$\mathcal{J} = \int_{\Gamma} \left(\psi \mathrm{d}x_2 - \mathbf{t} \cdot \frac{\partial \mathbf{u}}{\partial x_1} \mathrm{d}s \right) = \mathcal{D}\,, \tag{3.155}$$

where $\mathbf{n}$ denotes the normal to the path Γ encompassing the crack tip, see Fig. 3.52b. In addition, $\mathbf{u}$ and $\mathbf{t} = \boldsymbol{\sigma}\mathbf{n}$ denote the displacement and the Cauchy traction vector, respectively. The J-integral (3.155) is independent of the definition of Γ, as long as the path is closed [81, 460].

3.8.4.2 Cohesive Zone Modeling

Cohesive zone modeling introduces a *Traction Separation Law (TSL)* that serves as a surrogate measure of the failure process. The TSL defines the traction that acts in response to the opening of the crack faces, see Fig. 3.53. It may be derived from a cohesive potential that governs the material-dependent resistance against failure and thus specifies the properties of the cohesive zone *independently* from the properties of the bulk material.

We may introduce an isotropic cohesive potential $\Psi_{\mathrm{c}}(\mathbf{u}_{\mathrm{d}}, \zeta)$ per unit undeformed area of the failure surface [200, 413], where $\mathbf{u}_{\mathrm{d}}$ is the opening displacement, whilst ζ denotes a scalar internal variable that determines the state of damage. Coleman and Noll's procedure [97] allows us then to compute the cohesive traction $\mathbf{T} = \mathbf{PN}$ and the dissipation $\mathcal{D}$ according to

$$\mathbf{T} = \frac{\partial \Psi_{\mathrm{c}}}{\partial \mathbf{u}_{\mathrm{d}}} \quad ; \quad \mathcal{D} = -\frac{\partial \Psi_{\mathrm{c}}}{\partial \zeta}\dot{\zeta} \geq 0\,, \tag{3.156}$$

where $\mathbf{P}$ and $\mathbf{N}$ denote the first Piola–Kirchhoff stress and the normal to the failure zone in the reference configuration Ω_0, respectively. Cohesive zone modeling is a well-established concept [135] to describe failure in metals [26] and concrete [254], and it has also been used to describe the fracture of biological tissues [200, 233].

Fig. 3.53 Mechanical representation of a fracture by a cohesive zone model. A Traction Separation Law (TSL) serves as a surrogate description of the failure process and defines the traction $\mathbf{T}$ as a function of the opening of the crack faces $\mathbf{u}_{\mathrm{d}}$

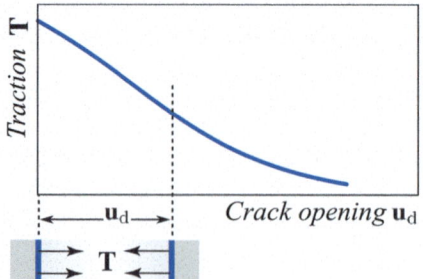

3.9   Summary and Conclusion

Continuum mechanics is a powerful tool in the investigation of biomechanical problems. It allows us to explore solid mechanical as well as fluid mechanical aspects of the vascular circulation. Factors, such as stress and strain, are not only of great importance to understand the physiology of the vascular system, but they are also common denominators of cardiovascular pathologies and thus factor of direct clinical relevance.

In continuum mechanics a solid or fluid is represented by an infinite number of material particles, each of which represents the material's internal structure through the RVE. The *stress* describes the mechanical loading of the RVE and may be regarded as the most fundamental quantity in continuum mechanics. It develops according to Cauchy's equation of motion and thus through Newton's laws of mechanics.

The material's *constitutive description* is needed to close the set of equations that describe a problem in continuum mechanics. It expresses the relation between stress and *strain* of the material particle. A constitutive description therefore represents a surrogate description of the RVE's mechanical properties and specifies how the material's internal structure interacts in the transmission of the mechanical load. It may explicitly represent such interactions, or simply using a phenomenological relation between the stress and deformation, eventually even based on Machine Learning (ML) technology. We may consider the deformation of a body as a thermodynamic process, an especially powerful concept in the description of finite deformation problems. Given the complexity of the vasculature, the related constitutive models are discussed in detail in the subsequent chapters.

The iBVP is the mathematical description of a problem in continuum mechanics and allows us to formulate many questions in biomechanics. The analytical solution of the iBVP is only possible for a very limited number of problems. The exploration of most vascular biomechanics problems therefore requires the numerical approximation of the iBVP. One such approach, the FEM, is discussed in the next chapter.

The Finite Element Method

4

This chapter introduces the linear and non-linear Finite Element Method (FEM). It begins with the spatial discretization of a continuum body, followed by the introduction of shape functions and gradient operators. We then discuss the Calculus of Variations, where problems such as diffusion, advection-diffusion, linear and non-linear solid mechanics, as well as incompressible flow are covered. The corresponding discretized sets of equations are provided in the next section. As vascular problems are inherently characterized by incompressible deformations, we therefore consider constraints at the finite element level, where Penalty, Lagrange and augmented-Lagrange methods are discussed. Next, the contributions of the individual finite elements are assembled into a global system of equations, and the stability and regularization of the discretized equations are explored. We then present a set of methods to solve the system of FEM equations- it covers the solution of sparse linear systems, non-linear systems, arc-length methods as well as explicit and implicit solution approaches. A number of case studies exemplify the solution of structural and fluid flow problems, and a summary concludes the chapter.

4.1 Introduction

The Finite Element Method (FEM) is the most widely used numerical method to solve partial differential equations (PDE) and thus also the first choice to compute solutions to problems in vascular biomechanics. A problem is subdivided into a large number of finite elements and then approximated through a discrete system of interconnected nodes. Time-marching or predictor–corrector methods

---

The original version of this chapter was revised: ESM has been added. The correction to this chapter is available at https://doi.org/10.1007/978-3-030-70966-2_8

Supplementary Information The online version contains supplementary material available at https://doi.org/10.1007/978-3-030-70966-2_4

© Springer Nature Switzerland AG 2021, corrected publication 2022
T. C. Gasser, *Vascular Biomechanics*, https://doi.org/10.1007/978-3-030-70966-2_4

are used to compute the independent nodal variables, such as the displacement **u** of a solid mechanical problem or the velocity **v** of a fluid mechanical problem. The FEM emerged from mathematical and engineering approaches, and the term "finite element" appears first in the 1960s [93]. Today excellent texts describe the FEM [30, 34, 53, 457, 617–619], and efficient commercial and open-source software support the solution of FEM problems for different applications.

Considerable insight in a problem, such as properties of the expected solution, can be gained by knowing the type of PDE that determines its physics. The classification of PDEs rests on whether lines or surfaces exist across which the derivation of the solution is discontinuous [34]. PDEs may be classified into three types:

- The *hyperbolic PDE* describes problems, such as wave propagation. Discontinuities in initial conditions propagate through the system. Given non-linear problems, even smooth data may localize towards shock waves.
- The *elliptic PDE* describes problems, such as the deformation of elastic bodies. The solutions are smooth, even for discontinues initial conditions. However, boundary data tend to affect the entire domain. A major difficulty is that acute corners in the boundary lead to singularities in the solution.
- The *parabolic PDE* describes problems, such as heat conduction and may be seen as somewhere in between the hyperbolic and elliptic problems.

This chapter aims at reviewing the foundation of the FEM towards targeted vascular biomechanics applications.

4.2   Spatial Discretization

The domain Ω is approximated by $\Omega^h = \sum_{i=1}^{n_e} \Omega_e$, which itself is split into n_e subdomains Ω_e, the so-called finite elements, see Fig. 4.1. The discretization of Ω^h allows us to focus on a single finite element Ω_e towards the implementation of the governing equations. Each finite element has n_{npe} nodes $\mathbf{x}_i$, and the element connectivity specifies their arrangement. This information is stored in the FEM data structure.

4.2.1   Shape Function

The n_{npe} finite element nodes store the *independent* problem variables, such as the displacements $\mathbf{u}_i$; $i = 1, \ldots, n_{\mathrm{npe}}$, of a structural mechanical problem. They are the *essential* variables of the FEM problem. The shape function $N_i(\boldsymbol{\xi})$ interpolates such variables over the finite element Ω_e, and the expression

$$\mathbf{u}(\boldsymbol{\xi}) = \sum_{i=1}^{n_{\mathrm{npe}}} N_i(\boldsymbol{\xi})\mathbf{u}_i \tag{4.1}$$

then specifies their value at the natural coordinate $\boldsymbol{\xi}$ inside the finite element.

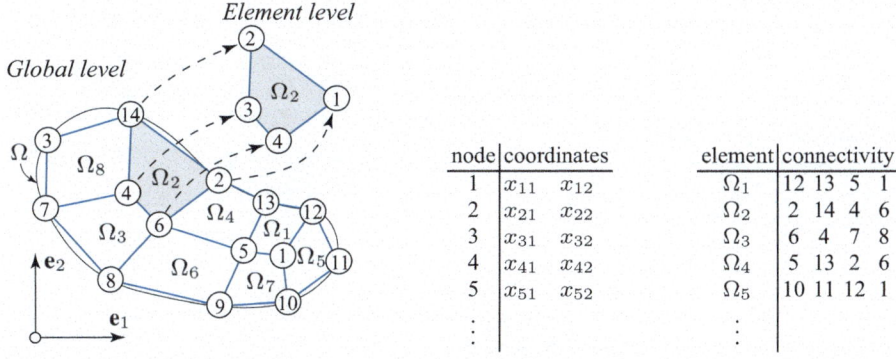

Fig. 4.1 Spatial discretization of the 2D domain Ω^{h} through the nodes $\mathbf{x}_i$; $i = 1, \ldots, 14$ and the quadrilateral finite elements Ω_e; $e = 1, \ldots, 8$. (left) The global node numbering $i = 1, \ldots, 14$ describes the global problem, whilst the local node numbering $i = 1, 2, 3, 4$ corresponds to the level of the finite element. (right) The FEM data structure stores the nodal coordinates and the element connectivity

We may also introduce $\mathbf{h}$, a n_{dof}-dimensional vector that stores the essential variables of one finite element. Given a continuum finite element with n_{dim} denoting the problem's number of dimensions, the vector $\mathbf{h}$ has the dimension $n_{\mathrm{dof}} = n_{\mathrm{npe}} n_{\mathrm{dim}}$. The interpolation (4.1) then reads

$$\mathbf{u} = \mathbf{N}\mathbf{h} \; ; \quad u_i = N_{ia}h_a \, , \tag{4.2}$$

where $\mathbf{N}$ is the $n_{\mathrm{dim}} \times n_{\mathrm{dof}}$-dimensional *interpolation matrix*. It reads

$$\mathbf{N} = \begin{bmatrix} N_1 & 0 & 0 & \ldots & N_{n_{\mathrm{npe}}} & 0 & 0 \\ 0 & N_1 & 0 & \ldots & 0 & N_{n_{\mathrm{npe}}} & 0 \\ 0 & 0 & N_1 & \ldots & 0 & 0 & N_{n_{\mathrm{npe}}} \end{bmatrix}$$

for a 3D continuum problem, whilst $\mathbf{N} = \begin{bmatrix} N_1 \ldots N_{n_{\mathrm{npe}}} \end{bmatrix}$ interpolates scalar essential variables.

Given an *isoparametric* finite element formulation, the interpolation of the spatial coordinate $\mathbf{x}$ uses the same shape functions as used to interpolate $\mathbf{u}$, and thus $\mathbf{x}(\boldsymbol{\xi}) = \sum_{i=1}^{n_{\mathrm{npe}}} N_i(\boldsymbol{\xi})\mathbf{x}_i$.

The shape functions are often defined within the finite element's natural coordinate $\boldsymbol{\xi}$, a domain linked to the finite element's global coordinate $\mathbf{x}$ through the *Jacobian transformation* $\mathbf{J} = \partial\mathbf{x}/\partial\boldsymbol{\xi}$. Whilst the introduction of the natural coordinate $\boldsymbol{\xi}$ is not necessary, it allows for an easier design of shape functions. Figure 4.2 illustrates the mapping of a quadrilateral finite element between its physical domain, described by the global coordinate $\mathbf{x}$, and its parent domain, the bi-square ($-1 \leq \xi_1 \leq 1$; $-1 \leq \xi_2 \leq 1$) and described by the natural coordinate $\boldsymbol{\xi} = [\xi_1 \ \xi_2]^{\mathrm{T}}$.

Fig. 4.2 Jacobian transformation $\mathbf{J} = \partial\mathbf{x}/\partial\boldsymbol{\xi}$ that maps the finite element from its physical domain into the parent domain. The example shows a quadrilateral finite element of nodes $\mathbf{x}_i$; $i = 1, 2, 3, 4$, that then transforms into the bi-square $(-1 \leq \xi_1 \leq 1; -1 \leq \xi_2 \leq 1)$

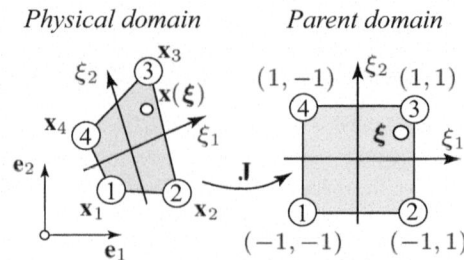

For the FEM to yield numerically stable and plausible results, the shape functions have to guarantee at least three properties:

- *Property of unity,* such that $\sum_{i=1}^{n_{\text{npe}}} N_i(\boldsymbol{\xi}) = 1$ holds for all $\boldsymbol{\xi}$
- *Delta property,* such that $N_i(\boldsymbol{\xi}_j) = \delta_{ij}$ holds, where $\boldsymbol{\xi}_j$ denotes the natural coordinates of the j-th finite element node, and δ_{ij} is the Kronecker delta
- *Linear independence,* such that $\sum_{i=1}^{n_{\text{npe}}} c_i N_i(\boldsymbol{\xi}) \neq 0$ holds for the non-zero coefficients c_i

Whilst these requirements are sufficient for continuum finite elements, structural finite elements, such as beams and plates, require addition properties of the shape functions to be satisfied [618]. Given FEM problems with moving meshes, such as finite deformation problems in Lagrange formulation, the shape functions are functions of the essential problem variables.

4.2.1.1 Shape Functions for 1D Problems
Given a two-noded line element that covers the domain $-1 \leq \xi \leq 1$ of the natural coordinate, the corresponding linear shape functions read

$$N_1(\xi) = (1 - \xi)/2 \; ; \; N_2(\xi) = (1 + \xi)/2, \tag{4.3}$$

which is illustrated in Fig. 4.3a. We may add a mid-point node, which then leads to a three-noded finite element and the quadratic shape functions

$$N_1(\xi) = (\xi^2 - \xi)/2 \; ; \; N_2(\xi) = (\xi^2 + \xi)/2 \; ; \; N_3(\xi) = 1 - \xi^2 \tag{4.4}$$

shown in Fig. 4.3b. It can easily be shown that the polynoms (4.3) and (4.4) satisfy the aforementioned requirements of shape functions.

4.2.1.2 Shape Functions for 2D Problems
Given the natural coordinates $-1 \leq \xi_1 \leq 1$ and $-1 \leq \xi_2 \leq 1$ of a quadrilateral finite element, 2D shape functions may be designed through the generalization of 1D shape functions. The bi-linear shape functions

$$N_1(\xi_1, \xi_2) = \frac{1}{4}(1 - \xi_1)(1 - \xi_2) \; ; \; N_2(\xi_1, \xi_2) = \frac{1}{4}(1 + \xi_1)(1 - \xi_2) \; ;$$

$$N_3(\xi_1, \xi_2) = \frac{1}{4}(1 + \xi_1)(1 + \xi_2) \; ; \; N_4(\xi_1, \xi_2) = \frac{1}{4}(1 - \xi_1)(1 + \xi_2) \tag{4.5}$$

Fig. 4.3 Shape functions for 1D finite elements plotted at their parent domain $-1 \leq \xi \leq 1$. (**a**) Two-noded finite element with linear shape functions $N_i(\xi)$; $i = 1, 2$. (**b**) Three-noded finite element with quadratic shape functions $N_i(\xi)$; $i = 1, 2, 3$

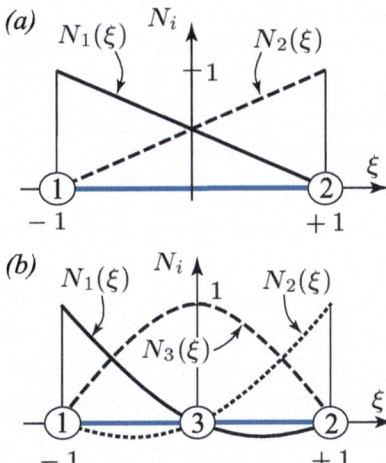

specify the interpolation of field variables, where the node numbering corresponds to the parent domain shown in Fig. 4.2.

An alternative representation of the bi-linear shape functions separates the constant, linear, and non-linear terms according to

$$
\begin{bmatrix} N_1 \\ N_2 \\ N_3 \\ N_4 \end{bmatrix} = \frac{1}{4} \begin{bmatrix} 1 \\ 1 \\ 1 \\ 1 \end{bmatrix} + \frac{1}{4} \begin{bmatrix} -1 \\ 1 \\ 1 \\ -1 \end{bmatrix} \xi_1 + \frac{1}{4} \begin{bmatrix} -1 \\ -1 \\ 1 \\ 1 \end{bmatrix} \xi_2 + \frac{1}{4} \begin{bmatrix} 1 \\ -1 \\ 1 \\ -1 \end{bmatrix} \xi_1 \xi_2 ,
$$

where the last term, the so-called *hourglass* term, hosts the non-linear contributions.

Whilst quadrilateral elements perform very good and are widely used, the generation of quadrilateral meshes may be challenging. Triangular elements often ease mesh generation and support straightforward *local mesh refinement*—one triangle is simply split into three sub-triangles. In contrary, quadrilaterals do not support local mesh refinement and mesh refinement always require complex *non-local* mesh manipulations.

The unit rectangular triangle is the parent domain of triangular finite elements, and the shape functions then read

$$
N_1 = \xi_1 \; ; \quad N_2 = \xi_2 \; ; \quad N_3 = 1 - \xi_1 - \xi_2 .
$$

They may also be interpreted as area coordinates, $N_i = A_i/A$; $i = 1, 2, 3$, where A_i denote the sub-areas, whilst $A = \sum_{i=1}^3 A_i$ is the area of the triangular finite element, see Fig. 4.4.

Whilst shape functions are often based on polynomial expressions, any other function may be used that satisfies the aforementioned requirements of shape

Fig. 4.4 Jacobian transformation **J** that maps a triangular finite element from its physical domain to its parent domain. The shape function may be seen as area coordinates

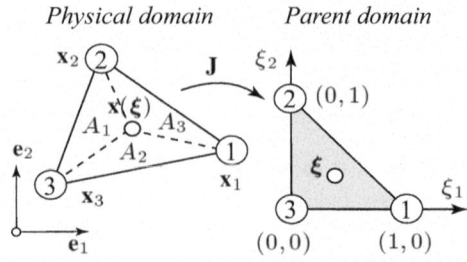

functions. The literature is rich on shape functions, and some of them are tailored to the solution of specific problems [619]. Even Non-Uniform Rational Basis Spline (NURBS) may be used to build finite elements, an approach known as isogeometric analysis [269]. In addition, the Virtual Element Method (VEM) does not explicitly introduce shape functions and allows to use arbitrarily shaped finite elements [12].

Example 4.1 (Transformation of the Quadrilateral Finite Element). The nodal coordinates

$$\begin{bmatrix} x_1 \\ y_1 \end{bmatrix} = \begin{bmatrix} 0.0 \\ 0.0 \end{bmatrix} \; ; \; \begin{bmatrix} x_2 \\ y_2 \end{bmatrix} = \begin{bmatrix} 3.5 \\ 0.5 \end{bmatrix} \; ; \; \begin{bmatrix} x_3 \\ y_3 \end{bmatrix} = \begin{bmatrix} 3.0 \\ 2.3 \end{bmatrix} \; ; \; \begin{bmatrix} x_4 \\ y_4 \end{bmatrix} = \begin{bmatrix} -0.5 \\ 2.4 \end{bmatrix} ,$$

specify a quadrilateral finite element with the dimensions given in centimeters.

(a) Provide the expressions for the interpolation of the global coordinates $\mathbf{x} = [x \; y]^\mathrm{T}$.
(b) Compute the Jacobian transformation matrix $\mathbf{J} = \partial \mathbf{x}/\partial \boldsymbol{\xi}$ that maps the finite element to its parent domain $-1 \le \xi \le 1$ and $-1 \le \eta \le 1$ with $\boldsymbol{\xi} = [\xi \; \eta]^\mathrm{T}$.
(c) Show that the expression $A = 4\det \mathbf{J}(\xi = 0, \eta = 0)$ denotes the area of the finite element. ∎

4.2.1.3 Shape Functions for 3D Problems

The extension of the quadrilateral finite element to the eight-noded hexahedral finite element yields the tri-linear shape functions

$$N_1 = \frac{1}{8}(1 - \xi_1)(1 - \xi_2)(1 - \xi_3) \; ; \; N_2 = \frac{1}{8}(1 + \xi_1)(1 - \xi_2)(1 - \xi_3) ;$$

$$N_3 = \frac{1}{8}(1 + \xi_1)(1 + \xi_2)(1 - \xi_3) \; ; \; N_4 = \frac{1}{8}(1 - \xi_1)(1 + \xi_2)(1 - \xi_3) ;$$

$$N_5 = \frac{1}{8}(1 - \xi_1)(1 - \xi_2)(1 + \xi_3) \; ; \; N_6 = \frac{1}{8}(1 + \xi_1)(1 - \xi_2)(1 + \xi_3) ;$$

$$N_7 = \frac{1}{8}(1 + \xi_1)(1 + \xi_2)(1 + \xi_3) \; ; \; N_8 = \frac{1}{8}(1 - \xi_1)(1 + \xi_2)(1 + \xi_3) ,$$

whilst the four-noded linear tetrahedral finite element is described by the shape functions

$$N_1 = \xi_1 \; ; \quad N_2 = \xi_2 \; ; \quad N_3 = \xi_3 \; ; \quad N_4 = 1 - \xi_1 - \xi_2 - \xi_3 \,.$$

Many other shape functions have been proposed and successfully used in the solution of FEM problems, see amongst others [619].

4.2.2   Gradient Interpolation

Given the interpolation (4.1), the symmetric spatial gradient of the essential variable reads

$$\text{grad}_s \mathbf{u}(\boldsymbol{\xi}) = \frac{1}{2} \sum_{i=1}^{n_{npe}} \mathbf{u}_i \otimes \text{grad} N_i(\boldsymbol{\xi}) + \text{grad} N_i(\boldsymbol{\xi}) \otimes \mathbf{u}_i \,. \tag{4.6}$$

It is convenient to introduce *Voigt notation* and to express the n_s independent components of $\text{grad}_s \mathbf{u}$ through a vector $\text{grad}_s^v \mathbf{u}$. Given a continuum finite element in 1D, 2D, and 3D, $n_s = 1, n_s = 4$ and $n_s = 6$, respectively. The gradient interpolation then reads

$$\text{grad}_s^v \mathbf{u} = \mathbf{B} \mathbf{h} \,, \tag{4.7}$$

where the *gradient interpolation matrix* $\mathbf{B}$ is a $n_s \times n_{dof} n_{npe}$-dimensional matrix that stores the spatial gradients of the shape functions. It reads

$$\mathbf{B} = \begin{bmatrix} \partial N_1/\partial x_1 & 0 & 0 & \ldots & \partial N_{n_{npe}}/\partial x_1 & 0 & 0 \\ 0 & \partial N_1/\partial x_2 & 0 & \ldots & 0 & \partial N_{n_{npe}}/\partial x_2 & 0 \\ 0 & 0 & \partial N_1/\partial x_3 & \ldots & 0 & 0 & \partial N_{n_{npe}}/\partial x_3 \\ \partial N_1/\partial x_2 & \partial N_1/\partial x_1 & 0 & \ldots & \partial N_{n_{npe}}/\partial x_2 & \partial N_{n_{npe}}/\partial x_1 & 0 \\ 0 & \partial N_1/\partial x_3 & \partial N_1/\partial x_2 & \ldots & 0 & \partial N_{n_{npe}}/\partial x_3 & \partial N_{n_{npe}}/\partial x_2 \\ \partial N_1/\partial x_3 & 0 & \partial N_1/\partial x_1 & \ldots & \partial N_{n_{npe}}/\partial x_3 & 0 & \partial N_{n_{npe}}/\partial x_1 \end{bmatrix}$$

for a 3D continuum problem, and we used the Voigt notation with the storage convention $[(\bullet)_{11} \; (\bullet)_{22} \; (\bullet)_{33} \; (\bullet)_{12} \; (\bullet)_{23} \; (\bullet)_{13}]^{\mathsf{T}}$. To interpolate the gradient of a scalar essential variable over the finite element, the gradient interpolation matrix in 3D reads

$$\mathbf{B} = \begin{bmatrix} \partial N_1/\partial x_1 & \ldots & \partial N_{n_{npe}}/\partial x_1 \\ \partial N_1/\partial x_2 & \ldots & \partial N_{n_{npe}}/\partial x_2 \\ \partial N_1/\partial x_3 & \ldots & \partial N_{n_{npe}}/\partial x_3 \end{bmatrix} \,.$$

The shape functions $N_i(\boldsymbol{\xi})$ are functions of the natural coordinate $\boldsymbol{\xi}$, and the computation of the gradient with respect to the global coordinate $\mathbf{x}$ requires the Jacobian matrix $\mathbf{J} = \partial \mathbf{x}/\partial \boldsymbol{\xi}$. Given its inverse $\mathbf{J}^{-1} = \partial \boldsymbol{\xi}/\partial \mathbf{x}$ exists, the expression

$$\frac{\partial N_i(\boldsymbol{\xi})}{\partial \mathbf{x}} = \frac{\partial N_i(\boldsymbol{\xi})}{\partial \boldsymbol{\xi}} : \frac{\partial \boldsymbol{\xi}}{\partial \mathbf{x}} = \frac{\partial N_i(\boldsymbol{\xi})}{\partial \boldsymbol{\xi}} : \mathbf{J}^{-1} \ ; \quad i = 1, \ldots, n_{\text{npe}}$$

allows us then to compute the coefficients of $\mathbf{B}$.

Finally, the divergence of a vector $\mathbf{u}$ may be interpolated by $\text{div}\,\mathbf{u} = \mathbf{G} \cdot \mathbf{h}$, where the n_{dof}-dimensional divergence interpolation matrix $\mathbf{G}$ has been introduced. It reads

$$\mathbf{G} = \left[\partial N_1/\partial x_1 \ \partial N_1/\partial x_2 \ \partial N_1/\partial x_3 \ \ldots \ \partial N_{n_{\text{npe}}}/\partial x_1 \ \partial N_{n_{\text{npe}}}/\partial x_2 \ \partial N_{n_{\text{npe}}}/\partial x_3 \right]$$

for a 3D continuum problem.

Example 4.2 (Displacement and Strain Interpolation). A quadrilateral finite element with the bi-linear shape function (4.5) approximates the displacement $\mathbf{u}$ of a solid mechanical problem.

(a) Assemble the displacement interpolation matrix $\mathbf{N}$.
(b) Consider the definition of the engineering strain $\boldsymbol{\varepsilon} = (\text{grad}\,\mathbf{u} + \text{grad}^{\text{T}}\mathbf{u})/2$ and derive the strain interpolation matrix $\mathbf{B}$, such that $\boldsymbol{\varepsilon}^{\text{v}} = \mathbf{B}\mathbf{h}$ holds.
(c) Given the finite element's global and natural coordinates are related by the Jacobian matrix

$$\mathbf{J} = \begin{bmatrix} 1.75 & -0.25 \\ 0.1 - 0.15\eta & 1.05 - 0.15\xi \end{bmatrix},$$

compute the numerical expressions of the matrices $\mathbf{N}$ and $\mathbf{B}$ at the natural coordinates $\xi = 0.3$ and $\eta = -0.1$.   ∎

4.2.3   Mixed and Hybrid Finite Elements

The interaction of physical processes often determines a biomechanical problem, where each process is described by an individual set of essential variables. The flow of an incompressible fluid would be such an example—the flow velocity $\mathbf{v}$ and the fluid pressure p are then essential problem variables. The velocity and pressure could then be stored at the finite element nodes by the two vectors $\mathbf{h}$ and $\mathbf{q}$ of the essential problem variables, respectively. The both fields may be interpolated at different interpolation orders, which then leads to a *hybrid* finite element, see Fig. 4.5.

Given a constraint problem, such as the incompressible deformation of the vessel wall, an *irreducible* finite element approach may show volume locking—the discretized problem is then (much) stiffer than the continuum problem. The irreducible FEM uses the minimum number and thus the irreducible set of problem variables to interpolate the solution over the finite element. In contrary, a *mixed*

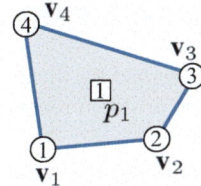

Fig. 4.5 Hybrid finite element with a bi-linear interpolation of the velocity **v**, and a constant interpolation of the pressure p. The velocity **v** is stored at the four vertex nodes, whilst the pressure p is stored at a single mid-point node

finite element approach introduces additional variables, more than necessary, to interpolate the solution over the finite element. At some points over the finite element domain, the solution fields are then connected through the problem's constraint equations. The mixed finite element also leads to a hybrid finite element, given the essential variables are prescribed at different interpolation orders.

4.3   Calculus of Variations

The description of a physical problem is often given by *local* governing equations, which then directly yields the *strong form* of the problem description. Whilst some numerical concepts, such as the Finite Difference Method (FDM), directly use the strong form, the FEM requires the *weak or integral form* of the problem description. The weak form of some engineering problems may be derived through concepts, such as the Principle of Virtual Work (PVW). However, such physics-motivated principles are not available for all engineering problems, and the more general machinery of *Calculus of Variations* allows then the derivation of the weak form from any strong form description of a problem. Calculus of Variations is a method that multiplies the governing equations with a *test function*, also known as *admissible variation*. The result is then integrated over the body's spatial domain, and the Dirichlet boundary condition is embedded—the problem's *weak form* has now been derived.

Given the strong form description of a problem has a solution, it is identical to the solution of the weak form problem description, a statement known as the *Fundamental Lemma of Calculus of Variations*. The weak form is also known as the *irreducible* description, and no additional reduction of the order of differentiation of the essential variable is therefore possible. The accuracy of a numerical method is directly compromised by the number of differentiations of the essential variables, whilst an integration over the variables does not contribute (much) to the numerical error.

4.3.1   Diffusion Boundary Value Problem

The diffusion of a substance is described by the *Poisson*[1] *equation*, and the strong form description of such a transport problem reads

$$
\left.
\begin{aligned}
\mathrm{div}(\mathrm{grad}c) + \alpha &= 0 \text{ in } \Omega, \\
c &= \bar{c} \text{ at } \partial\Omega_c, \\
\mathrm{grad}c \cdot \mathbf{n} &= \bar{q} \text{ at } \partial\Omega_q,
\end{aligned}
\right\}
\tag{4.8}
$$

where the concentration c is the *independent* or *essential* problem variable. The particular physical properties determine the constant α. Figure 4.6 illustrates the Boundary Value Problem (BVP) that describes the diffusion of a substance. The *Dirichlet* and *Neumann* boundary conditions are formulated along $\partial\Omega_c$ and $\partial\Omega_q$, respectively. Along $\partial\Omega_c$ the value of c is set to $\bar{c}$, whilst along $\partial\Omega_q$ the normal flux $q = \mathrm{grad}c \cdot \mathbf{n}$ is set to $\bar{q}$, where $\mathbf{n}$ denotes the outward-oriented normal vector to the boundary $\partial\Omega$.

Following Calculus of Variations, we first introduce a continues *test function δc*, also known as admissible variation of the essential variable c. The test function δc is arbitrary but vanishes along the Dirichlet boundary, and thus $\delta c = 0$ holds at $\partial\Omega_c$. The local governing equation $\mathrm{div}(\mathrm{grad}c) + \alpha = 0$ is then multiplied with δc and integrated over Ω, which yields

$$
\int_\Omega \delta c \, \mathrm{div}(\mathrm{grad}c) \mathrm{d}v = -\int_\Omega \delta c \, \alpha \mathrm{d}v \,.
$$

Integration by parts of the term at the left-hand side yields

$$
\int_\Omega \delta c \, \mathrm{div}(\mathrm{grad}c) \mathrm{d}v = \int_\Omega \mathrm{div}\left(\delta c \, \mathrm{grad}c\right) \mathrm{d}v - \int_\Omega \mathrm{grad}\delta c \cdot \mathrm{grad}c \, \mathrm{d}v,
$$

Fig. 4.6 Boundary Value Problem (BVP) that describes the diffusion of a substance, and thus the distribution of its concentration $c(\mathbf{x})$ within the body Ω. The normal flux $\mathrm{grad}c(\mathbf{x}) \cdot \mathbf{n} = \bar{q}$ is prescribed along the Neumann boundary $\partial\Omega_q$, whilst the concentration $c(\mathbf{x}) = \bar{c}$ is prescribed along the Dirichlet boundary $\partial\Omega_c$

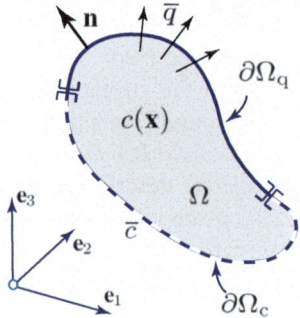

---

[1] Siméon Denis Poisson, French mathematician, engineer, and physicist, 1781–1840.

which, with the divergence theorem $\int_\Omega \text{div}\,(\delta c\,\text{grad}c)\,dv = \int_{\partial\Omega} \delta c\,\text{grad}\cdot\mathbf{n}ds = \int_{\partial\Omega} \delta c\,\overline{q}ds$, then results in the diffusion problem's weak form

$$\int_\Omega \text{grad}\delta c \cdot \text{grad}c\,dv = \int_{\partial\Omega_q} \delta c\,\overline{q}ds + \int_\Omega \delta c\,\alpha dv\,, \qquad (4.9)$$

where the condition $\delta c = 0$ at $\partial\Omega_c$ has been used. The weak form problem description (4.9) is irreducible and involves only the first derivative, whilst the strong form (4.8) requires the second derivative of c.

4.3.2   Advection–Diffusion Boundary Value Problem

A more general transport is described by the Advection–Diffusion (AD) problem. For simplicity we consider the steady-state AD problem, which strong form reads

$$\left.\begin{aligned} \mathbf{v}\cdot\text{grad}c - \nu\text{div}(\text{grad}c) + \alpha = 0 \text{ in } \Omega\,, \\ c = \overline{c} \text{ at } \partial\Omega_c\,, \\ \text{grad}c \cdot \mathbf{n} = \overline{q} \text{ at } \partial\Omega_q\,, \end{aligned}\right\} \qquad (4.10)$$

where c is the essential problem variable—it could be the concentration of a substance. In addition, ν describes the diffusivity of the substance within the body Ω, whilst $\mathbf{v}$ denotes the adventive velocity and thus the velocity of the medium that advects the substance. Dirichlet and Neumann boundary conditions are formulated along $\partial\Omega_c$ and $\partial\Omega_q$, respectively.

Calculus of Variations and thus the aforementioned manipulations lead to the weak form

$$\int_\Omega \delta c\,\mathbf{v}\cdot\text{grad}c\,dv + \int_\Omega \nu\,\text{grad}\delta c \cdot \text{grad}c\,dv = \int_{\partial\Omega_q} \delta c\,\overline{q}ds + \int_\Omega \delta c\alpha dv \qquad (4.11)$$

of the AD problem, where δc denotes the test function.

Example 4.3 (Heat Conduction Problem). Given a body of the density $\rho\,[\text{kg}\,\text{m}^{-3}]$ and the specific heat $c\,[\text{J}\,\text{kg}^{-1}\,\text{K}^{-1}]$, the partial differential equation

$$-\text{div}\mathbf{q} + r = \rho c\frac{\partial\theta}{\partial t} \qquad (4.12)$$

determines a transient heat conduction problem and describes how the temperature θ changes over the time t within the domain Ω. Here, $\text{div}(\bullet)$ is the spatial divergence operator, whilst $\mathbf{q}\,[\text{W}\,\text{m}^{-2}]$ and $r\,[\text{W}\,\text{m}^{-3}]$ denote the heat flux and the heat source, respectively. The partial differential equation (4.12) is complemented by the

Dirichlet boundary condition $\theta = \bar{\theta}$ at $\partial\Omega_\theta$ and the Neumann boundary condition $\mathbf{q} \cdot \mathbf{n} = \bar{q}$ at $\partial\Omega_q$, where $\mathbf{n}$ denotes the outward normal vector to the boundary $\partial\Omega$.

(a) Multiply the transient heat conduction problem (4.12) with the admissible variation of the temperature $\delta\theta$ and integrate then over the domain Ω.
(b) Embedded the Dirichlet boundary condition and use Fourier's law $q_i = -k\partial\theta/\partial x_i$ towards the derivation of the weak form of the transient heat conduction problem. Here, k [W m^{-1} K^{-1}] denotes the thermal conductivity of the isotropic thermal body. ∎

4.3.3   Linear Solid Mechanics Boundary Value Problem

We consider a solid body that occupies the spatial configuration Ω, and the displacement $\mathbf{u}$ is the essential variable in the description of the problem. The displacement $\mathbf{u}$, the velocity $\dot{\mathbf{u}}$, and the acceleration $\ddot{\mathbf{u}}$ then determine the state of motion. Given a linear-elastic isothermal body, the Cauchy stress reads $\boldsymbol{\sigma}(\mathbf{u}) = \mathbb{c} : \boldsymbol{\varepsilon}(\mathbf{u})$, where $\mathbb{c}$ and $\boldsymbol{\varepsilon}(\mathbf{u}) = \mathrm{grad}_s\mathbf{u}$ denote the elasticity tensor and the engineering strain, respectively. At the Dirichlet boundary $\partial\Omega_u$ the displacement $\mathbf{u} = \bar{\mathbf{u}}$ and at the Neumann boundary $\partial\Omega_t$ the traction $\mathbf{t} = \boldsymbol{\sigma}\mathbf{n} = \bar{\mathbf{t}}$ are prescribed. The outward-oriented normal vector to $\partial\Omega_t$ is denoted by $\mathbf{n}$. Given a transient problem, the Cauchy's momentum equation $\mathrm{div}\boldsymbol{\sigma} + \mathbf{b}_f = \rho\ddot{\mathbf{u}} + \eta\dot{\mathbf{u}}$ describes the motion of the particle, where $\mathbf{b}_f$ [N m^{-3}] denotes the body force per unit volume, whilst ρ [kg m^{-3}] is the density. In addition, η [N s m^{-4}] denotes a friction-like resistance per unit volume. The *strong form* of the BVP then reads

$$\left.\begin{array}{r} \mathrm{div}\boldsymbol{\sigma} + \mathbf{b}_f - \rho\ddot{\mathbf{u}} - \eta\dot{\mathbf{u}} = \mathbf{0} \text{ in } \Omega\,, \\[4pt] \mathbf{u} = \bar{\mathbf{u}} \text{ at } \partial\Omega_u\,, \\[4pt] \mathbf{t} = \boldsymbol{\sigma}\mathbf{n} = \bar{\mathbf{t}} \text{ at } \partial\Omega_t\,. \end{array}\right\} \qquad (4.15)$$

It is complemented with the initial conditions for the displacement $\mathbf{u} = \mathbf{u}_0$ and the velocity $\dot{\mathbf{u}} = \dot{\mathbf{u}}_0$ that describe the motion of the body at the time $t = t_0$, and thus at the beginning of the analysis. The set of Eqs. (4.15), together with the initial conditions, describes the *initial BVP* (iBVP).

Following Calculus of Variations, the Cauchy's momentum equation $\mathrm{div}\boldsymbol{\sigma} + \mathbf{b}_f - \rho\ddot{\mathbf{u}} - \eta\dot{\mathbf{u}} = \mathbf{0}$ is multiplied with $\delta\mathbf{u}$ and integrated over Ω, which then yields

$$\int_\Omega \delta\mathbf{u} \cdot (\mathrm{div}\boldsymbol{\sigma} + \mathbf{b}_f - \rho\ddot{\mathbf{u}} - \eta\dot{\mathbf{u}})\mathrm{d}v = 0\,, \qquad (4.16)$$

where the test function $\delta\mathbf{u}$ satisfies the condition $\delta\mathbf{u} = \mathbf{0}$ at $\partial\Omega_u$. Integration by parts

$$\int_\Omega \delta\mathbf{u} \cdot \mathrm{div}\boldsymbol{\sigma} \, \mathrm{d}v = \int_\Omega \mathrm{div}(\boldsymbol{\sigma}\,\delta\mathbf{u})\mathrm{d}v - \int_\Omega (\boldsymbol{\sigma} : \mathrm{grad}\delta\mathbf{u}) \, \mathrm{d}v$$

and the divergence theorem $\int_\Omega \mathrm{div}(\sigma\,\delta\mathbf{u})\mathrm{d}v = \int_{\partial\Omega} \delta\mathbf{u} \cdot \sigma\mathbf{n}\,\mathrm{d}s$ allow us to express (4.16) through

$$\int_{\partial\Omega} \delta\mathbf{u} \cdot \sigma\mathbf{n}\,\mathrm{d}s + \int_\Omega \left[\delta\mathbf{u} \cdot (\mathbf{b}_\mathrm{f} - \rho\ddot{\mathbf{u}} - \eta\dot{\mathbf{u}}) - \sigma : \mathrm{grad}_\mathrm{s}\delta\mathbf{u} \right] \mathrm{d}v = 0 \,,$$

where the symmetry $\sigma = \sigma^\mathrm{T}$ of the Cauchy stress has been used. With the Cauchy stress theorem $\mathbf{t} = \sigma\mathbf{n} = \bar{\mathbf{t}}$ and the Dirichlet condition $\delta\mathbf{u} = \mathbf{0}$ at $\partial\Omega_\mathrm{u}$, the iBVP (4.15) may be expressed by its weak form

$$\underbrace{\int_{\partial\Omega_\mathrm{t}} \bar{\mathbf{t}} \cdot \delta\mathbf{u}\mathrm{d}s + \int_\Omega \delta\mathbf{u} \cdot (\mathbf{b}_\mathrm{f} - \rho\ddot{\mathbf{u}} - \eta\dot{\mathbf{u}})\mathrm{d}v}_{\delta\Pi_\mathrm{ext}} - \underbrace{\int_\Omega \sigma : \mathrm{grad}_\mathrm{s}\delta\mathbf{u}\,\mathrm{d}v}_{\delta\Pi_\mathrm{int}} = 0 \,, \qquad (4.17)$$

where $\delta\Pi_\mathrm{ext}$ and $\delta\Pi_\mathrm{int}$ denote the external and internal work upon the variation $\delta\mathbf{u}$. Recalling the definition of the admissible engineering strain $\delta\boldsymbol{\varepsilon} = \mathrm{grad}_\mathrm{s}\delta\mathbf{u}$, the weak form (4.17) is identical to the PVW, see Sect. 3.7.4. It is irreducible and involves only the first derivative of the displacement, whilst the term $\mathrm{div}\sigma$ that appears in the strong form (4.15) involves the second derivative of $\mathbf{u}$.

4.3.4 Non-linear Solid Mechanics Boundary Value Problem

Large deformations, a non-linear constitutive model, and non-constant external forces lead to a non-linear problem in solid mechanics. Equation (4.17) is then to be generalized. Let us for simplicity consider a body at so-called dead loads, the external loads $\mathbf{t} = \mathbf{t}_0$ and $\mathbf{b}_\mathrm{f} = \mathbf{b}_{\mathrm{f}0}$ are then constant, and

$$\delta\Pi_\mathrm{int} = \int_\Omega \delta\mathbf{e} : \sigma\,\mathrm{d}v$$

is the only non-linear term in the generalized version of (4.17), where the variation of the Euler–Almansi strain $\delta\mathbf{e} = \mathrm{grad}_\mathrm{s}\delta\mathbf{u}$ has been introduced. It is convenient to express $\delta\Pi_\mathrm{int}$ with respect to the non-moving reference configuration Ω_0,

$$\delta\Pi_\mathrm{int} = \int_\Omega \delta\mathbf{e} : \sigma\,\mathrm{d}v = \int_{\Omega_0} \delta\mathbf{E} : \mathbf{S}\,\mathrm{d}V \,, \qquad (4.18)$$

where $\mathbf{S} = J\mathbf{F}^{-1}\sigma\mathbf{F}^{-\mathrm{T}}$ denotes the second Piola–Kirchhoff stress and $\delta\mathbf{E} = \mathrm{sym}(\mathbf{F}^\mathrm{T}\,\mathrm{Grad}\delta\mathbf{u})$ is the variation of the Green–Lagrange strain. The variation of kinematic quantities is listed in Appendix D.

Given the fixed configuration Ω_0, expression $(4.18)_2$ simplifies the derivation of the *directional derivative* $D_\mathbf{u}\delta\Pi_\mathrm{int} = \delta\Pi_\mathrm{int}(\mathbf{u}+\Delta\mathbf{u}) - \delta\Pi_\mathrm{int}(\mathbf{u})$ along the increment $\Delta\mathbf{u}$ of the displacement $\mathbf{u}$. The linearization in the reference configuration may be pushed forward to the spatial configuration Ω. It then yields

$$D_{\mathbf{u}}\delta\Pi_{\text{int}} = \int_{\Omega} \left(\text{grad}_s\delta\mathbf{u} : \text{grad}_s\Delta\mathbf{u}\boldsymbol{\sigma} + \text{grad}_s\delta\mathbf{u} : \mathbb{c} : \text{grad}_s\Delta\mathbf{u}\right)dv, \qquad (4.19)$$

where the spatial elasticity $\mathbb{c}$ expresses the relation between the stress and the strain according to $\Delta\boldsymbol{\sigma} = \mathbb{c} : \text{grad}_s\Delta\mathbf{u}$. Example 4.4 further details this derivation.

Given the linearization (4.19), the *geometrical* contribution

$$D_{\mathbf{u}}\delta\Pi_{\text{int geo}} = \int_{\Omega} \left(\text{grad}_s\delta\mathbf{u} : \text{grad}_s\Delta\mathbf{u}\boldsymbol{\sigma}\right)\, dv \qquad (4.20)$$

and the *material* contribution

$$D_{\mathbf{u}}\delta\Pi_{\text{int mat}} = \int_{\Omega} \left(\text{grad}_s\delta\mathbf{u} : \mathbb{c} : \text{grad}_s\Delta\mathbf{u}\right)\, dv \qquad (4.21)$$

are distinct parts of the linearized internal work.

Finite deformation problems in solid mechanics follow a Lagrangian description, where the finite element mesh follows the deformation of the continuum. The term $\text{grad}_s\delta\mathbf{u}$ therefore depends on the displacement and caused the geometric contribution (4.20). The material contribution (4.21) appeared already in the linear FEM problem, but now it is no longer a constant.

Example 4.4 (Linearization of a Spatial Variational Statement). The spatial configuration Ω moves in time and complicates the linearization of variational statements formulated with respect to Ω. It is therefore convenient to pull back the variational statement to the reference configuration Ω_0, linearize it at the fixed configuration Ω_0, and then push forward the linearized statement to the spatial configuration Ω.

(a) Linearize the relation $(4.18)_2$, and provide the directional derivative $D_{\mathbf{u}}\delta\Pi_{\text{int}}$ with respect to the reference configuration Ω_0.
(b) Use the Piola transform together with the transformation of the volume element and show that the expression (4.19) represents the linearization of variational statement (4.18).   ∎

4.3.5   Incompressible Flow Boundary Value Problem

Problems in fluid mechanics commonly use an Eulerian description, where the finite element mesh is fixed in space and the fluid flows through it. The fluid then occupies the fixed spatial domain Ω, and the velocity $\mathbf{v}$ is the essential variable in the description of such a problem. Given an incompressible, laminar, and isothermal flow, the set

$$\left.\begin{array}{c} \operatorname{div}\mathbf{v} = 0 \text{ in } \Omega \,, \\[2mm] \operatorname{div}\boldsymbol{\sigma} + \mathbf{b}_{\mathrm{f}} - \rho\dfrac{D\mathbf{v}}{Dt} = \mathbf{0} \text{ in } \Omega \,, \\[2mm] \mathbf{v} = \bar{\mathbf{v}} \text{ at } \partial\Omega_{\mathrm{u}} \,, \\[2mm] \mathbf{t} = \boldsymbol{\sigma}\mathbf{n} = \bar{\mathbf{t}} \text{ at } \partial\Omega_{\mathrm{t}} \end{array}\right\} \tag{4.24}$$

describes the strong form of the iBVP, where $D\mathbf{v}/Dt = \partial\mathbf{v}/\partial t + \mathbf{v} \cdot \operatorname{grad}\mathbf{v}$ denotes the material time derivative of the velocity and thus the acceleration felt by the fluid particle. The description expresses the continuity, the conservation of momentum, and the boundary conditions. The initial velocity $\mathbf{v} = \mathbf{v}_0$ describes the motion of the fluid at the time $t = t_0$ and complements (4.24).

Given the two local Eqs. (4.24)$_1$ and (4.24)$_2$, we introduce the test functions δp and $\delta\mathbf{v}$ and apply the Calculus of Variations independently to both expressions. Their weak forms then read

$$\int_{\Omega} \delta p \operatorname{div}\mathbf{v}dv = 0 \,,$$

$$\int_{\partial\Omega_{\mathrm{t}}} \delta\mathbf{v} \cdot \bar{\mathbf{t}}ds + \int_{\Omega} \delta\mathbf{v} \cdot \left(\mathbf{b}_{\mathrm{f}} - \rho\frac{D\mathbf{v}}{Dt}\right) dv - \int_{\Omega} \delta\mathbf{d} : \boldsymbol{\sigma} \, dv = 0 \,,$$

where $\delta\mathbf{d} = \operatorname{grad}_{\mathrm{s}}\delta\mathbf{v}$ denotes the symmetric virtual velocity gradient, and the divergence theorem has been used.

With the decoupled representation of the stress $\boldsymbol{\sigma} = \bar{\boldsymbol{\sigma}} - p\mathbf{I}$, we get

$$\left.\begin{array}{c} \displaystyle\int_{\Omega} \delta p \operatorname{div}\mathbf{v}dv = 0 \,, \\[4mm] \displaystyle\int_{\Omega} \delta\mathbf{v} \cdot \rho\left(\frac{\partial\mathbf{v}}{\partial t} + \mathbf{v} \cdot \operatorname{grad}\mathbf{v}\right) dv + \int_{\Omega} \delta\mathbf{d} : \bar{\boldsymbol{\sigma}} \, dv \\[4mm] \displaystyle - \int_{\Omega} \operatorname{div}\delta\mathbf{v} \, p \, dv - \int_{\partial\Omega_{\mathrm{t}}} \delta\mathbf{v} \cdot \bar{\mathbf{t}}ds - \int_{\Omega} \delta\mathbf{v} \cdot \mathbf{b}_{\mathrm{f}}dv = 0, \end{array}\right\} \tag{4.25}$$

where the deviatoric stress $\bar{\boldsymbol{\sigma}}(\mathbf{d})$ is defined by the constitutive model of the fluid. The continuity (4.24)$_1$ may also be seen as a constraint equation, and the test function δp would then be a Lagrange multiplier.

4.4 Finite Element Equations

Given the spatial interpolation of the essential variables as well as their test functions, the *discretized* weak form of the (initial) BVP may be derived. The

Galerkin[2] approach, first introduced by Ritz,[3] uses the same interpolations for the essential variables and their test functions, whilst the *Petrov*[4]*–Galerkin* approach uses different functions to interpolate the two fields.

4.4.1   Diffusion Problems

The concentration $c(\mathbf{x})$ and its gradient may be interpolated through $c = N_a h_a$ and grad$c = B_{ia} h_a$, where the n_{npe}-dimensional vector $\mathbf{h}$ stores the concentrations at the finite element nodes, whilst $\mathbf{N}$ and $\mathbf{B}$ are interpolation and gradient interpolation matrices, respectively. Following the Galerkin approach the same matrices are used to interpolate the test function $\delta c(\mathbf{x})$, and thus $\delta c = N_a \delta h_a$ and grad$\delta c = B_{ia} \delta h_a$ with $\delta\mathbf{h}$ denoting the nodal variations. The weak form (4.9) of the diffusion problem and the e-th finite element is then given by

$$\delta h_i \left[\underbrace{\int_{\Omega_e} B_{ai} B_{aj} \mathrm{d}v}_{D_{ij}} \; h_j - \underbrace{\left(\int_{\Omega_e} \alpha N_i \mathrm{d}v + \int_{\partial\Omega_{eq}} \overline{q} N_i \mathrm{d}s \right)}_{f_i} \right] = 0 \,, \qquad (4.26)$$

where $\mathbf{D}$ and $\mathbf{f}$ are the finite element's diffusion matrix and the force vector, a terminology borrowed from structural mechanics. Given arbitrary (admissible) nodal variation $\delta\mathbf{h}$, the statement (4.26) yields the set of algebraic linear equations

$$\mathbf{Dh} - \mathbf{f} = \mathbf{0} \,. \qquad (4.27)$$

It represents the discretized diffusion problem of the e-th finite element, where $\mathbf{D}$ is a symmetric $n_{\mathrm{npe}} \times n_{\mathrm{npe}}$ matrix, whilst $\mathbf{h}$ and $\mathbf{f}$ are n_{npe}-dimensional vectors, respectively.

4.4.2   Advection–Diffusion Problems

Following the Galerkin approach, the concentration $c(\mathbf{x})$ and the test function $\delta c(\mathbf{x})$ are interpolated according to $c = N_a h_a$ and $\delta c = N_a \delta h_a$. The gradients then read grad$c = B_{ia} h_a$ and grad$\delta c = B_{ia} \delta h_a$, where $\mathbf{h}$ and $\delta\mathbf{h}$ store the concentrations and the admissible variations at the n_{npe} finite element nodes, whilst $\mathbf{N}$ and $\mathbf{B}$ are interpolation and gradient interpolation matrices, respectively. The AD problem (4.11) of the e-th finite element is then expressed by the discrete weak form

---

[2]Boris Grigoryevich Galerkin, Soviet mathematician and an engineer, 1871–1945.
[3]Walther Heinrich Wilhelm Ritz, Swiss theoretical physicist, 1878–1909.
[4]Georgii Ivanovich Petrov, Russian scientist in the field of aerodynamics, gas dynamics, and space research, 1912–1987.

$$\delta h_i \left[\left(\underbrace{\int_{\Omega_e} N_i v_a B_{aj} dv}_{K_{ij}} + \underbrace{\int_{\Omega_e} B_{ai} v B_{aj} dv}_{D_{ij}} \right) h_j \right.$$

$$\left. - \underbrace{\left(\int_{\Omega_e} \alpha N_i dv + \int_{\partial \Omega_{eq}} \overline{q} N_i ds \right)}_{f_i} \right] = 0 , \tag{4.28}$$

where $\mathbf{K}$ and $\mathbf{D}$ are the advection and diffusion matrices, whilst $\mathbf{f}$ denotes the force vector. Given arbitrary (admissible) nodal variation $\delta\mathbf{h}$, the statement (4.28) yields the algebraic set of linear equations

$$(\mathbf{K} + \mathbf{D})\mathbf{h} - \mathbf{f} = \mathbf{0} , \tag{4.29}$$

where $\mathbf{K}$ is a non-symmetric and $\mathbf{D}$ is a symmetric $n_{npe} \times n_{npe}$ matrix, whilst $\mathbf{h}$ and $\mathbf{f}$ are n_{npe}-dimensional vectors, respectively.

Example 4.5 (The 1D Advection–Diffusion Finite Element). Consider a two-noded finite element of length $h[m]$ that models a 1D Advection–Diffusion (AD) problem (4.11).

(a) Use the linear shape functions N_1, N_2 given in (4.3) and provide the expressions of the advection matrix $\mathbf{K}$, the diffusion matrix $\mathbf{D}$, and the force vector $\mathbf{f}$ of a Galerkin finite element.
(b) Use the shape functions N_1, N_2 for the interpolation of c, and the shape functions $S_1 = (1 - \xi)/2 - 3\beta(1 - \xi^2)/4$ and $S_2 = (1 + \xi)/2 + 3\beta(1 - \xi^2)/4$ for the interpolation of the test function δc, where β denotes a constant that weights the non-linear terms in S_1 and S_2. Given these interpolations, provide the expressions of $\mathbf{K}$, $\mathbf{D}$, and $\mathbf{f}$ of the corresponding Petrov–Galerkin finite element. ∎

4.4.3 Linear Solid Mechanics Problems

Following the Galerkin approach, the displacement $\mathbf{u}(\mathbf{x})$ and the test function $\delta\mathbf{u}(\mathbf{x})$ are interpolated according to $u_i = N_{ia} h_a$ and $\delta u_i = N_{ia} \delta h_a$, where $\mathbf{h}$ and $\delta\mathbf{h}$ store the displacements and its admissible variations at the n_{npe} finite element nodes, whilst $\mathbf{N}$ denotes the $n_{dim} \times n_{dof}$-dimensional interpolation matrix. The same interpolation is used for the velocity $\dot{u}_i = N_{ia} \dot{h}_a$ and the acceleration $\ddot{u}_i = N_{ia} \ddot{h}_a$, where $\dot{\mathbf{h}}$ and $\ddot{\mathbf{h}}$ store the nodal velocities and accelerations, respectively. The symmetric gradients of $\mathbf{u}(\mathbf{x})$ and $\delta\mathbf{u}(\mathbf{x})$ are the engineering strain and the respective virtual engineering strain. They are expressed by $\varepsilon_i = B_{ia}h_a$ and $\delta\varepsilon_i = B_{ia}\delta h_a$, where $\mathbf{B}$ denotes the $n_s \times n_{\text{dof}}$ gradient interpolation matrix.

Given a linear-elastic material at small strains, Hooke's law $\boldsymbol{\sigma} = \mathbb{C} : \boldsymbol{\varepsilon}$ describes the stress $\boldsymbol{\sigma}$ as a function of the engineering strain $\boldsymbol{\varepsilon}$, where $\mathbb{C}$ denotes a constant stiffness tensor. In Voigt notation, Hooke's law (3.49) reads $\sigma_i = C_{ij}\varepsilon_j; i, j = 1, \ldots, n_s$, which together with the aforementioned interpolations, yields the weak form

$$\delta h_i \left[\underbrace{\int_{\Omega_e} N_{ai}\rho N_{aj}\,dv}_{M_{ij}} \ddot{h}_j + \underbrace{\int_{\Omega_e} N_{ai}\eta N_{aj}\,dv}_{C_{ij}} \dot{h}_j + \underbrace{\int_{\Omega_e} B_{ci}C_{ca}B_{aj}\,dv}_{K_{ij}} h_j \right.$$

$$\left. - \underbrace{\left(\int_{\Omega_e} b_{fa}N_{ai}\,dv + \int_{\partial\Omega_{et}} \bar{t}_a N_{ai}\,ds \right)}_{f_i} \right] = 0 \qquad (4.30)$$

of the linear solid mechanics problem (4.17) and the e-th finite element.

Given arbitrary (admissible) nodal variation $\delta\mathbf{h}$, the statement (4.30) yields the linear system of differential equations

$$\mathbf{M}\ddot{\mathbf{h}} + \mathbf{C}\dot{\mathbf{h}} + \mathbf{K}\mathbf{h} - \mathbf{f} = \mathbf{0}, \qquad (4.31)$$

where $\mathbf{f}$ denotes the n_{dof}-dimensional element nodal force vector, whilst $\mathbf{M}$, $\mathbf{C}$, and $\mathbf{K}$ are the finite element's $n_{\text{dof}} \times n_{\text{dof}}$-dimensional and symmetric mass, damping, and stiffness matrices, respectively.

4.4.3.1 The Linear Truss Finite Element Equations

Figure 4.7 shows a two-noded finite element that represents a truss of the cross-section A and length l. It represents a 1D problem, and the truss is loaded by the body force b_f along the x direction. The material has the density ρ, and the Young's modulus E describes linear-elastic properties.

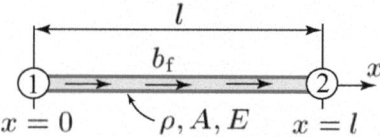

Fig. 4.7 A single two-noded finite element that represents a truss of cross-section A and length l. It is loaded by the body force b_f. Its linear-elastic material has the density ρ and the Young's modulus E

The mapping $x = (1 + \xi)l/2$ relates global and natural coordinates of the finite element and defines the Jacobian transformation $J = \partial x/\partial \xi = l/2$. We consider the linear shape functions (4.3) together with their spatial gradients (4.6), and the matrices

$$\mathbf{N} = \frac{1}{2}[(1 - \xi) \ (1 + \xi)] \ ; \ \mathbf{B} = \frac{1}{l}[-1 \ 1] \tag{4.32}$$

then define the interpolation of the displacements and strains, respectively. These matrices have a single row, and we will use the simplified notation $N_{1i} = N_i$ and $B_{1i} = B_i$ for $i = 1, 2$ in the following.

Given the displacement interpolation $\mathbf{N}$ together with the definition shown in (4.30), the truss element's nodal force vector reads

$$\mathbf{f} = \int_{\Omega_e} b_f N_i \, dv = \frac{b_f A l}{4} \int_{\xi=-1}^{\xi=1} \begin{bmatrix} 1 - \xi \\ 1 + \xi \end{bmatrix} d\xi = \frac{b_f A l}{2} \begin{bmatrix} 1 \\ 1 \end{bmatrix}, \tag{4.33}$$

where the volume element $dv = A dx = A J d\xi = (Al/2)d\xi$ has been used. The total external load $b_f A l$ is therefore split into half and then allocated to the two element nodes.

In addition, the definition shown in (4.30) specifies the truss element's mass matrix

$$\mathbf{M} = \int_{\Omega_e} N_i \rho N_j \, dv$$

$$= \frac{\rho A l}{4} \int_{\xi=-1}^{\xi=1} \begin{bmatrix} (1 - \xi)^2 & (1 - \xi)(1 + \xi) \\ (1 - \xi)(1 + \xi) & (1 + \xi)^2 \end{bmatrix} d\xi = \frac{m}{6} \begin{bmatrix} 2 & 1 \\ 1 & 2 \end{bmatrix}, \tag{4.34}$$

where $m = \rho A l$ denotes the mass of the truss element. Calculus of Variations has been used in the derivation of (4.34)—it is therefore known as the *variational consistent* mass matrix. It is symmetric but shows off-diagonal terms, which significantly increase the time of the *explicit solution* of the global set of equations, see Sect. 4.8.5. The mass matrix is therefore often approximated by a diagonal, or *lumped mass matrix*, where the sum of all coefficients in one row (or column) are lumped into the diagonal. Given the mass matrix (4.34), the corresponding lumped mass matrix then reads

$$\mathbf{M}_{lumped} = \frac{m}{2} \begin{bmatrix} 1 & 0 \\ 0 & 1 \end{bmatrix},$$

and each of the two nodes "receives" half of the element mass. A lumped mass matrix neglects *gyroscopic effects*, which can be important to consider in the analysis of structures that involve large finite elements.

The strain interpolation $\mathbf{B}$, and the definition shown in (4.30), allows us to compute the truss element's stiffness matrix

$$\mathbf{K} = \int_{\Omega_e} B_i E B_j \, dv = B_i B_j \frac{EA}{2l} \int_{\xi=-1}^{\xi=1} d\xi = \frac{EA}{l} \begin{bmatrix} 1 & -1 \\ -1 & 1 \end{bmatrix}, \tag{4.35}$$

where the Young's modulus E expresses the material's elasticity tensor C_{ij} of the 1D truss problem. The analyzed truss problem has neither a contribution from frictional forces, $\eta = 0$, nor a Neumann boundary term, $\bar{t} = 0$. All contributions of the weak form (4.30) have therefore been derived.

With the mass matrix (4.34), the stiffness (4.35), and the nodal forces (4.33), the system

$$\frac{m}{6}\begin{bmatrix} 2 & 1 \\ 1 & 2 \end{bmatrix}\begin{bmatrix} \ddot{u}_1 \\ \ddot{u}_2 \end{bmatrix} + \frac{EA}{l}\begin{bmatrix} 1 & -1 \\ -1 & 1 \end{bmatrix}\begin{bmatrix} u_1 \\ u_2 \end{bmatrix} = \frac{b_f A l}{2}\begin{bmatrix} 1 \\ 1 \end{bmatrix}$$

of second-order differential equations then expresses the equilibrium of the truss finite element.

Example 4.6 (Vessel Segment at Quasi-static Tension). The tensile force of $P = 0.3\,\text{N}$ loads a stiff vessel segment of the length $l = 1.2\,\text{cm}$, a problem that may be modeled by a single linear truss element, see Fig. 4.8. The Young's modulus $E = 223.5\,\text{kPa}$ describes the linear-elastic properties of the vessel wall, and $d_i = 3.2\,\text{mm}$ and $d_o = 5.8\,\text{mm}$ are the inner and outer vessel diameters, respectively.

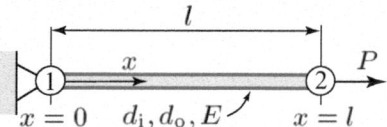

Fig. 4.8 A two-noded finite element represents a vessel segment of the length l. It is loaded by the force P and simply supported at the left end. The parameters d_i, d_o and E describe the geometry and the linear-elastic vessel wall properties, respectively

(a) Derive the set of algebraic equations that describes the finite element's quasi-static equilibrium.
(b) Solve the problem and compute the distribution of the displacement along the vessel's axial direction.   ∎

4.4.4   Non-linear Solid Mechanics Problems

The substitution of the test function gradient $\text{grad}^v_s \delta\mathbf{u} = \delta\mathbf{e}^v = \mathbf{B}\delta\mathbf{h}$ in the variation $(4.18)_1$ yields $\delta\Pi_{\text{int}} = f_i \delta h_i$ with

$$f_i = \int_{\Omega_e} B_{ai}\sigma_a dv \tag{4.36}$$

denoting the *nodal force vector*. For a non-linear solid mechanics problem, the gradient interpolation matrix $\mathbf{B}$ and the stress σ are non-linear functions of the displacements—the nodal force $\mathbf{f}$ is then also a non-linear function of the nodal displacement $\mathbf{h}$.

Given the gradient expressions $\mathrm{grad}_s^v \delta\mathbf{u} = \mathbf{B}\delta\mathbf{h}$ and $\mathrm{grad}_s^v \Delta\mathbf{u} = \mathbf{B}\Delta\mathbf{h}$, the linearized variational statement (4.21) yields $D_\mathbf{u} \delta \Pi_{\mathrm{int\,mat}} = \delta h_i K_{\mathrm{mat}\,ij} \Delta h_j$, where

$$K_{\mathrm{mat}\,ij} = \int_{\Omega_e} B_{ai} C_{ab} B_{bj} \,\mathrm{d}v \qquad (4.37)$$

is the *material stiffness matrix* of the finite element, and C_{ab} denotes the material's stiffness matrix in Voigt notation. Whilst C_{ab} is sparsely populated for a Hooke material, it is in general fully populated in the description of a non-linear material at finite deformations.

The substitution of the displacement $\delta\mathbf{u} = \mathbf{N}\delta\mathbf{h}$ and $\Delta\mathbf{u} = \mathbf{N}\Delta\mathbf{h}$ in the geometrical stress contribution (4.20) yields $D_\mathbf{u} \delta \Pi_{\mathrm{int\,geo}} = \delta h_i K_{\mathrm{geo}\,ij} \Delta h_j$, where

$$K_{\mathrm{geo}\,ij} = \int_{\Omega_e} \frac{\partial N_{ai}}{\partial x_c} \sigma_{ab} \frac{\partial N_{cj}}{\partial x_b} \,\mathrm{d}v \qquad (4.38)$$

denotes the *geometric stiffness matrix*, and σ_{ab} are the components of the Cauchy stress. The structure of the interpolation matrix results in a diagonally populated geometric stiffness matrix [53].

Both stiffness contributions, $\mathbf{K}_{\mathrm{geo}}$ and $\mathbf{K}_{\mathrm{mat}}$, are functions of the essential variable $\mathbf{h}$ and need to be computed at each iteration of the Newton–Raphson[5]-based solution method, see Sect. 4.8.6.1. Given an iterative solution schema to solve a finite element problem, the stress σ in (4.38) refers to the stress at the beginning of the iteration, and the geometric stiffness $K_{\mathrm{geo}\,ij}$ is therefore also called *initial stress stiffness*.

The nodal force vector (4.36) as well as the stiffness matrices (4.37) and (4.38) are integrals over the deformed finite element—they are commonly computed through *numerical quadrature*, see Sect. 4.4.6.

4.4.4.1 Pressure Boundary Condition

A pressure p_0 may be prescribed along the Neumann boundary $\partial\Omega_t$. The traction $\mathbf{t} = -p_0\mathbf{n}$ then acts at $\partial\Omega_t$, where $\mathbf{n}$ denotes the outward normal to the boundary. Given finite deformations, the boundary deforms and then changes the traction—the pressure load is therefore also called a *follower load*. The external contribution from the pressure boundary to the virtual work reads

$$\delta\Pi_{\mathrm{ext\,p}} = \int_{\partial\Omega_t} \delta\mathbf{u}\cdot\mathbf{t}\,\mathrm{d}s = -p_0 \int_{\partial\Omega_t} \delta\mathbf{u}\cdot\mathbf{n}\,\mathrm{d}s . \qquad (4.39)$$

---

[5]Joseph Raphson, English mathematician, 1648–1715.

With the displacement interpolation $\delta u_i = N_{ia}\delta h_a$, it may be expressed through $\delta \Pi_{\text{ext}\,p} = f_{p\,i}\delta h_i$, where

$$f_{p\,i} = -p_0 \int_{\partial\Omega_t} N_{ai}n_a\mathrm{d}s$$

denotes the contribution of the pressure boundary to the *nodal force vector*.

Towards the derivation of the external stiffness from the pressure boundary condition, we use Nanson's formula (3.5) $\mathbf{n}\mathrm{d}s = J\mathbf{F}^{-\mathrm{T}}\mathbf{N}\mathrm{d}S$ to pull back the expression (4.39) to the reference configuration. It then reads

$$\delta \Pi_{\text{ext}\,p} = -p_0 \int_{\partial\Omega_t} \delta\mathbf{u}\cdot\mathbf{n}\,\mathrm{d}s = -p_0 \int_{\partial\Omega_{0\,t}} \delta\mathbf{u}\cdot J\mathbf{F}^{-\mathrm{T}}\mathbf{N}\mathrm{d}S\,, \qquad (4.40)$$

and the integration is now taken over the fixed domain $\partial\Omega_{0\,t}$. Consequently,

$$
\begin{aligned}
D_{\mathbf{u}}\delta \Pi_{\text{ext}\,p} &= -p_0 \int_{\partial\Omega_{0\,t}} \delta\mathbf{u}\cdot D_{\mathbf{u}}\left(J\mathbf{F}^{-\mathrm{T}}\right)\mathbf{N}\mathrm{d}S \\
&= -p_0 \int_{\partial\Omega_{0\,t}} \delta\mathbf{u}\cdot\left(D_{\mathbf{u}}J\,\mathbf{F}^{-\mathrm{T}} + J\,D_{\mathbf{u}}\mathbf{F}^{-\mathrm{T}}\right)\mathbf{N}\mathrm{d}S \\
&= -p_0 \int_{\partial\Omega_{0\,t}} \delta\mathbf{u}\cdot\left(J\,\mathrm{div}\Delta\mathbf{u}\,\mathbf{F}^{-\mathrm{T}} - J\,\mathrm{grad}^{\mathrm{T}}\Delta\mathbf{u}\mathbf{F}^{-\mathrm{T}}\right)\mathbf{N}\mathrm{d}S \qquad (4.41)
\end{aligned}
$$

expresses the directional derivative of $\delta \Pi_{\text{ext}\,p}$, where the results $D_{\mathbf{u}}J = J\,\mathrm{div}\Delta\mathbf{u}$ and $D_{\mathbf{u}}\mathbf{F}^{-1} = -\mathbf{F}^{-1}\mathrm{grad}\Delta\mathbf{u}$ have been used, see Appendix D.

The push-forward of (4.41), and thus another application of Nanson's formula, yields

$$D_{\mathbf{u}}\delta \Pi_{\text{ext}\,p} = -p_0 \int_{\partial\Omega_t} \delta\mathbf{u}\cdot\left(\mathbf{I}\,\mathrm{div}\Delta\mathbf{u} - \mathrm{grad}^{\mathrm{T}}\Delta\mathbf{u}\right)\mathbf{n}\mathrm{d}s \qquad (4.42)$$

and expresses the linearization of the pressure loading. Given the displacement interpolations $\delta u_i = N_{ia}\delta h_a$ and $\Delta u_i = N_{ia}\Delta h_a$, this expression reads

$$D_{\mathbf{u}}\delta \Pi_{\text{ext}\,p} = \delta h_i\ (-p_0)\underbrace{\int_{\partial\Omega_t}\left(N_{ai}\frac{\partial N_{cj}}{\partial x_c}n_a - N_{ai}\frac{\partial N_{cj}}{\partial x_a}n_c\right)\mathrm{d}s}_{K_{p\,ij}}\ \Delta h_j\,,$$

where $K_{p\,ij}$ denotes the *stiffness matrix* of the pressure boundary condition. In general this stiffness matrix is *not symmetric* $\mathbf{K}_{\mathrm{p}} \neq \mathbf{K}_{\mathrm{p}}^{\mathrm{T}}$.

A different derivation, not based on the pull-back of the expression (4.39) to the reference configuration, has been reported elsewhere [53].

4.4.5 Incompressible Flow Problems

Given the interpolation matrix $\mathbf{N}$, the divergence interpolation matrix $\mathbf{G}$, and the gradient interpolation matrix $\mathbf{B}$, the expressions $v_i = N_{ia}h_a$, $\partial v_i/\partial x_i = G_a h_a$, and $d_i = B_{ia}h_a$ interpolate the velocity $\mathbf{v}(\mathbf{x})$, its divergence $\mathrm{div}\mathbf{v}(\mathbf{x})$, and the rate of deformation $\mathbf{d}(\mathbf{x})$ over the finite element. Here, the $n_{\mathrm{npe}}n_{\mathrm{dim}}$-dimensional vector $\mathbf{h}$ stores the nodal velocities. With the identity $\mathbf{v}\cdot\mathrm{grad}\mathbf{v} = \mathrm{grad}|\mathbf{v}|^2/2$, we may also express the advective acceleration by $\mathrm{grad}|\mathbf{v}|^2/2 = N_{ia}^2 h_a^2/2$.

In addition to the velocity $\mathbf{v}(\mathbf{x})$, the pressure $p(\mathbf{x})$ is also an independent variable of an incompressible flow. The expression $p = N_a q_a$ interpolates it over the finite element, where the n_{npe}-dimensional vector $\mathbf{q}$ holds the nodal pressures. The pressure p is a scalar, and thus the interpolation matrix differs from the interpolation matrix applied to the velocities. It is indicated by a single index of the coefficients N_a, whilst the interpolation matrix that interpolates the velocity has the coefficients N_{ia} with two indices.

Given a Galerkin approach, and thus using the same interpolations for the physical variables $\mathbf{v}$, p and their test functions $\delta\mathbf{v}$, δp, the two variational statements (4.25) may be discretized and result in

$$\delta q_i \underbrace{\int_{\Omega_e} N_i G_j \mathrm{d}v}_{D_{ij}} h_j = 0\,;$$

$$(4.43)$$

$$\delta h_i \left[\underbrace{\int_{\Omega_e} N_{ai}\rho N_{aj}\mathrm{d}v}_{M_{ij}} \frac{\partial h_j}{\partial t} + \frac{1}{2}\underbrace{\int_{\Omega_e} N_{ai}\rho N_{aj}^2\mathrm{d}v}_{A_{ij}} h_j^2 + \underbrace{\int_{\Omega_e} B_{ci}\overline{c}_{ca}B_{aj}\mathrm{d}v}_{K_{ij}} h_j \right.$$

$$\left. - \underbrace{\int_{\Omega_e} G_i N_j \mathrm{d}v}_{D_{ij}^{\mathsf{T}}} q_j - \underbrace{\left(\int_{\Omega_e} b_{fa} N_{ai}\mathrm{d}v + \int_{\partial\Omega_{e\,t}} \overline{t}_a N_{ai}\mathrm{d}s\right)}_{f_i} \right] = 0\,,$$

$$(4.44)$$

where the stiffness $\overline{c}_{ij}$ in (4.44) represents the Voigt notation of the stiffness tensor $\overline{\mathbb{c}} = \partial\overline{\boldsymbol{\sigma}}/\partial\mathbf{d}$ and thus the resistance of the fluid against shearing.

Given arbitrary (admissible) nodal variations $\delta\mathbf{h}$ and $\delta\mathbf{q}$, the statement (4.44) yields the non-symmetric system of first-order partial differential equations

$$\mathbf{M}\dot{\mathbf{h}} + \mathbf{A}\mathbf{h}^2 + \mathbf{K}\mathbf{h} - \mathbf{D}^{\mathsf{T}}\mathbf{q} - \mathbf{f} = \mathbf{0}\,. \qquad (4.45)$$

It is to be solved under the constraint $\mathbf{Dh} = \mathbf{0}$, which represents the continuity (4.43) of the incompressible fluid. In (4.45), $\mathbf{f}$ denotes the $n_{nep}n_{dim}$-dimensional nodal force vector, $\mathbf{M}$, $\mathbf{A}$, $\mathbf{K}$ are $n_{nep}n_{dim} \times n_{nep}n_{dim}$-dimensional square matrices, whilst $\mathbf{D}$ is of the dimension $n_{nep}n_{dim} \times n_{nep}$. Due to the advective acceleration term, the system (4.45) is non-linear, even for the case of constant matrices.

4.4.6   Numerical Quadrature

Finite elements frequently use polynomial shape functions, which allows the analytical computation of the integrals towards expressing the element force vector and the element stiffness matrices. However, a numerical quadrature of these integrals is often faster. The integrals are transformed from the element's physical domain Ω_e to its parent domain ω_e, and the quadrature is then performed over ω_e. The Jacobian transformation $\mathbf{J}$ connects Ω_e and ω_e, and given the relation $ds = \det\mathbf{J}ds_\omega$ between the area elements ds and ds_ω,

$$I = \int_{\partial\Omega_e} \mathcal{F}(\mathbf{x})ds = \int_{\partial\omega_e} \mathcal{F}(\boldsymbol{\xi})\det\mathbf{J}(\boldsymbol{\xi})ds_\omega \approx \sum_l^{l_{int}} \mathcal{F}(\boldsymbol{\xi}_l)\det\mathbf{J}(\boldsymbol{\xi}_l)w_l \qquad (4.46)$$

expresses a surface integral and thus allows for the computation of the Neumann contribution to the finite element nodal force vector. Here, $\boldsymbol{\xi}_l; l = 1, \ldots, l_{int}$ denotes the integration points' natural coordinates, and w_l are their weights. Given the relation $dv = \det\mathbf{J}dv_\omega$ of the volume elements dv and dv_ω,

$$I = \int_{\Omega_e} \mathcal{F}(\mathbf{x})dv \approx \sum_l^{l_{int}} \mathcal{F}(\boldsymbol{\xi}_l)\det\mathbf{J}(\boldsymbol{\xi}_l)w_l \qquad (4.47)$$

expresses a volume integral and allows us to compute the body force contribution to the finite element nodal force vector as well as the computation of the finite element stiffness matrices.

Whilst any quadrature rule may be applied, *Gauss[6]–Legendre[7] quadrature* is commonly used to compute the integrals of the FEM equations. Table 4.1 reports Gauss–Legendre integration point coordinates ξ_l and weights w_l in 1D.

The superposition of 1D quadrature is used to integrate over higher-dimensional functions, and thus

$$I = \int_{-1}^{+1} \int_{-1}^{+1} \mathcal{F}(\xi_1, \xi_2)d\xi_1 d\xi_2 \approx \sum_k^{k_{int}} \sum_l^{l_{int}} \mathcal{F}(\xi_{1k}\xi_{2l})w_k w_l \qquad (4.48)$$

---

[6]Johann Carl Friedrich Gauss, German mathematician and physicist, 1777–1855.
[7]Adrien-Marie Legendre, French mathematician, 1752–1833.

Table 4.1 Integration point coordinates ξ_l and weights w_l of 1D Gauss–Legendre quadrature

l_{int}	Integration point coordinates and weights (ξ_l, w_l)
1	$(0.0, 2.0)$
2	$(\pm 0.57735, 2.0)$
3	$(0.0, 0.88889)$; $(\pm 0.77460, 0.55556)$
4	$(\pm 0.86113, 0.34786)$; $(\pm 0.33998, 0.65214)$
5	$(0.0, 0.56889)$; $(\pm 0.90618, 0.23693)$; $(\pm 0.53847, 0.47863)$
6	$(\pm 0.66121, 0.36076)$; $(\pm 0.23862, 0.46791)$; $(\pm 0.93247, 0.17132)$
7	$(0.0, 0.41796)$; $(\pm 0.40585, 0.38183)$; $(\pm 0.74153, 0.27970)$; $(\pm 0.94911, 0.12948)$

represents the numerical integration of the 2D function $\mathcal{F}(\xi_1, \xi_2)$. Gauss–Legendre quadrature has the order $2n - 1$, such that n integration points along a spatial coordinate direction are required for the exact integration of the function $\mathcal{F}(\boldsymbol{\xi})$ of the polynomial degree n. Given triangular and tetrahedral elements, similar quadrature rules exist [619].

Example 4.7 (Nodal Forces and Stiffness of a Quadrilateral Finite Element). A small section of a vessel wall may be modeled as a 2D plane stress problem and should be described by a single quadrilateral finite element, see Fig. 4.9. The wall is $h = 1.0\,$mm thick and the finite element has the nodal coordinates

$$\begin{bmatrix} x_1 \\ y_1 \end{bmatrix} = \begin{bmatrix} 0.0 \\ 0.0 \end{bmatrix} \; ; \; \begin{bmatrix} x_2 \\ y_2 \end{bmatrix} = \begin{bmatrix} 3.0 \\ 0.3 \end{bmatrix} \; ; \; \begin{bmatrix} x_3 \\ y_3 \end{bmatrix} = \begin{bmatrix} 4.0 \\ 2.0 \end{bmatrix} \; ; \; \begin{bmatrix} x_4 \\ y_4 \end{bmatrix} = \begin{bmatrix} -0.5 \\ 3.4 \end{bmatrix} ,$$

where the numbers are given in millimeters. A linear-elastic constitutive description of the vessel wall may be used, and

$$\begin{bmatrix} \sigma_{11} \\ \sigma_{22} \\ \sigma_{12} \end{bmatrix} = \frac{E}{1 - \nu^2} \begin{bmatrix} 1 & \nu & 0 \\ \nu & 1 & 0 \\ 0 & 0 & 1 - \nu \end{bmatrix} \begin{bmatrix} \varepsilon_{11} \\ \varepsilon_{22} \\ \varepsilon_{12} \end{bmatrix} \tag{4.49}$$

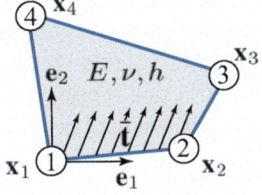

Fig. 4.9 Quadrilateral finite element at plane stress. The element represents a vessel segment made of linear-elastic material with the Young's modulus E and the Poisson's ratio ν. The element is h thick and loaded by the traction $\bar{\mathbf{t}}$ along the edge between node 1 and node 2

then describes the relation between stress and strain, where $E = 1.3\,\text{MPa}$ and $\nu = 0.45$ denote the Young's modulus and Poisson's ratio, respectively. The traction $\bar{\mathbf{t}} = [0.3\ 0.5]^{\mathsf{T}}\ [\text{N mm}^{-1}]$ acts along the edge that is formed by node 1 and node 2 of the finite element, see Fig. 4.9.

(a) Compute the strain interpolation matrix $\mathbf{B}$ in the center of the finite element.
(b) Identify the required number of Gauss–Legendre integration points to compute exact expressions of the nodal force vector $\mathbf{f}$ and the element stiffness matrix $\mathbf{K}$.
(c) Use Gauss–Legendre integration to compute exact expressions of $\mathbf{f}$ and $\mathbf{K}$, and investigate the eigenvalues of $\mathbf{K}$.
(d) Use one-point Gauss–Legendre integration to compute $\mathbf{K}$ and explain why this matrix is rank-deficit.   ∎

4.5   Constrained Problems

The essential variables of many biomechanical problems cannot "freely" develop—they have to satisfy constraints. Given the flow problem discussed in Sect. 4.4.5, the incompressibility of the fluid represents such a constraint. It led to the expression $\mathbf{Dh} = \mathbf{0}$ in addition to the momentum equation (4.45). A *constrained problem* may also be formulated by including the constraint directly at the finite element level, a concept followed in this section. We will outline such an approach in the description of an incompressible material and therefore split the Cauchy stress tensor $\boldsymbol{\sigma} = \bar{\boldsymbol{\sigma}} + \boldsymbol{\sigma}_{\text{vol}}$ into deviatoric $\bar{\boldsymbol{\sigma}}$ and volumetric $\boldsymbol{\sigma}_{\text{vol}}$ stress contributions. The deviatoric stress $\bar{\boldsymbol{\sigma}}$ represents constitutive information, whilst the volumetric stress $\boldsymbol{\sigma}_{\text{vol}}$ represents the constraint and *enforces the incompressibility*, see Sect. 3.3.9.

4.5.1   Penalty Constraint

The decoupled stress representation $\boldsymbol{\sigma} = \bar{\boldsymbol{\sigma}} + \boldsymbol{\sigma}_{\text{vol}}$ results in a decoupled elasticity tensor $\mathbb{c} = \bar{\mathbb{c}} + \mathbb{c}_{\text{vol}}$, where $\bar{\mathbb{c}}$ and $\mathbb{c}_{\text{vol}}$ denote the volumetric and isochoric contributions, respectively. The *Penalty method* implements the condition $\mathbb{c}_{\text{vol}} \gg \bar{\mathbb{c}}$ towards approximating an incompressible material. At the incompressible limit the coefficients of $\mathbb{c}_{\text{vol}}$ would then tend to infinity, and incompressibility holds exactly. However, the larger the coefficients of $\mathbb{c}_{\text{vol}}$ relative to the coefficients of $\bar{\mathbb{c}}$, the larger is the condition number of the matrix that represents $\mathbb{c}$. It is then more challenging to solve the penalty-constrained finite element problem.

A finite element problem may be seen as an *optimization problem*, where the solution represents the displacement $\mathbf{u}$ at the minimum of the potential, see Fig. 4.10. The Penalty method augments the isochoric potential $\overline{\Pi}(\mathbf{u})$ and adds a penalty contribution to deformation states that violate the constraint $C(\mathbf{u}) = 0$. It then results in a potential $\Pi_{\text{P}}(\mathbf{u})$ whose minimization is more challenging than the minimization of the unconstraint potential $\overline{\Pi}(\mathbf{u})$. Whilst the Penalty method is *easy to implement* and adds no additional degrees of freedom to the finite element

Fig. 4.10 Constrained optimization using the Penalty method towards approximately enforcing the constraint $C(u_1, u_2) = 0$. Adding the penalty contribution results in a more "valley-like" potential $\Pi_P(u_1, u_2)$. It is then more challenging to find the minimum $\Pi_P \rightarrow$ min, as compared to an unconstrained optimization $\overline{\Pi} \rightarrow$ min

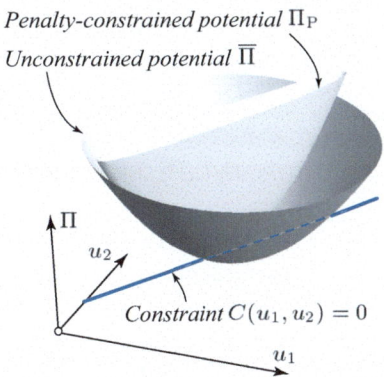

Penalty-constrained potential Π_P

Unconstrained potential $\overline{\Pi}$

Π

u_2

Constraint $C(u_1, u_2) = 0$

u_1

problem, it is always a *trade-off* between the error of violating the constraint and the difficulty (time) to solve the constraint problem.

4.5.2 Lagrange Constraint

The *Lagrange method* considers the constraint $C(\mathbf{u}) = 0$ by adding a *Lagrange contribution* to the isochoric potential $\overline{\Pi}(\mathbf{u})$. The Lagrange potential then reads

$$\Pi_L(\mathbf{u}, p) = \overline{\Pi}(\mathbf{u}) + \int_\Omega p\,(J(\mathbf{u}) - 1)\,dv\,, \tag{4.50}$$

where the displacement $\mathbf{u}$ and the Lagrange parameter p are the essential (independent) problem variables. Given an incompressible material, the Lagrange parameter p may be identified as the hydrostatic pressure $p = \mathrm{tr}\boldsymbol{\sigma}/n_{\mathrm{dim}}$ with n_{dim} denoting the numbers of dimension. The pressure is then no longer a function of $\mathbf{u}$, but an additional essential problem variable. Given the two independent variables $\mathbf{u}$ and p, the potential (4.50) represents a *two-field variational problem* and leads to two variational statements. They read

$$
\left.
\begin{aligned}
\delta_{\mathbf{u}}\Pi_L(\mathbf{u}, p) &= \int_\Omega \overline{\boldsymbol{\sigma}}(\mathbf{u}) : \mathrm{grad}_s\delta\mathbf{u}\,dv + \int_\Omega p\,J(\mathbf{u})\mathrm{div}\delta\mathbf{u}\,dv - \delta_{\mathbf{u}}\Pi_{\mathrm{ext}} \\
&= \int_\Omega (\overline{\boldsymbol{\sigma}}(\mathbf{u}) + p\mathbf{I}) : \mathrm{grad}_s\delta\mathbf{u}\,dv - \delta_{\mathbf{u}}\Pi_{\mathrm{ext}} = 0\,; \\
\delta_p\Pi_L(\mathbf{u}, p) &= \int_\Omega \delta p(J(\mathbf{u}) - 1)dv - \delta_p\Pi_{\mathrm{ext}} = 0\,,
\end{aligned}
\right\} \tag{4.51}
$$

where the variation of the volume ratio $\delta_{\mathbf{u}}J(\mathbf{u}) = J(\mathbf{u})\mathrm{div}\delta\mathbf{u}$ has been used, see Appendix D. Whilst the variational statement $(4.51)_1$ appears formally identical to the displacement-based finite element formulation (4.17), it is noted that p in $(4.51)_1$ is an independent variable.

The linearization of the variational statements (4.51) yields

$$
\left.\begin{aligned}
D_{\mathbf{u}}\delta_{\mathbf{u}}\Pi_{\mathrm{L}} &= \int_{\Omega} (\mathrm{grad}\delta\mathbf{u} : \mathrm{grad}\Delta\mathbf{u}\overline{\sigma} + \mathrm{grad}\delta\mathbf{u} : \overline{\mathbb{C}} : \mathrm{grad}\Delta\mathbf{u})\, dv\,, \\
D_{\mathbf{u}}\delta_{\mathrm{p}}\Pi_{\mathrm{L}} = D_{\mathrm{p}}\delta_{\mathbf{u}}\Pi_{\mathrm{L}} &= \int_{\Omega} (\mathrm{grad}\delta\mathbf{u} : \mathrm{grad}\Delta\mathbf{u}\, p\mathbf{I} + \mathrm{grad}\delta\mathbf{u} : \overline{\mathbb{C}}_p : \mathrm{grad}\Delta\mathbf{u})\, dv\,, \\
D_{\mathrm{p}}\delta_{\mathrm{p}}\Pi_{\mathrm{L}} &= 0\,,
\end{aligned}\right\}
$$

$$(4.52)$$

where $D_{\mathbf{b}}\delta_{\mathbf{a}}\Pi_{\mathrm{L}}$ denotes the directional derivative of $\delta_{\mathbf{a}}\Pi_{\mathrm{L}}$ along the direction b. These results may be derived by pulling back (4.51) to the reference configuration Ω_0, carrying out the actual linearization by using the expressions in Appendix D, and then pushing the results forward to the current configuration Ω.

The first and second terms in $(4.52)_{1,2}$ refer to *geometrical* and *material* contributions, respectively. The isochoric stiffness is denoted by $\overline{\mathbb{C}}$ and depends on the displacement $\mathbf{u}$, whilst the Lagrange parameter p determines the volumetric stiffness $\overline{\mathbb{C}}_p = p(\mathbf{I} \otimes \mathbf{I} - 2\mathbb{I})$ with $\mathbb{I} = (\delta_{ij}\delta_{kl} + \delta_{ik}\delta_{jl})/2$ denoting the fourth-order identity tensor. We may interpolate the essential variables $\mathbf{u}$, p and the corresponding test functions $\delta\mathbf{u}$, δp in (4.51) and (4.52), which then leads to the force vector and the stiffness matrix of the Lagrange-constrained finite element.

Whilst the Lagrange method always enforces the constraint exactly, it results in zeros in the diagonal of the finite element stiffness, the so-called *pivot* elements, see $(4.52)_3$. It may be seen as a general drawback and does not allow to solve a Lagrange-constrained problem with fast Gaussian elimination-based solvers.

4.5.3   Augmented-Lagrange Constraint

The *Augmented-Lagrange* method uses the potential

$$
\Pi_{\mathrm{AL}}(\mathbf{u}, p) = \Pi_{\mathrm{L}}(\mathbf{u}, p) - \frac{1}{2\kappa} \int_{\Omega} p^2 dv \tag{4.53}
$$

to implement the constraint $C(\mathbf{u}) = 0$, an approach that for $\kappa \to \infty$ approaches the Lagrange potential (4.50). It avoids the ill-conditioning of the finite element stiffness matrix as well as zero pivot elements, drawbacks known from the penalty-constrained and the Lagrange-constrained problems, respectively.

The Augmented-Lagrange potential (4.53) leads to the two variational statements

$$
\left.\begin{aligned}
\delta_{\mathbf{u}}\Pi_{\mathrm{AL}}(\mathbf{u}, p) &= \int_{\Omega} (\overline{\sigma}(\mathbf{u}) + p\mathbf{I}) : \mathrm{grad}_s\delta\mathbf{u}\, dv - \delta_{\mathbf{u}}\Pi_{\mathrm{ext}} = 0\,, \\
\delta_p\Pi_{\mathrm{AL}}(\mathbf{u}, p) &= \int_{\Omega} \delta p[(J(\mathbf{u}) - 1) - p/\kappa]dv - \delta_p\Pi_{\mathrm{ext}} = 0\,,
\end{aligned}\right\}
$$

$$(4.54)$$

where the results (4.51) have been used. Localization of $(4.54)_2$ yields the Euler equation $p = \kappa (J(\mathbf{u}) - 1)$ and allows to substitute the additional degree of freedom p. The finite element is then fully described by the displacement $\mathbf{u}$ and is known as *general displacement* finite element implementation [53].

Example 4.8 (Hu–Washizu Variational Principles). Material incompressibility may lead to volume looking of a structural problem. The solution of the finite element-approximated description is then (much) stiffer than the continuum problem. At the finite element level, locking is the direct consequence of the shape functions not being able to represent the constraint deformation all over the finite element. Volume locking may be resolved by a variational formulation that introduces the additional kinematic variable θ that represents the change of the element volume. It must not be confused with the volume ratio $J = \det \mathbf{F}(\mathbf{u})$, which is a function of the displacement $\mathbf{u}$. Given the displacement $\mathbf{u}$ and the pressure p, the potential

$$\Pi_{HW}(\mathbf{u}, p, \theta) = \overline{\Pi}(\mathbf{u}) + \int_\Omega U(\theta)\mathrm{d}v + \int_\Omega p(J - \theta)\mathrm{d}v - \Pi_{ext} = 0 \qquad (4.55)$$

then defies the three-field Hu–Washizu variational principle, where $\mathbf{u}, p, \theta$ are the essential variables.

(a) Derive the three variational statements that follow from the potential (4.55).
(b) Consider an augmented Hu–Washizu potential $\mathcal{L}(\mathbf{u}, p, \theta, \lambda) = \Pi_{HW}(\mathbf{u}, p, \theta)$ $+ \int_\Omega \lambda h(\theta)\mathrm{d}v$, where $h(\theta)$ denotes a function with the condition $h(1) = 0$, whilst λ is an additional essential variable. Derive the four variational statements that follow from $\mathcal{L}(\mathbf{u}, p, \theta, \lambda)$. ∎

4.6 Globalization

The finite element representation of an engineering problem requires the *assembly* of the global system of equations, and thus the allocation of the contributions from the individual finite elements to the respective global nodes. Figure 4.11 illustrates

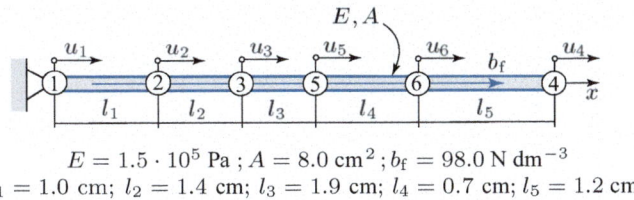

$$E = 1.5 \cdot 10^5 \text{ Pa} \; ; A = 8.0 \text{ cm}^2 \; ; b_{\mathrm{f}} = 98.0 \text{ N dm}^{-3}$$
$$l_1 = 1.0 \text{ cm}; \; l_2 = 1.4 \text{ cm}; \; l_3 = 1.9 \text{ cm}; \; l_4 = 0.7 \text{ cm}; \; l_5 = 1.2 \text{ cm}$$

Fig. 4.11 Truss structure discretized by five two-noded finite elements that is simply supported at node 1. The truss is loaded by the body force b_{f}, has the cross-section A and is made of linear-elastic material with the Young's modulus E

this process for a 1D truss that is discretized by $n_e = 5$ linear elements of the length l_e; $e = 1, \ldots, 5$. The truss has the cross-section A and is loaded by the body force b_f, whilst the Young's modulus E describes its linear-elastic material properties. The global force vector $\mathbf{f}$ is then assembled according to

$$\mathbf{f} = \mathop{\mathbf{A}}_{e=1}^{n_e} \mathbf{f}_e \,,$$

which, given the nodal force vector $\mathbf{f}_e = b_f A l_e [1 \ 1]^{\mathrm{T}}/2$ of the e-th finite element, reads

$$\mathbf{f} = \frac{b_f A l_1}{2}\begin{bmatrix} 1 \\ 1 \\ 0 \\ 0 \\ 0 \\ 0 \end{bmatrix} + \frac{b_f A l_2}{2}\begin{bmatrix} 0 \\ 1 \\ 1 \\ 0 \\ 0 \\ 0 \end{bmatrix} + \frac{b_f A l_3}{2}\begin{bmatrix} 0 \\ 0 \\ 1 \\ 1 \\ 0 \\ 0 \end{bmatrix} + \frac{b_f A l_4}{2}\begin{bmatrix} 0 \\ 0 \\ 0 \\ 0 \\ 1 \\ 1 \end{bmatrix} + \frac{b_f A l_5}{2}\begin{bmatrix} 0 \\ 0 \\ 0 \\ 1 \\ 0 \\ 1 \end{bmatrix} .$$

In addition to the nodal forces, we also have to allocate the finite element stiffness

$$\mathbf{K}_e = \frac{EA}{l_e}\begin{bmatrix} 1 & -1 \\ -1 & 1 \end{bmatrix}$$

of the e-th finite elements to the nodal stiffness of the respective global nodes. The operation

$$\mathbf{K} = \mathop{\mathbf{A}}_{e=1}^{n_e} \mathbf{K}_e$$

then explicitly reads

$$\mathbf{K} = \frac{EA}{l_1}\begin{bmatrix} 1 & -1 & 0 & 0 & 0 & 0 \\ -1 & 1 & 0 & 0 & 0 & 0 \\ 0 & 0 & 0 & 0 & 0 & 0 \\ 0 & 0 & 0 & 0 & 0 & 0 \\ 0 & 0 & 0 & 0 & 0 & 0 \\ 0 & 0 & 0 & 0 & 0 & 0 \end{bmatrix} + \frac{EA}{l_2}\begin{bmatrix} 0 & 0 & 0 & 0 & 0 & 0 \\ 0 & 1 & -1 & 0 & 0 & 0 \\ 0 & -1 & 1 & 0 & 0 & 0 \\ 0 & 0 & 0 & 0 & 0 & 0 \\ 0 & 0 & 0 & 0 & 0 & 0 \\ 0 & 0 & 0 & 0 & 0 & 0 \end{bmatrix}$$

$$+\frac{EA}{l_3}\begin{bmatrix} 0 & 0 & 0 & 0 & 0 & 0 \\ 0 & 0 & 0 & 0 & 0 & 0 \\ 0 & 0 & 1 & 0 & -1 & 0 \\ 0 & 0 & 0 & 0 & 0 & 0 \\ 0 & 0 & -1 & 0 & 1 & 0 \\ 0 & 0 & 0 & 0 & 0 & 0 \end{bmatrix} + \frac{EA}{l_4}\begin{bmatrix} 0 & 0 & 0 & 0 & 0 & 0 \\ 0 & 0 & 0 & 0 & 0 & 0 \\ 0 & 0 & 0 & 0 & 0 & 0 \\ 0 & 0 & 0 & 0 & 0 & 0 \\ 0 & 0 & 0 & 0 & 1 & -1 \\ 0 & 0 & 0 & 0 & -1 & 1 \end{bmatrix}$$

$$+\frac{EA}{l_5}\begin{bmatrix} 0 & 0 & 0 & 0 & 0 & 0 \\ 0 & 0 & 0 & 0 & 0 & 0 \\ 0 & 0 & 0 & 0 & 0 & 0 \\ 0 & 0 & 0 & 1 & 0 & -1 \\ 0 & 0 & 0 & 0 & 0 & 0 \\ 0 & 0 & 0 & -1 & 0 & 1 \end{bmatrix}.$$

Given the contributions of all finite elements have been assembled, we derive at the global system

$$\mathbf{Kh} - \mathbf{f} = \mathbf{0} \qquad (4.56)$$

of n_{dof} equations. The n_{dof}-dimensional vectors $\mathbf{h}$ and $\mathbf{f}$ store the essential variables and the nodal forces, respectively. The $n_{\text{dof}} \times n_{\text{dof}}$-dimensional global stiffness matrix $\mathbf{K}$ is sparsely populated, and, in many cases, also symmetric.

The allocation of the values in the stiffness matrix depends on the global node numbering. The narrower the non-zero terms are populated around the diagonal, and thus the smaller the bandwidth of $\mathbf{K}$, the faster and more robustly the system (4.56) may be solved. Automatic algorithms are used to renumber the global nodes towards minimizing the bandwidth of $\mathbf{K}$ [619]. In addition, sparse storage approaches avoid storing zero elements of $\mathbf{K}$ and optimize memory management of FEM problems [418].

The global system of Eqs. (4.56) of the truss problem and the parameters shown in Fig. 4.11 reads

$$\begin{bmatrix} 12000 & -12000 & 0 & 0 & 0 & 0 \\ -12000 & 20571 & -8571 & 0 & 0 & 0 \\ 0 & -8571 & 14887 & 0 & -6316 & 0 \\ 0 & 0 & 0 & 10000 & 0 & -10000 \\ 0 & 0 & -6316 & 0 & 23459 & 17143 \\ 0 & 0 & 0 & -10000 & 17143 & 27143 \end{bmatrix}\begin{bmatrix} u_1 \\ u_2 \\ u_3 \\ u_4 \\ u_5 \\ u_6 \end{bmatrix} = \begin{bmatrix} 0.392 \\ 0.941 \\ 1.294 \\ 0.47 \\ 1.019 \\ 0.745 \end{bmatrix}.$$

Because of $\det\mathbf{K} = 0$, it cannot be solved without embedding the Dirichlet boundary condition $u_1 = 0$. The system has otherwise a zero eigenvalue that represents the translation of the truss along the x direction.

The Dirichlet boundary condition $u_1 = 0$ removes the first row and column from the system. It then reads

$$
\begin{bmatrix}
20571 & -8571 & 0 & 0 & 0 \\
-8571 & 14887 & 0 & -6316 & 0 \\
0 & 0 & 10000 & 0 & -10000 \\
0 & -6316 & 0 & 23459 & 17143 \\
0 & 0 & -10000 & 17143 & 27143
\end{bmatrix}
\begin{bmatrix}
u_2 \\ u_3 \\ u_4 \\ u_5 \\ u_6
\end{bmatrix}
=
\begin{bmatrix}
0.941 \\ 1.294 \\ 0.47 \\ 1.019 \\ 0.745
\end{bmatrix} ,
$$

and the nodal displacements $u_2 = 0.37, u_3 = 0.78, u_4 = 1.14, u_5 = 1.21, u_6 = 1.26$ mm is the solution to the problem.

4.7 Stabilization

The stability of the solution of non-linear problems is of major interest, and a solution may become unstable because of physical and/or numerical reasons. In this section we will address *numerical stability* and thus instabilities that are linked to the discretization of a physically stable problem. The analysis is limited to linear problems and we "hope" that our conclusions may also apply to non-linear problems.

4.7.1 Positive Definiteness of the Finite Element Stiffness

One source of instabilities of a discretized structural problem can be linked to instantaneous spurious deformation modes at the finite element level. Let us consider a structural finite element with $\mathbf{Kh} = \mathbf{f}$ determining the equilibrium at the element level. Dirichlet boundary conditions have already been implemented and suppress any rigid body motion. Lyapunov[8] stability would then require that *small* perturbations in the load vector $\Delta \mathbf{f} = \mathbf{f}^* - \mathbf{f}$ at most result in *small* alterations of the displacements $\Delta \mathbf{h} = \mathbf{h}^* - \mathbf{h}$, where $(\bullet)^*$ denotes quantities of the perturbed system. The finite element would otherwise be unstable. Given a symmetric finite element stiffness $\mathbf{K}$, the relation $\mathbf{K}\Delta\mathbf{h} = \Delta\mathbf{f}$ implies that Lyapunov stability is ensured for any positive definite $\mathbf{K}$.

Figure 4.12 shows an *hourglass-instability*, an instability seen with quadrilateral or hexahedral finite elements and caused by the under-integration of the stiffness matrix. It is one cause of an instable finite element. Whilst under-integration is a common approach to avoid locking of finite elements or to speed up the computa-

---

[8] Aleksandr Mikhailovich Lyapunov, Russian mathematician, mechanician, and physicist, 1857–1918.

Fig. 4.12 Quadrilateral finite element showing an hourglass-instability. The quadrature by a single integration point leads to under-integration and a non-positive-definite element stiffness

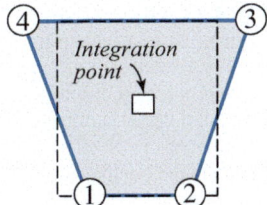

tions, it leads to rank-deficiency and therefore a non-positive definite $\mathbf{K}$. Methods for the stabilization of under-integrated finite elements are well reported [35, 266, 601].

4.7.2 Stabilization of the Advection–Diffusion Finite Element

The AD problem discussed in Sect. 4.3.2 shows another type of numerical instability. It is not related to the positive definiteness of the element stiffness matrix but appears in an *advection-dominated* problem. A high value of the advective term $\mathbf{v} \cdot \mathrm{grad}\,c$ in relation to the diffusive term $\nu\,\mathrm{div}(\mathrm{grad}\,c)$ of the governing equations (4.10) then results in an unstable solution. The same instability appears in the description of fluid mechanical problems (4.3.5) with a significant adventive velocity term $\mathbf{v} \cdot \mathrm{grad}\,\mathbf{v}$ and thus at high Reynolds number Re.

We may consider the 1D AD Galerkin finite element analyzed in Example 4.5 towards the exploration of such an instability. The governing system (4.29) then reads

$$\left(\frac{v}{2} \begin{bmatrix} -1 & 1 \\ -1 & 1 \end{bmatrix} + \frac{\nu}{h} \begin{bmatrix} 1 & -1 \\ -1 & 1 \end{bmatrix} \right) \begin{bmatrix} \bar{c}_1 \\ \bar{c}_2 \end{bmatrix} = \frac{\alpha h}{2} \begin{bmatrix} 1 \\ 1 \end{bmatrix}, \tag{4.57}$$

where v, ν, and α denote the advective velocity, the diffusivity, and a constant that describes the physics of the problem. In addition, the finite element is h long, and $\bar{c}_1$ and $\bar{c}_2$ are the essential problem variables at the two nodes. Figure 4.13 illustrates a 1D AD problem discretized by n such finite elements. The system

$$\left\{ Pe \begin{bmatrix} -1 & 1 & & & \\ -1 & 0 & 1 & & \\ & -1 & 0 & 1 & \\ & & -1 & 0 & 1 \\ & & & \ddots & \ddots & \ddots \end{bmatrix} + \begin{bmatrix} 1 & -1 & & & \\ -1 & 2 & -1 & & \\ & -1 & 2 & -1 & \\ & & -1 & 2 & -1 \\ & & & \ddots & \ddots & \ddots \end{bmatrix} \right\} \begin{bmatrix} c_1 \\ c_2 \\ c_3 \\ c_4 \\ \vdots \end{bmatrix} = k \begin{bmatrix} 1 \\ 2 \\ 2 \\ 2 \\ \vdots \end{bmatrix} \tag{4.58}$$

then determines the numerical problem and the unknowns are stored in the vector $\mathbf{h} = [c_1 \ldots c_{n+1}]^{\mathrm{T}}$. Towards a dimensionless study of the problem, the elemental Péclet number $Pe = vh/(2\nu)$ and the factor $k = \alpha h^2/(2\nu)$ have been introduced.

Fig. 4.13 Galerkin finite element-based predictions of the 1D Advection–Diffusion (AD) problem at different elemental Péclet numbers Pe. The advective velocity v points in the positive x direction, and $k = 0.005$ defines the right-hand side of the AD problem. The problem is discretized by $n = 10$ finite elements, and $c = 0.0$ describes Dirichlet boundary conditions at node 1 and node 11, respectively

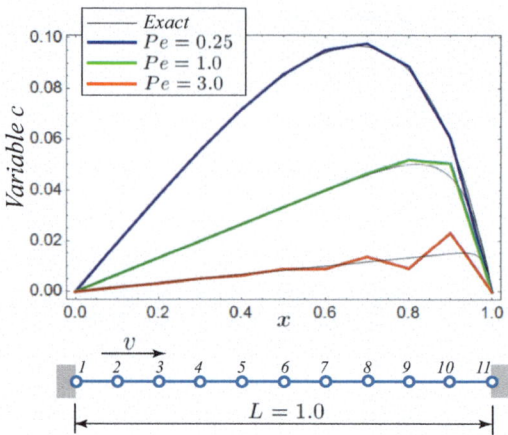

The elemental Péclet number specifies the ratio between advective and diffusive transport over one finite element.

Figure 4.13 shows the result from the solution of the system (4.58). Whilst at low Pe numbers the numerical solution approximates the exact solution

$$c_{\text{exact}} = \frac{1}{v} \left[x - \frac{1 - \exp(vx/v)}{1 - \exp(v/v)} \right]$$

of the problem, at larger Pe numbers, however, we observe spurious node-to-node oscillations around the exact solution. Given a high Pe number, a sharp boundary layer develops close to the right boundary of the problem and then "triggers" the oscillations in the solution.

4.7.2.1 Full Upwind Stabilization

Different approaches have been proposed to stabilize the AD problem and to suppress the aforementioned oscillations in the solution. At the i-th node, the algebraic relation

$$v \underbrace{\frac{c_{i+1} - c_{i-1}}{2h}}_{\approx \mathrm{d}c/\mathrm{d}x} - v \underbrace{\frac{c_{i+1} - 2c_i + c_{i-1}}{h^2}}_{\approx \mathrm{d}^2 c/\mathrm{d}x^2} = \alpha \tag{4.59}$$

derives directly from the system (4.58) and illustrates the approximations used to express the first and second derivatives of c by the Galerkin finite element. The approach introduces a truncation error and solves an AD problem with less diffusivity and thus less dissipation than the exact problem. We may therefore add *artificial diffusivity* to compensate for the truncation error.

In contrary to the Galerkin approach, the finite difference discretization at the i-th node leads to

$$v\frac{c_i - c_{i-1}}{h} - v\frac{c_{i+1} - 2c_i + c_{i-1}}{h^2} = \alpha \,, \tag{4.60}$$

where backward-Euler and mid-point rules have been used to approximate adventive and diffusion terms, respectively. This discretization of the AD problem does not show oscillations in the solution. The identical discretization is achieved through the alteration of (4.59) towards

$$v\frac{c_{i+1} - c_{i-1}}{2h} - \left(v + v^\star\right)\frac{c_{i+1} - 2c_i + c_{i-1}}{h^2} = \alpha \,,$$

where the diffusivity $v^\star = vh/2$ has been added to stabilize the problem. We may express the added diffusivity directly at the finite element level, and the system (4.57) then reads

$$\left(\frac{v}{2}\begin{bmatrix} -1 & 1 \\ -1 & 1 \end{bmatrix} + \frac{v + v^\star}{h}\begin{bmatrix} 1 & -1 \\ -1 & 1 \end{bmatrix}\right)\begin{bmatrix} \bar{c}_1 \\ \bar{c}_2 \end{bmatrix} = \frac{\alpha h}{2}\begin{bmatrix} 1 \\ 1 \end{bmatrix}, \tag{4.61}$$

an approach known as *full upwind stabilization*. Figure 4.14 shows the results achieved with this method. Whilst it prevents oscillations, the solution is overly diffusive and affects also the results at low Pe numbers.

4.7.2.2 Petrov–Galerkin Finite Elements

A Petrov–Galerkin finite element uses *different* shape functions to interpolate the physical quantity and its corresponding test function. It leads to a non-symmetric effect on the i-th node, as shown in Fig. 4.15. To achieve an *upwind stabilization*, the influence of the upwind side needs to be amplified. The bubble mode $3\beta(1 - \xi^2)/4$ may be considered in addition to the linear shape functions. The expressions

$$S_1 = (1 + \xi)/2 - 3\beta(1 - \xi^2)/4 \ ; \ \ S_2 = (1 - \xi)/2 + 3\beta(1 - \xi^2)/4 \tag{4.62}$$

then determine the interpolation $\delta c = S_i \delta h_i$ of the test function, where $\xi = 2x/h - 1$ denotes the natural coordinate, whilst β specifies the amount of upwinding. It is

Fig. 4.14 Numerical predictions of the full upwind stabilized 1D Advection–Diffusion (AD) problem at different elemental Péclet numbers Pe

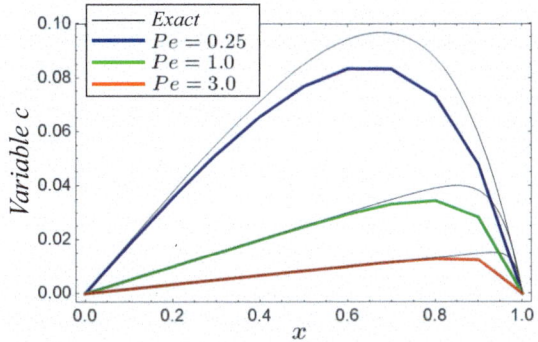

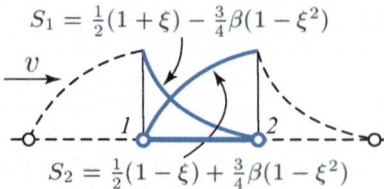

$$S_1 = \tfrac{1}{2}(1 + \xi) - \tfrac{3}{4}\beta(1 - \xi^2)$$

$$S_2 = \tfrac{1}{2}(1 - \xi) + \tfrac{3}{4}\beta(1 - \xi^2)$$

Fig. 4.15 Upwind stabilization of the 1D Advection–Diffusion (AD) problem through Petrov–Galerkin finite elements. Bubble modes are added to the linear shape functions and alter the interpolation of the test function δc. It results in a non-symmetric effect at the finite element nodes

only the test function's interpolation that is changed, and $c = N_i h_i$ still expresses the interpolation of the physical quantity c. The advection and diffusion matrices

$$\mathbf{K} = \frac{v}{2}\begin{bmatrix} -1 & 1 \\ -1 & 1 \end{bmatrix} + \underbrace{\beta\frac{v}{2}\begin{bmatrix} 1 & -1 \\ -1 & 1 \end{bmatrix}}_{\text{upwinding}} \; ; \quad \mathbf{D} = \frac{v}{h}\begin{bmatrix} 1 & -1 \\ -1 & 1 \end{bmatrix}$$

together with the force vector

$$\mathbf{f} = \frac{\alpha h}{2}\begin{bmatrix} 1 \\ 1 \end{bmatrix} + \underbrace{\beta\frac{\alpha h}{2}\begin{bmatrix} -1 \\ 1 \end{bmatrix}}_{\text{upwinding}}$$

then define the Petrov–Galerkin 1D AD finite element, see Example 4.5. The upwinding contribution of $\mathbf{f}$ affects only the domain's boundary and is neglected in many applications. The diffusion matrix $\mathbf{D}$ and the upwinding contribution to $\mathbf{K}$ have the same structure, which allows to express the finite element equations according to (4.61), where $v^\star = \beta v h / 2$ determines the added diffusivity. Here, the parameter β controls the amount of added diffusivity. It should be large enough to suppress oscillations, but at the same time as small as possible to prevent overly dissipative results.

4.7.2.3 Generalization for Multi-dimensional Problems

The aforementioned ideas of stabilizing the 1D AD problem may be extended to higher dimensions. The artificial diffusivity is then added through an anisotropic *diffusion tensor*

$$v^\star = v^\star \frac{\mathbf{v} \otimes \mathbf{v}}{|\mathbf{v}|^2} . \tag{4.63}$$

The only non-zero eigenvalue of $v^\star$ is $|\mathbf{v}|^2$ and points in the direction of the flow velocity $\mathbf{v}$. Along the flow direction the diffusivity $v^\star$ is then added towards

stabilizing the problem, whilst it remains unchanged perpendicular to the flow—
crosswind diffusion is therefore avoided.

Given the property $v\mathrm{div}(\mathrm{grad}c) = v\mathbf{I} : \mathrm{grad}(\mathrm{grad}c)$, the AD problem (4.10) may
be rewritten, and the artificial diffusivity $v^\star$ can be added to the physical diffusivity
$v\mathbf{I}$. The local expression $\mathbf{v} \cdot \mathrm{grad}c - (v\mathbf{I} + v^\star) : \mathrm{grad}(\mathrm{grad}c) + \alpha = 0$ then governs
the stabilized AD problem, and (4.63) leads to

$$\mathbf{v} \cdot \mathrm{grad}c - \frac{v^\star}{|\mathbf{v}|^2}(\mathbf{v} \otimes \mathbf{v}) : \mathrm{grad}(\mathrm{grad}c) - v\mathrm{div}(\mathrm{grad}c) + \alpha = 0 \,.$$

Calculus of Variations applied to the first three terms yields

$$\int_\Omega \left(\delta c + \frac{v^\star}{|\mathbf{v}|^2}\mathbf{v} \cdot \mathrm{grad}\,\delta c \right) \mathbf{v} \cdot \mathrm{grad}c \, dv + \int_\Omega v\,\mathrm{grad}\delta c \cdot \mathrm{grad}c \, dv \,,$$

where the surface terms are not of interest for the present analysis and have been
neglected. The Galerkin interpolation $c = N_i h_i$ and $\delta c = N_i \delta h_i$ then determines
the advection and diffusion matrices

$$K_{ij} = \int_{\Omega_e} \underbrace{\left(N_i + \frac{v^\star}{|\mathbf{v}|^2} v_a B_{ai} \right)}_{S_i} v_b B_{bj} \, dv \; ; \quad D_{ij} = \int_{\Omega_e} B_{ai} v \, B_{aj} \, dv \qquad (4.64)$$

of the e-th finite element. We may interpret S_i as a newly defined shape function
in the interpolation of the test function, $\delta c = S_i \delta h_i$, and $\mathbf{K}$ would then have
been derived from a Petrov–Galerkin approximation. Given $\mathbf{D}$ is still based on a
Galerkin approximation, the matrices (4.64) arise from an inconsistent variational
formulation, known as *inconsistent Streamline Upwind (SU)* stabilization.

4.7.2.4 Petrov–Galerkin Formulations
Petrov–Galerkin formulations for stabilized AD problems may be summarized by
the following weak form

$$\int_\Omega \delta c\, \mathbf{v} \cdot \mathrm{grad}c \, dv + \int_\Omega v\,\mathrm{grad}\delta c \cdot \mathrm{grad}c \, dv + \underbrace{\int_\Omega \mathcal{P}(\delta c)\tau R(c) \, dv}_{\text{Stabilization term}}$$

$$= \int_{\partial\Omega_q} \delta c\,\overline{q}\,ds - \int_\Omega \delta c \alpha dv \,, \qquad (4.65)$$

where τ denotes a user-selected stabilization parameter, and $R(c) = \mathbf{v} \cdot \mathrm{grad}c - v\mathrm{div}(\mathrm{grad}c) + \alpha$ is the residual error of the AD problem $(4.10)_1$. Given this residual
error term, the effect of the stabilization will be small for a low R. The operator
$\mathcal{P}(\delta c)$ is determined by the specific stabilization formulation.

The Streamline Upwind Petrov–Galerkin (SUPG) stabilization [59, 268] is one of the most widely used stabilization methods and implements

$$\mathcal{P}(\delta c) = \mathbf{v} \cdot \mathrm{grad}\delta c \text{ and } \tau = v^\star/|\mathbf{v}|^2 .$$

The Galerkin least-square (GLS) stabilization is another frequently used method [270].

Example 4.9 (SUPG-Stabilized 1D Advection–Diffusion Problem). To avoid oscillating solutions at coarse finite element meshes, the Advection–Diffusion (AD) problem requires stabilization. In this exercise the Streamline Upwind Petrov–Galerkin (SUPG)-stabilized finite element should be used to solve the 1D AD problem.

(a) Consider the linear shape functions (4.3), and provide the expressions of $\mathbf{K}$, $\mathbf{D}$, and $\mathbf{f}$ of the SUPG-stabilized finite element. Given the stabilization parameter β, the artificial diffusivity $v^\star = \beta v h/2$ may be considered, where h and v denote the finite element length and the advective velocity, respectively.

(b) Explain why the SUPG-stabilized finite element can be regarded as a consistent Petrov–Galerkin finite element.

(c) Consider the discretized problem shown in Fig. 4.13 and use the diffusivity $v = 1.0\,\mathrm{m}^2\,\mathrm{s}^{-1}$, the constant finite element length $h = 0.1\,\mathrm{m}$, and the parameter $\alpha = 1.0\,\mathrm{s}^{-1}$ of the AD problem. Compute the essential variable c for the Péclet number $Pe = 1.0$ and estimate the stabilization parameter β by trial and error that gives the closest result to the exact solution of this problem. ■

As with the AD problem, Galerkin-based FE formulations of the incompressible flow (4.43) and (4.44) may also result in spurious node-to-node oscillations of the velocity. It is known to appear in advection-dominated flows as well as flows with sharp boundary layers. The use of inappropriate combinations of interpolation functions for the velocity $\mathbf{v}$ and the pressure p fields is also known to be a source of a numerical instability. It results in oscillations that primarily effect the pressure field. Very similar approaches to the ones discussed to stabilize the AD problem have also been used to stabilize incompressible flows [546].

4.8 Solving the System of Finite Element Equations

The individual bioengineering application determines the system of equations to be solved. Whilst a linear static structural problem yields a system of linear algebraic equations, a non-linear transient problem requires the solution of a system of non-linear partial differential equations in time. The systems may or may not be symmetric, and in many cases the essential variables have to satisfy constraints. In this section some of the most popular approaches to solve systems of FEM equations are discussed. They can also be used to solve systems from many other discretization methods.

4.8.1 Solving Sparse Linear Systems

Direct or *iterative* solution stagiest may be used to solve the linear system

$$\mathbf{Kh} - \mathbf{f} = \mathbf{0} \tag{4.68}$$

of algebraic equations, and thus to compute the essential variables $\mathbf{h}$. Given FEM applications, the system of Eqs. (4.68) is in general very large, which makes its solution computationally demanding. The matrix $\mathbf{K}$, is constant, sparsely populated, and for many applications it is also symmetric. Storage schemas, such as *skyline* storage and *sparse* storage, greatly reduce the time needed to solve, as well as the memory to store the data. Sparse storage schemas store only the non-zero elements of $\mathbf{K}$, but in a way that still enables efficient computations, especially of matrix vector products.

4.8.1.1 Direct Solution Methods
Direct solution schemas, and thus variants of *Gaussian elimination*, are the method of choice for "small" systems of up to approximately 100k unknowns. Parallel computing splits the finite element problem into smaller sub-problems, and given enough processors, even large problems may be efficiently solved by direct methods.

LU factorization is a direct solution method and expresses the matrix through $\mathbf{K} = \mathbf{LU}$, where $\mathbf{L}$ and $\mathbf{U}$ are lower and upper triangular matrices, respectively. Here, $\mathbf{U}$ is the established notation in linear algebra and must not be mixed up with the right stretch tensor used in continuum mechanics. All Gaussian elimination-based methods are unstable in their pure form, and a *non-zero pivot* has to be ensured at every step of the elimination process. Given finite precision arithmetic, already small pivots can make the schema unstable or introduce considerable numerical errors in the solution. *Pivoting* is therefore used.

Pivoting interchanges rows of the equation system to ensure relatively large entries as pivot elements. Strategies, such as partial pivoting, complete pivoting, and rook pivoting, are known. The result of pivoting is the permutation matrix $\mathbf{I}_P$ that swaps rows or columns. The system of Eqs. (4.68) then reads

$$\overline{\mathbf{K}}\overline{\mathbf{h}} - \overline{\mathbf{f}} = \mathbf{0}$$

with $\overline{\mathbf{K}} = \mathbf{I}_P\mathbf{K}$, $\overline{\mathbf{f}} = \mathbf{I}_P\mathbf{f}$, and $\overline{\mathbf{h}} = \mathbf{I}_P\mathbf{h}$ representing the rearranged matrix and vectors, respectively.

LDU factorization is another commonly used approach to solve sparse linear systems that emerge from the FEM. It uses upper $\mathbf{U}$ and lower $\mathbf{L}$ triangular matrices of the property $\mathbf{L} = \mathbf{U}^T$ together with the decomposition $\mathbf{K} = \mathbf{LDU}$, where $\mathbf{D}$ is a nonsingular diagonal matrix. Consequently, the system (4.68) reads $\mathbf{LDUh} = \mathbf{f}$ and its solution involves the three consecutive steps

Forward reduction: set $\mathbf{DUh} = \mathbf{z}$ and solve $\mathbf{Lz} = \mathbf{f}$ for $\mathbf{z}$;

Diagonal scaling: set $\mathbf{Uh} = \mathbf{y}$ and solve $\mathbf{Dy} = \mathbf{z}$ for $\mathbf{y}$;

Back substitution: solve $\mathbf{Uh} = \mathbf{y}$ for $\mathbf{h}$.

Many other direct solution approaches are known, and the literature is rich of methods to solve sparse linear systems [113].

4.8.1.2 Iterative Solution Methods

The matrix of FEM problems is diagonally populated, a structure that supports the application of *iterative* solution methods, the strategy of choice to solve "large" systems of more than approximately 100k unknowns.

Change of sign and adding $\mathbf{Ah}$ to both sides of equation (4.68) yield

$$\mathbf{Ah} = \mathbf{Ah} + \mathbf{f} - \mathbf{Kh}, \qquad (4.69)$$

where $\mathbf{A}$ is a diagonal matrix to be specified later. The substitution $\mathbf{h} = \mathbf{h}_n$, applied only to the right side of (4.69), yields then the iteration rule

$$\mathbf{Ah} = \mathbf{Ah}_n + \mathbf{f} - \mathbf{Kh}_n. \qquad (4.70)$$

Given $\mathbf{A}$ approximates the finite element problem and therefore contains "similar" information than $\mathbf{K}$, the expression (4.70) yields a fixpoint iteration and converges to the solution of (4.68). This iteration is only efficient for a diagonal matrix $\mathbf{A}$, and the system of Eqs. (4.70) can then be solved row by row. The diagonal matrix $\mathbf{A}$ may be formed by the diagonal terms of $\mathbf{K}$, or by lumping all column elements of $\mathbf{K}$ into the corresponding diagonal term of $\mathbf{A}$.

The convergence of the iteration (4.70) can be enhanced through *preconditioning*. The system (4.68) is then multiplied with a problem-specific precondition matrix before the iteration (4.70) starts. Iterative solution methods that are based on many other approaches are known, and the literature is rich in the description of fast iterative solvers [140].

Example 4.10 (Solving a Linear System of Equations). The symmetric system of linear equations $\mathbf{Kh} = \mathbf{f}$ with

$$\mathbf{K} = \begin{bmatrix} 8 & 2 & 1 & 0 \\ 2 & 9 & 5 & 3 \\ 1 & 5 & 12 & 2 \\ 0 & 3 & 2 & 6 \end{bmatrix} \quad ; \quad \mathbf{f} = \begin{bmatrix} 0 \\ 0 \\ 1 \\ 2 \end{bmatrix}$$

represents a discretized structural problem.

(a) Use a LU factorization-based direct solution method to solve the system of equations and report **h**.
(b) Use an iterative solution method to solve the system of equations and report its convergence.   ∎

4.8.2   Time Integration

The solution of a transient FEM problem requires the integration of a system of partial differential equations in time. To explore such a problem, let us for simplicity consider the first-order differential equation $dy/dt = f(t, y)$, where y and t denote the essential variable and the time, respectively. An *explicit* or *implicit* approach may be used to integrate the function $f(t, y)$. Given the essential variable y_n at the time t_n, an explicit method allows for the explicit expression of the essential variable y_{n+1} at the time $t_{n+1} = t_n + \Delta t$, and thus at the time increment Δt later. In contrary, implicit methods require the solution of an implicit system of equations to calculate the essential variable y_{n+1}. In general, a time-marching iteration

$$y_{n+1} = y_n + \phi \Delta t$$

may be used for the integration—it is explicit for $\phi = \phi(t_n, y_n)$ and implicit for $\phi = \phi(t_{n+1}, y_{n+1})$.

The *Euler integration* uses the factor $\phi = dy/dt$ and leads either to explicit or implicit integrations, known as forward-Euler and respectively backward-Euler integrations, see Fig. 4.18. The Euler integration is *first-order* accurate and its *global* truncation error, the error at a given time t, is therefore proportional to the step size Δt. The error estimation

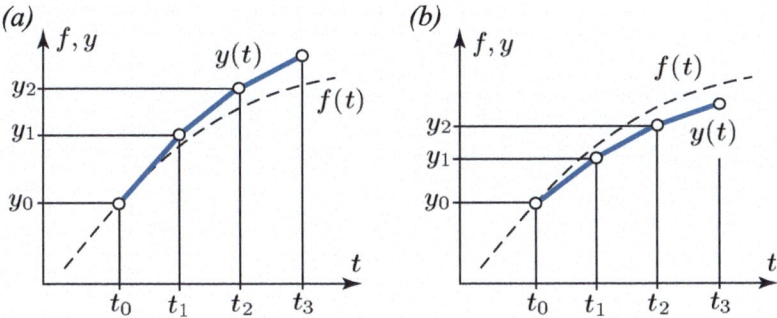

Fig. 4.18 Numerical integration of a first-order differential equation with the exact solution denoted by the function $f(t)$ (dashed curve). (**a**) Forward-Euler and (**b**) Backward-Euler integration yield the approximations $y(t)$ (solid curves)

$$\epsilon(\Delta t) = |f - y| < C\Delta t^q$$

has then the coefficient $q = 1$. The constant C depends on the function f. In contrary, the *local* truncation error, the error per step, is proportional to Δt^2. The Euler integration is the simplest case of a family of time-marching algorithms, known as *Runge[9]–Kutta[10]* methods.

Aside the accuracy, the stability of the time integration is crucial to the solution of problems. It defines conditions for the step size Δt that then lead to a converged numerical solution. It is especially important for *stiff* problems, where the solution of the problem shows sharp gradients in time, conditions that may trigger the formation of instabilities.

Example 4.11 (Stability of the Euler Integration). Consider the stiff differential equation

$$\mathrm{d}y/\mathrm{d}t = -yt \;\; ; \;\; y(0) = 1 \tag{4.72}$$

and test the stability of the Euler integration.

(a) Specify the iteration of forward-Euler and backward-Euler integrations, respectively.
(b) Use the step size $\Delta t = 9/10$ and compute the forward-Euler and backward-Euler integrations in the interval $0 \le t \le 9$. ∎

4.8.3   Non-linear Formulations

A number of bioengineering problems are described by non-linear governing equations. The essential variable $\mathbf{h}$ is then determined by the non-linear vector equation

$$\mathbf{g}(\mathbf{h}) - \mathbf{f}(\mathbf{h}) = \mathbf{0}\,, \tag{4.73}$$

where $\mathbf{g}(\mathbf{h})$ and $\mathbf{f}(\mathbf{h})$ denote the internal and external contributions to the nodal forces, respectively. Non-linearities may stem from factors, such as a non-linear constitutive description, finite deformations, the advective term in the material time derivative, and follower loads.

A first-order Taylor expansion of (4.73) around $\mathbf{h}_0$ and over the increment $\Delta\mathbf{h}$ yields the system

$$\mathbf{g}(\mathbf{h}_0) + \mathbf{K}(\mathbf{h}_0)\Delta\mathbf{h} - \mathbf{f}(\mathbf{h}_0) - \mathbf{K}_\mathrm{f}(\mathbf{h}_0)\Delta\mathbf{h} = \mathbf{K}(\mathbf{h}_0)\Delta\mathbf{h} - \mathbf{K}_\mathrm{f}(\mathbf{h}_0)\Delta\mathbf{h} = \mathbf{0} \tag{4.74}$$

---

[9]Carl David Tolmé Runge, German mathematician, physicist, and spectroscopist, 1856–1927.
[10]Martin Wilhelm Kutta, German mathematician, 1867–1944.

of linearized equations, where $\mathbf{K}(\mathbf{h}_0)$ and $\mathbf{K}_f(\mathbf{h}_0)$ are the internal and external contributions to the system stiffness, respectively. The derivation of $(4.74)_2$ used the equilibrium relation $\mathbf{g}(\mathbf{h}_0) - \mathbf{f}(\mathbf{h}_0) = \mathbf{0}$ at the state $\mathbf{h}_0$ of the essential variable.

The internal $\mathbf{K}$ and external $\mathbf{K}_f$ contributions depend on the particular application. Given the linear problem discussed in Sect. 4.6, the constant external load b_f results in $\mathbf{K}_f = \mathbf{0}$, which together with the constant stiffness $\mathbf{K}$ reduces the equilibrium (4.73) to $\mathbf{Kh} - \mathbf{f} = \mathbf{0}$ and thus to the linear system (4.68).

4.8.4 Incremental Formulation

Given the external $\mathbf{f}$ and internal $\mathbf{g}$ nodal forces as well as the nodal contributions from inertia $\mathbf{m}$ and damping $\mathbf{d}$,

$$\mathbf{m}(\ddot{\mathbf{h}}) + \mathbf{d}(\dot{\mathbf{h}}) + \mathbf{g}(\mathbf{h}) - \mathbf{f}(\mathbf{h}) = \mathbf{0} \tag{4.75}$$

determines the motion of a material particle. It describes how the particle's state of motion, characterized by the displacement $\mathbf{h}$, the velocity $\dot{\mathbf{h}}$, and the acceleration $\ddot{\mathbf{h}}$, changes over time.

Let us consider a small deviation from the equilibrium state. The incremental relation $\Delta\mathbf{m}(\ddot{\mathbf{h}}) + \Delta\mathbf{d}(\dot{\mathbf{h}}) + \Delta\mathbf{g}(\mathbf{h}) - \Delta\mathbf{f}(\mathbf{h}) = \mathbf{0}$ then follows from relation (4.75) and yields

$$\mathbf{M}(\mathbf{h})\Delta\ddot{\mathbf{h}} + \mathbf{D}(\mathbf{h})\Delta\dot{\mathbf{h}} + \mathbf{K}(\mathbf{h})\Delta\mathbf{h} - \mathbf{K}_f(\mathbf{h})\Delta\mathbf{h} = \mathbf{0}, \tag{4.76}$$

where the linear approximations $\Delta\mathbf{m}(\ddot{\mathbf{h}}) = \mathbf{M}\Delta\ddot{\mathbf{h}}$, $\Delta\mathbf{d}(\dot{\mathbf{h}}) = \mathbf{D}(\mathbf{h})\Delta\dot{\mathbf{h}}$, $\Delta\mathbf{g}(\mathbf{h}) = \mathbf{K}(\mathbf{h})\Delta\mathbf{h}$, and $\Delta\mathbf{f}(\mathbf{h}) = \mathbf{K}_f(\mathbf{h})\Delta\mathbf{h}$ have been used. The mass and damping matrices are denoted by $\mathbf{M}$ and $\mathbf{D}$, whilst $\mathbf{K}$ and $\mathbf{K}_f$ are the stiffness matrices upon internal and external nodal forces, respectively. The solution of the system (4.76) provides the increments $\Delta\mathbf{h}$, $\Delta\dot{\mathbf{h}}$, and $\Delta\ddot{\mathbf{h}}$ and allows us then to update the state $\mathbf{h} = \mathbf{h}_n + \Delta\mathbf{h}$, $\dot{\mathbf{h}} = \dot{\mathbf{h}}_n + \Delta\dot{\mathbf{h}}$, and $\ddot{\mathbf{h}} = \ddot{\mathbf{h}}_n + \Delta\ddot{\mathbf{h}}$, where $(\bullet)_n$ denotes state variables at the previous time t_n.

Whilst the *boundary conditions* have already been included in (4.76), the computation of $\mathbf{h}, \dot{\mathbf{h}}, \ddot{\mathbf{h}}$ over the time t also requires the specification of the state variables at the time $t = 0$—the *initial conditions* of the problem.

4.8.5 Explicit Solution

The *explicit solution* strategy solves the system (4.76) for the acceleration, which then reads

$$\mathbf{M}_n\Delta\ddot{\mathbf{h}}_{n+1} = \mathbf{K}_{fn}\Delta\mathbf{h}_n - \mathbf{D}_n\Delta\dot{\mathbf{h}}_n - \mathbf{K}_n\Delta\mathbf{h}_n, \tag{4.77}$$

where $(\bullet)_n$ denotes quantities known from the last time point, whilst $(\bullet)_{n+1}$ are the quantities to be calculated. Given a *lumped mass matrix*, the acceleration increment $\Delta\ddot{\mathbf{h}}_{n+1}$ is rapidly computed. The matrix $\mathbf{M}_n$ is then a diagonal matrix and (4.77) represents an iteration rule—it avoids solving a system of equations. The acceleration $\ddot{\mathbf{h}}_{n+1} = \ddot{\mathbf{h}}_n + \Delta\ddot{\mathbf{h}}_{n+1}$ and the time increment Δt allow then the update of the velocities $\dot{\mathbf{h}}_{n+1} = \dot{\mathbf{h}}_n + \ddot{\mathbf{h}}_{n+1}\Delta t$, as well as the displacements $\mathbf{h}_{n+1} = \mathbf{h}_n + \dot{\mathbf{h}}_{n+1}\Delta t$. A single update step therefore determines this time-marching algorithm.

The solution strategy uses exclusively information at the time t_n to compute the system state at t_{n+1}, it is therefore an *explicit* time integration. The Courant–Friedrichs–Lewy (CFL) criterion [100]

$$\Delta t < \alpha \min(h/c) \tag{4.78}$$

may be used to set the time step Δt and ensure the stability of the solution. The factor α is often set to one, whilst h [m] and c [m s^{-1}] denote the characteristic length of the smallest finite element in the problem and the speed of sound in the material. The CFL criterion ensures that a sound wave would not propagate through the smallest finite element during one time step.

An explicit solution is recommended for problems with significant inertia or viscose forces, where the first or second terms in (4.76) therefore contribute a fair amount to the nodal forces. Given the solution of non-linear problems with an explicit solution strategy, it is always difficult to control the drift from the exact solution.

4.8.6   Implicit Solution

The *implicit solution* strategy solves the system (4.76) for the displacements. It is then rewritten as

$$\mathbf{K}^\star(\mathbf{h})\Delta\mathbf{h} - \mathbf{f}^\star = \mathbf{0} \,, \tag{4.79}$$

where $\mathbf{K}^\star$ and $\mathbf{f}^\star$ are the algorithmic stiffness matrix and the algorithmic load vector, respectively.

Let us consider the first-order transient problem $\mathbf{D}\Delta\dot{\mathbf{h}} + \mathbf{K}\Delta\mathbf{h} - \mathbf{K}_\mathrm{f}\Delta\mathbf{h} = \mathbf{0}$ to exemplify the derivation of $\mathbf{K}^\star$ and $\mathbf{f}^\star$. Given $\Delta\dot{\mathbf{h}}_{n+1} = \dot{\mathbf{h}}_{n+1} - \dot{\mathbf{h}}_n$ and the backward-Euler approximation $\dot{\mathbf{h}}_{n+1} \approx \Delta\mathbf{h}_{n+1}/\Delta t$ of the velocity, the time-discretized system reads

$$\mathbf{D}_{n+1}\Delta\mathbf{h}_{n+1}/\Delta t - \mathbf{D}_{n+1}\dot{\mathbf{h}}_n + \mathbf{K}_{n+1}\Delta\mathbf{h}_{n+1} - \mathbf{K}_{\mathrm{f}n+1}\Delta\mathbf{h}_{n+1} = \mathbf{0}$$

$$\underbrace{\left(\mathbf{D}_{n+1}/\Delta t + \mathbf{K}_{n+1} - \mathbf{K}_{\mathrm{f}n+1}\right)}_{\mathbf{K}^\star_{n+1}}\Delta\mathbf{h}_{n+1} - \underbrace{\mathbf{D}_{n+1}\dot{\mathbf{h}}_n}_{\mathbf{f}^\star_{n+1}} = \mathbf{0} \,,$$

an expression according to (4.79).

The algorithmic stiffness and the load vector are in general given by

$$\mathbf{K}^\star = k_1 \mathbf{K} + k_2 \mathbf{D} + k_3 \mathbf{M} - \mathbf{K}_f \; ; \; \mathbf{f}^\star = (m_1 \mathbf{D} + m_2 \mathbf{M}) \ddot{\mathbf{h}}_n + (d_1 \mathbf{D} + d_2 \mathbf{M}) \dot{\mathbf{h}}_n \, ,$$

where the applied time-integration method, such as the Newmark method or the Hilber–Hughes–Taylor method [267], then determines the coefficients k_i, d_i, m_i. For many applications $\mathbf{D}$ and $\mathbf{M}$ are constant, and the algorithmic load vector $\mathbf{f}^\star$ is then independent from the current time. It may be regarded as a history vector, and the system

$$\mathbf{K}^\star_{n+1} \Delta \mathbf{h}_{n+1} - \mathbf{f}^\star_n = 0 \tag{4.80}$$

then describes the non-linear problem—its solution will be discussed in the following sections.

4.8.6.1 Newton–Raphson Method

A non-linear FEM problem may always be expressed by the system

$$\mathbf{K}^\star(\mathbf{h}) \Delta \mathbf{h} - \mathbf{f}^\star = \mathbf{r}(\mathbf{h}) \tag{4.81}$$

of equations, where $\mathbf{r} = \mathbf{0}$ denotes the residuum. The quasi-static problem with dead loads is a particular case: $\mathbf{K}^\star(\mathbf{h}) = \mathbf{K}(\mathbf{h})$ then denotes the stiffness from the internal contributions to the nodal forces, and $\mathbf{f}^\star(\mathbf{h}) = \mathbf{0}$. The non-linear system (4.81) is solved iteratively through the minimization of the residuum $\mathbf{r}$ until $|\mathbf{r}| < \epsilon$, where ϵ denotes a tolerance level. Given the solution $\mathbf{h}_n$ at the previous step, the current increment $\Delta \mathbf{h}$ can be computed and allows us to update the solution $\mathbf{h} = \mathbf{h}_n + \Delta \mathbf{h}$.

The *Newton–Raphson* fixpoint iteration $\Delta \mathbf{h} \leftarrow \Delta \mathbf{h} - [\partial \mathbf{r}/\partial \mathbf{h}]^{-1} \mathbf{r}$ represent a common approach to solve the system (4.81). Figure 4.20 exemplifies the computation of the increment $\Delta \mathbf{h}_{n+1}$, which then is used to update the essential variable $\mathbf{h}_{n+1} = \mathbf{h}_n + \Delta \mathbf{h}_{n+1}$. Given the large dimension of a FEM system, its stiffness matrix $\mathbf{K}^\star = \partial \mathbf{r}/\partial \mathbf{h}$ is not inverted and the allocation of memory to store $\Delta \mathbf{h}$ is also avoided. The practical implementation of the Newton–Raphson fixpoint iterations then reads

$$\mathbf{h} \leftarrow \mathbf{h} - \delta \mathbf{h} \text{ with } \delta \mathbf{h} \text{ from solving } \mathbf{K}^\star(\mathbf{h}) \delta \mathbf{h} = \mathbf{0} \text{ and } \mathbf{h}_0 = \mathbf{0} \, . \tag{4.82}$$

Given a continuous residuum $\mathbf{r}$ and a vector $\mathbf{h}$ that is close to the solution, the Newton–Raphson fixpoint iteration converges *quadratically*. The solution is then found within a small number (three to six) of iterations and satisfies the condition

$$\log(|\mathbf{r}|_{n+1}) / \log(|\mathbf{r}|_n) \approx 2 \tag{4.83}$$

between two consecutive iterations.

Fig. 4.20 Newton–Raphson fixpoint iteration to compute the essential variable **h** at the residuum **r** = **0**. Starting at **h**$_n$ and the residuum **r**$_n$, the image illustrates the iterations δ**h**$_i$, which then sum up and determine the increment Δ**h**$_{n+1}$. The update **h**$_{n+1}$ = **h**$_n$ + Δ**h**$_{n+1}$ finally yields the solution

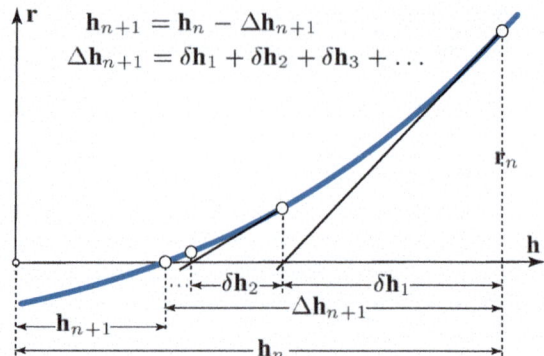

4.8.6.2 Load Incrementation

The Newton–Raphson iteration may not converge for large increments Δ**h** and thus for initializations **h**$_n$ that are too far away from the solution **h**$_{n+1}$. Whilst for transient problems the time step Δt can be reduced, *load incrementation* is commonly applied for quasi-static problems towards overcoming this drawback of the Newton–Raphson method. A load factor $0 \leq \lambda \leq 1$ is introduced, and instead of (4.73), **g**(**h**) $- \lambda$**f** = **0** is then solved, where we considered dead loads for simplicity. Given the solution is found, the load factor λ is increased, and the problem is solved again. It is equivalent to the solution of the system

$$\mathbf{K}(\mathbf{h})\Delta\mathbf{h} - \Delta\mathbf{f} = \mathbf{r}(\mathbf{h}) , \qquad (4.84)$$

at each load increment Δ**f** $= (\lambda_{n+1} - \lambda_n)$**f** and leads to two nested loops, see Table 4.4. The load factor λ may also be linked to the essential variables **h**, which then leads to a separate family of solution techniques, known as *arc-length* or *continuation methods*, see Sect. 4.8.6.4.

4.8.6.3 Dirichlet *Versus* Neumann Boundary Conditions

The description of either Dirichlet or Neumann boundary conditions can have important implication towards the ability to solve a non-linear problem. Let us consider a truss of Young's modulus E and cross-section A under tension. It is modeled by two finite elements of the lengths $L_a = \xi L$ and $L_b = (1 - \xi)L$ with $0 < \xi < 1$, see Fig. 4.21. Element b is linear-elastic and has the stiffness $k_b = EA/[(1 - \xi)L]$, whilst element a reflects softening and then failure of the truss. The referential stiffness $k_{a0} = EA/(\xi L)$ of element a changes with increasing engineering strain $\varepsilon = u_1/(\xi L)$ towards $k_a = -(k_{a0}/\varepsilon) \exp(-\varepsilon)[\exp(-\varepsilon) - 1]$.

The two equations $F = -k_{a0} \exp[-u_1/(\xi L)]\{\exp[-u_1/(\xi L)] - 1\}$ and $F = k_b(u_2 - u_1)$ describe then the internal equilibrium at node 1, and given the substitution $u_1 = u_2 - F/k_b$, the residuum equation

$$r(u_2, F) = k_{a0}\alpha(\alpha - 1) + F = 0 \quad \text{with} \quad \alpha = \exp[(F/k_b - u_2)/(\xi L)] \qquad (4.85)$$

Table 4.4 Algorithm to solve a non-linear FEM problem with load incrementation and Newton–Raphson iterations

(1) Set load increment $\Delta\lambda = 1/n_l$ with n_l denoting the number of load increments
(2) Initialize essential variables $\mathbf{h} = \mathbf{0}$ and load iteration $\lambda = 0$
(3) Set load increment $\Delta\mathbf{f} = \Delta\lambda\mathbf{f}$

(4) Loop over load incrementation
     Do While $\lambda \leq 1$
 	
         $\lambda \leftarrow \lambda + \Delta\lambda$

(5) Loop over Newton–Raphson iteration
     Do While $|\mathbf{r}| \geq 10^{-8}$
         Compute $\delta\mathbf{h}$ from the direct or iterative
         solution of $\mathbf{K}(\mathbf{h})\delta\mathbf{h} - \Delta\mathbf{f} = \mathbf{r}(\mathbf{h})$
         $\mathbf{h} \leftarrow \mathbf{h} - \delta\mathbf{h}$
     End Do
 	
     End Do

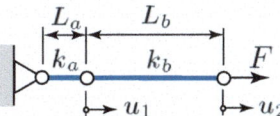

Fig. 4.21 Two finite elements to model the failure of a truss at tension. Element a captures strain softening, whilst element b has linear-elastic properties

represents the problem. We may prescribe either the load F or the displacement u_2, which would represent either a Neumann or a Dirichlet boundary condition at the right end of the truss. A Newton–Raphson method may then be used to solve for the non-prescribed variable.

Table 4.5 illustrates the prescription of the Dirichlet boundary condition or the *displacement-controlled* approach. The displacement u_2 is prescribed, whilst the iteration $F \leftarrow F - \delta F$ with $\delta F = r(\partial r/\partial F)^{-1}$ determines the force. In contrary, Table 4.6 illustrates the prescription of the Neumann boundary condition or the *load-controlled approach*, where the force F is prescribed and $u_2 \leftarrow u_2 - \delta u_2$ with $\delta u_2 = r(\partial r/\partial u_2)^{-1}$ determines the displacement. Given the distensibility $\partial r/\partial F \neq 0$, the displacement-controlled approach is able to compute strain softening, whilst the load-controlled approach terminates as soon as the ultimate load is reached. Here, the stiffness $\partial r/\partial u_2 = 0$ and can therefore not be inverted. Figure 4.23 shows the force *versus* displacement computed with the displacement-controlled and load-controlled solution approaches.

Table 4.5 Algorithm for the displacement-controlled solution of the residuum (4.85)

(1) Set Parameters

$\xi = 0.4$; $E = 70.0$ MPa; $A = 1.0$ cm^2; $L = 0.5$ m

$k_b = EA/[(1 - \xi)L]$; $k_{a0} = EA/(\xi L)$

(2) Set displacement increment $\Delta u_2 = 1/n$ with n denoting the number of increments

(3) Initialize variables $u_2 = 0$ and $F = 0$

(4) Loop over load incrementation

Do While $u_2 \leq n \Delta u_2$

$u_2 \leftarrow u_2 + \Delta u_2$

(5) Loop over Newton–Raphson iteration

$r = 1$

Do While $r \geq 10^{-8}$

$r = k_{a0}\alpha[\alpha - 1] + F$; $K = \frac{k_{a0}\alpha(2\alpha - 1)}{L\xi k_b} + 1$ with

$\alpha = \exp\left[\frac{F/k_b - u_2}{\xi L}\right]$

$F \leftarrow F - \delta F$ with $\delta F = r/K$

End Do

End Do

Table 4.6 Algorithm for the load-controlled solution of the residuum (4.85)

(1) Set Parameters

$\xi = 0.4$; $E = 70.0$ MPa; $A = 1.0$ cm^2; $L = 0.5$ m

$k_b = EA/[(1 - \xi)L]$; $k_{a0} = EA/(\xi L)$

(2) Set load increment $\Delta u_2 = 1/n$ with n denoting the number of increments

(3) Initialize variables $u_2 = 0$ and $F = 0$

(4) Loop over load incrementation

Do While $F \leq n\Delta F$

$F \leftarrow F + \Delta F$

(5) Loop over Newton–Raphson iteration

$r = 1$

Do While $r \geq 10^{-8}$

$r = k_{a0}\alpha[\alpha - 1] + F$; $K = -\frac{k_{a0}\alpha(2\alpha - 1)}{\xi L}$ with

$\alpha = \exp\left[\frac{F/k_b - u_2}{\xi L}\right]$

$u_2 \leftarrow u_2 - \delta u_2$ with $\delta u_2 = r/K$

End Do

End Do

Example 4.12 (Spring Lever Structure). Figure 4.22 shows a structure that is loaded by the force F. A rigid lever redirects the force into a spring of the stiffness $k = 1.2$ N mm^{-1}. The structure is weightless, and the dimensions $a = 38$ mm and

$b = 20$ mm define its unloaded configuration. The equilibrium determines it spatial configuration, which may be parameterized by the vertical displacement v.

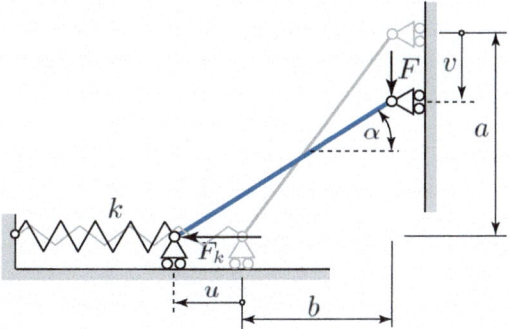

Fig. 4.22 Structure that is loaded by the force F, which is then redirected through a rigid lever into a spring. The displacement v defines the structure's spatial configuration

(a) Derive the non-linear governing equation of this problem.
(b) Use a Newton–Raphson algorithm to compute the force F upon a prescribed displacement v.
(c) Use a Newton–Raphson algorithm to compute the displacement v upon a prescribed force F.   ■

4.8.6.4 Arc-Length, or Continuation Methods

A structure may exhibit a so-called *snap-back* behavior. The force *versus* displacement path becomes then instable, and the structure quickly changes its configuration. Even a displacement-controlled solution approach does not allow to solve the problem. Whilst a transient solution approach, and thus considering the dynamics of the snap-back, would be fully feasible, it is much faster to compute the quasi-static solution with the arc-length method.

The *arc-length solution approach* introduces the load factor λ, an additional degree of freedom that weights the external load. It needs to be distinguished from Sect. 4.8.6.2, where the load factor has been imposed in the solution of the problem. Now, the nodal displacements $\mathbf{u}$ together with the load factor λ define the generalized variable $\mathbf{h} = [\mathbf{u}\ \lambda]^{\mathrm{T}}$, and the problem is solved along the path that is formed by the increments $\Delta\mathbf{h}$. Given a quasi-static problem with $\mathbf{g}$ and $\mathbf{f}$ denoting the internal and respective external nodal forces, the system

$$\mathbf{g} - \lambda\mathbf{f} = \mathbf{r}_{\mathbf{u}} \ ; \quad \Delta\mathbf{u} \cdot \Delta\mathbf{u} + \eta^2 \Delta\lambda^2 \mathbf{f} \cdot \mathbf{f} - \Delta h^2 = r_\lambda \qquad (4.89)$$

then defines the problem, where Δh is a predefined step size. It may be seen as the radius of a hypersphere in a space that is formed by the problem's degrees of freedom. The external force acting on the structure is $\lambda\mathbf{f}$, and η in $(4.89)_2$ serves as a scaling factor that weights the influence of the displacement and the load in determining the solution path. At the limit cases $\eta = 0$ and $\eta = \infty$, the arc-length

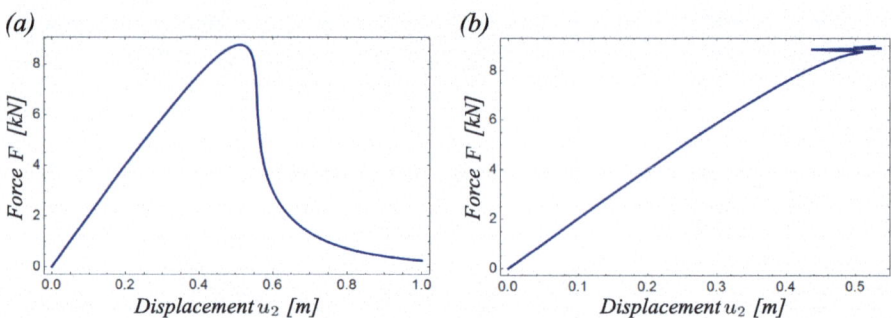

Fig. 4.23 Force displacement properties of the truss problem shown in Fig. 4.21 and the use of the parameters $E = 70.0\,\text{MPa}$, $A = 1.0\,\text{cm}^2$, $L = 0.5\,\text{m}$, and $\xi = 0.4$. (**a**) Displacement-controlled solution. (**b**) Load-controlled solution with non-converged results towards the ultimate load level of the structure

method resembles the displacement-controlled and force-controlled approaches, respectively.

Given a *trial* (or guess) of the generalized variable, the Newton–Raphson iteration $\mathbf{h} \leftarrow \mathbf{h} - \mathbf{K}^{-1}\mathbf{r}$ with

$$\mathbf{K} = \begin{bmatrix} \partial \mathbf{r}_{\mathrm{u}}/\partial \mathbf{u} & \partial \mathbf{r}_{\mathrm{u}}/\partial \lambda \\ \partial r_{\lambda}/\partial \mathbf{u} & \partial r_{\lambda}/\partial \lambda \end{bmatrix} \quad \text{and} \quad \mathbf{r} = \begin{bmatrix} \mathbf{r}_{\mathrm{u}} \\ r_{\lambda} \end{bmatrix} \tag{4.90}$$

serves then as the *corrector step* and determines $\mathbf{h}$. Even for the case the displacement-controlled stiffness $\partial \mathbf{r}_{\mathrm{u}}/\partial \mathbf{u}$ would be singular, $\mathbf{K}$ in $(4.90)_1$ is not singular and may be inverted. However, $\mathbf{K}$ is neither symmetric nor banded. Whilst the aforementioned concept founds the basics of all arc-length approaches, many different variants are known [106].

Let us solve the problem shown in Fig. 4.21 with a specific arc-length method, known as *pseudo-displacement control*. The displacement of a single node then controls the problem, and Δu_1^2 substitutes the term $\Delta \mathbf{u} \cdot \Delta \mathbf{u}$ in Eq. (4.89_2). Given u_1, u_2, and F, the non-linear system

$$r_{\mathrm{u}\,1} = k_{a0}\alpha(\alpha - 1) + \lambda F = 0\,;$$

$$r_{\mathrm{u}\,2} = k_b(u_2 - u_1) - \lambda F = 0\,;$$

$$r_{\lambda} = \Delta u_1^2 + \Delta \lambda^2 \eta^2 F^2 - \Delta h^2 = 0$$

then determines the problem, where the abbreviation $\alpha = \exp[-u_1/(\xi L)]$ has been used. In addition Δh defines the step size of the solution method, and Δu_1, Δu_2 and $\Delta \lambda$ denote the increments of the nodal displacements and load factor, respectively.

Given a trial state, the Newton–Raphson iteration

$$\begin{bmatrix} u_1 \\ u_2 \\ \lambda \end{bmatrix} \leftarrow \begin{bmatrix} u_1 \\ u_2 \\ \lambda \end{bmatrix} - \begin{bmatrix} -\frac{k_{a0}}{\xi L}(2\alpha^2 - \alpha) & 0 & F \\ -k_b & k_b & -F \\ 2\Delta u_1 & 0 & 2F^2\eta^2\Delta\lambda \end{bmatrix}^{-1} \begin{bmatrix} r_{\mathrm{u}\,1} \\ r_{\mathrm{u}\,2} \\ r_{\lambda} \end{bmatrix}$$

corrects the guess. Figure 4.25 illustrates results for different ξ values, and Table 4.8 reports the algorithm that has been used to compute said results.

Fig. 4.25 Force displacement properties of the truss problem shown in Fig. 4.21. The parameter ξ describes the discretization, and the example uses $E = 70.0\,\text{MPa}$, $A = 1.0\,\text{cm}^2$, and $L = 0.5\,\text{m}$

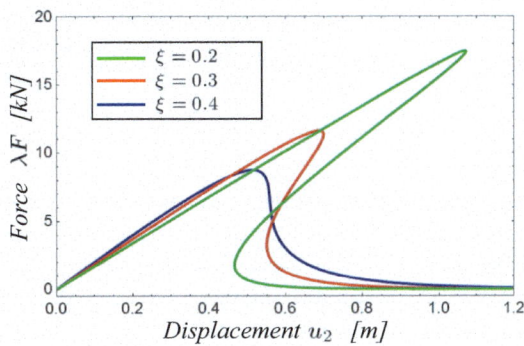

Table 4.8 Algorithm for the arc-length controlled solution of the problem shown in Fig. 4.21

(1) Set Parameters

$\xi = 0.4$; $E = 70.0$ MPa; $A = 1.0$ cm^2; $L = 0.5$ m

$k_b = EA/[(1-\xi)L]$; $k_{a0} = EA/(\xi L)$

(2) Set reference load $F = 1.0 \cdot 10^3$ N and scaling factor $\eta = 1.0 \cdot 10^{-4}$.

(3) Set step size $\Delta h = 0.01$ and the number of increments $n = 300$.

(4) Set initial conditions $\mathbf{h}_0 = \mathbf{0}$ and $\Delta \mathbf{h} = [\sqrt{\Delta s}\ 0\ \sqrt{\Delta s}]^\mathrm{T}$

(5) Loop over arc-length path increments

Do $i = 1, \ldots, n$

(6) Set trial state

$\mathbf{h} = \mathbf{h}_0 + \Delta \mathbf{h}$

(7) Loop over Newton–Raphson iteration

$|\mathbf{r}| = 1$

Do While $|\mathbf{r}| \geq 10^{-8}$

$\alpha = \exp[-h_1/(\xi L)]$; $\Delta h_1 = h_1 - h_{01}$; $\Delta h_3 = h_3 - h_{03}$

$$\mathbf{r} = \begin{bmatrix} k_{a0}\alpha(\alpha - 1) + h_3 F \\ k_b(h_2 - h_1) - h_3 F \\ \Delta h_1^2 + \Delta h_3^2 \eta^2 h_3^2 - \Delta h^2 \end{bmatrix} \quad ; \quad \mathbf{K} = \begin{bmatrix} -\frac{k_{a0}}{\xi L}(2\alpha^2 - \alpha) & 0 & F \\ -k_b & k_b & -F \\ 2\Delta h_1 & 0 & 2F^2\eta^2\Delta h_3 \end{bmatrix}$$

$\mathbf{h} \leftarrow \mathbf{h} - \mathbf{K}^{-1}\mathbf{r}$

End Do

(8) Update solution

$\Delta \mathbf{h} = \mathbf{h} - \mathbf{h}_0$; $\mathbf{h}_0 = \mathbf{h}$

End Do

4.8.6.5 Incompressible Flow Equation

The incompressible flow equation (4.45) and the flow incompressibility $\mathbf{Dh} = \mathbf{0}$ may be solved *monolytically*, and the entire system of equations

$$\begin{bmatrix} \mathbf{M} & \mathbf{0} \\ \mathbf{0} & \mathbf{0} \end{bmatrix} \begin{bmatrix} \dot{\mathbf{h}} \\ \dot{\mathbf{q}} \end{bmatrix} + \begin{bmatrix} \mathbf{Ah} + \mathbf{K} & -\mathbf{D}^{\mathsf{T}} \\ \mathbf{D} & \mathbf{0} \end{bmatrix} \begin{bmatrix} \mathbf{h} \\ \mathbf{q} \end{bmatrix} = \begin{bmatrix} \mathbf{f} \\ \mathbf{0} \end{bmatrix} \tag{4.91}$$

is then solved together towards computing the nodal velocity $\mathbf{h}$ and the nodal pressure $\mathbf{q}$.

The solution of the entire system (4.91) is challenging and time consuming—it may alternatively be decoupled into weakly interacting sub-systems. One of such methods is *Chorin's projection* method [87]. It assumes that viscous forces and pressure forces weakly interact, and we may therefore decouple the system (4.91). The momentum equation (4.45) is split into the two sub-systems

$$\mathbf{M}\dot{\mathbf{h}} + \mathbf{Ah}^2 + \mathbf{Kh} - \mathbf{f} = \mathbf{0} \,, \tag{4.92}$$

$$\mathbf{M}\dot{\mathbf{h}} - \mathbf{D}^{\mathsf{T}}\mathbf{q} = \mathbf{0} \tag{4.93}$$

and solved together with the incompressibility $\mathbf{Dh} = \mathbf{0}$.

We may now compute the intermediate velocity $\mathbf{h}^{\star}$ by solving (4.92) and then update the solution with (4.93) to compute the final velocity $\mathbf{h}$. However, the pressure momentum (4.93) requires the nodal pressure $\mathbf{q}$, which may be derived as follows. Given the derivation of the incompressible flow equations in Sect. 4.3.5, the pressure momentum (4.93) corresponds to the local expression

$$\operatorname{div} p\, \mathbf{I} = -\rho(\partial \mathbf{v})/\partial t \,, \tag{4.94}$$

where p and ρ denote the pressure and density of the fluid, respectively. Let us imagine the flow is already at its intermediate velocity $\mathbf{h}^{\star}$. The discretization of the pressure momentum (4.94) then reads $\operatorname{div} p\, \mathbf{I} = -\rho(\mathbf{v}_{n+1} - \mathbf{v}^{\star})/\Delta t$, where Δt and v_{n+1} denote the time increment and the velocity at the time t_{n+1}, respectively. Given the incompressibility $\operatorname{div}\mathbf{v}_{n+1} = 0$, the divergence of this expression yields the Poisson equation

$$\operatorname{div}(\operatorname{div} p) - \rho\operatorname{div}\mathbf{v}^{\star}/\Delta t = 0 \,. \tag{4.95}$$

The FEM approximation of Poisson's equation has been discussed in Sect. 4.8 and leads to the linear system $\mathbf{Dq} - \mathbf{f}_{\mathrm{q}} = \mathbf{0}$. Its solution then yields the nodal pressure $\mathbf{q}$. The time discretization of (4.93), $\mathbf{M}(\mathbf{h} - \mathbf{h}^{\star})/\Delta t - \mathbf{D}^{\mathsf{T}}\mathbf{q}$ may finally be used to update the nodal velocities $\mathbf{h}$. Table 4.9 summarizes the three steps of Chorin's projection method.

Table 4.9 Chorin's projection method to solve incompressible flow problems

(1) Compute the intermediate velocity $\mathbf{h}^\star$ from the solution of $\mathbf{M\dot{h}}^\star + \mathbf{Ah}^{\star 2} + \mathbf{Kh}^\star - \mathbf{f} = \mathbf{0}$.
(2) Compute the nodal pressure $\mathbf{q}$ from the solution of $\mathbf{Dq} - \mathbf{f_q} = \mathbf{0}$.
(3) Update the velocity and compute $\mathbf{h}$ from the solution of $\mathbf{M}(\mathbf{h} - \mathbf{h}^\star)/\Delta t - \mathbf{D}^T\mathbf{q}$.

4.9   Case Study: Planar Biaxial Testing

Planar biaxial testing is a commonly applied experimental protocol in the characterization of the vessel wall [482], see Fig. 4.26a. We consider a flat quadratic patch of the edge length $A = 16.0$ mm and the thickness $H = 2.3$ mm. At in total 20 anchor points the sample is hooked up to stiff wires that connect it to the actuators of the testing machine. The dimensions $a = 9.2$ mm and $b = A/12$ determine the spacing and positioning of the wires, and the FEM should be used in the exploration of the stress distribution within the test sample. Section 5.6 and Appendix F provide additional information about planar biaxial testing of vascular wall tissue.

4.9.1   Mechanical Problem Description

Given the *symmetry* of the problem, we model the quadrant $x_1 \geq 0$, $x_2 \geq 0$ and use the 2D plane stress assumption in the description of the mechanical problem, see Fig. 4.26b. The displacements $\bar{u}_1, \bar{u}_2$ are directly prescribed at the anchor points, whilst $u_1 = u_2 = 0$ holds at the symmetry axes, respectively. These settings determine the problem's Dirichlet boundary conditions, and the Neumann condition of zero traction applies to all other boundary conditions. Body forces are neglected in this analysis.

The vessel wall is described as an incompressible Yeoh material, where $\Psi(\mathbf{C}) = c_1(I_1 - 3) + c_2(I_1 - 3)^2$ denotes the strain energy per unit (undeformed) tissue volume. Here, $I_1 = \text{tr}\mathbf{C}$ denotes the first invariant of the right Cauchy–Green strain $\mathbf{C}$, whilst $c_1 = 1.1$ kPa and $c_2 = 15.5$ kPa are the material parameters characterizing the vessel wall sample.

The problem has been discretized by a quadratic triangular FEM mesh of approximately 69k degrees of freedom, and the Lagrange method enforced the incompressibility. We do not expect any limit points in the force *versus* displacement properties of the problem, and therefore a standard *displacement-controlled static analysis* is performed until the target displacements $\bar{u}_1 = 4.0$ mm and $\bar{u}_2 = 2.0$ mm will be reached.

4.9.2   Results and Verification

Figure 4.27 shows the computed distribution of the Cauchy stress components σ_{11} and σ_{22} superimposed on the deformed configuration Ω of the wall sample. The

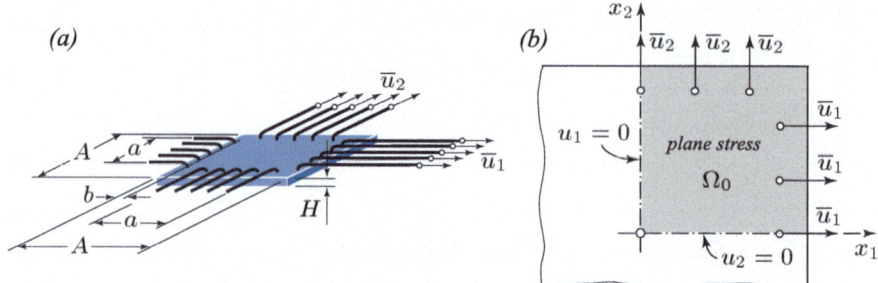

Fig. 4.26 Planar biaxial tension experiment towards the characterization of the vessel wall. (**a**) A rectangular patch of vessel wall is hooked up to stiff wires and then exposed to the displacements $\bar{u}_1$ and $\bar{u}_2$. (**b**) Mechanical problem description exploiting problem symmetries and using the 2D plane stress assumption

Fig. 4.27 Cauchy stress distribution in the vessel wall at planar biaxial testing. The deformation corresponds to $\bar{u}_1 = 4.0$ mm and $\bar{u}_2 = 2.0$ mm, and σ_{11} (left) and σ_{22} (right) are shown. For illustrative purposes, the analyzed domain is mirrored and the stress shown in half of the test sample

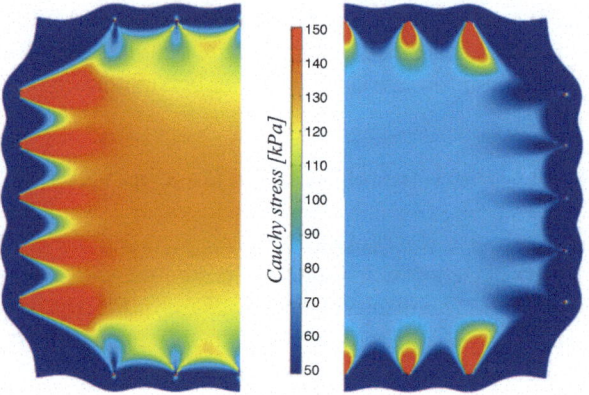

prescription of the displacements $\bar{u}_1$ and $\bar{u}_2$ directly at the finite element nodes results in stress concentrations and therefore highly inhomogeneous stress close to the anchor points. Away from the anchor points the stress state homogenizes and reaches the Cauchy stress components $\sigma_{11} = 131.85$ kPa and $\sigma_{22} = 75.36$ kPa at the center of the sample. The stretches $\lambda_1 = 1.513$ and $\lambda_2 = 1.200$ then correspond to this stress state. Given the analyzed quarter of the tissue sample, the prescription of the displacements at the anchor points results in the total reaction force of $F_1 = 1.5249$ N and $F_2 = 1.2011$ N along the x_1 and x_2 directions, respectively.

Towards the verification of these results, we consider an incompressible biaxial deformation in principal stress directions. Coleman and Noll's procedure (5.7) then yields the Cauchy stress

$$\sigma_1 = \alpha(\lambda_1^2 - \lambda_1^{-2}\lambda_2^{-2}) \; ; \; \sigma_2 = \alpha(\lambda_2^2 - \lambda_2^{-2}\lambda_1^{-2}) \, , \tag{4.96}$$

where the first invariant $I_1 = \lambda_1^2 + \lambda_2^2 + \lambda_1^{-2}\lambda_2^{-2}$ and the parameter $\alpha = 2c_1 + 4c_2(I_1 - 3)$ have been used. We may approximate the principal stretches by $\lambda_1 =$

$1 + 2\overline{u}_1/A = 1.5$ and $\lambda_2 = 1 + 2\overline{u}_2/A = 1.25$, values close to the FE predictions in the center of the sample. The expression (4.96) then yields $\sigma_1 = 138.0$ kPa and $\sigma_2 = 89.7$ kPa, values that again verify the FEM results. The stress is mainly distributed over the square of tissue that is inside of the anchor points, see Fig. 4.27. The force in the x_1 and x_2 directions then reads $F_1 = P_1 H(5A/12) = 1.411$ N and, $F_2 = P_2 H(5A/12) = 1.1$ N, where $P_i = \sigma_i/\lambda_i, i = 1, 2$ (no summation), denote the principal first Piola–Kirchhoff stress.

4.10 Case Study: Inflated Cylindrical Vessel

The stress and deformation of the vessel wall during the inflation experiment shown in Fig. 4.28a should be investigated. At the unloaded configuration the vessel segment is $L = 8.0$ cm long, and $R_i = 8.0$ mm and $R_o = 11.0$ mm are the respective inner and outer radii. We neglect residual stresses, and this configuration is therefore also the vessel's stress-free configuration. The vessel is cannulated, mounted at its reference length L in the testing device, and then inflated up to the pressure of $p_i = 26.0$ kPa.

The vessel wall is modeled as an incompressible Yeoh material, where $\Psi(\mathbf{C}) = c_1(I_1 - 3) + c_2(I_1 - 3)^2$ denotes the strain energy per unit (undeformed) tissue volume. Here, $I_1 = \mathrm{tr}\mathbf{C}$ denotes the first invariant of the right Cauchy–Green strain $\mathbf{C}$, whilst $c_1 = 18.1$ kPa and $c_2 = 138.0$ kPa are the material parameters characterizing the vessel wall sample.

The *symmetry* of the problem allows us to perform a *2D axisymmetric analysis*, where the domain $R_i \leq r \leq R_o, 0 \leq z \leq L/2$ determines the reference configuration Ω_0, see Fig. 4.28b. The displacements $u_r = u_z = 0$ are prescribed at the level of the cannulation $z = L/2$, and the symmetry condition $u_z = 0$ at $z = 0$ completes the description of the problem's Dirichlet boundary conditions. At the endothelium $r = R_i$ the pressure p_i acts and "follows" the structure during deformation, whilst the vessel's outer surface is traction-free. Neumann boundary

Fig. 4.28 Inflation of a cylindrical vessel segment. (a) The sample is cannulated at it reference length L and then inflated at the pressure p_i. (b) Axisymmetric structural analysis model

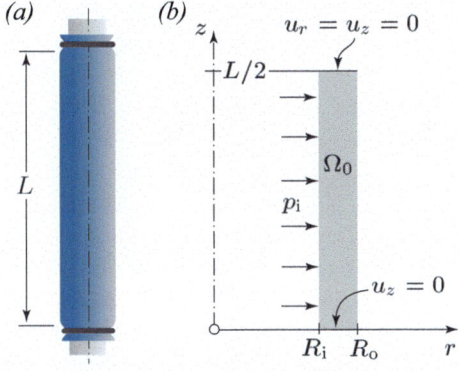

conditions are used in the prescription of these conditions. Body forces are neglected in this analysis.

4.10.1 Results and Verification

Figure 4.29a illustrates the distribution of the von Mises stress in the vessel wall at $p_i = 26.0$ kPa inflation. It represents the result from a *static FEM analysis* that used quadratic triangular finite elements of approximately 2.5k degrees of freedom with the Lagrange method enforcing the incompressibility.

We may use a membrane model to verify the FE results. As with the previous example, the vessel wall is at biaxial tension. Given $(4.96)_1$, the circumferential Cauchy stress in the membrane then reads

$$\sigma_\theta = \alpha(\lambda_\theta^2 - \lambda_\theta^{-2}) \ ; \quad \alpha = 2c_1 + 4c_2(\lambda_\theta^2 + \lambda_\theta^{-2} - 2), \tag{4.97}$$

where edge effects of the inflation experiment have been neglected, and $\lambda_z = 1$ specifies the axial stretch. Given the equilibrium in the circumferential direction $p_i(d - h) = 2\sigma_\theta h$ and the stress expression (4.97), the inflation pressure reads

$$p_i = 2\alpha(1 - \lambda_\theta^{-4})H/(D_m - H/\lambda_\theta^2), $$

where $D_m = R_i + R_o$ denotes the diameter of the membrane model. In addition, the relations $h = H/\lambda_\theta$ and $d_m = D_m\lambda_\theta$ have been used to express the wall thickness and the diameter of the deformed vessel. Given the circumferential stretch λ_θ, the inflation pressure p_i may be plotted over the vessel's outer diameter $d_o = d_m + h$, see Fig. 4.29(b). The plot compares the membrane-based solution to the FEM-predicted outer vessel diameter at $z = 0$. The non-linearity of the Yeoh material leads to a high stress gradient across the wall thickness, a *bending effect* that is not considered

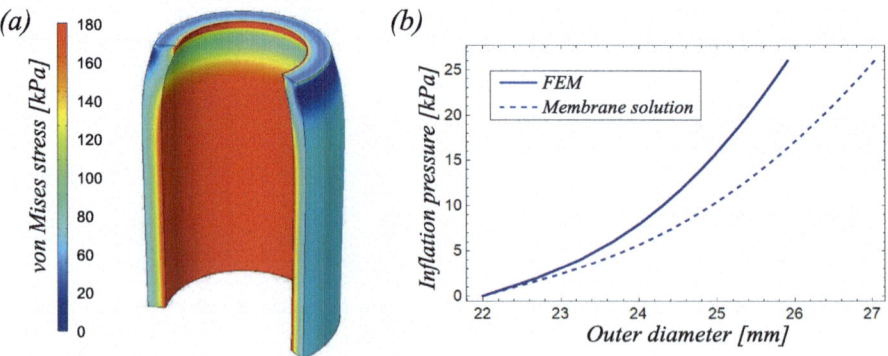

Fig. 4.29 (a) von Mises stress in the vessel wall at 26.0 kPa inflation. For a better visualization, the 2D axisymmetric solution has been rotated around the vessel axis. (b) Development of the outer diameter in the center of the vessel at increasing inflation pressure

by the membrane model, and then explains the significantly stiffer response of the FEM solution.

4.11 Case Study: Bulge Inflation

A bulge inflation experiment may be used to explore the elastic and failure properties of the vessel wall [476]. Given the present application, we consider a flat circular patch of the diameter $D = 4.0\,\text{cm}$ and the thickness $H = 1.5\,\text{mm}$ that is clamped along the circumference and inflated by the pressure p_i, see Fig. 4.30a. A FEM model should be used to compute the stress in the sample during the bulge inflation experiment.

4.11.1 Mechanical Problem Description

Body forces are neglected, and the vessel wall is described by an incompressible neoHookean material, where $\Psi(\mathbf{C}) = G(I_1 - 3)/2$ represents the strain energy per unit (undeformed) tissue volume. Here, $I_1 = \text{tr}\mathbf{C}$ denotes the first invariant of the right Cauchy–Green strain $\mathbf{C}$, and the referential shear modulus $G = 422.5\,\text{kPa}$ characterizes the vessel wall specimen.

Given the *isotropic* description of the vessel wall, the bulge inflation is *rotational symmetric*. The domain $0 \leq r \leq D/2$ and $0 \leq z \leq H$ then defines the reference configuration Ω_0 of the *2D axisymmetric structural analysis* problem, see Fig. 4.30b. Other than the pressure p_i that acts at $z = 0$, the boundary is traction-free, conditions that determine the problem's Neumann boundary conditions. The pressure p_i "follows" the surface and results in a non-symmetric stiffness of our problem, see Sect. 4.4.4.1.

The symmetry condition $u_r = 0$ at $r = 0$ specifies the problem's Dirichlet boundary conditions. We could also prescribe $u_r = u_z = 0$ at $r = D/2$, another Dirichlet boundary condition. However, this leads to severe mesh distortion and local buckling at large deformations, which eventually terminates the FEM computations, see Fig. 4.30c. At $r = D/2$, we instead prescribe a linear-elastic

Fig. 4.30 Bulge inflation experiment. (**a**) A circular patch of vessel wall is clamped and then exposed to the pressure p_i. (**b**) Axisymmetric structural analysis model. (**c**) Severe mesh distortion with the description of a Dirichlet boundary condition at $r = D/2$

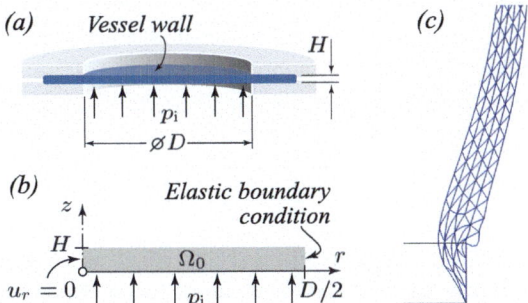

boundary condition with the spring stiffness $k_{bc} = 1.0\,\text{GPa m}^{-1}$. Alternative boundary conditions, including contact modeling that more accurately reflects the clamping of the tissue sample, would also be possible, but not further considered in the present case study.

4.11.2 Analytical Problem Assessment

We explore basic characteristics of the bulge inflation problem with a coarse analytical approximation prior to the FEM computation. The test specimen may then be approximated by a membrane that deforms into sphere, see Fig. 4.31a. The sphere has the radius $R = D^2/(8t) + t/2$, where t denotes the deflection in the center. The model assumes the stretch $\lambda = 2R\beta/D$ to be homogenously distributed all over the specimen, where the angle $\beta = \arcsin[D/(2R)]$ is shown in Fig. 4.31a. The kinematics result in an equi-biaxial deformation $\lambda = \lambda_\theta = \lambda_r$ with the Cauchy stress

$$\sigma = G(\lambda^2 - \lambda^{-4})$$

in the membrane. This expression follows from (4.96) and the incompressibility $\lambda_\theta \lambda_r \lambda_z = 1$, where θ, r and z denote the principal stress–stretch directions, the circumferential, radial, and axial directions, respectively.

The static equilibrium $D\pi h\sigma \sin\beta = p_i D^2 \pi/4$ relates the stress σ in the membrane to the inflation pressure p, where $h = H/\lambda^2$ denotes the thickness of the deformed membrane. We may use the substitution $\sin\beta = D/(2R)$, and the pressure then reads

$$p_i = \frac{2H\sigma}{R\lambda^2},$$

where σ, λ, and R are explicit functions of the deflection t. The pressure p_i may therefore be plotted over t, a graph shown by the dashed curve in Fig. 4.31b. We notice a *limit point* of the pressure-deflection properties, beyond which p_i decreases with increasing deflection t.

The above-outlined analytical assessment represents a very coarse structural description of the problem. Most important, the applied kinematics are *incompatible* with the boundary conditions of the problem—close to the clamping an equi-biaxial deformation cannot develop. The stress in the sample can therefore not be homogeneous, as it is illustrated by the FEM-based result in Fig. 4.32.

4.11.3 Solution Strategy and Results

As with the aforementioned analytical assessment, we also expect a limit point to appear in the FEM analysis of the problem—a standard static analysis cannot

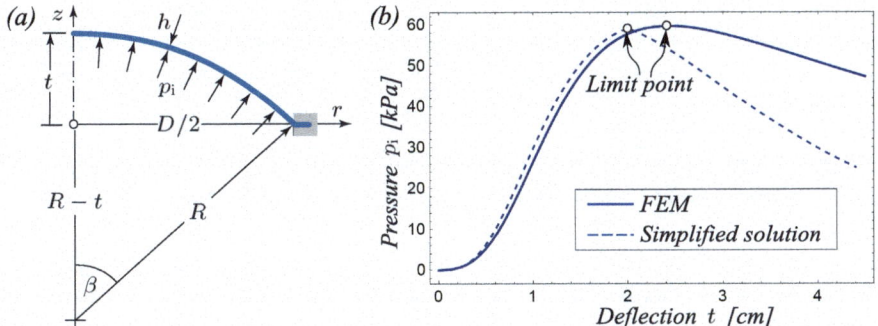

Fig. 4.31 Simplified solution of the bulge inflation experiment. (**a**) Spherical deformation-based kinematics. (**b**) Pressure p_i as a function of the deflection t with limit points

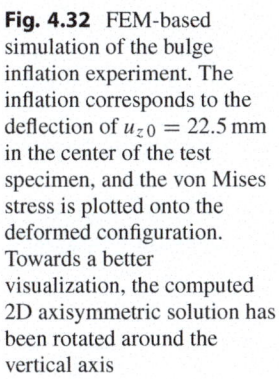

Fig. 4.32 FEM-based simulation of the bulge inflation experiment. The inflation corresponds to the deflection of $u_{z0} = 22.5$ mm in the center of the test specimen, and the von Mises stress is plotted onto the deformed configuration. Towards a better visualization, the computed 2D axisymmetric solution has been rotated around the vertical axis

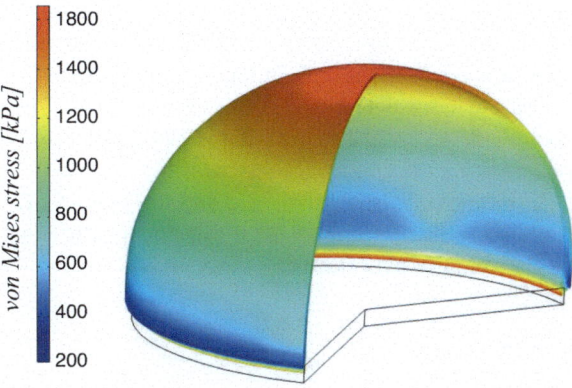

be used to solve the bulge inflation problem. We therefore express the inflation pressure by $p_i = \lambda p_0$, where λ and p_0 denote a load factor and a reference pressure, respectively. The global residuum equation

$$r_\lambda = \overline{u}_{z0} - u_{z0}(\lambda) = 0$$

is then added to the system of equations and couples the load factor λ, an additional degree of freedom, to the other degrees of freedom of our model. Here, u_{z0} denotes the axial deflection in the center, and thus at the finite element node located at $r = z = 0$, whilst $\overline{u}_{z0}$ is the prescribed deflection of this node. The outlined approach is equivalent to a *pseudo-displacement-controlled arc-length method* and supports computations beyond the limit point, and thus of an instable equilibrium. However, it avoids the access to the increments of solution variables as they have been used in Sect. 4.8.6.4. The problem has been discretized by 506 quadratic triangular finite elements and the Lagrange method enforced the incompressibility. Figure 4.32 illustrates the deformation of the test specimen and the corresponding von Mises stress.

4.12   Case Study: Flow Through a Network of Conduit Arteries

The FEM should be used to compute mean arterial flow in the major conduit arteries below the diaphragm. A network of rigid *cylindrical tubes* with circular cross-sections may be used in the description of the vascular tree, see Fig. 4.33. A Newtonian fluid with the dynamic viscosity of $\eta = 3.5\,\mathrm{mPa\ s}$ describes the rheological properties of blood, and we consider *Stokes flow* to describe the blood flow. Hagen–Poiseuille's law (2.31) then determines the resistance of a vessel segment. The aortic flow $q_{\mathrm{inl}} = 30\,\mathrm{ml\ s}^{-1}$ is prescribed at the level of the diaphragm, and the pressure of $p_{\mathrm{out}} = 13.0\,\mathrm{kPa}$ is set at all outlets. Vasoreactivity of the downstream tissue strongly influences the outlet pressures at the individual vessels, factors not considered in the present analysis. Gravitational effects are also not considered, and we assume ideal flow in all bifurcations—no pressure drops to appear across the connection points.

A *static analysis* is performed and the system of equations with 468 degrees of freedom is solved. Figure 4.34 illustrates the distribution of the pressure, flow rate, and the velocity over the arterial tree, respectively.

Fig. 4.33 Geometry and vessel dimensions to model the blood flow through a network of cylindrical tubes

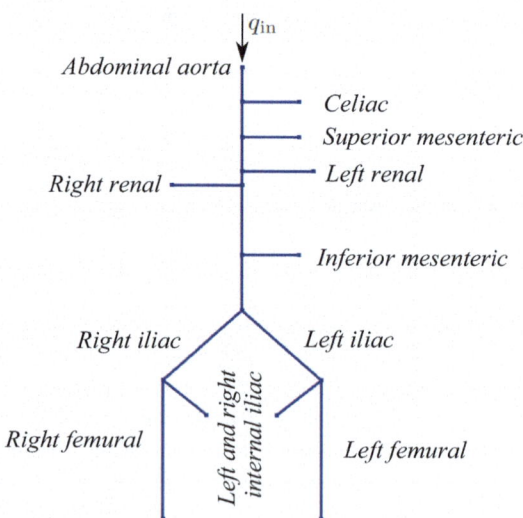

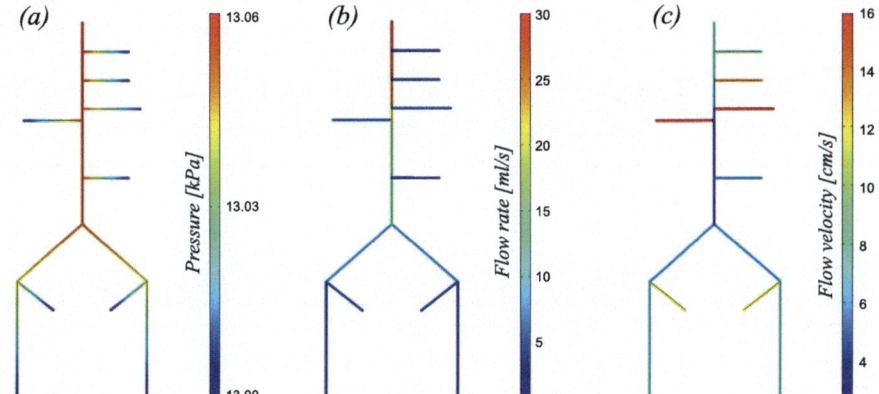

Fig. 4.34 Flow through a network of conduit arteries. The inflow of the abdominal aorta as well as the pressures at the individual vessel outlets is prescribed. (**a**) Pressure. (**b**) Flow rate. (**c**) Flow velocity

---

4.13   Case Study: Flow Through the Cylindrical Tube

The FEM should be used to explore fluid flow through a straight cylindrical tube, a widely used model to study blood flow in the vasculature. Blood may be described by a Newtonian fluid with the dynamic viscosity of $\eta = 3.5$ mPa s and the density $\rho = 1060$ kg m^{-3}. The vessel segment of the length $l = 30.0$ cm and the diameter $d = 2.4$ cm aims at modeling the human aorta. The blood flow in the aorta is complex, and *steady-state* as well as *steady-pulsatile* conditions should be explored. Regardless of the problem being rotational symmetric, a full *3D analysis* should be performed—it would support the development of any instabilities of the numerically predicted solution.

4.13.1   Steady-State Analysis

In addition to the no-slip boundary condition at the wall, we prescribe the inlet velocity $v_{\mathrm{inl}} = 40.0$ cm s^{-1} homogenously over the inlet cross-section, see Fig. 4.35. These conditions determine the problem's Dirichlet boundary, whilst the prescription of the outlet pressure $p_{\mathrm{out}} = 13.3$ kPa specifies the Neumann boundary. Given an incompressible fluid and a rigid vessel wall, the actual value of the outlet pressure *does not* influence the predicted blood flow. We also notice that the prescription of the inflow velocity of an incompressible fluid in the confined domain results in a highly constraint problem.

Figure 4.37a shows the development of the velocity profile of the steady-state tube flow. The result has been achieved with a tetrahedral finite element mesh of approximately 250k degrees of freedom. The mesh has been refined close to the wall

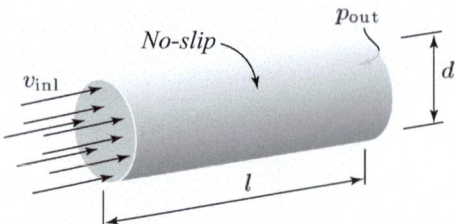

Fig. 4.35 Flow of an incompressible Newtonian fluid through the circular tube. The description of the inflow velocity v_{inl} together with the no-slip condition at the wall defines the problem's Dirichlet boundary, whilst the outlet pressure p_{out} specifies the problem's Neumann boundary

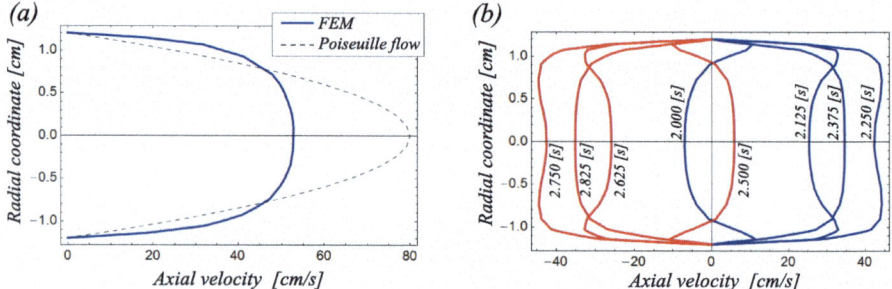

Fig. 4.36 Numerically predicted velocity profiles of Newtonian fluid flow in the circular tube. (**a**) Steady-state flow in comparison to the fully developed, and thus Poiseuille flow. (**b**) Pulsatile flow at a number of time points. Blue and red velocity profiles correspond to forward and backward flows, respectively

towards a better approximation of the strain rate gradient in the *boundary layer*. The tube is much shorter than the *hydrodynamic entrance length* of $l_h = 0.05 \, Re \, d = 345.6 \, cm$ that is needed to establish a Poiseuille flow, where $Re = v d \rho / \eta = 2880$ denotes the Reynolds number. Even at the end of the tube, the Poiseuille flow profile (6.34)

$$v = \frac{8q}{d^2 \pi} \left[1 - \left(\frac{2r}{d} \right) \right] \tag{4.98}$$

has therefore not yet been established, where $q = v_{inl} d^2 \pi / 4 = 181.0 \, ml \, s^{-1}$ denotes the flow rate through the vessel, see Fig. 4.36a.

4.13.2 Steady-Pulsatile Analysis

The inlet velocity $v_{inl} = v_0 \sin(2\pi t)$ with $v_0 = 40.0 \, cm \, s^{-1}$ and the no-slip condition at the wall describes the problem's Dirichlet boundary, whilst the outlet pressure $p_{out} = 0$ sets the Neumann boundary. A *transient analysis* with the

(a)

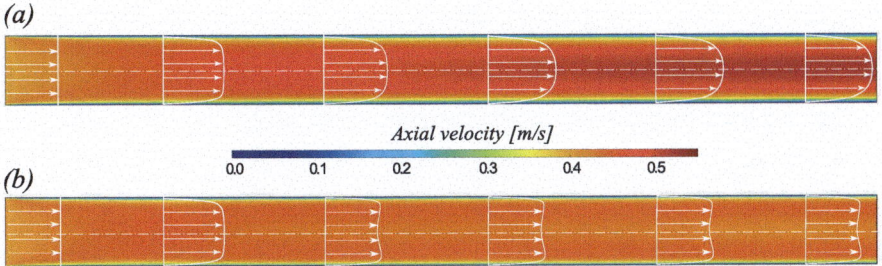

Axial velocity [m/s]

| 0.0 | 0.1 | 0.2 | 0.3 | 0.4 | 0.5 |

(b)

Fig. 4.37 Development of the velocity profile in the cylindrical tube. The inlet velocity is prescribed homogenously over the inlet cross-section and the flow develops against the constant outlet pressure. (**a**) Steady-state flow. (**b**) Pulsatile flow at peak inflow

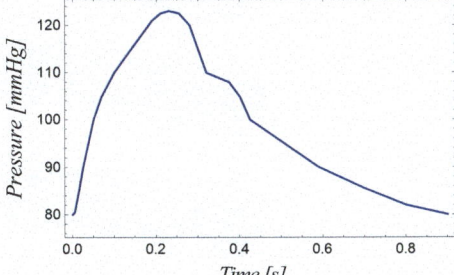

Time [s]	Pressure [kPa]	Time [s]	Pressure [kPa]
0.0	10.667	0.255	16.333
0.005	10.733	0.28	16.000
0.015	11.333	0.32	14.667
0.025	12.000	0.375	14.400
0.05	13.333	0.4	14.000
0.07	14.000	0.425	13.333
0.1	14.667	0.59	12.000
0.19	16.133	0.7	11.400
0.21	16.333	0.8	10.933
0.23	16.400	0.9	10.667

Fig. 4.38 Pressure waveform over the cardiac cycle time $T = 0.9\,\text{s}$ that represents normal ventricular function [119]

aforementioned tetrahedral finite element mesh has been used to compute the flow over the time period $0.0 \leq t \leq 3.0\,\text{s}$ and thus over three cycles.

Figure 4.37b shows the development of the velocity profile at the time $t = 2.25\,\text{s}$, conditions of peak inflow of the third cycle. The computed velocity profiles at a number of time points and $18.0\,\text{cm}$ downstream the inlet are shown in Fig. 4.36b. Given the inertia effects the velocity is very different from the steady-state flow. The flow changes its direction, and flow reversal starts at the layer close to the wall.

4.13.3 Transient Analysis with WindKessel Outlet Boundary Condition

We describe the pressure p_{inl} at the inlet by the profile shown in Fig. 4.38, whilst a three-element WindKessel (WK) model describes the systemic circuit at the outlet of the flow domain, see Sect. 2.3.1.5 and Fig. 4.39a. The differential equation (4.99)

$$C\frac{\mathrm{d}p_{out}(t)}{\mathrm{d}t} = \frac{R_1 + R_2}{R_2}q(t) - \frac{p_{out}(t)}{R_2} + R_1\frac{\mathrm{d}q(t)}{\mathrm{d}t}, \tag{4.99}$$

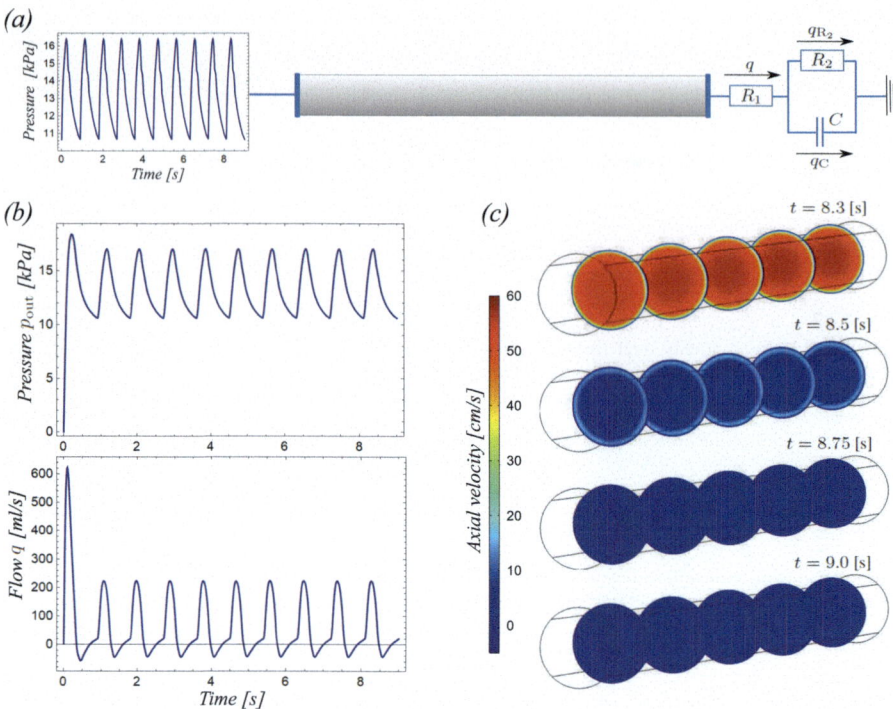

Fig. 4.39 (**a**) Pressure wave at the inlet and WindKessel (WK) at the outlet describe Neumann boundary conditions. (**b**) Development of the outlet pressure and system flow over a number of cardiac cycles. (**c**) Axial velocity at selective cross-sections and time points in the cardiac cycle

together with the initial condition $p_{out}(0) = 0$, then describes the relation between the time-dependent outlet pressure $p_{out}(t)$ and the system flow $q(t)$. Here, $R_1 = 0.1 \, \text{mmHg s ml}^{-1}$, $R_2 = 1.8 \, \text{mmHg s ml}^{-1}$, and $C = 0.9 \, \text{ml mmHg}^{-1}$ are the resistances and the capacitance of the systemic circuit, respectively. Given this set-up, a *transient analysis* computes the flow over a number of cardiac cycles. Figure 4.39b illustrates the development of the outlet pressure and system flow—already at the second cardiac cycle the system reached its steady-pulsatile response. In addition, Fig. 4.39c shows the axial velocity at selective cross-sections and time points in the cardiac cycle.

4.14   Summary and Conclusion

Over the past half century, computer simulation has become the third pillar of science alongside theory and experiment. The FEM has emerged as one of the greatest successes and has revolutionized engineering design and analysis. This chapter has been designed to equip the reader with knowledge for the targeted application of the FEM in the solution of vascular biomechanics problems. Whilst modern software eases the use of the FEM, the understanding of the algorithm running in the background is mandatory for efficient problem solving. This is especially true for the many non-linear problems in vascular biomechanics. A deeper understanding directly results in a higher success rate and less frustration in the application of the FEM. One is also advised to always start with a very simple model (linear, low dimensionality, robust boundary conditions, etc.) and then successively refine it until the required model fidelity has been reached.

The spatial discretization and the interpolation of problem variables, together with the weak form of the governing BVP, are the very foundation of the FEM. Whilst the local BVP of many bioengineering problems is readily available, the Calculus of Variations allowed us to derive its weak form. This procedure has been demonstrated for several classes of engineering problems. Many vascular biomechanics applications are characterized by constraints, and the essential variables can then not freely develop. Constraints can either be directly introduced at the finite element level or solved in addition to the set of governing equations—both approaches have been discussed.

The FEM-discretized problem results in a system of equations to be solved for the essential nodal variables. The nature of the engineering problem and the type of the investigation determine this system and different solution strategies have been reviewed. Given the non-linear character of most biomechanical problems, the numerical stability of the solution is important for an efficient numerical analysis. We exemplified the stabilization of the AD problem and discussed a number of approaches, methods that have also been successfully applied to other problems, such as the description of the incompressible flow.

A number of case studies concluded the chapter and demonstrated the application of the FEM in vascular biomechanics—more such problems will be provided in the forthcoming chapters.

Conduit Vessels

5

This chapter concerns the analysis of conduit vessels, and given their clinical relevance in vascular disease, it mainly relates to arteries. We begin with the histology of the vessel wall, with specific emphasis placed on collagen and elastin, the main structural proteins in the ExtraCellular Matrix (ECM). Following the general description of vessel wall properties, vessels, such as the aorta, carotid, coronary, iliac and femoral are addressed in greater detail. Atherosclerosis and aneurysm disease, two common vascular diseases are then discussed, where the influence of biomechanical factors in the disease progression is specially analyzed. A key section of this chapter concerns the constitutive description of the vessel wall, where phenomenological and histomechanical approaches are addressed. The inflated tube, a common loading example, is used to test the structural implications of the different constitutive models. In addition to elastic and viscoelastic descriptions, we also analyze damage and failure models in the description of the vessel wall. Experimental data from planar biaxial tissue characterization is then used to estimate material parameters for a number of constitutive descriptions. A case study uses the Finite Element Method (FEM) to predict the stress in the aneurysmatic aorta, and concluding remarks summarize the chapter.

5.1 Introduction

The (visco)-elastic properties of the *conduit vessels* play a critical role in the proper functioning of the cardiovascular system [394]. Large arteries determine the *compliance* of the vascular system, and the relation between *arterial stiffness*

The original version of this chapter was revised: ESM has been added. The correction to this chapter is available at https://doi.org/10.1007/978-3-030-70966-2_8

Supplementary Information The online version contains supplementary material available at https://doi.org/10.1007/978-3-030-70966-2_5

© Springer Nature Switzerland AG 2021, corrected publication 2022
T. C. Gasser, *Vascular Biomechanics*, https://doi.org/10.1007/978-3-030-70966-2_5

and cardiovascular morbidity and mortality is well documented [332]. In addition, large veins serve as *compartments* for blood and thus influence the long-term blood pressure. The mechanical properties of the veins are therefore linked to hypertension, a factor known to cause a number of cardiovascular diseases.

Aside from the aforementioned circulatory implications, arteries themselves are particularly prone to *mechanics-triggered injuries*, and histopathological changes of the artery wall show striking spatial correlation with biomechanical stress. The literature reports a number of such observations. The dilation of the ascending aorta occurs preferentially at its outer curvature [218] and thus at the site of the maximum axial stress. The alteration of blood flow, and thus of Wall Shear Stress (WSS), from above-knee amputations causes a five-fold higher prevalence for Abdominal Aortic Aneurysm (AAA) development [579]. Regions of the vessel wall exposed to complex blood flow are more likely to develop atherosclerotic lesions [176, 315, 612]. Many more studies may be listed, and there has therefore been an enormous motivation to assess the mechanical properties of vessels so as to understand cardiovascular physiology, and the role of tissue stress and strain in pathology. Besides being of a purely scientific interest, such information would improve clinical therapeutics and diagnostics, and may help to optimize the design of vascular devices, amongst many other applications.

A blood vessel should always be seen as a *dynamic biological system*. In addition to the *passive* deformation in response to external forces, the vessel also *adapts* through the action of contractile cells in the wall, as well as the continuous turnover of the wall's internal tissue structure. These mechanisms allow the vessels to change their mass and mechanical properties in time and in response to environmental cues. Whilst the present chapter discusses the passive mechanical properties of the vessel wall, Chap. 7 concerns mechanisms of chemo-mechanotransduction in the description of vessel wall adaptation mechanisms.

5.2   Histology and Morphology of the Vessel Wall

ECM components, such as elastin, collagen, ProteoGlycans (PGs), GlycosAmino-Glycan (GAG), fibronectin, and fibrillin ensure the vessel wall's *structural integrity*, whilst vascular cells, such as Endothelial Cells (EC), Smooth Muscle Cells (SMC), FibroBlasts (FB), and myofibroblasts maintain its *metabolism*. ECM components in the vessel wall are continuously produced and degraded—in the normal vessel these processes are at homeostatic equilibrium. Whilst the cells produce ECM components, it is primarily Matrix MetalloProteinases (MMPs) that degrades them [365].

The production and degradation of tissue constituents determine the continuous turnover of vascular tissue, a mechanism that allows the wall to *grow* and *remodel*. The vessel's geometrical, histological, and mechanical properties change along the vascular tree towards the maintenance of conditions for optimal mechanical operation [445]. The properties of the vessel wall alter also in response to many other factors, such as age, disease, and lifestyle.

5.2.1   Layered Vessel Wall Organization

The vessel wall is formed by *intima, media,* and *adventitia,* layers that are separated by the *Internal Elastica Lamina (IEL)* and the *External Elastic Lamina (EEL),* see Fig. 5.1.

Situated inner to the IEL, the intima is the innermost layer, see Fig. 5.1a. It is composed of the *endothelium,* a single layer of ECs that rests on a thin *basal membrane* and a subendothelial layer of ECM. Collagen type IV, laminin, and PG [611] form the basal membrane and provide physical support for ECs. It mechanically *decouples* shear deformation from other deformations of the vessel wall and allows the EC to sense and respond to Wall Shear Stress (WSS). Whilst WSS sensing plays an important role in vessel biomechanics, the structural–mechanical impact of the intima is often negligible in non-diseased conduit vessels. The endothelium also provides an anti-thrombogenic and low-resistance lining between blood and vessel tissue and has important barrier function.

Situated between the IEL and the EEL, the media is the middle layer of the vessel wall, see Fig. 5.1a. It consists of a complex 3D network of SMCs, elastin, collagen fibers, and fibrils, as well as other connective tissue. These structural components are predominantly aligned along the circumferential vessel direction [195, 400] and organized in repeating *Medial Lamellar Units (MLUs)* of 13 to 15 μm thickness [91, 126, 400], see Fig. 5.1b. The thickness of MLUs is independent of the radial location in the wall and the number of MLUs increases with increasing vessel diameter—in the human aorta up to 60 MLUs may be found. The tension that is carried by a single MLU in the normal wall remains constant at about 1.6 to 2.4 N m^{-1} [91]. The media's layered structure is gradually lost towards the periphery, and a discrete laminated architecture is hardly present in muscular arteries or smaller veins. The high SMC content equips the media with excellent *vasoactivity.* This property is especially important in smaller vessels to regulate bloodstreams, see Chap. 2.

Situated outside of the EEL, the adventitia is the outermost vessel wall layer, see Fig. 5.1a. It consists mainly of FBs embedded in an ECM of thick bundles of *collagen,* PGs, GAGs, and some other connective tissue, such as a few number of elastin fibers. The FBs continuously synthesize ECM compounds, mainly to maintain the high amount of collagen in the layer. The adventitia *shields* the vital medial and intimal layers from overstretching and *anchors* the vessel to its surrounding. It is perforated by *nerves* connecting the SMCs in the medial layer, and hosts the *vasa vasorum,* tiny blood vessels that perfuse not only the adventitia itself, but also the outer media. In contrary, the intima and the inner media are perfused by the radial convection of fluid that establishes through the pressure gradient between the circulation and the much lower interstitial pressure in the adventitia.

The thickness of the individual vessel wall layers depends on the blood vessel's physiological function and changes with age and diseases, such as atherosclerosis and aneurysm formation.

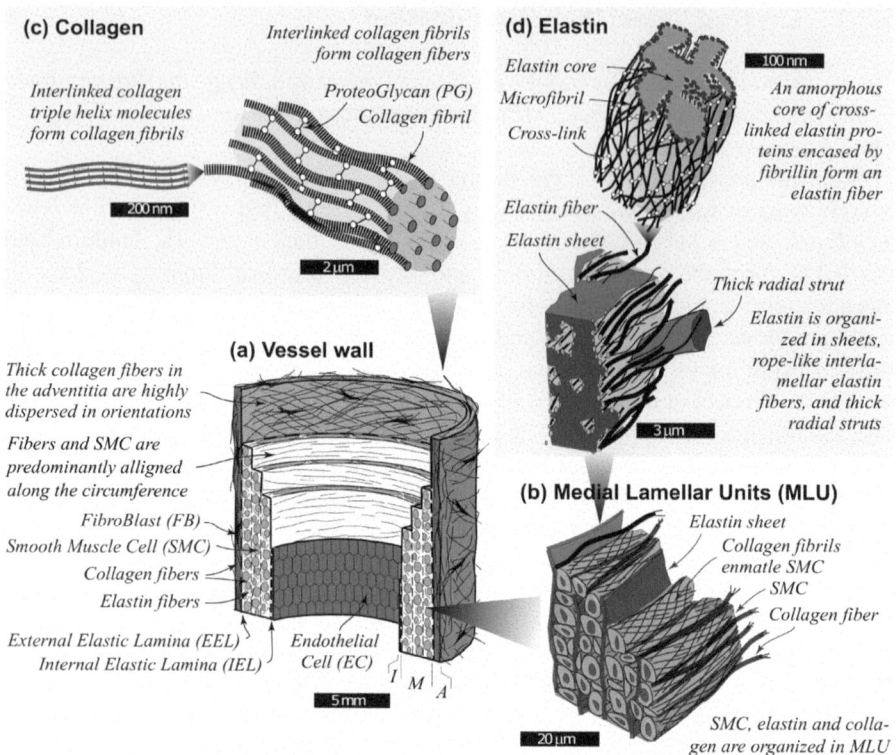

Fig. 5.1 Histological idealization of the normal aorta. (**a**) The vessel wall is composed of three layers: Intima (I), Media (M), and Adventitia (A). The intima is the innermost layer consisting of a single layer of endothelial cells, a thin basal membrane, and a subendothelial layer. (**b**) Smooth Muscle Cells (SMCs), elastin, and collagen are key mechanical constituents in the media, and are arranged in Medial Lamellar Units (MLUs) with the thickness of 13 to 15 μm. In the adventitia, the primary constituents are collagen fibers and FibroBlasts (FB). (**c**) Collagen fibers with a thickness in the range of micrometers are assembled by 50 to 300 nm thick collagen fibrils. Load transition between collagen fibrils is (likely) maintained by Proteoglycan (PG) bridges. (**d**) Elastin fibers with a thickness of hundreds of nanometers are formed by an amorphous core of highly cross-linked elastin protein that is encapsulated by 5 nm thick microfibrils. Elastin fibers present as thin concentric elastic sheets, rope-like interlamellar elastin fibers, and thick radial struts

5.2.2 Differences Between Arteries and Veins

Despite the structures of arteries and veins are very similar, some key differences may be noticed. Whilst the arterial wall is *thicker* to cope with the higher blood pressure in the arterial system, arteries also contain *more elastin* than veins to support recoil in response to the high-pressure pulsatility. Given the very low-pressure gradient along the venous path, some veins are equipped with passive *valves* to establish unidirectional blood flow. No such valves are needed in the arterial system.

5.2.3   Extra Cellular Matrix (ECM)

The ECM is a highly organized network of proteins that contain, amongst many others, collagen, elastin, and a loose network of PGs. See the schematic representation of the aorta in Fig. 5.1. The 3D organization of ECM components is vital to the vessel's proper physiological functioning—the ECM rather than being merely a system of scaffolding for the surrounding cells is a mechanical structure that controls the *micro-mechanical* and *macro-mechanical* environments of vascular tissue. It quantifies the amount of stress and strain that is transmitted to the individual cells, known to influence their metabolism [70] and factors, such as cell adhesion, proliferation, migration, differentiation, and gene expression.

5.2.3.1 Collagen Structure

Collagen of the *types I, III, IV, V, and VI* are found in the vessel walls. The fibrilar collagen types I and III make-up most of the collagen, out of which type I accounts for 50 to 70%. Type IV is mainly seen in the basal membrane and around SMCs. Veins tend to have in general a higher collagen content than arteries.

Collagen is at a continuous state of deposition and degradation at a normal half-life time of *60 to 70 days* [395]. The continuous maintenance of the collagen structure relies on a delicate (coupled) balance between production and degradation. Whilst cells, such as SMCs, FBs, and myofibroblasts [394] synthesize collagen, it is mainly MMPs, collectively called *collagenases*, that degrade it. Collagen synthesis lasts throughout the lifespan, and given a normal vessel, it leads to collagen of stable quality.

Collagen contributes *stiffness, strength*, and *toughness* to the vascular wall. Earlier observations indicated that the collagen-rich abdominal aorta is stiffer than the collagen-poor thoracic aorta [37, 323], and the regional variation of aortic properties has then later been specifically documented [514]. Apart from the amount of collagen in the wall, its spatial orientation [175], as well as the spread in orientations [202], strongly influences its macroscopic mechanical properties. At Mean Arterial Pressure (MAP), only approximately 6 to 7% of collagen fibers are mechanically engaged [14, 226].

Collagen *fibrils* range from fifty to a few hundreds of nanometers in diameter, and they may be seen as the basic building block of many collagenous tissues [175]. However, it is their organization into *suprafibrillar structures* that dominates the vessel wall's macroscopic mechanical properties. Within the MLU, and thus between the elastic lamellae of the media, collagen fibrils or bundles of fibrils (10 to 40 fibrils per bundle) run in parallel closely enveloping the SMCs [400]. The collagen fibers are *not* woven together but aligned in parallel, very much like in tendon or ligament, a structure designed to cope with high mechanical load [400], see Fig. 5.1b.

The collagen fibrils that build a collagen *fiber* are cross-linked towards the formation of a structure that is suitable to carry mechanical load. *PG bridges* [499, 500] could potentially support such interfibrillar cross-linking, see Fig. 5.1c. A

PG unit consists of a "core protein" with one or more covalently attached GAG chain. Small PGs, such as decorin, bind non-covalently but specifically to collagen fibrils and cross-link adjacent collagen fibrils at about 60 nm intervals [499]. Reversible deformability of the PG bridges is crucial to serve as shape-maintaining modules [499] and, *fast* and *slow* deformation mechanisms have been identified. The fast (elastic) deformation is supported by the sudden extension of about 10% of the L-iduronate (an elastic sugar) at a critical load of about 200 pN [246]. The slow (viscous) deformation is based on a sliding filament mechanism of the twofold helix of the glycan [499]. It could also explain the large portion of macroscopic viscoelasticity that characterizes collagen. PG-based cross-linking is supported by a number of experimental studies that highlight the role played by PG in inter-fibril load transmission [342,470,492,499], a mechanism that also has been verified through theoretical investigations [157,456,576]. However, the biomechanical role of PGs is still somewhat uncertain, and some data indicates minimal, if any, PG contribution to the tensile properties of the tissue [157,462,463].

5.2.3.2 Elastin Structure

Elastin functions in partnership with collagen. It mainly determines the mechanical properties of the vessel wall at *low* strain levels [468] and is important to *recoil* arteries during each pulse cycle. Whilst elastin in the adventitia plays a negligible role, it is a main structural component of the *media*. Elastin presents as 1.0 to 2.0 μm thick concentric sheets (71%), 100 to 500 nm thick rope-like interlamellar elastin fibers (27%), and approximately 1.5 μm thick radial struts (2%) [39,126,400], see Fig. 5.1b. The elastin sheets encapsulate the MLUs, and they are perforated and gusseted by elastin fibers, see Fig. 5.1d. In muscular arteries, the layered structure of the vessel wall is lost and most elastin appears in the EEL, a layer that is then thicker and formed by elastic fibers predominantly aligned along the vessel's axial direction. They are important to maintain the axial prestress of the vessel to avoid bucking of the vessels during, for example, limb motion [292].

Microscopy studies indicate that elastin is made-up of repeating self-similar structures at many length-scales [538]. An amorphous core of highly cross-linked elastin protein accounts for 90% of the mass, whilst 10% relate to a fibrillar mantle of about 5 nm thick microfibrils [92, 442], see Fig. 5.1d. A number of elastin molecules are cross-linked and connected to each other as well as to other molecules, such as microfibrils, fibulins, and collagen.

It is the shift from the low-pressurized to the high-pressurized circulation around the birth that triggers the production of elastin in central arteries [149]. Elastin is synthesized and secreted by vascular SMCs and FBs, a process that normally stops soon after puberty once the body reaches maturity. Although the dense lysyl cross-linking makes elastin fibrils extremely insoluble and stable that then explains the half-life times of *tens of years* [7], elastin may be degraded by selective MMPs, collectively called *elastases*. Elastases cause disruption of elastin fiber integrity and diminish mechanical tissue properties.

Elastin is a critical autocrine factor that maintains vascular homeostasis through a combination of biomechanical support and biologic signaling [11, 310]. Whilst elastin degradation has been related to diseases, such as atherosclerosis, Marfan syndrome, Cutis laxa, it is also important for physiological processes, such as growth, wound healing, pregnancy, and tissue remodeling [595]. The proteolytic degradation of elastin may therefore have also important consequences in normal elastogenesis and repair processes [585]. The repair of protease-damaged elastin does not appear to produce elastin of the same quality than originally laid down during primary vascular growth [518]. In addition, elastic fibers that are damaged or degraded with aging or disease, are often not repaired but replaced by collagen and PG that then stiffens the vessel wall [94].

Elastin and rubber share a number of mechanical similarities, such as high deformability, entropic-elasticity, and both materials go through a glassy transition. In contrary to rubber, elastin's hydrophobic interactions are a determining factor in its elasticity, and it is *only* elastic when swollen in water [585].

5.2.4 Cells

Vascular cells *sense* and *respond* to mechanical loads, a mechanism that allows the vessel to undergo many changes during normal development, ageing, and in response to diseases or implanted medical devices. The cells work in partnership with the ECM, and it is this orchestrated response that then leads to a vessel of optimal physiological function. In addition, cells produce a number of immune and inflammatory mediators that stimulate the migration of immune cells and inflammatory cells from the bloodstream into the vessel wall.

A single layer of ECs forms the endothelium, a *non-thrombotic* surface between blood and tissue. ECs have a half-life time of 1 to 3 years. They are constantly exposed to WSS, in response to which they secret *vasoactive agents* that are then transported into the wall and control the tonus of adjacent contractile SMCs.

SMCs are aligned with the circumferential direction and at a radial tilt of approximately 20 degrees [180, 400]. Aside from their physiological functions, abnormal SMC function is also involved in a number of vascular diseases, such as atherosclerosis, restenosis, hypertension, and the formation of aneurysms. SMCs are able to switch between phenotypes in response to *environmental cues*—pulsatility and pressure appear to determine the SMC phenotype in the arteries [52].

At the *contractile* phenotype, SMCs are quiescent and do not proliferate. The cells contract in response to electrical, chemical, or mechanical stimuli, factors that provide control over the vessel's diameter [375]. It is a key property for resistance vessels towards the diversion of bloodstreams. Whilst a conduit vessel is too large to divert bloodstreams, SMC contraction influences the vessel stiffness, and thus the pulse wave velocity.

SMCs at their *synthetic or dedifferentiated* phenotype can migrate, proliferate, and react with a number of proinflammatory and secretary responses to respond to injuries and diseases. As with FBs, SMC produce also ECM proteins and simultaneously secrete MMPs, both of which controls the remodeling of the vessel wall tissue.

5.3   Mechanical Properties and Experimental Observations

The biomechanical properties of the vessel wall have been investigated by a large number of experimental studies, and data of many vessels are available in the literature [2]. *In vitro* tensile testing displays pronounced stress softening during the first few loading cycles, until the tissue is *preconditioned* and a stable cyclic response is observed, see Fig. 5.2. Even after preconditioning, the vessel wall shows *strain-rate dependency*, and exhibits phenomena, such as creep, relaxation, and dissipation at cyclic loading, see Fig. 5.3a. The dissipation at cyclic loading over a frequency that ranges over five orders of magnitude does not change by a factor of more than two [540]. This observation led to the *pseudo-elastic* description of the vessel wall. The loading and unloading paths are rate-independent, but *different*, and the tissue then yields frequency-independent dissipation at cyclic loading.

The normal artery is highly deformable, and as with most soft biological tissues, exhibits a non-linear stress *versus* strain response [478]. The wall progressively stiffens at levels around the vessel's physiological *in vivo* strain, see Fig. 5.3b. The characterization of vascular tissue after the selective digestion of elastin or collagen [129, 231, 358, 468] helped to understand this phenomenon. The data suggested that the vessel wall stiffens in response to the gradual recruitment of the embedded *wavy* collagen fibers [468, 490, 599], a mechanism that explains not only the *non-linear elasticity* of the vessel wall, but also the vessel wall's *anisotropy* [421, 568], see Fig. 5.3b, c. The vessel wall appears often stiffest along the circumferential direction and thus along the direction most collagen fibers are aligned [195, 400]. Tissue stiffness may be seen as the mechanical manifestation of the vessel wall's histological composition, and the collagen-rich muscular arteries are regularly stiffer than the elastic arteries. The stiffness of the vessel wall depends on the deformation and changes during normal development, ageing, and in response to disease and many other factors.

The artery wall may be regarded as a mixture of solid components, such as elastin, collagen, and SMCs, that are immersed in water. Much of this water is *not* particularly mobile, but bound to the hydrophile PG, GAG, and elastin. The large amount of bounded water then explains that *in vivo* the volumetric

Fig. 5.2 Preconditioning of a vessel wall sample taken from the porcine descending thoracic aorta. The data has been recorded during the first three loading cycles of planar equi-biaxial testing. The plot shows the circumferential First Piola–Kirchhoff stress *versus* the circumferential stretch

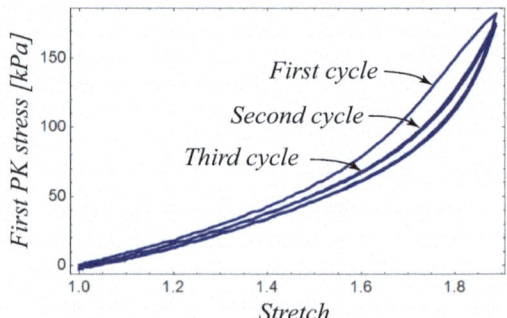

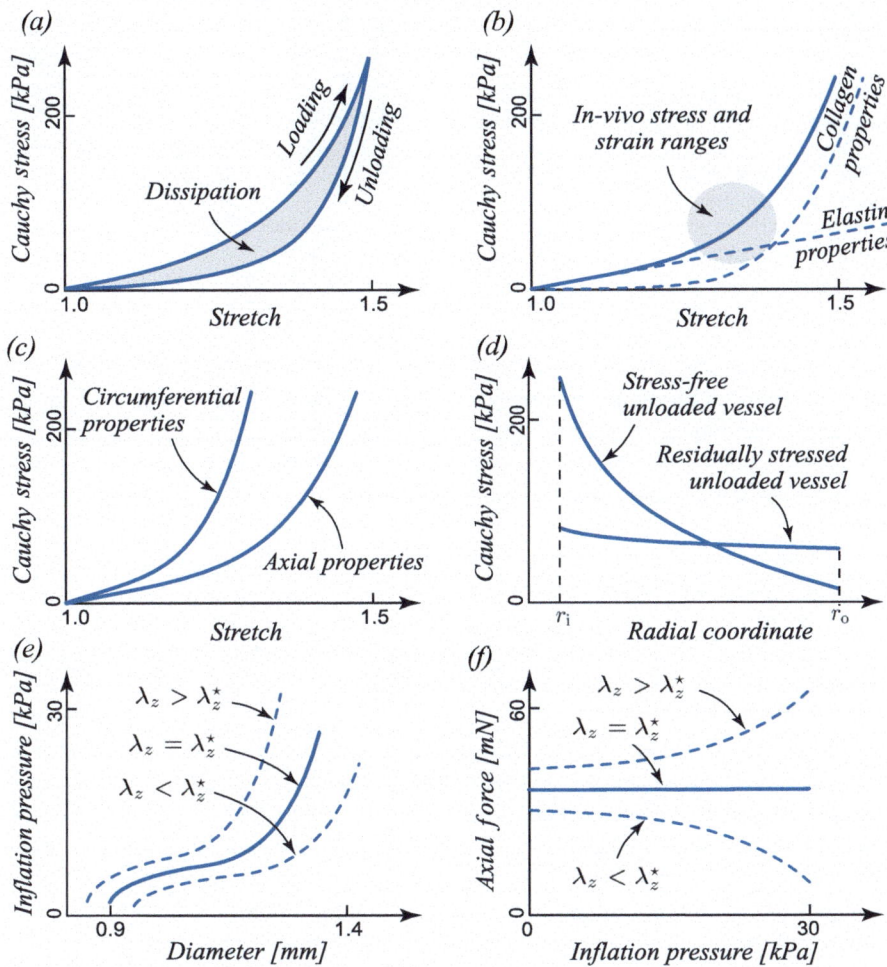

Fig. 5.3 Mechanical properties of the vessel wall with stress and strain magnitudes characteristic for a normal artery. (**a**) Stress-stretch properties of the vessel wall at cyclic loading. The area enclosed by the loading and unloading curves represents the specific energy that is dissipated during a single loading cycle. (**b**) Non-linear stress-stretch properties (solid line) of the vessel wall at simple tension. The dashed lines denote the individual contributions from collagen and elastin, respectively. At the *in vivo* deformation, the transition from the soft elastin-dominated to the stiff collagen-dominated properties appears. (**c**) Anisotropy of the vessel wall. The properties along the circumferential vessel direction are stiffer than along the axial direction. (**d**) Residual stress of the load-free vessel wall leads to approximately homogenous stress across the wall of the vessel that is inflated at Mean Arterial Pressure (MAP). (**e**, **f**) The axial pre-stretch λ_z influences the pressure-diameter properties and the axial force of a vessel segment. Given inflation at the *in vivo* axial pre-stretch $\lambda_z^{\star}$, the axial force is independent from the inflation pressure

Fig. 5.4 Stress-stretch properties of the vessel wall exposed to supra-physiological loading. Compared to the physiological loading cycle (**a**), the vessel wall shows remaining deformations and softening after the exposure to supra-physiological loadings (**b**)

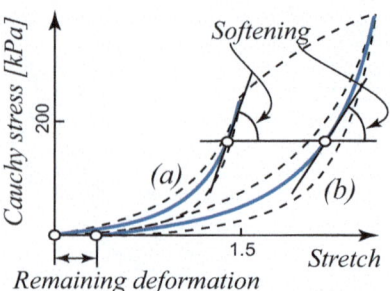

strain is three orders of magnitude lower than the circumferential strain [69]. For many mechanical problems the vessel wall may therefore been regarded as an *incompressible* homogenized mixture of solid constituents immersed in water. The affinity of water to PG, GAG, and elastin results in the development of water pressure within the tissue that is inhomogeneously distributed across the vessel wall [471].

The continuous turnover of tissue constituents tends to *homogenize* the stress across the vessel wall. The excision (and deflation) of a vessel segment from its *in vivo* configuration introduces therefore *residual stresses* in the vessel's *load-free* configuration. Residual stresses in arteries have been known for at least half a century [37], and their biomechanical consequences are well discussed in the literature [88, 183, 444, 558]. It is reported that both, circumferential [88, 558] and longitudinal [584] strips change their curvature when excised from the load-free artery. Residual stresses in the vascular wall are therefore multi-dimensional and disregarding residual stresses can be a severe limitation [114, 302, 349]. Given (passive) biomechanical simulations, the neglection of residual stresses then often leads to considerable stress gradients across the wall at the vessel's *in vivo* loading, see Fig. 5.3d. This is not physiological and in contradiction to the uniform stress hypothesis [183].

Within the body, the vessel wall is under a multi-axial stress state with σ_r, σ_θ, and σ_z denoting the principal Cauchy stresses in radial, circumferential, and axial directions, respectively. The radial stress is approximately one order of magnitude lower than the others, and *plane stress* is a commonly-used approximation of the loading of the vessel wall. Whilst the *in vivo* circumferential stress σ_θ is directly related to the blood pressure, the axial stress σ_z is also influenced by the vessel's axial *pre-stretch*. Experiments that pressurized arteries at their *in vivo* length indicated that the axial force within the vessel did not depend on the inflation pressure [593], see Fig. 5.3e, f. This observation suggests that arteries in the body pulsate in diameter, but not along their axial direction. The ascending thoracic aorta is an exception and shows in addition to circumferential also axial pulsation [155].

The exposure of vascular tissue to supra-physiological mechanical stress leads to (irreversible) rearrangements of the tissue's microstructure. Given vascular tissue, *damage-related* effects [142, 410] and *plasticity-related* effects [410, 489] have been documented, observations that somewhat remind on preconditioning, see Fig. 5.4.

Fig. 5.5 Some major arteries of the systemic circulation

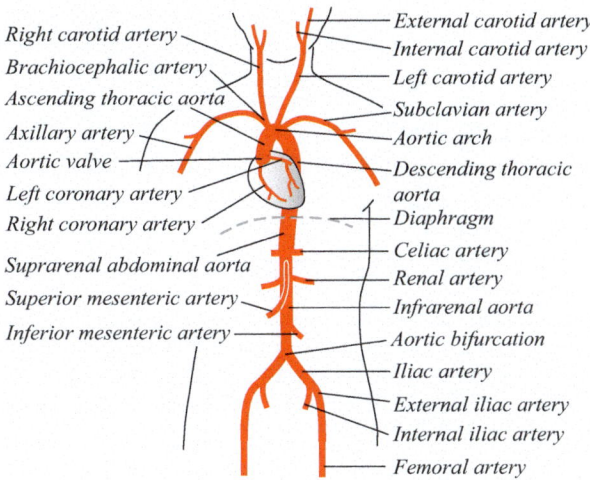

Right carotid artery
Brachiocephalic artery
Ascending thoracic aorta
Axillary artery
Aortic valve
Left coronary artery
Right coronary artery
Suprarenal abdominal aorta
Superior mesenteric artery
Inferior mesenteric artery

External carotid artery
Internal carotid artery
Left carotid artery
Subclavian artery
Aortic arch
Descending thoracic aorta
Diaphragm
Celiac artery
Renal artery
Infrarenal aorta
Aortic bifurcation
Iliac artery
External iliac artery
Internal iliac artery
Femoral artery

5.3.1   Aorta

The *aorta* is the first arterial segment of the systemic blood circulation, directly connected to the left ventricle through the aortic valve. It is the largest artery in the human body, and the diaphragm at the level of the twelfth thoracic vertebra separates it into the *thoracic* and *abdominal* aorta, see Fig. 5.5. The thoracic aorta is formed by the ascending thoracic aorta, the aortic arch, and the descending thoracic aorta. The level of the renal arteries separates the abdominal aorta and splits it into the suprarenal and infrarenal segments.

Given its prominent role, the aorta is one of the best explored vessels. At its origin it has a diameter of approximately 3.0 cm, which reduces to approximately 1.8 to 2.0 cm at the aortic bifurcation, situated at the level of the fourth lumbar vertebra. Different segments of the aorta have different embryologic origins. SMC in the aortic arch origins from the neural crest, in the descending thoracic aorta from the somites and in the infrarenal aorta from the splanchnic mesoderm [317].

The aorta's pressure-diameter property is of critical importance to the entire cardiovascular system and determines its pressure-flow characteristics. The compliance of the aorta is responsible for almost the entire capacity of the systemic circulation and defines its WindKessel (WK) properties. The aortic capacity is constant over a wide range of pressures, and the thoracic aorta alone contributes 85% to it [232]. Collagen fibers are undulated and they gradually engage with the stretch of the vessel wall. In the abdominal aorta approximately 10% and 30% of the collagen seem to be engaged at diastolic and systolic deformations, respectively [16].

As with other arteries, the aorta is axially pre-stretched in the body. In humans, the pre-stretch is approximately 5% [16], a property that remarkably reduces with age [264] and changes between the different aortic segments [232]. The aortic

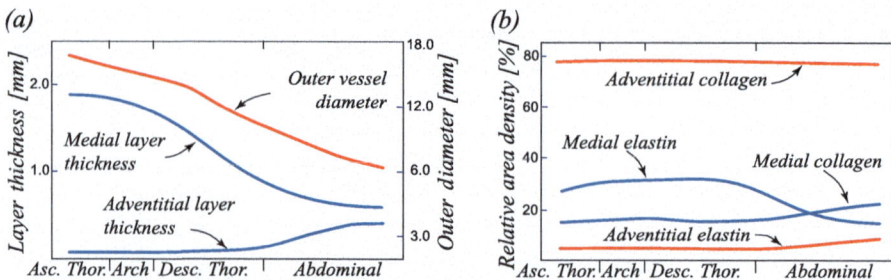

Fig. 5.6 Variation of the geometrical properties and the histological composition of the porcine aorta reported elsewhere [514]. (**a**) Change of the outer aortic diameter and thicknesses of the medial and the adventitial layers. The intimal layer thickness covers less than 1% of the total wall thickness. (**b**) Relative area density of elastin and collagen in the medial and the adventitial layers. The relative area density represents the area that is covered by the respective constituent in histological stains (Asc.Thor.-Ascending thoracic aorta; Arch.-Aortic arch; Desc.Thor.-Descending thoracic aorta; Abdominal-Abdominal aorta)

morphology varies remarkably from the root to the bifurcation, alterations that reflect the transition from an elastic vessel towards a muscular vessel, see Fig. 5.6. The abdominal aorta is equipped with a thick adventitia and then stiffer than the thoracic aorta. The circumferential stiffness seems to be highest at the level of the diaphragm [232, 540], and it might also be higher in males than in females [516]. Aortic stiffness and strength is further discussed in relation to aneurysmal disease in Sect. 5.4.6.

The aorta is highly vulnerable to *aging*. The amount of elastin in the wall decreases, whilst collagen and MMP-2 increase with age. The aged aortic wall is also thinner, shows split and fraying of MLUs, as well as an increased level of elastin glycation and collagen cross-linking. All these morphological changes may explain why the aorta's diameter and stiffness [16, 23, 284, 327, 394, 472] increase much faster over time than of any other artery. The thoracic aorta also grows faster in diameter than other vessels. However, it maintains its circumferential compliance, and the comparison of the vessel's load-free configurations does not show a faster widening in favor of the thoracic aorta [294]. In the thoracic aorta the stretch over a cardiac cycle decreases with age from approximately 14% to 6% [294].

5.3.2   Carotid Artery

The left and right common *carotid arteries* supply the head and the neck, see Fig. 5.5. Whilst the left common carotid artery originates directly from the aortic arch, the right common carotid artery is formed at the bifurcation of the brachiocephalic artery. The common carotid artery divides at the level of the fourth cervical vertebra and forms the *external* and *internal* carotid arteries. The internal carotid arteries supply the brain, whilst the external carotid arteries supply the face, scalp, and neck. The *carotid sinus* is a dilated area at the base of the internal carotid artery, right superior to the bifurcation of the internal and external carotid arteries. The carotid sinus hosts baroreceptors that continuously record blood pressure and feed the information to corresponding afferent neurons, see Sect. 2.1.7. The external and internal carotid arteries are approximately 6.5 mm and 5.0 mm in diameter, a measure that depends on many factors [311]. Given its clinical significance, the biomechanics of the carotid artery are well explored. The diseased (atherosclerotic) carotid artery is involved in many strokes, and the intima-media thickness of the carotid artery wall is a marker of subclinical atherosclerosis.

5.3.3   Coronary Artery

At the root of the ascending aorta, the lumen has three small pockets between the cusps of the aortic valve and the wall of the aorta—the *aortic sinuses*. The left and right aortic sinuses give rise to the left and right *coronary arteries*, vessels that supply together the cardiac muscle, see Fig. 5.5. The contraction of the heart muscle generates its own blood flow, and coronary flow differs therefore from all other arteries. During the systolic phase, the left ventricular contraction "throttles" or squeezes coronary blood flow and the majority (approximately 80%) of the flow appears during the *diastolic phase*, when the heart is relaxed, see the bottom panel in Fig. 5.7. The cross-talk between coronary flow and cyclic heart muscle contraction determines *unique* pulsatile characteristics. Pressure and flow waveforms as well as arterial and venous phasic differences distinguish remarkably from the blood flow elsewhere in the vasculature. During a single cardiac cycle, coronary flow passes through *two* flow pulsations, see Fig. 5.7. The exploration of the unique coronary flow characteristics has been the subject of a number of studies [589].

The luminal diameter at the origin of the coronary arteries is approximately 4.0 to 4.5 mm and depends on many factors [131]. Given its clinical significance, the biomechanics of the coronary artery are well explored [405]. The diseased (atherosclerotic) coronary artery is involved in clinical events, such as angina and heart attack. The oxygen exchange from the perfusion in the myocardium is very high. An increase of oxygen supply is only to be achieved by an increased coronary blood flow—a strong linear correlation between the oxygen demand of the cardiac muscle and the coronary flow therefore exists. It underlies the importance of the proper functioning of coronary autoregulation.

5.3.4   Iliac and Femoral Artery

The aortic bifurcation forms the left and right common *iliac arteries*, vessels that supply the lower limbs, see Fig. 5.5. After approximately 5.0 cm, the common iliac artery bifurcates and forms the *external* and *internal* iliac arteries. The internal iliac arteries supply the pelvis, the buttock, the reproductive organs, and the medial compartment of the thigh, whilst the external iliac arteries provide the main blood supply to the legs. In the lower part of the abdomen, the external iliac artery

Fig. 5.7 Coronary blood flow during the cardiac cycle. The majority of flow appears at the diastolic phase, and during a single cardiac cycle, coronary flow passes through two pulsations (ECG-ElectroCardioGram; LAD-Left Anterior Descending coronary; LCX-Left CircumfleX coronary; RCA-Right Coronary Artery)

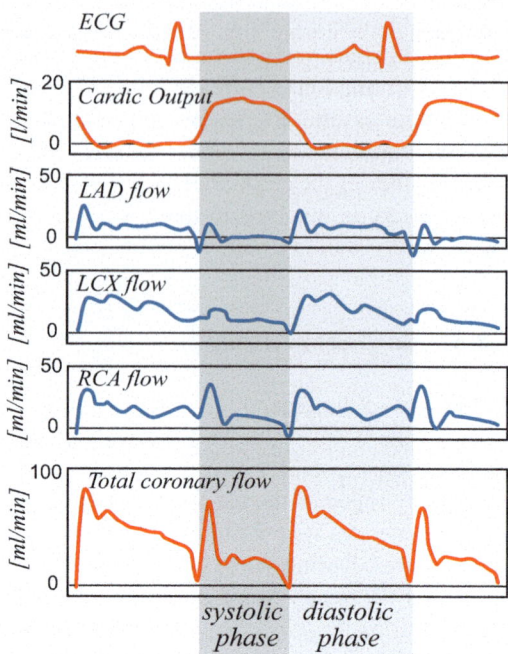

becomes the *femoral* artery. The luminal diameter of the normal iliac artery is 9.0 to 10.0 mm and correlates with age and sex [423]. As with other vessels, the axial pre-stretch of the iliac and femoral artery decreases with age [292]. However, the stretch over a cardiac cycle decreases with age only by approximately 1% [294]—muscular arteries are much less effected by age-related circumferential stiffening than elastic arteries. Both, iliac and femoral artery are vulnerable to the formation of atherosclerosis, whilst aneurysms are mainly seen in the iliac artery.

5.4   Vascular Diseases

Biomechanical factors, such a stress and strain are related to a number of vascular diseases. Whilst the scientific exploration of this influence is often driven by curiosity and very different study objectives, the following four questions are of main clinical relevance:

- *What are the hemodynamic implications of a vascular pathology, and how does it influence tissue perfusion?*
- *What is the likelihood of an acute cardiovascular event that may emerge from the diseased vessel?*
- *How will the disease progress and what are the expected implications?*

- *How does a specific treatment influence the cardiovascular system and how durable is the treatment expected to be?*

Vascular biomechanics may help to answer these questions and support clinicians in the selection of the optimal treatment for the individual patient.

Over the past decades, the level of sophistication with which vascular biomechanical analysis can be made has increased dramatically and nowadays, full patient-specific biomechanical investigations are possible. The progress of medical imaging, together with the advances in the mechanical characterization of vessels, led to biomechanical models that integrated more and more histological and mechanical features. Such models meanwhile play a considerable role in the analysis of vascular pathologies, some of which are discussed later in this chapter.

5.4.1 Diagnostic Examinations

A diagnostic examination is a test that is used to diagnose a disease or condition. In many cases, no single test can diagnose vascular disease, it is rather the conclusion from the outcome of a number of diagnostic tests. Whilst functional tests aim at detecting a *symptom* or *sign* of vascular dysfunction, a diagnostic examination may also acquire *structural, anatomical*, and *morphological* information related to vascular diseases. Very different methods are used in the acquisition and post-processing of diagnostic information, often based on medical imaging. Given the importance of the heart, a number of diagnostic tests examine the heart and its vasculature.

An *ElectroCardioGram (ECG)* is a simple test that records the heart's *electrical activity* and provides information regarding the beating and rhythm of the heart. An ECG can show coronary heart disease-related signs of heart damage as well as signs of previous or current heart attacks. ECG is often combined with a stress test, which can highlight possible signs and symptoms of coronary heart disease, such as shortness of breath, chest pain, and abnormal changes in the heart rhythm, blood pressure, or the heart's electrical activity. These would appear if plaque-narrowed coronary arteries fail to supply enough oxygen-rich blood to meet the heart's needs during exercise.

Echocardiography (echo) uses sound waves to *image* the heart. This imaging modality uses standard 2D, 3D, or Doppler ultrasound and provides anatomical, morphological, and functional information. A subsequent image analysis allows then for the determination of size and shape of the heart and how well the chambers and valves of the heart are working. It can also show areas of poor blood flow to the heart, areas of the heart muscle that are contracting abnormally, and injuries that appeared previously to the heart muscle.

Ultrasound examination is routinely used to examine many large vessels in the body, and Fig. 5.8a shows an ultrasound image of an Abdominal Aortic Aneurysm (AAA). *Intravascular Ultrasound (IVUS)* is a specific echocardiography modality that allows for a vessel to be seen from the *inside-out* and permitting the

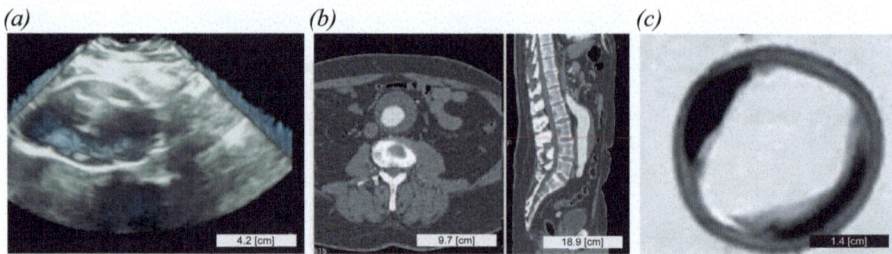

Fig. 5.8 Examples of images acquired by different vascular imaging modalities. (**a**) Vascular ultrasound and (**b**) Computed Tomography-Angiography (CT-A) of an Abdominal Aortic Aneurysm (AAA). (**c**) *In vitro* Magnetic Resonance-Angiography (MR-A) of an atherosclerotic carotid artery

visualization of atherosclerotic lesions. It is an *invasive* imaging modality, whereby a specially designed catheter with a miniaturized ultrasound probe attached to its distal end, moves inside the vessel.

Pulse Wave Imaging (PWI) is a non-invasive ultrasound-based imaging technique that allows to track the propagation of pulse waves along the aorta at high spatial and temporal resolutions. The acquired information allows to estimate the stiffness of the vessel wall [572], a factor linked to cardiovascular diseases.

Scintigraphy, also known as a Gamma scan, provides 2D anatomical, morphological, and functional information acquired from *gamma radiation* that is emitted by radioisotopes. The radioisotopes have been attached to drugs and traveled to the tissue of interest upon administration. Thallium-201 is a commonly used radioisotope, and its gamma radiation correlates with the tissue's blood supply. The very same physical principle is used by *Single-Photon Emission Computed Tomography (SPECT)*, an image modality that forms true 3D anatomical, morphological, and functional information.

Angiography uses *X-ray* based techniques, such as fluoroscopy, and provides transverse projections of the vascular lumen. It requires a radio-opaque contrast dye to be injected into a vein, where it travels to the tissue of interest and highlights the vessel lumen. In addition to anatomical information, angiography images also visualize obstructed vessel segments.

Computed Tomography-Angiography (CT-A) captures X-ray images of the body from many angles, and combines them into 2D or 3D images. Like an Angiography study, CT-A also requires a contrast dye to be injected into a vein. The degree of absorption of X-rays by a tissue component determines its appearance, and thus the grey value or *Hounsfield[1] Unit (HU)* in the image. Calcific deposits in atherosclerotic lesions have high absorption, and are clearly visible in X-ray images. However, the volume of calcific deposits is over-represented and quantitative studies should always adjust for this artifact. There is an inherent difficulty in X-ray-based

---

[1]Sir Godfrey Newbold Hounsfield, English electrical engineer, 1919–2004.

methods to discriminate between the individual components of soft tissues. The typical resolution in X-ray based studies is of the order of half a millimeter along each axis and total image acquisition can be performed within less than 10 s. With ECG-gated imaging, dynamic CT-A acquisitions are possible and allow the study of factors, such as vessel motion and myocardial function. Figure 5.8b illustrates traverse and sagittal CT-A images of a patient with an AAA.

Magnetic Resonance-Angiography (MR-A) is a *radio wave*-based imaging modality that can be performed with contrast dye to enhance the images. The acquisition is used to image vessel anatomy and morphology, blood flow, and vessel motion. MR-A shows better discrimination than CT-A amongst soft tissue components, but MR-A requires more time and has less spatial resolution than X-ray based modalities. MR-A imaging methods are reliant upon the detection of magnetization of the nucleus of hydrogen atoms in water. They aim to create high contrast between stationary and moving magnetic nuclear spins, using methods such as 2D and 3D time-of-flight, phase contrast MR-A, contrast-enhanced MR-A, and black blood MR-A [334]. Figure 5.8c illustrates a MR-A image acquired through *in vitro* acquisition of the carotid artery.

Whilst the aforementioned modalities are commonly used in the clinical examination of vascular diseases, many more modalities and post-processing approaches are available for vascular imaging, see [517, 531]. Recently also methods based on Machine Learning (ML) found their way into the processing of clinical images [145].

5.4.2 Atherosclerosis

Atherosclerosis is a slowly progressing disease of the intima that leads to the formation of *intimal plaques*—the accumulation of material within the intima that contain lipid, SMCs, inflammatory cells, connective tissue, and calcification. It should be distinguished from *arteriosclerosis*, which concerns age-related changes of the vessel wall and therefore represents a more general term including also other arteriopathies. Excellent reviews of atherosclerosis are available in the literature [54, 525] and numerous pathological studies investigated the morphologic characteristics of plaques, including features of *vulnerable* lesions that are prone to plaque rupture and atherothrombosis [298, 578].

Atherosclerosis may be seen as an ongoing *inflammation* in response to local endothelial dysfunction, a process that continuously weakens the vessel wall. The initiation and progression of atherosclerosis is strongly influenced by the local interaction of *biochemical* and *biomechanical factors*. In addition to flow-induced WSS and biomechanical stress of plaque tissue, atherosclerosis is determined by factors such as infection, oxidative stress, chronic hypertension, and most notably, elevated Low-Density Lipoprotein (LDL) levels. At the *dysfunctional endothelium*, LDL is allowed to pass into the vessel wall, where it is *oxidized* by reactive oxygen species, see Fig. 5.9b. The reactive oxygen species are present naturally, yet are observed in increased concentrations when the patient is exposed to one or

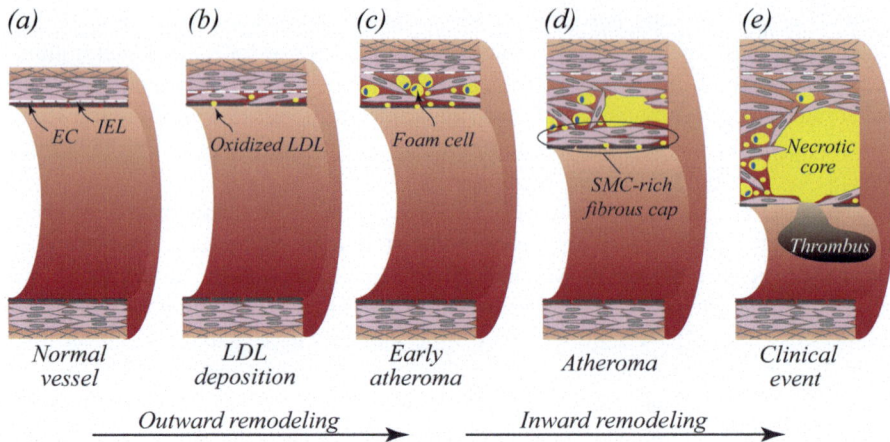

(a)   *(b)*   *(c)*   *(d)*   *(e)*

| *Normal* | *LDL* | *Early* | *Atheroma* | *Clinical* |
| *vessel* | *deposition* | *atheroma* | | *event* |

Outward remodeling   *Inward remodeling*

Fig. 5.9 Stages of atherosclerotic plaque formation. (**a**) Normal vessel: Intimal, medial, and adventitial layers form the wall. The intima is encapsulated by the Internal Elastic Lamina (IEL) and Endothelial Cells (EC). (**b**) Low-Density Lipoprotein (LDL) deposition: Dysfunctional endothelium allows inflammatory cells and apoB-containing (oxidized) LDL to accumulate in the vessel wall. (**c**) Early atheroma: Oxidized LDL and inflammatory cells continue to move into the vessel wall. Smooth Muscle Cells (SMC) are recruited and monocytes are activated to macrophages that phagocytize oxidized LDL. Following extensive phagocytic activity, the swollen macrophages are then termed foam cells. They accumulate and form together with other cellular and extracellular debris an atheromatous plaque. (**d**) Atheroma: SMC proliferate and recruit towards the formation of a SMC-rich fibrous cap that overlays the lipid and macrophage-rich plaque components. (**e**) Clinical event: Superficial plaque erosion and rupture of the weakened SMC-rich fibrotic cap causes plaque-associated thrombosis, culminating in a clinical event

more atherosclerosis *risk factors*. An inflammatory response to the oxidized LDL is mounted, and monocytes are recruited from the blood locally through the expression of inflammatory mediators and specific adhesion factors on the endothelium. Inside the intima, *monocytes* are activated to *macrophages* that engulf oxidized LDL. After extensive phagocytic activity, the swollen macrophages take on a foamy appearance and are termed *foam cells*, see Fig. 5.9c. It is the accumulation of foam cells, other inflammatory cells and SMCs together with collagen, elastin, fibrin, extracellular cholesterol, cellular debris, and eventually calcifications that forms an *atheromatous plaque*.

The earliest macro-scale manifestation of atherosclerosis is the presence of *fatty streaks*, which are caused by the accumulation of lipid-laden macrophages and T-lymphocytes under the intimal endothelium [335]. Whilst the atherosclerotic vessel still has the passive mechanical properties of the normal vessel, its active response may already have been compromised at this stage of the disease. As atherosclerosis progresses, the arterial wall adopts a structure that is quite different from a normal artery. The mechanical properties of the diseased vessel then clearly reflect such differences.

Earlier in the progression of a plaque, as *foam cells* accumulate in the intima, the vessel wall remodels and expands outward to preserve the patent lumen area, a process called *outward remodeling* [213], see Fig. 5.9b, c. The outer vessel wall then adopts a swollen appearance focally around the plaque, and little or no luminal narrowing is present during an angiographic study, a modality that images the vessel lumen. Given coronary arteries, this process continues until approximately 40% diameter stenosis is reached [213], beyond which *inward remodeling* and the narrowing of the lumen determines the development of atherosclerosis, see Fig. 5.9d, e.

As the inflammatory response continues, SMCs are signaled to proliferate and recruit from the media towards the intima, see Fig. 5.9d. These SMCs together with fibrous ECM components, then overlay the lipid and macrophage-rich plaque components and form the *SMC-rich fibrous cap*. In some atherosclerotic lesions, the fibrous cap is challenged—it weakens and erodes over time. It is the *rupture* or *ulceration* of such a fibrous cap that then exposes highly thrombogenic plaque contents to the flowing blood, see Fig. 5.9e, and eventually causes *acute thrombo-embolic events*, such as myocardial infarction and stroke. Smaller calcium hydroxyapatite deposits, so-called *micro calcifications*, may also play a critical role in the rupture or ulceration of the fibrous cap [68, 300].

Not every plaque disruption causes a thrombo-embolic event, but many of them heal with the *reformation* of the fibrous cap. A series of ruptures and repairs can eventually lead to a large plaque that grows inward into the vessel lumen. The plaque then determines a high-degree arterial stenosis with *hemodynamic significance*. These plaques have a more irregular appearance and structure. The fibrous cap is of non-uniform thickness, the lipid-rich necrotic core has very irregular shapes, and the plaque often contains bulk and micro calcifications.

A thickened intima contributes remarkable to the mechanical properties of the atherosclerotic vessel. The individual morphology and biomechanical properties of the plaque components then *dominate* the vessel wall properties of highly stenotic vessels.

5.4.3 Biomechanical Factors in Atherosclerosis

The atherosclerotic artery wall is *very* different from the normal artery in both, composition and mechanical properties. A coarse view of the atherosclerotic vessel would recognize the fibrous intima, SMC-rich fibrous cap, lipid-rich and macrophage-rich necrotic core, media, and the adventitia as distinct structural components of individual mechanical properties. Of course, at the histological scale, each one of these components has a highly heterogeneous microstructure of its own. Aside from the histological complexities of an atherosclerotic lesion, there are numerous ongoing *biochemical* processes that affect their mechanical properties. A host of MMPs present in the lesion is capable of *degrading* structural ECM proteins. The MMPs are largely *inflammatory mediated*, and the inflammatory state of the lesion and its macrophage/monocyte population influences then the stiffness and strength of the SMC-rich fibrous cap.

5.4.3.1 Atherosclerosis Development

The presence of atherosclerosis has been correlated with local *hemodynamic complexity*. Regions of the vessel wall experiencing *low WSS* below of approximately 0.2 Pa, *secondary flow*, and *oscillatory WSS* are more likely to develop atherosclerotic lesions [176,315,612]. In addition to plaque burden, low WSS is also a predictor of progressive plaque enlargement and lumen narrowing of atheromatous vessel segments [530].

Low mean WSS and oscillatory WSS appear at areas of flow recirculation or flow oscillation, and therefore at locations with elevated residence times for blood particles. Low WSS also alters the gene expression of the ECs and supports a pro-inflammatory state [394], see Sect. 7.3.2. It presumably increases the *permeability* of the endothelium [169, 461, 529], which together with the increase in residence time enhances the *mass transport* of LDL and inflammatory cells into the vessel wall [315]. In contrary to low WSS, the exposure of ECs to *constantly* normal *WSS* levels of above of approximately 1.0 Pa causes the expression of genes that then protect the wall and creates resistance to atherosclerosis.

In addition to the aforementioned factors, local *wall stiffness* could also play a role in the initiation of atherosclerotic lesions. It collocates with coronary plaques in patients with minimal coronary disease [406].

5.4.3.2 Plaque Stress and Strain

A number of biomechanical studies explored idealized geometries in the biomechanical study of atheromatous vessel segments. It eases the investigation of the mechanisms by which *plaque stresses and strains* are related to *morphological features*, such as fibrous cap thickness, lipid pool volume, calcification level, lumen eccentricity, and the remodeling index.

An earlier study [348] used a 2D model of an atherosclerotic coronary artery and found that *fibrous cap thickness*, and not stenosis severity alone, influenced the stress in the plaque cap. The thickness of the fibrous cap that covers a necrotic core may therefore be an important factor in plaque rupture. A fibrous cap that is thinner than 60 to 100 μm results in a peak stress that exceeds 300 kPa [280], a stress that may already appear at a diameter stenosis of as low as 10% [340]. Whilst the size of the lipid pool has *no* effect on the peak stress in the cap, bulk calcifications may *reduce* the stress [280] and stabilize the plaque [298]. These results have somewhat been confirmed by 3D [90] and Fluid Structure Interaction (FSI) [541] models.

Another study [404] concluded that the cap thickness alone was *not* able to discriminate between stable and instable coronary plaques, and factors, such as the remodeling index and the relative necrotic core thickness would be similarly important risk factor. The remodeling index represents the ratio between the area that is enclosed by the EEL at the most stenotic cross-section and a normal vessel cross-section, respectively.

Plaque tissue is highly heterogeneous, and *stiff inclusions* in a soft body are known to cause local stress peaks. Micro calcifications of approximately $10\,\mu$m in size that are closely spaced have therefore the potential to elevate cap stress by a factor of at least *two* [300]. It may explain the association of micro calcifications with more unstable and vulnerable lesions [390]. The *compliance mismatch* between a stiff stent and the much softer vessel wall also leads to local stress concentrations. It explains the development of peristrut micro-hemorrhages, cholesterol accumulation, and oxidation in coronary arteries, factors that may trigger in-stent neoatherosclerosis [544].

Whilst high WSS protects the vessel wall from the formation of atherosclerosis, plaque ruptures have been reported at locations of high WSS [560]. It may destabilize the structural integrity of the plaque cap [181, 227] and therefore contribute to plaque rupture at later stages of the disease. Given the correlation between WSS and tissue strain [212], it remains unclear whether WSS may be regarded as an independent rupture risk marker.

5.4.3.3   Clinical Relevance of Atherosclerosis

Atherosclerosis favored *some* larger arteries, but *not all*. Whilst arteries, such as the aorta, carotids, coronaries, and iliacs frequently show atherosclerosis, the disease spares arteries in the upper limbs, such as the mesenteric, renal, and internal mammary. Atherosclerosis has multiple clinical implications and may influence arterial blood flow and hemodynamics by at least six different factors [394].

- *Gradual stenosis formation.* Successive inward remodeling gradually creates a localized stenosis, which in turn limits the flow of oxygen-rich blood to downstream tissue.
- *Instant stenosis formation.* Enzymatic degradation of ECM components constantly weakens the fibrous cap. Over time, the cap erodes or even ruptures at a certain time point. Both events may form a thrombus in the lumen that creates a localized stenosis and/or entirely occludes the vessel. The thrombotic event then suddenly limits the flow of oxygen-rich blood to the downstream tissue.
- *Embolization.* Fibrous cap erosion or rupture may lead to the embolization of atheromatous material or the associated thrombus to smaller peripheral arteries. This occludes peripheral arteries and limits the flow of oxygen-rich blood to the downstream tissue.
- *Aneurysm formation.* Enzymatic degradation of ECM components diminishes the structural integrity of the vessel wall. Atherosclerotic plaque-based ulcerations can lead to dissections or the formation of an aneurysm in some vessels.
- *Systematic implication.* Atherosclerosis increases the stiffness of the vessel wall, and if wide spread, this decreases the capacity of the vascular system, a factor that augments left ventricular load. It may cause related cardiac complications, such as cardiomyopathy.
- *Endothelial dysfunction.* The endothelium plays a crucial role in the control of SMC tonus, platelet function, and fibrinolysis. Endothelial function in atheroscle-

rotic vessels is abnormal, which in turn may lead to abnormal vasoconstriction and limits the potential of a thrombus in the healing process.

Table 5.1 lists risk factors for arteriosclerosis and their potential disease mechanisms. Whilst some of the risk factors are well established, others are suspected and lack statistical evidence at the present.

5.4.4 Carotid Artery Disease

Carotid artery disease is a disease in which *plaque builds up* inside the carotid arteries. It commonly develops where the carotid divides into the internal and external carotid arteries, and thus at a location of complex blood flow. Both internal carotids feed the circle of Willis,[2] which then distributes blood to the brain. The formation of a stenosis in one of the carotids is therefore not of major concern to the delivery of oxygen-rich blood to the brain. However, the plaque in the carotid artery may *erode* or *rupture* and cause an *acute thrombo-embolic event*. The clot, or plaque

Table 5.1 Major risk factors for arteriosclerosis and the related disease mechanism

Diabetes	Insulin production and usage is impaired, which elevates the body's sugar level. Diabetes increases the prevalence for atherosclerosis by a factor of four
Genetic predisposition	Family history of atherosclerosis
Hypertension	High blood pressure leads to high mechanical stress in the vessel wall. This can damage the vessel tissue and enhance the development of atherosclerosis
Lack of physical activity	The lack of aerobic activity can worsen other risk factors relating to atherosclerosis disease, such as unhealthy blood cholesterol levels, high blood pressure, diabetes, and obesity
Smoking	Smoking leads to unhealthy cholesterol levels. It raises the blood pressure as a consequence of multiple proatherogenic mediators present in tobacco and cigarette smoke. This damages the vessel wall and hastens the development of atherosclerosis
Unhealthy cholesterol	Disturbed levels of blood lipids trigger endothelial injury/dysfunction, vessel wall inflammation, and promote atherosclerosis in all stages. This includes high Low-Density Lipoprotein (LDL) and low High-Density lipoprotein (HDL) cholesterol
Metabolic syndrome	The five metabolic risk factors are a large waistline (abdominal obesity), a high triglyceride level (a type of fat found in the blood), a low HDL cholesterol level, high blood pressure, and high blood sugar. Metabolic syndrome is diagnosed if three of the said metabolic risk factors are given

(continued)

---

[2]Thomas Willis, English physician, 1621–1675.

Table 5.1 (continued)

Older age	The risk for atherosclerosis increases in men starting at age 45, and in women at age 55
Unhealthy diet	Foods that are high in saturated and trans fats, cholesterol, sodium, and sugar can worsen other risk factors for atherosclerosis
Inflammation[a]	High levels of C-reactive protein in the blood is a sign of inflammation in the body and may raise the risk of atherosclerosis. Damage to the inner arterial wall segments may trigger inflammation and support the development of plaque
Sleep apnea[a]	Untreated sleep apnea can increase the risk for high blood pressure, diabetes, and even a heart attack or stroke
Mental stress[a]	Stress is a commonly reported "trigger" for a heart attack. It is an emotionally upsetting event, especially one involving anger

[a]Suspected risk factor

particles that have been broken away, then travels through the bloodstream and eventually occludes vital brain arteries [461]. Carotid artery disease has therefore the risk of causing an *ischemic stroke*. Given the brain tissue is not supplied by oxygen for more than a few minutes, the brain cells start to die. It then causes lasting brain damage with consequences, such as vision or speech problems, paralysis, and death. Carotid artery disease is a major cause of stroke in many industrialized countries, and 10 to 20% of all strokes emerge from carotid plaques [420].

An acute thrombo-embolic event may also appear in form of a *transient ischemic attack* and symptoms that usually disappear within 24 h. It is a serious clinical sign of carotid artery disease and patients need *urgent* clinical consideration to prevent them from a stroke and the associated consequences. Patients who experienced a stroke need immediate treatment to optimize the chances of full recovery.

Given the important clinical role of the diseased carotid artery, its biomechanics are well investigated and reported in the literature [405]. Table 5.2 summarizes a number of such studies, and Fig. 5.10 illustrates the elastic tensile properties of diseased carotid artery tissue.

5.4.5 Coronary Heart Disease

Coronary heart disease is a pathology where *plaque* builds up inside the coronary arteries. It narrows their lumen and forms a stenosis that gradually reduces the flow of oxygen-rich blood to the heart. An acute thrombo-embolic event can emerge from the coronaries, eventually resulting in *heart attack, angina,* or *arrhythmia*.

A heart attack occurs if the flow of oxygen-rich blood to a section of the heart muscle is cut off. Fluids then build up in the body causing shortness of breath, swelling in the ankles, feet, legs, stomach, and veins in the neck. Angina is chest pain or discomfort that may spread to the shoulders, arms, neck, jaw, or back, which disappears within a few minutes and with rest. Some symptoms of angina can be

Table 5.2 Constitutive properties of diseased carotid artery tissue acquired by *in vitro* tissue characterization. The constitutive models are further discussed in Sect. 5.5

Constitutive model	Constitutive parameters [kPa]	Reference
neoHookean (5.9)	Fibrous cap: $c = 500.0$	[403]
	Lipid core: $c = 5.0$	
	Normal vessel: $c = 150.0$	
Yeoh model (5.8) with $N = 3$	Hard: $c_1 = 302.1$; $c_2 = -228.0$; $c_3 = 261.0$	[333]
	Mixed: $c_1 = 23.5$; $c_2 = 126.0$; $c_3 = 112.0$	
	Soft: $c_1 = 29.6$; $c_2 = -33.2$; $c_3 = 128.0$	
Orthotropic linear-elastic	$E_r = 50.0$; $E_\theta = 1000.0$	[80]
Compressible neoHookean	$2c = 7.0$ to 100.0; $E = 6c$	[27]
Orthotropic linear-elastic	Normal vessel: $E_r = 10.0$; $E_\theta = E_z = 100.0$;	[404]
	$v_{r\theta} = 0.1^a$; $v_{\theta z} = 0.27^a$; $G_{r\theta} = 52.0$	
	Fibrosis: $E_r = 115.6$; $E_\theta = E_z = 2312$;	
	$v_{r\theta} = 0.07^a$; $v_{\theta z} = 0.27^a$; $G_{r\theta} = 1175.0$	
	Necrotic core: $E = 1.0$; $v = 0.49^a$	

[a]dimensionless parameter
r, θ, z denote the radial, circumferential, and axial vessel direction
E, G, v denote Young's modulus, shear modulus, and Poisson's ratio
c_i denotes material parameters used by the individual constitutive model

Fig. 5.10 Elastic properties of the diseased carotid wall at simple tension. The grey-shaded area reflects data reported in the literature and illustrates the high variability, whilst the curves characterize the properties of the soft, mixed, and hard carotid wall as reported elsewhere [333]

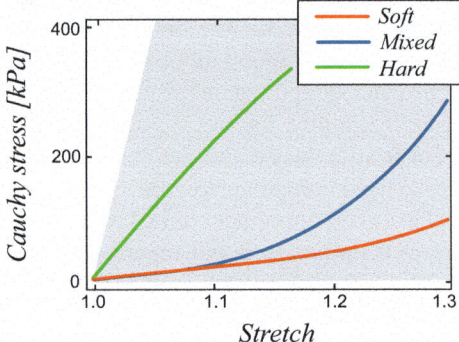

similar to the symptoms of a heart attack. In contrary, an arrhythmia is a problem with the rate or rhythm of the heartbeat. It bears the risk of *sudden cardiac arrest,* the heart then suddenly stops beating.

Given the blood supply to the cardiac tissue is cut off for more than a few minutes, the heart cells start to die, and normal heart tissue is replaced with scar tissue. Whilst the related heart damage may not be obvious to the patient, it can cause severe and long-lasting problems—over time, it can weaken the heart muscle leading to heart failure and arrhythmias.

5.4.6 Aneurysm Disease

An *aneurysm* is a local, slowly growing dilatation of an artery, typically defined by at least 1.5 times of its normal diameter. Although every artery can become aneurysmatic, the infrarenal abdominal aorta seems to be most vulnerable to this disease and frequently develops an AAA. The formation of AAAs is promoted by factors such as smoking, older age, male gender, unhealthy cholesterol, and high MAP; it is also over-represented in persons with a family history of AAA, patients with coronary heart disease and Chronic Obstructive Pulmonary Disease (COPD) [486], and aortic aneurysms are often seen in patients with rare genetic diseases, such as Ehlers–Danlos syndrome type IV, Marfan syndrome, Loeys–Dietz syndrome, and fibromuscular dysplasia. Whilst aneurysm disease and atherosclerosis have many common risk factors, both diseases have distinct gene expressions [49]. Given the clinical importance, AAA studies are well reported in the literature, concerning many different aspects of the disease [109, 343, 486].

If left untreated, aneurysms progress until the wall stress eventually exceeds the failure strength of the degenerated aortic wall—the artery *ruptures*. AAA rupture often leads to massive internal bleeding, and approximately 75% of patients die from such an event. AAA prevalence is much higher in males, whist rupture appears relatively more frequently in females, see Fig. 5.11.

In addition to the risk of internal bleedings and the drop of blood pressure, the rupture of a *cerebral aneurysm* may also lead to a *hemorrhagic stroke*. Blood then leaks into the brain causing cells to die from the increasing intracranial pressure.

5.4.6.1 Aneurysm Pathophysiology
An aneurysm is the end-result of irreversible pathological remodeling of the vessel wall [86, 112], and larger aneurysms show distinct pathological features [11, 86, 299, 373, 467], such as

- *Degradation and fragmentation of elastin fibers*
- *Apoptosis of vascular SMC*
- *Increased collagen content and collagen synthesis*
- *Excessive inflammatory response*

Fig. 5.11 Abdominal Aortic Aneurysm (AAA) rupture rate per year for men and females. Data represents estimates of the RESCAN study [459]

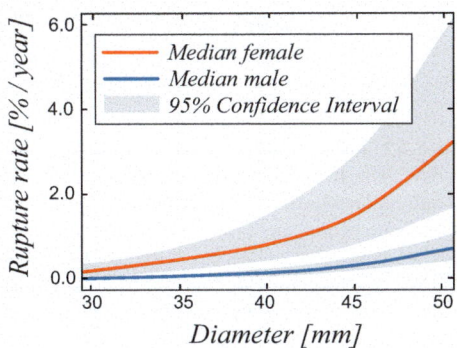

- *Increased oxidative stress*

Whilst the loss of elastin (and possibly increased SMC apoptosis) triggers the initial dilatation, it is the collagen turnover that determines the enlargement and local weakening of the wall, a mechanism that then eventually leads to the rupture [86].

Given its high clinical relevance, the aneurysmatic infrarenal aorta has been extensively studied and a number of pathophysiological processes have been proposed towards the understanding of AAA disease. Figure 5.12 summarizes some of them, ultimately resulting in the *destruction* of the well-defined organization of the normal vessel wall as it is illustrated in Fig. 5.1. The AAA wall exhibits a *degraded* media with few SMCs and fragmented elastin structures, and an inflammatory and/or fibrotic adventitia that can be *thicker* than normal. Mast cells in the adventitia trigger degranulation and release different vasoactive factors linked to *neovascularization* [364]. The "perforation" by the dense vasa vasorum may then diminishe the strength of the vessel wall [581]. It is often challenging to distinguish between medial and adventitial layers in large AAAs—the entire wall resembles a fibrous collagenous tissue that is similar to the adventitia in the normal aorta [195].

Almost all AAAs of clinically relevant size contain an *Intra-Luminal Thrombus (ILT)* [243], a pseudo-tissue that develops from coagulated blood. It has solid-like mechanical properties [196, 569] and its formation may be promoted by disturbed blood flow [42,43]. The ILT is composed of a fibrin mesh, traversed by a continuous network of interconnected canaliculi and contains blood particles, such as erythrocytes, neutrophils, aggregated platelets, blood proteins, and cellular debris [5, 205],

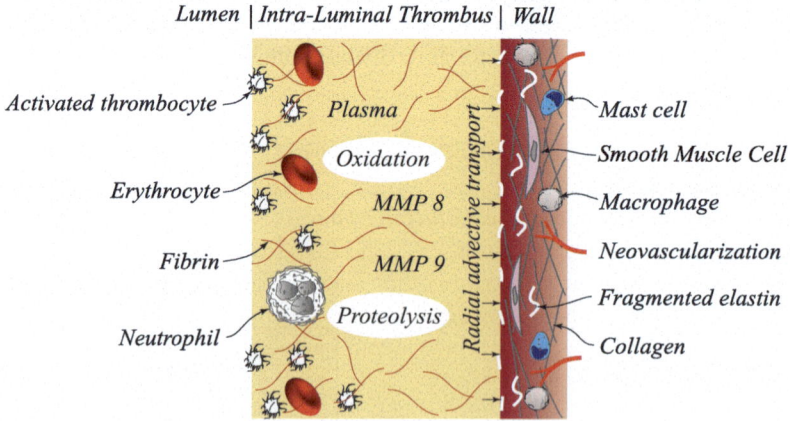

Fig. 5.12 Intra-luminal Thrombus (ILT) and vessel wall of an Abdominal Aortic Aneurysm (AAA). The ILT promotes proteolytic and oxidative activities and facilitates breakdown of the ExtraCellular Matrix (ECM), apoptosis of vascular Smooth Muscle Cells (SMCs), and activation of immune responses. This process also activates MMPs, such as MMP 8 and MMP 9, which together with other substances is transported into the vessel wall. The AAA wall shows depletion of SMC, fragmentation of elastic fibers, numerous inflammatory responses, and mast cells in the adventitia that promote neovascularization

see Fig. 5.12. The ILT creates an environment for increased *proteolytic* and *oxidative* activity [168, 533], possibly linked to the weakening [581] and thinning [299] of the vessel wall. The reaction products are convected towards and into the wall, where they contribute to ECM degradation and the adventitial immune response. Proteolysis that is catalyzed by protases can degrade fibrillar ECM and intermediate adhesive proteins. It then provokes SMC detachment and apoptosis [486], such that neither EC nor SMCs can spread and proliferate in close contact with the ILT. It may explain the absence of an endothelium layer of the ILT-covered AAA wall.

5.4.6.2 The Elastic Properties of the Infrarenal Aorta

The normal and aneurismatic aorta has been characterized by uniaxial and biaxial *in vitro* tissue characterization studies, and Table 5.3 reports some data of the vessel wall and the ILT. The AAA wall is remarkably stiffer than the aged, but normal infrarenal aorta, see Fig. 5.13a. As compared to the wall, the ILT is much softer and shows linear Cauchy stress-stretch properties, see Fig. 5.13b.

5.4.6.3 Strength of Aorta Tissue

As already indicated by the AAA wall's inhomogeneous patho-histology [143], AAA wall strength changes significantly within and across patients [458]. Whilst some of the strength-influencing factors have been identified [171, 172, 352, 357, 504], many are still unknown. Given this unknown, the acquired data is commonly highly dispersed. Table 5.4a summarizes some of the reported AAA strength and thickness measurements. In average the AAA wall withstands 865 (SD 390) kPa stress, and it is 1.6 (SD 0.6) mm thick. The aneurysmatic aorta

Table 5.3 Constitutive properties of (a) the wall and (b) the Intra-Luminal Thrombus (ILT) of Abdominal Aortic Aneurysm (AAA). Data has been acquired by *in vitro* tissue characterization, and the constitutive models are further discussed in Sect. 5.5

Constitutive description	Constitutive parameters [kPa]	Reference
(a) AAA wall		
Yeoh model (5.8) with $N = 2$	$c_1 = 177.0$; $c_2 = 1881.0$	[449]
Yeoh model (5.8) with $N = 5$	$c_1 = 5.0$; $c_2 = c_3 = 0.0$; $c_4 = 2.2 \cdot 10^3$; $c_5 = 13.741 \cdot 10^3$	[436]
Orthotropic hyperelastic (5.30)[b]	$c_0 = 0.14$; $c_1 = 477.0$[a]; $c_2 = 416.4$[a]; $c_3 = 408.3$[a]	[568]
(b) ILT		
Yeoh model (5.8) with $N = 2$	Luminal layer: $c_1 = 7.98$; $c_2 = 8.71$	[569]
Reduced Ogden model (5.11)	Luminal layer: $c_0 = 2.62$	[196]
	Medial layer: $c_0 = 1.98$	
	Abluminal layer: $c_0 = 1.73$	

c_i denotes material parameters used by the individual constitutive model
[a]Dimensionless parameter
[b]Limited to the description of a membrane problem

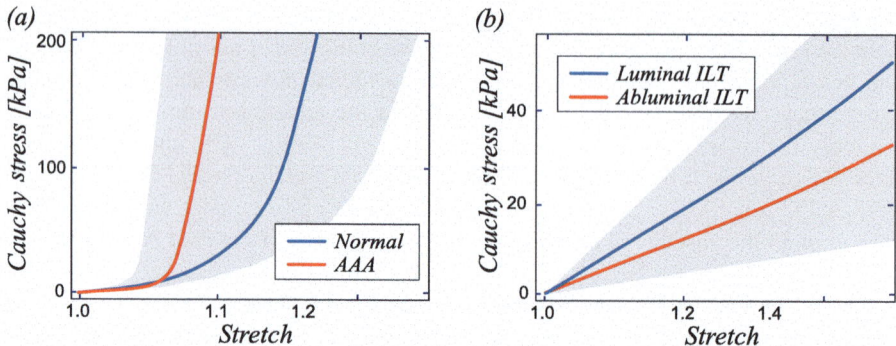

Fig. 5.13 Elastic properties of the infrarenal aortic tissue at equi-biaxial tension. (**a**) Aged, but normal abdominal aorta as compared to the Abdominal Aortic Aneurysm (AAA) wall. Properties are based on *in vitro* experimental tissue characterization [568]. (**b**) Intra-Luminal Thrombus (ILT) of the AAA. Data has been acquired from *in vitro* experimental tissue characterization of the luminal and abluminal ILT sections [196]

is therefore weaker than the normal infrarenal aorta, reported to have a strength of 1210 (SD 330) kPa [582] or even 1710 (SD 140) kPa [583].

Given the reported strength of 1950 (SD 600) kPa [4] and 1470 (SD 910) kPa [380], the normal thoracic aorta is slightly stronger than the normal infrarenal aorta. Less experimental data has been acquired from the Thoracic Aortic Aneurysm (TAA) wall, some of which is listed in Table 5.4(b).

As with ILT's elastic properties, its tensile strength changes with respect to the radial position, and luminal and abluminal ILT tissue fails at 156.5 (SD 57.9) kPa and 47.7 (SD 22.9) kPa, respectively [196].

Table 5.4a indicates the strong negative correlation of AAA strength and wall thickness, a property also reported previously [172, 458]. Given this correlation, a thinner wall is relatively stronger, and *wall tension* (wall stress multiplied with wall thickness) may be a more robust rupture risk predictor [451]. The inverse correlation also somewhat justifies a uniform wall thickness to be used in FEM models of the AAA.

Whilst the aorta wall is biaxially loaded in the body, most strength data, with very few exceptions [476], has been characterized by uniaxial tensile testing. The acquired data may therefore not adequately represent the threshold of the aorta against wall rupture in the body.

Table 5.4 Strength and thickness of the aneurysmatic aorta acquired from *in vitro* testing. (a) Abdominal Aortic Aneurysm (AAA) wall. (b) Thoracic Aortic Aneurysm (TAA) wall

Sample specification	Thickness [mm]	Strength [MPa]	Reference
(a) AAA wall			
$N = 31$; fibrous	1.2	1.2	[412]
$N = 38$; partly calcified	1.5	0.87	
$N = 28$	1.18	–	[411]
$N = 83$	–	0.81	[570]
$N = 26$	1.32	–	[568]
$N = 25$; $d < 55$ mm	1.53	0.77	[381]
$N = 65$; $d > 55$ mm	1.58	1.03	
$N = 76$	–	Female: 0.68	[566]
		Male: 0.88	
$N = 163$	1.57	1.42	[458]
$N = 374/48$	1.48	1.26	[448]
$N = 14$	1.5 to 1.9	Long.: 0.93	[605]
		Circ.: 1.15	
$N = 16$	2.06	0.57	[171]
Anterior: $N = 29$	2.73	Long.: 0.38	[550]
		Circ.: 0.52	
Lateral: $N = 9$	2.52	Long.: 0.51	
		Circ.: 0.73	
Posterior: $N = 9$	2.09	Long.: 0.47	
		Circ.: 0.45	
Intact AAA: $N = 26$	2.5	0.82	[122]
Ruptured AAA: $N = 13$	3.6	0.54	
Intact: $N = 278/56$	1.5	0.98	[451]
Ruptured: $N = 141/21$	1.7	0.95	
Long.: $N = 45$	–	0.86	[450]
Circ.: $N = 19$	–	1.02	
Long.: $N = 49$	1.57	0.715	[437]
Circ.: $N = 41$	1.58	1.1	
$h > 4$ mm: $N = 7$	–	1.38	[581]
$h < 4$ mm: $N = 7$	–	2.16	
(b) TAA wall			
$N = 163$	–	TAV: Long.: 0.54	[428]
	–	Circ.: 0.961	
	–	BAV: Long.: 0.698	
	–	Circ.: 1.656	
$N = 27$	1.99	TAV: 0.878	[172]
	1.7	BAV: 1.310	
$N = 26$	2.0	TAV: Long.: 0.88	[188]
	2.0	Circ.: 1.19	
	1.9	BAV: Long.: 0.84	
	1.9	Circ.: 1.23	

N- Number of samples; d- Aorta diameter at site of wall sample excision; h- Thickness of the Intra-luminal Thrombus (ILT) layer underneath the wall sample; Circ.- circumferential; Long.- longitudinal; ILT- Intra-Luminal Thrombus; TAV- Aorta with a Tricuspid Aortic Valve; BAV- Aorta with a Bicuspid Aortic Valve

5.5   Constitutive Descriptions

Constitutive modeling of vascular tissue is well documented in the literature, and numerous descriptions with application to blood vessels have been reported [408]. Whilst purely *phenomenological approaches* can successfully fit experimental data, they show limited robustness of predictions beyond the strain range within which the model parameters have been identified. In contrary, *structural constitutive descriptions* overcome such limitations and integrate histological and mechanical information of the vessel wall. Aside from being more robust, such modeling fosters our understanding concerning the load-carrying mechanisms in vascular tissue.

This section addresses the basic biomechanical properties of the passive vessel wall upon its description as homogenized *single phase incompressible solid*. Whilst much more complex frameworks are reported in the literature, it is the Intended Model Application (IMA) that determines the level of complexity. Examples show that even the highly porous ILT may be described as a single phase incompressible solid towards the prediction of the stress in the AAA wall [433].

5.5.1   Capacity of a Vessel Segment

The *capacity*, or volume compliance of a vessel segment is a property used by lumped parameter models, see Sect. 2.3.2. We approximate the vessel by a thin-walled tube at plane stress that is inflated at the blood pressure p_i. Given static equilibrium, $\sigma_\theta = p_i d/(2h)$ and $\sigma_z = \sigma_\theta/2$ describe the circumferential and respective axial stress in the vessel wall. A linear-elastic material with the Young's modulus E and the Poisson's ratio $v = 0.5$ may be used in the description of the vessel wall's elastic properties. At plane stress, Hooke's law (3.49) yields the circumferential strain $\varepsilon_\theta = \Delta d/d = 3\sigma_\theta/(4E)$, and with the aforementioned circumferential stress, results in the pressure increment $\Delta p_i = 8hE\Delta d/(3d^2)$, a function of the diameter increment Δd. The vessel's capacity therefore reads

$$C = \frac{\Delta V}{\Delta p_i} = \frac{3d^3\pi l}{16hE} \ \ [\text{m}^3 \ \text{Pa}^{-1}], \tag{5.1}$$

where the volume increment $\Delta V = d\pi l\Delta d/2$ has been used, and l denotes the length of the vessel segment. The capacity (5.1) is constant and thus independent of the blood pressure p_i.

Whilst the vessel capacity model (5.1) is often used, it is know that the vessel wall's stress–strain properties are non-linear, and Hooke's law reflects therefore a very coarse approximation. The description of the vessel's capacity (5.1) should for many applications then be modified towards the description of non-linear vessel wall properties. Let us consider a general non-linear material, where $D_{\theta\theta}$ and $D_{\theta z}$ are the coefficients of the (non-constant) compliance matrix. The increment of the circumferential wall strain then reads $\Delta\varepsilon_\theta = \Delta d/d = D_{\theta\theta}\Delta\sigma_\theta + D_{\theta z}\Delta\sigma_z$, and with $\sigma_z = \sigma_\theta/2$ we may express the strain increment $\Delta\varepsilon_\theta = (D_{\theta\theta} + D_{\theta z}/2)\Delta\sigma_\theta$ as a function of the stress increment $\Delta\sigma_\theta$. Given the stress increment $\Delta\sigma_\theta = \Delta p_i d/(2h)$ from the static equilibrium, we may express the pressure increment $\Delta p_i = 2h\Delta d/(d^2\alpha)$ as a function of the diameter increment Δd and the constitutive factor $\alpha = D_{\theta\theta} + D_{\theta z}/2$, respectively. The vessel's capacity then reads

$$C = \frac{\Delta V}{\Delta p_i} = \frac{d^3\pi l\alpha}{4h}, \tag{5.2}$$

where the volume increment $\Delta V = d\pi l\Delta d/2$ has been used. For a non-linear stress–strain law, the coefficient α depends on the strain. The capacity (5.2) is therefore not constant, but a function of the blood pressure p_i.

5.5.2 Hyperelasticity for Incompressible Solids

The vessel wall's stress–strain properties may be described within the framework of hyperelasticity for incompressible solids. The volume ratio $J = 1$ then holds, and Coleman and Noll's procedure for incompressible materials (3.131) allows us to derive the second Piola–Kirchhoff stress

$$\mathbf{S} = 2\frac{\partial\Psi(\mathbf{C})}{\partial\mathbf{C}} - \kappa\mathbf{C}^{-1} \tag{5.3}$$

from any given strain energy $\Psi(\mathbf{C})$ [J m^{-3}] per unit (reference) tissue volume, see Sect. 3.6.4. Here, $\mathbf{C} = \mathbf{F}^{\mathsf{T}}\mathbf{F}$ denotes the right Cauchy–Green strain, and $\mathbf{F}$ is the deformation gradient. The pressure κ, to be distinguished from the hydrostatic pressure p, acts at the inverse right Cauchy–Green strain $\mathbf{C}^{-1}$ and serves as a Lagrange parameter to *enforce incompressibility*. It is determined by the boundary conditions of the solid mechanics problem, whilst all constitutive information is captured by the particular form of the strain energy density—specific functions will be discussed later in this chapter.

The second Piola transform (3.32) allows us to compute the Cauchy stress

$$\boldsymbol{\sigma} = 2\mathbf{F}\frac{\partial\Psi(\mathbf{C})}{\partial\mathbf{C}}\mathbf{F}^{\mathsf{T}} - \kappa\mathbf{I}, \tag{5.4}$$

where the Lagrange pressure κ contributes to the hydrostatic stress and acts at the identity $\mathbf{I}$. Given the strain energy Ψ and the deformation gradient $\mathbf{F}$, the stress is determined up to the Lagrange parameter κ.

5.5.2.1 Stress and Strain Analysis in Principal Directions

We may consider the right Cauchy–Green strain $\mathbf{C} = \sum_{i=1}^{n_{\dim}} \lambda_i^2 \widehat{\mathbf{N}}_i \otimes \widehat{\mathbf{N}}_i$ in principal directions $\widehat{\mathbf{N}}_i$, where $n_{\dim}$ and λ_i denote the number of spatial dimensions and the principal stretches, respectively. Given an *incompressible* and *isotropic* material, the expression (5.3) then yields the principal second Piola–Kirchhoff stresses

$$S_i = \frac{1}{\lambda_i} \frac{\partial \Psi(\mathbf{C})}{\partial \lambda_i} - \frac{\kappa}{\lambda_i^2} \quad \text{(no sumation)}, \tag{5.5}$$

where κ denotes the Lagrange parameter and the relation $\partial \lambda_i / \partial C_{ii} = (2\lambda_i)^{-1}$ (no summation) has been used. It follows directly from the aforementioned eigenvalue representation of the right Cauchy–Green strain.

Given an isotropic material, the principal stress directions coincide with the principal strain directions, and (5.5) may always be used. In contrary, for an *anisotropic* material, the principal stress and strain directions are in general not equal, and (5.5) is therefore not generally valid. However, given the principal strain directions coincide with the *material's principal directions*, said directions are then also the principal stress directions. We may then use (5.5) even in the description of an anisotropic material. The material's principal directions are the directions along which the material exhibits extremal properties—it is stiffest or softest along them.

The first Piola transform of an incompressible material $P_i = \lambda_i S_i$ (no summation) and (5.5), then yields the first Piola–Kirchhoff stresses

$$P_i = \frac{\partial \Psi(\mathbf{C})}{\partial \lambda_i} - \kappa / \lambda_i, \tag{5.6}$$

whilst the second Piola transform $\sigma_i = \lambda_i^2 S_i$ determines the respective principal Cauchy stresses

$$\sigma_i = \lambda_i \frac{\partial \Psi(\mathbf{C})}{\partial \lambda_i} - \kappa, \tag{5.7}$$

where the boundary conditions of the solid mechanics problem determine the Lagrange parameter κ.

5.5.3 Purely Phenomenological Descriptions of the Vessel Wall

A constitutive description aims at the design of a strain energy function Ψ that is able to capture the vessel wall's mechanical properties. Following a purely phenomenological approach, Ψ is a mathematical function that lacks any information regarding the vessel wall's histology or the load-carrying mechanisms within the wall. Given no such "constraints," the function Ψ may be very "flexible" in the representation of experimental data, a property that makes it also unreliable in predictions *beyond* the range, within which its parameters have been calibrated.

5.5.3.1 Isotropic Models

Vascular tissue and rubber share some common mechanical properties, and models that originally have been proposed for rubber, are also frequently used to describe the vessel wall. One of them is the *Yeoh* strain energy function [608]

$$\Psi(\mathbf{C}) = \sum_{i=1}^{N} c_i (I_1 - 3)^i , \tag{5.8}$$

where $I_1 = \text{tr}\mathbf{C}$ denotes the first strain invariant, whilst c_i; $i = 1, \ldots, N$ [Pa] are stiffness-related material parameters of the N polynomial terms in the strain energy. Given vascular applications, most often two polynomial terms are used, as in the description of the aorta reported elsewhere [449, 450, 458, 569]. In addition, a three-term Yeoh model has been used to describe hard, mixed, and soft carotid plaques [333], whilst five terms have been proposed to capture the highly non-linear stress–strain properties of the aneurysmatic aortic wall [436].

Given the case $N = 1$, the Yeoh model reduces to the *neoHookean* model

$$\Psi(\mathbf{C}) = c(I_1 - 3) , \tag{5.9}$$

and $G = 2c$ [Pa] then denotes the small-strain shear modulus. It is frequently used in the description of atherosclerotic plaque components, such as the fibrous cap, the lipid core, and calcifications [403, 407].

The *Ogden* strain energy

$$\Psi(\mathbf{C}) = \Psi(\lambda_1, \lambda_2, \lambda_3) = \sum_{i=1}^{N} \frac{c_i}{k_i} \left(\lambda_1^{k_i} + \lambda_2^{k_i} + \lambda_3^{k_i} - 3 \right) \tag{5.10}$$

is another model to describe the 3D properties of vascular tissue, where $\lambda_1, \lambda_2, \lambda_3$ are the principal stretches. The material parameters are denoted by c_i [Pa] and k_i, where $c_0 = (\sum_i^N c_i k_i)/2$ [Pa] then determines the tissue's small-strain shear modulus. Whilst the model is frequently used with $N = 2$ terms, as in the description of normal and diseased carotid arteries [341], a reduced Ogden model

$$\Psi = c_0(\lambda_1^4 + \lambda_2^4 + \lambda_3^4 - 3) \tag{5.11}$$

captured the properties of ILT tissue with the single material parameter c_0 [Pa] [196].

In addition to models proposed for rubber, constitutive descriptions have also been developed to represent vascular tissues in particular. The *Demiray* strain energy [118]

$$\Psi(\mathbf{C}) = c_1 \exp[c_2(I_1 - 3) - 1] , \tag{5.12}$$

where c_1 [Pa] and c_2 denote material parameters, considers an *exponential* term to describe the progressive strain-stiffening of vascular tissue. The model has been calibrated to aortic wall tissue [436, 465], and used to analyze the carotid bifurcation [117].

Example 5.1 (Vessel Segment Characterization). We consider a femoral artery segment that is mounted in an experimental testing rig towards the characterization of its biomechanical properties. The vessel segment may be represented by the three-element lumped parameter model in Fig. 5.14. The vessel is perfused with a Newtonian fluid of the viscosity $\eta = 4.0$ mPa s and the density $\rho = 1025.0\,\text{kg m}^{-3}$, properties that mimics blood. The vessel segment may be regarded as a thin-walled membrane, and the diameter $D = 6.1$ mm and the wall thickness $H = 0.7$ mm describe its load-free configuration. The vessel length of $l = 42$ cm remains constant during the test cycle.

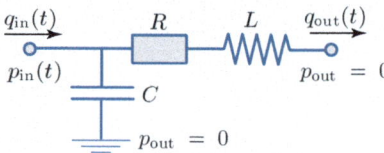

Fig. 5.14 Schematic representation of a three-element lumped parameter model of a vessel segment. The flows $q_{\text{in}}(t)$, $q_{\text{out}}(t)$ and the pressures $p_{\text{in}}(t)$, $p_{\text{out}} = 0$ describe the inlet and outlet conditions. The vessel's biomechanical properties are expressed by its capacity C, resistance R, and inductance L, respectively

(a) Compute the capacity C, the resistance R, and inductance L of the femoral artery segment. Given this task, small deformations may be assumed, and the vessel wall is described by a linear-elastic material with the Young's modulus $E = 300.0$ kPa and the Poisson's ration $v = 0.5$.

(b) In a refined analysis finite deformations should be considered, and the respective vessel's properties R, C, and L are to be computed at the Mean Arterial Pressure (MAP) of 100.0 mmHg. Given this task, non-linear wall properties are represented by the strain energy $\Psi(\mathbf{C}) = c(I_1 - 3)^2$ per unit tissue volume, where $c = 125.0$ kPa is a stiffness-related parameter, whilst $I_1 = \text{tr}\mathbf{C}$ denotes the first invariant of the right Cauchy–Green strain $\mathbf{C}$. Incompressibility may be assumed.

(c) Explore the response of the three-element lumped parameter model shown in Fig. 5.14 at the prescribed pressures of $p_{\text{in}} = \mathbf{P}\exp(i\omega t)$ and $p_{\text{out}} = 0$. Here, $\mathbf{P}$ denotes the complex pressure amplitude, and $i = \sqrt{-1}$ is the imaginary unit. At these boundary conditions, the flow $q(t)$ establishes in the femoral artery. Given R, C, and L through Tasks (a) and (b), compute the impedance module $|\mathbf{Z}|$ and phase shift $\arg(\mathbf{Z})$ as a function of the signal frequency $f = \omega/(2\pi)$. ■

5.5.4 Inflated and Axially Stretched Two-Layered Vessel

For a number of applications, a thin-walled cylinder of then media and adventitia may be used in the description of a normal, and thus non-diseased vessel. Figure 5.16 shows such a model, where $\Omega_{0\,m}$ and $\Omega_{0\,a}$ denote the stress-free reference configurations of the media and adventitia, respectively. Given the dimensions $D_m = 12\,mm$, $L_m = 23\,mm$, $H_m = 0.7\,mm$ and $D_a = 14\,mm$, $L_a = 27\,mm$, $H_a = 0.5\,mm$, the model aims at representing an infrarenal aorta. The strain energies

$$\Psi_m(\mathbf{C}) = c_m(I_{1\,m} - 3) \; ; \; \Psi_a(\mathbf{C}) = c_a\{\exp[b(I_{1\,a} - 3)] - 1\} \qquad (5.19)$$

describe the elastic properties of medial and adventitial tissue, where $I_{1\,m} = \mathrm{tr}\mathbf{C}$ and $I_{1\,a} = \mathrm{tr}\mathbf{C}$ denote the first invariants of the right Cauchy–Green strain $\mathbf{C}$ in the respective layers. The constitutive parameters $c_m = 18.5\,kPa$, $c_a = 3.6\,kPa$, and $b = 3.7$ have been identified from tissue characterization experiments.

We put together the two tissue layers, which then determines the vessel's load-free configuration $\Omega_{0\,ma}$, a configuration that is free from external loading. The individual configurations $\Omega_{0\,m}$ and $\Omega_{0\,a}$ will in general have different dimensions, which then leads to *residual stresses* in the load-free configuration $\Omega_{0\,ma}$.

Fig. 5.16 Configurations of the two-layered vessel model. The motions χ_a and χ_m map the stress-free reference configurations of the adventitia and media to the vessel's load-free configuration. The motion χ maps the vessel to its loaded configuration, whilst the motion χ_{pre} deforms the adventitia to fit the geometry of the media

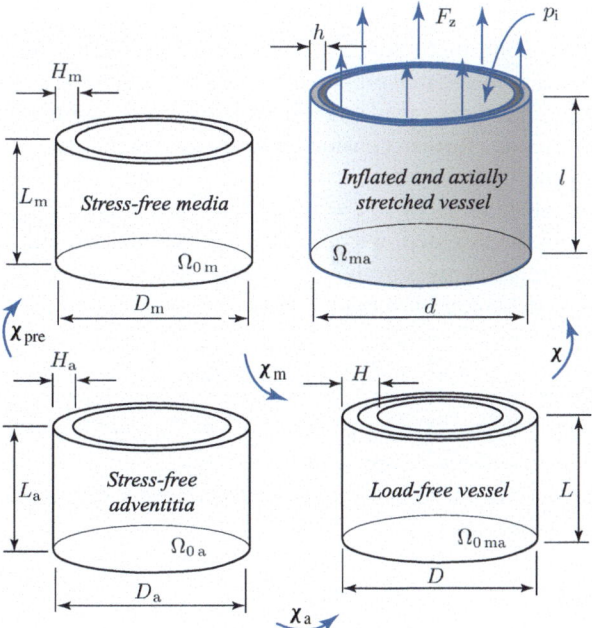

Towards the computation of the stress in the individual vessel layers, we start with the description of their deformations. Given the principal stretches $\lambda_{\theta i}, \lambda_{z i}, \lambda_{r i}$ in the media, $i = $ m, and the adventitia, $i = $ a, along the circumferential θ, axial z, and radial r vessel directions, $\mathbf{F}_i = \mathrm{Grad}(\boldsymbol{\chi}(\boldsymbol{\chi}_i)) = \mathrm{diag}[\lambda_{\theta i}, \lambda_{z i}, \lambda_{r i}]$ expresses the respective deformation gradient. The first invariant of the right Cauchy–Green strain $\mathbf{C}$ then reads $I_{1 i} = \lambda_{\theta i}^2 + \lambda_{z i}^2 + \lambda_{r i}^2 ; i = $ m, a, and the strain energy functions (5.19) together with the relation (5.7) allow us to compute the circumferential and axial Cauchy stresses

$$\sigma_{\theta i} = \xi_i(\lambda_{\theta i}^2 - \lambda_{r i}^2) \ ; \ \ \sigma_{z i} = \xi_i(\lambda_{z i}^2 - \lambda_{r i}^2) \ ; \ i = \mathrm{m, a} \ \ (\text{no summation}),$$

where the membrane condition $\sigma_{r a} = \sigma_{r m} = 0$ has been used. The abbreviations $\xi_a = 2bc_a \exp[b(\lambda_{\theta a}^2 + \lambda_{r a}^2 + \lambda_{z a}^2 - 3)]$ and $\xi_m = 2c_m$ have also been introduced.

The kinematic compatibility between the media and adventitia may now be addressed. We therefore introduce the pre-stretches

$$\lambda_{\theta \, \mathrm{pre}} = D_m/D_a \ ; \ \ \lambda_{z \, \mathrm{pre}} = L_m/L_a \tag{5.20}$$

that deform the adventitia to fit the load-free media $\Omega_{0 m}$, see Fig. 5.16. The stretches in the adventitia may then be expressed through

$$\lambda_{\theta \, a} = \lambda_{\theta \, m}\lambda_{\theta \, \mathrm{pre}} \ ; \ \ \lambda_{z \, a} = \lambda_{z \, m}\lambda_{z \, \mathrm{pre}} \ .$$

The definition $(5.20)_1$ used infinitesimally thin vessel wall layers, and alternatively the circumferential pre-stretch may also be defined by $\lambda_{\theta \, \mathrm{pre}} = (D_m - H_m/2)/(D_a + H_a/2)$. It then considers finite wall thicknesses of the respective layers, a refinement that is not further considered in this example. The kinematics description is then closed by the incompressibility of the vascular tissue. It allows us to express the radial stretches by $\lambda_{r a} = (\lambda_{\theta a}\lambda_{z a})^{-1}$ and $\lambda_{r m} = (\lambda_{\theta m}\lambda_{z m})^{-1}$, respectively.

In the last step, we introduce the external loading to the problem, and consider the inflation pressure p_i and the axial stress resultant N_z to act at the vessel segment. The static equilibrium then reads

$$p_i d = 2(\sigma_{\theta \, a}h_a + \sigma_{\theta \, m}h_m) \ ; \ \ N_z = (\sigma_{\theta \, a}h_a + \sigma_{\theta \, m}h_m)d\pi \ ,$$

where $h_m = H_m\lambda_{r m}$ and $h_a = H_a\lambda_{r a}$ are the deformed thicknesses of the medial and adventitial layers, whilst $d = D_m\lambda_{\theta \, m} = D_a\lambda_{\theta \, a}$ denotes the deformed diameter of the vessel. Given the geometrical and constitutive data of the problem, these equations read

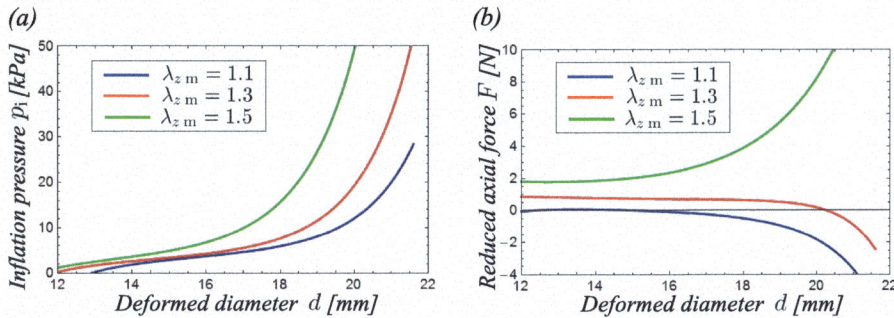

Fig. 5.17 Characteristics of the two-layered thin-walled composite tube during inflation at the fixed axial stretches $\lambda_{z\,\mathrm{m}}$ of the media. (**a**) Inflation pressure p_i *versus* the deformed vessel diameter d. (**b**) Reduced axial force F *versus* the deformed vessel diameter d

$$\left.\begin{aligned}
\frac{p_i}{\lambda_{\theta\,\mathrm{m}}^4\lambda_{z\,\mathrm{m}}^3} &= 4316.7(\lambda_{\theta\,\mathrm{m}}^4\lambda_{z\,\mathrm{m}}^2 - 1) + \zeta(3.376\lambda_{\theta\,\mathrm{m}}^4\lambda_{z\,\mathrm{m}}^2 - 8.619)\;; \\
\frac{N_z}{\lambda_{\theta\,\mathrm{m}}^2\lambda_{z\,\mathrm{m}}^3} &= 0.976(\lambda_{\theta\,\mathrm{m}}^2\lambda_{z\,\mathrm{m}}^4 - 1) + 10^{-4}\zeta(7.542\lambda_{\theta\,\mathrm{m}}^2\lambda_{z\,\mathrm{m}}^4 - 19.495)
\end{aligned}\right\} \tag{5.21}$$

with $\zeta = 10^2\exp[2.718\lambda_{\theta\,\mathrm{m}}^2 + 6.94(\lambda_{\theta\,\mathrm{m}}\lambda_{z\,\mathrm{m}})^{-2} + 2.685\lambda_{z\,\mathrm{m}}^2]$. In addition to the influence on the circumferential stress, the pressure p_i also acts in the axial direction—an inflation experiment that characterizes the vessel does not measure N_z, but the reduced axial force $F = N_z - p_i d^2\pi/4$.

Figure 5.17 shows the vessel's biomechanical characteristics as predicted by (5.21). It is in qualitative agreement with the experimental observations and illustrates the non-linearity of the inflation pressure p_i and the reduced axial force F.

Aside from being a function of the circumferential and axial stretches, the stress in the vessel wall is also different in the medial and adventitial layers. Figure 5.18a shows the circumferential stress in both layers during the inflation at the fixed axial stretch $\lambda_{zm} = 1.3$. The progressive increase of the stress in the adventitia at higher deformations is characteristic for blood vessels, a mechanism towards protecting the media and intima from overstretching. The stress in the vessel wall is biaxial, and the von Mises stress $\sigma_{\mathrm{Mises}} = \sqrt{\sigma_\theta^2 - \sigma_\theta\sigma_z + \sigma_z^2}$ may be used to express the stress state. Figure 5.18b shows it in both layers during inflation at $\lambda_{zm} = 1.3$, a response that is very similar to the circumferential stress *versus* diameter properties.

Towards the computation of the residual stress in the vessel's load-free configuration, we point out that neither pressure nor axial force act at $\Omega_{0\,\mathrm{ma}}$. The solution of the governing equations (5.21) at $p_i = F_z = 0$ therefore determines the residual stress in $\Omega_{0\,\mathrm{ma}}$. It results in a system of two equations, which solution is given by the stretches $\lambda_{\theta\,\mathrm{m}} = 1.0836$ and $\lambda_{z\,\mathrm{m}} = 1.0863$ of the medial layer. It determines the diameter $D = D_{\mathrm{m}}\lambda_{\theta\,\mathrm{m}} = 13.003\,\mathrm{mm}$, the length $L = L_{\mathrm{m}}\lambda_{z\,\mathrm{m}} = 24.985\,\mathrm{mm}$, and

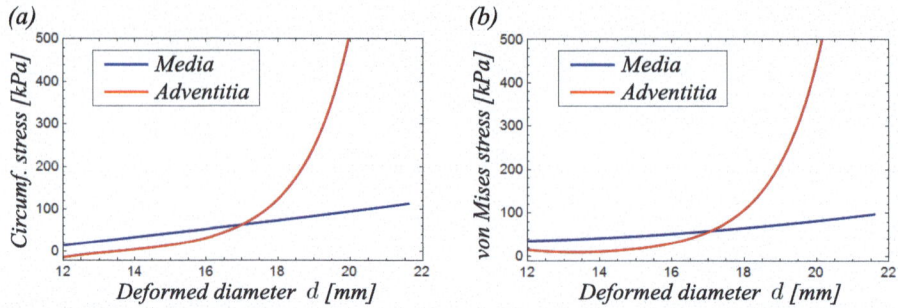

Fig. 5.18 Development of the stress in the medial and adventitial layers during inflation at the fixed axial stretches $\lambda_{z\,m} = 1.3$. (**a**) Circumferential stress and (**b**) von Mises stress *versus* the deformed vessel diameter d

the wall thickness $H = H_a\lambda_{r\,a} + H_m\lambda_{r\,a} = 1.176\,\text{mm}$ of the vessel at its load-free configuration $\Omega_{0\,\text{ma}}$.

5.5.5 Inflated and Axially Stretched Thick-Walled Vessel

The thickness of the normal vessel wall is approximately 10% of its diameter, and a thick-walled cylindrical tube could be a more accurate biomechanical representation as compared to the afore-described membrane model. We therefore consider the aorta segment shown in Fig. 5.19. It occupies the stress-free reference configuration Ω_0, where $R_i = 10\,\text{mm}$, $L = 23.0\,\text{mm}$, and $H = 2.0\,\text{mm}$ denote the inner radius, length, and wall thickness, respectively. At its *in vivo* configuration, the vessel is inflated at the pressure p_i, and stretched by $\lambda_z = l/L$ along its axial direction, such that r_i, l, and h determine the vessel's spatial configuration Ω.

The incompressible elastic deformation of the aorta wall may be described by the Yeoh strain energy density

$$\Psi(\mathbf{C}) = c_1(I_1 - 3) + c_4(I_1 - 3)^4, \tag{5.22}$$

where $I_1 = \text{tr}\mathbf{C}$ denotes the first invariant of the right Cauchy–Green strain tensor $\mathbf{C}$, whilst $c_1 = 12.7\,\text{kPa}$ and $c_4 = 1.4\,\text{kPa}$ are constitutive parameters.

Towards the description of the kinematics of this problem, we introduce the outer radius $R_o = R_i + H$ of the stress-free vessel and look at the ring segment that is formed between the inner radius R_i and the radius $R < R_o$. Given incompressibility, and thus volume-preserving deformation, $L(R^2 - R_i^2) = l(r^2 - r_i^2)$ holds, and

$$r = \sqrt{r_i^2 + (R^2 - R_i^2)/\lambda_z} \tag{5.23}$$

then describes the radial position of a material particle in the deformed vessel wall.

Fig. 5.19 Reference (left) and spatial (right) configurations of the thick-walled vessel model

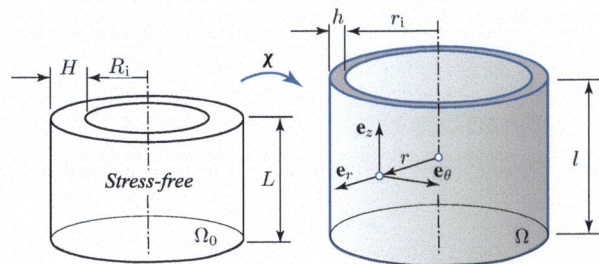

The strain energy function (5.22), together with relation (5.7) allows the determination of the principal Cauchy stresses $\sigma_i = \overline{\sigma}_i - \kappa$; $i = \theta, z, r$ along the circumferential θ, axial z, and radial r vessel directions. The tissue deformation determines the stress $\overline{\sigma}_i = 2\lambda_i^2[c_1 + 4c_4(I_1 - 3)^3]$, and the boundary conditions set the Lagrange pressure κ.

Cauchy's momentum equation (3.107) in cylindrical coordinates $\{\mathbf{e}_\theta, \mathbf{e}_z, \mathbf{e}_r\}$ determines the static equilibrium of the vessel wall. Given the axisymmetric problem, the only non-trivial component of Cauchy's momentum equation is the equilibrium along the radial directions $\mathbf{e}_r$ and reads $r\,d\sigma_r/dr = \sigma_\theta - \sigma_r$. Its integration at the boundary conditions $\sigma_r(r_i) = -p_i$ and $\sigma_r(r_o) = 0$, then yields the relation

$$p_i = \int_{r_i}^{r_o} \frac{\overline{\sigma}_\theta - \overline{\sigma}_r}{r}\,dr\,, \tag{5.24}$$

where the identity $\sigma_\theta - \sigma_r = \overline{\sigma}_\theta - \overline{\sigma}_r$ has been used, and $r_o = r_i + h$ denotes the outer radius of the deformed vessel.

With the kinematics relation (5.23), we may express the circumferential stretch by $\lambda_\theta(r_i, \lambda_z, R) = r/R = \sqrt{r_i^2/R^2 + (1 - R_i^2/R^2)/\lambda_z}$, and the incompressibility allows the substitution $\lambda_r(r_i, \lambda_z, R) = \lambda_z^{-1}\lambda_\theta^{-1}$ of the radial stretch. The stresses $\overline{\sigma}_\theta$ and $\overline{\sigma}_r$ may therefore be written as functions of r_i, λ_z, and R, and the equilibrium (5.24) then reads

$$p_i = \int_{R_i}^{R_o} \frac{\overline{\sigma}_\theta(r_i, \lambda_z, R) - \overline{\sigma}_r(r_i, \lambda_z, R)}{R}\,dR\,,$$

where $r = \lambda_r R$ and $dr = \lambda_r dR$ have been used. At prescribed axial stretch λ_z and inner radius r_i, this expression can be integrated—at least numerically. Figure 5.20a shows the inflation pressure p_i *versus* the deformed inner radius r_i at a fixed axial stretch λ_z.

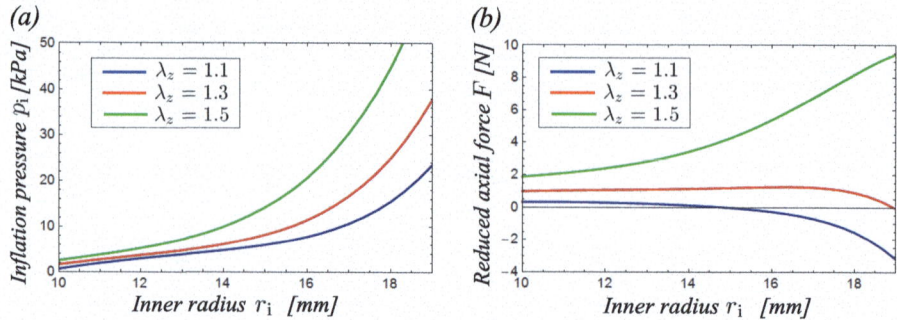

Fig. 5.20 Characteristics of the inflated thick-walled tube at fixed axial stretches λ_z. (**a**) Inflation pressure p_i *versus* the deformed inner radius r_i. (**b**) Reduced axial force F *versus* the deformed inner radius r_i

In addition to the stretching along the circumference, the vessel is loaded in the axial direction, where $N = 2\pi \int_{r_i}^{r_o} (\overline{\sigma}_z - \kappa) r \, dr$ denotes the axial stress resultant. Given the relation $\kappa(\xi) = \overline{\sigma}_r(\xi) - \sigma_r(\xi)$ for the Lagrange contribution at the radial position ξ, and the substitution $\sigma_r(\xi) = -p_i - \int_{r_i}^{\xi} (\overline{\sigma}_\theta - \overline{\sigma}_r)/r \, dr = \int_{r_o}^{\xi} (\overline{\sigma}_\theta - \overline{\sigma}_r)/r \, dr$ of the radial stress, we may express the axial stress resultant by

$$N = 2\pi \int_{r_i}^{r_o} \left(\overline{\sigma}_z - \overline{\sigma}_r + \int_{r_o}^{\xi} \frac{\overline{\sigma}_\theta - \overline{\sigma}_r}{r} \, dr \right) \xi \, d\xi$$

$$= 2\pi \int_{r_i}^{r_o} (\overline{\sigma}_z - \overline{\sigma}_r) r \, dr + 2\pi \int_{r_i}^{r_o} \int_{r_o}^{\xi} \frac{\overline{\sigma}_\theta - \overline{\sigma}_r}{r} \, dr \, \xi \, d\xi .$$

The reversion of the order of integration of the second integral then yields

$$N = 2\pi \int_{r_i}^{r_o} (\overline{\sigma}_z - \overline{\sigma}_r) r \, dr + 2\pi \int_{r_i}^{r_o} \frac{\overline{\sigma}_\theta - \overline{\sigma}_r}{r} \int_{r_i}^{r} \xi \, d\xi \, dr$$

$$= 2\pi \int_{r_i}^{r_o} (\overline{\sigma}_z - \overline{\sigma}_r) r \, dr - 2\pi \int_{r_i}^{r_o} \frac{\overline{\sigma}_\theta - \overline{\sigma}_r}{r} \left(\frac{r^2}{2} - \frac{r_i^2}{2} \right) dr$$

$$= \pi \int_{r_i}^{r_o} (2\overline{\sigma}_z - \overline{\sigma}_r - \overline{\sigma}_\theta) r \, dr + \pi r_i^2 \underbrace{\int_{r_i}^{r_o} \frac{\overline{\sigma}_\theta - \overline{\sigma}_r}{r} \, dr}_{0 + p_i} ,$$

and

$$F = N - p_i r_i^2 \pi = \pi \int\limits_{r_i}^{r_o} (2\overline{\sigma}_z - \overline{\sigma}_r - \overline{\sigma}_\theta) \, r \, dr \qquad (5.25)$$

expresses the reduced axial force. It is a function of the stresses $\overline{\sigma}_z, \overline{\sigma}_r, \overline{\sigma}_\theta$ and independent of the Lagrange pressure κ.

As with the inflation pressure p_i, the reduced axial force (5.25) may be expressed as a function of r_i, λ_z, and R. It then reads

$$F = \pi \int\limits_{R_i}^{R_o} [2\overline{\sigma}_z(r_i, \lambda_z, R) - \overline{\sigma}_r(r_i, \lambda_z, R) - \overline{\sigma}_\theta(r_i, \lambda_z, R)] \, R\lambda_r^2(r_i, R) dR \,,$$

where $r = \lambda_r R$ and $dr = \lambda_r dR$ have been used. At prescribed axial stretch λ_z and inner radius r_i, it can be integrated. Figure 5.20b shows the development of the reduced axial force F during vessel inflation at fixed λ_z.

Whilst the presented approach supports the computation of the vessels diameter and the reduced axial force, the Lagrange pressure κ remains unknown and permits from the computation of the stress in the inflated vessel. A discretization method, such as the one used in Sect. 7.6.4 or the FEM model discussed in Chap. 4, is needed to compute the stress across the vessel wall.

5.5.5.1 Anisotropic Models

The vascular wall is anisotropic, a property that can be captured by *anisotropic* hyperelastic models. Said models are often formulated with respect to the *principal material directions*, and thus within the circumferential $\mathbf{e}_\theta$, the axial $\mathbf{e}_z$, and the radial $\mathbf{e}_r$ direction of the vessel's reference configuration Ω_0. The *Fung-type* model [89] is such a description and expresses the strain energy

$$\Psi(\mathbf{E}, \mathbf{e}_\theta, \mathbf{e}_z, \mathbf{e}_r) = c_0[\exp(Q) - 1] \,, \qquad (5.26)$$

as a function of the Green–Lagrange strain $\mathbf{E} = (\mathbf{C} - \mathbf{I})/2$, where

$$Q = c_1 E_{\theta\theta}^2 + c_2 E_{zz}^2 + c_3 E_{rr}^2 + c_4 E_{\theta\theta} E_{zz} + c_5 E_{zz} E_{rr} \qquad (5.27)$$

$$+ c_6 E_{rr} E_{\theta\theta} + c_7 E_{\theta z}^2 + c_8 E_{rz}^2 + c_9 E_{r\theta}^2 \,, \qquad (5.28)$$

denotes a strain-dependent scalar. It is a function of the components of the Green–Lagrange strain $E_{ij} = \mathbf{e}_i \cdot (\mathbf{E}\mathbf{e}_j); i, j = \theta, z, r$ with respect to the vessel's circumferential θ, axial z, and radial r directions, respectively. The model uses the material parameters c_0 [Pa] and $c_i; i = 1, \ldots, 9$, and is widely used in vascular biomechanics, such as for the description of the aorta [620]. Towards the definition of a strictly convex strain energy, the material parameter c_0 has to be positive and the parameters $c_i; i = 1, \ldots, 9$ cannot be independent from each other [151], see Sect. 5.6.2.2.

A class of models expresses the strain energy in terms of the in-plane strain components $E_{\theta\theta}$, E_{zz}, and $E_{\theta z}$. Whilst such models may be used for a *membrane analysis* of the vessel wall, they do not support a general 3D stress analysis, see Section 4.2.1 in [261]. The Hayashi-type model [536] is such a description and uses the strain energy

$$\Psi(\mathbf{E}, \mathbf{e}_\theta, \mathbf{e}_z) = c_0[\ln(1 - Q)], \qquad (5.29)$$

where the scalar $Q = c_1 E_{\theta\theta}^2/2 + c_2 E_{zz}^2/2 + c_3 E_{\theta\theta} E_{zz}$ is a function of the in-plane strain components. The model uses the material parameters c_0 [Pa] and $c_i; i = 1, 2, 3$ and has been used for applications, such as the description of the dog carotid artery [536].

Another model [85] limited to a membrane analysis of the vessel wall, uses the strain energy

$$\Psi(\mathbf{E}, \mathbf{e}_\theta, \mathbf{e}_z) = c_0 \left[\exp[c_1 E_{\theta\theta}^2] + \exp(c_2 E_{zz}^2) + \exp(c_3 E_{\theta\theta} E_{zz}) - 3 \right], \qquad (5.30)$$

with the material parameters c_0 [Pa] and $c_i; i = 1, 2, 3$. It has been calibrated to vessels, such as the normal and aneurysmatic aorta [568].

The models discussed in this sections are all characterized by a high degree of phenomenology; they are essentially *ad hoc* proposals of mathematical functions. Many more such models have been proposed towards capturing the properties of the vessel wall, see amongst others [184, 263, 319, 408, 559].

5.5.6   Histo-mechanical Descriptions

The passive properties of the vessel wall may be seen as the superposition of the mechanical responses from *collagen* and *elastin* [261, 273, 409, 515]. This assumption is supported by results from selective digestion of these structural proteins [129, 468], experiments that also suggest that elastin and collagen are major *independent* determinants of the wall's mechanical properties at small and respective large deformations. Collagen and elastin may therefore be represented by *separate* terms in the strain energy function Ψ.

Given the fibrous structure of the vessel wall, a number of modeling concepts follow the description of a *fiber-reinforced composite*. Fibers are arranged in N families that are embedded in an otherwise isotropic matrix material. The fibers within the i-th family are parallel to each other, and their orientation in the vessel's reference configuration is determined by the unit direction vector $\mathbf{a}_{0i}$. One example is the HGO model [260] and materialized by the strain energy

$$\Psi(\mathbf{C}, \mathbf{a}_{0i}) = c_0(I_1 - 3) + \sum_{i=1}^{N} c_{1i}\{\exp[c_{2i}(I_{4i} - 1)^2] - 1\}, \tag{5.31}$$

where the neoHookean parameter c_0 [Pa] reflects the properties of the matrix material, whilst c_{1i} [Pa] and c_{2i}; $i = 1, \ldots, N$ are parameters related to the i-th family of fibers. The matrix deformation is captured by the first invariant $I_1 = \mathrm{tr}\mathbf{C}$, whilst the fiber stretch is introduced through the fourth invariant $I_4 = \mathbf{C} : \mathbf{A}_i = \lambda_a^2$, where $\mathbf{A}_i = \mathbf{a}_{0i} \otimes \mathbf{a}_{0i}$ denotes the *structural tensor* and represents the organization of the i-th family of fibers. The model has been frequently used in the description of vessels, such as the normal and aneurysmatic aorta, modeled with $N = 2$ [523, 614, 620] and $N = 4$ [20, 156] families of fibers, respectively. Another application to the aorta and a number of other arteries also used $N = 4$ families of fibers [293, 472].

The HGO model introduces the exponential function towards the phenomenological description of the successive engagement of collagen fibers at increasing tissue stretch—it captures the *progressive stiffening* of the vessel wall. A modified version of the model used a 6-th order polynomial term [28], and proposed the strain energy

$$\Psi(\mathbf{C}, \mathbf{a}_{0i}) = c_0(I_1 - 3) + \sum_{i=1}^{2} c_1(I_{4i} - 1)^6, \tag{5.32}$$

where c_0, c_1 [Pa] are material parameters that are related to the matrix and fibers, respectively. It has been calibrated to describe the biomechanical properties of the aorta [28, 568].

Another modification of the HGO model replaced the neoHookean-based description of the matrix with an exponential function [465]. The strain energy then reads

$$\Psi(\mathbf{C}, \mathbf{a}_{0i}) = c_0\{\exp[c_1(I_1 - 3)] - 1\} + c_2\{\exp[c_3(I_4 - 1)] - 1\}, \tag{5.33}$$

where c_0, c_2 [Pa] and c_1, c_3 are material parameters. It has been used to describe the mechanics of vessels, such as the aorta [465].

Yet another model proposed the strain energy [475]

$$\Psi(\mathbf{C}, \mathbf{a}_{0i}) = c_0(I_1 - 3) + \sum_{i=1}^{2} c_1[\exp(Q) - 1], \tag{5.34}$$

with $Q = c_2[c_3(I_{4i} - c_4)^2 + (1 - c_3)(I_1 - 3)^2]$. The model introduced the material parameters c_0, c_1 [Pa] and c_i; $i = 2, \ldots, 4$ and has also been used to describe the aorta [474, 475, 568].

The orientation of collagen fibers in the vessel wall, and especially in the adventitia, is dispersed [434, 496, 551], see for example data acquired from the normal [160] and the aneurysmatic [195] aorta. The basic assumption of two (or more) families of parallelly aligned collagen fibers seems therefore unrealistic [434] and led to the development of the GOH model. It represents an efficient way to include the dispersion of fiber orientations in the description of the strain energy function [202]. Given a fiber-reinforced composite with N families of (dispersed) fibers, the GOH model proposes the strain energy

$$\Psi(\mathbf{C}, \mathbf{a}_{0i}, \kappa_i) = c_0(I_1 - 3) + \sum_{i=1}^{N} c_1[\exp(c_2 E_i^2) - 1], \qquad (5.35)$$

where the neoHookean parameter c_0 [Pa] describes the matrix, whilst c_1 [Pa] and c_2 are related to the properties of the collagen fibers. The model considers all collagen fibers within the i-th (dispersed) family of fibers to be *homogenously* deformed at the strain $E_i = \mathbf{H}_i : \mathbf{C} - 1$, where $\mathbf{H}_i = \kappa_i \mathbf{I} + (1 - 3\kappa_i)(\mathbf{a}_{0i} \otimes \mathbf{a}_{0i})$ denotes a *general structural tensor*. It describes the structure of the i-th family of fibers, which orientation is transverse-isotropically dispersed. The i-th family of collagen fibers is then captured by its *mean orientation* $\mathbf{a}_{0i}$ and its *dispersion* κ_i, a parameter that determines how much the fibers are dispersed around the mean orientation $\mathbf{a}_{0i}$. The model is frequently used, such as in the description of the aorta [397] and the non-diseased parts of carotid plaques [407].

Fibers, such as collagen fibers can only contribute to the wall stress if they are at tension, and the contribution of fibers that are at axial compression should be excluded. Given the aforementioned constitutive description of dispersed fibers, the exclusion of individual fibers is challenging and different criteria [202, 367] have been proposed to neglect axially compressed fibers.

Given the Probability Density Function (PDF) $\rho(\mathbf{M})$ of the fiber orientation, the general structural tensor may be computed by the integration [202]

$$\mathbf{H} = \frac{1}{2\pi} \int_{-\frac{\pi}{2}}^{\frac{\pi}{2}} \int_{-\frac{\pi}{2}}^{\frac{\pi}{2}} \rho(\mathbf{M}) \mathbf{M} \otimes \mathbf{M} \cos\phi \, d\theta \, d\phi \qquad (5.36)$$

over the hemisphere, which then allows the application of the model (5.35) to any fiber orientation distribution $\rho(\mathbf{M})$. The factor $1/(2\pi)$ in (5.36) normalizes $\mathbf{H}$ over the hemisphere, and $\mathbf{M} = [\cos\phi\cos\theta \quad \cos\phi\sin\theta \quad \sin\phi]$ denotes an arbitrary orientation in the reference configuration Ω_0, expressed through the azimuthal θ and elevation ϕ angles, respectively.

The collagen in the vessel wall is continuously synthesized and degraded. The assumption of homogenously deformed fibers within the i-th family of fibers, as made by the GOH model, therefore appropriately reflects *in vivo* (physiological) vessel wall conditions, whilst for many other load cases the collagen fibers may not be homogeneously deformed.

5.5.7 General Theory of Fibrous Connective Tissue

The properties of the vessel wall may also be expressed by the *general theory of fibrous connective tissue* [325], an approach that results in an angular integration model. The stress in the wall is then the superposition of non-interacting fibers which orientations are captured by $\rho(\mathbf{M})$, a PDF that determines their orientation within the vessel wall. The vessel wall's Cauchy stress therefore reads

$$\boldsymbol{\sigma} = \frac{1}{2\pi} \int_{\phi=-\pi/2}^{\pi/2} \int_{\theta=-\pi/2}^{\pi/2} \rho(\mathbf{M})\sigma(\lambda)\mathrm{dev}(\mathbf{m} \otimes \mathbf{m}) \cos\phi \, \mathrm{d}\phi \mathrm{d}\theta - p\mathbf{I} \,, \qquad (5.37)$$

where $\mathbf{m} = \mathbf{F}\mathbf{M}$ denotes the push-forward of the referential fiber direction $\mathbf{M}$ with $|\mathbf{M}| = 1$, and $\mathrm{dev}(\bullet) = (\bullet) - \mathrm{tr}(\bullet)\mathbf{I}/3$ is the spatial deviator operator. Here, $\sigma(\lambda)$ expresses the Cauchy stress in a fiber as a function of the fiber stretch λ. It is a 1D constitutive model and therefore eases the description of the detailed load-carrying mechanisms at the fiber level. The hydrostatic pressure p acts at the identity $\mathbf{I}$ and serves as a Lagrange parameter to enforce incompressibility.

5.5.7.1 Collagen Fiber Models
The collagen fibers in the unloaded vessel wall are undulated and then *gradually engage (recruit)* upon loading, a mechanism that determines the strong stiffening of vascular tissue at increasing strain. Whilst the aforementioned strain energy functions implemented a phenomenological description of said mechanism, we may also explicitly model the *statistics* of the engagement of collagen fibers.

Almost 40 years ago, a tissue model based on wavy collagen fibers that engage during loading has been reported [116]. The model expressed the tissue stress by

$$\sigma(\varepsilon) = c_0 \int_{-\infty}^{\varepsilon} (\varepsilon - x)\rho(x)\mathrm{d}x \,, \qquad (5.38)$$

where c_0 [Pa] denotes the stiffness of the individual collagen fibers. The engagement PDF $\rho(\varepsilon)$ determines the amount of collagen fibers that *mechanically engages* at ε, the normal strain component of the tissue strain in direction of the fiber.

Different engagement PDFs $\rho(x)$ have been used to model the mechanics of the vessel wall according to (5.38). The Cauchy–Lorentz[3] PDF

$$\rho(x) = \frac{c_1}{\{2\pi[c_1^2/4 + (x - c_2)^2]\}}$$

was proposed to describe the collagen fiber engagement in the aorta [604], whilst the log-logistic PDF

---

[3] Hendrik Antoon Lorentz, Dutch physicist, 1853–1928.

$$\rho(x) = \begin{cases} \dfrac{c_1(x - c_3/c_2)^{c_1-1}}{c_2[1 + (x - c_3/c_2)^{c_2}]^2} & ; \, x \le c_3 \,, \\[4mm] 0 & ; \, x > c_3 \,, \end{cases}$$

has been applied to the carotid artery wall [620]. Given these PDFs, c_1, c_2, c_3 are dimensionless parameters that describe the engagement of collagen fibers. In addition, a triangular engagement PDF

$$\rho(x) = \begin{cases} 0 & ; \, 0 < x \le c_1 \,, \\[2mm] \dfrac{2(x-c_1)}{(c_2-c_1)[(c_1+c_2)/2-c_1]} & ; \, c_1 \le x \le \dfrac{c_2+c_1}{2} \,, \\[2mm] \dfrac{2(c_2-x)}{(c_2-c_1)[c_2-(c_1+c_2)/2]} & ; \, \dfrac{c_2+c_1}{2} \le x \le c_2 \,, \\[2mm] 0 & ; \, c_2 < x < \infty \,, \end{cases} \qquad (5.39)$$

has been reported, where c_1 and c_2 denote the onset and offset of the triangular PDF [360]. In addition to the symmetric triangular distribution (5.39), the general and thus unsymmetrical triangular distribution has also been used to more closely represent experimental data [373].

Most collagen fiber-engagement models assume the fibers to follow an *affine* transformation. The tissue stretch $\sqrt{\mathbf{C} : (\mathbf{M} \otimes \mathbf{M})}$ along $\mathbf{M}$ matches then the fiber stretch λ. Here, $\mathbf{C}$ is the right Cauchy–Green strain, and $\mathbf{M}$ denotes the unit direction vector of the fiber in the reference configuration Ω_0. Given the PDF $\rho(x)$, the Cumulative Density Function (CDF) $\Upsilon(\lambda) = \int_{-\infty}^{\lambda} \rho(x) \mathrm{d}x$ determines the amount of collagen fibers that are engaged at λ, and therefore all fibers that carry load. We may decompose the total stretch $\lambda = \lambda_e \lambda_s$ of a collagen fiber into the part λ_s that *straightens* the fiber and defines its *intermediate configuration*, and the part λ_e that *elastically deforms* it with respect to the intermediate configuration, see Fig. 5.21.

In addition to the kinematics, a constitutive model of the fibers is needed to close the system of equations. The description of the first Piola–Kirchhoff stress in the fiber's *intermediate configuration* is the most convenient approach to formulate such a model [242]. We demand the model to be independent from the fibers' intermediate configurations—it then avoids keeping track of the infinite large number of intermediate configurations. Given this access, the first Piola–Kirchhoff

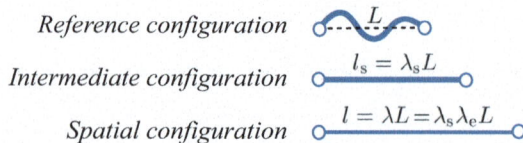

Fig. 5.21 Different configurations during the stretching of an undulated fiber. In the reference configuration, the fiber has the length L, whilst l_s denotes the straightened out but still unstressed fiber

stress $T = c_0 \log \lambda_e$ is the only possible function to describe the fiber [242], where c_0 [Pa] denotes the fiber stiffness. With the kinematics relation $\lambda = \lambda_e \lambda_s$, the model yields the differential

$$\frac{dT}{d\lambda} = \frac{c_0}{\lambda_s} \frac{d\lambda_s}{d\lambda} = \frac{c_0}{\lambda} \tag{5.40}$$

that is indeed independent from the fiber's intermediate configuration, and thus from the stretch λ_s. The expression

$$P(\lambda) = c_0 \int_{-\infty}^{\lambda} \frac{\Upsilon(x)}{x} dx \tag{5.41}$$

then determines the first Piola–Kirchhoff stress, and thus the sum from all fibers that are engaged, where $\Upsilon(x)$ denotes the CDF of $\rho(x)$.

The expression (5.41) may be integrated, and the application of the first Piola transform for incompressible solids $\sigma = \lambda P$ then determines the closed-form solution of the Cauchy stress

$$\sigma(\lambda) = \begin{cases} 0 & ; \lambda \le c_1 , \\ \frac{2c_0(\lambda-c_1)^3}{3(c_2-c_1)^2} & ; c_1 < \lambda \le \frac{c_2+c_1}{2} , \\ \frac{c_0(\lambda-c_1)(c_1^2-3c_2^2-2c_1\lambda+6c_2\lambda-2\lambda^2)}{3\lambda(c_2-c_1)^2} & ; \frac{c_2+c_1}{2} < \lambda \le c_2 , \\ c_0(\lambda - c_1) & ; c_2 < \lambda \le \infty , \end{cases} \tag{5.42}$$

where c_0 [Pa] and c_1, c_2 are parameters. Slightly different assumptions have been used elsewhere [360], and the model has been calibrated to aortic wall tissue [195, 360, 374, 434].

During the cardiac cycle, the artery wall deforms differently along its circumferential and axial directions, respectively. The collagen fibers that are aligned along these direction are then exposed to different stretches and may develop towards different homeostatic targets. Another model [195] therefore considered orientation-dependent collagen fiber properties and proposed the stress–stretch law

$$\sigma(\lambda) = \begin{cases} 0 & ; \lambda \le 1 , \\ 2c_0\lambda(1 + c_1 \sin \theta)(\lambda^2 - 1) \exp[c_2(\lambda^2 - 1)] & ; 1 < \lambda \le \infty , \end{cases} \tag{5.43}$$

where c_0 [Pa] and c_1, c_2 are material parameters. The angle between the fiber orientation and the vessel's circumferential direction is denoted by θ, and the model has been calibrated to capture the aneurysmatic aorta [195].

Yet another model [551] considered the stress

$$\sigma(\lambda) = \begin{cases} 0 & ; \lambda \le c_0 , \\ c_1(\lambda - c_0) & ; c_0 < \lambda \le \infty \end{cases} \tag{5.44}$$

in a collagen fiber that is stretched at λ. The material parameters are denoted by c_0 [Pa] and c_1, and the model has been calibrated to the aortic wall [378,623]. Many more models to describe the load-bearing of collagen have been proposed, including also the complex time-dependent properties of this structure [374].

5.5.7.2 Description of Collagen Fiber Orientations

The orientation of the collagen fibers in the vessel wall, and especially in the adventitia, is *dispersed*. A structural property that has a remarkable effect on the macroscopic mechanical properties of the vessel wall [202]. The dispersion of collagen fibers may be described by an orientation PDF $\rho(\mathbf{M}(\phi, \theta))$, where $\cos \phi d\theta d\phi$ determines the normalized amount of collagen fibers that are aligned within the infinitesimal small sector $\{[\phi, \phi + d\phi], [\theta, \theta + d\theta]\}$, see Fig. 5.22. The elevation and azimuthal angles are denoted by ϕ and respectively θ, and given the orientation PDF is *normalized*, the integral over all possible directions $\mathbf{M}(\phi, \theta)$ then yields the surface of the unit hemisphere,

$$\int_{-\pi/2}^{\pi/2} \int_{-\pi/2}^{\pi/2} \rho(\mathbf{M}) \cos \phi d\theta d\phi = 2\pi . \tag{5.45}$$

The orientation PDF $\rho(\mathbf{M})$ allows us to compute the general structural tensor $\mathbf{H}$ (5.36) of the GOH model (5.35), or it may be included directly in the stress

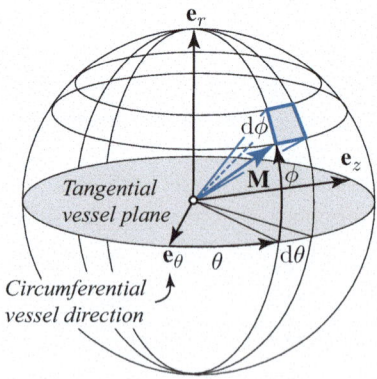

Fig. 5.22 Fiber orientation in the vessel wall. The Cartesian base vector $\mathbf{e}_\theta, \mathbf{e}_z, \mathbf{e}_r$ denote the circumferential, axial, and radial vessel wall directions, respectively. The tangential plane is formed by the vessel's circumferential $\mathbf{e}_\theta$ and axial $\mathbf{e}_z$ directions. Elevation ϕ and azimuthal θ angles determine the fiber direction $\mathbf{M}$ in the vessel wall. The infinitesimal small sector $\{[\phi, \phi + d\phi], [\theta, \theta + d\theta]\}$ is used to define the orientation Probability Density Function (PDF) $\rho(\mathbf{M})$ of fibers

computations according to the general theory of fibrous tissues (5.37). Whilst $\rho(\mathbf{M})$ could basically be interpolated between experimental data points [428, 551], such data are often noisy and sparse. The orientation data of the collagen is therefore commonly represented by a statistical (analytical) PDF that is then calibrated to the experimental data points.

Given the orientations of collagen fibers are transverse-isotropically dispersed around the circumferential direction $\mathbf{e}_\theta$, the orientation PDF

$$\rho(\mathbf{M}) = \rho(\theta) = \bar{c}\exp[c_0\cos(2\theta)] \tag{5.46}$$

describes the tissue structure, where c_0 denotes a dimensionless distribution parameter, whilst $\bar{c}$ is the dimensionless normalization parameter that ensures the condition (5.45). The transverse isotropic distribution of collagen fibers in the vessel wall is justified by earlier experimental data [160], and the PDF (5.46) has been used to describe the collagen distribution in the aorta [496].

More recent experimental data [195, 496] suggests a more complex orientation distribution of collagen fibers in the vessel wall. Collagen fibers exhibit a higher dispersion within the tangential plane than perpendicular to it—the dispersion of the azimuthal angle θ is then higher than the dispersion of the elevation angle ϕ. This structure cannot be captured by the transverse isotropic PDF (5.46) and the Bingham PDF [47]

$$\rho(\mathbf{M}) = \rho(\phi, \theta) = \bar{c}\exp[c_0(\cos\phi\cos\theta)^2 + c_1(\cos\phi\sin\theta)^2] \tag{5.47}$$

may be used instead, where c_0, c_1 denote dimensionless distribution parameters, whilst $\bar{c}$ ensures the condition (5.45). The Bingham PDF has been calibrated to collagen orientations of the carotid artery wall [483], and the aneurysmatic aortic wall [195].

Regardless collagen orientation is 3D, for some applications a planar distribution may be used. The von Mises PDF

$$\rho(\mathbf{M}) = \rho(\theta) = \bar{c}\exp[c_0\cos(2\theta)] \tag{5.48}$$

has therefore also been proposed to describe the collagen in the aortic wall [434]. The parameter c_0 then determines the shape of the PDF, and the 2D normalization condition

$$\int_{-\pi/2}^{\pi/2} \rho(\mathbf{M})\mathrm{d}\theta = \pi \tag{5.49}$$

determines the normalization parameter $\bar{c}$.

5.5.8 Residual Stress and Load-Free Configuration

The vessel's load-free configuration is not stress-free, and we cannot always directly use the aforementioned constitutive models—they are formulated with respect to the vessel wall's *stress-free* reference configuration. Some of these models have therefore been combined with techniques to estimate the residual strain field from the uniform stress hypothesis [431], through the back-calculation of the load-free configuration from the vessel's inflated geometry [114, 209, 465], or the use of a modified updated Lagrangian formulation [209].

The residual stress is multi-dimensional, and the dissection of a vessel into stress-free segments is in general not possible. However, a vessel may always been cut open which then results in the *opened-up configuration* Ω_0 shown in Fig. 5.23. The opened-up configuration is often regarded stress-free and then used as the reference configuration Ω_0 for stress and strain calculations. Whilst this approach is widely used, it neglects the existence of axial residual stresses and does not address the multi-dimensionality of the residual deformation in the vessel wall.

Example 5.2 (Residual Stresses of a Thick-Walled Artery). The load-free configuration Ω of an artery segment has the dimensions r_i, l, h, see Fig. 5.23. At this configuration the vessel contains residual stresses, and cutting it open then results in its opened-up configuration Ω_0 of the dimensions $R_i = 10\,\text{mm}$, $L = 23.0\,\text{mm}$, $H = 2.0\,\text{mm}$, and $\alpha = \pi/2$. Given this example, Ω_0 may be regarded as the stress-free configuration. The vessel tissue is incompressible and its constitutive properties may be modeled by the neoHookean strain energy

$$\Psi(\mathbf{C}) = c_1(I_1 - 3)\,, \tag{5.50}$$

per unit volume vessel wall tissue. Given incompressibility, we do not need to distinguish between unit volume of deformed or undeformed volume. The stiffens-related parameter $c_1 = 12.7\,\text{kPa}$ has been identified from mechanical tissue characterization, and $I_1 = \text{tr}\mathbf{C}$ denotes the first invariant of the right Cauchy–Green strain $\mathbf{C}$.

Fig. 5.23 Opened-up and stress-free configuration Ω_0 as well as load-free configuration Ω of a vessel segment. The stress-free configuration Ω_0 serves as reference configuration in the description of stress and strains

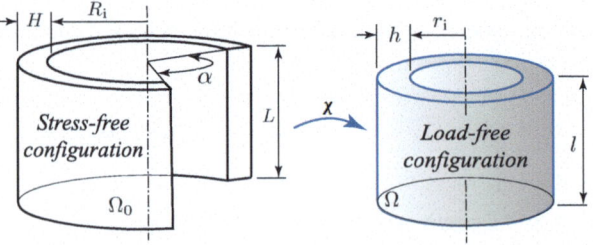

(a) Describe the deformation kinematics of this problem, and derive the Cauchy stress as a function of the principal stretches.

(b) Derive the equilibrium equation in the radial direction that follows from Cauchy's equation of motion $\text{div}\boldsymbol{\sigma} = \mathbf{0}$.

(c) Provide the relation of the axial force N, and thus the stress resultant in the vessel's axial direction.

(d) Show that the configuration Ω with $r_{\mathrm{i}} = 7.26\,\text{mm}$, $l = 23\,\text{mm}$, $h = 2.0\,\text{mm}$ satisfies the balance equations of Tasks (b) and (c).

(e) Plot the distribution of the Cauchy stress differences $\sigma_\theta - \sigma_r$ and $\sigma_\theta - \sigma_z$ across the wall thickness.

(f) Compute the circumferential Cauchy stress σ_θ at the inside and outside of the artery.   ∎

5.5.9 Visco-hyperelastic Descriptions

Vascular tissue shows *non-linear viscoelastic* properties [2, 182]. In addition to the time-dependent properties of cells and ECM constituents themselves, the motion of fluids within the tissue also contributes to the over-all time-dependent tissue properties. We may generalize the linear viscoelastic models discussed in Sect. 3.5.4 towards the description of soft biological tissues, an approach known as *quasi-linear viscoelasticity*. It is widely applied in the descriptions of vascular tissue, such as the coronary artery [262], carotic artery [248], and ventricular tissue [194]. As with the series approximation of a non-linear function, the *superposition* of a finite number of linear viscoelastic models results in non-linear viscoelastic properties.

We may describe the vascular wall with the *generalized Maxwell rheological model* shown in Fig. 5.25, a model that could also be called *generalized Standard Solid model*. Its elastic properties are represented by the Helmholtz free energy $\Psi(\mathbf{C}, \mathbf{C}_{\mathrm{M}1}, \mathbf{C}_{\mathrm{M}2}, \ldots, \mathbf{C}_{\mathrm{M}N})$ per unit vessel tissue, where $\mathbf{C}$ denotes the right Cauchy–Green strain, whilst $\mathbf{C}_{\mathrm{M}i}$ is the right Cauchy–Green strain of the i-th Maxwell body (spring) and serves as *hidden internal variable*. The elastic energy

$$\Psi_{\mathrm{iso}} = \Psi_{\mathrm{iso\,E}}(\mathbf{C}) + \sum_{i=1}^{N} \Psi_{\mathrm{iso\,M}i}(\mathbf{C}_{\mathrm{M}i}) \tag{5.54}$$

Fig. 5.25 Generalized Maxwell rheological element with $\Psi_{\mathrm{iso\,E}}$ and $\Psi_{\mathrm{iso\,M}i}$; $i = 1, \ldots, N$ representing the strain energies stored in the respective hyperelastic springs

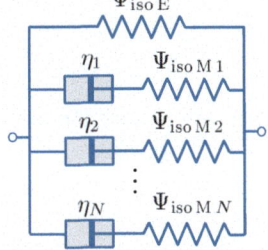

then represents the tissue, where $\Psi_{\mathrm{iso\,E}}(\mathbf{C})$ and $\Psi_{\mathrm{iso\,M}i}(\mathbf{C}_{\mathrm{M}i})$ denote the strain energies stored in the elastic body (mainspring) as well as in the spring of the i-th Maxwell body, respectively. Coleman and Noll's procedure for incompressible materials (3.131) yields therefore the second Piola–Kirchhoff stress

$$
\mathbf{S} = 2\underbrace{\frac{\partial \Psi_{\mathrm{iso\,E}}(\mathbf{C})}{\partial \mathbf{C}}}_{\overline{\mathbf{S}}_{\mathrm{E}}} + 2\underbrace{\sum_{i=1}^{N} \frac{\partial \Psi_{\mathrm{iso\,M}i}(\mathbf{C}_{\mathrm{M}i})}{\partial \mathbf{C}_{\mathrm{M}i}} : \frac{\partial \mathbf{C}_{\mathrm{M}i}}{\partial \mathbf{C}}}_{\overline{\mathbf{S}}_{\mathrm{M}}} - \kappa \mathbf{C}^{-1} \,, \tag{5.55}
$$

where $\overline{\mathbf{S}}_{\mathrm{E}}$ is the *elastic stress*, whilst $\overline{\mathbf{S}}_{\mathrm{M}} = \sum_{i=1}^{N} \overline{\mathbf{S}}_{\mathrm{M}i}$ denotes the *over stress*, a stress that appears as long as the material has not yet reached its thermodynamic equilibrium. As with an incompressible elastic description, the Lagrange pressure κ contributes to the hydrostatic pressure p and enforces the incompressibility.

The term $\partial \mathbf{C}_{\mathrm{M}i}/\partial \mathbf{C}$ in (5.55) establishes the relation between the strain increment $\Delta \mathbf{C}_{\mathrm{M}i}$ of the i-th Maxwell body and the total strain increment $\Delta \mathbf{C}$, a kinematic relation determined through *rate equations*, see Sect. 3.5.4.7. We may also implicitly specify this relation through the development of the over stresses $\overline{\mathbf{S}}_{\mathrm{M}i}$, and introduce the linear rate equations

$$
\dot{\overline{\mathbf{S}}}_{\mathrm{M}i} + \frac{1}{\tau_i}\overline{\mathbf{S}}_{\mathrm{M}i} = \beta_i \dot{\overline{\mathbf{S}}}_{\mathrm{E}} \text{ (no summation)} \,, \tag{5.56}
$$

where β_i and τ_i $[\mathrm{s}^{-1}]$ are the properties of the i-th Maxwell body. Whilst the scalar β_i relates the stiffness of the i-th Maxwell body to the stiffness of the elastic body, τ_i defines the relaxation time of the i-th Maxwell body. The over stress may then be expressed by the convolution integral

$$
\overline{\mathbf{S}}_{\mathrm{M}}(t) = \sum_{i=1}^{N} \overline{\mathbf{S}}_{\mathrm{M}i} = \sum_{i=1}^{N} \left\{ \beta_i \int_0^t \exp[-(t-x)/\tau_i]\dot{\overline{\mathbf{S}}}_{\mathrm{E}}\mathrm{d}x \right\} \,, \tag{5.57}
$$

the closed-form solution of the governing equation (5.56). The outlined viscoelastic model leads to an extremely efficient FEM implementation [222, 291, 506], and the observation that the *algorithmic tangent* $\mathbb{C} = 2\partial \mathbf{S}/\partial \mathbf{C}$ exhibits major symmetry, verifies that a Helmholtz free energy function Ψ_{iso} indeed exists.

5.5.10 Cyclic Deformation of the Visco-hyperelastic Thin-Walled Tube

We may describe the wall of the common iliac artery by the generalized Maxwell model shown in Fig. 5.25 with $N = 2$ Maxwell bodies (devices). Given its stress-free reference configuration Ω_0, the artery may be regarded as a cylindrical membrane of the diameter $D = 9.0\,\text{mm}$ and the wall thickness $H = 0.8\,\text{mm}$. The vessel tissue is incompressible, and the Yeoh strain energy

$$\Psi(\mathbf{C}) = c_3(I_1 - 3)^3 \tag{5.58}$$

describes its elastic properties per unit volume, where $I_1 = \text{tr}\mathbf{C}$ and $c_3 = 50.0\,\text{kPa}$ denotes the first invariant of the right Cauchy–Green strain and the stiffens-related parameter, respectively. The over stresses $\overline{S}_{Mi}; i = 1, 2$ in the vessel wall are determined by the linear rate equations (5.56), where the properties $\beta_1 = 0.3$, $\beta_2 = 0.2$ and $\tau_1 = 0.2\,\text{s}^{-1}$, $\tau_2 = 0.7\,\text{s}^{-1}$ describe the respective Maxwell bodies.

With the strain energy (5.58) and the relation (5.5), the principal second Piola–Kirchhoff stresses $S_{Ei} = 6c_3(I_1 - 3)^2 - \kappa/\lambda_i^2$ determine the vessel wall's elastic properties, where κ is the Lagrange contribution to the hydrostatic pressure, and $i = \theta, z, r$ denotes the circumferential, axial, and radial directions, respectively. Towards the computation of the over stress (5.57), we determine the isochoric elastic second Piola–Kirchhoff stress $\overline{\mathbf{S}}_E = \text{Dev}\mathbf{S}_E = \mathbf{S}_E - (\mathbf{C} : \mathbf{S}_E)\mathbf{C}^{-1}/3$, which then reads

$$\overline{\mathbf{S}}_E = 2c_3(I_1 - 3)^2 \text{diag}\left[2 - \frac{\lambda_r^2 + \lambda_z^2}{\lambda_\theta^2}, 2 - \frac{\lambda_\theta^2 + \lambda_r^2}{\lambda_z^2}, 2 - \frac{\lambda_\theta^2 + \lambda_z^2}{\lambda_r^2}\right],$$

where the right Cauchy–Green strain $\mathbf{C} = \mathbf{F}^{\mathrm{T}}\mathbf{F} = \text{diag}[\lambda_\theta^2, \lambda_z^2, \lambda_r^2]$ has been used.

The computation of the over stress requires the rate of the isochoric elastic stress $\dot{\overline{\mathbf{S}}}_E$, which itself is a function of the rate of stretches. Whilst these rates derive in general from the solution of the equilibrium, in the present example we directly prescribe them, and

$$\lambda_\theta = 1.4 + 0.1\sin(2\pi t) \quad ; \quad \lambda_z = 1.2 \quad ; \quad \lambda_r = (\lambda_\theta\lambda_z)^{-1}$$

determines the development of the stretches over time. We may now solve the convolution integral (5.57) and compute the over stresses $\overline{S}_M$, an exercise that in general requires a numerical schema [507], see Example 5.3. The equilibrium in the radial direction $\overline{S}_{Er} + \overline{S}_{Mr} - p/\lambda_r^2 = 0$ allows us then to express the hydrostatic pressure p (to be distinguished from the above introduced κ) and closes the description of the second Piola–Kirchhoff stress $\mathbf{S} = \overline{\mathbf{S}}_E + \overline{\mathbf{S}}_M - p\mathbf{C}^{-1}$.

The equilibrium of the inflated thin-walled tube determines the inflation pressure $p_i = 2\sigma_\theta h/d$ and the reduced axial force $F = \sigma_z d\pi h - pd^2\pi/4$, where $\sigma_\theta = S_\theta\lambda_\theta^2$ and $\sigma_z = S_z\lambda_z^2$ denote the circumferential and axial Cauchy stresses, respectively. The substitution of the deformed diameter $d = D\lambda_\theta$ and the deformed wall thickness $h = H\lambda_r = H(\lambda_\theta\lambda_z)^{-1}$ then yields

$$p_i = \frac{2S_\theta H}{\lambda_z D} \quad ; \quad F = S_z\lambda_z DH\pi - p_i\frac{D^2\lambda_\theta^2\pi}{4}.$$

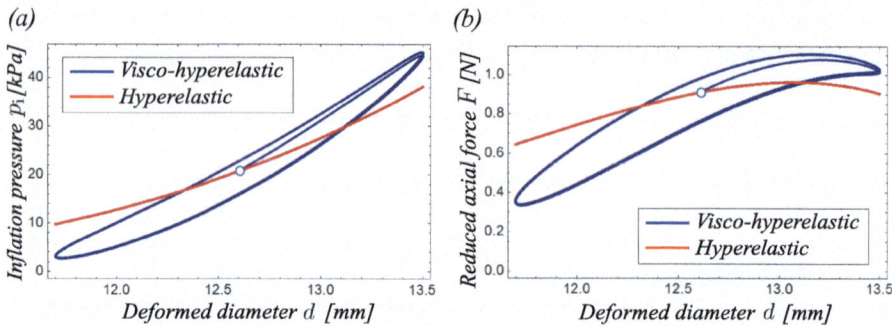

Fig. 5.26 Properties of the inflated visco-hyperelastic thin-walled tube (blue) at the prescribed circumferential and axial stretches of $\lambda_\theta = 1.4 + 0.1\sin(2\pi t)$ and $\lambda_z = 1.2$, respectively. (**a**) Inflation pressure p_i *versus* the deformed diameter d of the tube. (**b**) Reduced axial force F *versus* the deformed diameter d of the tube. The dot denotes the time point $t = 0$ and the curves cover the time interval $0 < t < 3.0$ s. For comparison, the hyperelastic solution is shown in red

Figure 5.26 illustrates the inflation pressure and the reduced axial force over the time interval $0 < t < 3.0$ s, where the aforementioned properties of the iliac artery have been used. Aside from a *hysteresis*, the graph shows that the visco-hyperelastic tube is *stiffer* than the hyperelastic solution. Given cyclic periodic conditions, $\mathcal{D} = 2\pi \oint p_i r \, \mathrm{d}r$ determines the dissipation per cycle, where $r = d/2$ denotes the radius of the deformed vessel.

Example 5.3 (Discretization of the Convolution Integral). The viscoelastic properties of a vessel wall may be described by a generalized Maxwell model, and the convolution integral (5.57) then determines the over stress.

(a) Use the mid-point integration rule and discretize the convolution integral (5.57) at the time t_{n+1}. The elastic stress and the over stress at the previous time point t_n are available in a history vector.
(b) Use the iteration rule of Task (a), compute the stress, and replicate the example discussed in Sect. 5.5.10. Investigate the influence of the time step Δt upon the accuracy of the results.
(c) Imagine a Finite Element Method (FEM) implementation of the model and provide the algorithmic tangent $\overline{\mathbb{C}}^{\text{algo}}_{\text{M}\,n+1}$ that follows from the linearization of the over stress, $\Delta \overline{\mathbf{S}}_{\text{M}} = 2\overline{\mathbb{C}}^{\text{algo}}_{\text{M}\,n+1} : \Delta \mathbf{C}$, where $\mathbf{C}$ denotes the right Cauchy–Green strain. ∎

5.5.11   Damage and Failure Descriptions

The mechanical stress throughout the different histological tissue constituents is inhomogeneous and results in local stress concentrations within the vascular wall. At high stress levels, *micro-defects*, such as the pull-out and breakage of collagen fibers, develop and then irreversibly rearrange the tissue's microstructure. Healthy vascular tissue at physiological stress levels continuously *repairs* such defects and maintains the structural integrity of the vessel wall. At supra-physiological stress levels or in diseased vascular tissue, healing cannot match up with the development of micro-defects and the tissue continues to accumulate weak links. Given the number of micro-defects per tissue volume exceeds a certain threshold, micro-defects coalesce towards the formation of *macro-defects*. A single macro-defect then eventually propagates and fractures the vessel wall.

The study of a number of problems in vascular biomechanics requires the constitutive description of vessel wall injury, and *damage-related* effects [141, 142, 410] as well as *plasticity-related* effects [410, 489] have been proposed. It stimulated the development of models that account for damage [24, 64, 258, 356, 397, 580], plasticity [197,539], and fracture [154,170,200,201,281] of the vascular wall. Most commonly a single-scale macroscopic framework has been proposed, which, however, fails to describe the experimentally reported localized irreversible deformation of individual collagen fibers [308,441].

There is still no clear definition of what constitutes vascular tissue injury, and conventional mechanical indicators such as visible failure and loss of stiffness may not adequately identify the tissue's tolerance level. A more complete definition of mechanical injury is therefore needed to adequately describe mechanical and physiological changes that then result in *anatomical* and *functional* damage [577].

5.5.11.1 Modeling Irreversible Properties of Collagen Fibers

At large deformations, *collagen* is the main load-carrying protein in the vessel wall, and its failure properties are therefore sensitive to the modeling of vessel wall damage and failure. A number of constitutive approaches have been proposed to describe the damage of collagen fibers, some of them are discussed in the following.

Continuum Damage Mechanics (CDM) Description A collagen fiber may be regarded as a 1D robe-like structure, and the first Piola–Kirchhoff stress is the most natural stress measure to specify the constitutive properties of such a fiber [242]. Given a *Kachanov-like* [290] damage variable $0 \leq d \leq 1$ that defines the state of damage or weakening of a collagen fiber,

$$P(\lambda, d) = (1 - d)\widetilde{P}(\lambda) \tag{5.60}$$

then determines the first Piola–Kirchhoff stress. Here, λ denotes the stretch, whilst $\widetilde{P}(\lambda)$ determines the properties of the undamaged collagen fiber, a function that may be captured by the descriptions discussed in Sect. 5.5.7.1. In addition to (5.60), a governing equation determines the *evolution* of the state of damage d and closes the constitutive description. Whilst such a damage-based approach is widely used in the description of vascular tissue [24,65,356], it is characterized by a high degree of phenomenology and cannot provide detailed structural insights into microstructural failure of vascular tissue.

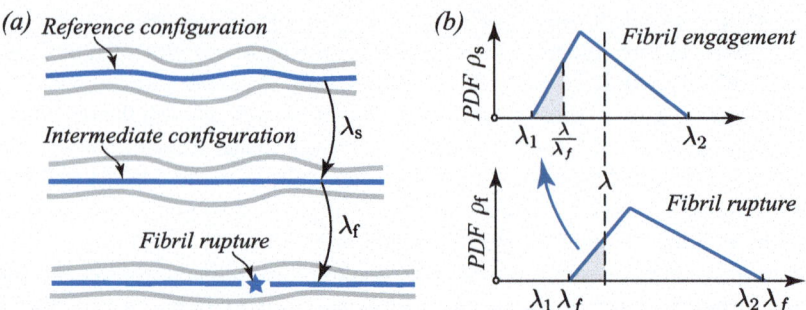

Fig. 5.28 Schematic load-carrying mechanisms of a collagen fiber that is assembled by undulated collagen fibrils. Collagen fibrils of different undulations, gradually engage until they eventually rupture. (**a**) Deformation of an individual fibril. The engagement stretch λ_s defines the fibril's intermediate configuration, a deformation at which the fibril mechanically engages and starts carrying load. The application of the failure stretch λ_f relative to the intermediate configuration defines the deformation at which the fibril ruptures. (**b**) The collagen fibrils' configurations are described by a triangular Probability Density Function (PDF). It describes the mechanical engagement (top), and "stretching" the PDF by λ_f, describes the rupture of collagen fibrils (bottom). The grey-shaded triangle denotes the portion of fibrils that has been ruptured at the stretch λ

Statistics-Based Description   A collagen fiber may be seen as the arrangement of a bundle of undulated collagen *fibrils* of negligible bending stiffness, see Fig. 5.28. An engagement PDF $\rho_s(\lambda)$ then determines the portion of the mechanically engaged collagen fibrils [360, 604, 620]. Given an individual fibril, λ_s defines its intermediate configuration, a state at which the fibril is straitened out but still stress-free. At stretches beyond λ_s, the collagen fibril is elastically stretched until it eventually fails at the failure stretch λ_f. The stretch λ_f is measured relative to the fibril's intermediate configuration, see Fig. 5.28a.

The portion of ruptured fibrils may be described by $\rho_f(\lambda)$, a PDF that results from "stretching" the engagement PDF $\rho_s(\lambda)$ by λ_f [243], see Fig. 5.28b. The portion of ruptured fibrils at the stretch λ is then directly given by $\Upsilon_f(\lambda) = \int_{\lambda_f \lambda_1}^{\lambda} \rho_f(x)\mathrm{d}x$, the CDF of $\rho_f(\lambda)$. It may also be illustrated in the graph of the engagement PDF $\rho_s(\lambda)$, where it covers the domain $\lambda_1 < x \leq (\lambda/\lambda_f)$, and thus the grey-shaded triangle in Fig. 5.28b. The triangular PDF [242] and the normal PDF [277] have been used in the statistics-based description of collagen fibril rupture.

Example 5.4 (Progressive Engagement and Rupture of Collagen Fibrils). The mechanical properties of a patch dissected from the media of the vascular wall may be modeled as a collagen fiber-reinforced composite. All collagen fibrils are aligned in the circumferential direction, and they are undulated in the tissue's unloaded reference configuration. The triangular Probability Density Function (PDF)

$$\rho_s(\lambda) = \begin{cases} 0 & ; 0 < \lambda \le \lambda_1 \, , \\ \frac{2(\lambda-\lambda_1)}{(\lambda_2-\lambda_1)(\lambda_3-\lambda_1)} & ; \lambda_1 \le \lambda \le \lambda_3 \, , \\ \frac{2(\lambda_2-\lambda)}{(\lambda_2-\lambda_1)(\lambda_2-\lambda_3)} & ; \lambda_3 \le \lambda \le \lambda_2 \, , \\ 0 & ; \lambda_2 < \lambda < \infty \end{cases}$$

describes the engagement of fibrils upon loading, where $\lambda_1 = 1.05$ and $\lambda_2 = 1.8$ denote the respective onset and offset of the triangular PDF, whilst $\lambda_3 = 1.3$ is the stretch at the peak probability. The tissue is incompressible, and $P_f = k(\lambda_e - 1)$ with $k = 3.8$ MPa determines the first Piola–Kirchhoff stress *versus* stretch properties of an individual collagen fibril. The stretch λ_e is measured relative to the fibril's intermediate configuration.

(a) Consider simple tension along the circumferential direction and compute the Cauchy stress *versus* stretch properties of the tissue.
(b) Consider the failure stretch $\lambda_f = 1.28$, a stretch measured relative to the fibril's intermediate configuration and at which a fibril ruptures. Given this assumption, compute the Cauchy stress that develops upon continuous elongation along its circumferential direction. Incompressibility may be assumed. ∎

Plastic-Like Description Collagen fibrils are *interconnected* by PG bridges as shown in Fig. 5.30, cross-linking that allows to a certain extent the sliding of adjacent collagen fibrils relative to each other. Let us assume the deformation of the PG-bridge defines *four* distinct mechanisms [191, 499]:

- *Elastic*—the PG-bridge instantaneously changes its configuration upon loading
- *Viscoelastic*—the sliding between the glycan chains develops over time, but the original PG-bridge configuration is fully recovered upon unloading
- *Plastic sliding*—the proteoglycan protein slides along the collagen fibril, and a permanent deformation remains even at complete unloading
- *Damage*—the glycan chains or proteoglycan protein have slid apart and the mechanical linkage is broken

These mechanisms motivate a "stretch-based" constitutive concept, where irreversible (plastic) sliding of the collagen fibrils not only defines the fiber's irreversible elongation but also its state of damage d. The damage variable d reflects the loss of fiber stiffness from non-functional (broken) PG bridges, and an exponential function $d = 1 - \exp(-a\varepsilon_{pl}^2)$ may be used to relate it to the plastic strain ε_{pl} of the collagen fiber, see [191]. Small and large values of the parameter a then define *ductile* and *brittle* failure properties, respectively. Full details, including the model's numerical implementation, are reported elsewhere [191].

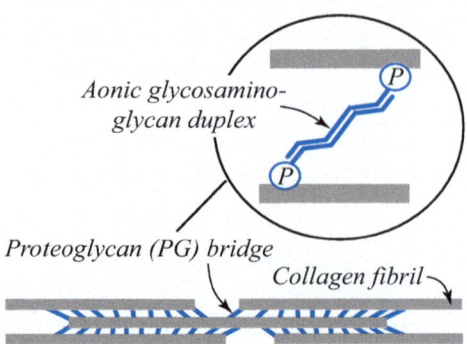

Fig. 5.30 Schematic load-carrying mechanisms of a collagen fiber assembled by a number of collagen fibrils. The model considers straight collagen fibrils, where Proteoglycan (PG) bridges support load transition between collagen fibrils. Antiparallel anionic glycosaminoglycan duplex binds non-covalently to the collagen fibrils at the proteoglycan protein (P)

5.5.11.2 Failure Represented by Cohesive Zone Models

The aforementioned injury models can be directly used in simulations, as long as the determinant of the material's stiffness tensor *remains positive definite* for all possible deformations, and thus $\det\mathbb{C} > 0$ holds [402]. Beyond this limit, a regularization is needed to avoid non-physical results [32]. Aside from the integration of an *internal length-scale* in the constitutive description that specifies the size of (collagen) failure, a number of other methods have also been proposed to describe the development of failure in a continuum, see Sect. 3.8.

Vascular tissue failure has also been modeled by a *cohesive zone*, see Sect. 3.8.4.2. The resistance of a crack against (further) opening is then described by the cohesive potentials $\psi = \psi(\mathbf{u}_d \otimes \mathbf{u}_d, \zeta)$, where $\mathbf{u}_d$ and ζ denote the crack opening displacement and the damage variable, respectively. The theory of invariants [520] allows us to express the potential by $\psi = \psi(i_1, i_2, i_3, \zeta)$, where i_1, i_2, i_3 denote the invariants of the symmetric second-order tensor $\mathbf{u}_d \otimes \mathbf{u}_d$ [198]. Given a formulation that considers only the first invariant $i_1 = \mathbf{u}_d \cdot \mathbf{u}_d$, a particular cohesive potential reads

$$\psi(i_1, \zeta) = \frac{t_0}{2\zeta} \exp(-a\delta^b)i_1 ,$$

where t_0 [Pa] denotes the cohesive tensile strength, whilst the non-negative parameters a and b capture the softening response, and thus the decay of cohesive traction with increasing crack opening displacement $|\mathbf{u}_d|$. The constitutive description is closed with the definition of the damage surface $\phi(\mathbf{u}_d, \delta) = |\mathbf{u}_d| - \zeta = 0$ and the evolution of the damage variable $\dot{\zeta} = \overline{|\mathbf{u}_d|}$ for $\phi > 0$. Coleman and Noll's procedure (3.156) allows us then to express the cohesive traction

$$\mathbf{T} = c\mathbf{u}_d \; ; \; c = t_0 \exp(-a\zeta^b)/\zeta \; ; \; \gamma = c(1 + ab\zeta^b)/\zeta ,$$

where $\partial i_1/\partial \mathbf{u}_d = 2\mathbf{u}_d$ has been used. A proof of non-negativeness of the dissipation, and thus $\mathcal{D}_{int} \geq 0$, is given elsewhere [198]

Given a FEM model, cohesive zones may either be *embedded* within the finite elements, or simply added to the faces of the finite elements. The former approach leads to the Partition of Unity FEM (PUFEM) [198, 594], also known as eXtended FEM (XFEM), a concept that also requires a crack-tracking algorithm [199] to handle the propagation of the failure. Attaching cohesive zones to the faces of the finite elements requires *a priori* knowledge of the failure surface and makes the generation of the FEM mesh more challenging. It also requires the prescription of an *artificial elastic stiffness* with eventually considerable impact upon the physics of the problem.

5.6 Identification of Constitutive Parameters

Vascular tissue models, such as the ones listed in Sect. 5.5, need *constitutive parameters* to be identified from experimental data. The parameters may either be intuitively adjusted and the model predictions then compared to the experimental data, or the parameters may be identified in a more consistent way through the use of *optimization methods*.

To exemplify optimization-based parameter identification, we consider the experimental data from *planar biaxial testing* of the infrarenal porcine aorta shown in Appendix F. Planar biaxial testing is a well-established laboratory exercise and allows the application of loads that are close to the vessel's *in vivo* conditions. A square-shaped vessel wall sample is fully immersed in physiological saline solution at 37 degrees Celsius, and two actuators apply the displacements u_θ and u_z along the vessel wall's circumferential θ and axial z directions, see Fig. 5.31. At the same time, two load cells measure the corresponding forces F_θ and F_z. Given the acquisition of the data reported in Appendix F, a *displacement-controlled* loading protocol was used that materialized the three different combinations of displacements $u_\theta/u_z = 2/1\,;\,1/1\,;\,1/2$.

Fig. 5.31 Infrarenal porcine aorta wall sample mounted in a planar biaxial testing machine. Sample is immersed in saline solution and the vessel's circumferential θ and axial z directions are illustrated

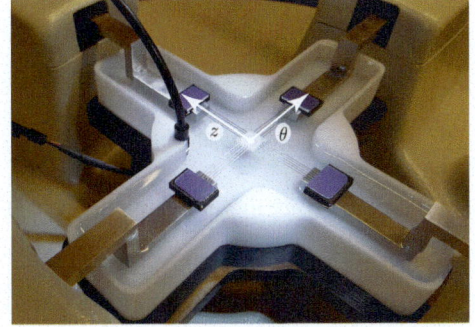

At its undeformed configuration, the vessel wall sample had the thickness $H = 1.9\,\text{mm}$ and the edge length $L = 9.7\,\text{mm}$, whilst $L_i = 6$ mm specified the square inside the gripping points. Edge effects of specimen mounting (pinching) may, for simplicity, been neglected. The prescribed displacements then directly yield the ratios $\varepsilon_\theta/\varepsilon_z = 2/1\,;1/1\,;1/2$ amongst the circumferential strain $\varepsilon_\theta = u_\theta/L_i$ and the axial strain $\varepsilon_z = u_z/L_i$, respectively. Given the sample dimensions, the experimentally recorded data can be translated into four-dimensional data points of the format $(\lambda_\theta^e, \lambda_z^e, P_\theta^e, P_z^e)$, see Tables F.1. Here, $\lambda_\theta^e = 1 + \varepsilon_\theta$ and $\lambda_z^e = 1 + \varepsilon_z$ are the stretches along the circumferential and axial vessel directions, whilst $P_\theta^e = F_\theta/(HL_i)$ and $P_z^e = F_z/(HL_i)$ are the corresponding first Piola–Kirchhoff stresses, respectively. This assumptions represent a severe simplification of the complex inhomogeneous stress field of planar biaxial testing, see the FEM analysis of the case study in Sect. 4.9. Towards the acquisition of reproducible experimental data, the vessel wall sample has been *preconditioned* during four loading cycles. The data from the fifth's cycle is reported in Tables F.1, as well as shown in Fig. 5.32. The data illustrates *finite deformation*, *non-linearity*, and *anisotropy* of the vessel wall sample.

5.6.1   Analytical Vessel Wall Models

Given (5.5) and (3.33), the first Piola–Kirchhoff stress reads $P_i = \partial\Psi(\mathbf{C})/\partial\lambda_i - \kappa/\lambda_i$, where κ denotes the Lagrange pressure that can be identified from the equilibrium along the radial direction, $P_r = \partial\Psi(\mathbf{C})/\partial\lambda_r - \kappa/\lambda_r = 0$. The two principal first Piola–Kirchhoff stresses then read

$$P_\theta = \frac{\partial\Psi(\mathbf{C})}{\partial\lambda_\theta} - \frac{\lambda_r}{\lambda_\theta}\frac{\partial\Psi(\mathbf{C})}{\partial\lambda_r}\;;\; P_z = \frac{\partial\psi(\mathbf{C})}{\partial\lambda_z} - \frac{\lambda_r}{\lambda_z}\frac{\partial\Psi(\mathbf{C})}{\partial\lambda_r} \qquad (5.63)$$

with $\Psi(\mathbf{C})$ denoting the vessel wall's Helmholtz free energy per unit volume.

5.6.1.1 Yeoh Model
Given the two-parameter Yeoh model $\Psi(\mathbf{C}) = c_1(I_1 - 3) + c_2(I_1 - 3)^2$, the stress relations (5.63) explicitly read

$$P_\theta(\lambda_\theta, \lambda_z) = \alpha(\lambda_\theta - \lambda_\theta^{-3}\lambda_z^{-2})\;;\; P_z(\lambda_\theta, \lambda_z) = \alpha(\lambda_z - \lambda_z^{-3}\lambda_\theta^{-2})\,, \qquad (5.64)$$

where the definition of the first invariant $I_1 = \lambda_\theta^2 + \lambda_z^2 + \lambda_r^2$ and the parameter $\alpha = 2c_1 + 4c_2(I_1 - 3)$ have been used. The Yeoh model is *isotropic*, and the stress relations (5.64) are therefore symmetric with respect to the stretches λ_θ and λ_z.

5.6.1.2 Fung Model
Planar biaxial testing along the vessel wall's principal directions suppresses shear deformations, and the off-diagonal components of the Green–Lagrange strain $\mathbf{E}$ are zero. The strain energy $\Psi(\mathbf{E}, \mathbf{e}_\theta, \mathbf{e}_z, \mathbf{e}_r) = c_0[\exp(Q) - 1]$ of Fung's model involves

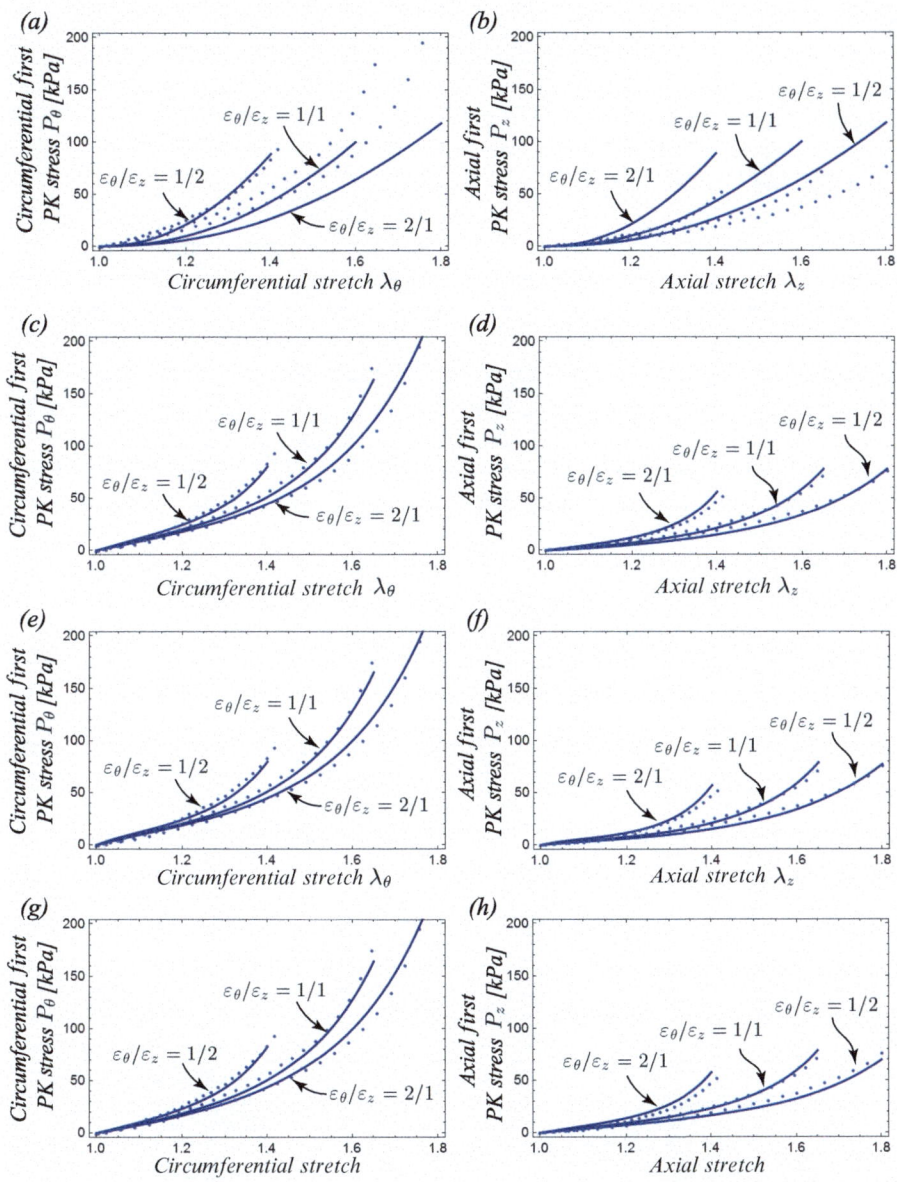

Fig. 5.32 Vessel wall properties at planar biaxial tension for the three different loading protocols $\varepsilon_\theta/\varepsilon_z = 2/1; 1/1; 1/2$. (Left column) Circumferential first Piola–Kirchhoff stress P_θ *versus* the circumferential stretch $\lambda_\theta = 1 + \varepsilon_\theta$. (Right column) Axial first Piola–Kirchhoff stress P_z *versus* the axial stretch $\lambda_z = 1 + \varepsilon_z$. Solid curves are model-based predictions, whilst the dots denote experimental acquisitions. (**a, b**) Predictions using the Yeoh strain energy (5.8) and least-square optimized parameters $c_i; i = 1, 2$. (**c, d**) Predictions using the Fung strain energy (5.26) and least-square optimized parameters $c_i; i = 0, \ldots, 6$. (**e, f**) Predictions using the Fung strain energy (5.26) and least-square optimized parameters $c_i; i = 0, \ldots, 6$ under the constraint of a convex strain energy. (**g, h**) Predictions using the GOH strain energy (5.35) and least-square optimized parameters $c_0, c_{1m}, c_{2m}, c_{3m}, c_{4m}$. PK–Piola–Kirchhoff

then the scalar $Q = c_1 E_{\theta\theta}^2 + c_2 E_{zz}^2 + c_3 E_{rr}^2 + c_4 E_{\theta\theta} E_{zz} + c_5 E_{zz} E_{rr} + c_6 E_{rr} E_{\theta\theta}$, where $c_i; i = 0, \dots, 6$ are the material parameters. The first Piola–Kirchhoff stresses derives then from the relations (5.63), where the lengthly expressions are not explicitly shown here.

5.6.1.3 GOH Model

The GOH strain energy (5.35) considers the collagen fibers' orientation in the vessel wall. Whilst the collagen in the media is coherently aligned along the circumferential direction, its orientation in the adventitia is highly dispersed. We therefore describe the collagen alignment by $\rho_m(\phi, \theta)$ and $\rho_a(\phi, \theta)$, Bingham PDFs (5.47) that are shown in Fig. 5.33, where the parameters $c_{0\,m} = 15.0$, $c_{1\,m} = 0.0$ and $c_{0\,a} = 10.0$, $c_{1\,a} = 10.0$ represent the media and adventitia, respectively. The medial collagen is concentrated along the circumferential direction and thus grouped around at the azimuthal angle $\phi = 0$ and the elevation angle $\theta = 0$, respectively. The adventitial collagen is isotropic in the tangential plane (along $\phi = 0$), whilst it has a moderate dispersion in radial direction. Aside from the parameters that determine the shape of the Bingham distribution, $\bar{c}_m = 1.769 \cdot 10^{-5}$ and $\bar{c}_a = 3.24 \cdot 10^{-5}$ have been used to normalize the distributions according to (5.45).

The definition (5.36) allows us now to express the generalized structural tensors by

$$\mathbf{H}_i = \frac{1}{2\pi} \int_{-\frac{\pi}{2}}^{\frac{\pi}{2}} \int_{-\frac{\pi}{2}}^{\frac{\pi}{2}} \rho(\phi, \theta) \mathbf{M}(\phi, \theta) \otimes \mathbf{M}(\phi, \theta) \cos\phi \, d\theta d\phi \; ; \; i = m, a \,, \qquad (5.65)$$

where the orientation $\mathbf{M} = [\cos\phi\cos\theta \;\; \cos\phi\sin\theta \;\; \sin\phi]^{\mathrm{T}}$ has been expressed by the azimuthal and the elevation angles ϕ and θ, respectively.

Given (5.65) and the densities $\rho_m(\phi, \theta)$ and $\rho_a(\phi, \theta)$, the general structural tensors $\mathbf{H}_m = \mathrm{diag}[0.9306, 0.0347, 0.0347]$ and $\mathbf{H}_a = \mathrm{diag}[0.475, 0.475, 0.05]$

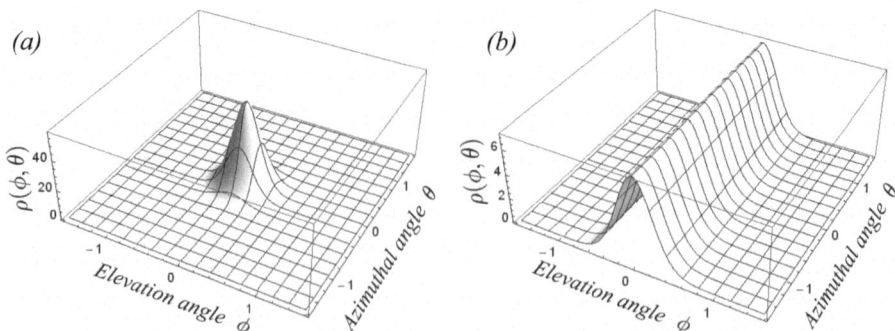

Fig. 5.33 Collagen orientation Probability Density Function (PDF) in the media (**a**) and the adventitia (**b**). Distributions are represented by the Bingham PDF (5.47). (**a**) Collagen in the media is transverse isotropic and coherently aligned with the circumferential direction. (**b**) Collagen in the adventitia is isotropic in the tangential plane, and moderately dispersion in radial direction

represent the collagen in the medial and adventitial layers, respectively. We note the condition $\text{tr}\mathbf{H}_\mathrm{m} = \text{tr}\mathbf{H}_\mathrm{a} = 1$, a direct consequence of the normalization (5.45).

The description of the media and the adventitia by the GOH model (5.35) and the use of membrane kinematics, then results in the strain energy

$$\Psi(\mathbf{C}, \mathbf{H}_\mathrm{m}, \mathbf{H}_\mathrm{a}) = c_0(I_1 - 3) + \sum_{i=\mathrm{m,a}} c_{1\,i}[\exp(c_{2\,i} E_i^2) - 1], \qquad (5.66)$$

where $E_\mathrm{m} = \mathbf{H}_\mathrm{m} : \mathbf{C} - 1$ and $E_\mathrm{a} = \mathbf{H}_\mathrm{a} : \mathbf{C} - 1$ represent the strain of medial and adventitial collagen, respectively. Whilst the material parameter c_0 describes the mechanics of the matrix (everything else but collagen), the parameters $c_{1\,\mathrm{m}}, c_{2\,\mathrm{m}}$, and $c_{1\,\mathrm{a}}, c_{2\,\mathrm{a}}$ capture the properties of medial and adventitial collagen, respectively. We may then compute the first Piola–Kirchhoff stresses from the relations (5.63), where the lengthly expressions are not explicitly shown here.

5.6.2   Optimization Problem

A least-square method may be used to identify the model parameters c_j from the experimental measurement points $(\lambda_{z\,i}^\mathrm{e}, \lambda_{\theta\,i}^\mathrm{e}, P_{z\,i}^\mathrm{e}, P_{\theta\,i}^\mathrm{e})$; $i = 1, \ldots, n$. The *objective function*

$$\Phi = \sum_{i=1}^n \left[\left(P_\theta(\lambda_{\theta\,i}^\mathrm{e}, \lambda_{z\,i}^\mathrm{e}) - P_{\theta\,i}^\mathrm{e} \right)^2 + \left(P_z(\lambda_{\theta\,i}^\mathrm{e}, \lambda_{z\,i}^\mathrm{e}) - P_{z\,i}^\mathrm{e} \right)^2 \right] \qquad (5.67)$$

is therefore introduced, and its minimization $\Phi \to \text{MIN}$ identifies the unknown material parameters c_j.

The definition (5.67) equally weights the stress differences $P_\theta - P_{\theta\,i}^\mathrm{e}$ and $P_z - P_{z\,i}^\mathrm{e}$ and therefore leads to parameters c_j that provide an optimized model representation over the stretch domain that is covered by *all* experimental data points. The use of application-specific weight functions $w_k(\lambda_\theta, \lambda_z)$; $k = \theta, z$ that are then multiplied with the stress differences $P_k - P_{k\,i}^\mathrm{e}$; $k = \theta, z$ in (5.67) allows for an improved model representation within a predefined stretch domain, however, at the cost of a worse representation outside this domain.

5.6.2.1 Yeoh Model
The objective function Φ of the two-parameter Yeoh model leads to a second-order polynomial. It is illustrated in Fig. 5.34 for the experimental data listed in Table F.1. The objective function Φ has a single *local minimum* ($\partial\Phi/\partial c_1 = \partial\Phi/\partial c_2 = 0$), which then is also the function's *global minimum*. It is at $c_1 = 1.054\,\text{kPa}$ and $c_2 = 7.068\,\text{kPa}$, and Fig. 5.32a, b shows the predicted model response on top of the experimental data points.

Fig. 5.34 Objective function
$\Phi(c_1, c_2)$ with the global
minimum defining the
least-square optimized
material parameters c_1 and c_2
of the two-parameter Yeoh
constitutive model

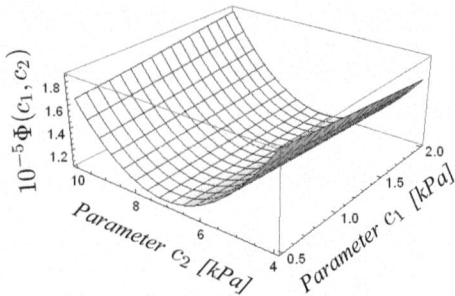

5.6.2.2 Fung Model

With the Fung model, Φ is a complex seven-dimensional function that has multiple
local minima. A numerical minimization algorithm identified the global minimum
at $c_0 = 22.329$ kPa, $c_1 = 0.3684$, $c_2 = 0.3394$, $c_3 = -0.482$, $c_4 = -0.02877$,
$c_5 = -0.1473$, and $c_6 = -1.7678$. Figure 5.32c, d shows the model's response for
these parameters together with the experimental data.

Towards the investigation of the convexity of Fung's strain energy (5.26), we note
that Ψ is strictly convex as long as the scalar Q remains positive for *all* possible
combinations of the principal strain components $E_{\theta\theta}$, E_{zz}, E_{rr}. The scalar may also
been written as

$$Q = \begin{bmatrix} E_{\theta\theta} & E_{zz} & E_{rr} \end{bmatrix} \underbrace{\begin{bmatrix} c_1 & c_4/2 & c_6/2 \\ c_4/2 & c_2 & c_5/2 \\ c_6/2 & c_5/2 & c_3 \end{bmatrix}}_{\mathbf{c}} \begin{bmatrix} E_{\theta\theta} \\ E_{zz} \\ E_{rr} \end{bmatrix}, \tag{5.68}$$

and the positive definiteness $\det\mathbf{c} > 0$ then ensures a strictly convex strain energy
Ψ [151]. With the afore-estimated parameters c_j; $j = 1, \ldots, 6$, the determinant
reads $\det\mathbf{c} = -0.329$, and Ψ is therefore not convex. A non-convex strain
energy may result in the prediction of non-physical stress states, and the *constraint*
$\det\mathbf{c} > 0$ should therefore be added to the minimization problem. Given the present
example, the constraint optimization led to the identification of the parameters
$c_0 = 31.682$ kPa, $c_1 = 0.4825$, $c_2 = 0.3271$, $c_3 = 0.3292$, $c_4 = 0.1666$,
$c_5 = 0.2922$, and $c_6 = -0.6235$, see Fig. 5.32e, f. Whilst these parameters
yield very similar results as compared to the unconstraint Fung's model shown in
Fig. 5.32c, d, they ensure the convexity of the vessel wall model.

Towards supporting the identification of "plausible" material parameters, it is
always recommended to add any available information to the minimization problem.
Aside from the convexity constraint, the uniform stress hypothesis [183] and other
plausible assumptions [593] have been used to constrain the parameter estimation
towards improving the reliability of the estimated model parameters [447, 522, 535,
536].

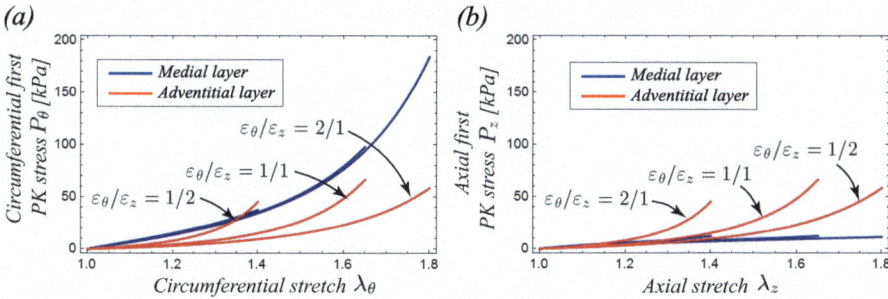

Fig. 5.35 Contributions of the media and the adventitia to the mechanical vessel wall properties at three different loading protocols $\varepsilon_\theta/\varepsilon_z = 2/1 ; 1/1 ; 1/2$. (**a**) Circumferential first Piola–Kirchhoff stress P_θ *versus* the circumferential stretch $\lambda_\theta = 1 + \varepsilon_\theta$. (**b**) Axial first Piola–Kirchhoff stress P_z *versus* the axial stretch $\lambda_z = 1 + \varepsilon_z$. Solid curves denote the two-layered analytical model, where each layer is modeled by the GOH strain energy (5.35). The model parameters have been estimated by least-square optimization from the experimental data listed in Table F.1. PK–Piola–Kirchhoff

5.6.2.3 GOH Model

With the GOH model, the objective function Φ is a complex five-dimensional function with multiple local minima. A numerical minimization algorithm identified the global minimum at $c_0 = 2.8973$ kPa, $c_{1\,\mathrm{m}} = 25.538$ kPa, $c_{2\,\mathrm{m}} = 0.20206$, $c_{3\,\mathrm{m}} = 12.357$ kPa, and $c_{4\,\mathrm{m}} = 0.39602$, and Fig. 5.32e, f illustrates the model in relation to the experimental data. This model considered a two-layered vessel wall description, and Fig. 5.35 shows the *individual* stress contributions of the media and adventitia, respectively.

5.6.2.4 Influence of Noise in the Experimental Data

Experimental data is always *noisy*, a factor that may influence the estimation of material parameters. Let us add random noise to the data listed in Tables F.1 towards the exploration of this effect. The randomly perturbed stretches and stresses then read

$$\left. \begin{array}{l} \lambda_\theta^{\star\mathrm{e}} = \lambda_\theta^{\mathrm{e}} + \alpha w_{\lambda_\theta}(\lambda_\theta^{\mathrm{e}} - 1) \; ; \; \lambda_z^{\star\mathrm{e}} = \lambda_z^{\mathrm{e}} + \alpha w_{\lambda_z}(\lambda_z^{\mathrm{e}} - 1) \; ; \\[2mm] P_\theta^{\star\mathrm{e}} = P_\theta^{\mathrm{e}} + \alpha w_{P_\theta} P_\theta^{\mathrm{e}} \; ; \; P_z^{\star\mathrm{e}} = P_z^{\mathrm{e}} + \alpha w_{P_z} P_z^{\mathrm{e}} \; , \end{array} \right\} \tag{5.69}$$

where α denotes the noise level, and $-\mathrm{mean}[(\bullet)] < w_{(\bullet)} < \mathrm{mean}[(\bullet)]$ with $(\bullet) = \{\lambda_\theta\}, \{\lambda_z\}, \{P_\theta\}, \{P_z\}$ are Gaussian-distributed pseudo-random numbers that are scaled by the magnitude of the experimental data.

Given the perturbed input data and Yeoh's as well as Fung's constitutive models, Table 5.6 reports the respective least-square identified material parameters. Whilst the models yield the best possible representation of the (noisy) experimental data, different nose levels resulted in very different model parameters. The graphical illustration of the objective function Φ in Fig. 5.34 of Yeoh's model helps us to understand this observation. The graph already indicates that Φ is more challenging

Table 5.6 Influence of noise in the identification of material parameters

Noise level α	0.0	0.05	0.1	0.15	0.2
(a) Yeoh model					
c_1 [kPa]	1.05424	1.42133	0.701512	1.68206	1.93777
c_2 [kPa]	7.06777	6.93361	7.06531	6.82147	6.75673
(b) Fung model					
c_0 [kPa]	31.6821	26.515	51.7766	41.0168	53.576
c_1	0.482481	0.5047	0.428366	0.406477	0.428047
c_2	0.327107	0.331239	0.211366	0.306464	0.172561
c_3	0.329228	0.380465	0.0618468	0.313789	0.0290524
c_4	0.166558	0.114796	0.149742	0.147198	0.104941
c_5	0.292222	0.162037	0.115555	0.335757	−0.0446757
c_6	−0.623488	−0.816747	−0.231142	−0.506722	−0.221245

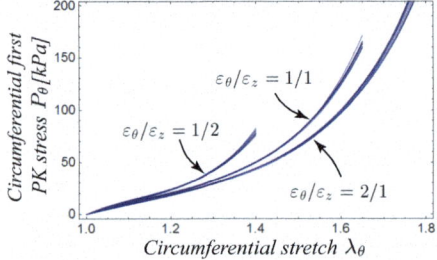 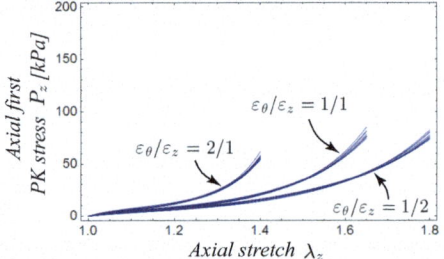

Fig. 5.36 Estimation of material parameters from noisy experimental data. Ten individual results are shown and the corresponding sets of parameter have been identified from randomly perturbed input data at the noise level of $\alpha = 0.1$. The identifications used Fung's model (5.26) that has been constraint to a strictly convex strain energy

to minimize for c_1 than for c_2—the curvature of Φ along c_1 is much smaller than along c_2. The inclusion of noise in the experimental data results then in a higher variability of c_1 than c_2, see the Yeoh model in Table 5.6.

Noise in the experimental data influences the estimation of material parameters, and the identified parameters are therefore no longer deterministic but probabilistic, a set of parameters has a certain probability of appearance. Given the noise level of $\alpha = 0.1$ and Fung's model, the identification results in the parameters $c_0 = 28.36\ (SD\ 4.319)$ kPa, $c_1 = 0.49\ (SD\ 0.01978)$, $c_2 = 0.3484\ (SD\ 0.0571)$, $c_3 = 0.4381\ (SD\ 0.2032)$, $c_4 = 0.1437\ (SD\ 0.05207)$, $c_5 = 0.2901\ (SD\ 0.177)$, $c_6 = -0.7498\ (SD\ 0.1849)$. These parameters are given by their mean and Standard Deviation (SD), information that has been identified through ten independent minimizations of the objective function Φ with the randomly perturbed input data according to (5.69). Whilst the individually estimated sets of parameters vary substantially, each set very nicely captures the experimental data, see Fig. 5.36. The examination of model parameter variations is therefore not conclusive in the assessment of a non-linear model.

Example 5.5 (Simple Shear Testing in Vessel Tissue Characterization). The experimental set-up shown in Fig. 5.37 is used to characterize the material properties
of a batch of the vascular wall. The tissue sample is exposed to a simple shear
deformation between two rigid plates.

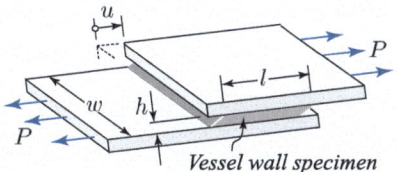

Fig. 5.37 Schematic illustration of simple shear testing of a sample of the vascular wall. The
actuator prescribes the displacement u, whilst the force P is measured by a load cell

(a) Derive the motion $\boldsymbol{\chi}(\mathbf{X})$ that maps the undeformed configuration Ω_0 into the
    deformed configuration Ω of the tissue batch.
(b) Derive the relation of the right Cauchy–Green strain $\mathbf{C}$ and the left Cauchy–
    Green strain $\mathbf{b}$.
(c) Provide the eigenvalue representation of $\mathbf{C}$ and $\mathbf{b}$.
(d) Derive the relation of the force P as a function of the displacement u. The
    vascular tissue may be regarded as an incompressible material with the strain
    energy

$$\Psi = c_1(I_1 - 3) + c_2(I_1 - 3)^2, \tag{5.70}$$

   where $I_1 = \mathrm{tr}\mathbf{C}$ denotes the first invariant of the right Cauchy–Green strain
   $\mathbf{C} = \mathbf{F}^{\mathsf{T}}\mathbf{F}$, and c_1, c_2 are material parameters.
(e) Consider the dimensions $l = 17\,\mathrm{mm}$, $w = 25\,\mathrm{mm}$, $h = 1.5\,\mathrm{mm}$, and use
    a least-square method to estimate the material parameters c_1 and c_2 from
    the experimentally recorded set $\{(0.5,0.3), (1.0,0.5), (1.5,1.2), (2.0,2.3)\}$ of
    measurement points, stored in the format (u [mm], P [N]).   ■

Example 5.6 (Pressure Inflation in Vessel Tissue Characterization). An *in vitro*
inflation test is used to characterize the material properties of a coronary artery
segment. The vessel segment of the referential diameter $D = 4.0\,\mathrm{mm}$, the length
$L = 70.0\,\mathrm{mm}$, and the wall thickness $H = 1.0\,\mathrm{mm}$ is mounted in a customized
tensile testing system, see Fig. 5.41. The vessel is fixed at its deformed length l, and
then slowly (quasi-statically) inflated up to the pressure p_i. A camera records the
deformed vessel diameter d, and a load cell measures the reduced axial force F_z,
and thus the reaction force that is conveyed from the vessel to the left clamp. The
vessel wall may be regarded as an incompressible material that is characterized by
the strain energy density

$$\Psi = c(I_1 - 3)^2, \tag{5.75}$$

where $I_1 = \mathrm{tr}\mathbf{C}$ denotes the first invariant of the Cauchy–Green strain $\mathbf{C}$. Membrane theory may be used in the analysis of this inflation experiment.

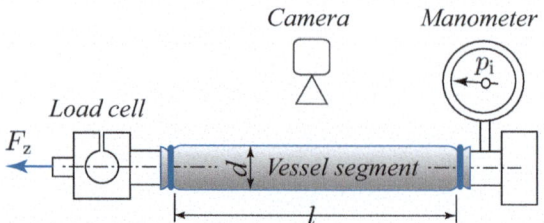

Fig. 5.41 Schematic illustration of an *in vitro* vessel inflation test. The vessel segment is kept at fixed axial length l and the inflation pressure p_i is measured by the manometer. The camera takes images to measure the vessel diameter d, and the load cell records the reduced axial force F_z

(a) Derive the relations of the circumferential σ_θ and the axial σ_z Cauchy stresses as a function of the circumferential λ_θ and the axial λ_z stretches.
(b) Use the equilibrium relation to express the inflation pressure p_i and the reduced axial force F_z as a function of the deformed vessel diameter d, the length l, and the wall thickness h, as well as the constitutive parameter c.
(c) Consider the set $\{(4.6,\ 73.5,\ 5.2,\ 21),\ (5.1,\ 73.5,\ 15.0,\ 3)\}$ of experimental measurement points and perform a least-square optimization to identify the constitutive parameter c. A measurement point is stored in the format (d [mm], l [mm], p_i [kPa], F_z [mN]). Consider the estimated parameter c, and plot the inflation pressure p_i and the reduced axial force F_z as a function of the deformed diameter d at the fixed vessel length of $l = 73.5$ mm. ∎

The parameter estimation approaches outlined in this section required the formulation of an analytical model that describes the mechanical properties of the test sample. Such a description is usually limited to deformations that are homogenous throughout the sample. However, constraints in sample size and shape often not support said assumption, and more generally applicable inverse parameter estimation approaches have been proposed [27, 33, 136, 248].

5.7 Case Study: Structural Analysis of the Aneurysmatic Infrarenal Aorta

Given the clinical relevance of an AAA, its biomechanical exploration is of enormous scientific and medical interest. Wall stress is known to be a marker for AAA rupture risk, see Sect. 1.6, and in this case study we therefore demonstrate all steps towards the computation of said wall stress. Whilst our case of 4.5 cm maximum transversal diameter is well below the clinical indication to perform AAA repair, significant peaks of wall stress may already have been developed.

5.7.1   Modeling Assumptions

The wall of the infrarenal aorta was reconstructed from clinically recorded CT-A images (A4clinics Research Edition, VASCOPS GmbH) and then saved in STereoLithography (STL) file format, see Fig. 5.43. CT-A scanning of the abdomen requires a few seconds, time during which the aorta pulsates between diastolic and systolic phases. Most protocols aim for the acquisition at the diastolic phase, and Fig. 5.43a shows such a CT-A recording. Given the AAA wall is much stiffer than the normal aorta, its deformation has been neglected and the diastolic geometry represented the stress-free reference configuration Ω_0 of our FEM model. In addition, the AAA wall thickness depends on a number of factors [299, 458] and CT-A images do not allow for its robust reconstruction. The model therefore used a predefined wall thickness between 1.13 mm and 1.5 mm, a value that depends on the thickness of the underlying ILT-layer [193].

The AAA wall has been modeled as an isotropic non-linear continuum at finite deformations. The contributions from the different vessel wall layers have therefore been homogenized and described by the Yeoh model (5.8) with the constitutive parameters $c_1 = 177.0\,\text{kPa}$ and $c_2 = 1881.0\,\text{kPa}$, information that has been identified from *in vitro* 1D tensile testing of the AAA wall [449]. Whilst our case contained an ILT, the structural implications of this pseudo-tissue have been neglected in the FEM model.

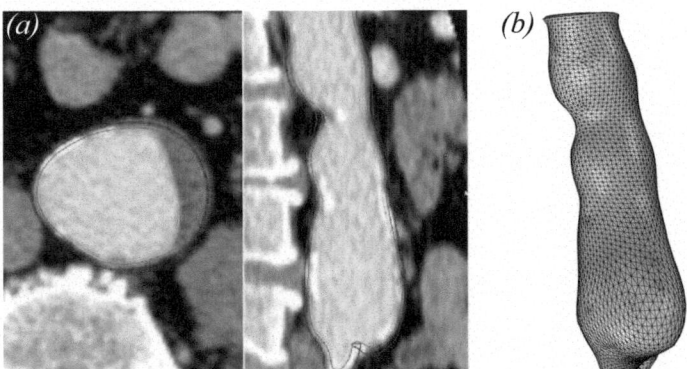

Fig. 5.43 (a) Computed Tomography-Angiography (CT-A) images to reconstruct the vascular geometry of an Abdominal Aortic Aneurysms (AAA). The yellow curve encapsulates the lumen, the domain between the yellow and green lines represents the Intra-luminal Thrombus (ILT), whilst the green and blue lines encapsulate the vessel wall. (b) Reconstructed vessel wall in STereoLithography (STL) file format

Given the individual systolic/diastolic blood pressure of 140/90 mmHg, the MAP $p_i = 14.2$ kPa was prescribed as a Neumann boundary condition at the inside of the vessel wall. The abdominal pressure was set to zero, and the outer vessel wall was therefore free of traction. In addition to the inflation by the arterial pressure, the aorta is stretched in axial direction. Whilst the axial *in vivo* stretch reaches approximately 40% in the young aorta, it decreases rapidly with age [264], and probably also with the development of aneurysm disease. The axial pre-stretch has therefore been neglected in our AAA model.

The aorta is anchored in the body, and assumptions regarding this boundary condition have to be made. Two locations seem to play an important role: the level of the renal arteries and the level of the aortic bifurcation, sites where the aorta is firmly connected to the surrounding tissues. All displacements at these positions have therefore been locked, and the prescription of this Dirichlet conditions concludes the description of the FEM model. We may now generate the FEM mesh and compute the stress in the AAA wall from the solution of the quasi-static equilibrium.

5.7.2 Results and Interpretation of the Computed Wall Stress

The FEM model used quadratic tetrahedral finite elements of approximately 100k degrees of freedom to approximate the structural problem. Figure 5.44 illustrates the distribution of the von Mises stress. Given the stress in the wall of inflated tube-like structures is determined by the local curvature, the stress is complexly distributed over the aneurysm wall. It reaches up to approximately 200 kPa and is therefore already much higher than the von Mises stress

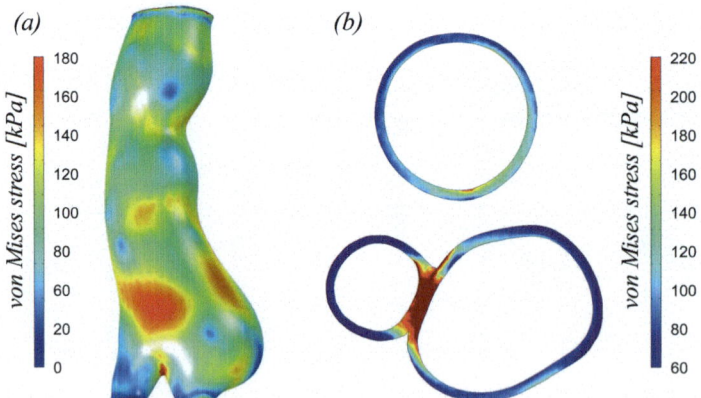

Fig. 5.44 Loading of the wall of an Abdominal Aortic Aneurysms (AAA) at the blood pressure of $p_i = 14.2$ kPa. The von Mises stress all over the AAA (**a**) and at a selective cross-section (**b**) is shown.

$$\sigma_M = \sqrt{\sigma_\theta^2 - 2\sigma_\theta\sigma_z + \sigma_z^2} = \frac{\sqrt{3}}{4}\sigma_\theta = 86.1 \text{ kPa} \qquad (5.77)$$

in the normal infrarenal aorta. In the derivation of this result, we used the membrane assumption and considered the vessel diameter $d = 21.0$ mm and the wall thickness $h = 1.5$ mm, figures that result in the circumferential stress $\sigma_\theta = p_i d/(2h) = 99.4$ kPa and the axial stress $\sigma_z = \sigma_\theta/2 = 49.7$ kPa in the vessel wall.

The highest stress appears in the aortic bifurcation, see Fig. 5.44b, at a site that, however, needs to be excluded from the analysis. At branching regions vascular tissue adapts towards a tendon-like collagenous tissue that is very different from the normal vessel wall [161]. The modeling assumptions concerning the wall thickness and the tissue properties are therefore not valid in the bifurcation. In addition, the assumption of a homogeneous continuum that represents the vessel wall permits from drawing conclusions regarding the local stress inside the wall. We may therefore only assess the stress in average across the wall. Our analysis did also not consider residual stress in the load-free configuration Ω_0, which otherwise would have decreased the stress gradient across the wall.

5.8 Summary and Conclusion

The biomechanical properties of conduit vessels are critical to the proper functioning of the cardiovascular system. In addition to the exploration of physiological mechanisms, the interaction of biomechanical, biochemical and clinical factors is of key importance to further our understanding of vascular pathologies [204].

The vessel wall's biomechanical macroscopic properties result from the complex 3D arrangement and interaction of microstructural constituents. Numerous constitutive descriptions have been proposed in the literature, and the complex vessel wall properties generally result in a large number of *structural* and *mechanical* model parameters. They need to be identified from appropriate experimental information. The *in vitro* characterization allows to load the vessel wall at well-defined conditions, and then supports the acquisition of rich data for the inverse identification of material parameters. However, the vessel needs to be dissected, an intervention that potentially changes its mechanical properties. In contrary, the *in vivo* characterization keeps the vessel intact, it tests it within its natural environment, but often acquires too less information for a robust parameter identification.

Despite encouraging progress in vascular tissue biomechanics, the variability of biomechanical predictions from the *uncertainty* of the input information remains a challenging limitation. Key input information, such as detailed vessel morphology and local biomechanical properties of vascular wall tissue often remains unknown, or their robust identification is difficult. It is also difficult to draw a clear border between distinct vessel components. Even if possible, the mechanical interaction across these borders remains unclear. The description of boundary conditions is another challenging task, and there exists very limited understanding of factors, such as the perivascular support [155], the mechanisms by which the adventitia is anchored to surrounding tissues. Owing to this lack of input information, homogeneous *mean population* input information is often used, and the extent to which this simplification influences model predictions remains to be explored for each individual application. *Probabilistic approaches* seem in this respect to be a promising way to deal with input uncertainty of vascular biomechanics [44, 45, 61, 432].

There usually exists a large intra-patient and inter-patient variation in material properties of vessel tissues. Currently *no* non-invasive technique is known that would be able to acquire individual biomechanical properties of vascular tissues that fully facilitate their biomechanical analysis. Even under well-defined laboratory conditions, the *in vitro* experimental characterization of vascular wall properties shows huge variability and the estimated tissue properties usually vary at least by one order of magnitude. Whilst the detailed causes of this variability remain largely unknown, several influential factors have already been identified. Chronic Kidney Disease (CKD) [458], COPD [171], bicuspid aortic valve anatomy [172, 504], Marfan syndrome [357], diabetes mellitus [352], results from blood tests and tissue metabolic activity [458], tobacco use [292] as well as the administration of drugs, such as beta blocker and Angiotensin-Converting-Enzyme (ACE) inhibitors [352], have shown to alter the biomechanical vessel wall properties. A better understanding of how such, and currently unknown, factors influence the mechanics of vascular tissue, would be of key importance to improve the reliability of biomechanical simulations. These factors often interact, which complicates the exploration of their isolated influence in biomechanics. Given sufficient experimental data, the application of Machine Learning-based (ML-based) approaches has been successfully demonstrated [61, 344] and may help to close this knowledge gap.

Hemodynamics

6

This chapter addresses the flow of blood in conduit vessels. We review the composition of blood, a suspension of different-sized particles in plasma, and investigate the forces that act upon said particles. It results in the description of the rheological properties of blood, where single-phasic and bi-phasic models are covered. We then explore blood damage mechanisms with focus on hemolysis and abnormal thrombocyte activation. A key section of this chapter concerns the description of incompressible flows by solving the Navier-Stokes equations for a number of 1D flows. It results in the description of steady-state and steady-periodic flows through circular tubes —the Poiseuille and respective Womersley flows. The exploration of the flow in elastic tubes reveals the expression of the wave speed, an important biomechanical property linked to the condition of the vascular system. Multidimensional flow phenomena, the characteristics of boundary layer flow and the difference between laminar, transitional and turbulent flow are then specified. Wall Shear Stress-related (WSS-related) and transport-related flow parameters, values used in the quantitative description of blood flows, are then addressed. A case study uses the Finite Element Method (FEM) to predict the blood flow in the aneurysmatic aorta, and concluding remarks summarize the chapter.

6.1 Introduction

Blood is a *suspension* that continuously delivers nutrients, oxygen, and other factors to the cells. At the same time, it transports away metabolic waste products. Besides these primary objectives, blood contributes also to factors, such immune response,

---

The original version of this chapter was revised: ESM has been added. The correction to this chapter is available at https://doi.org/10.1007/978-3-030-70966-2_8

Supplementary Information The online version contains supplementary material available at https://doi.org/10.1007/978-3-030-70966-2_6

wound healing, transportation of information through hormones, regulation of tissue temperature, and Potential of Hydrogen (pH) levels. All these tasks are accomplished by approximately *five* liters of blood in adults. The flow dynamics of blood changes remarkably along the vasculature. Whilst the blood velocity is *strongly oscillatory* in the larger arteries and shows in addition to forward flow also phases of backward flow, it develops towards *unidirectional* flow in smaller vessels and the veins. Blood flow is able to influence vascular biology through mechanical factors [11], such as WSS [78], blood pressure [498], and Vortical Structure (VS) dynamics [42]. They guide the adaptation of the normal vasculature towards homeostasis and allow the system to cope with environmental alterations. However, the very same mechanical factors are also involved in the development of cardiovascular pathologies.

6.2   Blood Composition

Blood is a suspension of *cells* and *macromolecules* in plasma. Erythrocytes, leukocytes, and thrombocytes are the most prominent cells, whilst globulin, albumin, fibrinogen, vitamins, and fatty acids are noticeable macromolecules in the blood. The mechanical, electrical, and molecular interactions amongst these particles define the macroscopic biomechanical properties of blood and determine blood flow characteristics.

6.2.1   Erythrocyte

The principle particles in blood are *erythrocytes* (or red blood cells) and their principal mean is the *delivery of oxygen*. One microliter blood contains approximately 5 million erythrocytes and their percentage by volume is called *hematocrit*. In males, the hematocrit ranges normally from 41% to 52%, whilst values from 36% to 48% are seen in females. Given their high volume fractions, erythrocytes are the main determinant of *blood viscosity*. Erythrocytes are highly *deformable*, and they have no nucleus. At rest or low shear rates, they appear as bi-concave-shaped discs, measuring approximately 6 to 8 μm in diameter and 2 to 4 μm in height, see Fig. 6.1a, b. Erythrocyte size and shape are remarkably similar across all mammals.

Erythrocytes carry oxygen from the lungs to the cells. They contain a special protein called *hemoglobin* that allows for the reversible binding of oxygen. A portion of 98% of the oxygen in blood binds to hemoglobin and only 2% remains unattached. Hemoglobin that is fully saturated with oxygen is called oxyhemoglobin and appears red. Hemoglobin that is free of oxygen is in contrast called de-oxyhemoglobin and appears purplish blue. In the lungs, where the partial oxygen pressure is *high*, oxygen binds to the hemoglobin of erythrocytes, and it is then transported away by the bloodstream. Given sites of tissue at low oxygen conditions, the partial oxygen pressure is *low*, and oxygen is then released from hemoglobin of nearby erythrocytes into the vessel lumen. The oxygen then travels further into tissue cells. Normal erythrocytes are highly deformable and their cytoskeleton

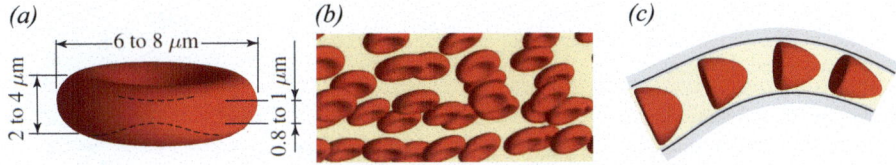

Fig. 6.1 Erythrocytes (or red blood cells) shape and dimensions. (**a**) Biconcave shape of a single erythrocyte at rest. (**b**) Erythrocytes in blood flow of low shear rate. (**c**) Erythrocytes appear at bullet-like shapes to pass through a microvessel

Fig. 6.2 Shape and dimensions of some leucocytes (or white blood cells), cells of the immune system that protect the body from infectious disease and foreign invaders

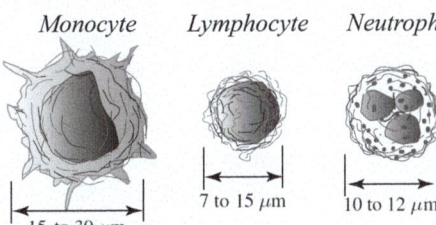

supports shape changes without destruction. They can easily adapt into bullet-liked shapes, a mechanism that allows them to pass micro-vessels much smaller in diameter than erythrocytes at rest, see Fig. 6.1c.

Erythrocytes have a lifespan of approximately 120 days, and the body continuously replaces old erythrocytes with newly formed ones—a process that appears in bone marrow and that is called *erythropoiesis*.

6.2.2 Leukocyte

Leukocytes (or white blood cells) are cells of the *immune system* and occupy approximately 0.7% of the blood volume. They reach a size of up to 30 μm in diameter and represent a family of cells with some of them illustrated in Fig. 6.2. Each leucocyte has a different role in protecting the body against *infectious disease* and foreign *invaders*. *Neutrophils* ingest and digest bacteria and fungi. *Eosinophils* and *basophils* attack larger parasites and are involved in allergy response. *Monocytes* are carried by blood to tissues and then transform into *macrophages* that destruct harmful organisms and activate other immune cells, such as T cells. *Lymphocytes* attack invading bacteria and viruses and help to destroy diseased or infected cells.

An increased volume fraction of leucocytes is the normal response to *infections*, whilst an abnormal increase relates to diseases such a leukemia. A decreased volume of leukocytes in contrast may indicate diseases that are linked to the impaired function of the immune system.

6.2.3 Thrombocyte

Thrombocytes (or platelets) are fragments of much larger cells called *megakaryocytes* and occupy approximately 0.3% of blood volume. Together with *coagulation*

Fig. 6.3 Thrombocyte (or
platelet) in its non-activated
(**a**) and activated (**b**) states.
Activated thrombocytes are
able to form a thrombus, the
first step in wound healing

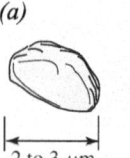 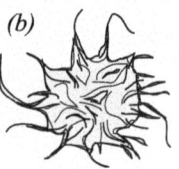

(a)   *(b)*

2 to 3 μm

factors they stop bleeding through the formation of a *blood clot*, a process that
appears within seconds to minutes. Thrombocytes have *no* nucleus, and may appear
in *non-activated* and *activated* states, respectively. The majority of thrombocytes
are non-activated. They then have a compact plate-like shape (hence they are named
platelets) and are approximately 2 to 3 μm large, see Fig. 6.3a. On activation,
platelets turn into (sticky) octopus-like shapes with multiple arms and legs, see
Fig. 6.3b. This transformation is important in enabling activated thrombocytes to
clump together and form blood clots. At a site of *endothelial damage*, the contact
with collagen activates thrombocytes and then allows them to stick to the collagen.
This process seals-off the damaged area and initializes the subsequent repair
steps [561]. *High shear stress* is also able to activate thrombocytes through the
stretching of *von Willebrands*[1] *Factor (vWF)*, a blood-borne adhesive protein. Given
enough time, vWF then binds to platelets and activates them [177, 479, 480].

Thrombocyte has a lifespan of approximately 8 to 10 days and a number of
hematological diseases prime the responsiveness of platelet populations [19]. A
low volume fraction of thrombocyte relates to a number of diseases and results
in impaired clotting ability and the risk of life-threatening blood loss from small
wounds. In contrary, high levels of thrombocytes increase the risk of thrombosis
and subsequent cardiovascular events, such as stroke and heart attack.

6.2.4 Plasma

Plasma is a straw-colored fluid that consists of 90% water, 9% various molecules,
and 1% electrolytes. *Electrolyte concentration* is important in the regulation of the
fluid content within cells, whilst *molecules* determine immune response and control
Colloid Osmotic Pressure (COP) as well as blood clotting.

The majority of molecules are *proteins*, such as globulin that plays a central role
in immune response, albumin that controls COP (see Sect. 2.1.6.2), and fibrinogen, a
key factor in blood clotting. Molecules in transit comprise, vitamins, carbon dioxide,
and oxygen, as well as waste products, such as urea and ammonia. Plasma also
contains a number of digestion products, such as fatty acids, amino acids, and
peptides. Low-Density Lipoprotein (LDL) and High-Density Lipoprotein (HDL)
are prominent fatty acids that have been related to atherosclerosis, see Sect. 5.4.2.

---

[1]Erik Adolf von Willebrand, Finish internist, 1870–1949.

A low volume fraction of plasma is caused by dehydration, salt depletion, or blood loss, whilst high plasma volume can occur as a result of inadequate salt excretion eventually linked to factors, such as kidney disease.

6.3 Forces Acting at Blood Particles

Blood particles in plasma are under the effect of *mechanical, electrical,* and *molecular* forces. Whilst they directly determine the transport of the particles in plasma, they also influence the macroscopic biomechanical properties of blood, and therefore the blood flow characteristics. We may relate the forces to drag effects, to gravitation and inertia effects, to inhomogeneous pressure and velocity fields, to collisions, and to chemical and electrical effects. Whilst the drag force moves particles *along* streamlines, the other forces may direct particles to *cross* the streamlines.

6.3.1 Drag Force

The *drag force* tends to carry immersed particles along streamlines, and thus along the direction of the flow velocity, see Fig. 6.4a. The drag force therefore pushes particles to move together with the fluid, a transport phenomenon known as *advection*. How closely a particle of diameter d follows the streamlines is determined by the *Stokes[2] number*

$$St = \frac{T|\mathbf{v}|}{d} , \tag{6.1}$$

where $\mathbf{v}$ denotes the velocity of the fluid, whilst T is the relaxation time of the particle. Let us imagine a resting particle is put into a moving fluid. The relaxation time T then denotes the time the particle needs to approach 63% of the flow velocity. A small St number determines a particle that closely follows streamlines, whilst the path of a particle with a high St number may substantially divert from the streamlines. At normal conditions, the St number is low in the vasculature and *Stokes drag* therefore applies—large particles related to thrombo-embolic events would be an exception. For *creep or Stokes flow*, the St number is the inverse of the Reynolds (Re) number, $St = Re^{-1}$.

---

[2]Sir George Gabriel Stokes, Irish physicist and mathematician, 1819–1903.

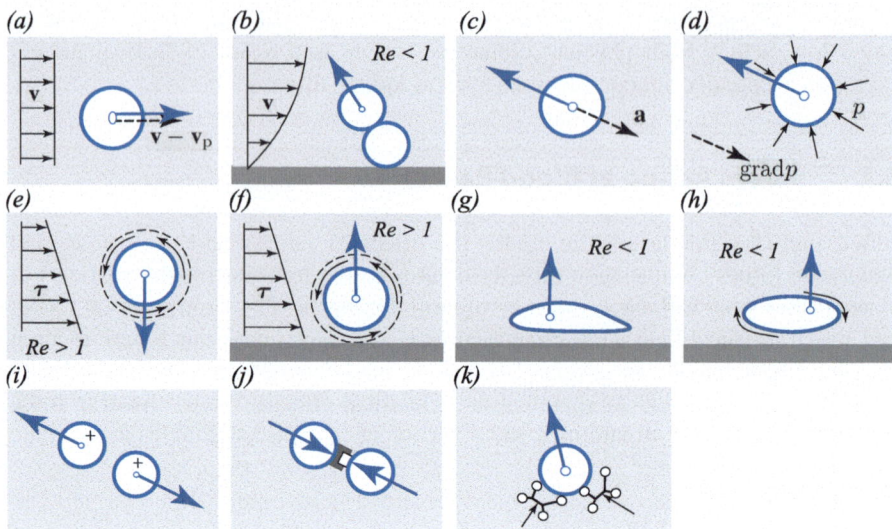

Fig. 6.4 Forces acting at blood particles in plasma. Plasma velocity, shear stress, and pressure are denoted by **v**, τ, and p, respectively. (**a**) Drag force acting at a particle of the velocity $\mathbf{v}_p$. (**b**) Impact forces from collision with other particles in the vicinity of the wall. (**c**) Gravitational or inertia force. (**d**) Hydrostatic force from an in-homogenous pressure p. (**e**) Inertia lift force, and (**f**) inertia wall lift force due to inertia effects of fluid rotating around a spinning particle. (**g**) Lift due to the loss of particle symmetry. (**h**) Lift due to tank-treading. (**i**) Electrostatic repulsion of similarly charged particles. (**j**) Molecular binding force. (**k**) Impacts with macromolecule results in erratic Brownian motion

6.3.2   Gravitational and Inertia Forces

Given the acceleration field **a**, the *inertia force* $\mathbf{F}_i = m\mathbf{a}$ acts at the particle of the mass m, see Fig. 6.4c. With the density of $1125 \ \mathrm{kg \, m^{-3}}$, erythrocytes are slightly heavier than plasma of the density $1025 \ \mathrm{kg \, m^{-3}}$, and the inertia force allows therefore the separation of erythrocytes from plasma in a centrifuge. The gravitational force is a special case of the inertia force and relates to the acceleration of $|\mathbf{a}| = 9.81 \ \mathrm{m \, s^{-2}}$.

6.3.3   Forces Related to Fluid Pressure

The *hydrostatic force* $\mathbf{F}_h = -\int_{\partial\Omega} p\mathbf{n}\mathrm{d}s$ acts on a particle that is immersed in fluid of the pressure p, where **n** denotes the outward normal vector to the particle surface, whilst $\mathrm{d}s$ is the area element. The integration is taken over the entire surface $\partial\Omega$ of the particle. Given an inhomogeneous pressure, this integral results in a hydrostatic force acting at the particle, see Fig. 6.4d.

6.3.4 Forces Related to Fluid Velocity and Shear Stress

A velocity field that represents a shear stress gradient results in *rotating* particles. Given a flow with a *high Re* number, the *inertia effects* of the fluid dominate the forces—the inertia of the fluid that surrounds the rotating circular particle then results in the *inertia lift force* [208], see Fig. 6.4e. The inertia lift force directs particles in the free flow towards the region of higher shear [255]. A number of inertia forces act on particles that are close to a wall, or even touch a wall. They are collectively called *inertia wall lift forces* and always direct away from the wall, see Fig. 6.4f. Lift forces result in particle migration, an effect that is already observable in laminar tube flow. Particles that are uniformly distributed at the inlet of a tube tend to locate in a ring of approximately 60% of the tube's diameter [501], an effect known as *tubular pinch effect* or *Segré–Silberberg effect*, see Fig. 6.5a.

In addition to the aforementioned inertia-dominated effects, *viscous forces* determine the migration of particles at *low Re* numbers. Erythrocytes are highly deformable and close to the wall they deform into *asymmetric* shapes. The forces acting on such particles direct them away from the wall [67], see Fig. 6.4g. Elastic particles close to the wall may also show *tank-treading* and move visually similar to tank treads, see Fig. 6.4h. It also leads to migration forces pointing away from the wall.

Given the high temporal and spatial resolution needed, the direct experimental observation of the aforementioned phenomena is challenging and simulations provided detailed insight into the motion and transport of blood particles in suspension [6, 152, 419, 455, 537].

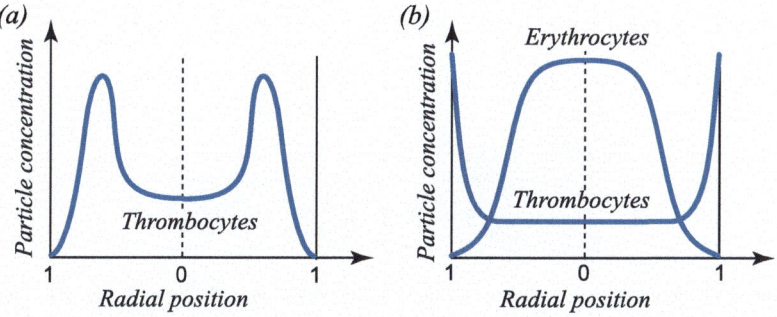

Fig. 6.5 Distribution of blood particles in laminar tube flow. Data stems from the flow in glass tubes of 3 mm in diameter and at the Wall Shear Stress (WSS) of approximately $1000\,\mathrm{s}^{-1}$ [1]. (**a**) Thrombocytes in saline solution show the Segré–Silberberg effect and group in a ring of approximately 60% of the tube's diameter [501]. (**b**) Erythrocyte ghosts and thrombocytes in saline solution. Erythrocyte ghosts are concentrated in the center, whilst thrombocytes are concentrated at the wall

6.3.5   Forces from Collisions

Random variations in the number of plasma molecules that collide with particles determine small random forces. The blood particles suspended in plasma therefore show random *erratic* movements known as *Brownian*[3] *motion*, see Fig. 6.4k.

With the increase in particle fraction, the collision between the particles themselves also increases, an effect that eventually dominates the distribution of blood particles and determines the local viscosity of the suspension. Given rigid spherical particles, the viscosity approaches *infinity* at a volume ratio of approximately 64% [219]. The particles are then no longer able to pass each other—the fluid stops flowing. With deformable particles, no such limit exists and erythrocytes suspensions of volume fractions as high as 80% still flow [217].

With the shear rate increasing, more particles have to pass each other, which then also results in more collisions. The collision rate of particles is therefore proportional to the shear rate in the fluid. The particles try to avoid collisions and migrate towards regions of lower collision rates, and thus towards lower shear rates, see Fig. 6.4b. The shear-induced migration of particles reduces the concentration of erythrocytes (or the hematocrit level) at the wall in vessels of small diameters [1], known as the *Fåhræus*[4] *effect* [146]. It also has a strong impact on the viscosity of blood and explains the decreased viscosity in small vessels, known as the *Fåhræus–Lindqvist effect* [147]. Both effects are observed in vessels of diameters below of approximately 0.2 mm, and thus vessels that are small enough that the thickness of the erythrocytes-depleted boundary plays an important role, see Fig. 6.6. Constitutive descriptions, such as the Phillips' model [427] allows for the quantitative description of shear-induced migration of particles.

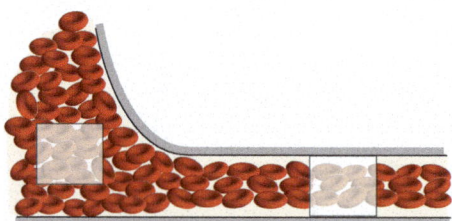

Fig. 6.6 Fåhræus effect. Shear induces migration of erythrocytes towards the center of the tube leading to a lower average hematocrit level in small vessels when compared to large vessels. It may be concluded from the number of erythrocytes within the two same-sized transparent squares in the figure

---

[3]Robert Brown, Scottish botanist and palaeobotanist, 1773–1858.
[4]Robert Fåhræus, Swedish clinical professor and pathologist, 1888–1968.

6.3.6 Chemical and Electrical Forces

Chemical and *electrical forces* operate at *close range* and appear through the interaction among blood particles rather than through the plasma. Electrically charged particles attract particles of the opposite charge and repel similarly charged particles, see Fig. 6.4i. Erythrocytes, leukocytes, and thrombocytes, all of them are *negatively* charged, which protects them from colliding and helps to prevent thrombus formation in the normal circulation. Electrostatic forces are also able to establish from electrical charges in the vessel wall and blood particles.

If two particles touch each other, *molecular bindings* may overcome electrostatic repulsion—the particles adhere, see Fig. 6.4j. Rouleaux formation [587], leukocyte adhesion, thrombocyte aggregation and adhesion, are prominent adhesion phenomena in blood.

Rouleaux formation appears at low shear rates, allowing erythrocytes to clump together face-to-face, see Fig. 6.7. The aggregation of erythrocytes is caused by a reversible aggregation–dissociation mechanism and requires the presence of albumin, fibrinogen, and globulin. Fibrinogen plays the most important role [29] and the aggregation first increases with the polymer concentration, reaches a maximum, and then decreases [165]. At very low shear rates, the rouleaux even align themselves in an end-to-end and side-to-end fashion, forming a secondary structure of 3D aggregates. At this configuration, blood shows solid-like properties.

Leukocyte adhesion establishes if a leukocyte comes close to the *endothelium*, and leukocyte ligands bind to selectin molecules on the endothelium cell. The leukocyte can then migrate into the vessel wall and deliver immune response.

Thrombocyte aggregation and *adhesion* develop through the binding of *activated* thrombocytes to each other via plasma macromolecules, such as fibrinogen and vWF [480]. In addition to self-adhesion, thrombocytes bind to a number of substances that are exposed to the bloodstream. Thrombocyte aggregation and adhesion is an important factor in the formation of a thrombus and therefore a critical step in wound healing.

Fig. 6.7 Rouleaux formation. Erythrocytes clump together face-to-face, a mechanism that appears at low shear rates in blood

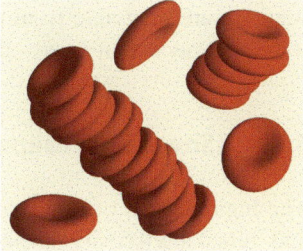

6.3.7   Segregation of Blood Particles

The smaller blood particles prevent in general the larger ones from contact—
macromolecules, such as albumin hinders particles, such as erythrocytes from
colliding with themselves. Blood contains many different-sized particles, and their
interactions in plasma lead to a number of *segregation* effects. Already in bimodal
suspensions, the distribution of particle evolves very differently from unimodal
suspensions. This has been illustrated by *in vitro* experiments in glass tubes that used
erythrocyte ghosts (erythrocytes made transparent by removing their hemoglobin)
and thrombocytes in saline solution [1]. The larger particles (erythrocyte ghosts)
are then concentrated in the center, whilst the smaller particles (thrombocytes)
are concentrated at the wall, see Fig. 6.5b. Particle segregation is always hindered
by *mixing*, a phenomenon that is inherent to all complex flows, such as swirl,
translational, and turbulent flow.

6.4   Blood Rheology Modeling

The *rheological properties* of blood are critical to proper *tissue perfusion*. Knowledge concerning the alterations in blood properties and their consequences for
the cardiovascular system is therefore important clinical information. Pathologies
with hematological origin, such as leukemia, hemolytic anemia, sickle-cell disease,
conditions associated with the risk for thrombosis and other cardiovascular events,
all relate to the *disturbance of local homeostasis*. In addition to experimental studies,
versatile mathematical descriptions of blood rheology may help to explore and
understand said circulatory implications [148, 469].

The rheological properties of a fluid are measured by a *rheometer*, where the
Couette[5] rheometer, the cone-and-plate rheometer, and the capillary rheometer are
the most commonly used experimental devices. They apply very different principles
to establish shear rates, which then also limits the cross-comparison of the acquired
experimental data. In addition to the experimental technique, vascular control
mechanisms such as metabolic autoregulation and/or modulation of endothelial
function are also able to modify blood rheology. It may explain that blood seems
to be less viscous at *in vivo* than *in vitro* conditions, see [365].

6.4.1   Shear Rate-Dependent Changes of Blood Microstructure

The high content of *erythrocytes* results in significant *non-Newtonian effects* [83],
and blood's viscosity is *not* constant. Given very low shear rates, erythrocyte
rouleaux join each other and form secondary structures of aggregates, see Fig. 6.8.
At this state, blood has solid-like properties with theoretically *infinite viscosity*.

---

[5]Maurice Marie Alfred Couette, French physicist, 1858–1943.

Fig. 6.8 Dependence of the viscosity of whole blood on the shear rate and the hematocrit level. Inlets show the shape and distribution of erythrocytes at different shear rates

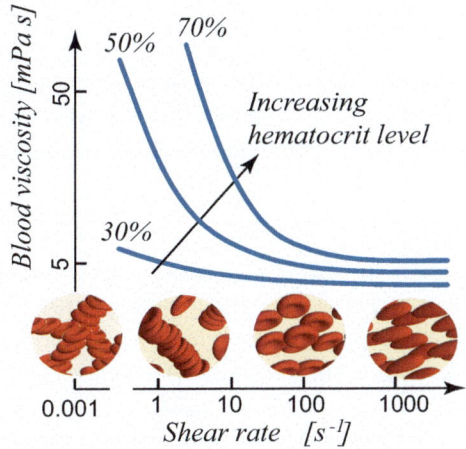

Exceeding the *yields stress*, solid-like blood breaks up into a network of aggregates. This process is time-dependent but reversible, and it is still unclear whether the yield stress may be regarded as a material property of blood. The aggregates then move independently within blood and their size reduces with increasing shear rates, a process that continues until all aggregates are broken down and the erythrocytes can move independently. Large solid-like obstacles severely increase the viscosity of a fluid, and breaking them down into smaller pieces therefore leads to the *fast drop* of blood viscosity already at low shear rates, see Fig. 6.8. At high shear rates, erythrocytes deform into ellipsoidal shapes that are aligned with the bloodstream, a mechanism that further reduces blood's flow resistance, see Fig. 6.8. At shear rates beyond of approximately 100 to 200 s^{-1}, blood viscosity becomes finally *insensitive* to the shear rate, and the Newtonian assumption of a *constant viscosity* is then justified.

The break-down and the formation of blood microstructure is *time-dependent*, and blood's rheological properties are *thixotropic*. Let us consider blood at a constant and high shear rate until its microstructure is equilibrated, and the blood established the viscosity η_{high}. Following a sudden drop to a lower and again constant shear rate, some time is required until blood's microstructure is again equilibrated, and the viscosity η_{low} establishes. The viscosity therefore lags always behind the shear rate, a property called *thixotropy*, which in some sense is similar to viscoelasticity in solids, see Sect. 3.5.4. The half-life time for aggregate formation is in the range of 3 to 5 s [495]—it may be shorter for pathological blood.

6.4.2 Modeling Generalized Newtonian Fluids

We consider an incompressible and isotropic fluid at the velocity **v**, and **l** = grad**v** denotes the spatial velocity gradient. The Cauchy stress may then be expressed by

$$\sigma = \tau(\mathbf{d}) - \kappa \mathbf{I} \,, \tag{6.2}$$

where $\mathbf{d} = (\mathbf{l} + \mathbf{l}^{\mathsf{T}})/2$ denotes the rate of deformation tensor, whilst κ is a Lagrange contribution to the hydrostatic stress that enforces the incompressibility. One possible particularization of the constitutive model (6.2) reads

$$\sigma = 2\eta(\mathbf{d})\mathbf{d} - p\mathbf{I} \,, \tag{6.3}$$

where the fluid's *dynamic viscosity* $\eta(\mathbf{d})$ [Pa s] is a function of the rate of deformation $\mathbf{d}$. The *Newtonian* fluid is a special case with a *constant* viscosity η. Given an incompressible fluid, the *continuity* reads $\mathrm{tr}\mathbf{d} = 0$, and p in (6.3) then denotes the hydrostatic pressure. This statement follows directly from the trace of (6.3), $\mathrm{tr}\sigma = -n_{\mathrm{dim}}p$, where n_{dim} denotes the number of spatial dimensions.

As with any function of a symmetric second-order tensor, the viscosity may be expressed by $\eta = \eta(\mathbf{d}) = \eta(I_{\mathbf{d}1}, I_{\mathbf{d}2}, I_{\mathbf{d}3})$, where

$$I_{\mathbf{d}1} = \mathrm{tr}\mathbf{d} \,;\; I_{\mathbf{d}2} = [(\mathrm{tr}\mathbf{d})^2 - \mathrm{tr}\mathbf{d}^2]/2 \,;\; I_{\mathbf{d}3} = \det\mathbf{d} \tag{6.4}$$

are the three invariants of the rate of deformation tensor $\mathbf{d}$. Given the continuity $\mathrm{tr}\mathbf{d} = 0 = I_{\mathbf{d}1}$, the first invariant contributes no information to the fluid's constitutive description. In addition, viscosity models almost always consider only the second invariant $I_{\mathbf{d}2} = (\mathrm{tr}\mathbf{d}^2)/2$, which is then substituted by the *scalar shear rate*

$$\dot{\gamma} = \sqrt{2\mathrm{tr}\mathbf{d}^2} \;\; [\mathrm{s}^{-1}] \,. \tag{6.5}$$

It leads to a non-constant viscosity model of the form $\eta(\dot{\gamma})$, called a *generalized Newtonian model*. A number of such models will be discussed in the following sections, and the description of blood viscosity is generally more important in smaller vessel, where the viscous forces dominate over the inertia forces in the momentum equation (3.107).

6.4.3 Single-Phase Viscosity Models for Blood

6.4.3.1 Power Law Model
The *Power Law model* expresses the fluid viscosity through

$$\eta(\dot{\gamma}) = \eta_0(\lambda\dot{\gamma})^{n-1}, \tag{6.6}$$

where η_0 [Pa s] denotes the viscosity at $\dot{\gamma} = 1/\lambda$, and n is the power law constant. Blood is a shear-thinning fluid and captured by $n < 1$. Given a shear-thickening fluid $n > 1$, whilst $n = 1$ describes a Newtonian fluid. The time constant λ [s] may be regarded as a model parameter and identified together with the other parameters from experimental data. However, it is commonly set at $\lambda = 1.0\,\text{s}$ and therefore only used to correct for dimensions in (6.6). The Power Law model approaches the non-physical viscosity $\eta = 0$ at the limit $\dot{\gamma} \to \infty$.

6.4.3.2 Carreau–Yasuda Model

The *Carreau–Yasuda model* [429] expresses the fluid viscosity through

$$\eta(\dot{\gamma}) = \eta_\infty + (\eta_0 - \eta_\infty)(1 + \lambda \dot{\gamma}^\alpha)^{(n-1)/\alpha}, \tag{6.7}$$

where λ [s] is a time constant, η_0 and η_∞ [Pa s] are fluid viscosity levels at low and high shear rates, respectively, whilst n denotes the power law constant. The parameter α determines the transition from the low to high viscosity levels. The low viscosity limit η_0 has only theoretical meaning, and it is unclear whether or not such limit would exist. Given the parameter $\alpha = 2$, the model is often referred to as *Carreau model*.

6.4.3.3 Casson Model

The *Casson model* [73] expresses the fluid viscosity through

$$\eta(\dot{\gamma}) = \eta_\infty \left[1 + \left(\frac{\tau_0}{\eta_\infty \dot{\gamma}} \right)^{1/2} \right]^2, \tag{6.8}$$

where η_∞ [Pa s] is the fluid viscosity level at high shear rates, whilst τ_0 [Pa] denotes the yield stress above which no erythrocytes rouleaux formation appears [95]. Such a limit is difficult to set, and may not even exist, such that τ_0 is more or less a phenomenological model parameter.

Figure 6.9 illustrates a comparison of different blood viscosity models. The model predictions are shown in relation to experimental data of canine blood at 37 °C [83] as well as human erythrocytes in homologous Acid Citrate Dextrose-

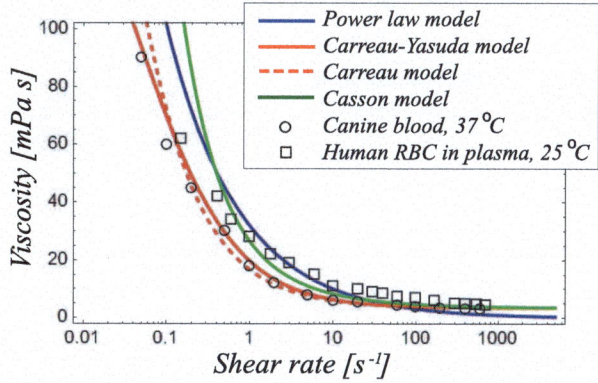

Fig. 6.9 Comparison of blood viscosity models in relation to experimental data of canine blood at 37 °C (○) and human Red Blood Cells (RBC) in homologous ACD-plasma at 25 °C and 48% hematocrit (□). Model predictions use the Power Law model (6.6), the Carreau–Yasuda model (6.7), the Casson model (6.8), and the parameters listed in Table 6.1

Table 6.1 Parameters used in the rheological description of blood

Power law model (6.6)	$\eta_0 = 31.9 \, \text{mPa s}$; $n = 0.5$.
Carreau–Yasuda model (6.7)	$\eta_0 = 200 \, \text{mPa s}$; $\eta_\infty = 3.5 \, \text{mPa s}$; $\lambda = 8.2 \, \text{s}$; $n = 0.2128$; $\alpha = 0.7$.
Carreau model (6.7)	$\eta_0 = 200 \, \text{mPa s}$; $\eta_\infty = 3.5 \, \text{mPa s}$; $\lambda = 1800 \, \text{s}$; $n = 0.28$; $\alpha = 2.0$.
Casson model (6.8)	$\eta_\infty = 3.5 \, \text{mPa s}$; $\tau_0 = 10.86 \, \text{mPa}$.
Krieger-based model (6.11)	$\eta_p = 1.2 \, \text{mPa s}$; $\phi_{\max} = 0.98$; $\lambda = 95 \, \text{s}$; $b = 6.1$; $c = 2.3$; $\beta = 8.23$; $\nu = 1.34$.

plasma (ACD-plasma) at 25 °C and 48% hematocrit. Model predictions are based on the parameters listed in Table 6.1.

6.4.4 Composition-Based Viscosity Models for Blood

A truly versatile model for blood viscosity includes the functional dependence of the interacting factors, such as shear rate, hematocrit, temperature, and plasma proteins. Whilst such descriptions often start with rational concepts and physical phenomena, empirical relations are then introduced to match experimental blood properties.

6.4.4.1 Walburn–Schneck Model
The Power Law viscosity model (6.6) may be generalized towards parameters that depend on the hematocrit level $0 < \phi < 1$, the volume density of erythrocytes. This generalization is then known as the *Walburn–Schneck model* [588] and uses the empirical relations

$$\eta_0 = c_1 \exp(c_2 \phi) \; ; \; n = 1 - c_3 \phi \, , \tag{6.9}$$

where c_1 [Pa s] and c_2, c_3 are model parameters to be identified from experimental data.

6.4.4.2 Quemada Model
The *Quemada model* [440] considers blood to be a *concentrated dispersion* of erythrocytes in plasma. It uses a semi-phenomenological approach that results in the viscosity

$$\eta = \eta_p \left[1 - \frac{\left(k_0 + k_\infty \sqrt{\dot{\gamma}/\dot{\gamma}_c} \right)}{2 \left(1 + \sqrt{\dot{\gamma}/\dot{\gamma}_c} \right)} \phi \right]^{-2} , \tag{6.10}$$

where η_p [Pa s] is the viscosity of blood plasma, whilst $0 < \phi < 1$ denotes the hematocrit level. The model parameters

$$k_0 = \exp\left(\sum_{i=0}^{3} a_i \phi^i\right) \; ; \; k_\infty = \exp\left(\sum_{i=0}^{3} b_i \phi^i\right) \; ; \; \dot{\gamma}_c = \exp\left(\sum_{i=0}^{3} c_i \phi^i\right)$$

depend on the hematocrit level ϕ, where a_i, b_i, and c_i $[\ln(s^{-1})]$ are constants. The Quemada model shows discontinuities when covering a large range of hematocrit levels ϕ [276], limit points that, however, do not concern hematocrit levels of most applications.

6.4.4.3 Krieger-Based Model

The *Krieger model* [312] is based on Eyring's[6] theory of rate processes applied to the suspension of solid spheres. In the description of blood, the framework has been empirically modified to cope with the deformation of erythrocytes [276, 416]. It yields the viscosity

$$\eta = \eta_p \left(1 - \frac{\phi}{\phi_{\max}}\right)^{-n} , \tag{6.11}$$

where $\phi_{\max}$ is the maximum possible hematocrit, and thus the hematocrit at which the suspension ceases to flow. Blood with normal erythrocytes flows up to hematocrit levels of 98% [624], and probably above. The maximum possible hematocrit is therefore set to $\phi_c = 0.98$ [276]—for pathologically stiffened erythrocytes, ϕ_c would be significantly lower.

Whilst for rigid spherical particles, the Krieger exponent $n = 1.66$, erythrocyte deformability alters this theoretical value, and n is no longer a constant but depends on the shear rate. The empirical relation

$$n = a + b\exp(-c\phi) + \begin{cases} 0 & ; \; \phi \leq 0.2 \\ n_{st} & ; \; \phi > 0.2 \end{cases} \tag{6.12}$$

has therefore been proposed [276], where a, b, and c are model parameters. Shear thinning of blood is experimentally only seen above a certain hematocrit level, and the contribution n_{st} is only added to the Krieger exponent for hematocrit levels $\phi > 0.2$. The shear-thinning contribution itself is determined by

$$n_{st} = \beta\left[1 + (\lambda\dot{\gamma})^2\right]^{-\nu} , \tag{6.13}$$

an empirical relation that aims at capturing effects from erythrocyte aggregation at low shear rates together with their elongation at high shear rates. Here, λ [s] and β are constants, whilst $\dot{\gamma}$ denotes the shear rate.

---

[6]Henry Eyring, Mexican-born American theoretical chemist, 1901–1981.

Fig. 6.10 Viscosity of a
suspension of human Red
Blood Cells (RBC) in
homologous ACD-plasma at
25 °C for 35.9% (□), 48%
(○), and 58.9% (△)
hematocrit levels. Model
predictions (solid lines) using
the Krieger model (6.11)
based on the parameters listed
in Table 6.1

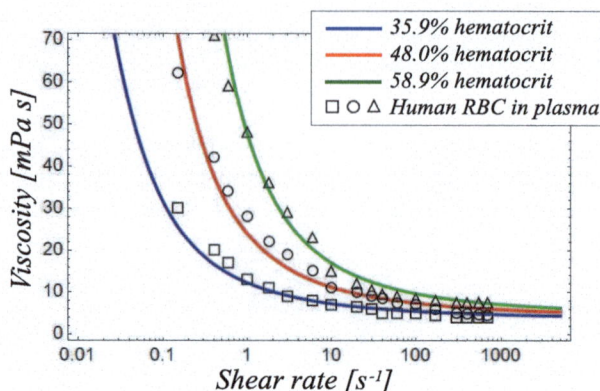

Figure 6.10 illustrates the model's ability to capture the experimentally measured viscosity of human erythrocyte suspensions in homologous ACD-plasma at
25 °C [60]. The applied model parameters are listed in Table 6.1. The Krieger-
based viscosity model can be further generalized to account for effects, such as
temperature and plasma protein concentrations [276].

Example 6.1 (Walburn–Schneck Viscosity Model). Table 6.2 lists the viscosity η of
erythrocytes (red blood cells) in Acid Citrate Dextrose-plasma (ACD-plasma), data
that has been measured at three different hematocrit levels ϕ_i, $i = 1, 2, 3$. It shows
typical shear thinning at increasing strain rate $\dot{\gamma}$. The measured viscosity should be
captured by the Walburn–Schneck viscosity model

$$\eta_{WS}(\dot{\gamma}, \phi) = \eta_0 (\lambda \dot{\gamma})^{n-1}, \tag{6.14}$$

where the expression (6.9) defines the parameter η_0.

Table 6.2 Viscosity η in
relation to the shear rate $\dot{\gamma}$ of
a suspension of human
erythrocytes in homologous
Acid Citrate Dextrose-plasma
(ACD-plasma) at 25 °C [60].
The hematocrit is denoted by
the three levels $\phi_1 = 0.359$,
$\phi_2 = 0.48$, and $\phi_3 = 0.589$,
respectively

$\dot{\gamma}$ [s⁻¹]	η [mPa s]			$\dot{\gamma}$ [s⁻¹]	η [mPa s]		
	ϕ_1	ϕ_2	ϕ_3		ϕ_1	ϕ_2	ϕ_3
0.15	30	62	–	30	6	9	10.5
0.41	20	42	71	40	5	8.5	9.5
0.6	17	34	59	60	5	7.5	9
1	13	28	48	100	4	7	8.5
1.8	11	22	36	170	4	6	8
3	9	19	29	300	4	5	7.5
5	8	15	23	400	4	4.9	7.5
10	7	11	15	550	4	4.7	7.5
20	6.5	10	12	700	4	4.5	7.5

(a) Use least-square optimization to identify the three Walburn–Schneck model
    parameters c_1, c_2, and c_3 from the experimental data listed in Table 6.2.

(b) Plot the predicted viscosity η_{WS} as a function of the strain rate $\dot{\gamma}$ and on top of the experimentally measured viscosities. Use a logarithmic scale of the shear rate $\dot{\gamma}$.   ∎

The Intended Model Application (IMA) of the individual simulation guides the selection of the blood's constitutive description. Different models will generally show different results, and the investigator has to decide upon the flow features of interest for a specific study. Figure 6.12 illustrates such differences in the computed blood flow in the aneurysmatic aorta using the Newtonian and the Carreau–Yasuda blood viscosity models, with otherwise identical modeling assumptions. The predicted WSS is almost identical between both models, whilst VS as well as the scalar shear rate $\dot{\gamma}$ differ remarkably in the core flow. As with the VS, also the secondary flow is strongly influenced by the blood's shear-thinning properties [82]. Given complex flows, blood flow indicators that are commonly used to explore the

Fig. 6.12 Comparison between Computational Fluid Dynamics (CFD) predictions based on the Newtonian (left column) and the Carreau–Yasuda (right column) blood viscosity models. The simulation represents blood flow in the aneurysmatic aorta at late systole in the cardiac cycle. (**a**) Wall Shear Stress (WSS). (**b**) Vortical Structure (VS). (**c**) Scalar shear rate $\dot{\gamma}$ at a single transversal plane

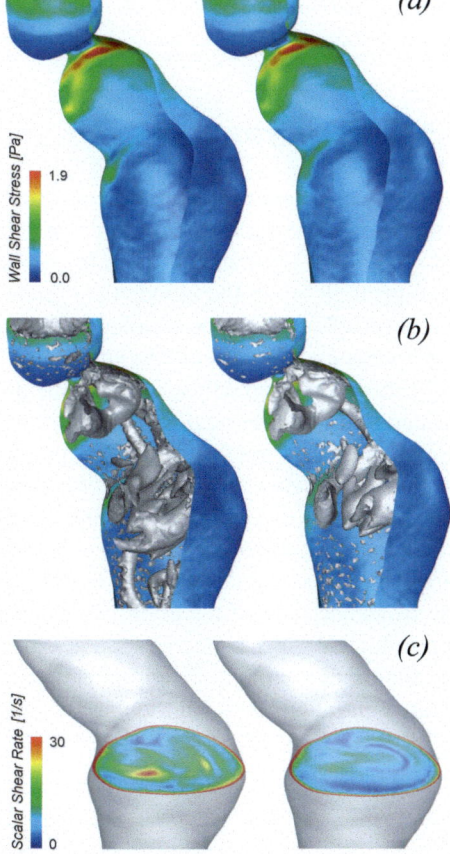

vasculature may change by 50% to 200% between Newtonian and non-Newtonian fluids, respectively [564, 565].

6.5   Blood Damage

Medical devices, such as ventricular assisting devices and artificial heart valves, introduce non-physiological aspects to the circulation and may result in *shear stress-induced* blood damage. Its minimization is one of the *most important* objectives in the development of medical devices. The minimization of blood damage allows to reduce the dose of anti-coagulants in patients with implanted medical devices, which then also greatly reduces the side effects of such drug therapy.

High mechanical shear stress can result in a number of blood-damage mechanisms, out of which *hemolysis* and (non-physiological) *thrombocyte activation* (also regarded as thrombocyte damage) are the most widely studied phenomena. Whilst blood damage is a major cause of failure of medical devices, the underlying damage mechanisms are poorly understood and current biomechanical models of blood damage are characterized by a very high degree of phenomenology.

6.5.1   Hemolysis

Mechanical shear stress is able to *rupture* or *damage* the membrane of erythrocytes, a blood-damage mechanism called *hemolysis*. It results in the release of *hemoglobin* into the blood plasma, and the ratio between the plasma-free hemoglobin and the total hemoglobin in the blood, a parameter called *Hemolysis Index (HI)*, quantifies the severity of this type of blood damage. In the normal (non-damaged) blood approximately 2% of hemoglobin appears freely in plasma, and higher levels would indicate erythrocyte damage. Experiments with erythrocytes indicate that damage-related release of hemoglobin appears only above a certain shear stress level—the literature reports thresholds in the range from $\overline{\tau} = 100$ to $250\,\mathrm{Pa}$ [320]. In addition to the *stress level*, the *exposure time* to stress influences hemolysis. Given the viscoelasticity of erythrocytes, time is needed to turn the stress level into cell membrane deformation, which then eventually results in hemolysis. Whilst the shear stress increases with increasing flow rate of the blood through a medical device, the residence time of the erythrocytes, and thus the time they are exposed to non-physiological aspects, decreases. The particular design of the medical device is then an important factor in HI minimization, see for example the analysis of different rotary continuous flow ventricular assisting devices [174].

Table 6.3 Parameters of the Power Law model (6.15) towards the description of hemolysis as a function of the shear stress τ [Pa] and the exposure time t [s]

C [Pa$^{-\alpha}$ s$^{-\beta}$]	α	β	Blood source	Reference
$3.63 \cdot 10^{-7}$	2.416	0.785	Human	[211, 603]
$1.8 \cdot 10^{-8}$	1.991	0.765	Porcine	[251]
$1.228 \cdot 10^{-7}$	1.9918	0.6606	Ovine	[125]
$6.701 \cdot 10^{-6}$	1.0981	0.2778	Porcine	[125]
$3.458 \cdot 10^{-8}$	2.0639	0.2777	Human	[125]
$9.772 \cdot 10^{-7}$	1.4445	0.2076	Bovine	[125]

Let us consider the damage-related release of hemoglobin, where ΔH_b denotes the amount of hemoglobin that is released into the bloodstream, whilst H_b is the total hemoglobin. We may then express the ratio between both values by the Power Law model

$$H(\tau, t) = \frac{\Delta H_b}{H_b} = \begin{cases} C\tau^\alpha t^\beta & \text{for } \tau \geq \overline{\tau}, \\ 0 & \text{for } \tau < \overline{\tau}, \end{cases} \qquad (6.15)$$

where C [Pa$^{-\alpha}$ s$^{-\beta}$] and α, β are constants to be identified from experimental data. A number of studies identified these parameters from shear flow experiments, and Table 6.3 lists parameters that have been reported in the literature. The generalization of the hemolysis model (6.15) towards multi-dimensional flows requires the introduction of a *scalar shear stress* τ, a scalar that represents the shear stress tensor $\boldsymbol{\tau}$. It should adequately capture the loading of erythrocytes and could be any function of the invariants of $\boldsymbol{\tau}$. As compared to the model parameters C, α, β, the definition of τ has a minor influence on the prediction of H [610]. Given this access, the von Mises stress

$$\tau = \sqrt{3(\tau_{12}^2 + \tau_{23}^2 + \tau_{13}^2)} = \sqrt{-3I_{\tau 2}}$$

has often been used in the study of hemolysis, where $I_{\tau 2} = [(\text{tr}\boldsymbol{\tau})^2 - \text{tr}\boldsymbol{\tau}^2]/2$ denotes the second invariant of the shear stress tensor. The derivation used the property $\text{tr}\boldsymbol{\tau} = 0$ that follows directly from the structure of the fluid stress tensor $\boldsymbol{\sigma} = \boldsymbol{\tau} - p\mathbf{I}$, where p denotes the hydrostatic pressure.

In addition to the Power Law model (6.15), many other approaches have been proposed to predict blood damage. A review of models together with a cross-comparison of their predictions has been reported elsewhere [610], and recently hemolysis has also been related to the energy dissipation rate [602]. The damage of blood components other than erythrocytes may also be modeled similarly to hemolysis. The same models, but different model constants, are then used.

The Power Law model (6.15) may be implemented in a Computational Fluid Dynamics (CFD) analysis using the *Eulerian* approach or the *Lagrangian* approach. Given the Eulerian approach, the model is rewritten in a standard transport equation, whilst in the Lagrangian approach it is integrated along pathlines.

6.5.1.1 Eulerian Implementation of the Hemolysis Power Law Model

The transformation $H_L = H^{1/\beta}$ maps the Power Law model (6.15) into $H_L = C^{1/\beta}\tau^{\alpha/\beta}t$, an equation that is linear in the time t. Given the material time derivative $DH_L/Dt = C^{1/\beta}\tau^{\alpha/\beta}$, (6.15) then leads to the *transport equation*

$$\frac{\partial H_L}{\partial t} + \mathbf{v} \cdot \operatorname{grad} H_L = C^{1/\beta}\tau^{\alpha/\beta}(1 - H_L), \tag{6.16}$$

where the term $(1 - H_L)$ serves as regularization term and prevents from unphysical solutions, such as $H_L > 1$, see [610]. The transport equation (6.16) is much more practical than the original model (6.15) and can be directly implemented in a CFD analysis.

6.5.1.2 Lagrangian Implementation of the Hemolysis Power Law Model

We may integrate the Power Law model (6.15) along pathlines, and

$$H(\tau, t) = H_0 + \int_0^t C\tau^\alpha t^\beta \mathrm{d}t \tag{6.17}$$

then represents the damage of erythrocytes that accumulates over the time t, where $H_0 = H(\tau, 0)$ denotes the initial erythrocyte damage. The simple approximation $H(\tau, t) \approx H_0 + \sum_{i=1}^n C\tau_i^\alpha t_i^\beta \Delta t_i$ of (6.17), where n denotes a number of discrete points in time, delivers a poor hemolysis prediction, and more advanced methods have therefore been proposed [221].

A pathline depends on its origin in space and time, and the accurate representation of the blood's damage potential relies on the adequate coverage of pathlines throughout the entire fluid domain. Whilst this is typically not difficult to attain for simple flow topologies without noticeable recirculation regions, it is challenging for more complex geometries with large stable VS. Pathlines seeded at the vessel inlet may not cover the entire flow domain, a factor that impacts hemolysis predictions negatively. The random seeding of pathlines all over the flow domain improves the situation, but it is seldom sufficient to fully resolve said issue—areas of high potential damage, such as recirculation zones and boundary layers, could again not have been passed by any pathline.

6.5.2 Thrombocyte Activation

In addition to *chemical agonists*, such as Adenosine DiphosPhate (ADP), thrombin, thromboxane A2, and serotonin, *mechanical shear stress* is also able to activate

Fig. 6.13 Thresholds of shear stress τ and exposure time t to activate thrombocytes (platelets). Data points are taken from experimental studies reported elsewhere [96, 249, 452, 487, 603, 615]. Least-square optimization leads to the regression line (6.18) that is shown in blue

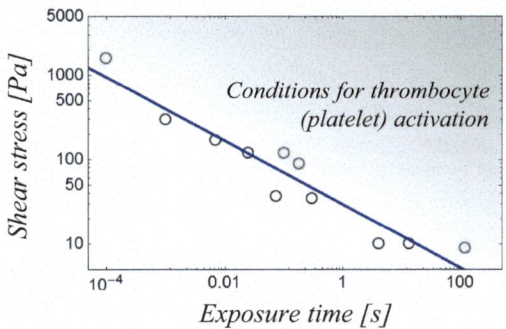

thrombocytes (platelets). Whilst this is a needed first step in wound healing, the activation of a large number of thrombocytes of blood that passes through medical devices, increases the risk for *device failure*. Clotted blood can occlude the device, or blood clots released in the bloodstream may occlude distal vessels and eventually lead to *acute thrombo-embolic events*. The activation of thrombocytes is influenced by the exposure time and the level of shear stress. Given a number of experimental observations [96, 249, 452, 487, 603, 615], the empirical expression

$$\tau = 29.598\, t^{-0.377} \tag{6.18}$$

determines conditions of shear stresses τ [Pa] and exposure times t [s] that results in the activation of thrombocytes, see Fig. 6.13.

6.6   Description of Incompressible Flows

6.6.1   Energy Conservation

For some applications, blood may be regarded as an *ideal fluid* without viscosity. Any dissipation of the blood flow is then neglected, and the flow may be studied entirely by the principle of *energy conservation*. At any point in a flow system, the sum of the specific potential energy $gh + p/\rho$ [J kg^{-1}] and the specific kinetic energy $v^2/2$ [J kg^{-1}] remains then constant, where v and p denote the velocity and pressure, whilst ρ is the density of the fluid. In addition, $g = 9.81$ m s^{-2} denotes the gravitation, and h is the height coordinate relative to a reference point of the flow system. The energy conservation then reads

$$\frac{v^2}{2} + gh + \frac{p}{\rho} = C, \tag{6.19}$$

Fig. 6.14 Blood flow
through a vessel segment,
where the flow path falls by
Δh from the inlet A to the
outlet B. The gravitation is
denoted by g

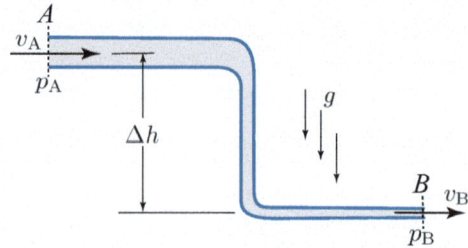

an expression known as *Bernoulli*[7] *equation* and used for a first estimation of the
average blood pressures that develops in the vascular system. In addition, the blood
flow velocities follow directly from the continuity of the blood volume.

6.6.1.1 Blood Flow Under the Action of Gravitation

The blood flow in the body is influenced by gravitation, and Fig. 6.14 shows an
arterial segment that allows us to investigate said mechanism. The vessel has circular
cross-sections, where $d_A = 1.9$ cm and $d_B = 1.5$ cm denote the diameter at the
inlet A and the outlet B, respectively. At the inlet the blood has the velocity $v_A =
12.4 \, \mathrm{cm \, s^{-1}}$ and the pressure $p_A = 98 \, \mathrm{mm \, Hg}$, it then drops by the height $\Delta h =
38.6$ cm and leaves the system through B. The blood then has the velocity v_B and
the pressure p_B, state variables to be determined hereinafter.

The continuity of a steady-state incompressible flow requires the inflow rate
$q_A = v_A d_A^2 \pi/4$ to match the outflow rate $q_B = v_B d_B^2 \pi/4$, a relation that allows
us to compute the outflow velocity

$$v_B = v_A (d_A/d_B)^2 = 19.9 \ \mathrm{cm \ s^{-1}} \, .$$

The outlet velocity v_B is therefore uniquely determined by the vessel's diameters
and independent from the gravitation as well as the drop in height Δh.

Given this example, blood may be regarded as an ideal (in-viscous) fluid with the
density $\rho = 1060 \, \mathrm{kg \, m^{-3}}$. Bernoulli's equation (6.19)

$$v_A^2/2 + g h_A + p_A/\rho = v_B^2/2 + g h_B + p_B/\rho$$

then allows us to express the outflow pressure

$$p_B = p_A + \rho(v_A^2 - v_B^2)/2 + \rho g \Delta h = 17.067 \ \mathrm{kPa} = 128.0 \ \mathrm{mm \ Hg} \, , \qquad (6.20)$$

where $\rho(v_A^2 - v_B^2)/2$ and $\rho g \Delta h$ denote *kinetic energy* and respective *potential energy*
contributions to the pressure. The potential energy contribution may be seen as
the gravitation's effect on blood flow. It often dominates over the kinetic energy

---

[7]Daniel Bernoulli, Swiss mathematician and physicist, 1700–1782.

contribution, and the vasculature actively reacts to gravitational effects through nervous system-controlled blood pressure compensation, see Sect. 2.1.7.2.

Given the special case $\Delta h = 0$, the gravitational effect in (6.20) vanishes and Bernoulli's equation (6.20) then results in the pressure difference

$$\Delta p = p_B - p_A = \rho(v_A^2 - v_B^2)/2 = -12.8 \text{ Pa} = -0.096 \text{ mm Hg}.$$

The *increase* in the flow velocity of 7.5 cm s^{-1} from the larger vessel diameter d_A towards the smaller vessel diameter d_B is then compensated by the *decrease* in pressure of 0.096 mm Hg. This is a counterintuitive result, and one could mistakenly think that blood at higher velocity would also have a higher pressure.

Example 6.2 (Collapse of a Constricted Vessel Segment). Figure 6.15 shows the blood flow through a constricted circular vessel, where d_i and d_c denote the diameters at the inlet and constriction, respectively. The blood has the density $\rho = 1060 \text{ kg m}^{-3}$ and may be regarded as an ideal (in-viscous) fluid with the velocity $v_i = 23.2 \text{ cm s}^{-1}$ and the pressure $p_i = 75 \text{ mm Hg}$ at the vessel's inlet. The vessel wall may be regarded as a membrane that has no bending stiffness, and $p_0 = 12.0 \text{ mm Hg}$ determines the ambient pressure outside the vessel.

Fig. 6.15 Blood flow through a vessel constriction

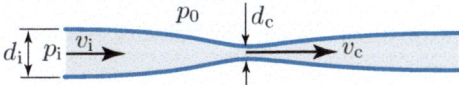

(a) Consider incompressible blood and express the flow velocity v_c in the constriction.
(b) Use Bernoulli's equation (6.19) and express the pressure p_i in the constriction.
(c) Specify the relative diameter stenosis $S_{\text{diameter}} = (d_i - d_c)/d_i$ as well as the relative area stenosis $S_{\text{area}} = (A_i - A_c)/A_i$, at which the vessel would collapse. Here, A_i and A_c denote the cross-sectional areas at the inlet and constriction, respectively.   ■

6.6.2 Linear Momentum Conservation

A 1D flow is a flow, which velocity $\mathbf{v}$ depends only on the time t and *one* spatial dimension. As with any other flow, it has to satisfy the linear momentum (3.107) $\rho(D\mathbf{v}/Dt) = \text{div}\boldsymbol{\sigma} + \mathbf{b}_f$, where $\boldsymbol{\sigma}$ and $\mathbf{b}_f$ denote the Cauchy stress and the body force per unit spatial volume, respectively. The material time derivative $D\mathbf{v}/Dt = \partial\mathbf{v}/\partial t + \mathbf{v} \cdot \text{grad}\mathbf{v}$ specifies the acceleration that is felt by the fluid particle, and ρ denotes fluid density per (spatial) unit volume. The continuity (3.104) of an incompressible fluid $\text{div}\mathbf{v} = 0$ is also part of the blood flow's mathematical description. The continuity and linear moment can then be combined into a *single*

Fig. 6.16 1D flows in (**a**) Cartesian and (**b**) cylindrical coordinates. The vectors denote the flow velocity $\mathbf{v}(x_3, t)$ and $\mathbf{v}(r, t)$ in Cartesian and cylindrical coordinates, respectively

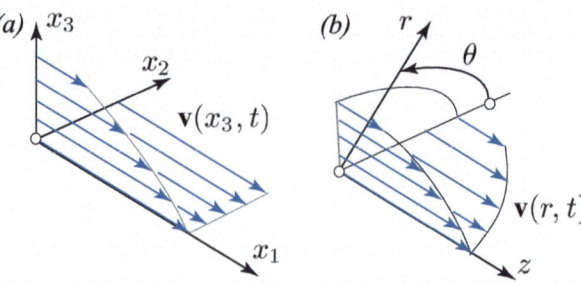

governing equation for incompressible 1D flows, a relation that looks differently in different coordinate systems.

6.6.2.1 Governing Equation for 1D Flows in Cartesian Coordinates

Without loss of generality, we may consider a fluid that flows along the x_1 coordinate direction, as it is illustrated in Fig. 6.16a. The velocity $\mathbf{v} = \mathbf{v}(x_3, t)$ then has the individual components $v_1(x_3, t)$ and $v_2 = v_3 = 0$, and the flow continuity

$$\mathrm{div}\,\mathbf{v} = \nabla \cdot \mathbf{v} = \frac{\partial v_1}{\partial x_1} + \frac{\partial v_2}{\partial x_2} + \frac{\partial v_3}{\partial x_3} = 0$$

is *a priori* satisfied. The velocity gradient of the 1D flow reads

$$\mathrm{grad}\,\mathbf{v} = \begin{bmatrix} \partial v_1/\partial x_1 & \partial v_1/\partial x_2 & \partial v_1/\partial x_3 \\ \partial v_2/\partial x_1 & \partial v_2/\partial x_2 & \partial v_2/\partial x_3 \\ \partial v_3/\partial x_1 & \partial v_3/\partial x_2 & \partial v_3/\partial x_3 \end{bmatrix} = \begin{bmatrix} 0 & 0 & \partial v_1/\partial x_3 \\ 0 & 0 & 0 \\ 0 & 0 & 0 \end{bmatrix},$$

and leads to the rate of deformation

$$\mathbf{d} = \frac{1}{2}\left(\mathrm{grad}\,\mathbf{v} + \mathrm{grad}^{\mathrm{T}}\mathbf{v}\right) = \frac{1}{2}\begin{bmatrix} 0 & 0 & \partial v_1/\partial x_3 \\ 0 & 0 & 0 \\ \partial v_1/\partial x_3 & 0 & 0 \end{bmatrix}.$$

The only non-vanishing shear stress component that acts at the fluid particle reads $\tau_{13} = 2\eta d_{13} = \eta(\partial v_1/\partial x_3)$, where η [Pa s] denotes the dynamic viscosity. Whilst η is constant for a Newtonian fluid, it depends on the shear rate for a general Newtonian fluid, see Sect. 6.4.2.

Given the decoupled stress representation $\boldsymbol{\sigma} = \boldsymbol{\tau} - p\mathbf{I}$ with the shear stress $\boldsymbol{\tau} = 2\eta\mathbf{d}$ and the pressure p, the linear momentum $\rho(\mathrm{D}\mathbf{v}/\mathrm{D}t) = \mathrm{div}\,\boldsymbol{\sigma} + \mathbf{b}_{\mathrm{f}}$ of the 1D incompressible flow reads

$$\rho\left(\begin{bmatrix} \partial v_1/\partial t \\ 0 \\ 0 \end{bmatrix} + \begin{bmatrix} 0 & 0 & \partial v_1/\partial x_3 \\ 0 & 0 & 0 \\ 0 & 0 & 0 \end{bmatrix}\begin{bmatrix} v_1 \\ 0 \\ 0 \end{bmatrix}\right) = \begin{bmatrix} \partial \tau_{13}/\partial x_3 - \partial p/\partial x_1 \\ -\partial p/\partial x_2 \\ -\partial p/\partial x_3 \end{bmatrix} + \begin{bmatrix} b_{f1} \\ b_{f2} \\ b_{f3} \end{bmatrix},$$

where the material time derivative $Dv/Dt = \partial v/\partial t + v \cdot \mathrm{grad} v$ and the stress divergence (A.18) in Cartesian coordinates have been used. Note that the *advective acceleration* vanishes, and $v \cdot \mathrm{grad} v = 0$ holds for the Cartesian 1D flow.

The shear stress may be expressed by $\tau_{13} = \eta(\partial v_1/\partial x_3)$, and the 1D fluid flow problem is then governed by the set

$$\rho\frac{\partial v_1}{\partial t} = \frac{\partial}{\partial x_3}\left(\eta\frac{\partial v_1}{\partial x_3}\right) - \frac{\partial p}{\partial x_1} + b_{f1}; \quad \frac{\partial p}{\partial x_2} = b_{f2}; \quad \frac{\partial p}{\partial x_3} = b_{f3} \qquad (6.22)$$

of equations, where $p = p(x_1, x_2, x_3)$ and $\mathbf{b}_f(x_1, x_2, x_3)$ express the fluid pressure and body force as a function of the Cartesian coordinates.

6.6.2.2 Fluid Flow Down the Inclined Plane

Whilst Cartesian 1D flow has limited applications in the vasculature, it is instrumental for the mechanical understanding of flow problems. One such example is shown in Fig. 6.17. A Newtonian fluid of viscosity $\eta = 37.0\,\mathrm{mPa\,s}$ and the density $\rho = 1.12\,\mathrm{kg\,dm^{-3}}$ flows down a plane that is inclined by the angle $\alpha = \pi/6$. The fluid is surrounded by air of negligible density, the flow is laminar, fully developed, and entrance effects may be neglected.

Let us start with the specification of the governing equations of this 1D steady-state problem. Given the angle $\beta = \pi/2 - \alpha$, the body force $\mathbf{b}_f = \rho g[\cos\beta - \sin\beta \ 0]^T$ acts, and the governing equations (6.22) then read

$$\frac{\partial \tau}{\partial y} - \frac{\partial p}{\partial x} + \rho g \cos\beta = 0\,; \quad -\frac{\partial p}{\partial y} - \rho g \sin\beta = 0\,, \qquad (6.23)$$

Fig. 6.17 Fluid flowing down a plane that is tilted by the angle α. The fluid layer has the thickness H, and the gravitation is denoted by g

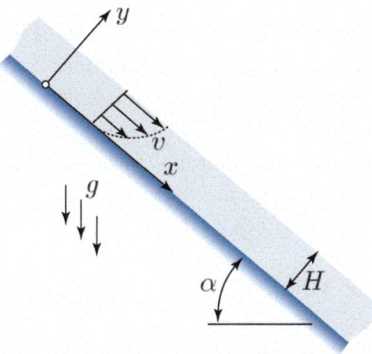

where $g = 9.81 \, \mathrm{m \, s^{-2}}$ denotes the gravitation. These equations determine the velocity profile v that establishes within the fluid layer.

The integration of $(6.23)_2$ yields the pressure

$$p = \rho g \sin \beta (H - y),$$

where the boundary condition $p(H) = 0$ has been used. Given the flow is fully developed, $\partial p / \partial x = 0$, and the governing equation $(6.23)_1$ then reads

$$\eta \frac{d^2 v}{dy^2} + \rho g \cos \beta = 0,$$

where a Newtonian fluid $\tau = \eta (dv/dy)$ has been used. The one-dimensionality of the problem allowed us to substitute the partial derivative with the ordinary derivative. Twice integration of this relation yields

$$v = -\frac{\rho g \cos \beta}{2\eta} y^2 + C_1 y + C_2,$$

where the integration constants $C_1 = \rho g \cos \beta H / \eta$ and $C_2 = 0$ can been identified from the *no-slip* condition at the wall $v(0) = 0$, as well as the *no-shear* condition at the free surface $\tau(H) = 0$. Given these conditions, the velocity of the fluid layer then reads

$$v = \frac{\rho g \cos \beta}{2\eta} \left(2yH - y^2 \right). \tag{6.24}$$

It is emphasized that the no-shear condition at the free surface $\tau(H) = 0$ implies $dv/dy = 0$ at $y = H$, and the velocity gradient vanishes at the free surface.

Let us consider 2.3 l of fluid flowing down per second and meter width of the inclined plane, information that allows us to identify the height H of the fluid layer. The integration of the velocity (6.24) over the layer thickness then yields the flow rate $q = \int_{y=0}^{H} v \, dy = \rho g \cos \beta H^3 / (3\eta) = 49491.9 H^3 \, \mathrm{m^2 \, s^{-1}}$ and results in the height $H = 3.595 \, \mathrm{mm}$ of the fluid layer. Given the height of the fluid layer, we are able to compute the shear stress $\tau(0) = \rho g \cos \beta H = 19.751 \, \mathrm{Pa}$ that acts at the wall, where the Newtonian fluid $\tau = \eta(dv/dy)$ and the velocity profile (6.24) have been used. This result follows also directly from the static equilibrium of the fluid layer. The gravitational force $G = \rho g H$ acts at the fluid layer per square meter of the inclined plane, and the tangential component of this force then determines the WSS $\tau = G \cos \beta$.

6.6.2.3 Fluid Layer at Oscillating Pressure Load
The contraction of the heart muscle exposes the arterial system to pulsatile pressure and flow. It leads to the dynamic exchange between the potential and kinetic energies of the blood and strongly influences the flow in large conduit arteries. The

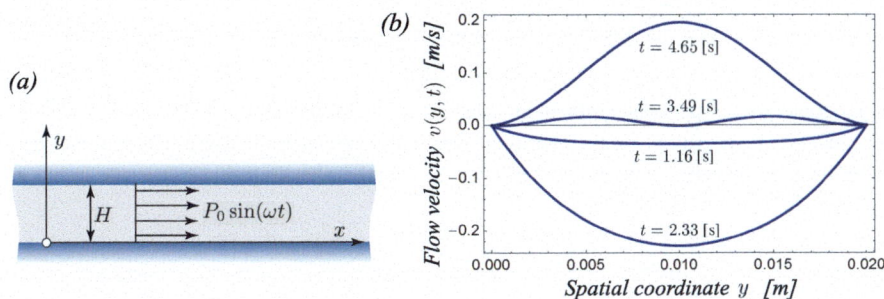

Fig. 6.18 Fluid layer exposed to oscillating pressure. (**a**) Problem definition. (**b**) Fluid velocity profiles $v(y, t)$ at a number of time points t

interaction of pressure and flow may already been studied by the 1D Cartesian flow problem shown in Fig. 6.18a. A fluid channel of the width $H = 2.0 \, \text{cm}$ is filled with a Newtonian fluid of the density $\rho = 1025.0 \, \text{kg m}^{-3}$ and the dynamic viscosity $\eta = 4.0 \, \text{mPa s}$. Over the time t, the fluid layer is exposed to the pressure $P_0 \sin(\omega t)$, where P_0 and ω denote the pressure amplitude and the angular velocity, respectively.

At steady-periodic conditions, the fluid velocity may be expressed by $v(y, t) = V(y) \sin(\omega t)$, where $V(y)$ denotes the amplitude of the velocity as a function of the channel depth coordinate y. Given (6.22) and the pressure $P_0 \sin(\omega t)$, the governing equation of this transient 1D flow problem then reads

$$\frac{\partial^2 V(y)}{\partial y^2} - \frac{\rho \omega V(y)}{\eta \tan(\omega t)} = P_0 \, ,$$

and its integration yields the velocity amplitude

$$V(y) = C_1 \exp(\alpha y) + C_2 \exp(-\alpha y) - \frac{P_0}{\alpha^2} \quad \text{with} \quad \alpha = \sqrt{\frac{\rho \omega}{\eta} \cot(\omega t)} \, . \qquad (6.25)$$

The two integration constants C_1 and C_2 may be identified from the two *no-slip* boundary conditions $V(0) = V(H) = 0$ and read

$$C_2(t) = P_0/\alpha^2 - C_1(t) \, ; \quad C_1(t) = \frac{P_0}{\alpha^2 [1 + \exp(\alpha H)]} \, .$$

The substitution of C_1 and C_2 in (6.25) then yields the velocity amplitude across the fluid layer

$$V(y) = \frac{[1 - \exp(-\alpha y)][-\exp(\alpha H) + \exp(\alpha y)] P_0}{\alpha^2 [1 + \exp(\alpha H)]} \, ,$$

which may also be expressed by

$$V(y) = \frac{P_0}{\alpha^2} [\cosh(\alpha y) - \sinh(\alpha y) \tanh(\alpha H/2) - 1] \sin(\omega t)$$

and completes the description of the flow velocity $v(y,t) = V(y)\sin(\omega t)$. Fig. 6.18b shows the velocity profile at the time points $t = 1.16; 2.33; 3.49; 4.65$ s. The plot considers an angular velocity $\omega = 0.6\,\text{s}^{-1}$ together with the pressure amplitude $P_0 = 10.0$ kPa.

Example 6.3 (Oscillating Plate on Top of a Fluid Layer). Figure 6.19 shows a rigid plate that moves at the velocity $V_0 \sin(\omega t)$ on top of a fluid layer of the thickness $H = 1.0$ cm, where t denotes the time. The velocity amplitude $V_0 = 1.0\,\text{cm}\,\text{s}^{-1}$ and the angular velocity $\omega = 0.6\,\text{s}^{-1}$ are given. Initial effects have already been fully dissipated, and the fluid layer moves at steady-state periodic conditions. The fluid has the density $\rho = 1025.0\,\text{kg}\,\text{m}^{-3}$ and the constant dynamic viscosity $\eta = 4.0\,\text{mPa}\,\text{s}$.

Fig. 6.19 Rigid plate that oscillates on top of a fluid layer of the thickness H

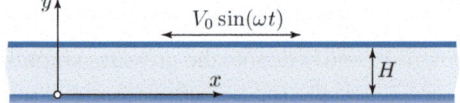

(a) Provide the governing equation for this transient 1D fluid problem. Assume the fluid velocity of the form $v(y,t) = V(y)\sin(\omega t)$ together with a constant pressure in the fluid layer. Integrate the governing equation, identify the integration constants, and express the flow velocity $v(y,t)$.
(b) Plot the fluid velocity profile $v(y,t)$ *versus* the fluid layer depth coordinate y at the time points $t = 1; 2; 3; 4$ s.   ∎

6.6.2.4 Governing Equation for 1D Flows in Cylindrical Coordinates
We consider fluid flow in a cylindrical coordinate system, and without loss of generality, along the z coordinate direction, see Fig. 6.16b. The velocity $\mathbf{v} = \mathbf{v}(r,t)$ then has the individual components $v_z(r,t)$ and $v_r = v_\theta = 0$, and flow continuity in cylindrical coordinates

$$\text{div}\mathbf{v} = \nabla \cdot \mathbf{v} = \frac{1}{r}\frac{\partial(rv_r)}{\partial r} + \frac{1}{r}\frac{\partial v_\theta}{\partial \theta} + \frac{\partial v_z}{\partial z} = 0$$

is *a priori* satisfied. In addition, the velocity gradient in cylindrical coordinates reads

$$\text{grad}\mathbf{v} = \begin{bmatrix} \partial v_r/\partial r & (\partial v_r/\partial \theta)/r - v_\theta/r & \partial v_r/\partial z \\ \partial v_\theta/\partial r & (\partial v_\theta/\partial \theta)/r + v_r/r & \partial v_\theta/\partial z \\ \partial v_z/\partial r & (\partial v_z/\partial \theta)/r & \partial v_z/\partial z \end{bmatrix} = \begin{bmatrix} 0 & 0 & 0 \\ 0 & 0 & 0 \\ \partial v_z/\partial r & 0 & 0 \end{bmatrix}, \quad (6.26)$$

where we considered rotational symmetry, and thus the condition $\partial(\bullet)/\partial\theta = 0$, in the derivation of this relation. The velocity gradient (6.26) yields the rate of deformation

$$\mathbf{d} = \frac{1}{2}\left(\mathrm{grad}\mathbf{v} + \mathrm{grad}^T\mathbf{v}\right) = \frac{1}{2}\begin{bmatrix} 0 & 0 & \partial v_z/\partial r \\ 0 & 0 & 0 \\ \partial v_z/\partial r & 0 & 0 \end{bmatrix}, \tag{6.27}$$

and $\tau_{rz} = 2\eta d_{rz} = \eta\partial v_z/\partial r$ is then the only non-vanishing shear stress component that acts at the fluid particle.

Given the decoupled stress representation $\boldsymbol{\sigma} = \boldsymbol{\tau} - p\mathbf{I}$ with the shear stress $\boldsymbol{\tau} = 2\eta\mathbf{d}$ and the pressure p, the linear momentum $\rho(D\mathbf{v}/Dt) = \mathrm{div}\boldsymbol{\sigma} + \mathbf{b}_f$ of the 1D incompressible flow in cylindrical coordinates reads

$$\rho\left(\begin{bmatrix} 0 \\ 0 \\ \frac{\partial v_z}{\partial t} \end{bmatrix} + \begin{bmatrix} 0 & 0 & 0 \\ 0 & 0 & 0 \\ \frac{\partial v_z}{\partial r} & 0 & 0 \end{bmatrix}\begin{bmatrix} 0 \\ 0 \\ v_z \end{bmatrix}\right) = \begin{bmatrix} \frac{\partial \tau_{rz}}{\partial z} - \frac{\partial p}{\partial r} \\ 0 \\ \frac{\partial \tau_{rz}}{\partial r} + \frac{\tau_{rz}}{r} - \frac{\partial p}{\partial z} \end{bmatrix} + \begin{bmatrix} b_{fr} \\ 0 \\ b_{fz} \end{bmatrix}, \tag{6.28}$$

where the material time derivative $D\mathbf{v}/Dt = \partial\mathbf{v}/\partial t + \mathbf{v}\cdot\mathrm{grad}\mathbf{v}$ has been used, and $\mathbf{b}_f = [b_{fr}\ 0\ b_{fz}]^T$ denotes the rotational symmetric body force vector per unit (spatial) volume. In addition,

$$\mathrm{div}\boldsymbol{\sigma} = \begin{bmatrix} \partial\sigma_{rr}/\partial r + (\sigma_{rr} - \sigma_{\theta\theta})/r + \partial\sigma_{rz}/\partial z \\ 0 \\ \partial\sigma_{rz}/\partial r + \sigma_{rz}/r + \partial\sigma_{zz}/\partial z \end{bmatrix}$$

expresses the stress divergence in cylindrical coordinates (A.19) at rotational symmetry, a condition upon which $\sigma_{r\theta} = \sigma_{\theta z} = 0$ and $\partial(\bullet)/\partial\theta = 0$ hold.

Given a fully developed flow, $\partial\tau_{rz}/\partial z = 0$ holds, and

$$\rho\frac{\partial v_z}{\partial t} = \frac{\partial}{\partial r}\left(\eta\frac{\partial v_z}{\partial r}\right) + \frac{\eta}{r}\frac{\partial v_z}{\partial r} - \frac{\partial p}{\partial z} + b_{fz}; \quad \frac{\partial p}{\partial r} = b_{fr} \tag{6.29}$$

governs the 1D flow problem. In the derivation of this result, the only non-vanishing shear stress component has been expressed by $\tau_{rz} = \eta(\partial v_z/\partial r)$.

With a Newtonian fluid, the viscosity η is constant, and

$$\rho\frac{\partial v_z}{\partial t} = \frac{\eta}{r}\frac{\partial}{\partial r}\left(r\frac{\partial v_z}{\partial r}\right) - \frac{\partial p}{\partial z} + b_{fz}; \quad \frac{\partial p}{\partial r} = b_{fr} \tag{6.30}$$

then governs the incompressible 1D fluid flow problem.

6.6.2.5 Steady-State Flow of a Newtonian Fluid Through a Circular Tube

Regardless the vessel changes its diameter over the cardiac cycle, for many applications the flow through a *rigid* cylindrical tube is an accurate model of blood flow. The flow through a straight circular tube is therefore a classical blood flow model and instrumental to the understanding of the vasculature.

Let us consider a fully developed steady-state flow at a flow rate that is high enough for blood to be modeled as a Newtonian fluid of the viscosity η. The conservation of linear momentum (6.28) then reads

$$
\begin{bmatrix} 0 \\ 0 \\ 0 \end{bmatrix} = \begin{bmatrix} \partial \tau_{rz}/\mathrm{d}z - \partial p/\partial r \\ 0 \\ \partial \tau_{rz}/\partial r + \tau_{rz}/r - \partial p/\partial z \end{bmatrix} ,
$$

where the steady-state condition $\partial(\bullet)/\partial t = 0$ has been used in the derivation of this expression. For a fully developed flow, $\partial \tau_{rz}/\partial z = 0$ holds, and the set

$$
\frac{\partial p}{\partial z} = \frac{1}{r}\frac{\partial(r\tau_{rz})}{\partial r} \; ; \quad \frac{\partial p}{\partial r} = 0 \tag{6.31}
$$

governs the tube flow problem, where $\tau_{rz} = \eta(\partial v_z/\partial r)$ is the only non-vanishing shear stress component that acts at the fluid particle.

Equation $(6.31)_2$ results in a constant pressure over a particular cross-section, $p(r, z) = p(z)$, and

$$
\frac{\mathrm{d}p}{\mathrm{d}z} = \frac{1}{r}\frac{\mathrm{d}(r\tau_{rz})}{\mathrm{d}r} \tag{6.32}
$$

remains the only no-trivial governing equation of the tube flow problem. The pressure and the shear stress are functions of single arguments, $p(z)$ and $\tau_{rz}(r)$, respectively—the partial derivatives have therefore been replaced by the ordinary derivatives.

The integration of (6.32) yields

$$
\frac{\mathrm{d}p}{\mathrm{d}z}\frac{r^2}{2} = r\tau_{rz} + C , \tag{6.33}
$$

where the integration constant $C = 0$ results from the *symmetry condition* $\tau_{rz}(r = 0) = 0$. The pressure gradient $\mathrm{d}p/\mathrm{d}z$ in (6.33) is a constant that "drives" the flow along the z-direction. We may express the shear stress by the Newtonian fluid model $\tau_{rz} = \eta(\mathrm{d}v_z/\mathrm{d}r)$ and use the no-slip boundary condition $v_z(r_0) = 0$ in the integration of (6.33) over r. It results in the *quadratic velocity profile*

$$v_z(r) = v_{\max}\left[1 - (r/r_0)^2\right] \quad \text{with} \quad v_{\max} = -\frac{dp}{dz}\frac{r_0^2}{4\eta} \; , \tag{6.34}$$

of the tube flow problem, where r_0 is the tube radius, whilst $v_{\max}$ denotes the velocity in the center—the highest fluid velocity in the tube. It is emphasized that a negative pressure gradient leads to a velocity in the positive z-direction. The velocity profile (6.34) is known as *Poiseuille flow profile*. Whilst it presents a good description of the blood flow in small arteries and veins, in large arteries said velocity profile only appears during the systolic phase of the cardiac cycle.

The integrating of (6.34) over the cross-section determines the flow rate

$$q = 2\pi \int_{r=0}^{r_0} v_z r \, dr = \frac{\pi r_0^4}{8\eta}\left(-\frac{dp}{dz}\right) \tag{6.35}$$

and provides the relation between the pressure gradient and the flow through the vessel. Given (6.33), the shear stress that is felt by the fluid particle then reads

$$\tau_{rz} = \frac{r}{2}\frac{dp}{dz} = -\frac{4\eta q r}{\pi r_0^4} \; , \tag{6.36}$$

which at the wall $r = r_0$, then results in the WSS of $\tau_{rz} = 4\eta q/(\pi r_0^3)$. The WSS is the stress that is felt by the endothelium, not by the blood particle, and we therefore changed the sign.

Example 6.4 (Steady-State Flow in a Vessel Segment).   Let us consider Poiseuille flow in the description of systolic blood flow through the iliac artery. The vessel has the diameter $d = 4.2\,\text{mm}$, and blood of the viscosity $\eta = 4.0\,\text{mPa s}$ flows at the rate of $q = 3.8\,\text{ml s}^{-1}$.

(a) Compute the fluid velocity $v(r)$ that establishes in the iliac artery, where r denotes the radial coordinate.
(b) Compute the shear stress $\tau(r)$ that acts at the fluid particles.
(c) Compute the drop of pressure Δp that appears over a length of $l = 10.0\,\text{cm}$ of the iliac artery.   ∎

Example 6.5 (Power Law Fluid at Steady-State Tube Flow).   The viscosity of blood is not constant, a factor that influences the evolution of the flow velocity. Given this example, blood is represented by the Power Law model (6.6) and the flow through a rigid circular tube is exemplified.

(a) Derive the governing equation and express the scalar shear rate $\dot{\gamma}$ for steady-state tube flow.
(b) Integrate the governing equation towards the derivation of the flow velocity. *Hint: Identify integration constants as early as possible to simplify the subsequent integration.*
(c) Given the power law constants $n = 5.0; 1.0; 0.5; 0.2$, plot the normalized velocity $v_z/v_{z\,\text{max}}$ as a function of the normalized radius r/r_0, where $v_{z\,\text{max}}$ and r_0 denote the maximum flow velocity and the tube radius, respectively.
(d) Compute the flow rate q that establishes in a tube of the diameter $r_0 = 2.0\,\text{mm}$. The pressure gradient $\mathrm{d}p/\mathrm{d}z = -7.0\,\text{mmHg}\,\text{m}^{-1}$ as well as the Power Law model parameters $\eta_0 = 5\,\text{mPa}\,\text{s}$ and $n = 0.8$ may be used. ∎

6.6.2.6 Resistance of a Vessel Segment

A vessel segment of radius r_0 and length l presents the resistance $R\,[\text{Pa}\,\text{m}^{-3}\,\text{s}]$ to the flow of the rate $q\,[\text{m}^3\,\text{s}^{-1}]$. Given Poiseuille flow, the resistance R of the vessel segment reads

$$R = \frac{\Delta p}{q} = \frac{8\eta l}{\pi r_0^4}, \tag{6.42}$$

where η denotes the blood's viscosity, and $\Delta p = p_{\text{inlet}} - p_{\text{outlet}} = -(\mathrm{d}p/\mathrm{d}z)l$ is the pressure drop between the inlet and outlet of the vessel segment. This relation is known as the *law of Hagen–Poiseuille* and directly follows from the expression (6.36).

Figure 6.22 illustrates the normalized resistance $R/(\eta l)$ of a vessel segment. It tends to infinity for $r_0 \to 0$, and the plot indicates that only the small vessels, the so-called resistance vessels, are able to build up resistance in the circulation. The peripheral resistance then determines the Mean Arterial Pressure (MAP) of the vascular system.

Fig. 6.22 Resistance of a circular tube to the flow of a Newtonian fluid. The graph shows the relation between the normalized flow resistance $R/(\eta l)$ and the radius r_0 of the tube

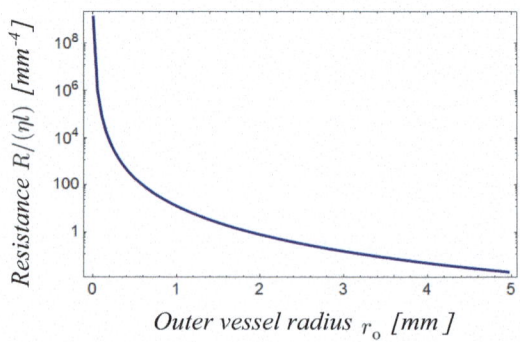

Example 6.6 (Stented Vessel). The development of atherosclerosis correlates with locations of low Wall Shear Stress (WSS), an observation that led to a number of non-conventional vascular stent designs. Figure 6.23 shows such a design, a stent that aims at increasing the WSS in a stenotic vessel segment. We consider a steady-state flow $q = 3.81\,\mathrm{min}^{-1}$ through a vessel of the diameter $d_\mathrm{v} = 4.9\,\mathrm{mm}$ and blood as a Newtonian fluid of the viscosity $\eta = 4.0\,\mathrm{mPa\,s}$. The flow is fully developed, and the stent of the length $l = 5.6\,\mathrm{cm}$ is long enough to neglect inlet and outlet effects. Given these assumptions, the hemodynamic effect of the stent should be assessed.

Fig. 6.23 Schematic illustration of a stent design that increases the Wall Shear Stress (WSS) in atherosclerotic lesions

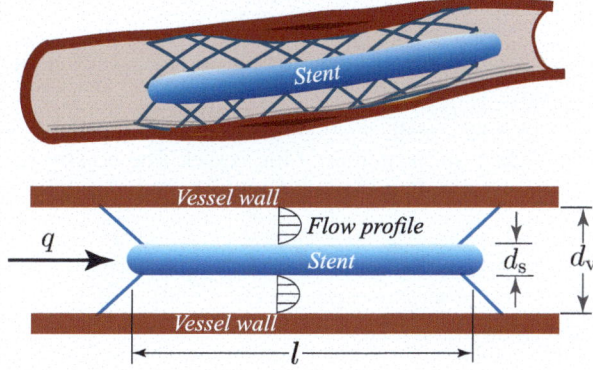

(a) Consider the native vessel, the vessel without the stent, and compute the blood flow velocity v as well as the WSS.
(b) Given the hub diameter $d_\mathrm{s} = 0.7\,\mathrm{mm}$, compute the blood flow velocity v and WSS of the stented vessel.
(c) Compute the resistance R of the stent to blood flow.   ∎

6.6.2.7 Pulsatile Newtonian Fluid Flow Through a Circular Tube
The rhythmic contraction of the heart muscle maintains the cyclic pulsatile flow in the arterial system, and the flow continuously exchanges potential and kinetic energies. The cyclic flow of a Newtonian fluid of viscosity η and density ρ in a circular and straight tube aims at modeling such conditions. At steady-periodic conditions, the pressure gradient may be expressed by $\mathrm{d}p/\mathrm{d}z = P\exp(i\omega t)$, where $i = \sqrt{-1}$ denotes the imaginary unit, whilst ω and P are the angular velocity and the pressure amplitude, respectively. In response to the cyclic pressure gradient, the cyclic flow velocity $v(r,t) = V(r)\exp(i\omega t)$ establishes, where the amplitude $V(r)$ is a function of the radius r. In a rigid tube, no phase lag between velocity and pressure can develop.

Given said forms of the velocity and pressure, the governing equation of 1D flows in cylindrical coordinates (6.30) yields the Bessel[8] *differential equation*

$$\frac{d^2 V}{dr^2} + \frac{1}{r}\frac{dV}{dr} - \frac{i\omega}{\nu}V = \frac{P}{\nu\rho},$$

where $\nu = \eta/\rho \, [\text{m}^2\,\text{s}^{-1}]$ denotes the kinematic viscosity of blood. The differential equation has the general solution

$$V(r) = \frac{-P}{i\rho\omega} + C_1 J_0\left(i^{3/2}\sqrt{\omega/\nu}\,r\right) + C_2 Y_0\left(i^{3/2}\sqrt{\omega/\nu}\,r\right),$$

where $J_0(x)$ and $Y_0(x)$ denote first-order Bessel functions of first and second kinds, respectively.

We notice $Y_0(0) = -\infty$, and $C_2 = 0$ is then required to bound the velocity amplitude V in the center of the vessel. The no-slip boundary condition at the wall $V(r = r_o) = 0$ allows us to identify the remaining integration constant

$$C_1 = \frac{-iP}{\rho\omega J_0\left(i^{3/2}\sqrt{\omega/\nu}\,r\right)}.$$

The velocity then reads

$$v(r,t) = \frac{iP\exp(i\omega t)}{\rho\omega}\left[1 - \frac{J_0\left(i^{3/2}\,Wo\,r/r_o\right)}{J_0\left(i^{3/2}\,Wo\right)}\right], \tag{6.46}$$

where $Wo = r_o\sqrt{\omega/\nu}$ denotes the *Womersley*[9] *number*. It may also be seen as

$$Wo = \sqrt{\frac{\text{transient inertial force}}{\text{viscouse force}}}, \tag{6.47}$$

and the Womersley number therefore represents the relative impact of transient and viscous effects on the flow.

Example 6.7 (Pulsatile Blood Flow in a Vessel Segment). We consider a steady-state pulsatile blood flow in an artery of the diameter $d = 6.0\,\text{mm}$. The flow is the response of blood to the pressure gradient $dp/dz = P\exp(i\omega t)$, where $P = 10.0\,\text{kPa}\,\text{m}^{-1}$ and $\omega = 2.0\pi$ denote the pressure amplitude and the angular velocity, respectively. The imaginary unit is denoted by $i = \sqrt{-1}$, and blood may be regarded as a Newtonian fluid with the dynamic viscosity $\eta = 3.5\,\text{mPa}\,\text{s}$ and the density $\rho = 1.06\,\text{kg}\,\text{dm}^{-3}$.

---

[8]Friedrich Wilhelm Bessel, German astronomer, mathematician, physicist, and geodesist, 1784–1846.

[9]John Ronald Womersley, British mathematician and computer scientist, 1907–1958.

(a) Compute the Womersley number that characterizes the blood flow.
(b) Compute velocity profiles that establish in the vessel at the time points $t =$ 0.0; 0.1; ...; 0.5 s over the cardiac cycle.   ■

6.6.3 Flow in the Elastic Tube

Blood flow is influenced by waves that propagate and reflect along the vascular tree. The flow of an incompressible and *inviscid* fluid in an *elastic* tube of circular cross-section is considered a first model system to study the propagation of such waves. The vessel is formed by a thin elastic wall and its diameter d is small as compared to the length of the tube. The wave length is also *much longer* than the tube diameter, and leads to a *small* disturbance Δd of the diameter. It results in the small, but important radial velocity v_r on top of the axial velocity v_z—the flow can therefore no longer been considered 1D. Given the time t and the axial position z, the vessel wall expands at the velocity $(\partial d/\partial t)/2$, see Fig. 6.25. The fluid follows the motion of the vessel wall, and

$$v_r = \frac{\partial d}{\partial t}\frac{r}{d}$$

determines the radial velocity as a function of the radial coordinate r. The continuity $\mathrm{div}\mathbf{v} = 0$ of the incompressible flow in cylindrical coordinates then reads

$$\mathrm{div}\mathbf{v} = \frac{1}{r}\frac{\partial(rv_r)}{\partial r} + \frac{\partial v_z}{\partial z} = \frac{2}{d}\frac{\partial d}{\partial t} + \frac{\partial v_z}{\partial z} = 0, \qquad (6.48)$$

where rotational symmetry $v_\theta = 0$ and $\partial(\bullet)/\partial\theta = 0$ has been used.

With the *diameter stiffness* $\alpha\,[\mathrm{m\,Pa^{-1}}]$ of the vessel wall, the incremental constitutive model

$$\alpha\Delta p_\mathrm{i} = \Delta d \qquad (6.49)$$

describes the elastic properties of the vessel wall. It relates the diameter d to the pressure p_i in the vessel, and allows us to substitute the rate of diameter change $\partial d/\partial t$ in (6.48). The expression

Fig. 6.25 Velocity components v_r and v_z in the radial and respective axial directions of a flow in an elastic circular tube of the diameter d. The vessel wall moves in the radial direction at the velocity $(\partial d/\partial t)/2$

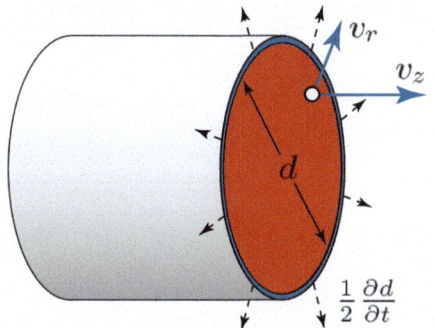

$$\alpha \frac{\partial p_i}{\partial t} + \frac{d}{2} \frac{\partial v_z}{\partial z} = 0 \tag{6.50}$$

then relates the change of pressure and the axial velocity gradient.

The balance of linear momentum closes the description, an expression that requires the velocity gradient in cylindrical coordinates

$$\mathrm{grad}\mathbf{v} = \begin{bmatrix} \partial v_r/\partial r & (\partial v_r/\partial \theta)/r - v_\theta/r & \partial v_r/\partial z \\ \partial v_\theta/\partial r & (\partial v_\theta/\partial \theta)/r + v_r/r & \partial v_\theta/\partial z \\ \partial v_z/\partial r & (\partial v_z/\partial \theta)/r & \partial v_z/\partial z \end{bmatrix} = \begin{bmatrix} 0 & 0 & 0 \\ 0 & 0 & 0 \\ \partial v_z/\partial r & 0 & 0 \end{bmatrix}$$

to be computed. It used the long-wave approximation $\partial(\bullet)/\partial z \approx 0$ that implies the condition $v_r \ll v_z$ and results in the velocity gradient (6.26) already known from the 1D flow.

The inviscid fluid is always at hydrostatic stress $\sigma = -p\mathbf{I}$, and the linear momentum $\rho(\mathrm{D}\mathbf{v}/\mathrm{D}t) = -\mathrm{div}p\mathbf{I}$ of the 1D incompressible flow in cylindrical coordinates then reads

$$\rho \left(\begin{bmatrix} 0 \\ 0 \\ \partial v_z/\partial t \end{bmatrix} + \begin{bmatrix} 0 & 0 & 0 \\ 0 & 0 & 0 \\ \partial v_z/\partial r & 0 & 0 \end{bmatrix} \begin{bmatrix} v_r \\ 0 \\ v_z \end{bmatrix} \right) = \begin{bmatrix} -\partial p/\partial r \\ 0 \\ -\partial p/\partial z \end{bmatrix}, \tag{6.51}$$

where the material time derivative $\mathrm{D}\mathbf{v}/\mathrm{D}t = \partial \mathbf{v}/\partial t + \mathbf{v} \cdot \mathrm{grad}\mathbf{v}$ has been used, whilst body forces have been neglected. Given $v_r \ll v_z$, the linear momentum yields the only non-trial equation

$$\rho \frac{\partial v_z}{\partial t} = -\frac{\partial p}{\partial z} \tag{6.52}$$

that then governs the flow problem.

We may now derive (6.50) with respect to the time t, and (6.52) with respect to the axial coordinate z, manipulations that allow us to substitute the axial velocity v_z. The flow in the elastic tube is then governed by the wave equation

$$\frac{\partial^2 p_i}{\partial t^2} - c^2 \frac{\partial^2 p_i}{\partial z^2} = 0,$$

and the *wave speed* c [m s^{-1}] reads

$$c = \sqrt{\frac{d}{2\rho\alpha}}. \tag{6.53}$$

Instead of the diameter stiffness α, the *area distensibility* $D = (\mathrm{d}A/\mathrm{d}p)/A = 2\alpha/d\,[\mathrm{Pa}^{-1}]$ is frequently used to describe the propagation of waves, where A denotes the vessel's cross-section. The wave speed (6.53) then reads $c = 1/\sqrt{\rho D}$, an expression previously introduced in Sect. 2.2.1. Another version of the wave equation reads

$$ c = \sqrt{\frac{A}{\rho(\mathrm{d}A/\mathrm{d}p)}} $$

and is referred to as *Frank/Bramwell*[10]*–Hill*[11] *equation*.

The vessel wall has non-linear stress–strain properties and is at biaxial stress in the body. However, in a very first approximation we may describe it by an incremental linear-elastic material at uniaxial circumferential stress. The diameter stiffness then reads $\alpha = d^2/(2Eh)$, where E and h denote the incremental Young's modulus and the (constant) wall thickness, respectively. The wave speed then reads

$$ c = \sqrt{\frac{Eh}{\rho d}}, \tag{6.54} $$

an expression known as the *Moens*[12]*–Korteweg*[13] *equation* [379] and widely used in the extraction of the vessel's stiffness from pulse wave velocity measurements.

6.7 Blood Flow Phenomena

The blood flow in the vasculature is influenced by factors, such as the geometry of the vessel segment, the flow rate, and the boundary conditions. Whilst the aforementioned analytical expressions are able to provide some basic insights, they are not general enough to explore a number of flow mechanisms. Already at steady-state, blood flow is complex and CFD simulations or direct experimental measurements such as Particle Image Velocimetry (PIV) are needed to acquire a comprehensive picture of blood flow phenomena.

6.7.1 Laminar, Transitional, and Turbulent Flow

Fluid flow in parallel layers without disruptions between said layers determines *laminar flow*. The adjacent fluid layers then slide past one another like "playing

---

[10] John Crighton Bramwell, British cardiologist, 1889–1976.

[11] Archibald Vivian Hill, British physiologist, 1886–1977.

[12] Adriaan Isebree Moens, Dutch physiologist, 1846–1891.

[13] Diederik Johannes Korteweg, Dutch mathematician, 1848–1941.

cards". There are neither cross-currents perpendicular to the direction of flow, nor eddies (vortices) or swirls that span over several small length-scales.

Laminar flow establishes if the fluid's viscosity is sufficient to *dissipate* the flow's kinetic energy without breaking-up into smaller flow structures. Otherwise the flow becomes *turbulent* and causes the formation of eddies (vortices) at many different length-scales—flow regimes show then chaotic particle velocities. By an essentially inviscid mechanism, large-scale structures of high turbulent kinetic energy transmit their energy to smaller and even smaller structures [438]. It produces a cascade of eddies (vortices) along which energy is transmitted all the way down to the smallest possible length-scale, the so-called *Kolmogorov*[14] *length-scale*, where viscosity dominates and the (remaining) kinetic energy is finally dissipated [309]. In a CFD problem, the computational mesh may be refined towards resolving the Kolmogorov length-scale, an approach that results in a Direct Numerical Simulation (DNS). The Kolmogorov length-scale is problem-specific, and of the order of $100\,\mu$m in the ventricle, for example [84].

Given the length-scale of the computational mesh is larger than the Kolmogorov length-scale, the *Reynolds decomposition* is used. The governing equations are then split into time-averaged and fluctuating quantities, and Reynolds-averaged Navier–Stokes equations (RANS equations) describe the fluid problem. The apparent stress that represents the fluctuating velocity field is referred to as the *Reynolds stress*. It is described by *turbulence models* and captures the chaotic nature of turbulent flows. They are based on statistical methods, and up to date no model is generally accepted in the description of turbulent blood flow.

The chaotic nature of turbulent flows strongly promotes the mixing of fluid particles and hinders their segregation. At physiological conditions, transitional and turbulent flows are rare in the cardiovascular system. Such conditions appear during the filling of the ventricle, at diastole in the first aortic segment [84] or determine small portions of the blood flow in the ascending as well as the descending aorta [521]. Transitional flow and turbulence are more pronounced in the diseased vasculature and easily establish in stenotic vessel segments or artificial heart valves, for example.

6.7.2 Boundary Layer Flow

A *boundary layer flow* represents the fluid flow in the immediate vicinity of a bounding surface. Due to the high shear rates in the boundary layer, *viscous effects* dominate over inertia effects and determine the motion of fluid particles. The boundary layer can separate into a broader wake and trigger the formation of a VS. It appears as soon as the flow in the layer closest to the bounding surface reverses its direction. The WSS is then zero and the boundary layer suddenly increases its thickness.

---

[14]Andrey Nikolaevich Kolmogorov, Russian mathematician, 1903–1987.

In blood vessels, the near-wall region is occupied by a *cell-depleted plasma layer* of the thickness of the size of the erythrocytes. The erythrocytes migrate away from the wall (see Sect. 6.3) and the cell-depleted layer contains only small and rheologically irrelevant particles, such as thrombocytes and macromolecules. The shear rate at the vessel wall is therefore a function of plasma viscosity [234], whilst erythrocytes strongly influence the shear rate in the bulk flow.

6.7.3 Blood Flow Through Circular Tubes

Vascular segments may be approximated by cylindrical tubes, a flow model of fundamental vascular biomechanical interest. Given the diameter distensibility is small and the vessel segment length is much larger than its diameter, a 1D flow with the aforementioned velocity profiles establishes, see Sects. 6.6.2.5 and 6.6.2.7.

For some applications blood may be described as a steady-state Newtonian fluid, and the quadratic velocity profile (6.34) of a *Poiseuille flow* builds-up, see the solid line in Fig. 6.26a. The velocity is then directly proportional to the negative pressure gradient $-\mathrm{d}p/\mathrm{d}z$, and thus how fast the pressure decreases along the flow path. Given laminar flow, the pressure gradient may be substituted by the flow rate q according to (6.36), which then leads to the law of *Hagen–Poiseuille* (6.42). It nicely illustrates that only in the resistance vessels of the vascular bed a significant pressure drop $\Delta p = (\mathrm{d}p/\mathrm{d}z)l$ over the vessel length l, is able to establish. The physiological relevance of this mechanism has been discussed in Chap. 2. Given a plug profile at the entrance and a flow of the Reynolds number Re, the hydrodynamic entrance length of $0.1r_0 Re$ is needed to approximate a Poiseuille flow, conditions that are challenging to meet in the vasculature.

Given a *non-Newtonian* fluid, the viscosity η depends on the shear rate $\dot{\gamma}$, a dependence that influences the velocity profile and the WSS. Blood is a shear-thinning fluid, and the linear momentum (6.28) then results in a more plug-like velocity profile as compared to Poiseuille flow, see the dashed line in Fig. 6.26a.

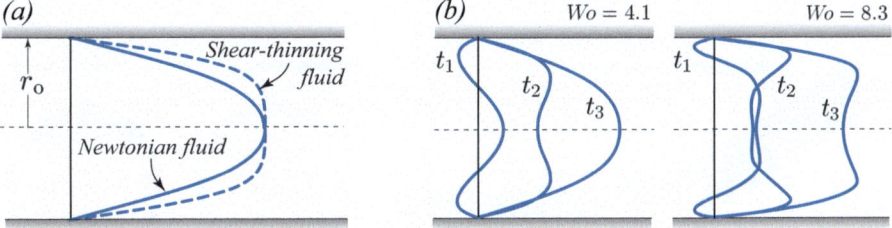

Fig. 6.26 Velocity profiles that establish within a circular tube at laminar flow. (**a**) Steady-state flows. The Newtonian fluid determines a quadratic velocity profile (Poiseuille flow), whilst a shear-thinning fluid leads to a more plug-like profile. (**b**) Pulsatile flows. Flow profiles at different times t_i during one cardiac cycle are shown, where *Wo* denotes the Womersley number

The blood flow in the larger arteries is not steady-state but pulsatile—it develops in response to inertia *and* viscous forces. Phases of forward and backward flow characterize then the blood motion, and the *Womersley number Wo* [600] determines the shape of the velocity profile, see Sect. 6.6.2.7. Womersley numbers of the base excitation angular velocity of $\omega = 2\pi$, range from approximately 2 to 20 in large arteries, and Fig. 6.26b illustrates typical velocity profiles.

6.7.4   Multi-dimensional Flow Phenomena

The analysis of multi-dimensional flow phenomena by analytical methods is challenging. CFD is often a more convenient approach, and the flow problems discussed in this section have been analyzed with the FEM, as discussed in Chap. 4.

6.7.4.1 Secondary Flow

The flow velocity in straight tubes changes across the tube's cross-section, and given such a non-uniform flow is forced to turn, the moment equation (3.107) predicts the development of *rotational* or *swirl* velocity components [115, 350, 389, 424, 512, 563]. Right-handed helical flow is a feature that is observed during late systole in most normal upper aortic arches, and retrograde flow along inner wall curvatures appears at end systole [306]. In addition to bends, many other conditions, such as jets can result in secondary flows.

Secondary flow significantly influences the blood flow [71] in large vessels—the WSS then decreases at the inside and increases at the outside of the bend, see Fig. 6.27. A detailed analysis of shape effects of steady-state and pulsatile flows in geometries, such as branches, anastomoses, and stenosis is reported elsewhere [132].

6.7.4.2 Vortex Flow and Vortex Structures

A VS is a region, within which blood is essentially at swirling or rotating motion, information that is also used in the clinical assessment of cardiovascular

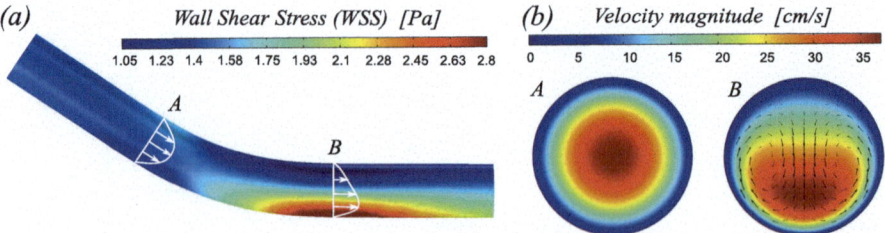

Fig. 6.27 Flow through a bent circular tube. (**a**) Wall Shear Stress (WSS) distribution with two characteristic velocity profiles. (**b**) Magnitude of the blood flow velocity at two cross-sections with arrows denoting the secondary flow

pathologies [303]. Rapid turning of the flow direction (see Fig. 6.27), or flow separation are two mechanisms that are able to trigger the formation of a VS—a number of other flow conditions have also been reported [318]. VSs are able to move, develop, and interact with each other [278], and then result in complex, but still laminar 3D flows. A VS may appear relatively weak in comparison to the mean flow and is therefore not always easy to detect.

A VS encapsulates blood and prevents it from the mixing with surrounding blood. Inside the VS, conditions that favor chemical reactions, such as thrombocyte activation and thrombin formation [42, 43, 51, 178] may establish. A VS has the potential to transport pockets of blood over long distances, until the VS finally breaks up and the entrapped blood mixes again with surrounding blood.

6.7.4.3 Jet Flow

A *jet flow* is a fast stream of fluid that develops in flow constrictions, such as stenotic vessels and pathological heart valves. The jet has a high momentum, which allows it to travel long distances without significant energy dissipation. Some distance away from the constriction, the break-up of (hairpin) VS may lead to transition from laminar to turbulent flow. A jet leads to highly inhomogeneous WSS. Whilst in the constriction the WSS exceeds upstream levels by several folds, it is very low in the flow separation zones immediately downstream the constriction [573, 574]. At the point of flow separation, the WSS is per definition zero.

6.7.4.4 Creep Flow

Given the advective inertia forces are small in comparison to the other forces in the fluid, its motion is described by a *creep* or *Stokes flow*. The *Re* number is then low, conditions that appear in the small blood vessels, where fluid velocities are low and the length-scales are small. Creep flow is described by the linearization

$$\rho \frac{\partial \mathbf{v}}{\partial t} = \text{div}\boldsymbol{\sigma} + \mathbf{b}_\text{f} \tag{6.55}$$

of the momentum equation (3.107), where $\boldsymbol{\sigma}$ denotes the Cauchy stress, whilst ρ and $\mathbf{b}_\text{f}$ are the density and the body force per unit (spatial) volume, respectively. Given the small velocity $\mathbf{v}$, the material time derivative $D\mathbf{v}/Dt = \partial \mathbf{v}/\partial t + \mathbf{v} \cdot \text{grad}\mathbf{v}$ has been approximated by the partial derivative $\partial \mathbf{v}/\partial t$. In many applications even $\partial \mathbf{v}/\partial t$ is negligible, and the linear momentum $\text{div}\boldsymbol{\sigma} + \mathbf{b}_\text{f} = \mathbf{0}$ then describes a creep or Stokes flow.

6.7.4.5 Inviscid Flow

Given the viscous forces are small in comparison to the inertia forces, the fluid motion is described by an *inviscid flow*. The *Re* number is high, and the fluid behaves like an ideal fluid. The stress $\boldsymbol{\sigma} = -p\mathbf{I}$ is then purely hydrostatic, and the momentum equation (3.107) reads

$$\rho \frac{D\mathbf{v}}{Dt} = -\mathrm{grad}\, p + \mathbf{b}_\mathrm{f}\,, \tag{6.56}$$

where p denotes the pressure, whilst ρ and $\mathbf{b}_\mathrm{f}$ are the density and the body force per unit (spatial) volume, respectively. In many applications even $\mathbf{b}_\mathrm{f}$ is negligible, and the linear momentum $\rho D\mathbf{v}/Dt = -\mathrm{grad}\, p$ then describes an inviscid flow.

Given inertia effects are small, the momentum equation (6.56) reads $\mathrm{grad}\, p = \mathbf{b}_\mathrm{f}$ and describes the *hydrostatic* equilibrium.

6.8   Description of Blood Flow Properties

Hemodynamics is tightly linked to the biochemical activity of the vasculature, and a number of flow parameters have been introduced to study the implications of blood flow in vascular physiology and pathology. Aside from *WSS-related* mechanisms, blood flow brings reactive species into close vicinity and influences chemical reactions through *transport-related* mechanisms. Both principles determined the introduction of a number of parameters, some of which are discussed hereinafter.

6.8.1   Wall Shear Stress-Based Parameters

Let us consider the vessel wall and denote the normal vector that points into the vessel lumen by $\mathbf{n}$. The endothelium is then exposed to the shear stress vector

$$\boldsymbol{\tau}_\mathrm{w}(t) = \boldsymbol{\sigma}(t)\mathbf{n} - (\mathbf{n} \cdot \boldsymbol{\sigma}(t)\mathbf{n})\,\mathbf{n}\,,$$

where $\boldsymbol{\sigma}(t)$ denotes the Cauchy stress of the fluid particle adjacent to the endothelium. We used Cauchy stress theorem (3.20), vector summation, and the equilibrium at the interface between fluid and endothelium in the derivation of this expression. The vector $\boldsymbol{\tau}_\mathrm{w}$ depends on the spatial position and the time t, and permits the derivation of a number of scalar blood flow parameters.

The *maximum shear stress* or the *time-averaged shear stress*

$$WSS_\mathrm{max} = \max(|\boldsymbol{\tau}_\mathrm{w}(t)|)\ ;\quad WSS_\mathrm{mean} = \frac{1}{T}\int_0^T |\boldsymbol{\tau}_\mathrm{w}(t)|\mathrm{d}t$$

over the cardiac cycle $0 < t < T$, as well as the *Oscillatory Shear Index*

$$OSI = \frac{1}{2}\left(1 - \frac{|\mathbf{WSS}_\mathrm{mean}|}{WSS_\mathrm{mean}}\right)\ \text{with}\ \mathbf{WSS}_\mathrm{mean} = \frac{1}{T}\int_0^T \boldsymbol{\tau}_\mathrm{w}(t)\mathrm{d}t$$

have been extensively used in the literature to express the implications of blood flow on the vessel wall. A sensitive endothelium is however a basic requirement to link said parameters to biology.

Given steady-state unidirectional flow, $|\mathbf{WSS}_{mean}| = WSS_{mean}$ and results in $OSI = 0$, whilst a purely oscillating flow with $|\mathbf{WSS}_{mean}| = 0$ leads to $OSI = 0.5$. The OSI is also linked to the *particle residence time*. At $OSI = 0.5$, blood particles close to the wall move back and forth at the same position, which in turn provides time for chemical reactions between blood particles and the endothelium. Given the viscosity η, the *Relative Residence Time* $RRT = \eta/[(1 - 2\,OSI)WSS_{mean}]$ is another parameter to express the time a particle spends close to a spatial position at the wall [187].

In an attempt to localize regions of the wall exposed to both, high OSI and low WSS_{mean}, the ratio

$$ECAP = \frac{OSI}{WSS_{mean}}$$

has also been proposed and referred to as the Endothelial Cell Activation Potential (ECAP) [121].

Whilst the aforementioned parameters addressed the interaction between blood and the endothelium, shear stress also plays a critical role in mechanisms, such as platelet activation, rouleaux formation, and erythrocyte damage—processes that may appear within the bloodstream and not necessarily at the wall. The *maximum shear stress* and the *time-averaged shear stress*

$$SS_{max} = \max\left(\frac{\sigma_{max}(t) - \sigma_{min}(t)}{2}\right) \;\; ; \;\; SS_{mean} = \frac{1}{T}\int_0^T \frac{\sigma_{max}(t) - \sigma_{min}(t)}{2}dt$$

over the cardiac cycle $0 < t < T$ are then reasonable metrics to study the implications of blood flow, where σ_{max} and σ_{min} are the largest and smallest principal Cauchy stresses, respectively.

Given a Newtonian fluid, the shear stress is proportional to the shear rate $\dot{\gamma}(t) = \sqrt{2\mathrm{tr}\mathbf{d}^2}$, where $\mathbf{d}$ denotes the rate of deformation tensor. The maximum scalar shear rate

$$\dot{\gamma}_{max} = \max\left(\dot{\gamma}(t)\right),$$

presents then as surrogate measure of the shear force that acts at blood particles.

Blood in the vasculature is exposed to highly inhomogeneous shear stress, and blood particles have already experienced shear stress when arriving at the time t at the position $\mathbf{x}$ of the flow domain. A local analysis of the flow at the position $\mathbf{x}$ therefore misses the *history* of the blood particle and may not adequately reflect its potential for chemical reactions. In addition to the shear stress level, the exposure time to shear stress may then also be an important factor. Given *small* particles, and thus problems of low *St numbers* (6.1), the particles closely follow streamlines, resulting in problem formulations discussed in Sect. 6.5.1. *Large* particles may not follow the flow and their inertia allow them to cross streamlines. We then have to solve the equation of motion of an individual particle to track its movement. Whilst in a number of applications the problem is one-way coupled and the particles do not influence the flow, *very* large particles, such a thrombi occupying a considerable part of the lumen, however influence the flow—the problem is then two-way-coupled.

6.8.2 Transport-Based Parameters

A chemical reaction can only appear if the reactive partners are in close proximity, and *mixing* a flow therefore enhances the probability for chemical reactions to appear. A *high vorticity* determines a mixing flow and a large value of $|\mathbf{w}|$ then relates to high chemical reactivity of blood. Here, $\mathbf{w} = (\mathbf{l} - \mathbf{l}^T)/2$ denotes the spin tensor, the skew-symmetric part of the spatial velocity gradient $\mathbf{l} = \text{grad}\mathbf{v}$.

VS encapsulate pockets of mixing flow that may outlast much longer than the cardiac cycle. VS therefore provide ideal conditions for chemical reactions, and VS visualization methods [286] have been used towards the qualitative study of the reactivity of blood flow [42].

6.9 Case Study: Blood Flow in the Aneurysmatic Infrarenal Aorta

Given the clinical relevance of an Abdominal Aortic Aneurysm (AAA), its hemodynamics have been intensively explored, a factor also to be investigated in this case study. We continue the analysis of the AAA discussed in Sect. 5.7, which lumen has been reconstructed from clinically recorded CT-A images (A4clinics Research Edition, VASCOPS GmbH) and then processed to allow for the computation of the blood flow velocity (COMSOL Multiphysics, COMSOL AB).

6.9.1 Modeling Assumptions

The AAA wall is known to be very stiff [542], a property that allowed us to use a *rigid-wall* FEM model in the computation of the blood flow over the cardiac cycle. Dirichlet boundaries were used to prescribe the no-slip condition at the blood–wall interface as well as to model the pulsatile inflow condition. The distribution of the flow velocity over the cross-section of the infrarenal aorta is unknown and influenced by many factors, see Sect. 6.7.3. Whilst methods, such as Doppler ultrasound imaging supports the acquisition of such information, it is not part of the regular patient examination. We therefore prescribe the velocity $v_{in}(t) = q(t)/A$ homogenously over the inlet of the cross-section $A = 3.8\,\text{cm}^2$, where the flow rate $q(t)$ represents the average infrarenal flow in AAA patients [336], see Fig. 6.28. Given our case study, we considered the cardiac cycle time $T = 0.8\,\text{s}$ and solved the transient flow equations over three cardiac cycles.

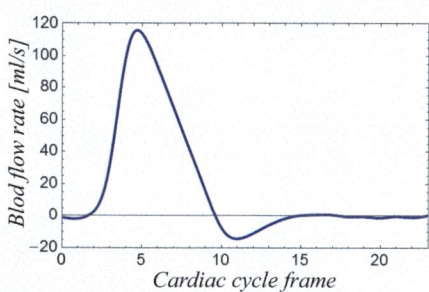

Time frame	Flow [ml s^{-1}]	Time frame	Flow [ml s^{-1}]
0	-1.08	12	-10.9
1	-1.8	13	-5.71
2	2.88	14	-1.67
3	32.4	15	0.333
4	99.6	16	0.702
5	114	17	0.594
6	92.8	18	-0.497
7	66.7	19	-0.481
8	41.1	20	-1.17
9	15.4	21	-0.306
10	-7.55	22	-0.951
11	-14.2	23	0.418

Fig. 6.28 Blood flow rate over the cardiac cycle to prescribe Abdominal Aortic Aneurysms (AAA) inflow. The data reflects the average infrarenal aortic flow in 36 patients with small ($d < 50$ mm) AAA, acquired through Magnetic Resonance-Angiography (MR-A) [336]

At the outlet the Neumann boundary condition of the constant pressure $p = 13.3$ kPa described the problem's pressure level. Given an incompressible fluid and a rigid vessel wall, the actual value of the outlet pressure does *not* influence the predicted blood flow.

The flow in the aorta is complex, and blood is exposed to phases of high and low shear rates over each cardiac cycle. The Carreau–Yasuda model (6.7) with the parameters $\eta_0 = 200$ mPa s, $\eta_\infty = 3.5$ mPa s, $\lambda = 1800$ s, $n = 0.28$ and $\alpha = 2.0$ (see Table 6.1, and Fig. 6.9) has been used to capture the rheological properties of blood over such a wide spectrum of shear rates. Thixotropic properties were neglected, and we therefore assumed that blood's viscosity changes instantaneously in response to the change of the shear rate.

6.9.2 Results

A tetrahedral mesh with quadratic finite elements of approximately 180k degrees of freedom was used to solve the hemodynamic problem, and Fig. 6.29 illustrates selective results taken from the third computed cardiac cycle. At peak systole, the flow is well organized and shows parallel streamlines. During the deceleration at the late-systolic phase the flow breaks up, and the blood is then strongly mixed during the entire diastolic phase. The blood–wall interface is exposed to complex loads, and the WSS is highly inhomogeneous in space and over the cardiac cycle. At peak systole the scalar shear rate at the aneurysmatic wall is between approximately $\dot{\gamma}_{min} = 40$ s^{-1} and $\dot{\gamma}_{max} = 1200$ s^{-1}, see Fig. 6.29b. The Carreau–Yasuda model predicts then the blood viscosity of $\eta_{min} = 4.42$ mPa s and $\eta_{max} = 3.58$ mPa s and the wall experiences therefore WSS between $\tau_{min} = \eta_{min}\dot{\gamma}_{min} = 0.18$ Pa and $\tau_{max} = \eta_{max}\dot{\gamma}_{max} = 4.30$ Pa at peak systole. These values may be compared to the normal infrarenal aorta that reaches the peak WSS of 2.72 Pa [414]. The prescribed

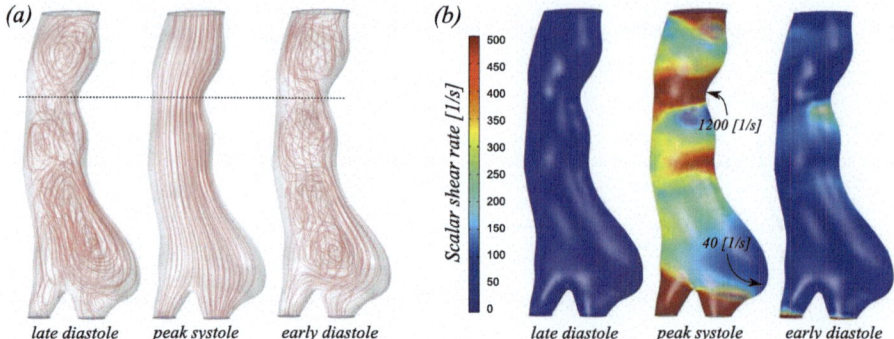

Fig. 6.29 Abdominal Aortic Aneurysm (AAA) hemodynamics at late diastole, peak systole, and early diastole. (**a**) Streamlines indicate strong mixing of blood during diastole and well organized flow at peak systole. The dotted line indicates the cross-section that is further analyzed in Fig. 6.30. (**b**) Distribution of the scalar shear rate $\dot{\gamma}$ at the blood–wall interface. The results represent the third cardiac cycle of the computations and relate to the time of 1.6 s (late diastole), 1.76 s (peak systole), and 2.0 s (early diastole), respectively

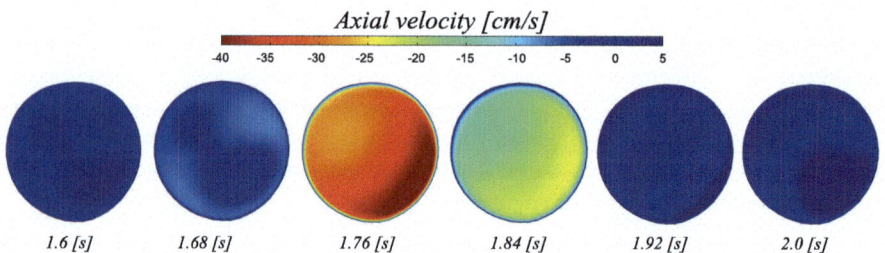

Fig. 6.30 Development of the velocity component v_z in axial vessel direction at the cross-section that is indicated by the dotted line in Fig. 6.29a. The plots show the velocity between the late and early diastolic phases

inlet velocity $v_{in}(t)$ results in unrealistic shear rate at the inlet, and the aortic segment close to the inlet has therefore been excluded from this analysis.

Figure 6.30 shows the development of the velocity component v_z in axial vessel direction between late diastole and early diastole. The plots correspond to the cross-section that is indicated by the dotted line in Fig. 6.29a and illustrate phases and domains of forward and backward flows, respectively. The hemodynamics at this cross-section is similar to blood flow in the normal aorta, whilst the aneurysm-related pathological flow develops further downstream.

6.10   Summary and Conclusion

Blood is a suspension with particles covering approximately 50% of its volume. Following continuum mechanics, blood properties are averaged over the Representative Volume Element (RVE), a domain that contains a *sufficiently large* number of blood particles. The largest blood particles reach the size of 30 μm, and the consistent application of continuum mechanics is therefore limited to the analysis of blood flow in vessels that are larger than approximately half a millimeter in diameter.

A number of vascular studies assumed *Poiseuille* flow. Whilst it provides insights in vascular function, such a model cannot be rigorously applied to the circulation. It requires a steady-state flow of a *Newtonian* fluid in rigid, long, and straight tubes to establish a Poiseuille flow. The simplification of a rigid wall may be acceptable in a number of arteries, but veins particularly depart from this assumption and change their diameter remarkably with the pressure—some of them even collapse during the cardiac cycle. In addition, the flow in large arteries is strongly influenced by inertia effects, and the blood flow velocity develops towards a *Womersley* profile. As with the Poiseuille flow, also the Womersley flow is based on a Newtonian fluid that flows in rigid, long, and straight tubes. These conditions barely apply to the circulation, and already in the normal vasculature the motion of blood is influenced by factors, such as a non-Newtonian viscosity, vessel curvature, vessel bifurcations, and edge effects.

Erythrocytes represent by far the largest proportion of blood particles and dominate its *shear-thinning* rheological behavior, a property that is not only important to describe the flow in small vessels, but also in vessels as large as the aorta—core flow dynamics differ remarkably between Newtonian and non-Newtonian fluids, see Fig. 6.12. Whilst many models are available to capture the rheological properties of blood, the identification of model parameters may be challenging for some of them [186].

Non-physiological aspects to the circulation may result in shear stress-induced *blood damage*, with hemolysis and thrombocyte activation being of most concern to medical device developers. The acquisition of experimental data led to a number of purely phenomenological blood-damage models, and an uncertainty assessment should always be considered [178] in the deployment of such models.

Mixing counteracts the segregation of blood particles and contributes to the chemical reactivity of blood. A VS represents an isolated pocket of blood at high vorticity and may therefore play a significant role in the biology of the blood flow. In addition to *transport-related* effects, also *shear stress* has well-known implications on vascular physiology and pathology. Both mechanisms led to the introduction of a number of blood flow parameters and post-processing techniques to assess the biological impact of blood flow on the vasculature.

Whilst the study of blood flow in the normal vasculature already revealed many physiological mechanisms, it is the investigation of diseased vessels that is of much more clinical relevance. The computation of the 3D blood flow within such vessel segments almost always requires CFD methods, sometimes even the access

to High Performance Computing (HPC) facilities. The results of CFD simulations
are sensitive to factors, such as geometry, boundary conditions, loading, and blood
rheology. Even under well-defined laboratory conditions, the acquisition of input
information is challenging and limits the applicability of CFD simulations. An
uncertainty analysis may therefore be foreseen to assess the robustness of CFD-
based results [164, 453]. As with many biomechanical models, the uncertainty of
(clinically) acquired input information is almost always the limiting factor and
determines the robustness of the results. Modern image modalities provide detailed
anatomical information of the vasculature, but it is the boundary conditions that
influence CFD results most [36, 56, 179, 466]—model personalization requires more
than anatomical personalization [265].

The Vascular Wall, an Active Entity

7

This chapter addresses the active properties of blood vessels with a focus on vasoreactivity and arteriogenesis. We review the structure and function of Smooth Muscle Cells (SMC) and their interaction with Endothelium Cells (EC) in the control of blood vessels. The discussion of collagen and elastin synthesis then aims at providing a fundamental understanding of arteriogenesis. A key section of this chapter concerns the constitutive description of vasoreactivity, where phenomenological, structural-based, and calcium concentration-based descriptions are addressed. We also provide a chemo-mechanical description of SMC and discuss the related thermodynamical aspects. Another key section concerns the modeling of arteriogenesis within the framework of open-system thermodynamics. Kinematics-based and continuous turnover-based growth descriptions are introduced—and showcased through simple tension and tube inflation examples. We illustrate the spatial distribution of the synthesized mass with respect to the growth descriptions and investigate the prescription of a homeostatic target toward which the vessel adapts. A review of multiphasic and miscellaneous vessel descriptions follows, and concluding remarks summarize the chapter.

7.1 Introduction

The vessel is a *dynamic structure*, and a number of mechanisms allow vascular tissue to *adapt* to environmental changes in an effort to optimize tissue perfusion. *Homeostasis* directs the adaption in normal vessels and keeps target biomechanical properties, such as Wall Shear Stress (WSS) [74, 236], circumferential wall

The original version of this chapter was revised: ESM has been added. The correction to this chapter is available at https://doi.org/10.1007/978-3-030-70966-2_8

Supplementary Information The online version contains supplementary material available at https://doi.org/10.1007/978-3-030-70966-2_7

stress [363,598], and axial stretch [215,283] relatively constant over time [275,586]. The failure to reach homeostasis may however result in vascular *pathologies*. Given an aneurysm for example, the diminished biological activity of the aneurysmatic vessel wall [362, 391] then leads to its continuous expansion in size and the development of life-threatening high levels of wall stress. Tissue adaptation occurs through a wide range of mechanisms, including SMC activation/relaxation, cell proliferation, apoptosis, pattern formation and synthesis and/or degradation of ExtraCellular Matrix (ECM).

Vascular cell function is tightly linked to *mechanics*, and ECM synthesis, cell proliferation, and vasoreactivity have been correlated with the onset of pressure in the embryonic circulation. Through chemo-mechanotransduction, external stimuli influence cell function at the level of gene expression and thereby contribute to the overall control of the structure and function of vascular tissues. A number of cell complexes are able to convert mechanical signals into chemical responses, and therefore activate intercellular signaling pathways. We may distinguish four major families of sensors:

- *Molecules* in the nuclear envelope, such as nuclear pores
- *Molecules* in the cell membrane, such as ion channels, receptors, adhesion molecules, and the glycocalyx
- Membrane *micro-domains*, such as the primary cilia and the caveola
- *Cell-supporting structures*, such as the cytoskeleton and the lipid bilayer plasma membrane

These sensors constantly monitor the mechanical state of the cells, information in response to which the vessel wall then undergoes many changes during normal development, aging, and in response to disease or implanted devices. At the local tissue level, it results in

- *Contractility*—the change of muscle tonus
- *Remodeling*—the change of macroscopic mechanical tissue properties
- *Growth or atrophy*—the change of tissue mass

The interaction of these factors then also results in the development of residual stresses in the vessels' load-free configuration and determines vessel morphogenesis, the evolution of vessel shape.

In addition to vascular cells that actively sense and respond to mechanical loads, the ECM also plays an important role in the adaptation of the vessel wall. It transmits the load to the cells and controls their micro-mechanical environment. The ECM may therefore not only be seen as a passive structure that carries wall stress, but actively contributes to vessel wall biology. Vessel wall dynamics may be classified as *vasoreactivity, arteriogenesis*, and *angiogenesis*. Whilst the first two factors are discussed in this chapter, the latter will only be touched upon.

7.2   Vasoreactivity

The vascular wall is equipped with contractile cells—pericytes in the capillaries and SMC in the other vessels. It allows the vessel to control its caliber and to meet demands within seconds, a property called *vasoreactivity*. In the microvasculature, vasoreactivity allows to divert the blood flow and to control the Mean Arterial Pressure (MAP), see Chap. 2. In contrary, in medium and large vessel, it mainly influences pulse wave velocity through the alteration of the elasticity of the vessel wall.

The physiology and function of general SMC [153, 592] and vascular SMC [52] are well explored, but less is known from pericytes. Whilst their involvement in processes, such as blood vessel formation (arteriogenesis) and maintenance [38] has been reported, the contractile function of pericytes remains still somewhat debated [241, 253].

7.2.1   SMC Phenotypic Modulation

A unique feature of SMCs is their ability to switch the *phenotype* between *contractile or differentiated* and *synthetic or dedifferentiated* states—nuclei, cell shape, and the expression of contractile markers, all is linked to SMC phenotype. The switch is *reversible* and appears in response to mechanical forces and chemical factors. The change between the two phenotypes equips vascular SMC with a wide range of functions. At the synthetic phenotype, the cell synthesizes/secretes ECM constituents, proliferates and migrates. At the contractile phenotype, the SMC is able to contract and shows neither proliferation nor migration. SMC in the (elastic) thoracic aorta appears at both, synthetic and contractile phenotypes [426], whilst muscular arteries contain mainly contractile SMC. The switching among phenotypes is not an instant event, but involves several stages [52], and recently it has also been suggested that SMC can also appear in a *degradative* phenotype, accomplishing tasks similar to macrophages [339].

To maintain the right balance between the two phenotypes, SMCs communicate with each other and with ECs to acquire environmental information. SMC phenotype composition is known to change in response to mechanical factors. In addition to forces directly applied to the SMC, also WSS at the endothelium modulates the phenotype of neighboring SMCs [439]. The ECs release Nitric Oxide (NO), which together with Reactive Oxygen Species (ROS) results in phenotypic modulation. ROS is a strong modulator of NO concentration, and an increase in ROS causes a decrease in NO concentration and then promotes SMCs to switch into their synthetic phenotype. Breakdown of NO is also a key mechanism of endothelium-dependent vasorelaxation, see Sect. 7.2.2.1.

With age the concentration of ROS in the vessel wall increases, and aging is therefore linked to a progressive shift from contractile to synthetic phenotypes. The synthetic phenotype promotes the production of ECM (especially collagen) and may

therefore explain vessel wall stiffening in the elderly. In addition, SMCs in the aged
vessel wall are smaller, and they often show senescence and/or apoptosis.

7.2.2 Structure and Contraction Machinery of Contractile SMC

SMCs are *elongated spindle-shaped* cells that are integrated in the vessel wall. They
are 2 to 10 μm in diameter, 50 to 400 μm long, and organized in Medial Lamellar
Unit (MLU). SMCs are aligned with the circumferential direction and at a radial
tilt of approximately 20 degrees [180, 400]. They have a centrally located nucleus
and filaments that are arranged in *branching* bundles, see Fig. 7.1. *Myofilament* are
formed by the interaction of *actin* (thin filament) and *myosin* (thick filament), whilst
intermediate filaments consist largely of vimentin and desmin. Filaments branch
at *dense bodies* and *dense plaques*, parts of the cell that appear darker under an
electron microscope, they are "electron-dense". Dense bodies are found inside the
SMC, whilst dense plaques, also called *focal adhesions*, are integrated in the cell
membrane. They connect cells with each other or anchor them to ECM constituents,
see Fig. 7.1a. Focal adhesions are macromolecular protein complexes able to
transmit and *sense* the forces entering the SMC. These *mechanosensors* translate
mechanical load into biochemical signals, which then results in gene-expression
patterns and influence factors, such as SMC phenotype, myofilament contraction,
and ECM synthesis. Whilst myofilaments actively contract, intermediate filaments
and microtubules are passively integrated in the SMC contraction kinematics. In
addition to the mechanical coupling, *gap junctions* allow for electrical and chemical
communication between neighboring SMCs.

In contrast to skeleton muscle, SMCs do not appear striated under light
microscopy, and the contractile proteins are not arranged in myofibrils. Actin
is cross-linked to myosin and contracts through an *antiparallel cross-bridge*
mechanism, see Fig. 7.1c. The actin filament of a muscle fiber connects to dense
bodies or dense plaques, and therefore transmits the contraction to the cell body.

7.2.2.1 Contraction and Relaxation
SMC *contraction* is regulated predominantly through the concentration of cytosolic
(intracellular) *free* Ca^{2+}. The exchange between the cell and the extracellular
space, and thus the transport through the cell membrane, as well as the exchange
with the sarcoplasmic reticulum (an intracellular storage of Ca^{2+}) influences the
concentration of free Ca^{2+}. At resting levels the concentration of free Ca^{2+} is
approximately 80 to 140 nmol l^{-1}, a level that changes in response to a number
of pro-contractive agonists, some of which are shown in Fig. 7.3 and will be further
discussed in Sect. 7.2.2.2. An agonist is a chemical that binds to a receptor and
activates the receptor to produce a biological response.

Given the activation concentration of 500 to 700 nmol l^{-1}, Ca^{2+} binds to
a molecule, called *calmodulin*, and forms a calcium-calmodulin complex. It is
this complex that then binds to the *Myosin Light Chain Kinase (MLCK)* and
activates it. Upon activation the regulatory Myosin Light Chain (MLC), a complex
located on the myosin motor is phosphorylated, and allows the chain of reactions

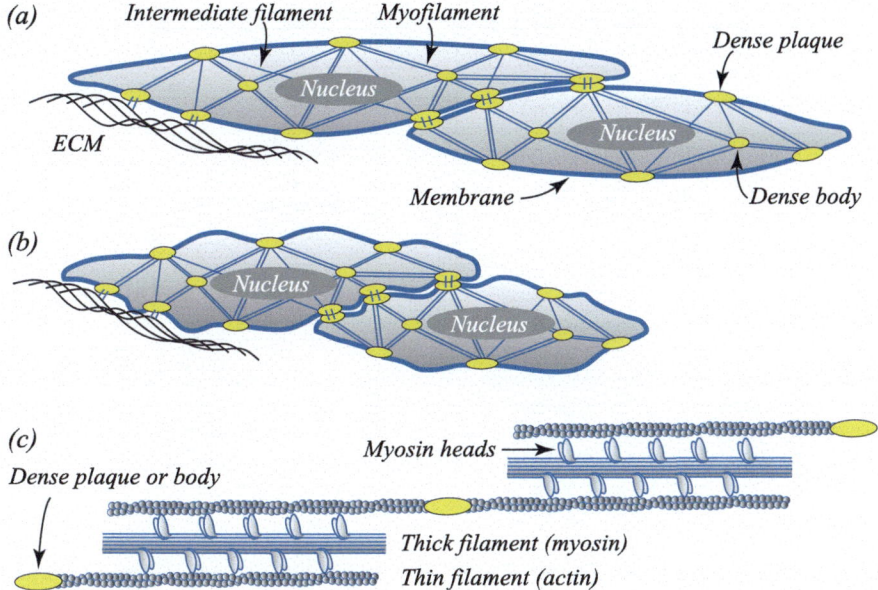

Fig. 7.1 Contraction mechanism of Smooth Muscle Cells (SMC). (**a**) Bundles of myofilaments (actin stress fibers) and intermediate filaments connect at dense bodies and dense plaques. Dense plaques connect across SMCs and to ExtraCellular Matrix (ECM) constituents. (**b**) On activation the myofilaments shorten, and the SMC shrinks predominantly along its longitudinal direction. (**c**) An antiparallel cross-bridge mechanism moves two adjacent filaments relative to each other. It shortens myofilaments and reduces the distance between two dense bodies/plaques

towards contraction. The actin and myosin then interact and the *cross-bridge cycles* establish—the myofilament contracts, and the SMC shortens. The energy for the cross-bridge cycle stroke is provided by *Adenosine Triphosphate (ATP)*, which splits into Adenosine Diphosphate (ADP) and a Phosphate ion (Pi), a reaction that produces a *molecular conformational change* in the neck domain of the myosin heavy chain. The myosin head tilts and drags along the actin filament a small distance of approximately 10 to 12 nm.

In contrary to contraction, the *Myosin Light Chain Phosphatase (MLCP)* catalyzes the *dephosphorylation* of the MLC, which then prevents from the formation of new actin–myosin cross-bridges—the SMC relaxes. Figure 7.2 summarizes the individual steps of SMC activation and relaxation, and the regulatory pathways of MLCK and MLCP activation and deactivation are well documented [386], see also Fig. 7.3.

7.2.2.2 Regulation of SMC Contraction and Relaxation

SMC *contraction* and *relaxation* is regulated through the *phosphorylation* and respective *dephosphorylation* of the regulatory *MLC*. *MLC phosphorylation* is a requirement for the interaction of myosin and actin. It allows *cross-bridge cycles* to establish and the SMC to contract and/or to generate stress. The state of MLC

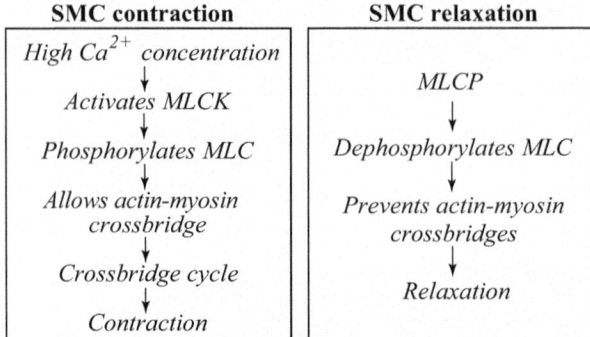

Fig. 7.2 Smooth Muscle Cell (SMC) contraction: Ca^{2+} activates the Myosin Light Chain Kinase (MLCK) and phosphorates the regulatory Myosin Light Chain (MLC). It then allows for the formation of cross-bridge cycles. SMC relaxation: Myosin Light Chain Phosphatase (MLCP) results in the dephosphorylation of the MLC, and then prevents from the formation of cross-bridge cycles

phosphorylation results from the balance between MLCK and MLCP activities and is regulated by a number of pathways.

Vasculature endothelium plays a central role in the regulation of SMC contraction and relaxation. Through the degradation, conversion, or uptake of *vasoactive signaling molecules*, the EC is able to *passively* regulate the contractile state of nearby SMCs. The EC can also *actively* regulate SMC through the formation and release of endothelium-1, NO, prostacyclin, and endothelium-derived hyperpolarization factors. The latter are collectively called vasoactive autacoids.

Endothelium-1 is an amino acid with well-known *vasoconstrictor* properties. The different regional receptor expressions of endothelium-1 support selective responses in different blood vessels [616]. The thoracic aorta therefore responds weakly to endothelium-1, whilst it is a powerful constrictor in the abdominal aorta. Endothelium-1 activates an enzyme that liberates *Inositol TriPhosphate (IP3)* and *DiAcylGlycerol (DAG)* from a membrane lipid, see Fig. 7.3a. Whilst IP3 activates Receptor-Operated Calcium Channels (ROCC) and increases the release of Ca^{2+} from sarcoplasmic reticulum, DAG stimulates Protein Kinase C (PKC), which depolarizes the cell and allows Ca^{2+} entry over Voltage-Gated Ca^{2+} Channels (VGCC). Both factors *increase* the concentration of cytosolic free Ca^{2+}, and *activates* MLCK. It phosphorates the regulatory MLC and leads to cross-bridge cycling. In addition to endothelin-1, a variety of *other agonists*, such as norepinephrine, histamine, leukotrienes, and thromboxane A2 activate SMC contraction through the very same pathways.

Another pathway of SMC contraction is through *rho kinase*, see Fig. 7.3c. Several agonists, some of which are EC-derived, activate the small monomeric G-protein RhoA through binding receptors in the cell membrane. It activates rho kinase and *inhibits* MLCP. The phosphorylation of the regulatory MLC is then preserved and maintains force development, even in the absence of a sustained

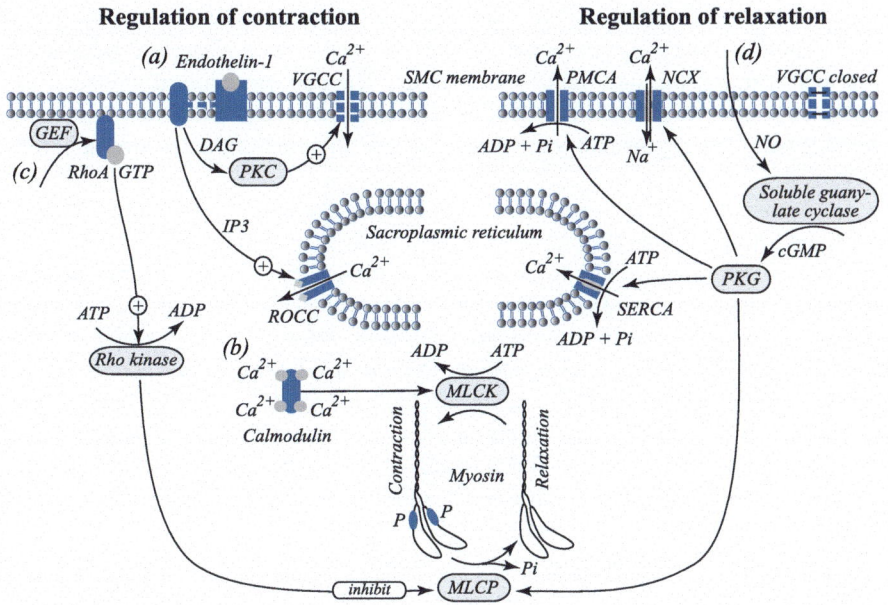

Fig. 7.3 Regulation of contraction (left panel) and relaxation (right panel) of Smooth Muscle Cells (SMC). (**a**) Endothelin-1 activates the release of Inositol TriPhosphate (IP3) and DiAcyl-Glycerol (DAG) from the membrane. IP3 upregulates Ca^{2+} release from sarcoplasmic reticulum through Receptor-Operated Calcium Channels (ROCC). DAG stimulates Protein Kinase C (PKC) and upregulates intercellular Ca^{2+} flux through Voltage-Gated Calcium Channels (VGCC). (**b**) Cytosolic free Ca^{2+} binds to calmodulin and activates the Myosin Light Chain Kinase (MLCK). It phosphorates the regulatory Myosin Light Chain (MLC), such that actin and myosin cross-bridges establish, and the SMC contracts. (**c**) Agonists bind to receptors in the cell membrane and activate rho kinase. It inhibits Myosin Light Chain Phosphatase (MLCP) and preserves the phosphorylation of the regulatory MLC and maintains SMC contraction. (**d**) Nitric Oxide (NO) diffuses into cell membrane and activates soluble guanylate cyclase. This increases cytosolic cyclic Guanylyl MonophosPhate (cGMP) and activates Protein Kinase G (PKG). PKG activates MLCP and prevents from MLC phosphorylation. It also upregulates Ca^{2+} transport into the sarcoplasmic reticulum through Sarco/Endoplasmic Reticulum Ca^{2+}- ATPase (SERCA) calcium pumps, and downregulates Ca^{2+} entry over voltage-gated Ca^{2+} channels in the surface membrane by the use of Na^{+}- Ca^{2+} eXchanger (NCX) and Plasma Membrane Ca^{2+} ATPase (PMCA)

Ca^{2+} concentration. Overall MLCP inhibition increases the sensitivity of MLCK to the Ca^{2+} concentration, and SMC contraction therefore becomes more sensitive to changes in Ca^{2+}, a property called *calcium sensitization*.

In contrary to the aforementioned vasoconstrictors, NO is the most important *vasodilator* produced by the endothelium. The formation of NO is catalyzed by the *endothelial NO synthase (eNOS)*; it is continuously expressed by numerous factors and affects the level of eNOS expression and activity. In addition to mechanical forces, neurotransmitters, hormones, autacoids, and coagulation factors modulate eNOS expression and activity [163]. The endothelium-based NO diffuses into the SMC, where it activates soluble guanylate cyclase and increases cytosolic cyclic guanylyl monophosphate (cGMP), see Fig. 7.3d. This activates Protein Kinase G (PKG) and promotes the transport of Ca^{2+} into the sarcoplasmic reticulum, as well as its outflux through the cell membrane. The concentration of cytosolic free Ca^{2+} therefore reduces, and the SMC *relaxes*. PKG also activates MLCP, and thus prevents from MLC phosphorylation and the establishment of actin–myosin cross-bridges.

The transport of Ca^{2+} into the sarcoplasmic reticulum is facilitated through sarco/endoplasmic reticulum Ca^{2+}-ATPase (SERCA) calcium pumps, where two Ca^{2+} are transported per hydrolyzed ATP. ATPases are a class of enzymes that catalyze the decomposition of ATP into ADP and a free Pi, and the respective inverse reaction. This dephosphorylation reaction releases energy, which in most cases activates other chemical reactions. In contrary, Ca^{2+} outflux across the cell membrane is mediated by Na^+ - Ca^{2+} exchanger (NCX), which exchanges three Na^+ for one Ca^{2+}, and Plasma Membrane Ca^{2+} ATPase (PMCA), which transports one Ca^{2+} out of the cell per ATP molecule hydrolyzed. NO also leads to SMC hyperpolarization, and thus inhibits Ca^{2+} entry over voltage-gated Ca^{2+} channels (VGCC) in the surface membrane.

The aforementioned pathways are differently active in elastic, muscular, and resistance arteries [386], and endothelial dysfunction leads to the imbalance resulting in pathologies, such as hypertension. A detailed knowledge of these pathways is also important in the *in vitro* experimental characterization of active vessel wall properties.

7.2.3   SMC Tone and Vessel Wall Properties

The level of SMC activation or vascular tone is determined by factors, such as neurotransmitters released from autonomic nerves, circulating vasoactive compounds, tissue metabolites, and endothelium-derived autacoids. They change in response to biomechanical stimuli, such as flow [111] or pressure [282, 498], hormonal stimuli, neural stimuli, and drugs. SMC's contractile apparatus is also activated through stretch, leading to an autonomous contraction known as *myogenic response*. The level of SMC activation depends not only on the circumferential stretch [282], but also on the axial stretch [75, 621].

SMCs are organized in MLU and predominantly aligned in the circumferential direction. SMC contraction therefore adds up to the circumferential wall stress and has very minor influence on the other stress components. In arteries, the activation of SMC is able to reduce the vessel diameter by 20 to 50% [25, 104, 130, 385]. The stress from SMC contraction is maximal at the physiological blood pressure, and thus at the vessel's *in vivo* diameter (see Fig. 7.4a) with the SMC being at the homeostatic length [387].

The activation of SMCs influences the stiffness of the vessel wall, and an increase [103] as well as a decrease [223] in stiffness upon vessel activation has been reported. The related experimental conditions may explain this contradiction, see Fig. 7.4. It illustrates data from *in vitro* inflation experiments of rat carotid

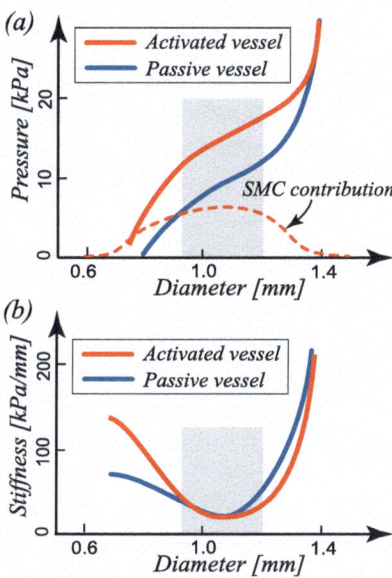

Fig. 7.4 Mechanical vessel properties with (activated vessel) and without (passive vessel) the contribution of Smooth Muscle Cells (SMC). The data represents the properties from *in vitro* testing of rat carotid arteries [622], where the gray-shaded area indicates the vessel's *in vivo* deformation. (**a**) Pressure *p versus* diameter *d* properties of the activated (red) and passive (blue) vessel, respectively. The dashed curve illustrates the SMC-related contribution and represents the difference between the pressure-diameter response of activated and passive vessels. (**b**) Diameter stiffness $\Delta p / \Delta d$ of the activated (red) and passive (blue) vessel

arteries [622], where p denotes the transmural pressure. The vessel's diameter stiffness $\Delta p / \Delta d$ is clearly a function of the vessel diameter d. Whilst at smaller diameters the stiffness of the activated vessel is higher than of the passive vessel, the opposite is valid at larger diameters, see Fig. 7.4b.

Given sufficient cytosolic free Ca^{2+}, the shortening velocity of SMC correlates well with the level of the MLC phosphorylation. Within a few minutes upon contraction initiation the Ca^{2+} level markedly decreases, such that MLC phosphorylation, and therefore energy utilization, decreases. Whilst this function is similar to skeletal muscle, SMC is in addition able to sustain contraction and maintain force for a prolonged time even at low levels of Ca^{2+}. SMC is then very economic and maintains the contraction force at minimal energy utilization and low rates of ATP hydrolysis. This sustained phase has been attributed to specific myosin cross-bridges, termed *latch bridges*. They are dephosphorylated but still attached cross-bridges [124]. Regardless the unknown molecular basis, latch bridges show very slow and Ca^{2+} level-independent MLC dephosphorylation. It results in slow cycling rates of latch bridges and the maintenance of force at low energy cost. In SMC, normalized force and shortening velocities are therefore regulated functions of cross-bridge phosphorylation [384], which is in clear contrast to skeletal muscle.

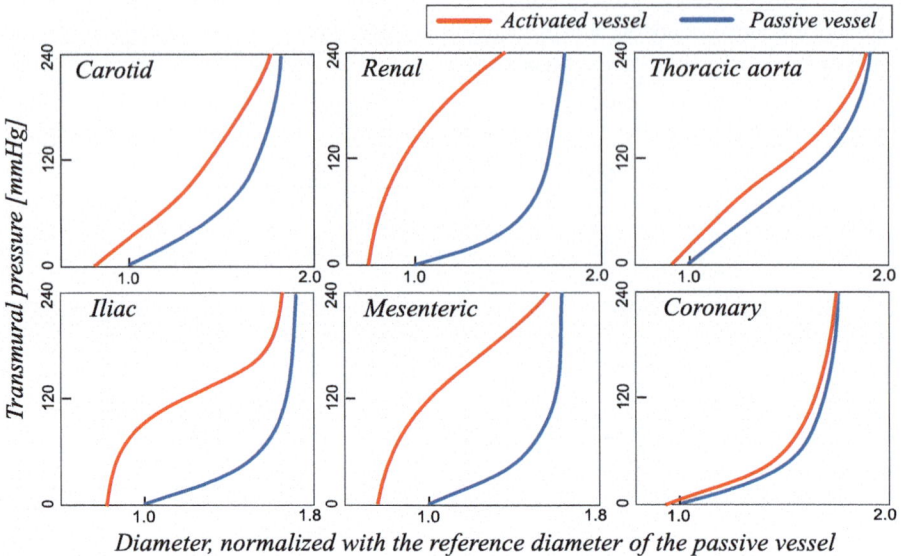

Fig. 7.5 Diameter *versus* pressure of activated (red) and passive (blue) arteries. Data represent results from *in vitro* cyclic inflation at *in vivo* length of canine arteries and used potassium (K) as vasoconstrictor agent [105]

Another factor that determines the amount of active wall stress relates to whether the SMCs contract at *isometric* or *isotonic* conditions. The stretch or respectively the stress in the vessel wall is then kept constant upon activation. As with skeleton muscle, SMC also generates the maximal contraction force at isometric conditions.

According to their physiological function, an artery may be classified as an *elastic vessel* that contributes to the capacity of the vascular system, or as a *muscular vessel* that is able to contract and divert the blood flow. Vasoconstriction alters the pressure-diameter properties much more in muscular than elastic arteries, see Fig. 7.5. This figure shows the pressure-diameter properties of activated vessel relative to their passive response for a number of canine arteries [105]. The data was acquired by *in vitro* cyclic inflation at the vessels' *in vivo* lengths and potassium (K) in physiological salt solution was the vasoconstrictor agent. Whilst the coronaries are commonly regarded as muscular arteries, the data shows minimal SMC-related contribution to its pressure-diameter properties and underlines the exceptional properties of the coronaries.

In addition to the modification of the vessel's pressure-diameter properties, the SMC stress may also contribute to the reduction of the transmural stress gradient at physiological loading [274, 446].

7.3   Arteriogenesis

Whilst vasoreactivity allows the vessel to react within seconds to environmental cues through the modulation of contractile cells in the wall, *arteriogenesis* operates at the timescale of days and weeks through the ability to add or remove vessel wall tissue. It relies on a delicate (coupled) balance between degradation and synthesis of vessel wall proteins, collectively called *tissue turnover*. Vascular cells, such as macrophages, SMCs, and fibroblasts not only synthesize ECM and contractile proteins, but secrete also Matrix MetalloProteinases (MMPs) that then constantly degrades ECM. It allows the vessel wall to undergo many changes during normal development and in response to disease. Whilst vasoactivity is the first functional response of the vessel to the tissue's sudden demand of oxygen and nutrient, vasoreactivity may be seen as the continuation of such response.

Mechanical forces play a central role in arteriogenesis and effect both, the synthesis and the degradation of vessel wall tissue. Vascular cells are anchored to ECM proteins, mechanical communications that allow the cells to sense forces and to adjust the synthesis of ECM proteins, MMPs, and other compounds [41, 50, 207, 228, 388, 502, 553]. The forces may have very different origins, referring to the WSS from the blood flowing over the endothelium, cyclic wall stress, and the stress induced by the transmural interstitial flow through the vessel wall, for example. The transmural interstitial flow also advects a number of mediators towards the cells that are able to influence cell function [375]. As with the synthesis, also the degradation of ECM proteins is linked to mechanical factors. Given a laboratory experiment, it is therefore often challenging to attribute a mechanical factor either to the synthesis or the degradation of ECM compounds.

The *reaction kinetics*, and thus the rates at which individual ECM proteins are synthesized or degraded, are very different and introduce very different time scales in the maintenance of the vessel wall. In the homeostatic and mature vascular wall, the collagen has a half-life time of approximately two months [395]. Elastin in contrary is extremely insoluble and stable—it has half-life times in the order of tens of years [7]. Elastin may, however, been degraded by selective MMPs, a process that is important for physiological processes, such as elastogenesis and repair [585].

7.3.1   Interplay of Endothelium Cells and Smooth Muscle Cells

The involvement of mechanical factors in arteriogenesis has been known for a long time: cyclic stretching of vascular SMCs increases their MMP production [228], which then enhances collagen degradation and therefore results in the upregulation of collagen turnover [337]. In addition to such direct cellular response to mechanical factors, SMC function is also indirectly controlled through the endothelium. ECs are constantly exposed to WSS, loading that has been associated with changes in gene-expression patterns through positive and negative WSS-responsive elements in their promoter regions [79]. WSS sensing has been attributed to different mechanosensors, including G protein-coupled receptors, glycocalyx, primary cilium, ion channels, and the endothelial-specific junctional complex comprising Platelet Endothelial Cell Adhesion Molecule-1 (PECAM1), Vascular Endothelial cadherin (VE-cadherin), and Vascular Endothelial Growth Factors (VEGFs) [22,238]. It is known for a long time that alterations of the flow rate result in proportional vessel diameter changes, an effort to restore the initial WSS level [295,296,324,549,554,556]. These observations suggested that ECs modulate arteriogenesis to maintain WSS. Given homeostatic WSS levels, pathways that promote blood vessel stabilization are activated, whilst arteriogenesis-characteristic pathways are triggered away from homeostasis [22] and modulate MMP in the vessel wall [99,237,372].

7.3.2 WSS Profile and Inflammation

Blood flow changes remarkably along the vascular tree and is especially complex at vessel segments, such as bifurcations, constrictions, and dilations. All these complexity is sensed and processed by the endothelium. EC's mechanosensors are able to decode subsecond-frequency characteristics. The exposure of ECs to WSS of different frequency spectra yields very different Nuclear Factor kappa-light-chain-enhancer of activated B cells (NF-κB) expression [150], see Fig. 7.6. NF-κB is a protein complex that controls transcription of DeoxyriboNucleic Acid (DNA), cytokine production, and cell survival. It is found in almost all animal cell types, and it is an important regulator of inflammation. Given inflammation plays a central role in atherosclerosis (see Sect. 5.4.2), NF-κB expression has been extensively studied in atherosclerotic vessels.

The implication of blood flow on the endothelium might roughly be classified as *static WSS*, *pulsatile WSS*, and *oscillatory WSS*. Static WSS is constant in time, pulsatile WSS changes over the cardiac cycle but points always into the same direction, whilst oscillatory WSS changes its direction with phases of forward and

Fig. 7.6 *In vitro* expression of Nuclear Factor kappa-light-chain-enhancer of activated B cells (NF-κB) of Endothelial Cells (ECs) in response to Wall Shear Stress (WSS) [150]. Data marked by the star represent the mechanical *in vivo* environment of carotid artery segments that are protective (red) and prone (green) to the formation of atherosclerosis, respectively

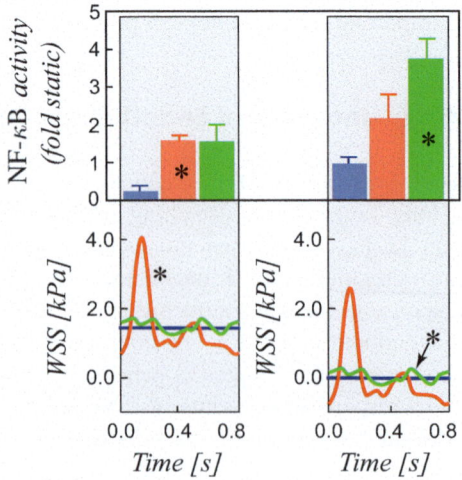

backward flows. Vascular EC responds very differently to such WSS characteristics, see Fig. 7.6. The stars in the plots denote carotid artery segments that are protective (red curve; pulsatile WSS) and prone (green curve; oscillatory WSS) to the formation of atherosclerosis. A high expression of NF-κB appears in EC that is exposed to the oscillatory WSS profile seen in the internal carotid artery. This vessel segment is known to be very prone to atherosclerosis. Atherosclerosis is an inherently inflammatory disease, and the data in Fig. 7.6 correlates well with the key role played by NF-κB in the control of many genes involved in inflammation.

7.3.3 Collagen Synthesis

In addition to synthetic SMC, most synthesis of collagen appears through *fibroblasts*, a cell type that is specialized for collagen synthesis. Figure 7.7 illustrates the

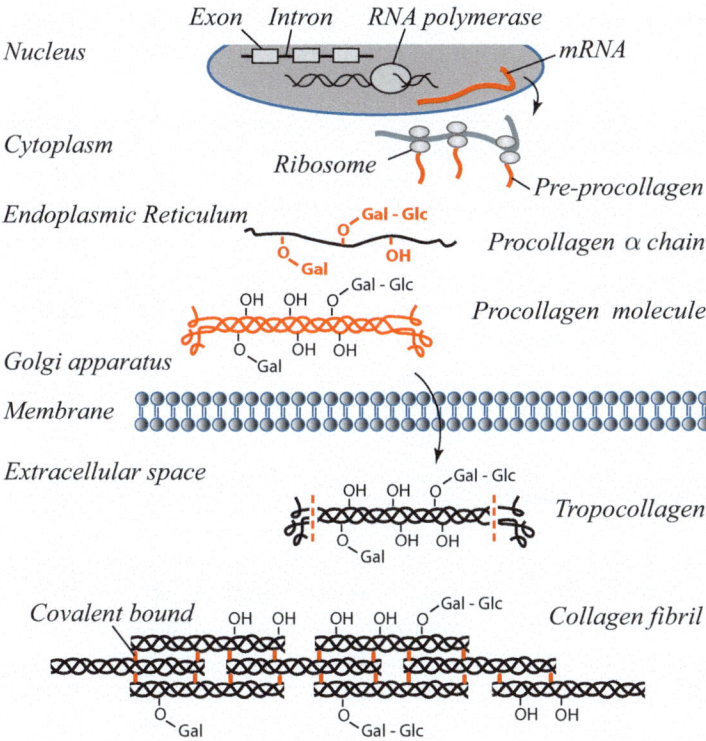

Fig. 7.7 Collagen synthesis. Transcription and translation lead to pre-procollagen, which is then modified into a procollagen α chain. Three such chains form a procollagen molecule and then moved to the extravascular space. Removing both ends from procollagen forms tropocollagen. A stack of tropocollagen molecules finally forms a collagen fibril. mRNA–messenger RiboNucleic Acid; HO–hydroxyl group; Gal–galactose; Glc–glucose

pathway of collagen synthesis, involving *intracellular* and *extracellular* processes. Collagen synthesis starts in the cell nucleus and ends in the extracellular space.

7.3.3.1 Intracellular Processes

In the cell nucleus the first step of DNA-based gene expression, the *transcription* of messenger RiboNucleic Acid (mRNA), takes place. The mRNA moves then into the cytoplasm and interacts with free ribosomes for *translation*. The compound is now called *pre-procollagen* and travels further into the endoplasmic reticulum for post-translational modification, steps resulting in a *procollagen α chain*. The modifications include the removal of the signal peptide on the N-terminal, adding hydroxyl groups (OH), and glycosylation of the selected hydroxyl groups with galactose (Gal) and glucose (Glc). The N-terminal refers to NH_2, a free amine group located at one end of a protein, whilst the C-terminal, a free carboxyl group COOH, forms the other end. The translation of a protein from mRNA starts with the N-terminal and ends with the C-terminal. Three procollagen α chains then form a *procollagen molecule* and mark the last step of intracellular collagen synthesis. The molecule is then moved to the Golgi apparatus, where it is assembled into secretory vesicles and transported through the cell membrane into the extracellular space, a process called exocytosis.

7.3.3.2 Extracellular Processes

Entering the extracellular space, propeptide cleavage forms *tropocollagen*. Here, both ends (N-terminal and C-terminal) of the procollagen molecule are removed, by enzymes known as collagen peptidases. Lysyl oxidase, a copper-dependent enzyme then acts on lysine and hydroxylysines, and covalent bonding between tropocollagen molecules results in a *collagen fibril*. Collagen fibrils may be regarded as the basic structural unit of collagenous tissues—a large number of supra-fibril structures emerge from it. In blood vessels, collagen fibrils form *collagen fibers*, rope-like bundles of fibrils that reinforce the vessel wall, see Sect. 5.2.1.

7.3.4   Elastin Synthesis

In the mature and normal vessel, elastin synthesis is often negligible. However, proteolytic degradation and repair of elastin are important factors in growth, wound healing, pregnancy, and tissue remodeling, all of which require upregulated elastin syntheses. As with collagen, also elastin is a remarkable hierarchical structure, and a number of small *tropoelastin* protein molecules form the finally very stable elastin complex. The tropoelastin molecules are cross-linked via their lysine residues with desmosine and isodesmosine cross-linking molecules. Lysyl oxidase establishes the cross-linking and results in a very stable structure.

7.4   Angiogenesis

Angiogenesis is another vascular process towards the optimization of tissue perfusion. It leads to the formation of new capillaries sprouting from the pre-existing vessels. Whilst arteriogenesis alters existing collateral vessels, angiogenesis expands the vascular tree to secure tissue perfusion. Hypoxia is the primary cue. Angiogenesis is therefore *ischemia-driven* and new capillaries form in response to the lack of oxygen in the surrounding tissue.

In addition to the normal development, angiogenesis is also a key step in tumor growth and stimulated by proteins, such as integrins and prostaglandins as well as growth factors, such as VEGF. The discussion of mechanical factors in angiogenesis is beyond the scope of this book, and the reader is referred to the literature [505].

7.5   Modeling Vasoreactivity

Vasoreactivity appears at time scales that are too short for the exchange of tissue mass, and the classical *closed system* governing laws therefore apply, see Sect. 3.6. Within this framework, a number of mathematical descriptions have been proposed towards modeling the stress contribution from contractile cells in the vessel wall—some of them are discussed hereafter.

7.5.1   Hill's Three-Parameter Muscle Model

Hill's three-parameter model for tetanized muscle contraction expresses the tension P as a function of the contraction velocity v. It follows from the *energy balance* and results in the hyperbolic relation

$$P = \frac{P_0(1 - v/v_0)}{1 - (P_0/a)(v/v_0)} , \tag{7.1}$$

where P_0 [N] and v_0 [m s^{-1}] denote the maximum isometric tension and the maximum contraction velocity of the muscle, respectively. The shape of the force-velocity curve is described by the parameter a. The muscle generates the maximum force P_0 at $v_0 = 0$, and the maximum contraction velocity is reached at $P_0 = 0$. Hill's model (7.1) predicts a decreasing force (stress) at increasing contraction velocity (strain rate). It is therefore in direct contrast to viscoelasticity, where the force (stress) increases at increasing strain rate. Whilst the model (7.1) has been widely applied in the description of skeletal muscles, it is barely used for SMC.

Table 7.1 Data points in the format $(\lambda_\theta, F$ [N]) that characterize the active and passive vessel wall, respectively

Active wall	(1, 0.08)	(1.3, 0.6)	(1.6, 0.8)	(1.9, 0.95)
Passive wall	(1, 0.01)	(1.3, 0.17)	(1.6, 0.3)	(1.9, 0.89)

7.5.2   Phenomenological Descriptions

The total stress in the active vessel wall results from the superposition of the ECM-based passive stress and the cell-based active stress. Whilst many external factors are able to activate contractile cells in the wall, the description of the *myogenic response* attracted most biomechanical attention. The cells are then activated through the stretch in the vessel wall.

Let us consider an example of a purely phenomenological description of vasoreactivity. A flat vessel wall sample of stress-free length $L = 54.0$ mm, width $W = 5.4$ mm, and thickness $H = 2.2$ mm is mounted in a tensile testing machine and stretched in the circumferential direction. The testing protocol foresees two acquisition cycles—in the first cycle the SMC in the vessel wall are active, whilst a drug suppresses their contractility in the second cycle. Table 7.1 reports the wall properties after preconditioning from both acquisition cycles. The data refers to the properties with and without SMC contribution, respectively. In the table λ_θ and F denote the stretch and force recorded from tensile testing.

The vessel wall's passive response represents ECM properties and is expressed by the first Piola–Kirchhoff stress

$$P_{\mathrm{ECM}} = c_{\mathrm{ECM}}(\lambda_\theta - 1)\exp[(\lambda_\theta - 1)^2] \,,$$

where λ_θ denotes the tissue stretch in circumferential direction, whilst c_{ECM} [Pa] is a stiffens-related material parameter. We may identify c_{ECM} from the minimization of the objective function

$$\Phi = \sum_{k=1}^{4} \left[P_{\mathrm{ECM}}(\lambda_{\theta\,k}) W H - F_{\mathrm{ECM}\,k}^{\mathrm{exp}} \right]^2 \,, \tag{7.2}$$

where WH [m^2] denotes the referential cross-section of the vessel wall specimen, and $F_{\mathrm{ECM}\,k}^{\mathrm{exp}}$ is the force that has been recorded at the k-th measurement point. The data refer to the properties of the passive wall, as listed in second row of Table 7.1. The minimization $\Phi \to$ MIN of (7.2) then yields the parameter $c_{\mathrm{ECM}} = 36.025$ kPa, and the passive force *versus* stretch properties are plotted in Fig. 7.8.

Following a purely phenomenological modeling, we describe the contribution from SMC contraction by the first Piola–Kirchhoff stress

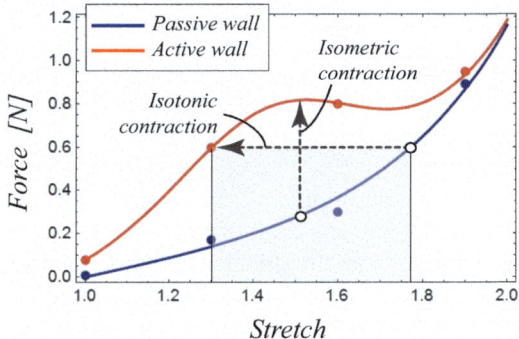

Fig. 7.8 Force *versus* stretch properties of the passive (blue) and respective active (red) vessel wall. Model parameters have been identified from the illustrated measurement points. Dashed arrows represent isotonic and isometric contractions, respectively. The shaded area illustrates the specific mechanical work done by isotonic contraction against 0.6 N

$$P_{\text{SMC}} = c_{\text{SMC}} \exp\left[-\frac{(\lambda_\theta - \lambda_m)^2}{2\lambda_d} \right] \tag{7.3}$$

with c_{SMC} [Pa] and λ_m, λ_d denoting material parameters to be identified from the experimental recordings of the activated vessel wall. Given the sparse experimental data, the simultaneous identification of these parameters was not possible, and we therefore fixed the dispersion parameter at $\lambda_d = 0.05$. It then results in a well-posed minimization problem $\Phi \rightarrow \text{MIN}$ and the objective function

$$\Phi = \sum_{k=1}^{4} \left\{ [P_{\text{ECM}}(\lambda_{\theta\,k}) + P_{\text{SMC}}(\lambda_{\theta\,k})] W H - F_{\text{active}\,k}^{\text{exp}} \right\}^2 ,$$

identified the parameter $c_{\text{SMC}} = 47.141$ kPa and $\lambda_m = 1.4422$, where $P_{\text{ECM}} + P_{\text{SMC}}$ and $F_{\text{active}\,k}^{\text{exp}}$ denote the analytical stress and the experimental force measurements, respectively. Figure 7.8 illustrates the force-stretch properties of the active vessel wall.

We may now explore the external mechanical work done by SMC, and consider the two limit cases of *isotonic* and *isometric* contractions, respectively. Given isotonic contraction, SMC contracts against a constant force, whilst at isometric contraction the SMC stretch remains constant upon contraction. The path lines of both contractions are shown in Fig. 7.8, and the area under the path line represents the product of force and stretch. It is therefore the specific external work $\mathcal{W}/L$ per unit reference length L of the tissue sample. Whilst an isotonic contraction exhibits external work, this is not the case for an isometric contraction. Given the isotonic contraction against the force of $F = 0.6$ N, it results in the work

$$\mathcal{W}_{\text{isotonic}} = 0.6 W H L (\lambda_{\theta\,\text{passive}} - \lambda_{\theta\,\text{active}}) = 3.02237 \cdot 10^{-7} \text{ J},$$

where the stretches $\lambda_{\theta\,\mathrm{passive}} = 1.7722$ and $\lambda_{\theta\,\mathrm{active}} = 1.3011$ have been identified from

$$P_{\mathrm{ECM}}(\lambda_{\theta\,\mathrm{passive}})WH = 0.6\,;$$
$$(P_{\mathrm{ECM}}(\lambda_{\theta\,\mathrm{active}}) + P_{\mathrm{SMC}}(\lambda_{\theta\,\mathrm{active}}))WH = 0.6\,.$$

We may further refine the description of the active vessel wall and introduce a scalar $0 \leq \alpha \leq 1$ that represents the level of SMC contractile activity, a property known as *muscle tonus* [446]. The first Piola–Kirchhoff stress then reads $P_{\mathrm{SMC}} = \alpha P^{\star}_{\mathrm{SMC}}$, where $P^{\star}_{\mathrm{SMC}}$ expresses the stretch-tension relationship of the fully activated SMC.

Example 7.1 (Biaxially Loaded Vessel Wall Patch). A batch of aortic wall tissue is mounted in a planar biaxial testing machine to explore the change of stress upon Smooth Muscle Cell (SMC) activation. For this experiment, the spatial configuration Ω is fixed and determined by the edge length $a = 2.3\,\mathrm{cm}$ and the thickness $h = 1.7\,\mathrm{mm}$ of the tissue specimen, see Fig. 7.9. Incompressibility of the vessel wall may be assumed, and the SMC fibers are aligned along the direction $\mathbf{a} = [\sqrt{3}/2\;\;1/2\;\;0]^{\mathrm{T}}$.

Fig. 7.9 Vessel wall specimen mounted in a planar biaxial testing machine. The specimen covers the spatial domain Ω, and the unit direction vector **a** determines the alignment of Smooth Muscle Cell (SMC) fibers.

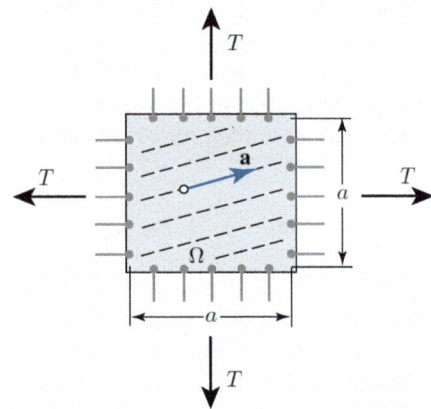

(a) Given the passive wall specimen is at equi-biaxial tension $T = 1.2\,\mathrm{N}$, compute the corresponding deformation and the elastic energy that is stored in the tissue sample. The ExtraCellular Matrix (ECM) may be described by a neoHookean strain energy $\psi_{\mathrm{ECM}} = c_{\mathrm{ECM}}(I_1 - 1)/2$, where $c_{\mathrm{ECM}} = 35.38\,\mathrm{kPa}$ denotes its referential stiffness, and $I_1 = \mathrm{tr}\mathbf{C}$ is the first invariant of the right Cauchy–Green strain tensor $\mathbf{C}$.

(b) Compute the first Piola–Kirchhoff stress of the fully activated vessel wall sample. The activated SMC fibers may be described by expression (7.3), where the material parameters $c_{\mathrm{SMC}} = 28.15\,\mathrm{kPa}$, $\lambda_{\mathrm{m}} = 1.15$, and $\lambda_{\mathrm{d}} = 0.01$ model the myogenic response. The stretch λ along the SMC fiber is taken with respect to specimen's stress-free configuration Ω_0. ∎

7.5.3 Structural-Based Descriptions

SMCs are predominately aligned along the circumferential direction and may be regarded as *active stress fibers* that carry load in parallel to the vessel wall's ECM. The rheology model in Fig. 7.10 represents such a structural view. It is known as *Hill's three-element model* (to be distinguished from Hill's three-parameter model [252]) and has originally been proposed to model skeleton muscle. In this section, Hill's model is generalized to hyperelasticity and towards modeling the nonlinear properties of the vessel wall.

We represent the vessel wall's circumferential stretch by $\lambda_\theta = \sqrt{\mathbf{C} : (\mathbf{e}_\theta \otimes \mathbf{e}_\theta)}$, where $\mathbf{e}_\theta$ is the unit direction vector along the vessel's circumference, and $\mathbf{C} = \mathbf{F}^\mathsf{T}\mathbf{F}$ is the right Cauchy–Green strain with respect to the vessel wall's stress-free reference configuration Ω_0. SMC fibers are aligned along the circumference and stretched at

$$\lambda_{SMC} = \lambda_\theta \lambda_{pre},$$

an expression that uses finite strain kinematics, where λ_{pre} denotes the pre-stretch of SMC relative to the vessel wall's reference configuration, see Fig. 7.10.

We regard the vessel wall as a mixture of ECM and SMC, and to be characterized by the Helmholtz free energy functions Ψ_{ECM} and Ψ_{SMC}, respectively. SMC appears at different activation levels, and the here introduced Ψ_{SMC} denotes the strain energy upon maximum activation. SMC activation level, or *tonus* is expressed by the scalar $0 \le \alpha \le 1$, and

$$\Psi(\mathbf{C}) = \Psi_{ECM}(\mathbf{C}) + \alpha(\mathbf{C})\Psi_{SMC}(\lambda_{SMC}) \qquad (7.5)$$

then describes the elastic energy stored in the vessel wall per unit (reference) tissue volume. With the intention to describe the *myogenic* response, the deformation $\mathbf{C}$ is

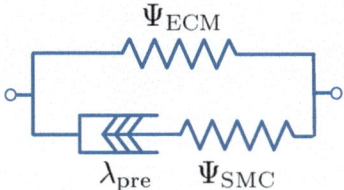

Fig. 7.10 Hill's three-element rheological model, where $\Psi_{iso\,ECM}$ and $\Psi_{iso\,SMC}$ represent the strain energies stored in the ExtraCellular Matrix (ECM) and the Smooth Muscle Cells (SMCs), respectively. SMC is pre-stretched at λ_{pre}, relative to the vessel wall's reference configuration

the most important argument to be considered in the tonus function α. In addition, any model from Sect. 5.5.2 may be used to describe the ECM contribution Ψ_{ECM} in (7.5).

A linear stress strain law has been proposed to model vascular SMC [622], and the strain energy

$$\Psi_{SMC}(\lambda_{SMC}) = c_{SMC}[\lambda_{SMC} - \ln(\lambda_{SMC}) - 1] \qquad (7.6)$$

then captures its elastic properties, where c_{SMC} [Pa] denotes the stiffness of the SMC tissue in the vessel wall.

To close the description of the vessel wall, the tonus α is specified as a function of the deformation $\mathbf{C}$. SMC is only able to contract in the range $\lambda_{\theta\,min} < \lambda_\theta < \lambda_{\theta\,max}$, where $\lambda_\theta = \lambda_{SMC}/\lambda_{pre}$ denotes the circumferential stretch of the vessel wall. Given $\lambda_\theta > \lambda_{\theta\,max}$, the actin–myosin overlap is too short for myofilaments to contract, whilst at $\lambda_\theta < \lambda_{\theta\,min}$, myofilaments are too long and the relative motion between actin and myosin is not able to generate tension. The stretch limits of $\lambda_{\theta\,min} = 0.68$ and $\lambda_{\theta\,max} = 1.505$ have been proposed for the rat carotid artery, for example [622].

Given the normal vessel wall and deformations in the range of $\lambda_{\theta\,min} < \lambda_\theta < \lambda_{\theta\,max}$,

$$\alpha(\mathbf{C}) = \alpha_{basal} + \frac{1 - \alpha_{basal}}{2} \exp\left[-\frac{(Q - Q_m)^2}{2Q_d}\right] \qquad (7.7)$$

may be used to express SMC tonus. Here, $Q(\mathbf{C})$ is a scalar that describes the deformation, and α_{basal} denotes the *basal tone* contraction, a tonus that is present even in the absence of any tissue deformation. In addition, the parameters Q_m and Q_d determine the median and variance of the tonus distribution. The aforementioned model is able to predict myogenic contraction in response to local stretch, and further details are reported elsewhere [622].

Example 7.2 (Ring Test to Characterize a Vessel Segment). The mechanical properties of a vessel segment of the dimensions D, L, and H are to be investigated by ring testing, see Fig. 7.11. Two steel pins are placed inside the vessel ring and then pulled apart in an effort to identify the segment's force F *versus* displacement δ properties. The experiment is conducted with active and passive vessel wall rings. A single-layer membrane model may be used in the analysis of the ring test, and the limit cases of simple tension and pure shear kinematics should be investigated. Friction-less contact between the sample and the steel pins, as well as the sample dimensions $L/D << 1$ would result in simple tension kinematics, whilst pure shear kinematics establishes for $L/D >> 1$ and stick contact.

Fig. 7.11 Schematic illustration of ring testing to characterize the mechanical properties of a vessel segment

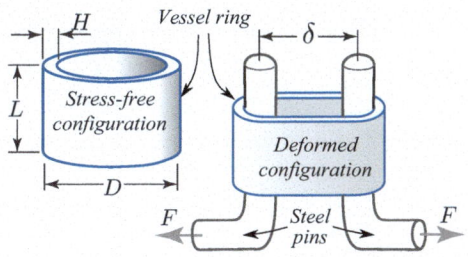

(a) Derive the Cauchy stress contributions of the passive vessel wall at simple tension and pure shear deformation kinematics. The vessel wall's ECM may be described by a neoHookean strain energy function $\Psi_{ECM}(\mathbf{C}) = c_{ECM}(I_1 - 3)/2$, where $I_1 = \mathrm{tr}\mathbf{C}$ denotes the first invariant of the right Cauchy–Green strain tensor $\mathbf{C}$.

(b) Derive the Cauchy stress contributions that follow from SMC contraction, where simple tension as well as pure shear is assumed to determine the deformation kinematics. SMCs are aligned along the circumferential direction, their elasticity may be described by the strain energy (7.6), and

$$\alpha(\mathbf{C}) = \exp\left[-\frac{(\lambda_{SMC} - \lambda_m)^2}{2\lambda_d}\right], \tag{7.8}$$

expresses their tonus, where λ_m and λ_d are material parameters. As compared to the model (7.7), this description uses the simplifications $Q = \lambda_{SMC}$ and $\alpha_{basal} = 0$.

(c) Consider ring tests with active and passive vessel rings and quantify the factor $r = F_{st}/F_{ps}$, where F_{st} and F_{ps} denote the forces upon simple tension and pure shear deformation kinematics, respectively.   ∎

7.5.4   Calcium Concentration-Based Descriptions

SMC contraction is determined by $\beta = [Ca^{2+}]$ [mol m^{-3}], the concentration of cytosolic (intracellular) free calcium—it is modulated by the flux across the cell membrane as well as the transport in and out of the sarcoplasmic reticulum. Electrochemical models, such as the *Hodgkin–Huxley-type electrical equivalent* [256] of cell membranes, may be used to describe the cytosolic calcium concentration. They explicitly express Ca^{2+} fluxes through VGCC, NCX, and PMCA (see Fig. 7.3) and other channels, and they have also been used to describe vascular SMC [607].

In addition to the cytosolic calcium concentration, we need to describe the actin–myosin interactions and therefore introduce four different states of myosin [240]:

α_1 free unphosphorylated myosin,
α_2 phosphorylated myosin,

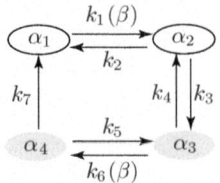

Fig. 7.13 Schematic representation of the four myosin functional states α_1, α_2, α_3, and α_4 of a Smooth Muscle Cell (SMC) [240]. Gray-shaded states are "force-generating" states. The different states are connected through the kinetic rates $k_1, \ldots, k_7$, where k_1 and k_6 depend on the concentration $\beta = [\mathrm{Ca}^{2+}]$ of cytosolic (intracellular) free calcium, whilst the others are constant

α_3 phosphorylated cross-bridges attached to actin,
α_4 dephosphorylated cross-bridges attached to actin.

The two functional states α_3 and α_4 refer to "force-generating" states and link the functional myosin state to the mechanical SMC description. The α_4 state corresponds to the latch bridges, which existence has been postulated [240] to capture the very slow MLC dephosphorylation, thereby maintaining the force at low energy costs.

The four different functional myosin states are connected through seven rates $k_1, \ldots, k_7$, see Fig. 7.13. The constants k_1 and k_6 represent the rates for the phosphorylation of myosin, whilst k_2 and k_5 define the dephosphorylation of myosin. The rates k_3 and k_4 relate to the attachment and detachment of the *(fast) cycling cross-bridges*, and k_7 is the rate constant for *latch bridge* detachment. Given these seven constants, the coupled system

$$\frac{\mathrm{d}}{\mathrm{d}t} \begin{bmatrix} \alpha_1 \\ \alpha_2 \\ \alpha_3 \\ \alpha_4 \end{bmatrix} = \begin{bmatrix} -k_1(\beta) & k_2 & 0 & k_7 \\ k_1(\beta) & -k_2 - k_3 & k_4 & 0 \\ 0 & k_3 & -k_4 - k_5 & k_6(\beta) \\ 0 & 0 & k_5 & -k_6(\beta) - k_7 \end{bmatrix} \begin{bmatrix} \alpha_1 \\ \alpha_2 \\ \alpha_3 \\ \alpha_4 \end{bmatrix} \tag{7.11}$$

of first-order differential equations then connects the four myosin states and thus functional SMC states. The sum of all fractions remains one, $\alpha_1 + \alpha_2 + \alpha_3 + \alpha_4 = 1$, which implies $\mathrm{d}(\alpha_1 + \alpha_2 + \alpha_3 + \alpha_4)/\mathrm{d}t = 0$, a property that is fulfilled by the kinetics relation (7.11).

The free Ca^{2+} concentration has different implications on the SMC kinetics; most noticeably it binds to calmodulin, activates the MLCK, and phosphorates the MLC. It therefore determines the proportions of phosphorylated and unphosphorylated myosin, and thus the rates k_1 and k_6. All the other rates are assumed to be constant. The reaction $\alpha_3 \rightarrow \alpha_4$ from phosphorylated cross-bridges to dephosphorylated cross-bridges is irreversible, and myosin cannot attach to actin and form "force-generating" states unless it has first been phosphorylated. It is also worth mentioning that a revised SMC kinetics model [239] extends the originally

Table 7.2 Typical coefficients $k_1, \ldots, k_7$ [s^{-1}] that govern the functional myosin state $\boldsymbol{\alpha} = [\alpha_1 \, \alpha_2 \, \alpha_3 \, \alpha_4]^T$ through the first-order differential equations (7.11). Here, sg(β) denotes the sigmoid function (7.12) and $\beta = [\text{Ca}^{2+}]$ [nmol l^{-1}] is the cytosolic (intracellular) free calcium concentration

k_1	k_2	k_3	k_4	k_5	k_6	k_7
0.3sg(β)	0.5	0.4	0.1	0.5	0.3sg(β)	0.01

proposed four states and adds an *ultra-slow cross-bridge cycle*. It suggest that thin-filament-based regulatory proteins may modulate actomyosin ATPase activity, which would then allow a SMC to have *two* discrete cross-bridge cycles.

Typical values of the coefficients $k_1, \ldots, k_7$ are listed in Table 7.2, where the sigmoid function

$$\text{sg}(\beta) = \frac{\exp\left[(\beta - \overline{\beta})/\gamma\right]}{\exp\left[(\beta - \overline{\beta})/\gamma\right] + 1} \tag{7.12}$$

has been used to describe k_1 and k_6, respectively. Here, $\overline{\beta}$ [mol l^{-1}] denotes the threshold of calcium concentration that activates the MLCK and then phosphorates the MLC, whilst γ specifies the sensitivity with respect to β of said activation.

Given the fractions α_3 and α_4, the related "force-generating" contributions may be modeled through individual constitutive descriptions. Several such descriptions have been reported in the literature [387, 503, 607].

Example 7.3 (Calcium Determines the Functional Myosin States). The cell membrane of Smooth Muscle Cells (SMC) is at the holding potential of -60.0 mV. A 1.6 s long step-depolarization to 0.0 mV leads then to the concentration $\beta = [\text{Ca}^{2+}]$ [nmol l^{-1}] of free calcium given by

$$\beta = \begin{cases} 150 & ; \ t < 0, \\ 630 \exp(-t/9.0) + 150 & ; \ t \geq 0. \end{cases} \tag{7.13}$$

The functional myosin states $\boldsymbol{\alpha} = [\alpha_1 \, \alpha_2 \, \alpha_3 \, \alpha_4]^T$ are governed by the first-order differential equations (7.11) with the coefficients in Table 7.2. The concentration threshold $\overline{\beta} = 400$ nmol l^{-1} and the sensation parameter $\gamma = 100$ may be used.

(a) Compute the functional myosin state $\boldsymbol{\alpha}$ that is reached at steady state upon the holding potential.
(b) Use a forward-Euler discretization of the first-order differential equations (7.11) and compute iteratively the transient change of $\boldsymbol{\alpha}$ that is determined by the calcium concentration profile (7.13). Consider the functional myosin state derived at Task (a) as initial condition. ∎

7.5.5   Thermodynamics of SMC Contraction

Given a tissue at homogenous temperature $\theta \neq \theta(\mathbf{X})$, the heat flux disappears, $\mathbf{q}_h = \mathbf{0}$, and the *first law of thermodynamics* (3.121) yields the balance equation

$$\dot{u} - r_h - \boldsymbol{\sigma} : \mathbf{d} = 0. \tag{7.19}$$

Here, u denotes the internal energy per unit volume, r_h is a heat source, and the term $\boldsymbol{\sigma} : \mathbf{d}$ expresses the stress power with $\boldsymbol{\sigma}$ and $\mathbf{d}$ denoting the Cauchy stress and the rate of deformation, respectively. These energies have been introduced with respect to unit volume of the deformed tissue. Let us consider a chemo-mechanical description, where the SMC changes said internal energy u through $\beta = [\mathrm{Ca}^{2+}]$ [mol m^{-3}], the calcium concentration per unit volume. The rate of the internal energy may then been expressed by

$$\dot{u} = \underbrace{b\dot{\beta}}_{\text{Chemical power}} + \underbrace{c_v\dot{\theta}}_{\text{Thermal power}} ,$$

where b [J mol^{-1} m^{-3}] is work-conjugate to the calcium concentration β. It represents the energy that is needed to change the Ca^{2+} concentration by one mole per unit SMC volume. In addition, c_v [J K^{-1} m^{-3}] denotes the heat capacity at constant volume of the unit volume SMC.

The energy balance (7.19) then reads $b\dot{\beta} + c_v\dot{\theta} - r_h - \boldsymbol{\sigma} : \mathbf{d} = 0$, and Piola transform (3.31) allows us to express it by

$$B\dot{\beta} + c_{0v}\dot{\theta} - R_h - \mathbf{P} : \dot{\mathbf{F}} = 0, \tag{7.20}$$

where $\mathbf{P}$ and $\mathbf{F}$ denote the first Piola–Kirchhoff stress and the deformation gradient, respectively. The term $B = Jb$ expresses the energy needed to change the calcium concentration per unit undeformed SMC volume, whilst $J = \det\mathbf{F}$ denotes the volume ratio. The heat capacity and the heat source per unit undeformed SMC volume are denoted by $c_{0v} = Jc_v$ and $R_h = Jr_h$, respectively.

In addition to the energy balance (7.20), an *admissible thermodynamic process* has to obey the *second law of thermodynamics*. Without the heat flux term, the Clausius–Duhem inequality (3.125) reads

$$-\dot{\Psi} - S\dot{\theta} + \mathbf{P} : \dot{\mathbf{F}} \geq 0, \tag{7.21}$$

where S denotes the entropy per undeformed SMC volume.

The standard deformation gradient $\mathbf{F}$ together with the active deformation gradient $\mathbf{F}_a$, a deformation that reflects the sliding between the actin and myosin filaments, describe the tissue deformation. Given the functional myosin/cross-bridge state α (see Sect. 7.5.4), these gradients define the thermodynamical state of SMC. We note that the free calcium concentration β is implicitly considered through the myosin-actin configurations $\boldsymbol{\alpha}$. The Helmholtz free energy then reads $\Psi(\mathbf{F}, \mathbf{F}_a, \boldsymbol{\alpha})$, and the Clausius–Duhem inequality (7.21)

$$\left(\mathbf{P} - \frac{\partial \Psi}{\partial \mathbf{F}}\right) : \dot{\mathbf{F}} - \frac{\partial \Psi}{\partial \mathbf{F}_a} : \dot{\mathbf{F}}_a - S\dot{\theta} - \frac{\partial \Psi}{\partial \boldsymbol{\alpha}} \cdot \dot{\boldsymbol{\alpha}} \geq 0$$

describes the thermodynamical admissible process. It holds for any arbitrary process and implies the constitutive law

$$\mathbf{P} = \frac{\partial \Psi}{\partial \mathbf{F}}, \tag{7.22}$$

which together with the inequality

$$-\frac{\partial \Psi}{\partial \mathbf{F}_a} : \dot{\mathbf{F}}_a - S\dot{\theta} - \frac{\partial \Psi}{\partial \boldsymbol{\alpha}} \cdot \dot{\boldsymbol{\alpha}} \geq 0$$

determine an admissible thermodynamical process.

At physiological conditions, the SMC may be described by an incompressible deformation. The constitutive relation (7.22) is then to be substituted by

$$\mathbf{P} = \frac{\partial \Psi}{\partial \mathbf{F}} - \kappa \mathbf{F}^{-1}, \tag{7.23}$$

where the Lagrange contribution κ has been introduced—it contributes to the hydrostatic pressure and enforces the incompressibility, see Sect. 3.6.4.

SMC are elongated cells, and for many applications they may be modeled as 1D active fibers in the vascular wall. Fibers are not able to build up stress perpendicular to the fiber direction, and $\kappa = 0$ therefore holds. The relation

$$B\dot{\beta} + c_{0v}\dot{\theta} - R_h - P\dot{\lambda} = 0,$$

then expresses the conservation of energy, whilst

$$P = \frac{\partial \Psi}{\partial \lambda} \;\; ; \;\; -\frac{\partial \Psi}{\partial \lambda_a}\dot{\lambda}_a - S\dot{\theta} - \frac{\partial \Psi}{\partial \boldsymbol{\alpha}} \cdot \dot{\boldsymbol{\alpha}} \geq 0, \tag{7.24}$$

determine the constitutive law and the respective dissipation inequality. Here, λ is the total SMC stretch, and λ_a expresses the active stretch, the kinematics of the sliding between actin and myosin filaments.

7.5.6 A Chemo-mechanical Description of SMC

Figure 7.15 shows a rheological model of a chemo-mechanical SMC description. It proposes two parallel elements, and the total first Piola–Kirchhoff stress then reads

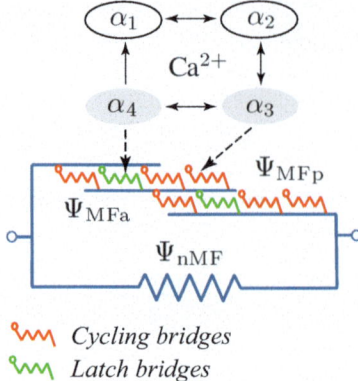

$\textcolor{red}{\text{\wavy}}$ *Cycling bridges*
$\textcolor{orange}{\text{\wavy}}$ *Latch bridges*

Fig. 7.15 Schematic representation of a chemo-mechanical description of Smooth Muscle Cells (SMC). The chemical potential Ψ_β determines the concentration β of free Ca^{2+}, and thus the development of functional SMC states. The two "force-generating" functional states α_3 and α_4 are then linked to the amount of latch bridges and respective cycling cross-bridges. The individual mechanical structures are expressed by the energy Ψ_{nMF} and Ψ_{MFp} from passive deformation as well as Ψ_{MFa} from active deformation, respectively

$$P = P_{nMF} + P_{MF}, \qquad\qquad (7.25)$$

where P_{nMF} denotes the stress from structures *other than* myofilaments, whilst the stress P_{MF} relates to the *myofilaments*. Cross-bridges transfer the myofilament stress across actin and myosin filaments, and $P_{MF} = P_{cb} + P_{lb}$ holds, where P_{cb} and P_{lb} denote the stress from *cyclic cross-bridges* and *latch bridges*, respectively. SMCs are slender fiber-like structures, and the condition $P = \max(0, P)$ complements the model—SMC fibers are therefore not able to carry any compressive load.

The individual stress contributions may be modeled directly at the microstructural level, as demonstrated elsewhere [607]. Given the thermodynamics framework in Sect. 7.5.5, we may derive the stress from the Helmholtz free energy

$$\Psi(\lambda, \lambda_a, \beta, \alpha_3, \alpha_4) = \Psi_{nMF}(\lambda) + \Psi_{MFp}(\lambda, \lambda_a, \alpha_3, \alpha_4) + \Psi_{MFa}(\lambda_a, \alpha_3, \alpha_4),$$

where Ψ_{nMF} denotes elastic energy that is stored in structures other than myofilaments. In addition, Ψ_{MFp} is the elastic energy stored upon passive deformation of the cross-bridges, whilst Ψ_{MFa} denotes the work done by (active) sliding of myosin against actin.

The reported *progressive* increase of SMC stress at increasing strain is attributed to structures other than the myofilaments. It is captured by the strain energy

$$\Psi_{nMF}(\lambda) = \frac{k_a}{2k_b} \left\{ \exp\left[k_b \left(\lambda - 1 \right)^2 \right] - 1 \right\}, \qquad\qquad (7.26)$$

where k_a [Pa] and k_b are material parameters. The potential (7.26) follows directly from a proposal made elsewhere [607].

Given myofilaments at tension, the cross-bridges are passively deformed and store elastic strain energy. It may be expressed by

$$\Psi_{\mathrm{MFp}}(\lambda, \lambda_a, \alpha_3, \alpha_4) = \frac{k_{cb}\alpha_3 + k_{lb}\alpha_4}{2} \lambda_a \left(\lambda/\lambda_a - 1\right)^2 \rho(\lambda_a), \tag{7.27}$$

where k_{cb}, k_{lb} [Pa] denote the stiffness of *cycling bridges* and *latch bridges*, respectively. They are multiplied by the corresponding portions of phosphorylated cross-bridges α_3 and dephosphorylated cross-bridges α_4, the portions of functional myosin states that are related to cycling bridges and latch bridges, respectively.

In (7.27) multiplicative kinematics $\lambda_b = \lambda/\lambda_a$ has been used, and the passive stretch λ_b of the cross-bridges has been substituted by the total stretch λ and the respective active stretch λ_a. The active stretch λ_a represents the motion between actin and myosin filaments and is taken in average over all actin–myosin overlaps of the cell. SMC contraction and relaxation correspond then to $\dot{\lambda}_a = \mathrm{d}\lambda_a/\mathrm{d}t < 0$ and $\dot{\lambda}_a > 0$, respectively.

A SMC hosts a large number myofibrils, and expression (7.27) therefore introduced the Probability Density Function (PDF) $\rho(\lambda_a)$ in the description of the actin–myosin overlap. The normal distribution (A.2) may be used [607], where the median corresponds to the optimal SMC length. It denotes the stretch $\bar{\lambda}_a$ at which the maximum overlap between myosin and actin filaments appears, the configuration at which the SMC develops the maximum tension.

We may assume only cycling cross-bridges, and thus phosphorylated cross-bridges, are able to produce active translational motion between actin and myosin filaments. The sliding appears at the stretch rate $\dot{\lambda}_c$, a material parameter that specifies how fast cyclic cross-bridges contract an unloaded SMC in the absence of latch bridges, $\alpha_4 = 0$. The expression $\dot{u}_c = L\dot{\lambda}_c$ [m s^{-1}] relates it to the cycling velocity $\dot{u}_c$ and specifies how fast cycling bridges slide against actin filaments in a SMC of the referential length L. In contrary to cycling bridges, latch bridges cannot contract the cell, but they are still able to generate resistance (force) against the sliding between actin and myosin filaments.

Given this access, the potential

$$\Psi_{\mathrm{MFa}}(\lambda_a, \alpha_3, \alpha_4) = \left[r_{cb}\alpha_3(\dot{\lambda}_c - \dot{\lambda}_a) + r_{lb}\alpha_4\dot{\lambda}_a\right] \Upsilon(\lambda_a), \tag{7.28}$$

captures the active SMC properties. The second term in the brackets describes the latch bridges, where α_4 denotes the portions of dephosphorylated cross-bridges. Latch bridges provide resistance against sliding, and r_{lb} [Pa s] may be seen as the frictional constant against actin–myosin filament sliding.

The first term in the bracket of (7.28) describes the contribution from cycling bridges, where α_3 denotes the portions of phosphorylated cross-bridges, and

$$r_{cb} = b(\dot{\lambda}_c - b\dot{\lambda}_a/a)^{-1} \text{ [Pa s]} \tag{7.29}$$

describes the friction of cycling bridges against actin–myosin filament sliding. The friction is somewhat motivated by Hill's three-parameter model (7.1) and determined by $\dot{\lambda}_c$ [s^{-1}] and a, b [Pa]. In (7.28), $\Upsilon(\lambda_a) = \int_{-\infty}^{\lambda_a} \rho(X)dX$ denotes the Cumulative Density Function (CDF) of $\rho(X)$ and accounts for the non-constant actin–myosin overlap. At steady-state condition $\dot{\lambda}_a = 0$, the term in the brackets of (7.28) then reduces to $P_0 = b\alpha_3$.

Our formulation has also to satisfy the internal force compatibility relation

$$P_{MF} = \underbrace{\frac{\partial \Psi_{MFa}}{\partial \lambda_a}}_{P_{MFa}} = \underbrace{\frac{\partial \Psi_{MFp}}{\partial \lambda}}_{P_{MFp}} \qquad (7.30)$$

to ensure the equilibrium among active P_{MFa} and passive P_{MFp} stress of the myofilament, a condition that then closes the constitutive SMC description.

Figure 7.16 illustrates the response of a micro vessel to a square wave of Ca^{2+} concentration using said modeling concept. The Ca^{2+} concentration determines the portions α_3 and α_4 of "force-generating" functional myosin states and results in the contractile stress P_{MF} of the myofilaments in the vessel wall. The diameter d of the vessel then responds accordingly, see also Example 7.4.

7.5.6.1 Stress *Versus* Stretch Properties

We consider an isometric experiment with SMC contracting at a fixed stretch $\lambda = \lambda_a\lambda_b$, where λ_a denotes the stretch from the actin–myosin overlap, and λ_b reflects the deformation of the cross-bridges. Upon contraction, the actin–myosin overlap λ_a changes until it reaches the steady-state condition $\dot{\lambda}_a = 0$, a state at which the SMC generates the stress P.

For simplicity, we neglect the stress from all non-myofibril-related structures in the present analysis. Coleman and Noll's procedure $(7.24)_1$ applied to the Helmholtz free-energies (7.27) and (7.28) then yields the first Piola–Kirchhoff stresses

$$P_{MFp} = \partial \Psi_{MFp}/\partial \lambda = (\alpha_3 k_{cb} + \alpha_4 k_{lb})(\lambda_b - 1)\rho(\lambda_a); \qquad (7.31)$$

$$P_{MFa} = \partial \Psi_{MFa}/\partial \lambda_a = \alpha_3 b\rho(\lambda_a). \qquad (7.32)$$

In the derivation of expression $(7.31)_2$ the steady-state condition $\dot{\lambda}_a = 0$ and the kinematics relation $\lambda = \lambda_a\lambda_b$ have been used, together with the relation $\partial(dX/dt)/\partial X = d(\partial X/\partial X)/dt = 0$ and the definition of the CDF $\Upsilon(X) = \int \rho(X)dX$.

Fig. 7.16 Development of the vessel diameter in response to a square wave of Ca^{2+} concentration. (**a**) Ca^{2+} concentration wave. (**b**) The two "force-generating" functional myosin states α_3 and α_4. (**c**) Vessel diameter *versus* time response with contraction and relaxation starting at the time $t = 60$ s and $t = 240$ s, respectively. Given the high amount of latch bridges α_4, the vessel relaxes slightly slower than it contracts. The simulation uses the parameters reported in Example 7.4

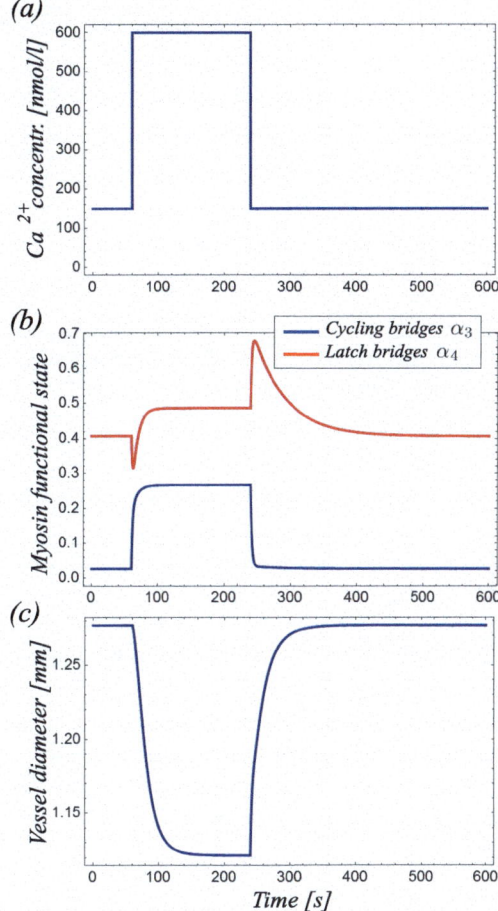

The internal equilibrium $P_{MF} = P_{MFp} = P_{MFa}$ allows us to compute the cross-bridge stretch

$$\lambda_b = (\alpha_3 k_{cb} + \alpha_4 k_{lb} + \alpha_3 b)/(\alpha_3 k_{cb} + \alpha_4 k_{lb}), \qquad (7.33)$$

and the myofilament stress P_{MF} can then be expressed as a function of the total stretch λ. Figure 7.17a shows the normalized stress P_{MF}/P_0, where $P_0 = \alpha_3 b/(\sqrt{2\pi}\sigma_a)$ denotes the maximum first Piola–Kirchhoff stress. The computation used the properties listed in Table 7.4.

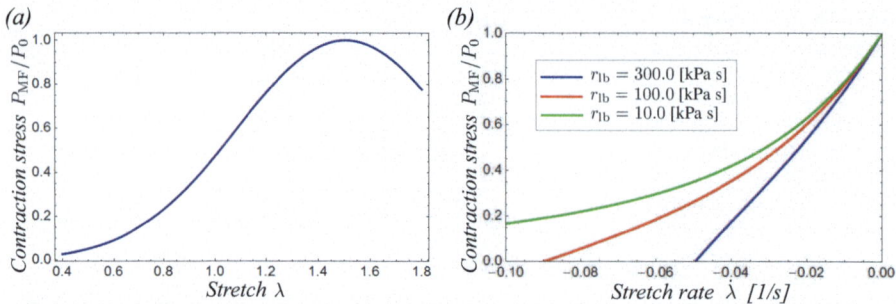

Fig. 7.17 (**a**) Normalized first Piola–Kirchhoff stress P_{MF}/P_0 at isometric contraction at the stretch λ. (**b**) Normalized first Piola–Kirchhoff stress P_{MF}/P_0 *versus* stretch rate $\dot{\lambda}$ at isotonic contraction. The maximum contraction stress is denoted by P_0 and the model parameters are reported in Table 7.4

7.5.6.2 Stress *versus* Stretch Rate Properties

A Quick-Release experiment that releases the vessel wall from the maximum contraction may be used to explore the stress *versus* stretch rate properties. The SMC is fixed at the stretch λ that results in $\lambda_a = \bar{\lambda}_a$, and the maximum contractile stress is then reached. At the end of the isometric contraction, the tissue is instantaneously released and contracts against a constant stress P. The SMC follows then an isotonic contraction at the stretch rate $\dot{\lambda}$. Myosin kinetics cannot follow the temporal changes at the point of release and allows us to assume a constant myosin state across the switch from isometric to isotonic conditions.

In addition to P_{MFp} according to (7.31), the Helmholtz free energy (7.28) yields the active first Piola–Kirchhoff stress

$$P = P_{MF} = \frac{\partial \Psi_{MFa}}{\partial \lambda_a} = \left[\frac{\alpha_3 b (\dot{\lambda}_c - \dot{\lambda}_a)}{\dot{\lambda}_c - (b/a)\dot{\lambda}_a} + r_{lb}\alpha_4 \dot{\lambda}_a \right] \left(\sqrt{2\pi}\sigma_a \right)^{-1} \qquad (7.34)$$

of the muscle contraction, where $\rho(\bar{\lambda}_a) = (\sqrt{2\pi}\sigma_a)^{-1}$ has been used.

The internal equilibrium $P_{MF} = P_{MFp} = P_{MFa}$ allows us then to express the cross-bridge stretch

$$\lambda_b = \frac{b\dot{\lambda}_a[\alpha_3 k_{cb} + \alpha_4(k_{lb} + \dot{\lambda}_a r_{lb})] + a[\alpha_3 b\dot{\lambda}_a - \alpha_3(b + k_{cb})\dot{\lambda}_c - \alpha_4\dot{\lambda}_c(k_{lb} + \dot{\lambda}_a r_{lb})]}{(\alpha_3 k_{cb} + \alpha_4 k_{lb})(b\dot{\lambda}_a - a\dot{\lambda}_c)},$$

a property that is constant during isotonic contraction. Given $\dot{\lambda}_a$ and the parameters listed in Table 7.4, the normalized stress P_{MF}/P_0 *versus* the stretch rate $\dot{\lambda} = \dot{\lambda}_a\lambda_b$ may then be plotted, see Fig. 7.17b. Here, $P_0 = \alpha_3 b/(\sqrt{2\pi}\sigma_a)$ denotes the maximum first Piola–Kirchhoff stress, the stress at which the tissue has been released in the quick-release experiment. As shown in Fig. 7.17b, the increase of the latch bridge friction r_{lb} changes the contraction properties from hyperbolic towards linear P_{MF}/P_0 *versus* $\dot{\lambda}$ properties.

Table 7.4 Algorithm to
compute the stretch λ of a
vessel wall specimen during a
Quick-Release experiment

(1) Set time discretization and problem parameters

$\Delta t = 0.01\,\text{s}$; $t_{\max} = 60\,\text{s}$; $n = t_{\max}/\Delta t$

$k_a = 67\,\text{kPa}$; $k_b = 3\,\text{kPa}$; $k_{lb} = 10\,\text{kPa}$; $k_{cb} = 28\,\text{kPa}$

$r_{lb} = 300\,\text{kPa s}$; $a = -1\,\text{kPa}$; $b = 30\,\text{kPa}$; $\lambda_c = -0.7$

$\overline{\lambda}_a = 0.8$; $\sigma_a = 0.22$; $\alpha_3 = 0.5$; $\alpha_4 = 0.3$

(2) Set initial conditions

$i = 0$; $\lambda = 1.0$; $\lambda_a = \overline{\lambda}_a$; $\dot{\lambda}_a = 0$

Do While $i \leq n$

$\quad t_n = i\,\Delta t$

(3) Compute stress contributions

$\rho = \exp[-(\lambda_a - \overline{\lambda}_a)^2/(2\sigma_a^2)]/(\sqrt{2\pi}\sigma_a)$

$P_{\text{nMF}} = k_a(\lambda - 1)\exp[k_b(\lambda - 1)^2]$

$P_{\text{MFp}} = (\alpha_3 k_{cb} + \alpha_4 k_{lb})(\lambda/\lambda_a - 1)\rho$

$P_{\text{MFa}} = \left[\frac{\alpha_3 b(\lambda_c - \dot{\lambda}_a)}{\lambda_c - (b/a)\dot{\lambda}_a} + r_{lb}\alpha_4\dot{\lambda}_a\right]\rho$

(4) Set stress level

$P_0 = \alpha_3 b/(\sqrt{2\pi}\sigma_a)$

If $[t_i > 20,\ P = 0.2/0.5/0.8\,P_0,\ P = P_0]$

(5) Solve internal and external equilibrium

$j = 0$; $f_1 = 1$; $f_2 = 1$

Do While $(f_1^2 + f_2^2 > 10^{-12})$

$\quad f_1 = P_{\text{MFa}} - P_{\text{MFp}}$; $f_2 = P_{\text{MFp}} - P$

$\quad \begin{bmatrix} \Delta\dot{\lambda}_a \\ \Delta\lambda \end{bmatrix} = \begin{bmatrix} \partial f_1/\partial\dot{\lambda}_a & \partial f_1/\partial\lambda \\ \partial f_2/\partial\dot{\lambda}_a & \partial f_2/\partial\lambda \end{bmatrix}^{-1} \begin{bmatrix} f_1 \\ f_2 \end{bmatrix}$

$\quad \dot{\lambda}_a \leftarrow \dot{\lambda}_a - \Delta\dot{\lambda}_a$; $\lambda \leftarrow \lambda - \Delta\lambda$; $j \leftarrow j + 1$

$\quad$ If $j = 15$, solution not found: terminate

End Do

Store λ_i for plotting

$\lambda_a \leftarrow \lambda_a + \dot{\lambda}_a\Delta t$; $i \leftarrow i + 1$

End Do

7.5.6.3 Quick-Release Experiment

A Quick-Release experiment begins with a phase of isometric contraction that is
then followed by isotonic contraction. The vessel wall specimen contracts over
several seconds, and the computation of the deformation characteristics requires
a discretization method.

We used the forward-Euler method to discretize the governing equations of
the Quick-Release experiment, and Table 7.4 lists the applied iterative algorithm.
Given the specification and initialization of parameters, the wall stress contributions
P_{nMF}, P_{MFp}, and P_{MFa} may be computed. The stress level P is then set, and
the rate $\dot{\lambda}_a$ of actin–myosin stretch, as well as the total circumferential stretch
λ are computed by a Newton–Raphson fixpoint iteration. It solves the internal

Fig. 7.18 Development of the stretch λ during a quick-release experiment. The dashed lines indicate the stretch rate at the beginning of the isotonic contraction. The algorithm that has been used to compute the results is given in Table 7.4

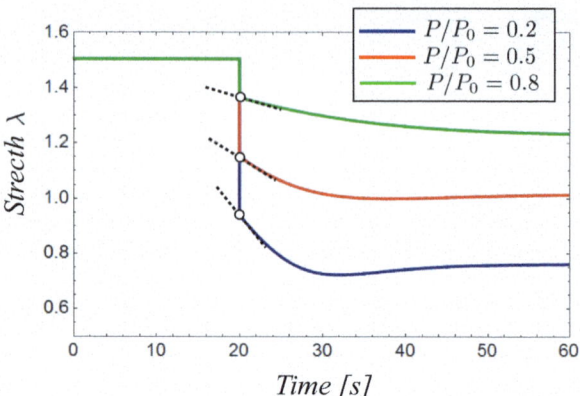

$f_1 = P_{\text{MFa}} - P_{\text{MFp}} = 0$ and external $f_2 = P_{\text{MFp}} - P = 0$ equilibrium. Figure 7.18 illustrates the evolution of the sample stretch λ over the time t with the sample being released from the stress levels of $P/P_0 = 0.2 \, ; 0.5 \, ; 0.8$. At the point of stress release, the stretch λ instantaneously drops due to elastic unloading, and the SMC then contracts at individual stretch rates, see the dashed lines in Fig. 7.18. For the release from $P/P_0 = 0.2$, the SMC model predicts also a small amount of stress recovery—the stress first drops and then recovers by a small amount.

Example 7.4 (Active Micro Vessel). We consider an arteriole segment of the stress-free reference diameter $D = 1.0 \, \text{mm}$ and the wall thickness $H = 0.08 \, \text{mm}$. The vessel is inflated at the pressure p_i and fixed in axial direction. The vessel wall is incompressible and entirely formed by Smooth Muscle Cell (SMC), all at the contractile phenotype. We may use the chemo-mechanical description of Sect. 7.5.6 to model the arteriole.

(a) Consider the fully relaxed vessel, a state at which the strain energy (7.26) with $k_a = 67.0 \, \text{kPa}$ and $k_b = 3.0$ describes the wall. Use membrane theory and compute the inflation pressure p_i that results in the diameter d that is 50% larger than the vessel's reference diameter D.

(b) The vessel wall is now inflated at the fixed circumferential stretch $\lambda = 1.2$, and the states $\alpha_1 = 0.549207, \alpha_2 = 0.0168519, \alpha_3 = 0.0266816$, and $\alpha_4 = 0.40726$ determine the equilibrated myosin in response to the Ca^{2+} concentration of $\beta = 150 \, \text{nmol}$. The parameters $k_{cb} = 28.0 \, \text{kPa}$ and $k_{lb} = 10.0 \, \text{kPa}$ describe cycling bridges and latch bridges, respectively, whilst $\lambda_a = 0.9$ determines the actin–myosin stretch. The overlap between actin and myosin may be modeled with the normal Probability Density Function (PDF)

$$\rho(\lambda_a) = \frac{\exp\left[-(\lambda_a - \overline{\lambda_a})^2/(2\sigma^2)\right]}{\sqrt{2\pi}\,\sigma}, \tag{7.35}$$

where $\overline{\lambda}_a = 0.8$ and $\sigma = 0.22$ denote the median and the Standard Deviation (SD), respectively. Use membrane theory and compute the inflation pressure p_i that relates to this state of the arteriole.

(c) Given the state described in Task (b), compute the rate of active stretch $\dot{\lambda}_a = d\lambda_a/dt$. The parameters $a = -1.0\,\text{kPa}$, $b = 30.0\,\text{kPa}$ together with the cycling stretch rate $\dot{\lambda}_c = -0.7\,\text{s}^{-1}$ describe the cycling bridges, whilst the latch bridges are determined by the frictional coefficient $r_{lb} = 2.1\,\text{MPa s}$.

(d) Use a forward-Euler discretization of the problem's governing equation and compute the vessel diameter *versus* time response. The square profile

$$\beta = \begin{cases} 150 \; ; & 0 \leq t < 60\,\text{s} \\ 600 \; ; & 60 \leq t < 240\,\text{s} \\ 150 \; ; & 240 \leq t \leq 600\,\text{s} \end{cases} \tag{7.36}$$

describes the concentration $\beta = [\text{Ca}^{2+}]\,[\text{nmol}\,l^{-1}]$ over the time t [s], and the functional myosin states α_i from Task (b) may be used as initial conditions. In addition, $\lambda = 1.27743$, $\lambda_a = 1.09549$, and $\dot{\lambda}_a = 0$ describe the problem.

(e) Express the change of the vessel wall's temperature $\dot{\theta}$ that is required to keep the entropy S at a constant level. ∎

7.6 Modeling Arteriogenesis

The vessel wall is at continuous mass turnover, conditions that result in *growth* and *remodeling* at a local tissue level. Whilst growth changes the tissue's stress-free configuration, remodeling modifies the tissue's internal structure and therefore changes its mechanical properties. Both mechanisms are interlinked and result in vascular morphogenesis, the change of the *in vivo* vessel geometry.

Arteriogenesis is determined by the synthesis and removal of tissue mass, and in contrary to vasoreactivity, an *open system* governing framework is to be used in its mathematical description. Arteriogenesis is also determined by *volume growth*, not surface growth or tip growth, mechanisms commonly linked to plant growth [220]. Mass turn-over and its implications may either be modeled at the macroscopic length-scale of the tissue, or at the length-scale of its constituents, such as collagen, elastin, and SMC.

Growth is a fundamental property of all soft biological tissues, and its mathematical description is well documented [10, 101, 220, 271, 275, 316, 369]. Given the clinical relevance of vascular diseases, much of the work relates to aneurysms [210, 314, 331, 361, 383, 580, 590, 597, 613], carotid disease [430], and coronary atherosclerosis [167].

7.6.1 Open System Governing Laws

We follow a Lagrange approach in the description of the growing continuum and adapt the description of Chap. 3. An *open system* describes then the material point and allows material to enter and to leave the system. The exchange of material has its own time scale τ, and given vascular tissue, it is in the range of days and weeks. It is much longer than the time scale t of elastic loading by the cardiac cycle. The *slow growth assumption* then applies, and the two time scales t and τ may be separated in the description of the growing vascular tissue.

7.6.1.1 Mass Balance
Given an open system, the right-hand side of the mass balance (3.102) does not disappear but describes the change in mass of the material point. It may either appear through a mass source or a mass flux across the boundaries of the open system. Whilst a flux of mass, such as the migration of SMC appears in vascular tissue, for many applications the flux term may be neglected over the source term. The open system mass balance then reads

$$\frac{\partial \rho}{\partial t} + \rho \mathrm{div}\mathbf{v} = \rho \varsigma_\mathrm{v} , \qquad (7.41)$$

where $\mathbf{v}$ is the velocity, and ρ denotes the spatial density of the tissue. In addition, $\varsigma_\mathrm{v}(\mathbf{x}, \tau)$ [s^{-1}] is a volume source field that expresses the rate of volume change. The term $\partial(\mathrm{d}m)/\partial t = \rho\varsigma_\mathrm{v}\mathrm{d}v$ then describes the mass that is added to material point of the volume $\mathrm{d}v$ per unit time. With the transformation of the volume element $\mathrm{d}v = J\mathrm{d}V$, it may also be expressed by $\partial(\mathrm{d}m)/\partial t = \rho_0\varsigma_\mathrm{v}\mathrm{d}V$, where $\rho_0 = J\rho$ denotes the tissue's density in the reference configuration.

7.6.1.2 Balance of Linear and Angular Momentum
The growing vascular tissue has to follow Newton's second law of mechanics. Given the *slow growth assumption*, tissue growth can neither influence linear nor angular momentum. Cauchy's momentum equations (3.108) and (3.19),

$$\rho \frac{\partial \mathbf{v}}{\partial t} = \mathrm{div}\boldsymbol{\sigma} + \mathbf{b}_\mathrm{f} \; ; \quad \boldsymbol{\sigma} = \boldsymbol{\sigma}^\mathrm{T} , \qquad (7.42)$$

therefore hold without modifications, where $\mathbf{b}_\mathrm{f}$ denotes body forces per unit spatial volume, see Sect. 3.6.2. The non-growth-related tissue deformation then allows us to compute the stress $\boldsymbol{\sigma}$, where the concepts, such as hyperelasticity, general theory of fibrous connective tissue, and visco-hyperelasticity may be used, see Sect. 5.5.

7.6.1.3 The First Law of Thermodynamics
The growth-related exchange of mass results in heat exchange across the boundary of the open system. Let us consider a material point at the internal energy u and the kinetic energy $\rho|\mathbf{v}^2|/2$. The exchange of material at the same energy, and thus the

exchange of *compatible* heat, does not change the system energy per unit volume—the energy balance (3.121) holds. However, the exchange of *non-compatible* heat gradually changes the tissue's energy and requires us to revise the balance towards

$$\frac{\partial u}{\partial t} + \text{div} \mathbf{q}_h - r_h - \boldsymbol{\sigma} : \mathbf{d} = \varsigma_h, \qquad (7.43)$$

where $\varsigma_h(\mathbf{x}, \tau)$ denotes the non-compatible heat source per unit spatial volume. It is linked to the time scale τ of tissue mass exchange. In (7.43), $\mathbf{q}_h$ and r_h denote the spatial heat flux and the spatial source term, whilst $\mathbf{d}$ is the rate of deformation, see Sect. 3.6.3.1.

7.6.1.4 The Second Law of Thermodynamics

Given the exchange of *compatible* entropy s, the Clausius–Duhem inequality (3.124) holds also for the open system. However, the exchange of *non-compatible* entropy gradually changes the tissue's entropy, and the inequality then reads

$$\gamma\theta = -\frac{\partial \psi}{\partial t} - s\frac{\partial \theta}{\partial t} + \boldsymbol{\sigma} : \mathbf{d} - \frac{\mathbf{q}_h \cdot \text{grad}\theta}{\theta} + \varsigma_s \geq 0, \qquad (7.44)$$

where $\varsigma_s(\mathbf{x}, \tau)$ denotes the source of non-compatible entropy per spatial unit volume. It enters the system through tissue growth at the time scale τ. The synthesis of tissue fibers at an undulation that is different to the undulation of existing fibers in the material point is an example of a *non-compatible* entropy source. See Sect. 3.6.3 for a detailed description of the other parameters used in (7.44).

7.6.2 Kinematics-Based Growth Description

Whilst *growth-related* and *non-growth-related* motions appear together in the adaptation of vascular tissue, both are very different in nature. The mathematical description therefore separates both motions through the introduction of the *intermediate reference configuration* Ω_0, see Fig. 7.20. It is stress-free and serves as the reference configuration of the non-growth-related motion $\chi(\mathbf{X}, t)$, whilst $\chi_g(\tilde{\mathbf{X}}, \tau)$ describes the growth-related motion relative to the *initial reference configuration* $\tilde{\Omega}_0$, a purely hypothetical entity. The decomposition of the motion shown in Fig. 7.20 reflects the *slow growth assumption* and separates the growth-related time scale τ from the non-growth-related time scale t. The referential gradients $\mathbf{G}(\tilde{\mathbf{X}}, \tau) = \partial\chi_g(\tilde{\mathbf{X}}, \tau)/\partial\tilde{\mathbf{X}}$ and $\mathbf{F}(\mathbf{X}, t) = \partial\chi(\mathbf{X}, t)/\partial\mathbf{X}$ then represent growth-related and non-growth-related deformations, respectively.

Regardless the kinematics framework introduced in Fig. 7.20, growth physically appears always in the spatial configuration Ω, the tissue's natural configuration. As a consequence, the stress-free intermediate reference configuration Ω_0 is incompatible—it can in general not be stress-free and compatible at the same time.

Fig. 7.20 Multiplicative kinematics of a growing tissue. The incompatible and stress-free intermediate reference configuration Ω_0 separates the growth-related and non-growth-related motions $\boldsymbol{\chi}_{\mathrm{g}}(\widetilde{\mathbf{X}}, \tau)$ and $\boldsymbol{\chi}(\mathbf{X}, t)$, respectively

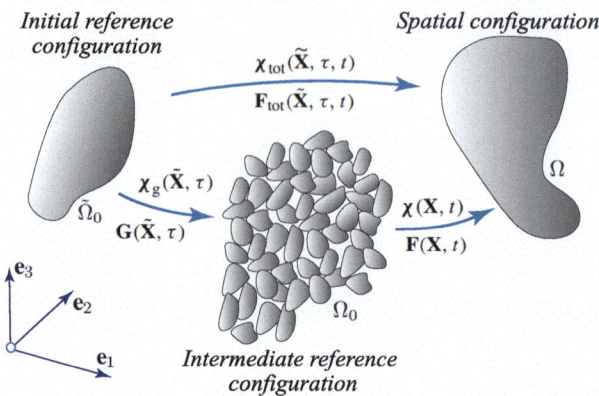

Each material point $\mathbf{X}$ therefore has its own reference configuration, independent from the reference configurations of the neighboring points [510]. Figure 7.20 aims at illustrating the incompatibility of Ω_0. The put-together of a compatible intermediate configuration (not shown in Fig. 7.20) requires tissue deformation, and even in the absence of external loading $\bar{\mathbf{t}}$ and $\mathbf{b}_{\mathrm{f}}$, it cannot be stress-free but contains residual stresses.

Given the kinematics framework introduced by Fig. 7.20, the deformation gradient of the total motion $\boldsymbol{\chi}_{\mathrm{tot}}(\widetilde{\mathbf{X}}, \tau, t)$ is multiplicatively decomposed and reads

$$\mathbf{F}_{\mathrm{tot}}(\widetilde{\mathbf{X}}, \tau, t) = \partial \boldsymbol{\chi}_{\mathrm{tot}}(\widetilde{\mathbf{X}}, \tau, t)/\partial \widetilde{\mathbf{X}} = \mathbf{F}(\mathbf{X}, t)\mathbf{G}(\widetilde{\mathbf{X}}, \tau). \tag{7.45}$$

The deformation gradient $\mathbf{G}(\widetilde{\mathbf{X}}, \tau)$ specifies the growth with respect to $\widetilde{\Omega}_0$, and the deformation gradient $\mathbf{F}(\mathbf{X}, t)$ records the non-growth-related deformation with respect to Ω_0. Multiplicative decomposition of the deformation gradient through the introduction of a non-compatible intermediate configuration is a well-established concept in the description of problems, such as elasto-plasticity [313], polymer swelling [166], thermoelasticity [528], and soft biological tissue growth [473, 511]. A detailed discussion of this conception and its limitations in the context of growth modeling is given elsewhere [10, 220].

Example 7.5 (Evolution of Residual Stresses Through Vessel Growth).   Figure 7.21 shows a 2D axisymmetric vessel segment, which spatial configuration Ω is inflated at the pressure p_{i}. The initial reference configuration $\widetilde{\Omega}_0$ is stress-free, and $\widetilde{R}_{\mathrm{i}} = 3.0$ mm and $\widetilde{R}_{\mathrm{o}} = 4.0$ mm denote the corresponding inner and outer radii, respectively. The growth deformation $\mathbf{G}(\widetilde{R})$ describes vessel wall growth relative to $\widetilde{\Omega}_0$. It is applied to infinitesimally thin vessel rings, and the mapping results in an infinite number of non-connected rings that form the stress-free intermediate reference configuration Ω_0. Figure 7.21 shows only one of these rings. An individual elastic deformation $\mathbf{F}(R)$ is then applied to each said stress-free rings and rejoins them into the spatial configuration Ω.

The vessel wall is at plane strain and described by an incompressible neoHookean material with the strain energy density $\Psi(\mathbf{C}) = c(\mathrm{tr}\mathbf{C} - 3)/2$, where $c = 120\,\text{kPa}$ is the referential stiffness, and $\mathbf{C}$ denotes the right Cauchy–Green strain tensor.

Fig. 7.21 Multiplicative kinematics of a growing thick-walled vessel in 2D. The growth deformation $\mathbf{G}(\widetilde{R})$ maps an infinitesimally thin ring into the respective intermediate ring. All such stress-free rings form the intermediate reference configuration Ω_0. It separates the growth deformation $\mathbf{G}(\widetilde{R})$ from the elastic deformation $\mathbf{F}(R)$

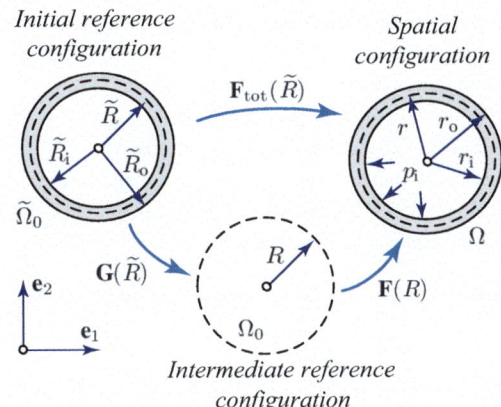

(a) Given the principal components of the growth deformation

$$G_\theta(\widetilde{R}) = 1.2 - 0.4(\widetilde{R} - \widetilde{R}_i)/(\widetilde{R}_o - \widetilde{R}_i) \ ; \quad G_R = 1.0\,, \tag{7.46}$$

express the radius r as a function of the radius $\widetilde{R}$ and the inner radius r_i.
(b) Solve the equilibrium of the thick-walled cylindrical tube problem and determine the relation between the pressure p_i and the radius r_i. Compute p_i for a number of radii r_i.
(c) Show that the pressure-free vessel, and thus the spatial configuration Ω at $p_i = 0$, contains residual strains. ∎

7.6.3 Spatial Distribution of Volume Growth

To close the description of the growth kinematics, the volume rate $\varsigma_v \mathrm{d}V$ introduced in Sect. 7.6.1.1 is to be linked to the growth-related deformation $\mathbf{G}$. We therefore consider the time scale τ, represented by the mapping between the initial reference configuration $\widetilde{\Omega}_0$ and the reference configuration Ω_0. The mass balance (7.41) then reads

$$\frac{\partial \rho_0}{\partial \tau} + \rho_0 \mathrm{div}\mathbf{v}_g = \rho_0 \varsigma_v\,,$$

where $\mathbf{v}_g$ is the growth-related velocity of the material particle, and $\rho_0 = J\rho$ denotes its density in the reference configuration Ω_0. With the kinematics relation $\mathbf{I} : \dot{\mathbf{G}}\mathbf{G}^{-1} = \mathrm{div}\mathbf{v}_g$ (see the equivalent expression (3.38) with the deformation gradient **F**), the mass balance may be expressed by

$$\frac{\partial \rho_0}{\partial \tau} + \rho_0 \mathrm{tr} \left(\dot{\mathbf{G}} \mathbf{G}^{-1} \right) = \rho_0 \varsigma_\mathrm{v} . \tag{7.49}$$

Given the rate of volume change ς_v and the density ρ_0 in Ω_0, it is a scalar equation in the nine unknowns G_{ij} in 3D—the identification of **G** therefore needs further kinematics assumptions to be made.

7.6.3.1 Constant-Volume Growth
Let us consider the exchange of mass across the boundary of the open system without net change in tissue volume. The condition $\dot{\mathbf{G}} = \mathbf{0}$ then holds, and the mass balance (7.49) reduces to

$$\frac{\partial \rho_0}{\partial \tau} = \rho_0 \varsigma_\mathrm{v} .$$

Whilst the vascular tissue then maintains its reference shape, the mechanical properties may change through the change of tissue density—the tissue remodels.

7.6.3.2 Constant-Density Growth
Given the exchange of mass across the boundary of the open system without the change in tissue density, $\dot{\rho}_0 = 0$ holds. The mass balance (7.49) then reduces to

$$\mathrm{tr} \left(\dot{\mathbf{G}} \mathbf{G}^{-1} \right) = \varsigma_\mathrm{v} , \tag{7.50}$$

and additional kinematics assumptions are needed to specify **G** upon the growth rate ς_v.

Tissue growth may be isotropic and therefore equally fast along all directions. The growth tensor $\mathbf{G} = \alpha \mathbf{I}$ is then the scaled spherical tensor **I**, and in 3D the mass balance (7.50) reads $\dot{\alpha} = \alpha \varsigma_\mathrm{v}/3$. The forward-Euler integration over the time increment $\Delta \tau$ then determines the growth tensor increment $\Delta \mathbf{G} = (\alpha/3)\varsigma_\mathrm{v} \Delta \tau \mathbf{I}$, and the iteration $\mathbf{G} \leftarrow \mathbf{G} + \Delta \mathbf{G}$ defines the growth kinematics.

Vascular tissue growth is often anisotropic, and a more general kinematics description is therefore needed. Let us consider the eigenvalue representation

$$\mathbf{G} = G_1 (\mathbf{E}_1 \otimes \tilde{\mathbf{E}}_1) + G_2 (\mathbf{E}_2 \otimes \tilde{\mathbf{E}}_2) + G_3 (\mathbf{E}_3 \otimes \tilde{\mathbf{E}}_3)$$

in 3D, where $G_i; i = 1, 2, 3$ denotes the eigenvalues of **G**, the growth-related stretches along the principal growth directions. Given **G** is a two-point tensor, $\mathbf{E}_i$ and $\tilde{\mathbf{E}}_i$ are eigenvectors in the reference configuration Ω_0 and the initial reference configuration $\tilde{\Omega}_0$, respectively. The balance (7.50) then yields the governing equation

$$\dot{G}_1 G_1^{-1} + \dot{G}_2 G_2^{-1} + \dot{G}_3 G_3^{-1} = \varsigma_v,$$

which integration

$$\ln G_1 + \ln G_2 + \ln G_3 = \int_{\tau_0}^{\tau} \varsigma_v(\bar{\tau}) d\bar{\tau}$$

links the rate of volume change and the growth along the principal directions. Here, the initial conditions $G_1 = G_2 = G_3 = 1$ at $\tau = 0$ have been used.

Biological tissue growth is known to be sensitive to chemical cues, such as concentrations of nutrient, hormone, and growth factors, as well as to mechanical cues, such as strain and stress. The tissue's histology also influences the growth kinematics. Experimental observations from cardiac dilatations, for example, indicate a clear difference between fiber and cross-fiber components of tissue growth [216]. A widely used approach [316] therefore considers the growth along predefined (anatomical) tissue directions, whilst chemical and/or mechanical cues determine the rate of growth and therefore the eigenvalues G_i.

7.6.3.3 Homeostatic Growth

Homeostasis is a common concept in the description of vascular adaption models. The rate equations are then formulated in a way that allows the field variables to approach homeostatic levels [443, 447, 534]. Whilst most of the models implement homeostasis at a local level, non-local targets may also be used. It allows for the homogenization of the stress across the vessel wall [436] according to the homogeneous stress hypothesis.

Example 7.6 (Growth that Maintains a Homeostatic Wall Stress Level). Let us consider a membrane model of a cylindrical vessel segment of the diameter d and the length l. The vessel is at the axial stretch λ_z, inflated at the pressure p_i, and its wall thickness h grows in an effort to maintain the circumferential stress σ_θ at a homeostatic (constant) level. In addition, constant-density volumetric growth may be assumed.

(a) Consider the growth tensor **G** within the principal coordinate system $\{\theta, z, r\}$ and express its eigenvalues G_θ, G_z, G_r as a function of the rate of volume change ς_v.

(b) Consider isotropic growth and compute ς_v that is required to maintain homeostasis in response to the rate $\dot{\lambda}_\theta$ of the circumferential stretch. ∎

7.6.4 The Thick-Walled Elastic Tube at Homeostatic Growth

We consider a 2D axisymmetric vessel segment which spatial configuration Ω is inflated at the pressure $p_i = 13.33\,\text{kPa}$, see Fig. 7.21. At the time $\tau = 0$ the vessel occupies its initial (and stress-free) reference configuration $\widetilde{\Omega}_0$, where $\widetilde{R}_i$ and $\widetilde{R}_o$ denote the inner and outer radii, respectively. The (incompatible) isotropic growth deformation $\mathbf{G} = \alpha \mathbf{I}$ is applied to $\widetilde{\Omega}_0$ and then followed by the elastic deformation $\mathbf{F}(R)$, mappings that then result in the spatial configuration Ω.

The ring of radius $\widetilde{R}$ and thickness $\mathrm{d}\widetilde{R}$ in $\widetilde{\Omega}_0$ maps into the ring of radius r and thickness $\mathrm{d}r$ in Ω, and allows us to express $2\pi r\mathrm{d}r = 2\pi\widetilde{R}\det\mathbf{G}(\widetilde{R})\det\mathbf{F}(R(\widetilde{R}))\mathrm{d}\widetilde{R}$. With the incompressibility of the elastic deformation $\det\mathbf{F} = 1$, the integration yields

$$
r^2 - r_i^2 = 2\int_{\widetilde{R}_i}^{\widetilde{R}} x\det\mathbf{G}(x)\mathrm{d}x\,,
$$

and the spatial radius then reads $r = \sqrt{r_i^2\alpha^2(\widetilde{R}^2 - \widetilde{R}_i^2)}$, where $\det\mathbf{G}(x) = \alpha^2$ has been used. Given the mapping $R = \alpha\widetilde{R}$, the elastic circumferential stretch

$$
\lambda_\theta(\widetilde{R}) = \frac{2\pi r}{2\pi R} = \frac{\sqrt{r_i^2\alpha^2(\widetilde{R}^2 - \widetilde{R}_i^2)}}{\alpha\widetilde{R}}\,, \tag{7.53}
$$

may be computed. Note that the intermediate reference configuration $\widetilde{\Omega}_0$ is incompatible and therefore $\lambda_r \neq r/R$.

We consider the vessel wall at plane strain and to be an incompressible neoHookean material. The strain energy density $\Psi(\mathbf{C}) = c(\mathrm{tr}\mathbf{C} - 3)/2$ per unit (undeformed) material then represents the vascular tissue, where $\mathbf{C}(R) = \mathbf{F}^\mathrm{T}(R)\mathbf{F}(R)$ denotes the right Cauchy–Green strain tensor. For this problem, we assume constant-density growth, and

$$
\varsigma_v = \zeta(\lambda_\theta - \overline{\lambda}_\theta) \tag{7.54}
$$

describes the rate of volume change. Here, ζ $[\text{s}^{-1}]$ denotes a time constant, and $\overline{\lambda}_\theta$ is a homeostatic target value for the circumferential stretch λ_θ. The exchange of mass scales then with the "deviation" from homeostasis.

With isotropic growth, $\mathbf{G}(\widetilde{R}) = \alpha(\widetilde{R})\mathbf{I}$, the forward-Euler integration of (7.50) over the time increment $\Delta\tau$ then defines the increment of the growth-related stretches

$$
\Delta G_\theta = \Delta G_r = \Delta\alpha(\widetilde{R}) = \alpha\varsigma_v(\widetilde{R})\Delta\tau/2 = \alpha\zeta\left[\lambda_\theta(\widetilde{R}) - \overline{\lambda}_\theta\right]\Delta\tau/2\,, \tag{7.55}
$$

where the elastic stretch $\lambda_\theta(\widetilde{R})$ follows from the solution of the structural equilibrium.

Given an axisymmetric problem in cylindrical coordinates, the only non-trivial equilibrium relation reads

$$r \mathrm{d}\sigma_r / \mathrm{d}r = \sigma_\theta - \sigma_r = \overline{\sigma}_\theta - \overline{\sigma}_r , \qquad (7.56)$$

where the principal Cauchy stresses $\sigma_i = \overline{\sigma}_i - \kappa; i = \theta, r$ in circumferential and respective radial directions have been used. The stresses $\overline{\sigma}_i = \lambda_i \partial \Psi / \partial \lambda_i = \lambda_i^2 c$ are determined by the strain energy, and κ denotes the Lagrange pressure that enforces the elastic incompressibility. The integration of (7.56) between the boundaries $\sigma_r(r_\mathrm{i}) = -p_\mathrm{i}$ and $\sigma_r(r_\mathrm{o}) = 0$ then results in

$$p_\mathrm{i} = \int_{r_\mathrm{i}}^{r_\mathrm{o}} \frac{\overline{\sigma}_\theta - \overline{\sigma}_r}{r} \mathrm{d}r = \int_{\widetilde{R}_\mathrm{i}}^{\widetilde{R}_\mathrm{o}} \frac{\overline{\sigma}_\theta - \overline{\sigma}_r}{\widetilde{R}} \mathrm{d}\widetilde{R} , \qquad (7.57)$$

where the kinematics relation $r = \alpha \lambda_r \widetilde{R}$ has been used with $\alpha \lambda_r$ denoting the total radial stretch relative to $\widetilde{\Omega}_0$.

We may use a spatial discretization of the vessel with m layers to approximate the integral in Eq. (7.57). A fixpoint iteration can then be applied to identify the discrete circumferential stretches $\lambda_{\theta j}$ from the equilibrium

$$p_\mathrm{i} = c \sum_{j=1}^{m} \frac{\lambda_{\theta j}^2 - \lambda_{\theta j}^{-2}}{\widetilde{R}_j} \Delta \widetilde{R} ,$$

where $\Delta \widetilde{R} = (\widetilde{R}_\mathrm{o} - \widetilde{R}_\mathrm{i})/m$ denotes the thickness of the discrete vessel layers, and the elastic incompressibility $\lambda_{r j} = \lambda_{\theta j}^{-1}$ has been used. Given $\lambda_{\theta j}$, the balance relation (7.55) defines $\Delta \alpha_j$ and allows us to update the growth parameters $\alpha_j \leftarrow \alpha_j + \Delta \alpha_j$ across all discretization layers j. The full iterative schema is listed in Table 7.6.

Figure 7.23a shows the development of the inner radius r_i over time, where $\overline{\lambda}_\theta = 1/m \sum_{j=1}^{m} \lambda_{\theta j}$ determined the target stretch. It represents a *non-local* target towards which the vessel adapts. Given this problem, the homeostatic circumferential stretch approached $\lambda_\theta = 1.10072$, constantly distributed throughout the vessel wall. The growth-related stretch changed then from $\alpha = 1.03046$ at the inside towards $\alpha = 0.968924$ at the outside of the vessel wall. The vessel's radius and wall thickness changed insignificantly. Whilst the growth towards the aforementioned target stretch stabilized the system after approximately 40 weeks, the prescription of a predefined and constant target stretch $\overline{\lambda}_\theta$ led to an overly constraint problem. The vessel kept then either expanding or shrinking, dependent on the value set for $\overline{\lambda}_\theta$, see Fig. 7.23b. Stress and strain are linked through the constitutive model of the vessel wall. Given the inflation pressure p_i, the structural equilibrium determines the wall stress and results in a unique circumferential strain field. Setting $\overline{\lambda}_\theta$ as the growth target then results in the non-plausible system response shown in Fig. 7.23b.

Table 7.6 Algorithm to
solve the thick-walled elastic
tube at isotropic growth over
a time increment of $\Delta\tau$

(1) Specify parameters

$\tilde{R}_i = 3.0\,\text{mm}$; $\tilde{R}_o = 4.0\,\text{mm}$; $m = 30$

$c = 120\,\text{kPa}$; $\zeta = 0.25\,\text{weeks}^{-1}$; $p_i = 13.33\,\text{kPa}$

(2) Set spatial discretization

$\Delta\tilde{R} = (\tilde{R}_o - \tilde{R}_i)/m$

$\tilde{R}_j = \tilde{R}_i + (j - 0.5)\Delta\tilde{R}$ for $j = 1, \ldots, m$

(3) Solve equilibrium

$r_i = 1.2\tilde{R}_i$; $r_{in} = \tilde{R}_i$; $p_n = 0$; $k = 1$; $k_{max} = 10$

Do While $k < k_{max}$

$\qquad p = 0$

$\qquad$ Do $j = 1, \ldots, m$

$$\lambda_{\theta\,j} = \frac{\sqrt{r_i^2 + \alpha_j^2(\tilde{R}_j^2 - \tilde{R}_i^2)}}{\alpha_j \tilde{R}_j}$$

$$p \leftarrow p + c\frac{\lambda_{\theta\,j}^2 - \lambda_{\theta\,j}^{-2}}{\tilde{R}_j}\Delta\tilde{R}$$

$\qquad$ End Do

$\qquad r_i \leftarrow r_i - \beta(p - p_i)$ with $\beta = \frac{r_i - r_{in}}{p - p_n}$

$\qquad r_{in} = r_i$; $p_n = p$

$\qquad$ If $|p - p_i| < 10^{-6}$, $k = k_{max} + 1$

$\qquad$ If $k = k_{max}$, solution not found: terminate

$\qquad k \leftarrow k + 1$

End Do

(4) Update growth parameter

$\bar{\lambda}_\theta = 1/m \sum_{j=1}^{m} \lambda_{\theta\,j}$ or $\bar{\lambda}_\theta = const$

Do $j = 1, \ldots, m$

$\qquad \alpha_j \leftarrow \alpha_j + \alpha_j\zeta(\lambda_{\theta\,j} - \bar{\lambda}_\theta)\Delta\tau/2$

End Do

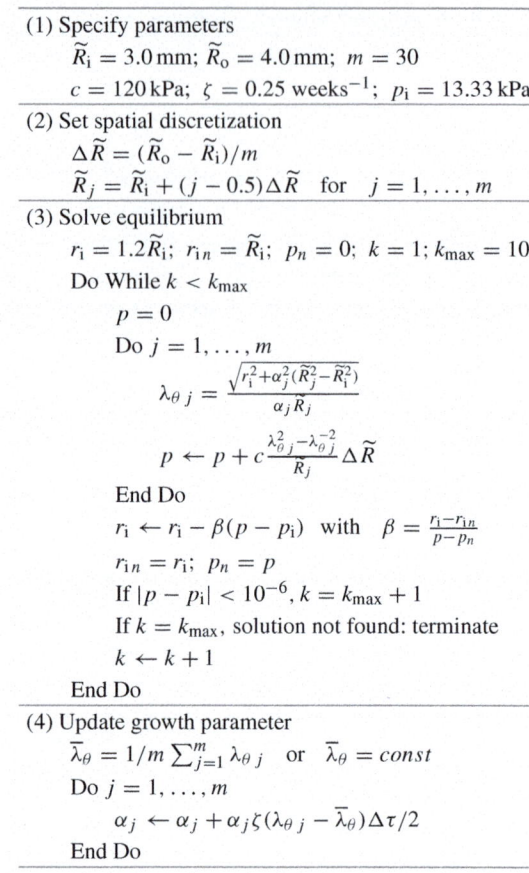

(a)

Inner vessel radius r_i [mm]

Time [weeks]

(b)

Inner vessel radius r_i [mm]

$\bar{\lambda}_\theta = 1.00$

$\bar{\lambda}_\theta = 1.05$

$\bar{\lambda}_\theta = 1.10$

$\bar{\lambda}_\theta = 1.15$

$\bar{\lambda}_\theta = 1.2$

Time [weeks]

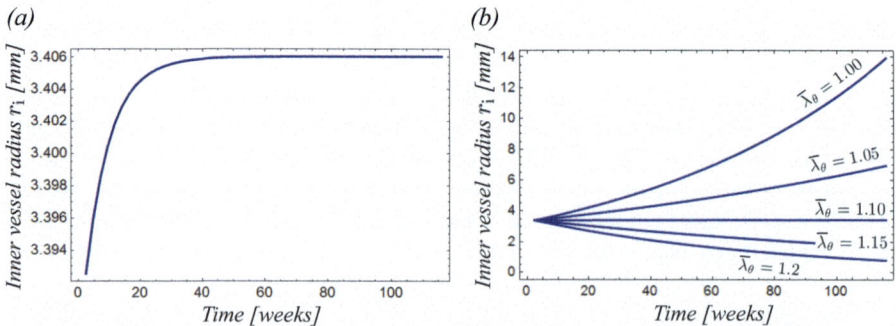

Fig. 7.23 Development of the inner radius r_i over time τ of the thick-walled tube with isotropic growth towards a stretch-based homeostatic target $\bar{\lambda}_\theta$. (**a**) The average circumferential stretch across the entire wall sets the homeostatic target. (**b**) A constant circumferential stretch is the homeostatic target. Table 7.6 lists the parameters that have been used for the computations

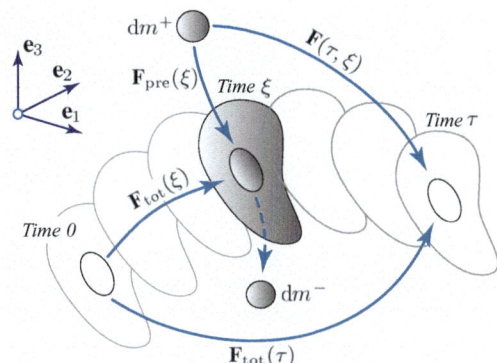

Fig. 7.24 Description of tissue turnover through the continuous addition of dm^+ and removal of dm^- of tissue mass. It leads to the development of natural configurations over time $0 < \xi < \tau$, and the corresponding deformation gradients

7.6.5 Continuous Turnover-Based Growth Description

Regardless the kinematics-based growth description shows physically reasonable results, it is not based on the representation of the continuous production and degradation of vascular tissue constituents. Vascular cells produce tissue constituents that are then integrated in the vessel wall, whilst chemicals, mostly MMPs, constantly degrade them. A process that results in the tissue turnover as illustrated in Fig. 7.24. At the time $\tau = \xi$, the tissue "package" of mass

$$dm^+ = \rho \varsigma_v^+ dv = \rho_0 \varsigma_v^+ dV = \varsigma_m^+ dv = \varsigma_M^+ dV$$

is produced during the time interval $d\tau$, *pre-stressed* by the deformation gradient $\mathbf{F}_{pre}$, and finally integrated in the continuum body Ω. The superscript "+" symbolizes influx into the system, and "$-$" denotes outflux. The rate of volume change is denoted by ς_v and given by the mass balance equation (7.41), whilst ς_m and ς_M are the rates of mass density per unit spatial and referenial volume, respectively.

We may also introduce the survival function $S(\tau - \tau_0)$ that quantifies the portion that remains at the time τ from a tissue "package" that has been deposited at the time τ, and has *not yet* been degraded by MMPs. The survival function $S(x)$ is defined for $x \geq 0$ and satisfies the conditions $S(0) = 1$ and $dS/dx < 0$. The rate of mass density and the survival function depend on the time, but also on factors, such as strain, stress, and chemical concentrations.

The referential mass density at the time τ from the continuous tissue turnover is then described by

$$\rho_0(\tau) = \rho_0(0)S(\tau) + \int_{\xi=0}^{\tau} \varsigma_M^+(\xi)S(\tau - \xi)d\xi \,, \tag{7.58}$$

an expression that involves the solution of a time convolution similar to linear viscoelasticity, see Sect. 3.5.4.2.

Given a hyperelastic description, the vascular tissue "package" of the mass dm^+ that is deformed by $\mathbf{F}$ stores the strain energy density $\Psi(\mathbf{F})/\rho_0$, where Ψ denotes the strain energy density per unit reference volume. The strain energy

$$\Psi(\tau) = \frac{\rho_0(0)S(\tau)}{\rho_0(\tau)}\Psi(\mathbf{F}(\tau,0)) + \int\limits_{\xi=0}^{\tau} \frac{\varsigma_{\mathrm{M}}^+(\xi)S(\tau-\xi)}{\rho_0(\xi)}\Psi(\mathbf{F}(\tau,\xi))\mathrm{d}\xi \qquad (7.59)$$

then expresses the energy that is elastically stored in the tissue at the time τ. Here, $\mathbf{F}(\tau,\xi)$ denotes the deformation gradient at the time τ, applied to the tissue "package" that has been deposited at the time ξ and thus

$$\mathbf{F}(\tau,\xi) = \mathbf{F}_{\mathrm{tot}}(\tau)\mathbf{F}_{\mathrm{tot}}^{-1}(\xi)\mathbf{F}_{\mathrm{pre}}(\xi), \qquad (7.60)$$

see Fig. 7.24.

As with the spatial distribution of tissue volume to the components of the growth tensor $\mathbf{G}$ discussed in Sect. 7.6.3, the continuous turnover-based growth description has also to specify how the mass dm^+ is deformed upon integration into the body and thus to specify the components of $\mathbf{F}_{\mathrm{pre}}(\xi)$.

The continuous turnover-based growth description has been combined with the mixture representation of vascular tissue [273], where tissue components follow an affine transformation. The deformation of the tissue constituents is then constraint, and the description therefore called *constrained mixture model*. Up to date, several versions and applications of this idea have been reported in the literature [21, 214, 272, 314, 331, 383]. The mixture model has been linked to the theory of volumetric growth [8], the microstructural description of collagen remodeling [351], and SMC basal tone variation [210]. Recently, also a homogenized constrained mixture model has been proposed towards the reduction of computational costs whilst preserving the key features of the theory [107, 329, 330]. Given the i-th tissue constituent, the growth deformation may be decomposed in the remodeling part $\mathbf{F}_{\mathrm{r}}$ and the growth part $\mathbf{F}_{\mathrm{g}}$. The change of the tissue's reference configuration is then described by $\mathbf{G} = \mathbf{F}_{\mathrm{r}}\mathbf{F}_{\mathrm{g}}$, where isotropic and anisotropic growth has been investigated [57].

7.6.6 Tissue at Simple Tension and Continuous Mass Turnover

We consider a tissue batch that is characterized by the neoHookean strain energy density $\Psi(\mathbf{C}) = c(\mathrm{tr}\mathbf{C} - 3)/2$ per unit (undeformed) material, where $\mathbf{C}$ denotes the right Cauchy–Green strain tensor and $c = 120\,\mathrm{kPa}$ defines the tissue's referential shear modulus. At the time $\tau = 0$ the tissue occupies its initial (and stress-free) reference configuration Ω_0, with respect to which it is loaded at simple tension. The stretch $\lambda_{\mathrm{tot}} = 1 + k\tau$ determines the tissue's deformation over the time τ, where $k = 10\ \mathrm{week}^{-1}$ is a constant. The tissue is at continuous deposition and removal, and the added tissue pre-stretched at $\lambda_{\mathrm{pre}} = 1.5$ relative to its unloaded configuration.

The initial referential density $\rho_0 = 0.6\,\mathrm{kg\,dm}^{-3}$ defines the mass density of the mechanically relevant structural proteins in the wall, and the constant rate of mass production $\varsigma_M^+ = 0.1\,\mathrm{kg\,dm}^{-3}\,\mathrm{week}^{-1}$ together with the survival function $S(\tau) = \exp(-\tau/\zeta)$ characterize the kinetics of the turnover process. Here, $\zeta = 2$ weeks denotes a time constant. The mass density (7.58) may be integrated, and

$$\rho_0(\tau) = \rho_0(0)\exp(-\tau/\zeta) + \varsigma_M^+\zeta[1 - \exp(-\tau/\zeta)] \tag{7.61}$$

then yields an analytic expression of the mass density per unit reference volume at the time τ, see Fig. 7.25a. In addition, the survival function $S(\tau)$ defines the removal of tissue mass $\mathrm{d}m^-$, a portion of tissue that has, however, no influence on the mechanics of the vessel and is therefore not further considered.

Given the free energy density (7.59), Coleman and Noll's procedure for incompressible materials (3.131) together with the second Piola transform yields the $i = 1, 2, 3$ principal Cauchy stresses

$$
\begin{aligned}
\sigma_i(\tau) = {} & \frac{\rho_0(0)S(\tau)}{\rho_0(\tau)}\lambda_i(\tau)\frac{\partial\Psi(\mathbf{F}(\tau,0))}{\partial\lambda_i(\tau)} \\
& + \int_{\xi=0}^{\tau}\frac{\varsigma_M^+ S(\tau-\xi)}{\rho_0(\xi)}\lambda_i(\tau,\xi)\frac{\partial\Psi(\mathbf{F}(\tau,\xi))}{\partial\lambda_i(\tau,\xi)}\mathrm{d}\xi - \kappa ,
\end{aligned} \tag{7.62}
$$

where κ denotes a Lagrange contribution to the hydrostatic stress that enforces the incompressibility.

The principal stretches of the incompressible material at simple tension read

$$\lambda_1 = \frac{\lambda_{\mathrm{tot}}(\tau)}{\lambda_{\mathrm{tot}}(\xi)}\lambda_{\mathrm{pre}} , \quad \lambda_2 = \lambda_3 = \sqrt{\lambda_1} , \tag{7.63}$$

where the multiplicative kinematics of the turnover-based growth description have been used, see Fig. 7.24. The substitution of said principal stretches in (7.62) and the result $\partial\Psi/\partial\lambda_i = c\lambda_i$, then yields the Cauchy stress in tensile direction

$$
\begin{aligned}
\sigma_1 = c\Bigg\{ & \frac{\rho_0(0)\exp\left(\frac{-\tau}{\zeta}\right)}{\rho_0(\tau)}\left(\lambda_{\mathrm{pre}}^2 - \lambda_{\mathrm{pre}}^{-1}\right) \\
& + \int_{\xi=0}^{\tau}\frac{\varsigma_M^+\exp\left(\frac{\xi-\tau}{\zeta}\right)}{\rho_0(\xi)}\left[\left(\frac{\lambda_{\mathrm{tot}}(\tau)\lambda_{\mathrm{pre}}}{\lambda_{\mathrm{tot}}(\xi)}\right)^2 - \frac{\lambda_{\mathrm{tot}}(\xi)}{\lambda_{\mathrm{tot}}(\tau)\lambda_{\mathrm{pre}}}\right]\mathrm{d}\xi \Bigg\} ,
\end{aligned} \tag{7.64}
$$

where the equilibrium $\sigma_2 = \sigma_3 = 0$ in cross-tension direction identified κ. The convolution integral in (7.64) can be numerically solved, and the model's stress *versus* time response is shown in Fig. 7.25b. Following an excessive stress deviation

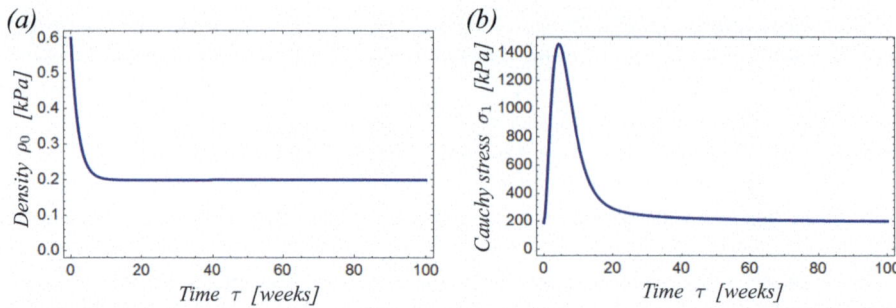

Fig. 7.25 Continuous turnover-based description of tissue at simple tension. Evolution of the density per unit reference volume (**a**) and the tensile Cauchy stress (**b**) over the time τ are shown

at approximately 5 weeks, a constant stress in the continuously elongating batch is predicted beyond the time of approximately 40 weeks.

Example 7.7 (The Thin-Walled Vessel at Tissue Turnover). Let us consider a vessel of diameter $D = 10.0\,\text{mm}$ and wall thickness $H = 1.0\,\text{mm}$ at its stress-free reference configuration Ω_0. At the time $\tau = 0$, the undeformed structural vessel wall proteins are described by the mass density $\rho_0 = 0.6\,\text{kg dm}^{-3}$. The vessel is incompressible and modeled with the neoHookean strain energy density $\Psi(\mathbf{C}) = c(\text{tr}\mathbf{C} - 3)/2$ per unit undeformed material, where $\mathbf{C}$ denotes the right Cauchy–Green strain tensor and $c = 120\,\text{kPa}$. The vessel is inflated by the pressure p_i and does not deform along the axial direction. The vessel wall is at continuous deposition and removal of tissue, where $\lambda_{\text{pre}} = 1.17$ defines the pre-stretch of added tissue. The kinetics are determined by the constant rate of mass production $\varsigma_M^+ = 0.1\,\text{kg dm}^{-3}\,\text{week}^{-1}$ together with the survival function $S(\tau) = \exp(-\tau/\zeta)$, where $\zeta = 2$ weeks denotes a time constant.

(a) Provide the analytical expression of the referential mass density.
(b) Derive the expression of the circumferential Cauchy stress σ from the strain energy (7.59).
(c) Split the time domain into n constant intervals $\Delta\tau$ and derive a time-discretized representation of the circumferential Cauchy stress in the vessel wall.
(d) Use the equilibrium of the inflated tube to compute the inflation pressure p_i over time. Investigate the limit case $\lambda_{\text{tot}} = \lambda_{\text{pre}}$ and show that the model approaches the inflation pressure p_i that is predicted by the thin-walled tube of neoHookean material. ∎

Example 7.8 (Vessel Wall Growth Through Tissue Turnover). We consider a 2D axisymmetric vessel segment, where Ω denotes its spatial configuration that is inflated at the pressure $p_i = 13.33\,\text{kPa}$, see Fig. 7.27. At the time $\tau = 0$ the vessel occupies its initial (and stress-free) reference configuration Ω_0, where the inner and outer radii are denoted by $R_i = 3.0\,\text{mm}$ and $R_o = 4.0\,\text{mm}$, respectively. The vessel wall tissue is at continuous tissue turnover, and Eq. (7.59) describes the vessel wall's strain energy $\Psi(\tau)$ per unit reference volume in Ω_0. We consider the vessel wall at plane strain and to be described by an incompressible neoHooken material of the strain energy density $\Psi(\mathbf{C}) = c(\text{tr}\mathbf{C} - 3)/2$ per unit undeformed material. Here, $\mathbf{C}$ is the right Cauchy–Green strain tensor, and $c = 120\,\text{kPa}$ denotes the referential shear modulus. The constant rate of mass production $\varsigma_M^+ = 0.1\,\text{kg}\,\text{dm}^{-3}\,\text{week}^{-1}$ together with the survival function $S(\tau) = \exp(-\tau/\zeta)$ define the kinetics of the turnover process, where $\zeta = 6$ weeks is a time constant.

Fig. 7.27 Continuous turnover-based growth description of a growing thick-walled vessel in 2D

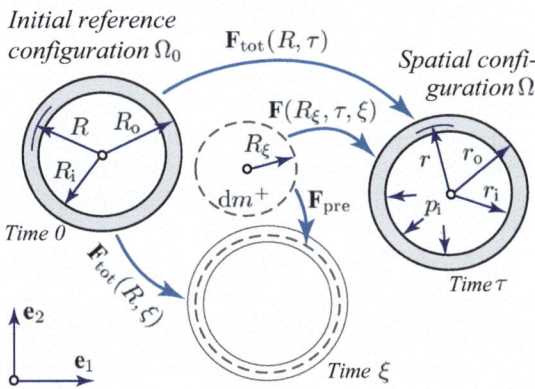

(a) Derive the Cauchy stress $\boldsymbol{\sigma}$ from the strain energy (7.59).
(b) Express the principal stretches and derive the principal Cauchy stresses. Discretize the time interval $0 < t < \tau$ by n equidistant increments $\Delta\tau$ and derive the time-discretized representation of the principal stresses.
(c) Consider the equilibrium of the thick-walled tube problem and discretize the governing equation in time and space.
(d) Adapt the algorithm shown in Table 7.6, and compute the development of the spatial configuration over time for different numbers n of time increments. ∎

7.6.7 Multiphasic and Miscellaneous Descriptions

The vascular wall may be regarded as a *mixture* of solid components, such as elastin, collagen, and SMCs, that are immersed in fluid. Whilst the convection of fluid within the tissues is essential for a number of physiological processes, in the description of many vascular biomechanical problems the wall may be regarded as an *incompressible homogenized solid*. The Cauchy stress then reads $\boldsymbol{\sigma} =$

$\sum_{i=1}^{n} \Phi_i \mathrm{dev}(\boldsymbol{\sigma}_i) - p\mathbf{I}$. The deviatoric stress is averaged over the deviatoric stress contributions $\mathrm{dev}(\boldsymbol{\sigma}_i)$ of the n constituents and weighted by their volume fractions Φ_i, whilst the hydrostatic pressure p enforces the incompressibility and links to pore-fluid flow. In addition to the *affine* deformation of their structural constituents, most mixture models also assume newly formed constituents to be deposited (integrated) at a *predefined stress or strain* into the existing structure. Vascular cells therefore not only sense and respond to stress or strain, but they would also have the ability to *pre-stretch* the secreted and newly integrated tissue constituents. The deposition stretch differs across tissues [137,245], and it is reasonable to assume that newly formed material deposits at the *homeostatic* stretch [139]. Aging, and thus shrinking of the deposited tissue components would be an alternative explanation for the development of the pre-stretch.

A specific mixture model [591] introduced separate variables in the description of the density and the respective undulation of collagen fibers. Both variables are governed by strain-mediated rate equations and then linked to the (passive) histology-based HGO model [261]. Several variations and applications of this approach have been reported [494, 590], and recently [13] the model has also been coupled to signaling pathways of collagen synthesis and degradation [108]. The change in normalized tissue densities, determines the volumetric growth and links to the theory of volumetric growth [473]. As with single phase models, the change of tissue volume is to be translated into the growth rate $\dot{\mathbf{G}}$ (see Sect. 7.6.3), and the related modeling assumption strongly influences vessel wall growth over time [229].

Another modeling approach [203, 361] follows the general theory of fibrous connective tissue, see Sect. 5.5.7. It links the multi-structural description of collagen (5.39) to a stretch-based rate equation, and the newly-synthesized collagen fibrils are integrated at a predefined and triangularly-distributed deposition stretch.

Yet another model [326] is based on a microstructural theory of soft tissue adaptation and considers elastin, collagen, and cells to be immersed in fluid. The model uses a strain-based stimulus and describes the constituents' turnover at a *fixed* fiber-matrix volume fraction. The deformation-dependent fiber degradation is lowest at the tissue's homeostatic stretch, and the model assumes a normally distributed deposition stretch of the tissue fibers.

Yet another model [345,346] addresses cell-mediated tissue compaction in combination with collagen fiber remodeling. In addition to many of the aforementioned mechanisms and concepts, it also introduces *active* stress fibers to cope with tissue compaction, an important aspect in tissue engineering. The cell's phenotype and the strain at the tissue level is assumed to influence the stress exerted by the stress fibers.

Some aspects of vascular wall adaptation can also be captured by the evolution of material constants used in constitutive formulations that have originally been developed to describe the passive vessel wall. Aside from *ad hoc* assumptions, the evolution of model parameters may consider factors, such as the mechanical stress or strain, and the concentration or flux of chemicals. This idea has been materialized as a bounded elastic strain energy [580], use of internal thermodynamic variables [368], through Continuum Damage Mechanics (CDM) [98], or the remodeling of collagen fibers [133, 134, 244].

7.7 Summary and Conclusion

The vascular wall is equipped with mechanisms to cope with environmental changes and to evolve towards optimal mechanical performance. Whilst the ECM determines the strain level of the individual cells, it is the cells themselves that sense and respond to mechanical load. Situated at the interface between tissue and blood, the *endothelium* plays a central role in this feed-back loop and links vessel wall biology to hemodynamic forces.

This chapter focused on the description of *vasoreactivity*, a mechanism that controls the vessel caliber, as well as *arteriogenesis*, the delicate balance between degradation and synthesis of vascular tissue constituents. The biochemical mechanisms that explain said vessel properties have been detailed, and modeling frameworks at different levels of complexity, including thermodynamics arguments, have been discussed. The kinematics-based and the continuous turnover-based descriptions of growth are two fundamentally different frameworks to model arteriogenesis. Whilst the latter one closely addresses tissue turn-over, it is computationally extremely demanding. Recent developments therefore combined elements of the kinematics-based and the continuous turnover-based descriptions [57, 107, 192, 210, 230, 314, 329, 361, 383, 580, 591]. One open issue concerns the relation between tissue mass and the actual growth kinematics. Different assumptions result in very different simulation outcomes [230]. Up to date no general conclusions can be made, and recent analytical insights suggest growth-related stretch to appear mainly in the direction(s) of lowest stiffness [58]. Ongoing research is expected to enlighten these aspects of vessel wall biomechanics.

The adaptation of the vascular wall has to obey a number of physical principles, and even within these constraints, very different approaches have been proposed. Given the scarcity of experiment data, the biological "forces" that govern the adaptation of the vascular wall are largely unknown. Models have to make a number of *ad hoc* assumptions that may or may not be appropriate. The thorough validation of these assumptions with respect to clearly specified Intended Model Applications (IMA) would be critically important to increase confidence in vascular adaptation models, a fundamental requirement towards their rigorous application. In addition to tailored *in vitro* experiments, modern image modalities allow the acquisition of *functional and biological* information of the vessel wall, data that could directly be used in the development and validation of adaption models. Regardless the current premature levels, the better understanding and modeling of vascular adaptation will have fundamental implications in vascular biomechanics.

Correction to: Vascular Biomechanics: Concepts, Models, and Applications

T. Christian Gasser

Correction to:
T. C. Gasser, *Vascular Biomechanics*,
https://doi.org/10.1007/978-3-030-70966-2

The book was inadvertently published with:

1. Low-resolution images in the online book PDF.
2. Incorrect rendering of numbers in some of the equations presented in the chapters on SpringerLink.
3. Missing electronic supplementary material.

   This has now been corrected.

The updated version of this book can be found at
https://doi.org/10.1007/978-3-030-70966-2

© Springer Nature Switzerland AG 2022
T. C. Gasser, *Vascular Biomechanics*, https://doi.org/10.1007/978-3-030-70966-2_8

Mathematical Preliminaries

A.1 Statistics

A.1.1 Definitions and Terminology

Correlation coefficient	Quantifies the correlation (dependence) between two sets of data.
Coefficient of determination	Quantifies the predictability of data.
Confidence Interval (CI)	Measure of the degree of uncertainty associated with a sample statistic.
Degrees of freedom	The number of independent observations in a sample minus the number of population parameters that must be estimated from sample data.
False positive	Mistakenly predicted positive outcome.
False negative	Mistakenly predicted negative outcome.
Independence	Two events are independent when the occurrence of one does not affect the probability of the occurrence of the other.
Interquartile Range (IQR)	A measure of variability, based on dividing a data set into quartiles. Alternatively: The difference between the largest and smallest values in the middle 50% of a set of data.
Interval scale	Numeric scales with order and exact differences between the values.
Mean	The arithmetic average of all values in an observation.
Median	The middle value of observations. If there is an even number of observations, the median is the average of the two middle values.
Nominal scale	Scale used for the labeling variables.

© Springer Nature Switzerland AG 2021
T. C. Gasser, *Vascular Biomechanics*, https://doi.org/10.1007/978-3-030-70966-2

One-tailed significance test	Significance test involving one tail of the probability distribution.
Ordinal scale	Scale that has an order but no specifical numerical value to it.
Outlier	An extreme value that differs greatly from other values of observations.
Population	The total set of observations that can be made.
Probability	A measure of occurrence of an observation.
Probability distribution	A rule that links each outcome of a statistical experiment with its probability of occurrence.
Quantile	Quantiles divide a rank-ordered data set into a number of equal parts.
Quartile	Quartiles divide a rank-ordered data set into four equal parts. The values that divide each part are called the first Q_1, second Q_2 (or median), and third Q_3 quartile.
Ratio scale	Numeric scales that can take any value.
Sample	Observations drawn from a population.
Significance level	The probability of committing a Type I error.
Size	The number of observations in a set.
Set	A well-defined collection of objects.
Standard error	A measure of the variability of a statistic.
Standard Deviation (SD)	A numerical value used to indicate how widely individuals in a group vary.
Statistic	Characteristic of a sample. Generally, a statistic is used to estimate the value of a population parameter.
Two-tailed significance test	Significance test involving both tails of the probability distribution.
Type I error	Error that occurs when the Null Hypothesis is wrongly rejected, and thus a false positive is predicted. The probability of committing a Type I error is called the significance level.
Type II error	Error that occurs when the Null Hypothesis fails to reject a negative observation, and thus a false negative is predicted. The probability of committing a Type II error is called the power.
Variance	A measure of the variability of a statistic.

A.1.2   Probability Distributions

The *Probability Density Function (PDF)* $\rho(X)$ is used to specify the distribution of a random variable X. Such a random variable may represent observations of a sample or a population. Variables describing samples are denoted by lower case letters (such as x), whilst upper case letters (such as X) describe observations of a population. The probability $0 \leq p \leq 1$ that such a random variable falls into the interval $X_1 \leq X \leq X_2$ reads

$$p = \int_{X_1}^{X_2} \rho(X)\mathrm{d}X \,. \tag{A.1}$$

The integration $\Upsilon(X) = \int_{-\infty}^{X} \rho(\xi)\mathrm{d}\xi$ yields the *Cumulative Density Function (CDF)* with the properties $\Upsilon(-\infty) = 0$ and $\Upsilon(\infty) = 1$. Fig. A.1a, b shows ρ and Υ of a *normal distribution*.

A random variable may be visualized by a *box-and-whisker plot*, see Fig. A.1c. The box is formed by the InterQuartile Range (IQR), and thus the range between the third quartile Q_3 and the first quartile Q_1, within which 50% of data falls. A box-and-whisker plot also shows the sample medium (or the second quartile Q_2) together with a bar covering 75% of the data—the whisker.

The *quartiles* (Q_1, median, Q_3) split the data into four equal portions, whilst splitting it into m equal portions, defines the $m - 1$ data *quantiles*.

Variables representing observations in samples or populations may follow different PDFs. The PDF of the *normal* (or *Gaussian*) *distribution* reads

$$\rho_{\mathrm{n}}(X) = \frac{\exp[-(X - \overline{X})^2/(2\sigma^2)]}{\sqrt{2\pi\sigma^2}} \,, \tag{A.2}$$

Fig. A.1 Representations of the random variable X that describes observations in a population. (**a**) Probability Density Function (PDF) $\rho(X)$, (**b**) Cumulative Density Function (CDF) $\Upsilon(X)$, and (**c**) box-and-whisker plot of X. The example shows a normal distribution of zero median and the standard deviation σ. In total 25% of data falls below the first quartile Q_1, 50% below the median, and 75% below the third quartile Q_3. The InterQuartile Range (IQR) is between Q_1 and Q_3 and covers 50% of the data

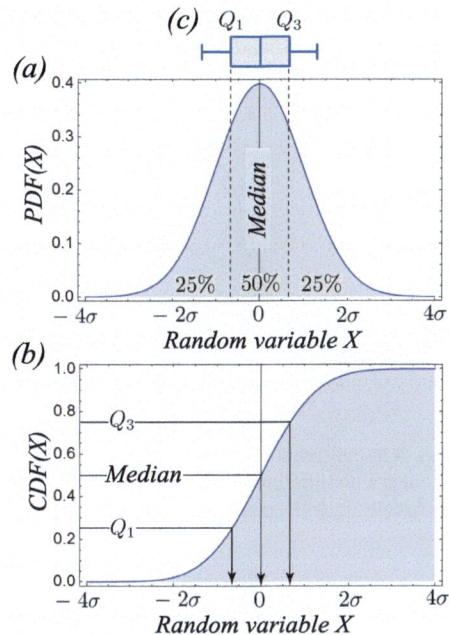

where $\overline{X}$ is the median, and σ^2 denotes the *variance*. It is the square of the distribution's *Standard Deviation (SD)* σ. Figure A.1 shows such a normal distribution.

The distribution of a sample of size n taken from a normal distributed population will *not* follow a normal distribution but a *student t-distribution* instead. The difference between both distributions diminishes with increasing sample size n. The sample size n also defines the degrees of freedom $\nu = n - 1$ of the student t-distribution. The student t-distribution's PDF reads

$$\rho_t(x) = \frac{\Gamma(\frac{\nu+1}{2})}{\sqrt{\nu\pi}\,\Gamma(\frac{\nu}{2})}\left(1 + \frac{x^2}{\nu}\right)^{-\frac{\nu+1}{2}}, \tag{A.3}$$

where Γ is the gamma function, and x denotes the sample variable. Figure A.2a shows student t-distributions of $\nu = 3$ and $\nu = \infty$ degrees of freedom, respectively. The difference of the tails between the normal distribution and the student t-distribution is especially emphasized, see Fig. A.2a. The student t-distribution appears often in hypothesis testing, see Sect. 1.4.3.

Some PDFs capture bounded data, where the random variable is zero beyond some bounds. The *beta-distribution* is such an example. It is bounded at 0 and 1, and its PDF reads

$$\rho_\beta(x) = X^{\alpha-1}(1-x)^{\beta-1}\frac{\Gamma(\alpha)\Gamma(\beta)}{\Gamma(\alpha+\beta)}, \tag{A.4}$$

where the parameters α and β determine the shape of the PDF. Figure A.2b illustrates the beta-distribution for some parameter combinations.

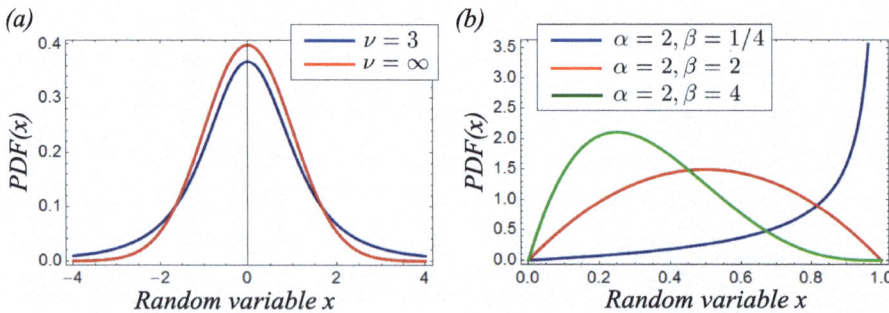

Fig. A.2 Different Probability Density Functions (PDF) that represent the random variable x. (**a**) Student t-distribution for different degrees of freedom ν. The case $\nu = \infty$ represents the normal distribution. (**b**) Beta-distribution for different shape parameters α and β

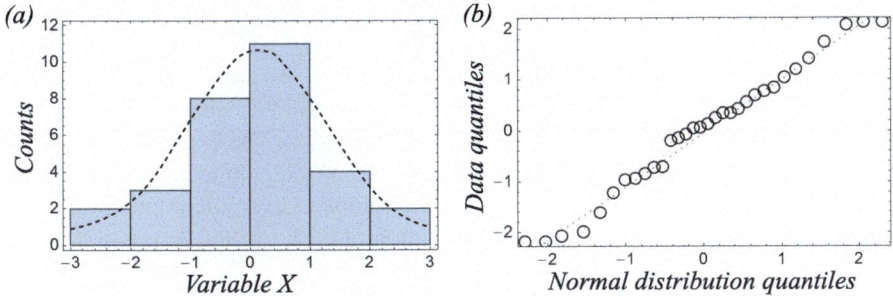

Fig. A.3 Graphical methods to determine whether a data set is well-modeled by a normal distribution. (**a**) Histogram plot. (**b**) Quantile-Quantile plot (QQ-plot)

A.1.3   Data Distribution Testing

Statistical analysis methods often demand a particular distribution of the data within a sample or population. Normally distributed data is the most relevant case. One of the first steps in data analysis is therefore the assessment of how well a normal PDF resembles the histogram of the data set, see Fig. A.3a. Given a low number of data points, a Quantile-Quantile plot (QQ-plot) may be used. A QQ-plot represents the data quantiles *versus* the quantiles of the normal distribution. Normality therefore holds if the data points fall around the diagonal of the QQ-plot, see Fig. A.3b. Any distribution other than a normal distribution, may be tested by a QQ-plot.

The distribution of the data within a set may also be tested against the hypothesis (see Sect. 1.4.3) that it follows a certain distribution. Such tests are called *frequentist inference tests*, and draw conclusions from sample data by emphasizing the frequency or proportion of the data.

A.1.4   Confidence Interval

Confidence Intervals (CI) describe the amount of uncertainty associated with a *sample estimate* of a *population parameter*. The estimation of the *population mean* μ from a sample is such an example; different samples will naturally yield different population means. Let us consider a sample that has been drawn from a normal distributed population, and sample mean $\overline{x}$ and SD s would be

$$\overline{x} = \frac{\sum_{i=1}^{n} x_i}{n} \quad ; \quad s = \sqrt{\frac{\sum_{i=1}^{n}(x_i - \overline{x})^2}{n-1}},$$

where x_i and n denote the observations in the sample and the sample size, respectively.

We may specify the *confidence level* c, the percentage we expect from the population mean μ to fall within the CI $\mu_{\min} \leq \mu_{\max}$. Given the confidence level, the critical probability $p^\star = (1 + c/100)/2$ follows, and the interval $1 - p^\star \leq p \leq p^\star$ covers the probability of the confidence level c. We now use the *student t-distribution* of $v = n - 1$ degrees of freedom to describe the sample. The corresponding statistic $t^\star$ follows then from the solution of the equation $p^\star = \Upsilon(t^\star)$, where $\Upsilon(x)$ denotes the CDF of the student t-distribution. The expressions $\overline{\mu}_{\min} = \overline{x} - t^\star e_s$ and $\overline{\mu}_{\max} = \overline{x} + t^\star e_s$ define then the CI of the population mean, where $e_s = s/\sqrt{n}$ denotes the *standard error* of the sample. Given the population's SD would also be known, a more accurate estimation of the CI would be possible.

Example A.1 (Population Mean of the Vessel Wall Strength). Let us consider an *in vitro* tissue characterization study that uses tensile testing to measure the strength x of the vessel wall. The experiment acquired data from vessel wall specimens of, in total, $n = 19$ animals; Table A.1 lists the recordings. Given this information, conclusions regarding the population mean μ of the vessel wall strength should be drawn.

Table A.1 Tensile strength x of vessel wall samples acquired from $n = 19$ animals

Specimen	Strength [kPa]	Specimen	Strength [kPa]	Specimen	Strength [kPa]
1	1140.5	8	764.8	15	1691.3
2	759.2	9	1221.9	16	1089.8
3	465.2	10	886.6	17	1361.0
4	855.1	11	1289.3	18	1023.5
5	1646.4	12	1205.4	19	641.2
6	1090.7	13	976.0		
7	578.3	14	376.0		

(a) Investigate wether or not the sample is normal distributed.
(b) Compute the sample mean $\overline{x}$, the sample Standard Deviation (SD) s, and the sample standard error e_s.
(c) Given the confidence levels of 90% and 95%, compute the respective Confidence Intervals (CI) of the population mean μ.   ■

A.2   Complex Numbers

A complex number may be expressed by $c = a + bi$, where a and b are real numbers, whilst $i = \sqrt{-1}$ denotes the imaginary unit. A complex number can be formally represented by a *vector* $\mathbf{c} = a\mathbf{e}_R + b\mathbf{e}_I$ in the complex plane, also known as Argand's

Fig. A.4 The complex number $a + bi$ represented by the vector **c** in the complex plane

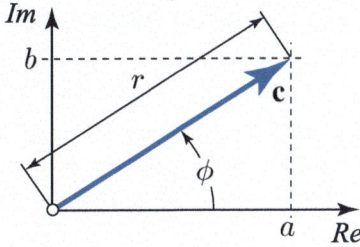

diagram, see Fig. A.4. The complex plane is defined by the one-dimensional base vectors $\mathbf{e}_R = 1$ and $\mathbf{e}_I = i$ pointing into the real and the imaginary directions, respectively.

In polar coordinates $\{r, \phi\}$, the vector has the coordinates $a = r \cos \phi$ and $b = r \sin \phi$, and the complex number then reads $\mathbf{c} = r \cos \phi \, \mathbf{e}_R + r \sin \phi \, \mathbf{e}_I = r(\cos \phi + i \sin \phi)$. Here, $r = |\mathbf{c}| = \sqrt{a^2 + b^2}$ and

$$\phi = \arg \mathbf{c} = \begin{cases} \arctan(b/a) & \text{if } a > 0 \\ \arctan(b/a) + \pi & \text{if } a < 0 \text{ and } b \geq 0 \\ \arctan(b/a) - \pi & \text{if } a < 0 \text{ and } b < 0 \\ \pi/2 & \text{if } a = 0 \text{ and } b > 0 \\ -\pi/2 & \text{if } a = 0 \text{ and } b < 0 \\ \text{indeterminate} & \text{if } a = 0 \text{ and } b = 0 \end{cases} \tag{A.5}$$

denotes its *norm* and *argument*, respectively. Given Euler's formula

$$\exp(i\phi) = \sum_{k=0}^{\infty} \frac{(i\phi)^k}{k!}$$

$$= \left(1 - \frac{\phi^2}{2!} + \frac{\phi^4}{4!} - \cdots \right) + i \left(\phi - \frac{\phi^3}{3!} + \frac{\phi^5}{5!} - \cdots \right)$$

$$= \cos \phi + i \sin \phi \,,$$

the complex number may be expressed by $\mathbf{c} = |\mathbf{c}| \exp(i\phi)$. All algebraic manipulations known of the exponential functions are then applicable.

Given a steady-state periodic problem, the vector representing the complex number rotates around the pole at the angular velocity ω. It then reads $\mathbf{c} \exp(i\omega t) = |\mathbf{c}| \exp(i\phi) \exp(i\omega t) = |\mathbf{c}| \exp[i(\omega t + \phi)]$, where t denotes the time, whilst $\phi = \arg[\mathbf{c}]_{t=0}$ is the angle between the real axis and **c** at the time $t = 0$.

A.3   Fourier Series Approximation

Any signal $s(t)$ that is *integrable* and *periodic* over a period p may be approximated by

$$\widetilde{s}(t) = \sum_{k=-N}^{N} c_k \exp(i\omega_k t); \; c_k = \frac{1}{p} \int_{t=0}^{p} s(t) \exp(i\omega_k t)\, dt \qquad (A.6)$$

with $\omega_k = 2\pi k/p$. Here, c_k are the Fourier coefficients and the approximation (A.6) converges towards the exact representation of $s(t)$ at $N \to \infty$. The expression (A.6) is then a Fourier series representation of $s(t)$.

A.4   Laplace and Fourier Transforms

The Laplace transform $\widehat{x}(s)$ and the Fourier transform $\widetilde{x}(s)$ of a time function $x(t)$ read

$$\widehat{x}(s) = \int_{0}^{\infty} x(t)\exp(-st)dt; \; \widetilde{x}(\omega) = \int_{-\infty}^{\infty} x(t)\exp(-i\omega t)dt. \qquad (A.7)$$

Both are linear operators and some of their properties are listed below:

- A convolution $I(t) = \int_{-\infty}^{\infty} x(\tau)y(t-\tau)d\tau$ of the functions $x(t)$ and $y(t)$ in the time domain is equivalent to the product $\widehat{I}(s) = \widehat{x}(s)\widehat{y}(s)$ in the Laplace domain, or the product $\widetilde{I}(\omega) = \widetilde{x}(\omega)\widetilde{y}(\omega)$ in the Fourier domain.
- In the Laplace domain, the time derivative of the continuous function $x(t)$ reads $\widehat{\dot{x}}(s) = s\widehat{x}(s) - x(0)$, where $x(0)$ denotes the value of $x(t)$ at $t = 0$. In the Fourier domain the time derivative reads $\widetilde{\dot{x}}(\omega) = i\omega\widetilde{x}(\omega)$.
- Given the function $x(t)$ has the Laplace transform $\widehat{x}(s)$, the Laplace transform of the function $t^n x(t)$; $n = 1, 2, 3 \ldots$ then reads $\widehat{t^n x(t)} = (-1)^n (d^n\widehat{x}(s)/ds^n)$.
- Given the function $x(t)$ has the Fourier transform $\widetilde{x}(\omega)$, the Fourier transform of the function $t^n x(t)$; $n = 1, 2, 3 \ldots$ then reads $\widetilde{t^n x(t)} = i^n (d^n\widetilde{x}(\omega)/d\omega^n)$.

A.5   Matrix Algebra

A *matrix* of the *dimension* $m \times n$

$$\mathbf{A} = A_{ij} = \begin{bmatrix} A_{11} & A_{12} & \ldots & A_{1n} \\ A_{21} & A_{22} & \ldots & A_{2n} \\ \vdots & \vdots & \ddots & \vdots \\ A_{m1} & A_{m2} & \ldots & A_{mn} \end{bmatrix}$$

stores information, such as the components of a second-order tensor. Given $n = m$, it is a *square matrix* and said matrix could also be *symmetric* $\mathbf{A} = \mathbf{A}^{\mathrm{T}}$; $A_{ij} = A_{ji}$ or *skew-symmetric* $\mathbf{A} = -\mathbf{A}^{\mathrm{T}}$; $A_{ij} = -A_{ji}$. A matrix $\mathbf{A}$ may be multiplied by a scalar α, resulting in the scaled matrix $\mathbf{C} = \alpha \mathbf{A}$; $C_{ij} = \alpha A_{ij}$; $i = 1, \ldots, m$; $j = 1, \ldots, n$.

The matrix $\mathbf{A}$ of dimension $m \times n$ may be multiplied by the matrix $\mathbf{B}$ of dimension $n \times o$, which then reads $\mathbf{C} = \mathbf{AB}$; $C_{ij} = \sum_{k=1}^{n} A_{ik} B_{kj}$. Following Einstein's summation convention, the summation symbol is not explicitly written. The multiplication then reads $C_{ij} = A_{ik} B_{kj}$, where k denotes the summation (or dummy) index, whilst i and j are the free indices, respectively.

A.5.1   Trace of a Matrix

The sum over the diagonal terms of a square matrix forms its *trace* $\mathrm{tr}\mathbf{A} = \mathbf{I} : \mathbf{A} = \delta_{kl} A_{kl} = A_{kk}$, where

$$\delta_{ij} = \begin{cases} 1 \text{ if } i = j, \\ 0 \text{ if } i \neq j \end{cases}$$

denotes the Kronecker delta.

A.5.2   Identity Matrix

The *identity matrix*

$$\mathbf{I} = \delta_{ij} = \begin{bmatrix} 1 & 0 & 0 \\ 0 & 1 & 0 \\ 0 & 0 & 1 \end{bmatrix}$$

is a particular square matrix, and its trace $\mathrm{tr}\mathbf{I} = \mathbf{I} : \mathbf{I} = \delta_{kl}\delta_{kl} = \delta_{kk} = n$ is equal to its dimension n.

A.5.3   Determinant of a Matrix

Given a square matrix of dimension $n = 3$, its *determinant* reads

$$\det\mathbf{A} = \frac{1}{6} e_{ijk} e_{lmn} A_{il} A_{jm} A_{kn},$$

where the *alternating symbol*

$$e_{ijk} = \begin{cases} 1 & \text{if } (i, j, k) \text{ is an even permutation, i.e. } e_{123} = e_{231} = e_{312} = 1, \\ -1 & \text{if } (i, j, k) \text{ is an odd permutation, i.e. } e_{321} = e_{213} = e_{132} = -1, \\ 0 & \text{if any of } i, j, k \text{ are equal, i.e. } e_{122} = e_{222} = e_{323} = 0, \end{cases}$$

(A.8)

has been introduced. The determinant then reads

$$\det \begin{bmatrix} A_{11} & A_{12} & A_{13} \\ A_{21} & A_{22} & A_{23} \\ A_{31} & A_{32} & A_{33} \end{bmatrix} = - A_{12}A_{22}A_{31} + A_{12}A_{23}A_{31} + A_{13}A_{21}A_{32}$$

$$- A_{11}A_{23}A_{32} - A_{12}A_{21}A_{33} + A_{11}A_{22}A_{33}.$$

The determinant of matrices of dimensions other than $n = 3$ dimensions may be similarly defined.

Example A.2 (Trace and Determinant of a Matrix). Given the matrix

$$\mathbf{A} = \begin{bmatrix} 1 & -2 & 3 \\ 4 & 5 & -6 \\ -7 & 8 & 9 \end{bmatrix},$$

∎

compute its trace and determinant.

A.5.4   Inverse and Orthogonal Matrix

Given $\det\mathbf{A} \neq 0$, the square matrix $\mathbf{A}$ can be *inverted* towards $\mathbf{A}^{-1}$, such that $\mathbf{A}^{-1}\mathbf{A} = \mathbf{A}\mathbf{A}^{-1} = \mathbf{I}$ holds.

The *orthogonal matrix* $\mathbf{R}$, is a particulary interesting matrix that satisfies $\mathbf{R}^{-1} = \mathbf{R}^{\mathrm{T}}$ and $\det\mathbf{R} = \pm 1$. Given $\det\mathbf{R} = +1$, $\mathbf{R}$ is a proper orthogonal matrix and represents a *rigid body rotation*.

A.5.5   Linear Vector Transform

The *linear vector transform* $\mathbf{b} = \mathbf{A}\mathbf{a}$; $b_i = A_{ij}a_j$ defines the multiplication of the matrix $\mathbf{A}$ with the vector $\mathbf{a}$. It maps the vector $\mathbf{a}$ into another vector $\mathbf{b}$. Using the inverse $\mathbf{A}^{-1}$, the transformation is performed in the "opposite direction", and thus $\mathbf{a} = \mathbf{A}^{-1}\mathbf{b}$. The orthogonal transformations $\mathbf{b} = \mathbf{R}\mathbf{a}$; $\mathbf{a} = \mathbf{R}^{\mathrm{T}}\mathbf{b}$ is a particular case of a linear vector transform.

Example A.3 (Linear Vector Transform). Given the vector $\mathbf{a} = [1 \ 2 \ 3]^{\mathrm{T}}$ and the matrix

$$\mathbf{A} = \begin{bmatrix} 7 & 2 & -1 \\ 4 & 3 & 4 \\ 5 & 6 & 9 \end{bmatrix},$$

compute the linear vector transform $\mathbf{b} = \mathbf{A}\mathbf{a}$.   ∎

A.5.6 Eigenvalue Problem

The linear vector transformation $\mathbf{A}\mathbf{x}$ that maintains the direction of the vector $\mathbf{x}$ and scales it by a factor λ appears frequently in engineering mechanics. It is called the *eigenvalue problem* and reads

$$\mathbf{A}\mathbf{x} = \lambda\mathbf{x} \quad \text{or alternatively} \quad (\mathbf{A} - \lambda\mathbf{I})\mathbf{x} = \mathbf{0}. \tag{A.9}$$

Its non-trivial solution satisfies $\det(\mathbf{A} - \lambda\mathbf{I}) = 0$, and in 3D, it leads to the *characteristic equation*

$$\lambda^3 - I_1\lambda^2 + I_2\lambda - I_3 = 0 ,$$

where $I_1 = \text{tr}\mathbf{A}$, $I_2 = 1/2[(\text{tr}\mathbf{A})^2 - \text{tr}\mathbf{A}^2]$, and $I_3 = \det\mathbf{A}$ are coefficients known as *invariants*.

Given $\mathbf{A} = \mathbf{A}^{\text{T}}$, (A.9) defines a *symmetric* eigenvalue problem. Its solution has n *real* eigenvalues λ_i ; $i = 1, \ldots, n$, and the corresponding eigenvectors $\mathbf{x}_i$ are *perpendicular* to each other, and thus $\mathbf{x}_i \cdot \mathbf{x}_j = \delta_{ij}$ for $i, j = 1, \ldots, n$.

The eigenvectors form the proper orthogonal matrix $\mathbf{R}^{\text{T}} = [\mathbf{x}_1, \ldots .\mathbf{x}_n]$ that rotates the matrix $\mathbf{A}$ into its diagonal form. In 3D

$$\mathbf{R}\mathbf{A}\mathbf{R}^{\text{T}} = \begin{bmatrix} \mathbf{x}_1^{\text{T}} \\ \mathbf{x}_2^{\text{T}} \\ \mathbf{x}_3^{\text{T}} \end{bmatrix} \underbrace{[\mathbf{A}\mathbf{x}_1 \quad \mathbf{A}\mathbf{x}_2 \quad \mathbf{A}\mathbf{x}_3]}_{[\lambda_1\mathbf{x}_1 \quad \lambda_2\mathbf{x}_2 \quad \lambda_3\mathbf{x}_3]} = \begin{bmatrix} \lambda_1 & 0 & 0 \\ 0 & \lambda_2 & 0 \\ 0 & 0 & \lambda_3 \end{bmatrix} \tag{A.10}$$

then holds, where the eigenvalues λ_i ; $i = 1, \ldots, 3$ appear along the diagonal.

Example A.4 (Eigenvalue Problem). Given the matrix

$$\mathbf{A} = \begin{bmatrix} 0.8 & 0.3 \\ 0.2 & 0.7 \end{bmatrix},$$

compute the eigenvalues λ_i and the respective eigenvectors $\mathbf{x}_i$.   ∎

A.5.7   Relation Between the Trace and the Eigenvalues of a Matrix

With the property $\mathbf{RR}^\mathrm{T} = \mathbf{I}$ of the rotation matrix, the relation $\mathrm{tr}(\mathbf{RAR}^\mathrm{T}) = R_{ik}A_{kl}R_{il} = \delta_{kl}A_{kl} = A_{kk} = \mathrm{tr}\mathbf{A}$ may be derived, which together with $\mathrm{tr}(\mathbf{RAR}^\mathrm{T}) = \lambda_1 + \lambda_2 + \lambda_3$ from (A.10) then leads to the relation

$$\mathrm{tr}\mathbf{A} = \lambda_1 + \lambda_2 + \lambda_3 \qquad (A.11)$$

between the eigenvalues λ_i and the trace of $\mathbf{A}$. In addition, the relations

$$\mathrm{tr}\mathbf{A}^2 = \lambda_1^2 + \lambda_2^2 + \lambda_3^2 \quad \text{and} \quad \mathrm{tr}\mathbf{A}^3 = \lambda_1^3 + \lambda_2^3 + \lambda_3^2 \qquad (A.12)$$

may be derived using the same arguments.

A.5.8   Cayley–Hamilton Theorem

The Cayley[1]–Hamilton[2] theorem states that every square matrix satisfies its own characteristic equation, and thus

$$\mathbf{A}^3 - I_1\mathbf{A}^2 + I_2\mathbf{A} - I_3\mathbf{I} = \mathbf{0} \qquad (A.13)$$

holds, where the definitions of the invariants $I_1 = \mathrm{tr}\mathbf{A}$, $I_2 = [(\mathrm{tr}\mathbf{A})^2 - \mathrm{tr}\mathbf{A}^2]/2$, $I_3 = \det\mathbf{A}$ have been used. The computation of the trace of (A.13) together with the use of (A.11) and (A.12) allows the derivation of (A.13).

A.6   Vector Algebra

A *vector* is a mathematical object that has a *magnitude* and a *direction*. Given the Cartesian coordinate system $\{\mathbf{e}_1, \ldots, \mathbf{e}_n\}$ of n orthonormal base vectors $\mathbf{e}_i$, the vector $\mathbf{a}$ is expressed by

$$\mathbf{a} = a_i\mathbf{e}_i \; ; \; i = 1, \ldots, n \,,$$

where a_i are the vector *components*. In 3D, it reads $\mathbf{a} = a_1\mathbf{e}_1 + a_2\mathbf{e}_2 + a_3\mathbf{e}_3$, and Fig. A.5 illustrates the vector $\mathbf{a}$ and its components $a_i; i = 1, 2, 3$.

---

[1] Arthur Cayley, British mathematician, 1821–1895.
[2] Sir William Rowan Hamilton, Irish physicist, astronomer, and mathematician, 1805–1865.

Fig. A.5 Illustration of the vector **a** with the components a_1, a_2, a_3 within the 3D Cartesian coordinate system $\{\mathbf{e}_1, \mathbf{e}_2, \mathbf{e}_3\}$

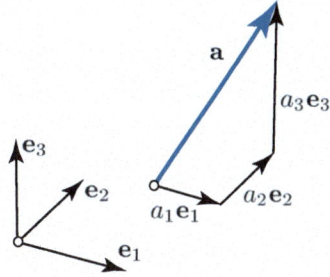

A.6.1   Basic Vector Operations

The *magnitude* of a n-dimensional vector is given by its Euclidian[3] norm $|\mathbf{a}| = \sqrt{\mathbf{a} \cdot \mathbf{a}} = \sqrt{a_i a_i}$; $i = 1, \ldots, n$. Two vectors may be added, and $\mathbf{c} = \mathbf{a} + \mathbf{b}$; $c_i = a_i + b_i$ then denotes the *sum* of the n-dimensional vectors **a** and **b**. A vector **a** may also be multiplied by a scalar α, which then yields the scaled vector $\mathbf{b} = \alpha \mathbf{a}$; $b_i = \alpha a_i$.

The *dot (or inner) product* $c = \mathbf{a} \cdot \mathbf{b} = |\mathbf{a}||\mathbf{b}| \cos \theta$; $c = a_i b_i$ of the vectors **a** and **b** results in the scalar c, where θ is the angle between both vectors, see Fig. A.6a. The dot product of a vector with itself yields the square of its norm $\mathbf{a} \cdot \mathbf{a} = |\mathbf{a}|^2$.

The *cross (or outer) product* of the 3D vectors **a** and **b** maps them into a 3D (pseudo) vector **c**. It reads

$$\mathbf{c} = \mathbf{a} \times \mathbf{b} = e_{ijk} a_j b_k \mathbf{e}_i ,$$

where e_{ijk} denotes the alternating symbol, see Sect. A.5.3. The norm $|\mathbf{c}| = |\mathbf{a} \times \mathbf{b}|$ is the area that is formed by the two vectors **a** and **b**, see Fig. A.6b. The cross product may formally also been written as

$$\mathbf{c} = \det \begin{bmatrix} \mathbf{e}_1 & \mathbf{e}_2 & \mathbf{e}_3 \\ a_1 & a_2 & a_3 \\ b_1 & b_2 & b_3 \end{bmatrix} = \begin{bmatrix} -a_3 b_2 + a_2 b_3 \\ a_3 b_1 - a_1 b_3 \\ -a_2 b_1 + a_1 b_2 \end{bmatrix} . \tag{A.14}$$

Another vector product that is formed by the 3D vectors **a**, **b**, and **c** is called the *triple scalar product*. It reads

$$d = (\mathbf{a} \times \mathbf{b}) \cdot \mathbf{c} = \det \begin{bmatrix} a_1 & a_2 & a_3 \\ b_1 & b_2 & b_3 \\ c_1 & c_2 & c_3 \end{bmatrix} = e_{ijk} a_i b_j c_k ,$$

---

[3] Euclid of Alexandria, Greek mathematician, 300 BCE.

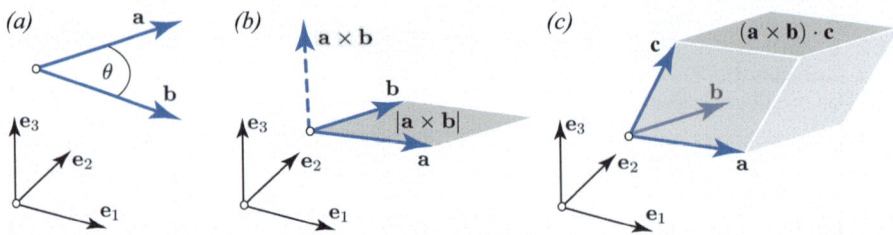

Fig. A.6 Definition of (**a**) the dot product $\mathbf{a} \cdot \mathbf{b}$, (**b**) the cross product $\mathbf{a} \times \mathbf{b}$, and (**c**) the triple scalar product $(\mathbf{a} \times \mathbf{b}) \cdot \mathbf{c}$

where e_{ijk} denotes the alternating symbol. The scalar d represents the volume of the parallelepiped that is formed by the three vectors $\mathbf{a}$, $\mathbf{b}$, and $\mathbf{c}$, see Fig. A.6c. Given the definition of the alternating symbol e_{ijk}, the triple scalar product explicitly reads

$$d = \det \begin{bmatrix} A_{11} & A_{12} & A_{13} \\ A_{21} & A_{22} & A_{23} \\ A_{31} & A_{32} & A_{33} \end{bmatrix} = -A_{13}A_{22}A_{31} + A_{12}A_{23}A_{31}$$

$$+A_{13}A_{21}A_{32} - A_{11}A_{23}A_{32} - A_{12}A_{21}A_{33} + A_{11}A_{22}A_{33}.$$

Another vector product is called the *dyadic product*. Given the n-dimensional vectors $\mathbf{a}$ and $\mathbf{b}$, the dyadic product maps them into the n-dimensional second-order tensor

$$\mathbf{A} = \mathbf{a} \otimes \mathbf{b} = a_i b_j \mathbf{e}_i \otimes \mathbf{e}_j \; ; \; i, j = 1, \ldots, n,$$

where the components of $\mathbf{A}$ may be presented by a n-dimensional matrix. In 3D, such a matrix reads

$$\mathbf{A} = \begin{bmatrix} a_1 b_1 & a_1 b_2 & a_1 b_3 \\ a_2 b_1 & a_2 b_2 & a_2 b_3 \\ a_3 b_1 & a_3 b_2 & a_3 b_3 \end{bmatrix} .$$

The dyadic product is a linear, but non-commutative operation, and the expressions

$$\mathbf{a} \otimes \mathbf{b} \neq \mathbf{b} \otimes \mathbf{a} ,$$

$$(\alpha \mathbf{a}) \otimes \mathbf{b} = \mathbf{a} \otimes (\alpha \mathbf{b}) = \alpha (\mathbf{a} \otimes \mathbf{b}) ,$$

$$\mathbf{a} \otimes (\mathbf{b} + \mathbf{c}) = \mathbf{a} \otimes \mathbf{b} + \mathbf{a} \otimes \mathbf{c} ,$$

$$(\mathbf{a} \otimes \mathbf{b}) \cdot \mathbf{c} = \mathbf{a}(\mathbf{b} \cdot \mathbf{c})$$

illustrate some properties of the dyadic product.

Example A.5 (Norm of a Vector and Vector Products). Consider the vectors $\mathbf{a} = [1\ 2\ 3]^{\mathrm{T}}$ and $\mathbf{b} = [4\ 5\ -6]^{\mathrm{T}}$ and compute their norms. In addition, compute the dot product, the cross product, and the dyadic product of $\mathbf{a}$ and $\mathbf{b}$, as well as the triple scalar product $(\mathbf{a} \times \mathbf{b}) \cdot \mathbf{c}$, where $\mathbf{c} = [-7\ 8\ 9]^{\mathrm{T}}$. ■

A.6.2   Coordinate Transformation

Whilst a vector $\mathbf{a} = a_i \mathbf{e}_i = \widetilde{a}_i \widetilde{\mathbf{e}}_i\ ; i = 1, \dots, n$ is *independent* of the coordinate system, its components $a_i, \widetilde{a}_i$ *depend* on the particular coordinate system. Let us consider a proper orthogonal transformation $\mathbf{R}$ that maps the Cartesian coordinate system $\{\mathbf{e}_1, \dots, \mathbf{e}_n\}$ into another Cartesian coordinate system $\{\widetilde{\mathbf{e}}_1, \dots, \widetilde{\mathbf{e}}_n\}$, a transformation shown in Fig. A.7. The proper orthogonal matrix

$$\mathbf{R} = \mathbf{e}_i \cdot \widetilde{\mathbf{e}}_j\ ; i = 1, \dots, n\,,$$

connects both Cartesian coordinate systems, and the linear transform $\widetilde{\mathbf{e}}_i = \mathbf{R}\mathbf{e}_i$ determines the transformation amongst their base vectors. The coefficients of $R_{ij} = \mathbf{e}_i \cdot \widetilde{\mathbf{e}}_j$ are the *direction cosines*, and thus the cosines of the angle between the base vectors $\mathbf{e}_j$ and $\widetilde{\mathbf{e}}_i$, respectively. Given a 3D problem, the coefficients of $\mathbf{R}$ therefore read

$$\mathbf{R} = \begin{bmatrix} \mathbf{e}_1 \cdot \widetilde{\mathbf{e}}_1 & \mathbf{e}_1 \cdot \widetilde{\mathbf{e}}_2 & \mathbf{e}_1 \cdot \widetilde{\mathbf{e}}_3 \\ \mathbf{e}_2 \cdot \widetilde{\mathbf{e}}_1 & \mathbf{e}_2 \cdot \widetilde{\mathbf{e}}_2 & \mathbf{e}_2 \cdot \widetilde{\mathbf{e}}_3 \\ \mathbf{e}_3 \cdot \widetilde{\mathbf{e}}_1 & \mathbf{e}_3 \cdot \widetilde{\mathbf{e}}_2 & \mathbf{e}_3 \cdot \widetilde{\mathbf{e}}_3 \end{bmatrix}\,.$$

The base vectors $\mathbf{e}_i$ have usually the components $e_{ij} = \delta_{ij}$, and the components of the rotated base vectors $\widetilde{\mathbf{e}}_i$ then read $\widetilde{e}_{ij} = R_{ji}$.

A.6.2.1 Vector Components undergoing Coordinate Transformation
We consider the transformed coordinate system $\{\widetilde{\mathbf{e}}_1, \dots, \widetilde{\mathbf{e}}_n\}$, within which

$$\widetilde{a}_i = \mathbf{a} \cdot \widetilde{\mathbf{e}}_i = \mathbf{a} \cdot (\mathbf{R}\mathbf{e}_i) = \mathbf{R}^{\mathrm{T}}\mathbf{a} \cdot \mathbf{e}_i\quad i = 1, \dots, n$$

are the components of the vector $\mathbf{a}$. Given the components $e_{ij} = \delta_{ij}$ of the base vectors $\mathbf{e}_i$ of the non-transformed coordinate system, the vector components in

Fig. A.7 The rotation matrix $\mathbf{R}$ determines the mapping between the Cartesian coordinate systems $\{\mathbf{e}_1, \mathbf{e}_2, \mathbf{e}_3\}$ and $\{\widetilde{\mathbf{e}}_1, \widetilde{\mathbf{e}}_2, \widetilde{\mathbf{e}}_3\}$, respectively

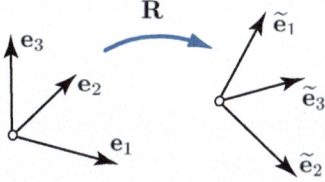

$\{\widetilde{\mathbf{e}}_1, \ldots, \widetilde{\mathbf{e}}_n\}$ then read $\widetilde{a}_i = R_{ai}a_a$. Consequently, the matrix relation

$$\begin{bmatrix} \widetilde{a}_1 \\ \widetilde{a}_2 \\ \widetilde{a}_3 \end{bmatrix} = \begin{bmatrix} \mathbf{R}^T \end{bmatrix} \begin{bmatrix} a_1 \\ a_2 \\ a_3 \end{bmatrix}$$

describes the change of the coefficients of the vector $\mathbf{a}$ in response to the rotation of the coordinate system in 3D.

Example A.6 (Coordinate Transformation). Figure A.8 illustrates the rotation between the coordinate systems $\{\mathbf{e}_1, \mathbf{e}_2, \mathbf{e}_3\}$ and $\{\widetilde{\mathbf{e}}_1, \widetilde{\mathbf{e}}_2, \widetilde{\mathbf{e}}_3\}$, where $\mathbf{e}_1 = [1 \ 0 \ 0]^T$, $\mathbf{e}_2 = [0 \ 1 \ 0]^T$, $\mathbf{e}_3 = [0 \ 0 \ 1]^T$ and $\widetilde{\mathbf{e}}_1 = [-1/\sqrt{2} \ 1/\sqrt{2} \ 0]^T$, $\widetilde{\mathbf{e}}_2 = [1/\sqrt{2} \ 1/\sqrt{2} \ 0]^T$, $\widetilde{\mathbf{e}}_3 = [0 \ 0 \ -1]^T$ denote the respective base vectors. Prove that both systems are Cartesian coordinate systems, and compute the rotation matrix $\mathbf{R}$ that determines the transformation between them. Provide a graphical sketch towards the geometrical interpretation of the coefficients of $\mathbf{R}$.

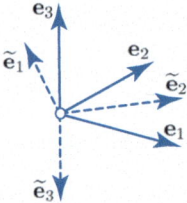

Fig. A.8 Transformation between the Cartesian coordinate systems $\{\mathbf{e}_1, \mathbf{e}_2, \mathbf{e}_3\}$ and $\{\widetilde{\mathbf{e}}_1, \widetilde{\mathbf{e}}_2, \widetilde{\mathbf{e}}_3\}$, respectively   ∎

Example A.7 (Vector Components in Different Coordinate Systems). The components of the vector $\mathbf{a} = [1 \ 1]^T$ are given with respect to the 2D Cartesian coordinate system of the base vectors $\mathbf{e}_1 = [1 \ 0]^T$ and $\mathbf{e}_2 = [0 \ 1]^T$, respectively. The system $\{\mathbf{e}_1, \mathbf{e}_2\}$ is then rotated by the angle $\alpha = \pi/6$ and thereby transformed into another Cartesian coordinate system $\{\widetilde{\mathbf{e}}_1, \widetilde{\mathbf{e}}_2\}$, see Fig. A.9. Derive the rotation matrix $\mathbf{R}$ that rotates the system $\{\mathbf{e}_1, \mathbf{e}_2\}$ into the system $\{\widetilde{\mathbf{e}}_1, \widetilde{\mathbf{e}}_2\}$, and compute the components of the vector $\mathbf{a}$ within the Cartesian coordinate system $\{\widetilde{\mathbf{e}}_1, \widetilde{\mathbf{e}}_2\}$.

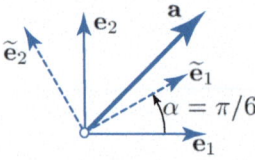

Fig. A.9 The vector $\mathbf{a}$ is described with respect to the two Cartesian coordinate systems $\{\mathbf{e}_1, \mathbf{e}_2\}$ and $\{\widetilde{\mathbf{e}}_1, \widetilde{\mathbf{e}}_2\}$, respectively. The systems are rotated to each other by the angle $\alpha = \pi/6$ ∎

Example A.8 (Objectivity of Vectors). Given a Cartesian coordinate system of base vectors $\mathbf{e}_1 = [1\ 0\ 0]^T$, $\mathbf{e}_2 = [0\ 1\ 0]^T$, and $\mathbf{e}_3 = [0\ 0\ 1]^T$, it should be verified that the identity $a_i \mathbf{e}_i = \widetilde{a}_i \widetilde{\mathbf{e}}_i$ holds upon coordinate system rotation. Here, $\mathbf{a}$ denotes a vector, whilst $\widetilde{\mathbf{e}}_i$ are the base vectors of the rotated Cartesian coordinate system. ∎

A.7   Tensor Algebra

Given the Cartesian coordinate system $\{\mathbf{e}_1, \ldots, \mathbf{e}_n\}$ of base vectors $\mathbf{e}_i$, a *second-order tensor* $\mathbf{A}$ may be defined by the dyadic product $\mathbf{A} = A_{ij}\mathbf{e}_i \otimes \mathbf{e}_j; i, j = 1, \ldots, n$. Here, A_{ij} and $\mathbf{e}_i \otimes \mathbf{e}_j$ denotes the tensor *components* and the corresponding second-order tensor *base*, respectively. As with vectors, tensors are *invariant* with respect to rotations of the coordinate system, and thus

$$\mathbf{A} = A_{ij}\mathbf{e}_i \otimes \mathbf{e}_j = \widetilde{A}_{ij}\widetilde{\mathbf{e}}_i \otimes \widetilde{\mathbf{e}}_j = \widetilde{A}_{ij}\mathbf{R}\mathbf{e}_i \otimes \mathbf{R}\mathbf{e}_j \qquad (A.15)$$

holds, where $\widetilde{\mathbf{e}}_i = \mathbf{R}\mathbf{e}_i$ denotes the rotated i-th base vector. The transformation (A.15) reads $A_{ij}e_{ik}e_{jl} = \widetilde{A}_{ij}R_{ik}e_{ik}R_{jl}e_{jl}$ in index notation, and $A_{ij} = \widetilde{A}_{ij}R_{ik}R_{jl}$ transforms therefore the tensor components among the two Cartesian coordinate systems $\{\mathbf{e}_1, \ldots, \mathbf{e}_n\}$ and $\{\widetilde{\mathbf{e}}_1, \ldots, \widetilde{\mathbf{e}}_n\}$, respectively. Given the matrix representations $\mathbf{A} = [A_{ij}]$ and $\widetilde{\mathbf{A}} = [\widetilde{A}_{ij}]$ of the tensor components, the symbolic relations

$$\mathbf{A} = \mathbf{R}^T\widetilde{\mathbf{A}}\mathbf{R} \quad \text{or} \quad \widetilde{\mathbf{A}} = \mathbf{R}\widetilde{\mathbf{A}}\mathbf{R}^T \qquad (A.16)$$

express the coordinate transformation. The base $\mathbf{e}_i \otimes \mathbf{e}_j$ of a second-order tensor is formed by two base vectors, and the coordinate transformation of a second-order tensor requires therefore *two* multiplications with $\mathbf{R}$.

Similarly to the definition of a second-order tensor, a tensor

$$\mathbf{A} = \underbrace{A_{ab\ldots l}}_{m \text{ indices}}\ \underbrace{\mathbf{e}_a \otimes \mathbf{e}_b \otimes \ldots \otimes \mathbf{e}_l}_{m \text{ base vectors}}$$

of m-th order may be defined through the dyadic vector product. Given n spatial dimensions, a tensor of m-th order has n^m components, and

$$\widetilde{A}_{pq\ldots t} = R_{pa}R_{qb}\ldots A_{ab\ldots l}$$

determines the coordinate transformation of the tensor components. Consistent with the introduced concept of tensors, a vector may be seen as a *first-order* tensor.

A.7.1   Spherical Tensor

The Kronecker delta δ_{ij} forms the components of the *spherical tensor* $\mathbf{I} = \delta_{ij}\mathbf{e}_i \otimes \mathbf{e}_j$, also called isotropic tensor or identity tensor. Its components remain *unchanged* upon coordinate transformation, and thus $\tilde{\delta}_{ij} = R_{ia}\delta_{ab}R_{jb} = R_{ib}R_{jb} = \delta_{ij}$, where the transforation (A.16) has been used.

A.7.2   Tensor Operations

The summation $\mathbf{C} = \mathbf{A} + \mathbf{B}$ is defined for tensors $\mathbf{A}$ and $\mathbf{B}$ of the *same* order and dimension. Given second-order tensors, this operation reads $C_{ij} = A_{ij} + B_{ij}$; $i, j = 1, \ldots, n$, where n denotes the spatial dimension.

As with vectors, different tensor products may be defined. The multiplication $\mathbf{B} = \alpha\mathbf{A}$; $B_{ij} = \alpha A_{ij}; i, j = 1, \ldots, n$ of a tensor $\mathbf{A}$ by the scalar α, scales each of the tensor components by α. The *outer tensor product* $\underline{\mathbf{C}} = \mathbf{a} \otimes \mathbf{A} = a_i A_{jk} \mathbf{e}_i \otimes \mathbf{e}_j \otimes \mathbf{e}_k$ of the second-order tensor $\mathbf{A}$ and the vector $\mathbf{a}$ forms the third-order tensor $\underline{\mathbf{C}}$ with the components $C_{ijk} = a_i A_{jk}$. The outer tensor product $\mathbb{C} = \mathbf{A} \otimes \mathbf{B} = A_{ij}A_{kl}\mathbf{e}_i \otimes \mathbf{e}_j \otimes \mathbf{e}_k \otimes \mathbf{e}_l$ of the two second-order tensors $\mathbf{A}$ and $\mathbf{B}$ forms the fourth-order tensor $\mathbb{C}$ with the components $C_{ijkl} = A_{ij}A_{kl}$. The *contraction* $\mathbf{c} = \mathbf{a}\mathbf{A} = \mathbf{A}^{\mathrm{T}}\mathbf{a} = a_k A_{ki}\mathbf{e}_i$ of the second-order tensor $\mathbf{A}$ and the vector $\mathbf{a}$ forms the vector $\mathbf{c}$ with the components $c_i = a_k A_{ki}$. The *single contraction* $\mathbf{C} = \mathbf{A}\mathbf{B} = \mathbf{B}^{\mathrm{T}}\mathbf{A}^{\mathrm{T}} = A_{ik}A_{kj}\mathbf{e}_i \otimes \mathbf{e}_j$ of the two second-order tensors $\mathbf{A}$ and $\mathbf{B}$ forms the second-order tensor $\mathbf{C}$ with the components $C_{ij} = A_{ik}A_{kj}$. Finally, the *double contraction* $c = \mathbf{A} : \mathbf{B} = \mathrm{tr}(\mathbf{A}\mathbf{B}^{\mathrm{T}}) = A_{kl}B_{kl}$ of the two second-order tensors $\mathbf{A}$ and $\mathbf{B}$ forms the scalar c. All above-mentioned operations require tensors and vectors of the same dimension.

Table A.2 list some multiplications of tensors, vectors, and scalars. Each line in the table shows identical expressions in symbolic, index, and matrix notations, respectively. The aforementioned operations may also be generalized to higher-order tensors.

Table A.2 Some multiplications with second-order tensors, vectors, and scalars. Each line presents identical expressions in symbolic, index, and matrix notations, respectively

Symbolic notation	Index notation	Matrix notation
$c = \mathbf{a} \cdot \mathbf{b}$	$c = a_i b_i$	$[c] = [\mathbf{a}]^{\mathrm{T}}[\mathbf{b}]$
$\mathbf{A} = \mathbf{a} \otimes \mathbf{b}$	$A_{ij} = a_i b_j$	$[\mathbf{A}] = [\mathbf{a}][\mathbf{b}]^{\mathrm{T}}$
$\mathbf{b} = \mathbf{A}\mathbf{a}$	$b_i = A_{ik}a_k$	$[\mathbf{b}] = [\mathbf{A}][\mathbf{a}]$
$\mathbf{b} = \mathbf{a}\mathbf{A}$	$b_i = a_k A_{ki}$	$[\mathbf{b}]^{\mathrm{T}} = [\mathbf{a}]^{\mathrm{T}}[\mathbf{A}]$
$c = \mathbf{a}\mathbf{A}\mathbf{b}$	$c = a_k A_{kl} b_l$	$[c] = [\mathbf{a}]^{\mathrm{T}}[\mathbf{A}][\mathbf{b}]$
$\mathbf{C} = \mathbf{A}\mathbf{B}$	$C_{ij} = A_{ik}B_{kl}$	$[\mathbf{C}] = [\mathbf{A}][\mathbf{B}]$
$\mathbf{C} = \mathbf{A}\mathbf{B}^{\mathrm{T}}$	$C_{ij} = A_{ik}B_{lk}$	$[\mathbf{C}] = [\mathbf{A}][\mathbf{B}]^{\mathrm{T}}$
$c = \mathbf{A} : \mathbf{B}$	$c = A_{kl}B_{kl}$	$[c] = \mathrm{tr}\left([\mathbf{A}][\mathbf{B}]^{\mathrm{T}}\right)$
$c = \mathrm{tr}(\mathbf{A}\mathbf{B})$	$c = A_{kl}B_{lk}$	$[c] = \mathrm{tr}\left([\mathbf{A}][\mathbf{B}]\right)$
$\mathbf{D} = \mathbf{A}\mathbf{B}\mathbf{C}$	$D_{ij} = A_{ik}B_{kl}C_{lj}$	$[\mathbf{D}] = [\mathbf{A}][\mathbf{B}][\mathbf{C}]$

A.7.3   Invariants of Second-Order Tensors

Whilst tensor components *depend* on the coordinate system, some combinations amongst them define *invariants* that are independent from the coordinate system. In 3D, the scalars $\mathrm{tr}\mathbf{A}$, $\mathrm{tr}\mathbf{A}^2$, $\mathrm{tr}\mathbf{A}^3$, and any combination amongst them, are independent from the coordinate system. In engineering mechanics, the linear combinations

$$I_1 = \mathrm{tr}\mathbf{A} \; ; \; I_2 = 1/2[(\mathrm{tr}\mathbf{A})^2 - \mathrm{tr}\mathbf{A}^2] \; ; \; I_3 = \det\mathbf{A} \; , \tag{A.17}$$

are commonly used as the invariants of a second-order tensor.

Example A.9 (Objectivity of Tensors). Show that the invariants I_1, I_2, and I_3 of a symmetric second-order tensor $\mathbf{A} = \mathbf{a} \otimes \mathbf{a}$ are invariant. The definition (A.17) of invariants may be used, and the coordinate transformation is determined by the rotation tensor $\mathbf{R}$.   ■

A.8   Vector and Tensor Calculus

A.8.1   Local Changes of Field Variables

A.8.1.1 Gradient

The *gradient* gradϕ of the *scalar* function $\phi(\mathbf{x})$ is a measure of how fast ϕ changes along the directions of the coordinate system's base vectors $\mathbf{e}_i \; ; i = 1, \ldots, n$. Figure A.10 provides a geometrical interpretation of the gradient in 2D. The gradient gradϕ is a n-dimensional *vector* and in Cartesian coordinates it reads

$$\mathbf{a} = \mathrm{grad}\phi = \frac{\partial \phi}{\partial x_i}\mathbf{e}_i \; ; i = 1, \ldots, n \; ,$$

where $\partial\phi/\partial x_i$ denotes the respective vector components. The components of the gradient depend on the coordinate system, and

$$[a_r \; a_\theta \; a_z]^{\mathrm{T}} = \left[\frac{\partial \phi}{\partial r} \; \frac{1}{r}\frac{\partial \phi}{\partial \theta} \; \frac{\partial \phi}{\partial z}\right]^{\mathrm{T}}$$

expresses the gradient in the cylindrical coordinate system $\{\mathbf{e}_r, \mathbf{e}_\theta, \mathbf{e}_z\}$, where r, θ, z denote the radial, circumferential and axial coordinate, respectively.

Fig. A.10 Geometrical
interpretation of the gradient
gradϕ of the scalar function
$\phi(\mathbf{x})$ in 2D

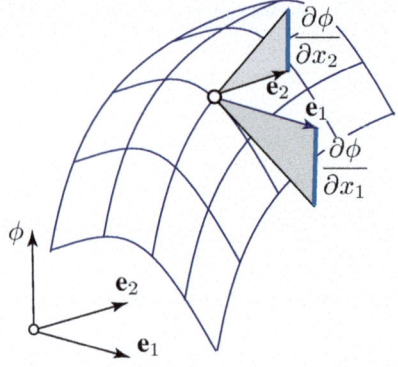

The gradient grad$\mathbf{a}$ of the n-dimensional *vector* function $\mathbf{a}(\mathbf{x}) = a_i(\mathbf{x})\mathbf{e}_i; i = 1, \ldots, n$ is a n-dimensional *second-order tensor* and reads

$$\mathbf{A} = \mathrm{grad}\,\mathbf{a} = \frac{\partial a_i}{\partial x_j}\mathbf{e}_i \otimes \mathbf{e}_j .$$

In 3D Cartesian coordinates, they read

$$\begin{bmatrix} A_{11} & A_{12} & A_{13} \\ A_{21} & A_{22} & A_{23} \\ A_{31} & A_{32} & A_{33} \end{bmatrix} = \begin{bmatrix} \frac{\partial a_1}{\partial x_1} & \frac{\partial a_1}{\partial x_2} & \frac{\partial a_1}{\partial x_3} \\ \frac{\partial a_2}{\partial x_1} & \frac{\partial a_2}{\partial x_2} & \frac{\partial a_2}{\partial x_3} \\ \frac{\partial a_3}{\partial x_1} & \frac{\partial a_3}{\partial x_2} & \frac{\partial a_3}{\partial x_3} \end{bmatrix} .$$

Given the cylindrical coordinate system $\{\mathbf{e}_r, \mathbf{e}_\theta, \mathbf{e}_z\}$, the gradient grad$\mathbf{a}$ of the vector function $a_i(r, \theta, z)\mathbf{e}_i; i = r, \theta, z$ has the components

$$\begin{bmatrix} A_{rr} & A_{r\theta} & A_{rz} \\ A_{\theta r} & A_{\theta\theta} & A_{\theta z} \\ A_{zr} & A_{z\theta} & A_{zz} \end{bmatrix} = \begin{bmatrix} \frac{\partial a_r}{\partial r} & \frac{1}{r}\frac{\partial a_r}{\partial \theta} - \frac{a_\theta}{r} & \frac{\partial a_r}{\partial z} \\ \frac{\partial a_\theta}{\partial r} & \frac{1}{r}\frac{\partial a_\theta}{\partial \theta} + \frac{a_r}{r} & \frac{\partial a_\theta}{\partial z} \\ \frac{\partial a_z}{\partial r} & \frac{1}{r}\frac{\partial a_z}{\partial \theta} & \frac{\partial a_z}{\partial z} \end{bmatrix} .$$

A.8.1.2 Divergence

The *divergence* of the n-dimensional *vector* function $\mathbf{a}(\mathbf{x}) = a_i(\mathbf{x})\mathbf{e}_i; i = 1, \ldots, n$ is a *scalar* and reads

$$a = \mathrm{div}\,\mathbf{a} = \frac{\partial a_i}{\partial x_i} .$$

The divergence materializes differently in different coordinate systems. Given 3D Cartesian coordinates $\{\mathbf{e}_1, \mathbf{e}_2, \mathbf{e}_3\}$, it reads div$\mathbf{a} = \partial a_1/\partial x_1 + \partial a_2/\partial x_2 + \partial a_3/\partial x_3$, whilst in cylindrical coordinates $\{\mathbf{e}_r, \mathbf{e}_\theta, \mathbf{e}_z\}$, it reads div$\mathbf{a} = [\partial(r a_r)/\partial r + \partial a_\theta/\partial \theta]/r + \partial a_z/\partial z$.

The divergence of the n-dimensional *second-order tensor* function $\mathbf{A}(\mathbf{x}) = A_{ij}(\mathbf{x})\mathbf{e}_i \otimes \mathbf{e}_j; i, j = 1, \ldots, n$ results in a n-dimensional *vector* and reads

$$\mathbf{a} = \text{div}\mathbf{A} = \frac{\partial A_{ik}}{\partial x_k}\mathbf{e}_i \ .$$

It has the components

$$
\begin{bmatrix} a_1 \\ a_2 \\ a_3 \end{bmatrix} = \begin{bmatrix} \frac{\partial A_{11}}{\partial x_1} + \frac{\partial A_{12}}{\partial x_2} + \frac{\partial A_{13}}{\partial x_3} \\ \frac{\partial A_{21}}{\partial x_1} + \frac{\partial A_{22}}{\partial x_2} + \frac{\partial A_{23}}{\partial x_3} \\ \frac{\partial A_{31}}{\partial x_1} + \frac{\partial A_{32}}{\partial x_2} + \frac{\partial A_{33}}{\partial x_3} \end{bmatrix}
\tag{A.18}
$$

with respect to the 3D Cartesian coordinate system $\{\mathbf{e}_1, \mathbf{e}_2, \mathbf{e}_3\}$. Given cylindrical coordinates $\{\mathbf{e}_r, \mathbf{e}_\theta, \mathbf{e}_z\}$,

$$
\begin{bmatrix} a_r \\ a_\theta \\ a_z \end{bmatrix} = \begin{bmatrix} \frac{\partial A_{rr}}{\partial r} + \frac{1}{r}\frac{\partial A_{r\theta}}{\partial \theta} + \frac{\partial A_{rz}}{\partial z} + \frac{A_{rr} - A_{\theta\theta}}{r} \\ \frac{\partial A_{r\theta}}{\partial r} + \frac{1}{r}\frac{\partial A_{\theta\theta}}{\partial \theta} + \frac{\partial A_{\theta z}}{\partial z} + \frac{2A_{r\theta}}{r} \\ \frac{\partial A_{rz}}{\partial r} + \frac{1}{r}\frac{\partial A_{\theta z}}{\partial \theta} + \frac{\partial A_{zz}}{\partial z} + \frac{A_{rz}}{r} \end{bmatrix}
\tag{A.19}
$$

are the components of the divergence $\mathbf{a} = \text{div}\mathbf{A}$ of the second-order tensor function $A_{ij}(r, \theta, z)\mathbf{e}_i \otimes \mathbf{e}_j; i, j = r, \theta, z.$

A.8.2   Divergence Theorem

The *divergence theorem*, or Gauss theorem, links the integral over a *domain* Ω to the integral over its *surface* $\partial\Omega$. Given the divergence div$\mathbf{a}$ of a vector function $\mathbf{a}(\mathbf{x}) = a_i(\mathbf{x})\mathbf{e}_i; i = 1, \ldots, n$, the divergence theorem reads

$$\int_\Omega \text{div}\mathbf{a}\,dv = \int_{\partial\Omega} \mathbf{a} \cdot \mathbf{n}ds \ ,$$

where $\mathbf{n}(\mathbf{x})$ denotes the outward unit normal vector onto the surface $\partial\Omega$, whilst

$$\int_\Omega \text{div}\mathbf{A}\,dv = \int_{\partial\Omega} \mathbf{A}\mathbf{n}ds$$

express the divergence theorem applied to the divergence of the second-order tensor function $\mathbf{A}(\mathbf{x}) = A_{ij}(\mathbf{x})\mathbf{e}_i \otimes \mathbf{e}_j, i, j = 1, \ldots, n.$

Some Useful Laplace and Fourier Transforms

B

B.1 Laplace Transforms

Dirac Delta Function $\delta(t)$

Using the definition $(A.7)_1$, the Laplace transform of the Dirac delta function $\delta(t)$ reads

$$\widehat{\delta}(s) = \int_0^\infty \delta(t)\exp(-st)\mathrm{d}t = [\mathcal{H}(t)]_0^\infty = 1 \, .$$

Here, the property

$$\int f(t)\delta(t)\mathrm{d}t = [f(t)]_{t=0}\int \delta(t)\mathrm{d}t = [f(t)]_{t=0}\mathcal{H}(t) \tag{B.1}$$

has been used, with $f(t)$ denoting an arbitrary function of t.

Heaviside Function $\mathcal{H}(t)$

Using the definition $(A.7)_1$ and the property $\delta(t) = \mathrm{d}\mathcal{H}(t)/\mathrm{d}t$, integration by parts together with the property (B.1), yields

$$\widehat{\mathcal{H}}(s) = \int_0^\infty \mathcal{H}(t)\exp(-st)\mathrm{d}t = -\frac{1}{s}\left[\mathcal{H}(t)\exp(-st)\right]_0^\infty + \frac{1}{s}\underbrace{\int_0^\infty \delta(t)\exp(-st)\mathrm{d}t}_{[\mathcal{H}(t)]_0^\infty}$$

$$= \frac{1}{s}\left\{\left[1 - \exp(-st)\right]\mathcal{H}(t)\right\}_0^\infty = \frac{1}{s}$$

for the Laplace transform of the Heaviside function $\mathcal{H}(t)$

© Springer Nature Switzerland AG 2021
T. C. Gasser, *Vascular Biomechanics*, https://doi.org/10.1007/978-3-030-70966-2

The Function $t\mathcal{H}(t)$

Using the property $\widehat{t^n x(t)} = (-1)^n (\mathrm{d}^n \widehat{x}(s)/\mathrm{d}s^n)$; $n = 1, 2, 3, \ldots$ of the Laplace transform together with the Laplace transform of the Heaviside function $\widehat{\mathcal{H}}(s) = s^{-1}$, yields

$$\widehat{t\mathcal{H}}(s) = -\frac{\mathrm{d}s^{-1}}{\mathrm{d}s} = \frac{1}{s^2}$$

for the Laplace transform of the function $t\mathcal{H}(t)$.

The Function $\sin(at)\mathcal{H}(t)$

Using the properties $\int \sin(at) \exp(-st)\mathrm{d}t = -\exp(-st)[a\cos(at) + s\sin(at)]/(a^2 + s^2)$ and $\delta(t) = \mathrm{d}\mathcal{H}(t)/\mathrm{d}t$, integration by parts together with the property (B.1) yields

$$\widehat{\sin(at)\mathcal{H}}(s) = \int_0^\infty \sin(at)\mathcal{H}(t)\exp(-st)\mathrm{d}t$$

$$= \frac{1}{a^2 + s^2} \underbrace{\{-\mathcal{H}(t)\exp(-st)[a\cos(at) + s\sin(at)]\}_0^\infty}_{0}$$

$$+ \frac{1}{a^2 + s^2} \underbrace{\int_0^\infty \delta(t)\exp(-st)[a\cos(at) + s\sin(at)]\mathrm{d}t}_{a[\mathcal{H}(t)]_0^\infty}$$

$$= \frac{a}{a^2 + s^2}$$

for the Laplace transform of the function $\sin(at)\mathcal{H}(t)$.

The Function $\cos(at)\mathcal{H}(t)$

Using the properties $\int \cos(at) \exp(-st)\mathrm{d}t = \exp(-st)[-s\cos(at) + a\sin(at)]/(a^2 + s^2)$ and $\delta(t) = \mathrm{d}\mathcal{H}(t)/\mathrm{d}t$, integration by parts together with the property (B.1) yields

$$\widehat{\cos(at)\mathcal{H}}(s) = \int_0^\infty \cos(at)\mathcal{H}(t)\exp(-st)\mathrm{d}t$$

$$= \frac{1}{a^2 + s^2} \underbrace{\{\mathcal{H}(t)\exp(-st)[-s\cos(at) + a\sin(at)]\}_0^\infty}_{0}$$

$$-\frac{1}{a^2+s^2}\underbrace{\int_0^\infty \delta(t)\exp(-st)[-s\cos(at)+a\sin(at)]dt}_{-s[\mathcal{H}(t)]_0^\infty}$$

$$=\frac{s}{a^2+s^2}$$

for the Laplace transform of the function $\cos(at)\mathcal{H}(t)$.

B.2   Fourier Transforms

Dirac Delta Function $\delta(t)$

Using the definition (A.7)$_2$ together with the property (B.1), the Fourier transform of the Dirac delta function $\delta(t)$ reads

$$\tilde{\delta}(\omega)=\int_{-\infty}^\infty \delta(t)\exp[(i\omega-a)t]dt=\frac{1}{i\omega-a}[\mathcal{H}(t)\exp[(i\omega-a)t]]_{-\infty}^\infty$$

$$-\frac{1}{i\omega-a}\underbrace{\int_{-\infty}^\infty \delta(t)\exp[(i\omega-a)t]dt}_{[\mathcal{H}(t)]_{-\infty}^\infty}$$

$$=\frac{1}{a-i\omega}.$$

The Function $\exp(-at)\mathcal{H}(t)$

Using the definition (A.7)$_2$ together with the property (B.1), the Fourier transform of the function $\exp(-at)\mathcal{H}(t)$ reads

$$\tilde{\delta}(\omega)=\int_{-\infty}^\infty \mathcal{H}(t)\exp(-i\omega t)dt=[\mathcal{H}(t)]_{-\infty}^\infty=1.$$

The Function $\sin(at)\mathcal{H}(t)$

Using the properties $\int \sin(at)\exp(i\omega t)dt=\exp(i\omega t)[a\cos(at)-i\omega\sin(at)]/(a^2-\omega^2)$ and $\delta(t)=d\mathcal{H}(t)/dt$, integration by parts together with the property (B.1) yields

$$\widetilde{\sin(at)\mathcal{H}}(\omega) = \int_{-\infty}^{\infty} \sin(at)\mathcal{H}(t)\exp(i\omega t)dt$$

$$= \frac{1}{a^2 - \omega^2} \underbrace{\{-\mathcal{H}(t)\exp(i\omega t)[a\cos(at) - i\omega\sin(at)]\}_{-\infty}^{\infty}}_{0}$$

$$+ \frac{1}{a^2 - \omega^2} \underbrace{\int_{-\infty}^{\infty} \delta(t)\exp(i\omega t)[a\cos(at) - i\omega\sin(at)]dt}_{a[\mathcal{H}(t)]_{-\infty}^{\infty}}$$

$$= \frac{a}{a^2 - \omega^2}$$

for the Fourier transform of the function $\sin(at)\mathcal{H}(t)$. Here, the symmetry of the cosinus function $\cos(\alpha) - \cos(-\alpha) = 0$ and the skew-symmetry of sinus function $\sin(\alpha) + \sin(-\alpha) = 0$ have been used.

The Function $\cos(at)\mathcal{H}(t)$

Using the properties $\int \cos(at)\exp(i\omega t)dt = i\exp(i\omega t)[\omega\cos(at) - ia\sin(at)]/(a^2 - \omega^2)$ and $\delta(t) = d\mathcal{H}(t)/dt$, integration by parts together with the property (B.1) yields

$$\widetilde{\cos(at)\mathcal{H}}(\omega) = \int_{-\infty}^{\infty} \cos(at)\mathcal{H}(t)\exp(i\omega t)dt$$

$$= \frac{i}{a^2 - \omega^2} \underbrace{\{\mathcal{H}(t)\exp(i\omega t)[\omega\cos(at) - ia\sin(at)]\}_{-\infty}^{\infty}}_{0}$$

$$- \frac{i}{a^2 - \omega^2} \underbrace{\int_{-\infty}^{\infty} \delta(t)\exp(i\omega t)[\omega\cos(at) - ia\sin(at)]dt}_{\omega[\mathcal{H}(t)]_{-\infty}^{\infty}}$$

$$= \frac{i\omega}{a^2 - \omega^2}$$

for the Fourier transform of the function $\cos(at)\mathcal{H}(t)$. Here, the symmetry of the cosine function $\cos(\alpha) - \cos(-\alpha) = 0$ and the skew-symmetry of sine function $\sin(\alpha) + \sin(-\alpha) = 0$ have been used.

Some Useful Tensor Relations

The Identity $(\partial \mathbf{C}/\partial \mathbf{F}) : \mathbf{F}^{-1} = 2\mathbf{I}$

Using the definition $\mathbf{C} = \mathbf{F}^{\mathrm{T}}\mathbf{F}$ of the right Cauchy–Green strain and the product rule, the relation $(\partial \mathbf{C}/\partial \mathbf{F}) : \mathbf{F}^{-1}$ reads

$$
\begin{aligned}
\partial(F_{aI}F_{aJ}/\partial F_{kL})F_{kL}^{-1} &= (\delta_{ak}\delta_{IL}F_{aJ} + F_{aI}\delta_{ak}\delta_{JL})F_{kL}^{-1} \\
&= (\delta_{IL}F_{kJ} + F_{kI}\delta_{JL})F_{kL}^{-1} \\
&= \delta_{IL}\delta_{LJ} + \delta_{LI}\delta_{JL} \\
&= 2\delta_{IJ} \ .
\end{aligned}
$$

The Time Derivative of $\mathbf{C}^{-1}$

The relation $\mathbf{C}\mathbf{C}^{-1} = \mathbf{I}$ together with the product rule yields $\dot{\mathbf{C}}\mathbf{C}^{-1} + \mathbf{C}\overline{\dot{\mathbf{C}^{-1}}} = \mathbf{0}$, such that $\overline{\dot{\mathbf{C}^{-1}}} = -\mathbf{C}^{-1}\dot{\mathbf{C}}\mathbf{C}^{-1}$ holds.

The Identity $\partial J/\partial \mathbf{C} = J\mathbf{C}^{-1}/2$

Using the definition of the volume ratio $\det\mathbf{C} = \det(\mathbf{F}^{\mathrm{T}}\mathbf{F}) = J^2$ together with Jacobi's formula $\partial\det\mathbf{A}/\partial\mathbf{A} = \det\mathbf{A}\mathbf{A}^{-\mathrm{T}}$ of the second-order tensor $\mathbf{A}$, the identity

$$
\frac{\partial J}{\partial \mathbf{C}} = \frac{\partial \sqrt{\det\mathbf{C}}}{\partial \mathbf{C}} = \frac{J}{2}\mathbf{C}^{-1}
$$

holds.

© Springer Nature Switzerland AG 2021
T. C. Gasser, *Vascular Biomechanics*, https://doi.org/10.1007/978-3-030-70966-2

The Identity $\partial I_1/\partial \mathbf{C} = \mathbf{I}$

Using the definition of the first strain invariant $I_1 = \mathrm{tr}\mathbf{C} = \mathbf{I} : \mathbf{C}$, the relation $\partial I_1/\partial \mathbf{C} = \mathbf{I}$ derives.

The Identity $\partial I_2/\partial \mathbf{C} = I_1 \mathbf{I} - \mathbf{C}$

Using the definition of the second strain invariant $I_2 = (I_1^2 - \mathrm{tr}\mathbf{C}^2)/2$, together with the definition of the first strain invariant $I_1 = \mathrm{tr}\mathbf{C}$, the relation $\partial I_2/\partial \mathbf{C} = I_1 \mathbf{I} - \mathbf{C}$ derives.

The Identity $\partial I_3/\partial \mathbf{C} = J^2 \mathbf{C}^{-1}$

Using the definition of the third strain invariant $I_3 = \det\mathbf{C}$, together with Jacobi's formula $\partial \det\mathbf{A}/\partial \mathbf{A} = \det\mathbf{A}\mathbf{A}^{-T}$, the property $\partial I_3/\partial \mathbf{C} = J^2 \mathbf{C}^{-1}$ with $J = \det\mathbf{F}$ denoting the volume ratio, holds.

The Identity $\partial \overline{\mathbf{C}}/\partial \mathbf{C} = J^{-2/3}[\mathbb{I} - (\mathbf{C} \otimes \mathbf{C}^{-1})/3]$

With the definition of the isochoric right Cauchy–Green strain $\overline{\mathbf{C}} = J^{-2/3}\mathbf{C}$, the product rule yields

$$\frac{\partial \overline{\mathbf{C}}}{\partial \mathbf{C}} = J^{-2/3}\frac{\partial \overline{\mathbf{C}}}{\partial \overline{\mathbf{C}}} + \mathbf{C} \otimes \frac{-2J^{-5/3}}{3}\frac{\partial J}{\partial \mathbf{C}} \, ,$$

which gives

$$\frac{\partial \overline{\mathbf{C}}}{\partial \mathbf{C}} = J^{-2/3}\left(\mathbb{I} - \frac{1}{3}\mathbf{C} \otimes \mathbf{C}^{-1}\right) \, ,$$

where the relation $\partial J/\partial \mathbf{C} = J\mathbf{C}^{-1}/2$ has been used. Here, $\mathbb{I}$ denotes the fourth-order symmetric identity tensor with the components $I_{ijkl} = (\delta_{ik}\delta_{jl} + \delta_{il}\delta_{jk})/2$.

The Identity $\mathbf{C}^{-1} : \dot{\mathbf{E}} = \mathrm{tr}\mathbf{d}$

Using the definitions of the right Cauchy–Green strain $\mathbf{C} = \mathbf{F}^T\mathbf{F}$ and the Green–Lagrange strain $\mathbf{E} = (\mathbf{F}^T\mathbf{F} - \mathbf{I})/2$, the term $\mathbf{C}^{-1} : \dot{\mathbf{E}}$ may be expressed by

$$\mathbf{C}^{-1} : \dot{\mathbf{E}} = \mathrm{tr}(\mathbf{C}^{-1}\dot{\mathbf{E}}) = \mathrm{tr}[\mathbf{F}^{-1}\mathbf{F}^{-T}(\dot{\mathbf{F}}^T\mathbf{F} + \mathbf{F}^T\dot{\mathbf{F}})/2]$$

$$= [\mathrm{tr}(\mathbf{F}^{-1}\mathbf{F}^{-T}\dot{\mathbf{F}}^T\mathbf{F}) + \mathrm{tr}(\mathbf{F}^{-1}\mathbf{F}^{-T}\mathbf{F}^T\dot{\mathbf{F}})]/2$$

$$= [\mathrm{tr}(\mathbf{F}^{-T}\dot{\mathbf{F}}^T) + \mathrm{tr}(\mathbf{F}^{-1}\dot{\mathbf{F}})]/2$$

$$= [\mathrm{tr}(\dot{\mathbf{F}}\mathbf{F}^{-1})^T + \mathrm{tr}(\dot{\mathbf{F}}\mathbf{F}^{-1})]/2 = (\mathrm{tr}\mathbf{l}^T + \mathrm{tr}\mathbf{l})/2 = \mathrm{tr}\mathbf{d} \, ,$$

where $\mathbf{d} = (\mathbf{l} + \mathbf{l}^T)/2$ denotes the rate of deformation tensor. Here, the property $\mathrm{tr}(\mathbf{ABC}) = \mathrm{tr}(\mathbf{BCA})$ of the trace operator has been used.

The Identity $\mathrm{dev}[\mathbf{FAF}^T] = \mathbf{F}\{\mathrm{Dev}[\mathbf{A}]\}\mathbf{F}^T$

This identity derives from the definitions of the spatial deviator operator $\mathrm{dev}(\bullet) = (\bullet) - [(\bullet) : \mathbf{I}]\,\mathbf{I}/3$ and the material deviator operator $\mathrm{Dev}(\bullet) = (\bullet) - [\mathbf{C} : (\bullet)]\,\mathbf{C}^{-1}$ according to

$$\mathbf{F}\{\mathrm{Dev}[\mathbf{A}]\}\mathbf{F}^T = \mathbf{FAF}^T - \frac{1}{3}\,(\mathbf{A} : \mathbf{C})\,\mathbf{FC}^{-1}\mathbf{F}^T$$

$$= \mathbf{FAF}^T - \frac{1}{3}\left[\mathbf{I} : (\mathbf{FAF}^T)\right]\mathbf{I} = \mathrm{dev}[\mathbf{FAF}^T]\,,$$

where the right Cauchy–Green strain $\mathbf{C} = \mathbf{F}^T\mathbf{F}$ and its inverse $\mathbf{C}^{-1} = \mathbf{F}^{-1}\mathbf{F}^{-T}$ have been used.

Some Useful Variations and Directional Derivatives $\quad$ D

The Relations $D_{\mathbf{u}}\mathbf{F} = \text{Grad}\Delta\mathbf{u}$ and $\delta\mathbf{F} = \text{Grad}\delta\mathbf{u}$

Given the deformation gradient $\mathbf{F} = \mathbf{I} + \text{Grad}(\mathbf{u})$, its directional derivative $D_{\mathbf{u}}\mathbf{F} = \text{Grad}(\mathbf{u} + \Delta\mathbf{u}) - \text{Grad}(\mathbf{u}) = \text{Grad}\Delta\mathbf{u}$ directly follows. The same arguments lead to the variation of the deformation gradient $\delta\mathbf{F} = \text{Grad}\,\delta\mathbf{u}$.

The Relations $D_{\mathbf{u}}\mathbf{E} = \text{sym}(\mathbf{F}^{\text{T}}\,\text{Grad}\Delta\mathbf{u})$ and $\delta\mathbf{E} = \text{sym}(\mathbf{F}^{\text{T}}\,\text{Grad}\delta\mathbf{u})$

Given the Green–Lagrange strain $\mathbf{E} = (\mathbf{F}^{\text{T}}\mathbf{F} - \mathbf{I})/2$, its directional derivative reads $D_{\mathbf{u}}\mathbf{E} = (D_{\mathbf{u}}\mathbf{F}^{\text{T}}\mathbf{F} + \mathbf{F}^{\text{T}}D_{\mathbf{u}}\mathbf{F})/2$, which with the property $D_{\mathbf{u}}\mathbf{F} = \text{Grad}\Delta\mathbf{u}$ yields $D_{\mathbf{u}}\mathbf{E} = (\text{Grad}^{\text{T}}\Delta\mathbf{u}\,\mathbf{F} + \mathbf{F}^{\text{T}}\,\text{Grad}\Delta\mathbf{u})/2$. The same arguments lead to the variation of the Green–Lagrange strain $\delta\mathbf{E} = (\text{Grad}^{\text{T}}\delta\mathbf{u}\,\mathbf{F} + \mathbf{F}^{\text{T}}\,\text{Grad}\delta\mathbf{u})/2$.

The Relation $D_{\mathbf{u}}\delta\mathbf{E} = \text{sym}(\text{Grad}^{\text{T}}\Delta\mathbf{u}\,\text{Grad}\delta\mathbf{u})$

Given the variation $\delta\mathbf{E} = (\text{Grad}^{\text{T}}\delta\mathbf{u}\,\mathbf{F} + \mathbf{F}^{\text{T}}\,\text{Grad}\delta\mathbf{u})/2$ of the Green–Lagrange strain and the directional derivative $D_{\mathbf{u}}\mathbf{F} = \text{Grad}\Delta\mathbf{u}$ of the deformation gradient, the directional derivative of $\delta\mathbf{E}$ then reads $D_{\mathbf{u}}\delta\mathbf{E} = (\text{Grad}^{\text{T}}\delta\mathbf{u}\,\text{Grad}\Delta\mathbf{u} + \text{Grad}^{\text{T}}\Delta\mathbf{u}\,\text{Grad}\delta\mathbf{u})/2$.

The Relations $D_{\mathbf{u}}\mathbf{e} = \text{grad}_{\text{s}}\Delta\mathbf{u}$ and $\delta\mathbf{e} = \text{grad}_{\text{s}}\delta\mathbf{u}$

Given the Euler–Almansi strain as the push-forward of the Green–Lagrange strain $\mathbf{e} = (\mathbf{I} - \mathbf{b}^{-1})/2 = \mathbf{F}^{-\text{T}}\mathbf{E}\mathbf{F}^{-1}$, its directional derivative yields $D_{\mathbf{u}}\mathbf{e} = \mathbf{F}^{-\text{T}}D_{\mathbf{u}}\mathbf{E}\,\mathbf{F}^{-1} = \text{grad}_{\text{s}}\Delta\mathbf{u}$, where $D_{\mathbf{u}}\mathbf{E} = (D_{\mathbf{u}}\mathbf{F}^{\text{T}}\,\mathbf{F} + \mathbf{F}^{\text{T}}D_{\mathbf{u}}\mathbf{F})/2$ denotes the directional derivative of the Green–Lagrange strain, and the properties

© Springer Nature Switzerland AG 2021
T. C. Gasser, *Vascular Biomechanics*, https://doi.org/10.1007/978-3-030-70966-2

$\mathrm{grad}_s\Delta\mathbf{u} = \mathrm{Grad}_s\Delta\mathbf{u}\mathbf{F}^{-1}$ and $D_\mathbf{u}\mathbf{F} = \mathrm{Grad}\Delta\mathbf{u}$ have been used. The same arguments lead to the variation of the Euler–Almansi strain $\delta\mathbf{e} = \mathbf{F}^{-T}\delta\mathbf{E}\mathbf{F}^{-1} = \mathrm{grad}_s\delta\mathbf{u}$.

The Relations $D_\mathbf{u}\mathbf{F}^{-1} = -\mathbf{F}^{-1}\mathrm{grad}\Delta\mathbf{u}$ and $\delta\mathbf{F}^{-1} = -\mathbf{F}^{-1}\mathrm{grad}\,\delta\mathbf{u}$

The directional derivative of the relation $\mathbf{F}\mathbf{F}^{-1} = \mathbf{I}$ yields $D_\mathbf{u}(\mathbf{F}\mathbf{F}^{-1}) = D_\mathbf{u}\mathbf{F}\,\mathbf{F}^{-1} + \mathbf{F}D_\mathbf{u}\mathbf{F}^{-1} = \mathbf{0}$. With $D_\mathbf{u}\mathbf{F} = \mathrm{Grad}\Delta\mathbf{u}$, the directional derivative of the inverse deformation gradient $\mathbf{F}^{-1}$ then reads $D_\mathbf{u}\mathbf{F}^{-1} = -\mathbf{F}^{-1}\mathrm{Grad}\,\Delta\mathbf{u}\,\mathbf{F}^{-1} = -\mathbf{F}^{-1}\mathrm{grad}\Delta\mathbf{u}$. The same arguments lead to the variation of the inverse deformation gradient $\delta\mathbf{F}^{-1} = -\mathbf{F}^{-1}\mathrm{grad}\,\delta\mathbf{u}$.

The Relations $D_\mathbf{u}\mathrm{grad}\delta\mathbf{u} = -\mathrm{grad}\delta\mathbf{u}\,\mathrm{grad}\Delta\mathbf{u}$

Given the property $\mathrm{grad}\delta\mathbf{u} = \mathrm{Grad}\delta\mathbf{u}\,\mathbf{F}^{-1}$, the directional derivative of the spatial gradient of the displacement variation reads

$$D_\mathbf{u}\mathrm{grad}\delta\mathbf{u} = D_\mathbf{u}\mathrm{Grad}\delta\mathbf{u}\,\mathbf{F}^{-1} + \mathrm{Grad}\delta\mathbf{u}\,D_\mathbf{u}\mathbf{F}^{-1}$$

$$= -\mathrm{Grad}\delta\mathbf{u}\,\mathbf{F}^{-1}\mathrm{grad}\Delta\mathbf{u} = -\mathrm{grad}\delta\mathbf{u}\,\mathrm{grad}\Delta\mathbf{u}\,,$$

where the properties $D_\mathbf{u}\mathrm{Grad}\delta\mathbf{u} = \mathbf{0}$ and $D_\mathbf{u}\mathbf{F}^{-1} = -\mathbf{F}^{-1}\mathrm{grad}\Delta\mathbf{u}$ have been used.

The Relations $D_\mathbf{u}J = J\mathrm{div}\Delta\mathbf{u}$ and $\delta J = J\mathrm{div}\delta\mathbf{u}$

Given the definition $J = \mathrm{det}\mathbf{F}$ of the volume ratio, the chain rule together with Jacobi's formula $\partial\mathrm{det}\mathbf{A}/\partial\mathbf{A} = \mathrm{det}\mathbf{A}\mathbf{A}^{-T}$ for the second-order tensor $\mathbf{A}$ yields $D_\mathbf{u}J = (\partial\mathrm{det}\mathbf{F}/\partial\mathbf{F}) : D_\mathbf{u}\mathbf{F} = J\mathbf{F}^{-T} : D_\mathbf{u}\mathbf{F}$. The property $\mathrm{div}(\bullet) = \mathrm{tr}(\mathrm{grad}(\bullet))$ and the directional derivative of the deformation gradient $D_\mathbf{u}\mathbf{F} = \mathrm{Grad}\Delta\mathbf{u}$ then yield $D_\mathbf{u}J = J\mathrm{div}\Delta\mathbf{u}$. The same arguments lead to the variation of the volume ratio $\delta J = J\mathrm{div}\delta\mathbf{u}$.

E.1 Basic Circuit Elements

E.1.1 Resistor Element

Given the flow at the rate $q(t)$ [m^3 s^{-1}] that passes a resistor of the *resistance* R [Pa s m^{-3}], the pressure $p(t)$ [Pa] decreases, and Ohm's[1] law

$$p(t) = Rq(t) \qquad \text{(E.1)}$$

provides the relation between these quantities. Figure E.1a shows the representation of the resistor in an electrical network chart together with the flow vector **q** and the pressure vector **p** in the complex plane. Passing a resistor, the flow and pressure *stay in phase*, and thus the phase angle $\phi = 0$ establishes between **p** and **q**.

E.1.2 Capacitor Element

Given the pressure $p(t)$ [Pa] that acts on a capacitor of the *capacity* C [m^3 Pa^{-1}], the flow at the rate $q(t)$ [m^3 s^{-1}] establishes, and

$$q(t) = C\frac{dp(t)}{dt} \qquad \text{(E.2)}$$

provides the relation between these quantities. Figure E.1b shows the representation of the capacitor in an electrical network chart together with the flow vector **q** and the pressure vector **p** in the complex plane. Passing a capacitor, the flow is $\pi/2$ *ahead* the pressure, and thus the phase angle $\phi = \pi/2$ establishes between **p** and **q**.

---

[1]Georg Simon Ohm, German physicist and mathematician (1789–1854).

© Springer Nature Switzerland AG 2021
T. C. Gasser, *Vascular Biomechanics*, https://doi.org/10.1007/978-3-030-70966-2

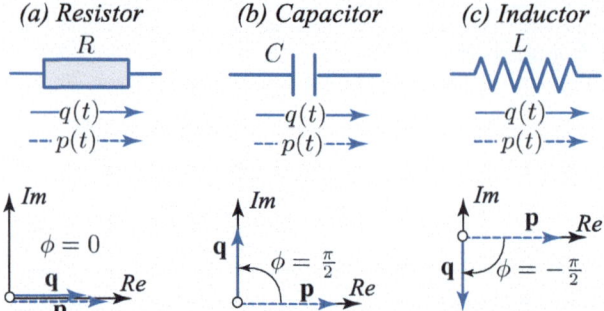

Fig. E.1 Basic circuit elements used to analyze vascular flow systems. Top row: representation used in electrical network charts. Bottom row: flow vector **q** and pressure vector **p** in the complex plane

E.1.3   Inductor Element

Given the flow at the rate $q(t)$ [m^3 s^{-1}] that passes through an inductor of the *inductance* L [Pa s^2 m^{-3}], the pressure $p(t)$ [Pa] establishes, and

$$p(t) = L \frac{dq(t)}{dt} \tag{E.3}$$

provides the relation between these quantities. Figure E.1c shows the representation of the inductance in an electrical network chart together with the flow vector **q** and the pressure vector **p** in the complex plane. Passing an inductor, the flow is $\pi/2$ *behind* the pressure, and thus the phase angle $\phi = -\pi/2$ establishes between **p** and **q**.

E.2   Transport Mechanisms

E.2.1   Diffusion

Diffusion is the movement of a solute (substance) from a region of high concentration towards a region of low concentration *relative* to any motion of the bulk material. Given the molal solute concentration c [mol m^{-3}], Fick's[2] laws of diffusion [158]

$$\mathbf{J_s} = -D \operatorname{grad} c \tag{E.4}$$

---

[2] Adolf Eugen Fick, German-born physician and physiologist, 1829–1901.

expresses the molar diffusion flux $\mathbf{J}_s$ [mol s^{-1}m^{-2}] that establishes in the system, where grad$c$ and D [s^{-1}m^{-1}] denote the gradient of the substance concentration and the *diffusion constant*, respectively. The diffusion flux $\mathbf{J}_s$ expresses how many moles of solute per second diffuse through the unit area perpendicular to the unit direction vector $\mathbf{n} = \mathbf{J}_s/|\mathbf{J}_s|$.

Equation (E.4) describes *isotropic* bulk materials, and thus materials without preferential directions for solute diffusion. Given *anisotropic* materials, the diffusion constant is a symmetric second-order tensor. It can also depend on the deformation of the bulk material. Fick's law is a purely phenomenological description of diffusion, and more detailed approaches use statistical descriptions of the diffusing particles.

E.2.2   Flow Through Porous Media

In analogy to Fick's law, the flow rate $\mathbf{q}_f$ [m s^{-1}] of fluid through the unit area of a *porous medium* is described by Darcy's law

$$\mathbf{q}_f = -\frac{k}{\eta}\mathrm{grad}\,p\,. \tag{E.5}$$

It states that $\mathbf{q}_f$ is proportional to the negative pressure gradient $-\mathrm{grad}\,p$ [Pa m^{-1}] and the *intrinsic permeability* k [m^2], whilst it is indirectly proportional to the *viscosity* η [Pa s] of the fluid. Equation (E.5) relates to an *isotropic* porous medium, and for an *anisotropic* material the permeability is a symmetric second-order tensor. Darcy's law may also be derived from the homogenization of the Navier–Stokes equations.

Given the flow through a porous wall of the thickness L [m], the pressure gradient is expressed by grad$p = (\Delta p/L)\mathbf{n}$, where $\mathbf{n}$ denotes the unit normal vector to the wall and Δp [Pa] is the pressure difference between the both sides of the wall. Darcy's law (E.5) then reads

$$\mathbf{q}_f = -L_p\Delta p\mathbf{n}\,, \tag{E.6}$$

where $L_p = k/(\eta L)$ [m Pa^{-1}s^{-1}] denotes the hydraulic conductivity of the wall. At finite deformations, the intrinsic permeability k and the hydraulic conductivity L_p depend on the deformation of the porous medium.

E.2.3   Advection

Advection is the movement of a solute (substance) *by* bulk motion, such as the transport of proteins together with blood plasma. Given the bulk velocity $\mathbf{v}$ [m s^{-1}] and the molal solute concentration c [mol m^{-3}], the transport equation

$$\mathbf{J}_s = c\mathbf{v} \tag{E.7}$$

expresses the molar advection flux $\mathbf{J}_s$ [mol $s^{-1}m^{-2}$]. It describes how many moles of solute per second advect in the direction of the bulk motion, and thus through the unit area perpendicular to the flow direction $\mathbf{n} = \mathbf{v}/|\mathbf{v}|$.

Advection is sometimes confused with the more encompassing process of convection, which is the combination of advective transport and diffusive transport.

E.3   Osmosis

Osmosis is the spontaneous net movement of fluid (solvent) through a semiperme-able membrane into a region of *higher* solute concentration. The flow points into the direction that tends to equalize the solute concentrations on the two sides, see Fig. E.2. Osmosis is a vital process and governs the exchange of water (solvent) and solutes in the vascular microcirculation.

E.3.1   Osmotic Pressure

The *osmotic pressure* is defined as the external pressure that would be required to prevent from any net movement of fluid (solvent) across a semipermeable

Fig. E.2 Development of osmotic pressure difference $\Delta \Pi$ across a semipermeable membrane by fluid (solvent) motion at the velocity (or flux) v_f from low solute concentration c_{low} to high solute concentration c_{high}

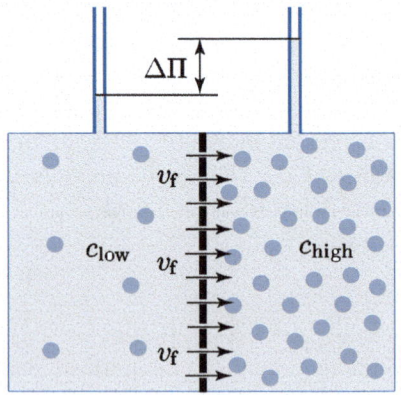

membrane, see Fig. E.2. The osmotic pressure Π [Pa] is related to the solute's molar concentration c [mol m^{-3}] through Hoff's law

$$\Pi = \phi \overline{R} \theta c , \tag{E.8}$$

where $\overline{R} = 8.3145$ J K^{-1}mol^{-1} and θ [K] denotes the ideal gas constant and the temperature, respectively. The osmotic coefficient ϕ adjusts the relation to non-idealized conditions, such as high solute concentration.

Most solutes are partially permeable and *leak* across a semipermeable membrane; they advect with the fluid (solvent) or diffuse across the membrane. The theoretically possible osmotic pressure difference $\Delta \Pi_{max}$ can then *not* be reached and the lower $\Delta \Pi$ is measured. Solutions are often classified as effective or ineffective osmoles on the basis of their ability to generate osmotic pressure.

Staverman used linear irreversible thermodynamics to investigate the permeability in such "leaky" membranes [526] and introduced a *reflection coefficient*

$$\sigma = \frac{\Delta \Pi}{\Delta \Pi_{max}} \tag{E.9}$$

that relates both osmotic pressure differences. Staverman's osmotic reflection coefficient σ ranges from zero to one towards the description of *free* movement and *full reflection* of the solute by the membrane, respectively. Given $\sigma = 1$, no solute leaks across the membrane and the maximum theoretically possible osmotic pressure difference $\Delta \Pi_{max}$ develops, whilst at $\sigma = 0$ the solutes can freely move across the membrane and the solution fails to produce any osmotic pressure difference and thus $\Delta \Pi = 0$. The establishment of osmotic fluid flux or pressure requires therefore (i) a solute concentration difference Δc, and (ii) a sizeable reflection coefficient $\sigma > 0$.

Solutions that contain several solutes exert an osmotic pressure that is the sum of the osmotic pressures by all solutes, $\Pi = \sum \Pi_j$. The osmotic pressure in the vasculature is exerted by proteins (noticeable by albumin) and called *oncotic pressure*, or *Colloid Osmotic Pressure (COP)*. The COP is simply osmotic pressure with protein colloids as effective osmoles and small solutes as ineffective osmoles. Charged proteins generate COP not only as dissolved molecules but also through electrostatic attraction of oppositely charged small counter-ions known as Gibbs[3]– Donnan[4] effect.

---

[3] Josiah Willard Gibbs, American mathematical physicist, 1839–1903.

[4] Frederick George Donnan, Irish physical chemist, 1870–1956.

E.3.2   Transport Across Semipermeable Membranes

The properties of the permeable membrane together with the ones of the solute particles determine the value of Staverman's *osmotic reflection coefficient* σ. It describes the selectivity of the membrane to a specific solute and may also been seen as the ratio $\sigma = L_{pf}/L_{ps}$ of the membrane's hydraulic conductivities L_{pf} and L_{ps} for fluid (solvent) and solute, respectively. Given the pressure gradient $\mathrm{grad}\,p$, fluid moves at the velocity (or the flux) $q_f = -L_{pf}\mathrm{grad}\,p$ across the semipermeable membrane. In contrary, the solute moves at the velocity $v_s = -L_{ps}\mathrm{grad}\,p$ across the membrane, and

$$v_s = (1 - \sigma)q_f \qquad\qquad (E.10)$$

therefore relates both velocities (or fluxes), see also Sect. E.2.2.

The sum of *diffusive* and *adventive* transports governs the solute flux

$$J_s = -D\mathrm{grad}\,c + (1 - \sigma)q_f\bar{c} \qquad\qquad (E.11)$$

of non-charged solutes across a semipermeable membrane, where q_f denotes the fluid (solvent) flux, see Sects. E.2.1 and E.2.3. Here, $\mathrm{grad}\,c = (c_{high} - c_{low})/h$ is the gradient of the molal solute concentration, whilst $\bar{c} = (c_{high} + c_{low})/2$ denotes the average molal solute concentration within the membrane of the thickness h.

Biaxial Experimental Vessel Wall Testing

F

F.1 Tissue Harvesting and Sample Preparation

An intact porcine aorta is harvested in the slaughterhouse, put into physiological saline solution (0.9% NaCl), and cooled during the transportation to the laboratory facilities. Figure F.1a shows the thoracic aortic segment with the surrounding connective tissue having already been carefully dissected. The aorta is opened along the intercostal arteries to acquire a tissue patch that allows for the preparation of homogenous rectangular wall patches. The residual stresses in the vessel contribute to the flattening of the tissue patch and ease the cutting-out of test specimens. Given this study, a quadratic-shaped tensile specimen is cut-out using a 18×18 mm large plexiglas template, see Fig. F.1b. The specimen is aligned with the vessel's circumferential and axial directions, and then marked and labeled in order to

(a) Facilitate strain measurements with a video extensiometer by tracking the motion of four dots (optical markers) in the center of the specimen,
(b) Assist the mounting of the specimen in the testing machine, and
(c) Uniquely identify the specimen's history and the alignment within the testing machine.

Whilst the specimen's edge length has already been set by the plexiglas template, we now measure its thickness. The specimen is placed between two glass plates, such as microscope slides, and with a caliper we measure the thickness of the three layers together. Given the thickness of the glass plates, the average thickness H of the vessel wall specimen may be calculated. It is crucial that the tissue is continuously moisturized and prevented from drying-out at any point during the specimen preparation.

© Springer Nature Switzerland AG 2021
T. C. Gasser, *Vascular Biomechanics*, https://doi.org/10.1007/978-3-030-70966-2

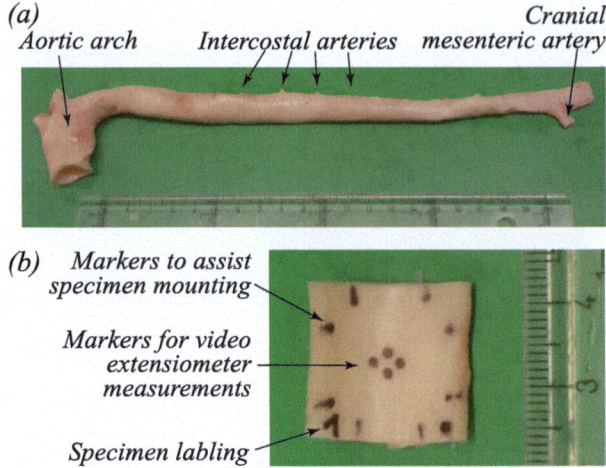

Fig. F.1 Tissue specimen preparation for the mechanical characterization of the porcine aortic wall. A ruler shows the dimensions in millimeters. (**a**) Thoracic aorta dissected from surrounding connective tissue. (**b**) Square-shaped wall specimen for planar biaxial tissue testing. The vertical direction in the image represents the axial vessel direction

F.2   Test Protocol Definition and Data Recording

A Biotester BT-50MM-01 (World Precision Instruments Ltd., UK) with electro-mechanical actuators is used to acquire the mechanical properties of the aortic wall, see Fig. 5.31. A displacement-controlled experiment is performed, and the displacement u_θ and u_z along the circumferential and respective axial vessel direction are prescribed by the machine's actuators. The displacements are applied symmetrically with respect to the specimen center, and the corresponding loads F_θ and F_z in axial and circumferential directions are then measured by load cells.

The vessel wall sample is biaxially stretched at three different displacement combinations. The targets for the prescribed displacements are set as follows:

- Protocol (a): $u_\theta = u_z = 6.5$ mm
- Protocol (b): $u_\theta = 3.25$ mm and $u_z = 6.5$ mm
- Protocol (c): $u_\theta = 6.5$ mm and $u_z = 3.25$ mm

We also have to specify the displacement rates, and thus the time to reach said targets. The present study aims at exploring the quasi-static properties of the vessel wall, an objective that determines the displacement rates. Given the specimen size, displacement rates in the range of 1 to 5 mm min^{-1} would result in the acquisition of quasi-static mechanical properties. Preliminary mechanical tests at different displacement rates may be used to validate this setting.

The vessel wall sample is tested in physiological saline solution at 37 °C to mimic thermal *in vivo* conditions. The zero-load level $F_\theta = F_z = 0$ is set, the specimen then mounted to the actuators, and the displacement-controlled test sequence activated. The acquisition of reproducible results requires tissue preconditioning, and each of the loading protocols (a), (b), and (c) is therefore applied five times in a row, respectively. Whilst the fifth cycle is used to characterize the tissue (see Sect. 5.6), the others precondition the test specimen.

F.3   Acquired Test Data

Planar biaxial testing results in a complex distributions of stress and strain over the tensile specimen, see the result from a Finite Element Method (FEM) analysis in Fig. 4.27. Regardless this complexity, we use a simplified analysis and assume a homogeneous biaxial deformation to be present inside the gripping hooks, see Fig. F.2. The domain $L_i \times L_i = 12.0 \times 12.0$ mm is then thought to be exposed to homogenous biaxial deformation. The aforementioned testing protocols result therefore in averaged target stretches of (a) $\lambda_\theta = \lambda_z = 1.54$, (b) $\lambda_\theta = 1.27$ and $\lambda_z = 1.54$, as well as (c) $\lambda_\theta = 1.54$ and $\lambda_z = 1.27$, respectively. The applied forces are also homogeneously distributed over the domain $L_i \times L_i$, and $P_\theta = F_\theta/(L_i H)$ and $P_z = F_z/(L_i H)$ characterizes the first Piola–Kirchhoff stress in the circumferential and axial directions, respectively.

At each time point, a biaxial test acquires the two displacements u_θ, u_z (or $\overline{u}_\theta$, $\overline{u}_z$ of the optical markers if the video extensiometer option is used) and the two forces F_θ, F_z. Given the test specimen's dimensions L_i, H, we may therefore compute the four-dimensional data points

$$\{\lambda_\theta = 1 + u_\theta/L_i; \ \lambda_z = 1 + u_z/L_i; \ P_\theta = F_\theta/(L_i H); \ P_z = F_z/(L_i H)\},$$

where λ_θ and λ_z denote the average circumferential and axial stretches, whilst P_θ and P_z are the corresponding first Piola–Kirchhoff stresses, respectively. Given video extensiometer measurements are used, u_θ, u_z is to be replaced by $\overline{u}_\theta$, $\overline{u}_z$.

Fig. F.2 Tissue specimen mounted in the tensile testing machine and immersed in physiological saline solution. The dimension L_o denotes the outer sample dimension, whilst L_i is measured inside the gripping hooks. The vertical direction in the image represents the axial vessel direction

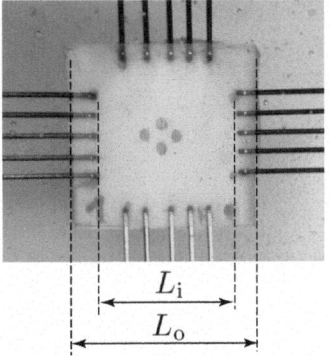

Figure F.3 illustrates the first Piola–Kirchhoff stress *versus* stretch properties of the loading protocols (a), (b), and (c). The data represents the experimental test data after preconditioning, and thus the recordings of the last loading–unloading cycle. The data show non-linear stretch–stress properties of the vessel wall and some degree of viscoelastic dissipation during the cycles, see Sect. 5.3. The experiment covers the transition from the elastin-dominated soft response at low strains, towards the collagen-dominated stiff response at larger strains, respectively. In addition, Table F.1 lists the acquired data, input information that has been used to identify model parameters of constitutive models in Sect. 5.6.

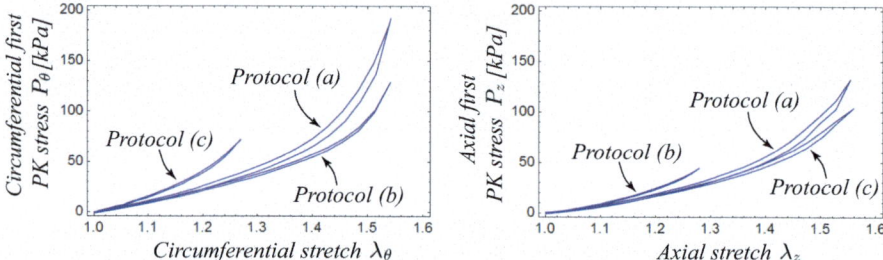

Fig. F.3 First Piola–Kirchhoff (PK) stress *versus* stretch properties of the porcine aorta after preconditioning. The data has been acquired through planar biaxial tension at the three different loading protocols (a), (b), and (c), see Sect. F.2

Table F.1 Experimental data from planar biaxial tensile testing of the porcine aorta wall. Stretches and first Piola–Kirchhoff stresses in circumferential and axial vessel wall directions are denoted by λ_θ^e, λ_z^e, and P_θ^e, P_z^e, respectively. The tissue has been displacement-controlled stretched following the protocols $u_\theta/u_z = 1/1$; $2/1$; $1/2$ with u_θ and u_z denoting the actuator displacements in axial and circumferential directions, respectively

$u_\theta/u_z = 1/1$				$u_\theta/u_z = 2/1$				$u_\theta/u_z = 1/2$			
λ_θ^e	λ_z^e	P_θ^e [kPa]	P_z^e [kPa]	λ_θ^e	λ_z^e	P_θ^e [kPa]	P_z^e [kPa]	λ_θ^e	λ_z^e	P_θ^e [kPa]	P_z^e [kPa]
1.0050	1.0060	−0.86	0.00	1.0000	1.0000	−1.09	−0.46	1.0225	1.0120	0.34	0.00
1.0317	1.0327	2.23	1.54	1.0157	1.0323	0.57	1.09	1.0558	1.0287	3.20	0.86
1.0585	1.0595	3.89	1.54	1.0323	1.0658	2.46	1.31	1.0893	1.0453	4.86	1.54
1.0852	1.0862	6.23	2.00	1.0490	1.0992	3.66	2.23	1.1227	1.0620	8.17	2.00
1.1118	1.1128	9.60	3.54	1.0657	1.1327	6.23	3.54	1.1562	1.0788	11.71	2.23
1.1387	1.1397	12.40	5.31	1.0823	1.1660	7.94	5.31	1.1895	1.0955	14.80	3.54
1.1653	1.1663	15.94	6.40	1.0992	1.1995	9.83	6.63	1.2230	1.1122	19.03	4.40
1.1922	1.1932	19.54	8.40	1.1158	1.2328	12.91	8.40	1.2563	1.1288	22.11	5.54
1.2188	1.2198	23.31	9.71	1.1325	1.2663	14.29	9.71	1.2898	1.1457	26.86	6.63
1.2455	1.2465	27.09	10.86	1.1492	1.2997	16.91	11.94	1.3232	1.1623	31.60	7.94
1.2723	1.2733	31.37	13.03	1.1660	1.3332	19.54	13.94	1.3567	1.1790	36.34	9.94
1.2990	1.3000	36.34	15.71	1.1827	1.3667	22.34	16.34	1.3900	1.1957	41.77	10.17
1.3257	1.3268	40.86	17.71	1.1993	1.4000	25.20	18.34	1.4235	1.2125	47.20	11.94
1.3525	1.3535	45.83	19.49	1.2160	1.4335	28.97	21.26	1.4568	1.2292	53.14	13.94
1.3792	1.3802	51.26	22.34	1.2328	1.4668	31.60	24.57	1.4903	1.2458	60.23	15.49
1.4060	1.4070	57.43	25.20	1.2495	1.5002	35.89	26.97	1.5237	1.2625	67.37	17.49
1.4327	1.4337	64.74	28.11	1.2662	1.5337	38.91	30.29	1.5572	1.2793	76.34	19.49
1.4593	1.4605	71.14	31.66	1.2828	1.5670	43.43	34.29	1.5905	1.2960	86.06	21.66
1.4862	1.4872	78.97	34.74	1.2997	1.6005	46.74	37.37	1.6240	1.3127	98.86	24.57
1.5128	1.5138	87.71	38.69	1.3163	1.6338	52.46	41.83	1.6573	1.3293	113.77	26.97
1.5397	1.5407	98.63	43.14	1.3330	1.6673	57.43	46.91	1.6908	1.3460	133.20	30.74
1.5663	1.5673	110.91	48.46	1.3497	1.7007	62.86	52.00	1.7242	1.3628	159.71	34.74
1.5930	1.5942	127.03	54.40	1.3665	1.7342	68.80	58.17	1.7577	1.3795	194.51	40.29
1.6198	1.6208	147.37	61.03	1.3832	1.7675	75.14	65.71	1.7910	1.3962	238.11	45.77
1.6465	1.6477	174.17	70.34	1.3998	1.8010	82.74	75.43	1.8245	1.4128	282.86	51.54
				1.4165	1.8333	92.46	90.29				

Definitions of Symbols, Functions, and Operators

G

Frequently used symbols

$\mathbf{a}_0$	Unit direction vector in the reference configuration
$\mathbf{b}$	Left Cauchy–Green strains
$\mathbf{B}$	Gradient interpolation matrix
$\mathbf{B}_f, \mathbf{b}_f$	Body force per reference and spatial volume
c	Molar concentration; wave speed
$\mathbf{C}$	Right Cauchy–Green strains; damping matrix
C	Capacity of a flow circuit
$\overline{\mathbf{C}}, \overline{\mathbf{b}}$	Isochoric right and left Cauchy–Green strains
$\mathbb{C}, \mathbb{c}$	Elasticity tensor in reference and spatial configuration
D	Distensibility; diameter
dS, ds	Area element in reference and spatial configuration
$d\mathbf{S}, d\mathbf{s}$	Area vector in reference and spatial configuration
d	Damage parameter; diameter
dV, dv	Volume element in reference and spatial configuration
dL, dl	Line element in reference and spatial configuration
$\mathbf{d}$	Rate of deformation tensor; Damping-related nodal forces
$\mathbf{D}$	Damping matrix
$\mathbf{E}_i, \mathbf{e}_i$	Cartesian base vector in reference and spatial configuration
E	Young's modulus
$\mathbf{E}, \mathbf{e}$	Green–Lagrange and Euler–Almansi strains
F	Reduced axial force
f	Frequency
$\mathbf{F}$	Deformation gradient
g	Gravitation
$\mathbf{g}$	Internal nodal force vector

(continued)

© Springer Nature Switzerland AG 2021

T. C. Gasser, *Vascular Biomechanics*, https://doi.org/10.1007/978-3-030-70966-2

$\mathbf{f}$	Finite element external nodal force vector
G	Shear modulus
G_i	Principal components of the growth-related deformation
$\mathbf{G}$	Divergence interpolation matrix; growth-related deformation
$\mathbf{H}$	General structural tensor
H	Hemolysis factor
H_0, H_a	Null hypothesis, alternative hypothesis
$\mathbf{h}, \dot{\mathbf{h}}, \ddot{\mathbf{h}}$	Essential variable vector
I_i	Tensor invariant
$\overline{I}_i$	Invariant of an isochoric tensor
$\mathbf{I}, \mathbb{I}$	Second-order and fourth-order identity tensor
$\mathbf{J}$	Jacobian transformation matrix
J	Volume ratio
$J_s, \mathbf{J}_s$	Molar substance flux
K	Bulk modulus
k	Permeability
$\mathbf{K}$	Stiffness matrix
$\mathbf{K}_f$	Stiffness matrix from external forces
$\mathbf{l}$	Velocity gradient tensor
l_p	Conductivity
L	Inertance of a flow circuit
$\mathbf{M}$	Mass matrix
$\mathbf{m}$	Inertia nodal forces
N_i	Shape functions
$\mathbf{N}, \mathbf{n}$	Unit normal vector in reference and spatial configuration; interpolation matrix
n_{dim}	Number of spatial dimensions
n_{dof}	Number of degrees of freedom
n_e	Number of elements
n_{npe}	Number of nodes per element
n_s	Number of independent stress or strain components
$\widehat{\mathbf{N}}_i, \widehat{\mathbf{n}}_i$	Eigenvector
p	Pressure; hydrostatic pressure; p-value
$p_{syst}, p_{dias}, p_{mean}$	Diastolic, systolic and mean pressure
$\mathbf{p}$	Complex pressure
$\mathbf{P}$	First Piola–Kirchhoff stresses
q	Flow
q_f	Filtration flux
$\mathbf{q}$	Complex flow; essential variable vector
$\mathbf{Q}$	Complex flow amplitude; proper orthogonal rotation tensor
$\mathbf{Q}_h, \mathbf{q}_h$	Heat flux per reference and spatial area
R	Resistance of a flow circuit
r	Pearson product-moment correlation coefficient
$\mathbf{r}$	Residuum

(continued)

r_s	Spearman rank correlation coefficient
$\overline{R}$	Ideal gas constant
$\mathbf{r}$	Residuum vector
$\mathbf{R}$	Proper orthogonal rotation tensor
R_h, r_h	Heat source per reference and spatial volume
R^2	Coefficient of determination
S, s	Entropy per reference and spatial volume
$\mathbf{S}$	Second Piola–Kirchhoff stresses
t	Time
T	Time of a cardiac cycle
$\mathbf{T}, \mathbf{t}$	Traction vector in reference and spatial configuration
U, u	Internal energies per reference and spatial volume
$\mathbf{u}, \dot{\mathbf{u}}, \ddot{\mathbf{u}}$	Displacement, velocity, and acceleration
$\mathbf{U}$	Right stretch tensor
v	Linear velocity
$\overline{\mathbf{v}}$	Left stretch tensor
$\mathbf{v}$	Velocity vector
V	Volume
$\mathbf{w}$	Spin tensor
$\mathbf{X}, \mathbf{x}$	Reference and spatial positions of a material particle
Z	Impedance of a flow circuit
$\mathbf{Z}$	Complex Impedance
$\beta = [Ca^{2+}]$	Calcium concentration
$\boldsymbol{\varepsilon}$	Engineering strain
δW	Virtual work
η	Viscosity
κ	Lagrange pressure to enforce incompressibility
θ	Absolute temperature
λ_i	Eigenvalue
λ	Load factor
$\lambda_{\mathbf{a}}$	Stretch along the direction $\mathbf{a}_0$
μ	Mean
ν	Poisson's ratio; kinematic fluid viscosity
ω	Angular velocity
$\boldsymbol{\omega}$	Angular velocity vector
Ω_0, Ω	Reference and spatial configuration of a body
$\partial\Omega_0, \partial\Omega$	Reference and spatial boundary of a body
Π	Osmotic pressure; Colloid Osmotic Pressure (COP); energy potential

(continued)

ρ, ρ_0	Probability Density Function (PDF); Density in the spatial and reference configuration
σ	Cauchy stress
σ	Cauchy stress; Standard Deviation (SD); Staverman's osmotic reflection coefficient
τ	Time related to tissue growth and remodeling
τ_w	Wall shear stress vector
ϕ	Phase angle; Hematocrit level
Φ	Objective function
χ	Motion or deformation function
Ψ, ψ	Helmholtz free energies per reference and spatial volume
Υ	Cumulative Density Function (CDF)
ς_v	Rate of volume change
ς_h	Non-compatible heat source
ς_s	Non-compatible entropy source

Mathematical operators and functions

$\arccos(x)$	Inverse cosinus of x
$\arcsin(x)$	Inverse sinus of x
$\cos(x)$	Cosinus of x
$\cot(x)$	Cotangents of x
$\mathrm{d}(\bullet)/\mathrm{d}t$	Total time derivative
$\mathrm{D}(\bullet)/\mathrm{D}t$	Material time derivative
$\mathrm{Dev}(\bullet), \mathrm{dev}(\bullet)$	Deviator in material and spatial description
$\det(\bullet)$	Determinant
$\mathrm{diag}[\bullet]$	Short-hand notation of a diagonal second-order tensor
$\mathrm{div}[\mathrm{grad}(\bullet)]$	Laplace operator
$\mathrm{Div}(\bullet), \mathrm{div}(\bullet)$	Divergence in the reference and spatial configuration
$\exp(x)$	Exponential function of x
$\mathrm{grad}_\mathrm{s}(\bullet)$	Symmetric gradient in the spatial configuration
$\mathrm{Grad}(\bullet), \mathrm{grad}(\bullet)$	Gradient in the reference and spatial configuration
$\ln(x)$	Natural logarithm of x
$\min(x)$	Minimum of x
$n!$	Factorial of n
$\mathrm{P}(A)$	Probability that event A will occur
$\mathrm{P}(A\|B)$	Probability that event A occurs, given that event B has occurred
$\mathrm{rg}(\bullet)$	Rank
$Re[(\bullet)], Im[(\bullet)]$	Real and imaginary parts
$\mathrm{sg}(x)$	Sigmoid function of x
$\mathrm{sign}(x)$	Sign of x
$\sin(x)$	Sinus of x
$\tan(x)$	Tangents of x
$\mathrm{Tr}(\bullet), \mathrm{tr}(\bullet)$	Trace in reference and spatial configuration
$\mathrm{Var}(X)$	Variance of X
$\mathrm{Var}(Y\|X)$	Variance of Y due to variability of X
$x(SD\ y)$	Sample with mean x and Standard Deviation (SD) y
$\delta(\bullet)$	Virtual variation; iteration increment
$\Delta(\bullet)$	Increment
$\partial(\bullet)/\partial t$	Partial time derivative
$D_\mathbf{u}(\bullet)$	Directional derivative along the increment $\Delta\mathbf{u}$
$\binom{n}{k}$	Binomial coefficient $n!/[k!(n-k)!]$
$\int(\bullet)\mathrm{d}x$	Integral over $(\bullet)$
$\int_a^b(\bullet)\mathrm{d}x$	Definite integral over $(\bullet)$ between a and b
$\|(\bullet)\|$	Norm
$\overline{(\bullet)}$	Mean value

Solutions

Example 1.1 (Sensitivity of the Resistance of a Blood Vessel).

(a) The sensitivity vector

$$\mathbf{s} = \frac{\partial R}{\partial \mathbf{x}} = R \left[l^{-1} \quad -4d^{-1} \quad \eta^{-1} \right]^{\mathrm{T}}$$

determines the local sensitivity of the Hagen–Poiseuille law.

(b) Given the sensitivity vector and the expression (1.1), the absolute and relative resistance errors read $\Delta R = \mathbf{s} \cdot \Delta \mathbf{x} = R(\Delta l/l - 4\Delta d/d + \Delta \eta/\eta)$ and $\Delta R/R = \Delta l/l - 4\Delta d/d + \Delta \eta/\eta$, respectively.

Let us consider the variation of one parameter at the time. The domains

$$|\Delta l/l| \leq 10.0\% \; ; \quad |\Delta d/d| \leq 2.5\% \; ; \quad |\Delta \eta/\eta| \leq 10.0\% \, ,$$

would then ensure that the resistance error remains below $\pm 10\%$. The relative resistance error $\Delta R/R$ of the Hagen–Poiseuille law is constant and does therefore not depend on the model parameter vector $\mathbf{x}$.

(c) The worst-case scenario results in the relative error $\Delta R/R = 10\% + 4 \cdot 2.5\% + 10\% = 30\%$. ∎

Example 1.2 (Global Versus Local Sensitivity Measures).

(a) The minimization of the objective function

$$\Phi(a_0, a_1, a_2, b_1, b_2, c_1, c_2, c_3, c_4) = \sum_{i=1}^{48} [r(h_i, s_i) - r_i]^2 \rightarrow \mathrm{MIN}$$

yields the set

$$a_0 = 0.3776 \, , \; a_1 = 0.08583 \, , \; a_2 = -0.045 \, , \; b_1 = 0.00125 \, ,$$

$$b_2 = 3.8954 \cdot 10^{-7} \, , \; c_1 = 0 \, , \; c_2 = 0 \, , \; c_3 = 0 \, , \; c_4 = 6.6667 \cdot 10^{-8}$$

© Springer Nature Switzerland AG 2021
T. C. Gasser, *Vascular Biomechanics*, https://doi.org/10.1007/978-3-030-70966-2

of surrogate model parameters. Here, r_i denotes the risk indices in Table 1.1, whilst the function $r(h_i, s_i)$ represents the surrogate model (1.3) at the data points h_i, s_i.

(b) Partial derivation of the model (1.3) yields the local sensitivity

$$\partial r/\partial h = a_1 + 2a_2 h + c_1 s + 2c_3 hs + c_2 s^2 + 2c_4 hs^2 ,$$

$$\partial r/\partial s = b_1 + c_1 h + c_3 h^2 + 2b_2 s + 2c_2 hs + 2c_4 h^2 s .$$

Given the identified parameters a_i, b_j, c_k, the sensitivity vector then reads

$$\begin{bmatrix} \partial r/\partial h \\ \partial r/\partial s \end{bmatrix} = \begin{bmatrix} 0.1218\,\text{mm}^{-1} \\ 2.43134 \cdot 10^{-3}\,\text{kPa}^{-1} \end{bmatrix}$$

at the point of question in the parameter space. Therefore, changing the wall thickness by 0.6 mm alters the risk index by 0.0731, and changing the tissue stiffness by 250.0 kPa alters the risk index by 0.6078, and thus almost an order of magnitude more.

(c) The provided probabilities of h and s yield the bi-normal probability distribution shown in Fig. 1.4a. Monte Carlo simulation has been used to sample the parameter domain $0.0 < h < 4.0$ mm and $0.0 < s < 1600.0$ kPa, using in total 10,000 points. Figure 1.4b illustrates the risk index predictions by the parameterized surrogate model, a sample that may be expressed by $r = 2.0633(SD\ 0.6268)$.

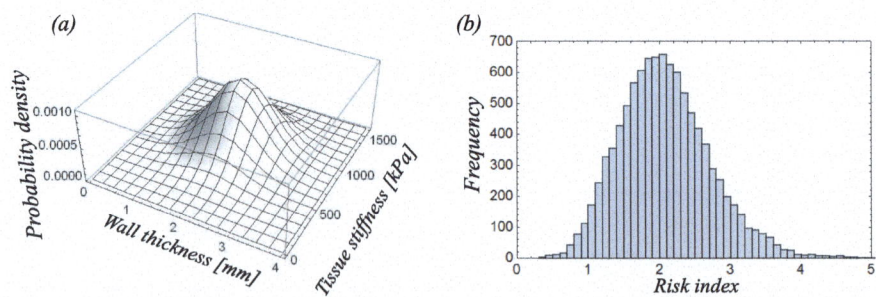

Fig. 1.4 (a) Bi-normal probability distribution to sample the parameter domain. (b) Distribution of the risk index computed by Monte Carlo simulation and the parameterized surrogate model ∎

Example 1.3 (Sobol's Variance-Based Sensitivity Analysis).

(a) Given an ANOVA-representation, the functions f_h, f_s, f_{hs} are orthogonal and

$$\int_0^1 f_h dh = \int_0^1 f_s ds = \int_0^1 f_{hs} dh = \int_0^1 f_{hs} ds = 0 \qquad (1.7)$$

holds, conditions that are satisfied by the model (1.6).

(b) The ANOVA-representation yields the decomposed variance expression

$$V = \text{Var}(r) = \int_0^1 r^2 dh ds - f_0^2$$

$$= \underbrace{\int_0^1 f_h^2 dh}_{V_h = \text{Var}(r|h)} + \underbrace{\int_0^1 f_s^2 ds}_{V_s = \text{Var}(r|s)} + \underbrace{\int_0^1 f_{hs}^2 dh ds}_{V_{hs} = \text{Var}(r|h,s) - V_h - V_s} \quad .$$

Given $V = 6.01 \cdot 10^{-3}$, $V_h = 0.732 \cdot 10^{-3}$, $V_s = 5.21 \cdot 10^{-3}$, and $V_{hs} = 0.0694 \cdot 10^{-3}$, Sobol's sensitivity indices are

$$S_h = \frac{V_h}{V} = 0.122 \ , \quad S_s = \frac{V_s}{V} = 0.867 \ , \quad S_{hs} = \frac{V_{hs}}{V} = 0.0116 \ .$$

Here, S_h and S_s are the first-order sensitivities, whilst S_{hs} represents a second-order sensitivity, or mixed effect. The different sensitivities are illustrated in Fig. 1.5, and the model output is by far most sensitive to the tissue strength.

(c) The variance of the wall thickness $\text{Var}(h) = 0.2$ results in the output variance $\text{Var}(r|h) = 0.2 S_h = 0.0244$, and the variance of the tissue strength of $\text{Var}(h) = 0.1$ in the output variance of $\text{Var}(r|s) = 0.1 S_s = 0.0867$, respectively.

Fig. 1.5 Sensitivity of the risk factor r with respect to the input information. The indices S_h and S_s denote the first-order sensitivities of wall thickness and tissue strength, whilst S_{hs} is the mixed effect from thickness and strength, respectively ∎

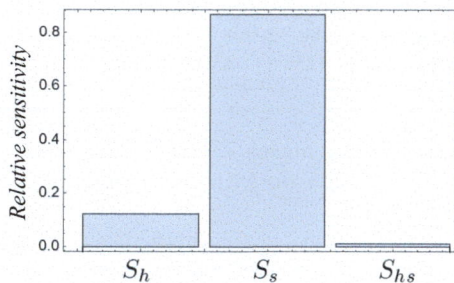

Example 1.4 (Correlation of Vessel Wall Stiffness and Strength).

(a) We start with the assumption of a linear correlation amongst the stiffness parameter x and the strength y. A linear regression model reads

$$y_i = b_0 + b_1 x_i + e_i \ ; \quad i = 1, \ldots, 20 , \qquad (1.14)$$

where e_i denotes the residuum. The coefficients b_0 and b_1 are identified through least-square optimization,

$$\sum_{i=1}^{n} e_i^2 = \sum_{i=1}^{n} (y_i - b_0 - b_1 x_i)^2 \rightarrow \text{MIN}.$$

Given the data listed in Table 1.2, the minimization problem yields $b_0 = 747.526$ kPa and $b_1 = 4.86802$, and with (1.14) we may compute the residuum $e_i = y_i - b_0 - b_1 x_i$.

To support the assumption of a linear correlation, e_i must be normal distributed. A Quantile-Quantile plot (QQ-plot) is used to investigate the normality of e_i. The QQ-plot shown in Fig. 1.9a illustrates that e_i falls around the diagonal, which fully supports the assumption of a linear correlation between x and y. Figure 1.9b shows the regression model.

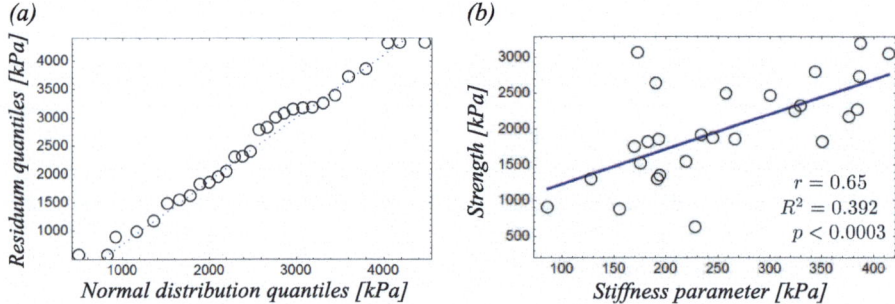

Fig. 1.9 Correlation between the vessel wall stiffness parameter and the vessel wall strength. (**a**) Quantile-Quantile plot (QQ-plot) illustrating normal distribution of the residuum e_i. (**b**) Linear regression with the Pearson's correlation coefficient r, the coefficient of determination R^2, and the significance p of the regression

(b) Given the means $\bar{x} = (\sum_{i=1}^{20} x_i)/20 = 255.1$ kPa and $\bar{y} = (\sum_{i=1}^{20} y_i)/20 = 1989.3$ kPa, the Pearson's product-moment correlation coefficient (1.8) reads

$$r = \frac{\sum_{i=1}^{20} (x_i - 255.1)(y_i - 1989.3)}{\sqrt{\sum_{i=1}^{20} (x_i - 255.1)^2 \sum_{i=1}^{n} (y_i - 1989.3)^2}} = 0.65.$$

The Pearson's correlation coefficient adequately quantifies a linear correlation, and $r = 0.65$ suggests a moderate and positive correlation between the stiffness parameter x and the strength y.

Given the Pearson's correlation coefficient, the expression (1.11) yields the coefficient of determination

$$R^2 = [r(n-1)/n]^2 = (0.6519/20)^2 = 0.392,\qquad(1.15)$$

which indicates rather poor predictability of the vessel wall strength.

(c) In order to test the significance of the regression, the Null Hypothesis that x_i and y_i are uncorrelated, is explored. With b_1 denoting the slope of the linear regression, a suitable Null Hypothesis reads $H_0 : b_1 = 0$. Given (1.13), the statistic reads

$$t = \frac{0.65\sqrt{18}}{\sqrt{1 - 0.65^2}} = 4.27836,\qquad(1.16)$$

and rejecting H_0 has the probability of $p = 2\int_t^\infty \rho_t(x)\mathrm{d}x = 0.000242$. Here, $\rho_t(x)$ denotes the PDF of the student t-distribution (A.3) of $n = 18$ degrees of freedom of the regression problem. The probability of 0.0242% is well below the significance level of 5%, and the regression can therefore be considered to be statistically significant.   ■

Example 1.5 (Testing for Clairvoyance).

(a) The Null Hypothesis H_0 assumes the person is simply guessing, and under the Alternative Hypothesis H_a, the person would be a clairvoyant. Given the probability $p = 1/4$ of guessing a suit correctly, said hypotheses read

$$H_0 : p = 1/4 \;;\; H_a : p > 1/4.$$

(b) The tree diagram shown in Fig. 1.11 illustrates the development of the probability of the test up to $n = 3$ trials. Given the answers would always be correct, we follow the most right path in Fig. 1.11. Rejecting H_0 for each of the $n = 25$ cards yields then the false positive probability of

$$\mathrm{P}(\text{reject } H_0 | H_0 \text{ is valid}) = \mathrm{P}(x = 25 | p = 1/4) = (1/4)^{25} < 10^{-15}.$$

Here, x denotes the number of correct answers, and this extremely low probability reflects answering by chance correctly 25 times in a row. Hence, out of 10^{15} persons, not a single one would have been associated mistakenly by a Type I error.

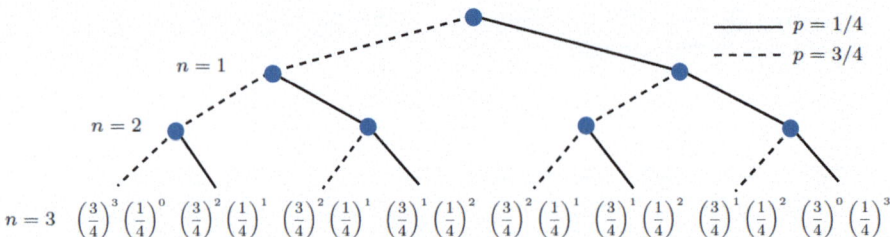

Fig. 1.11 Development of the probability for clairvoyant testing. The solid line denotes the probability $p = 1/4$ of answering correctly, and the dashed line the probability $p = 3/4$ of answering wrongly

(c) Looking at Fig. 1.11, the probability of giving at least two correct answers after three trials reads

$$P(x \geq 2 | p = 1/4) = 2 \left(\frac{3}{4} \right)^1 \left(\frac{1}{4} \right)^2 + \left(\frac{3}{4} \right)^1 \left(\frac{1}{4} \right)^2 + \left(\frac{3}{4} \right)^0 \left(\frac{1}{4} \right)^3 ,$$

a result that may be generalized towards

$$P(x \geq a | p = 1/4) = \sum_{k=a}^{n} \binom{n}{k} \left(\frac{3}{4} \right)^{n-k} \left(\frac{1}{4} \right)^k ,$$

where at least a correct answers after n trials have been given. The probability of rejecting H_0 for $a = 9$ cards then leads to

$$P(\text{reject } H_0 | H_0 \text{ is valid}) = \sum_{k=9}^{25} \binom{25}{k} \left(\frac{3}{4} \right)^{25-k} \left(\frac{1}{4} \right)^k = 0.149438$$

of false positives, i.e. roughly 15% of persons would have been associated mistakenly by a Type I error.

(d) The probability of rejecting H_0 for m correct answers reads

$$P(x \geq m | p = 1/4) = \sum_{k=m}^{25} \binom{25}{k} \left(\frac{3}{4} \right)^{25-k} \left(\frac{1}{4} \right)^k < 0.05 ,$$

an expression that allows us to determine $m = 11$. The related significance level would then be 2.97%. ∎

Example 1.6 (Conclusions from Vessel Wall Stiffness Data).

(a) Figure 1.12 illustrated the data by Quantile-Quantile plots (QQ-plots). The data are populated around the diagonals, and therefore normal distributed.

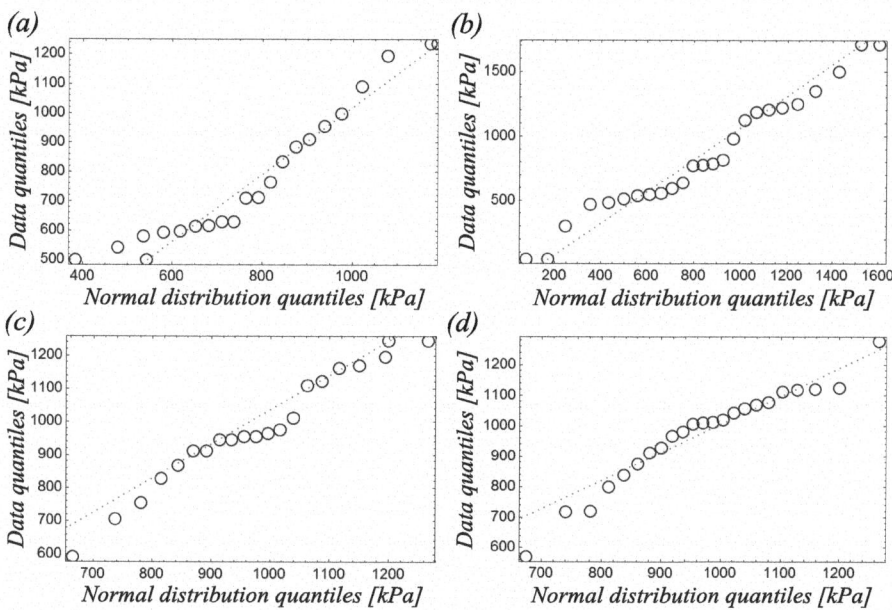

Fig. 1.12 Quantile-Quantile plots (QQ-plots) illustrating the normal distribution of the data listed in Table 1.3. (**a, b**) Stiffness data prior medication acquired by experimentalist A and experimentalist B, respectively. (**c, d**) Stiffness data post medication acquired by experimentalist A and experimentalist B, respectively

(b) We perform a mean difference test to test whether or not the data is influenced by the operator. As the data has been acquired from different rats, we may assume independence among the sample acquired by experimentalist A and experimentalist B. However, we do not know whether the medication influences the data and therefore test the baseline stiffness interdependently from the hypertensive stiffness.

Two-Sample t-test of the Baseline Stiffness The baseline stiffnesses, $\bar{x}_A = 778.3$ kPa and $\bar{x}_B = 839.7$ kPa denote the means of the data acquired by experimentalists A and B, respectively. The respective Standard Deviations (SDs) are $s_A = 222.3$ kPa and $s_B = 414.4$ kPa, and the combined distribution's standard error yields $\bar{s} = \sqrt{s_A^2/n_A + s_B^2/n_B} = 99.7$ kPa. The statistic $t = (\bar{x}_A - \bar{x}_B)/\bar{s} = -0.615$ then determines this problem, and the degrees of freedom of the student t-distribution is $\nu = 35$, given by the nearest integer of the expression

$$(s_A^2/n_A + s_B^2/n_B)^2 / \left[\frac{(s_A^2/n_A)^2}{n_A - 1} + \frac{(s_B^2/n_B)^2}{n_B - 1} \right].$$

The probability that the samples acquired by the two experimentalists A and B would describe the same population is then $p = 2\int_{-\infty}^{t} \rho_t(x)\mathrm{d}x = 0.542$, where $\rho_t(x)$ denotes the PDF of the student t-distribution (A.3). The Null Hypothesis that both samples describe the same population can therefore not be rejected.

Two-Sample t-test of the Hypertensive Stiffness The same analysis of the hypertensive stiffness data results in the statistic $t = -0.08705$ and the probability $p = 0.931$ that the samples acquired by the two experimentalists would describe different populations. It provides again counter-evidence that they would describe different populations.

(c) We test whether or not the difference Δx of the stiffness before and after medication would be statistically significant. A one-sample t-test is used, and we pool all samples together, an approach that is justified by the conclusion drawn from Task (b). The sample has the size $n = 43$, the mean $\overline{\Delta x} = 157.1$ kPa, and the SD $\Delta s = 378.2$ kPa, respectively. Its statistic reads

$$t = \frac{\overline{\Delta x} - \mu_0}{s/\sqrt{n}} = 2.724, \tag{1.21}$$

and rejecting the Null Hypothesis, i.e. the stiffness would not change by medication, has the probability of $p = 2\int_{-\infty}^{t} \rho_t(x)\mathrm{d}x = 0.009$, where $\rho_t(x)$ denotes the PDF of the student t-distribution (A.3). It is below the predefined significance level of 0.05, and the influence of medication is therefore statistically significant. Figure A.13 shows a box-and-whiskers plot of the difference in aorta stiffness through medication.

Fig. A.13 Box-and-whiskers plot illustrating the influence of medication on the wall stiffness. Statistical significance is denoted by p

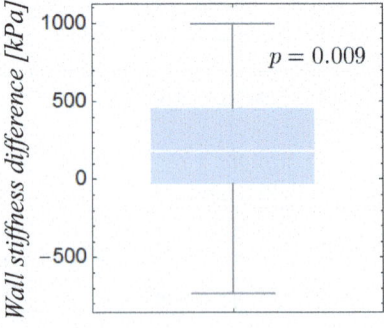

Example 1.7 (Training an Artificial Neural Network).

(a) We introduce the least-square error $\Phi = (y - \widetilde{y})^2$, and substitute y by the expression (1.22). The minimization of

$$\Phi = \left\{ \left[1 + \exp\left(-\sum_{i=1}^{2} x_i w_i \right) \right]^{-1} - \widetilde{y} \right\}^2 \rightarrow \text{MIN} \qquad (1.23)$$

then expresses the "learning" of the ANN. In the context of Artificial Intelligence (AI), Φ is commonly called loss function—only in the case $\Phi = 0$ we do not lose any information by the minimization.

(b) Given the data in Table 1.4, the minimization problem (1.23) yields $w_1 = 1.1848$ and $w_2 = -0.0813323$, and $\Phi = 0.00386171$ expresses the amount of information that has been lost. ∎

Example 2.1 (Upstream Pressure Wave Propagation).

(a) Given $c \gg v$, the control volume moves at the wave speed c in upstream direction, such that the mass flow rate ρcA passes through it.

(b) The hydrostatic force $F = -A\Delta p$ acts at the control volume, where Δp denotes the pressure difference between its right and left borders, respectively. When the pressure wave passes, the blood of the mass m in the control volume is accelerated by a. Given the inertia $ma = \rho cA\Delta v$, Newton's second law then yields $F = ma = -A\Delta p = \rho cA\Delta v$. The relation

$$\Delta p = -\rho c \Delta v, \qquad (2.2)$$

known as the water hammer equation, expresses then the change in pressure Δp as a function of the blood density ρ, the wave speed c, and the velocity change Δv, respectively.

(c) Given the relation (2.1), a wave propagates in the vessel at $c = (\rho D)^{-1/2} = 5.6 \text{ m s}^{-1}$.

(d) The water hammer equation (2.2) results in the velocity change of $\Delta v = -\Delta p/(\rho c) = -3.88 \text{ cm s}^{-1}$ in response to the pressure wave. ∎

Example 2.2 (Two-Element WindKessel Model Predictions).

(a) The cardiac cycle is split into k equidistant time increments $\Delta t = t_i - t_{i-1} = T/k$, and the backward-Euler discretization

$$\frac{\mathrm{d}p(t)}{\mathrm{d}t} \approx \frac{p_i - p_{i-1}}{\Delta t}$$

approximates the time derivative of the pressure, where p_i and p_{i-1} denote the pressure at the times t_i and t_{i-1}, respectively. Given the flow q_i at the time t_i, the discretized governing equation (2.6) reads

$$q_i = \frac{p_i}{R} + C \frac{p_i - p_{i-1}}{\Delta t}$$

and provides the explicit expression

$$p_i = \frac{\alpha p_{i-1} + R q_i}{1 + \alpha} \text{ with } \alpha = CR/\Delta t \tag{2.8}$$

for the pressure p_i at the time t_i.

(b) With the initial pressure $p_0 = 0$, the recursive application of Eq. (2.8) yields the pressure p_i at the discrete time points t_i. After a sufficient number of cardiac cycles, the steady-state periodic pressure response shown in Fig. 2.16 is reached. The figure also illustrates the convergence towards the exact solution for $k = 10; 50; 1000$ time increments over the cardiac cycle.

Fig. 2.16 Numerically predicted pressure over the cardiac cycle. Three different time discretizations k illustrate the convergence towards the exact solution. The simulation used $p_0 = 0$ as initial condition and the plot illustrates the pressure at the 10-th cardiac cycle ■

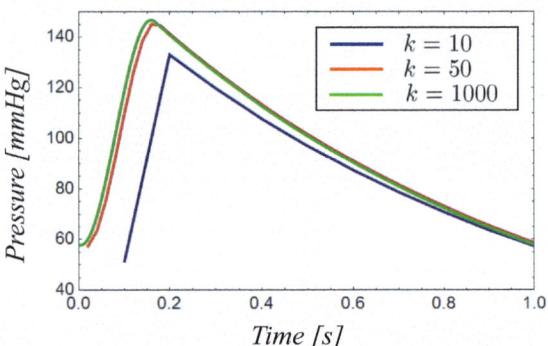

Example 2.3 (Systemic Implication of EVAR Treatment).

(a) Given the assumption the stent-graft's radial stiffness is much larger than the radial stiffness of the normal thoracic aorta, the capacity $C_{EVAR} = 0.3 C_n + 0.7[\alpha C_{sg} + (1 - \alpha) C_n]$ determines the systemic capacity of the treated patient. The stent-graft's radial stiffness k_{sg} determines it capacity

$$C_{sg} = \frac{\Delta V}{\Delta p} = \frac{1}{2} \pi d_{sg} l_{sg} k_{sg} = 0.165 \, \alpha \, [\text{cm}^3 \text{kPa}^{-1}],$$

such that

$$C_{EVAR} = 9.7 - 6.79 \, \alpha + 0.1155 \, \alpha^2 \, [\text{cm}^3 \text{kPa}^{-1}]$$

characterizes the treated patient.

(b) The governing equation of the two-element WK model reads

$$\dot{p}(t) + a p(t) = b(t) \quad \text{with} \quad a = (RC_{EVAR})^{-1} \quad \text{and} \quad b = Q \sin(\omega t)^2 / C_{EVAR},$$

where $\omega = \pi$. It is a linear first-order differential equation with the closed-form solution $p(t) = (\int \mu b dt + A)/\mu$, where $\mu = \exp\left(\int a dt\right) = \exp\left[t/(RC_{EVAR})\right]$ denotes the integrating factor, and A is an integration constant. The pressure is therefore given by

$$p(t) = \frac{\frac{Q}{C_{EVAR}} \int \exp[a\,t] \sin(\omega t)^2 dt + A}{\exp[a\,t]},$$

and has the closed-form solution

$$p(t) = A \exp[-a\,x] + \frac{QR}{2} - \frac{QR[\cos(2\omega t) + 2C_{EVAR} R\omega \sin(2\omega t)]}{2 + 8C_{EVAR}^2 R^2 \omega^2}.$$

The identification of the integration constant A from the initial condition $p(0) = p_0$, then yields

$$p(t) = 13.5 - 0.08875 \exp(-0.5727t) - 0.1112 \cos(2\pi x) - 1.22 \sin(2\pi x);$$

$$p(t) = 13.5 + 0.8623 \exp(-1.836t) - 1.062 \cos(2\pi t) - 3.635 \sin(2\pi t)$$

for the parameters $\alpha = 0$ and $\alpha = 1$, respectively. Figure 2.18 illustrates these solutions, and shows the pressure in the normal ($\alpha = 0$) and fully stent-graft-covered ($\alpha = 1$) thoracic aorta.

Fig. 2.18 Pressure that is predicted by the two-element WK model in the normal ($\alpha = 0$) and fully stent-graft-covered ($\alpha = 1$) thoracic aorta

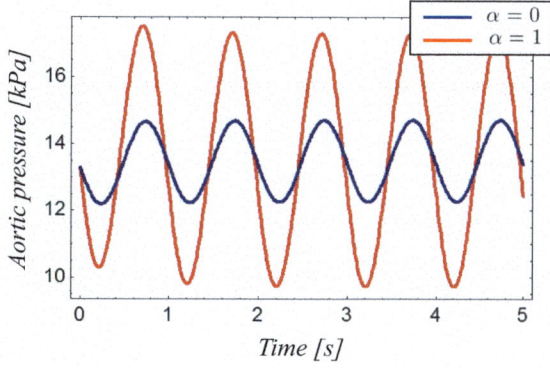

Example 2.4 (Impedance of the Vascular System).

(a) The cardiac cycle $T = 0.375$ s may be split into $N = 30$ (number of data points minus one) equidistant time intervals $\Delta t = T/N$. Given the trapezial integration rule, the Fourier coefficients

$$\mathbf{c}_{jk} = \frac{1}{T} \int\limits_{t=0}^{T} s_j(t) \exp\left(-i\omega_k t\right) \mathrm{d}t$$

$$\approx \frac{1}{T} \sum_{k=1}^{N} \frac{s_j(k) + s_j(k+1)}{2} \exp\left[i2\pi k(k-0.5)\frac{\Delta t}{p}\right] \; ; \; k = 1, \ldots, M$$

can be computed from the experimental data. Here, the index $j = \mathrm{q}$ and $j = \mathrm{p}$ relate to the flow and pressure, whilst k denotes the number of the individual Fourier series term, also known as the harmonics or modes. Table 2.3 lists harmonics of the flow and pressure waves, respectively.

Table 2.3 Fourier coefficients $\mathbf{c}_{\mathrm{q}k}$ and $\mathbf{c}_{\mathrm{p}k}$ approximating flow $q(t)$ and pressure p(t) waves, respectively

Mode k	Angular velocity ω_k	$\mathbf{c}_{\mathrm{q}k}$ [mls^{-1}]	$\mathbf{c}_{\mathrm{p}k}$ [mmHg]
10	−167.552	$0.0136667 - 0.00144338i$	$2.41288 \cdot 10^{-14} - 0.0721688i$
9	−150.796	$0.00722562 - 0.0347752i$	$-0.0813342 - 0.200008i$
8	−134.041	$-0.0593984 - 0.0217432i$	$-0.339697 + 0.0694108i$
7	−117.286	$-0.0233298 + 0.0691505i$	$-0.0208851 + 0.408154i$
6	−100.531	$0.0250477 + 0.0231058i$	$0.18498 + 0.136162i$
5	−83.7758	$0.0369504 + 0.0725i$	$0.259808 + 0.125i$
4	−67.0206	$0.216635 - 0.0557698i$	$0.981977 - 0.312774i$
3	−50.2655	$-0.0896779 - 0.408525i$	$-0.716424 - 1.97084i$
2	−33.5103	$-0.555816 + 0.112961i$	$-1.76808 + 0.691707i$
1	−16.7552	$0.0276938 + 0.646413i$	$-2.12634 + 1.60786i$
0	0	0.786333	85.25
1	16.7552	$0.0276938 - 0.646413i$	$-2.12634 - 1.60786i$
2	33.5103	$-0.555816 - 0.112961i$	$-1.76808 - 0.691707i$
3	50.2655	$-0.0896779 + 0.408525i$	$-0.716424 + 1.97084i$
4	67.0206	$0.216635 + 0.0557698i$	$0.981977 + 0.312774i$
5	83.7758	$0.0369504 - 0.0725i$	$0.259808 - 0.125i$
6	100.531	$0.0250477 - 0.0231058i$	$0.18498 - 0.136162i$
7	117.286	$-0.0233298 - 0.0691505i$	$-0.0208851 - 0.408154i$
8	134.041	$-0.0593984 + 0.0217432i$	$-0.339697 - 0.0694108i$
9	150.796	$0.00722562 + 0.0347752i$	$-0.0813342 + 0.200008i$
10	167.552	$0.0136667 + 0.00144338i$	$2.41288 \cdot 10^{-14} + 0.0721688i$

The Fourier-approximated signal then reads

$$\tilde{s}_j(t) = \mathrm{Re}\left[\sum_{k=-M}^{M} \mathbf{c}_{jk}\exp\left(i\omega_k t\right)\right] \; ; \; j = \mathrm{q,p} \qquad (2.13)$$

with $\omega_k = 2\pi k/T$ denoting the angular velocity of the complex vector. The flow and pressure waves are shown in Fig. 2.21a, b.

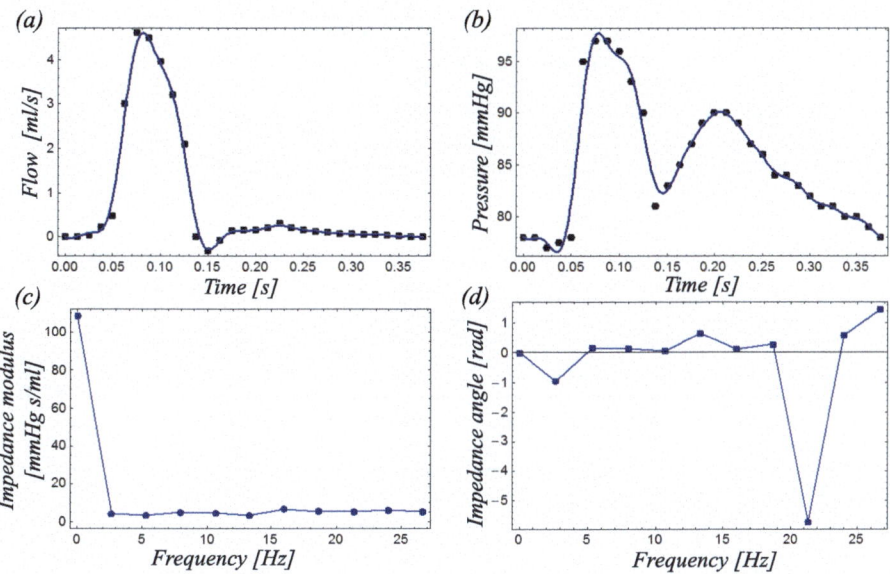

Fig. 2.21 Fourier series approximation (solid line) of (**a**) $q(t)$ and (**b**) $p(t)$ on top of the experimentally measured points (dots). (**c**) Impedance modulus Z_k and (**d**) impedance angle ϕ_k as a function of signal frequency f_k.

(b) With the complex Fourier coefficients $\mathbf{c}_{jk} = a_{jk} + ib_{jk}; k = 1\ldots, M$ and $j = \mathrm{q,p}$ of the flow and pressure waves, the impedance modulus $|\mathbf{Z}_k| = Z_k = |\mathbf{c}_{pk}|/|\mathbf{c}_{qk}|$ and impedance angle $\phi_k = \arg\mathbf{c}_{pk} - \arg\mathbf{c}_{qk}$ are defined. Figure 2.21c, d plots these data *versus* the frequency $f_k = k/T$. ∎

Example 2.5 (Decay Method to Estimate the Vascular Resistance).

(a) Figure 2.23 shows the lumped parameter model that represents the experimental set-up. Given $q_{\mathrm{m}} = 0$, the equation $q_{\mathrm{in}}(t) = 0 = p_{\mathrm{in}}(t)/R + C\dot{p}_{\mathrm{in}}(t)$ governs the problem, and the relation $\ln(p_1/p_0) = -(t_1 - t_0)/(RC)$ then determines the vascular bed resistance

$$R = -\frac{t_1 - t_0}{C\ln(p_1/p_0)} = 166.247 \text{ mmHg s ml}^{-1},$$

where the two pressure measurements p_1 and p_2 have been used.

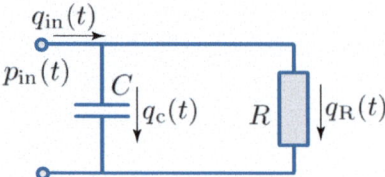

Fig. 2.23 Electrical representations of the two-element lumped parameter model that captures the experimental set-up. The flow $q(t)$ and the pressure $p(t)$ describe the system state, and R and C denote the organ's vascular bed resistance and arterial capacity, respectively

(b) Given the inflow q_{in} into the manometer, the problem's governing equation reads $q_{in}(t) = -\xi\,\dot{p}_{in}(t) = p_{in}(t)/R + C\,\dot{p}_{in}(t)$, and $\ln(p_1/p_0) = -(t_1 - t_0)/[R(C + \xi)]$ allows to compute the vascular bed resistance

$$R = -\frac{t_1 - t_0}{(C + \xi)\ln(p_1/p_0)} = \frac{1.9946}{0.012 + \xi} \text{ [mmHg s ml}^{-1}]. \qquad (2.16)$$

Figure 2.24 illustrates the development of the relative error e as a function of the device parameter ξ.

Fig. 2.24 Relative error $100(R - R_{\text{exact}})/R_{\text{exact}}$ as a function of the device-dependent parameter ξ of the manometer

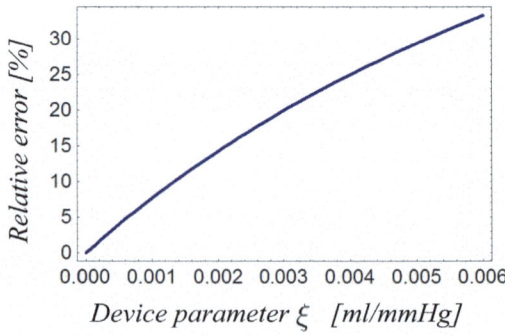

(c) With A and h denoting the cross-section and height of the water column in the uptake tube, the relations $q_m = A(dh/dt)$ and $p_{in} = \rho g h$ specify the inflow

$$q_m(t) = \xi\frac{dp_{in}}{dt} = \xi\rho g\frac{dh}{dt} = \frac{d_i^2\pi}{4}\frac{dh}{dt}$$

into the uptake tube. The device-dependent parameter

$$\xi = \frac{d_i^2 \pi}{4 \rho g} = 8.0061 \cdot 10^{-11} \mathrm{m}^3 \, \mathrm{Pa}^{-1} = 0.0106745 \, \mathrm{ml} \, \mathrm{mmHg}^{-1},$$

then determines with (2.16) the resistance $R = 87.98 \, \mathrm{mmHg} \, \mathrm{s} \, \mathrm{ml}^{-1}$ of the organ in question. ■

Example 2.6 (Two-Element Versus Three-Element WK Models).

(a) The peripheral resistance R is given by the relation

$$R = \frac{p_{\mathrm{mean}}}{q_{\mathrm{mean}}} = \frac{\int_0^T p(t) \mathrm{d}t}{\int_0^T q(t) \mathrm{d}t} \approx \frac{\sum_{k=1}^{N} (p_k + p_{k+1})}{\sum_{k=1}^{N} (q_k + q_{k+1})} = 108.42 \, \mathrm{mmHg} \, \mathrm{s} \, \mathrm{ml}^{-1},$$

where the integration over the cardiac cycle of $T = 0.375 \, \mathrm{s}$ has been approximated by the trapezial rule over $N = 30$ (number of data points minus one) time intervals. It represents the resistance of the two-element WK model, whilst R needs to be adjusted for the three-element WK model, see Task (c).

(b) At the late diastolic phase the flow $q(t) = 0$ and the system is governed by $p(t)/R + C\mathrm{d}p(t)/\mathrm{d}t = 0$. It determines the capacity

$$C = \frac{t_1 - t_0}{R \ln \left(\frac{p_0}{p_1} \right)} = 1.1808 \cdot 10^{-2} \, \mathrm{ml} \, \mathrm{mmHg}^{-1},$$

where the two late diastolic time points $t_0 = 0.25 \, \mathrm{s}$ and $t_1 = 0.375 \, \mathrm{s}$ together with the corresponding pressures $p(t_0) = 86 \, \mathrm{mmHg}$ and $p(t_1) = 78 \, \mathrm{mmHg}$ have been used.

(c) Given the aortic cross-section $A = d^2 \pi/4 = 3.801 \, \mathrm{mm}^2$, the aorta's characteristic impedance reads $Z_a = v_{\mathrm{pw}} \rho / A = 2.36186 \cdot 10^8 \, \mathrm{Pa} \, \mathrm{s} \, \mathrm{m}^{-3}$ or $1.7714 \, \mathrm{mmHg} \, \mathrm{s} \, \mathrm{ml}^{-1}$. For the three-element WK model, Z_a influence the total system resistance, and

$$R = \frac{p_{\mathrm{mean}}}{q_{\mathrm{mean}}} - Z_a = 108.42 - 1.7714 = 106.66 \, \mathrm{mmHg}$$

determines its peripheral resistance.

(d) The predicted pressure profiles of the two-element and three-element WK models are shown in Fig. 2.26b. These results use the prescribed flow wave shown in Fig. 2.26a together with the parameters estimated by Tasks (a) to (c).

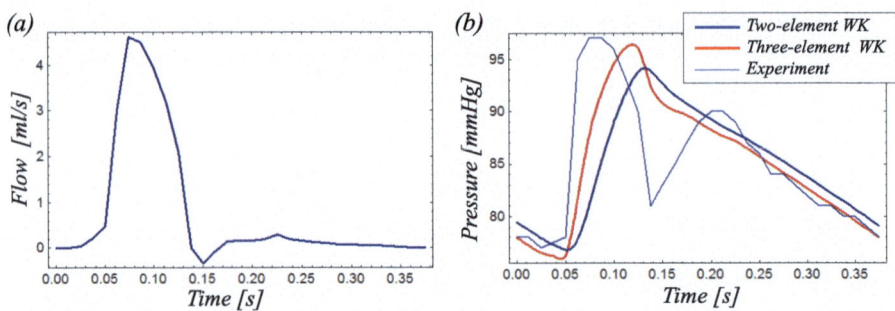

Fig. 2.26 (**a**) Prescribed flow profile. (**b**) Predicted pressure using two-element and three-element WindKessel (WK) models on top of the experimental data ■

Example 2.7 (Impedance-Based Estimation of WK Parameters).

(a) The objective function

$$\Phi(R, C; \omega) = \sum_{k=1}^{N} \alpha \left[Z(R, C; \omega_k) - Z_{\exp k} \right]^2 + \left[\phi(R, C; \omega_k) - \phi_{\exp k} \right]^2 ,$$

of the two-element WK model and

$$\Phi(R, C, Z_a; \omega) = \sum_{k=1}^{N} \alpha \left[Z(R, C, Z_a; \omega_k) - Z_{\exp k} \right]^2 + \left[\phi(R, C, Z_a; \omega_k) - \phi_{\exp k} \right]^2$$

of the respective three-element WK model may be introduced. Here, $Z_{\exp k}$ and $\phi_{\exp k}$ denote the $N = 5$ impedance moduli and angles reported in Table 2.4, whilst the relations (2.12) and (2.23) determine the analytical expressions for Z and ϕ of the two-element and three-element WK models, respectively. The analytical expressions are evaluated at $\omega_k = 2\pi f_k$, where f_k denotes the frequencies listed in Table 2.4. For simplicity we set $\alpha = 1$, and the minimization of the objective functions then yields the following least-square optimized parameters:

- Two-element WK model: peripheral resistance $R = 108.41$ mmHg s ml^{-1} and arterial capacity $C = 9.8143 \cdot 10^{-3}$ ml mmHg^{-1}
- Three-element WK model: peripheral resistance $R = 104.04$ mmHg s ml^{-1}, arterial capacity $C = 84.787 \cdot 10^{-3}$ ml mmHg^{-1}, and aortic characteristic impedance $Z_a = 4.37$ mmHg s ml^{-1}

Remark The use of $\alpha = 0.01$ scales the two contributions in the objective function more adequately and then results in a better approximation of the experimental data. The attended electronic Mathematica script allows the reader to recalculate the problem with said scaling parameter.

(b) Figure 2.27a, b shows the impedance modulus Z and the impedance angle ϕ predicted by the respective WK model.

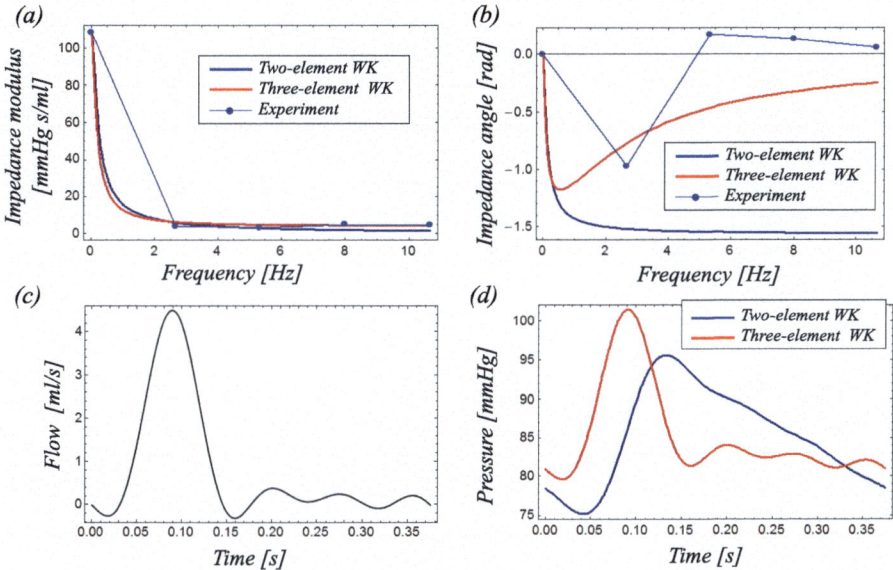

Fig. 2.27 (**a, b**) WindKessel (WK) model-predicted *versus* the measured impedance modulus Z and impedance angle ϕ of the system. (**c**) Flow wave $q(t)$ represented by the Fourier coefficients in Table 2.5. (**d**) WK model-predicted pressure waves $p(t)$ of a steady-state periodic analysis

(c) The Fourier coefficients $\mathbf{c}_{qk} = \mathbf{c}_{q-k}$ given in Table 2.5 determine the flow profile

$$\mathbf{q} = \sum_{k=-N}^{N} \mathbf{c}_{qk} \exp(i\omega_k t),$$

where $\omega_k = 2\pi k/T$ denotes the angular velocity of the complex vector. Figure 2.27c shows the real component $q(t) = Re[\mathbf{q}]$.

Given the Fourier coefficient of the flow $\mathbf{c}_{qk}$, the transformation (2.12) of the two-element WK model, and (2.23) of the three-element WK model, then determine the corresponding Fourier coefficients $\mathbf{c}_{pk}$ of the pressure waves

$$|\mathbf{c}_{pk}| = Z(\omega_k)|\mathbf{c}_{qk}| \; ; \quad \alpha_k = \phi(\omega_k) + \arg(\mathbf{c}_{qk}). \tag{2.24}$$

The superposition of all Fourier coefficients yields then the pressure wave

$$\mathbf{p} = \sum_{k=-N}^{N} |\mathbf{c}_{\mathrm{p}\,k}| \exp\left[i\left(\omega_k t + \alpha_k\right)\right], \qquad (2.25)$$

and Fig. 2.27d shows the real component $p(t) = Re[\mathbf{p}]$.

The numerical solution of the governing equation of the two-element WK model (2.6) as well as the three-element WK model (2.20) at the initial condition $p(t = 0) = 80$ mmHg, determine the pressure cycles shown in Fig. 2.28a, c. At the sixth cycle the solution is practically periodic, and the pressure cycle compares very well to the result of the steady-state analysis—compare Figs. 2.27d and 2.28b, d.

The Fourier coefficients $\mathbf{c}_{\mathrm{q}\,k}$ have been determined from the flow reported elsewhere [62], and Fig. 2.29 compares the WK model-predicted and the respective measured pressure waves. For further details see Example 2.4.

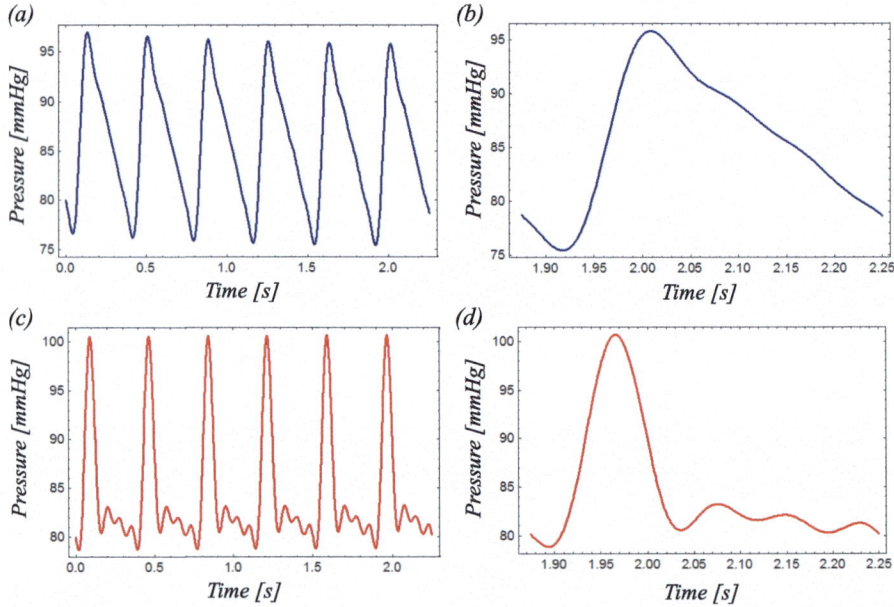

Fig. 2.28 Pressure pulse wave $p(t)$ predicted by the two-element WK model (top row) and the three-element WK model (bottom row). The waves are based on the numerical solution of the governing equations. The results over the first six cycles (**a, c**) and the sixth cycle (**b, d**) are shown separately

Fig. 2.29 WindKessel (WK)
model-predicted pressure
wave $p(t)$ on top of the
measured pressure wave
reported elsewhere [62]

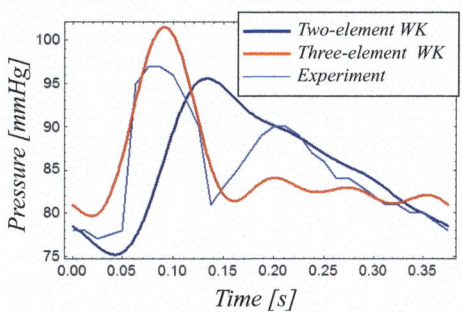

Example 2.8 (Renal Artery Adaptation to Partial Nephrectomy).

(a) Upon the pressure difference Δp between the renal artery and the renal vein,
    the mean flow $q = \Delta p/R$ passes the kidney, where R denotes its vascular bed
    resistance. The augmented resistance αR then yields the mean flow $q(\alpha) = \Delta p/(\alpha R)$.

(b) We consider Poiseuille flow, and thus a fully established steady-state laminar
    flow of a Newtonian fluid of viscosity η in a cylindrical tube of radius r. The
    blood flow velocity

$$v_z = -\frac{r^2}{4\eta}\frac{\mathrm{d}p}{\mathrm{d}z}\left[1 - \left(\frac{\xi}{r}\right)^2\right] \tag{2.34}$$

appears then at a parabolic profile, where $0 \leq \xi \leq r$ and $\mathrm{d}p/\mathrm{d}z$ denote the
radial coordinate and the pressure gradient along the tube's axial direction z,
respectively. A detailed derivation of this expression is given in Chapter 6. The
profile (2.34) determines the flow $q = 2\pi\int_0^r v_z r\mathrm{d}r = -\pi r^4(\mathrm{d}p/\mathrm{d}z)/(8\eta)$ and
results in the WSS $\tau_\mathrm{w} = \eta(\mathrm{d}v_z/\mathrm{d}\xi)_{\xi=r} = r(\mathrm{d}p/\mathrm{d}z)/(2\eta)$. The elimination of
the pressure gradient $\mathrm{d}p/\mathrm{d}z$ from these two relations then yields

$$q = -r^3\pi\tau_\mathrm{w}/(4\eta)\,, \tag{2.35}$$

an expression that relates the WSS to the flow in the renal artery.

(c) Given a thin-walled tube of wall thickness h that is inflated at the pressure p,
    the equilibrium in the circumferential direction $2\sigma_\theta h = 2pr$ yields the law of
    Laplace

$$\sigma_\theta = pr/h \tag{2.36}$$

and expresses the circumferential Cauchy stress σ_θ in the wall of the renal artery.

(d) According to Task (a), the flow q is inversely proportional to the augmentation factor α. The expression $q/q_0 = 1/\alpha$ and (2.35) at constant τ_{w}, then results in the homeostatic vessel radius $r = r_0/\sqrt[3]{\alpha}$.

Given (2.36) at constant σ_θ and p, the change from r_0 towards r results in the wall thickness $h = h_0 r/r_0$. Figure 2.33 illustrates the relative change of the radius r/r_0 (or the wall thickness h/h_0) in response to the augmentation factor α.

Fig. 2.33 Relative change of renal artery radius r/r_0 (or the wall thickness h/h_0) as a function of the augmentation factor α

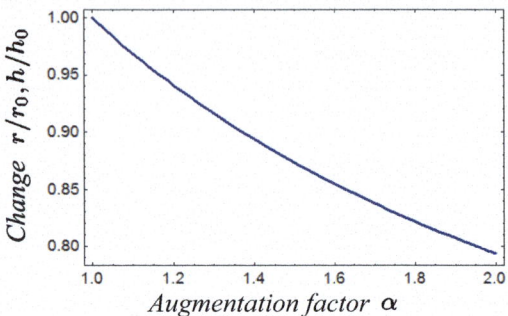

Example 2.9 (Two-Element Vessel Segment Model).

(a) The relations $\mathbf{p}_{\mathrm{in}} - R\mathbf{q}_{\mathrm{in}} = \mathbf{p}_{\mathrm{out}}$ and $\mathbf{q}_{\mathrm{in}} - \mathbf{q}_{\mathrm{c}} = \mathbf{q}_{\mathrm{out}}$ yield the system

$$\begin{bmatrix} \mathbf{p}_{\mathrm{out}} \\ \mathbf{q}_{\mathrm{out}} \end{bmatrix} = \begin{bmatrix} 1 & -R \\ 0 & 1 \end{bmatrix} \begin{bmatrix} \mathbf{p}_{\mathrm{in}} \\ \mathbf{q}_{\mathrm{in}} \end{bmatrix} + \begin{bmatrix} 0 & 0 \\ -C & RC \end{bmatrix} \begin{bmatrix} \dot{\mathbf{p}}_{\mathrm{in}} \\ \dot{\mathbf{q}}_{\mathrm{in}} \end{bmatrix} \tag{2.44}$$

of governing equations, where the definition $\mathbf{q}_{\mathrm{c}} = C\dot{\mathbf{p}}_{\mathrm{out}}$ of the capacitor has been used.

(b) With (2.31) and (2.32), the vessel segment's resistance and capacity read

$$R = \frac{128\eta l}{\pi d^4} = 2.83081 \cdot 10^6 \ \mathrm{Pa\ s\ m^{-3}} \,,$$

$$C = \frac{3\pi d^3 l}{16 E h} = 6.18751 \cdot 10^{-10} \ \mathrm{m^3\ Pa^{-1}} \,,$$

where the data in Fig. 2.25 has been used.

(c) The first equation of the system (2.44) directly allows us to compute the inlet pressure

$$\mathbf{p}_{\mathrm{in}} = \mathbf{p}_{\mathrm{out}} + R\mathbf{q}_{\mathrm{in}}$$

$$= 12.5 \exp[i(\omega t + \pi/6)] + 2.83081 \cdot 4.3 \exp[i\omega t]$$

$$= |\mathbf{p}_{\mathrm{in}}| \exp[i(\omega t + \alpha)]$$

with the amplitude $|\mathbf{p}_{\mathrm{in}}| = 23.83$ Pa and the phase angle $\alpha = 0.2653$ rad. Figure 2.35a illustrates $\mathbf{p}_{\mathrm{in}}$ in the complex plane.

The second equation of the system (2.44) yields the outlet flow

$$\begin{aligned}\mathbf{q}_{\rm out} &= \mathbf{q}_{\rm in} - C(\mathrm{d}\mathbf{p}_{\rm out}/\mathrm{d}t)\\ &= 4.3\cdot 10^{-6}\exp[i\omega t] - 6.18751\cdot 12.5\cdot\omega\cdot 10^{-10}\exp[i(\omega t + \pi/6 + \pi/2)]\\ &= |\mathbf{q}_{\rm out}|\exp[i(\omega t + \beta)].\end{aligned}$$

Given $\omega = 73\pi$, the flow $\mathbf{q}_{\rm out}$ has the amplitude $|\mathbf{q}_{\rm out}| = 5.41$ ml s^{-1} and the phase angle $\beta = -0.2879$ rad, respectively. Figure 2.35b illustrates $\mathbf{q}_{\rm out}$ in the complex plane.

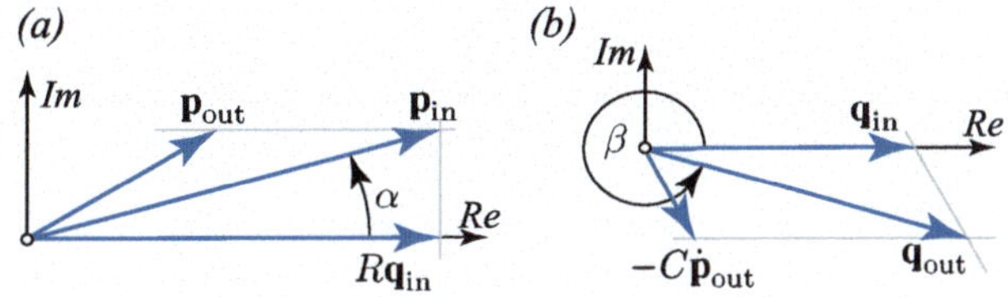

Fig. 2.35 Argand's diagrams to illustrate (**a**) the pressure $\mathbf{p}_{\rm in}$ at the inlet and (**b**) the flow $\mathbf{q}_{\rm out}$ at the outlet ■

Example 2.10 (Three-Element Vessel Segment Model).

(a) The relations $p_{\rm in} - \Delta p_{\rm R} - \Delta p_{\rm L} = p_{\rm out}$ and $q_{\rm in} - q_{\rm c} = q_{\rm out}$ lead to the set

$$\begin{bmatrix} p_{\rm out}\\ q_{\rm out}\end{bmatrix} = \begin{bmatrix} 1 & -R\\ 0 & 1\end{bmatrix}\begin{bmatrix} p_{\rm in}\\ q_{\rm in}\end{bmatrix} + \begin{bmatrix} 0 & -L\\ -C & RC\end{bmatrix}\begin{bmatrix} \dot p_{\rm in}\\ \dot q_{\rm in}\end{bmatrix} + \begin{bmatrix} LC & -RCL\\ 0 & 0\end{bmatrix}\begin{bmatrix} \ddot p_{\rm in}\\ \ddot q_{\rm in}\end{bmatrix} \tag{2.45}$$

of governing equations, where the expressions $\Delta p_{\rm R} = Rq_{\rm in}$, $\Delta p_{\rm L} = L\dot q_{\rm out}$ and $q_{\rm C} = C\dot p_{\rm C}$ with $p_{\rm C} = p_{\rm in} - \Delta p_{\rm R}$ describe the properties of resistor, inductor, and capacitor, respectively.

(b) Given the steady-state periodic inflow $\mathbf{q}_{\rm in} = |\mathbf{q}_{\rm in}|\exp[i\omega t]$, we may introduce the complex inflow pressure $\mathbf{p}_{\rm in} = |\mathbf{p}_{\rm in}|\exp[i(\omega t + \alpha)]$, and $(2.45)_1$ then yields the complex outlet pressure

$$\begin{aligned}\mathbf{p}_{\rm out} &= \mathbf{p}_{\rm in} - R\mathbf{q}_{\rm in} - L\dot{\mathbf{q}}_{\rm in} + LC\ddot{\mathbf{p}}_{\rm in} - RCL\ddot{\mathbf{q}}_{\rm in}\\ &= -7.89445 - 167.094i + 0.826159|\mathbf{p}_{\rm in}|\exp(i\alpha)\,.\end{aligned}$$

The implementation of the outlet pressure boundary condition $\mathrm{Re}[\mathbf{p}_{\rm out}] = 1000.0$ Pa and $\mathrm{Im}[\mathbf{p}_{\rm out}] = 0$ determines then the inlet pressure at the magnitude $|\mathbf{p}_{\rm in}| = 1236.63$ Pa and phase angle $\alpha = 0.16429$, respectively. Figure 2.37a illustrates the construction of $\mathbf{p}_{\rm out}$ in the complex plane.

Given $(2.45)_2$, the complex outflow vector reads

$$\begin{aligned}\mathbf{q}_{\rm out} &= \mathbf{q}_{\rm in} - C\dot{\mathbf{p}}_{\rm in} + RC\dot{\mathbf{q}}_{\rm in}\\ &= 5.20481\cdot 10^{-6} - i5.415\cdot 10^{-6}\,.\end{aligned}$$

It has the amplitude $|\mathbf{q}_{\text{out}}| = 7.51081$ ml s^{-1} and the phase angle arg$\mathbf{q}_{\text{out}} = -0.805188$ rad. Figure 2.37b illustrates the construction of $\mathbf{q}_{\text{out}}$ in the complex plane.

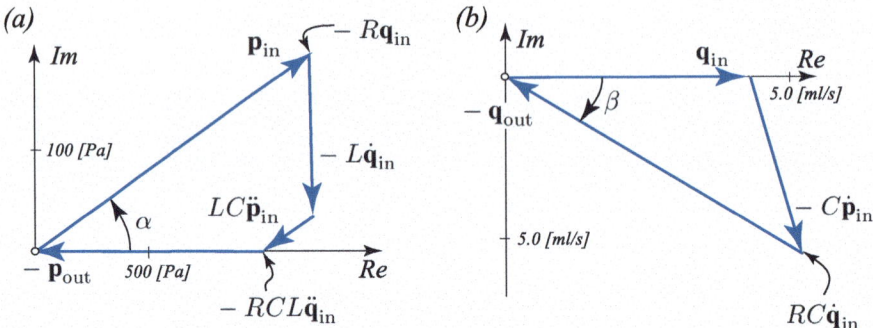

Fig. 2.37 Argand's diagrams illustrate the construction of (**a**) the outlet pressure $\mathbf{p}_{\text{out}}$ and (**b**) the outlet flow $\mathbf{q}_{\text{out}}$ of the three-element lumped parameter model ∎

Example 2.11 (Connected Vessel Segments).

(a) Table 2.7 reports the resistance, capacity, and inductance of the vessel segments, where the relations (2.31), (2.32), and (2.33) have been used to compute these figures.

Table 2.7 Resistance R, capacity C, and inductance L of vessel segments

	Vessel 1	Vessel 2	Vessel 3
Resistance R [Pa s m^{-3}]	$7.84316 \cdot 10^9$	$1.84119 \cdot 10^{10}$	$1.3369 \cdot 10^{11}$
Capacity C [m^3 s^{-1}]	$3.23977 \cdot 10^{-12}$	$7.15812 \cdot 10^{-13}$	$3.01593 \cdot 10^{-13}$
Inductance L [Pa s^2 m^{-3}]	$7.42299 \cdot 10^7$	$8.5385 \cdot 10^7$	$2.02445 \cdot 10^8$

(b) Given the pressure drops $\Delta p_{\text{R}} = Rq_{\text{out}}$ and $\Delta p_{\text{L}} = L\dot{q}_{\text{out}}$ over the resistance R and the inductance L, the relation

$$p_{\text{in}} - p_{\text{out}} = Rq_{\text{out}} + L\dot{q}_{\text{out}}$$

may be derived. The flow continuity

$$q_{\text{in}} - q_{\text{out}} = q_{\text{C}}$$

with the capacitor relation $q_{\text{C}} = C\dot{p}_{\text{in}}$ closes the mathematical description, and the system

$$\begin{bmatrix} p_{\text{out}} \\ q_{\text{out}} \end{bmatrix} = \begin{bmatrix} 1 & -R \\ 0 & 1 \end{bmatrix} \begin{bmatrix} p_{\text{in}} \\ q_{\text{in}} \end{bmatrix} + \begin{bmatrix} RC & -L \\ -C & 0 \end{bmatrix} \begin{bmatrix} \dot{p}_{\text{in}} \\ \dot{q}_{\text{in}} \end{bmatrix} + \begin{bmatrix} LC & 0 \\ 0 & 0 \end{bmatrix} \begin{bmatrix} \ddot{p}_{\text{in}} \\ \ddot{q}_{\text{in}} \end{bmatrix} \qquad (2.46)$$

of differential equations then governs the three-element vessel model.

(c) The substitution of the time derivatives in (2.46) through the backward-Euler discretization (2.41) allows us to derive the algebraic set

$$
\begin{bmatrix} p_{\text{out}} \\ q_{\text{out}} \end{bmatrix} = \begin{bmatrix} 1 + \frac{RC}{\Delta t} + \frac{LC}{\Delta t^2} & -R - \frac{L}{\Delta t} \\ -\frac{C}{\Delta t} & 1 \end{bmatrix} \begin{bmatrix} p_{\text{in}} \\ q_{\text{in}} \end{bmatrix} + \mathbf{H}
\tag{2.47}
$$

of equations, where

$$
\mathbf{H} = \begin{bmatrix} -\frac{RC}{\Delta t} - \frac{LC}{\Delta t^2} & \frac{L}{\Delta t} \\ \frac{C}{\Delta t} & 0 \end{bmatrix} \begin{bmatrix} p_{\text{in}\,n} \\ q_{\text{in}\,n} \end{bmatrix} - \begin{bmatrix} \frac{LC}{\Delta t} & 0 \\ 0 & 0 \end{bmatrix} \begin{bmatrix} \dot{p}_{\text{in}\,n} \\ \dot{q}_{\text{in}\,n} \end{bmatrix}
$$

denotes a history vector, and thus information from the previous time step. For the i-th vascular segment, $\mathbf{d}_{\text{out}\,i} = \mathbf{K}_i \mathbf{d}_{\text{in}\,i} + \mathbf{H}_i$ expresses the governing equation (2.47) in symbolic notation.

(d,e) Figure 2.39 shows the development of the system variables over the first three seconds, results that have been computed with the algorithm in Table 2.8.

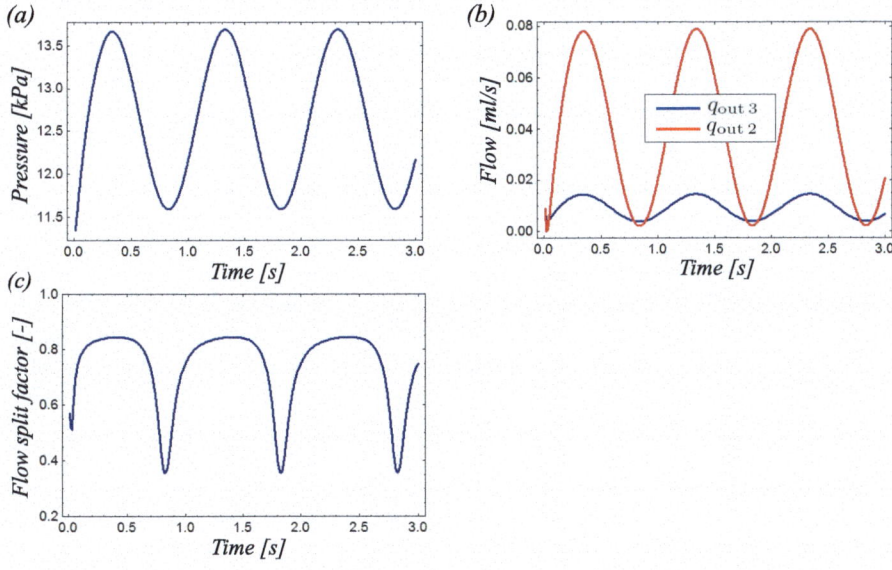

Fig. 2.39 Development of system variables over time. (**a**) Inlet pressure $p_{\text{in}\,1}$. (**b**) Outlet flows $q_{\text{out}\,2}$ and $q_{\text{out}\,3}$. (**c**) Flow split factor ξ

Table 2.8 Algorithm for the iterative solution of three connected vessel segments

(1) Initialize pressure and flow
$p_{\text{in}10} = (p_2 + p_3)/2$ as well as $q_{\text{in}i0} = q_0$ and $\dot{p}_{\text{in}i0} = \dot{q}_{\text{in}i0} = 0$ for $i = 1, 2, 3$

(2) Set the number of iterations $n = 300$ and the time step size $\Delta t = 0.01$ s

(3) Do For $k = 1, \ldots, n$
    (a) Compute the history vectors $\mathbf{H}_i$; $i = 1, 2, 3$
    (b) Prescribe the inflow $p_{\text{in}1} = p_0[1 + \sin(\omega k \Delta t)]$
    (c) Compute outflow conditions $\mathbf{d}_{\text{out}1} = \mathbf{K}_1 \mathbf{d}_{\text{in}1} + \mathbf{H}_1$ of the first vessel, where the inlet pressure $p_{\text{in}1}$ in the vector $\mathbf{d}_{\text{out}1}$ is unknown
    (d) Specify the compatibility conditions at the bifurcation
        $p_{\text{in}2} = p_{\text{in}3} = p_{\text{out}1}$; $q_{\text{in}2} = \xi q_{\text{out}1}$; $q_{\text{in}2} = (1 - \xi)q_{\text{out}1}$
    (e) Compute outflow conditions $\mathbf{d}_{\text{out}2} = \mathbf{K}_2 \mathbf{d}_{\text{in}2} + \mathbf{H}_2$ of the second vessel, where the inlet pressure $p_{\text{in}1}$ and the flow split factor ξ in $\mathbf{d}_{\text{out}2}$ are unknown
    (f) Compute outflow conditions $\mathbf{d}_{\text{out}3} = \mathbf{K}_2 \mathbf{d}_{\text{in}3} + \mathbf{H}_3$ of the third vessel, where the inlet pressure $p_{\text{in}1}$ and the flow split factor ξ in $\mathbf{d}_{\text{out}3}$ are unknown
    (g) Use the prescribed outlet pressures $p_{\text{out}2}(p_{\text{in}1}, \xi) = p_2$ and $p_{\text{out}3}(p_{\text{in}1}, \xi) = p_3$ to solve for $p_{\text{in}1}$ and ξ from these two equations
    End Do

∎

Example 2.12 (LDL Transport Through a Micro-channel).

(a) We may substitute the expression $v = q_f(1 - \sigma)$ in (2.49) and

$$J_s = c_b q_f(1 - \sigma) - \frac{q_f(1 - \sigma)(c_w - c_b)}{\exp(Pe\, L/d) - 1}$$

expresses then the LDL flux.
(b) Figure 2.41 illustrates J_s as a function of the normalized channel length L/d. A small Péclet number results in diffusion-dominated transport, whilst a large one determines an advection-dominated process. The transport is then entirely determined by the velocity v.

Example 2.13 (Transport Across the Ascending Vasa Recta wall).

(a) According to the law of Hagen–Poiseuille, the pressure gradient of

$$\Delta p/l = 128 q \eta/(\pi d^4) = 117.6 \text{ kPa m}^{-1} \tag{2.52}$$

develops in the capillary, and $p_v = p_{v\,\text{art}} - (\Delta p/l)x = 1.04 - 117.6x$ [kPa] describes the hydrostatic pressure along the AVR segment. Here, x [m] denotes the axial coordinate measured from the inlet. The hydrostatic pressure changes

Fig. 2.41 Low-Density Lipoprotein (LDL) flux J through a micro-channel as a function of the normalized channel length L/d. $Pe = vd/D$ denotes the Péclet number

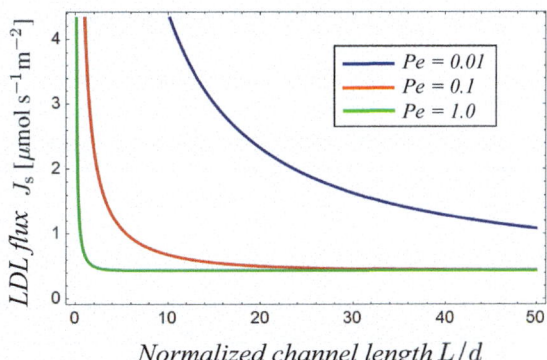

then linearly from 1.04 kPa at the AVR's arterial end towards 0.9 kPa at its venous end.

(b) Given $\Delta p_{art} = p_{v\,art} - p_i = 0.24$ kPa and $\Delta p_{ven} = p_{v\,ven} - p_i = 0.1$ kPa, the relation $\Delta p = 0.24 - 117.6x$ [kPa] describes the transcapillary hydrostatic pressure, where x [m] denotes the axial coordinate along the AVR segment. In addition, $\Delta\Pi = 2.97 - 1033.3x$ [kPa] describes the transcapillary COP along the AVR segment. Figure 2.44a shows these transcapillary pressures, and Starling's filtration law (2.51) then predicts the filtration flux $q_f = -23.4 + 7744.4x$ [nm s^{-1}]. Along the entire vessel an inward flux is therefore predicted (see Fig. 2.44b), and the fluid from the medullary interstitium is continuously absorbed into the AVR.

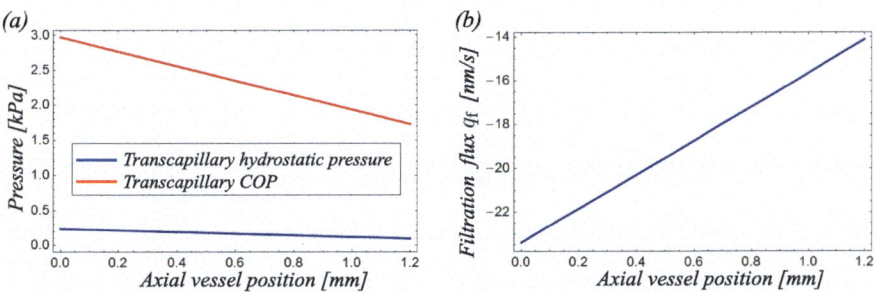

Fig. 2.44 Properties along the Ascending Vasa Recta (AVR). (a) Transcapillary hydrostatic pressure Δp and transcapillary Colloid Osmotic Pressures (COP) $\Delta\Pi$. (b) Filtration flux across the AVR wall

Example 2.14 (Glycocalyx-Cleft Model).

(a) Given $\Delta p_{art} = p_{v\,art} - p_i = 1.387$ kPa and $\Delta p_{ven} = p_{v\,ven} - p_i = 1.026$ kPa, the transcapillary hydrostatic pressure reads $\Delta p = 1.387 - 102.86x$ [kPa], where x [m] denotes the axial vessel position. In addition, $\Delta\Pi = 1.613 - 72.381x$ [kPa] describes the transcapillary COP along the capillary segment.

Starling's model (2.51) predicts then the filtration flux

$$q_f = L_p \left(\Delta p - \sigma \Delta \Pi \right) = 1.656 - 775.429x \quad [\text{nm s}^{-1}] \, ,$$

resulting in influx along the venous side, see Fig. 2.47 (case $\alpha = 1.0$).

(b) The filtration flux reads

$$q_f = L_p \left(\Delta p - \sigma \Delta \Pi_{gc} \right) \, , \tag{2.56}$$

where $\Delta \Pi_{gc} = \Pi_v - \Pi_{gc}$ denotes the COP difference between the vascular space and the position underneath the glycocalyx layer. It determines the effective transcapillary COP of the glycocalyx-cleft model. Hoff's relation allows us to translate the molar concentration into COP, and thus $\Pi_{gc} = (1 - \alpha)\Pi_v + \alpha \Pi_i$ holds. The effective transcapillary COP then reads $\Delta \Pi_{gc} = \Pi_v - \Pi_{gc} = \alpha(1.613 - 72.381x)$ [kPa] and the glycocalyx-cleft model (2.56) predicts the filtration flux $q_f = 23.92 + \alpha(-22.264 + 998.857x) - 1774.29x$ [nm s^{-1}], where the transcapillary hydrostatic pressure $\Delta p = 1.387 - 1.0286x$ [kPa] has been used. Figure 2.47 illustrates the filtration flux along the vessel and as a function of the factor α.

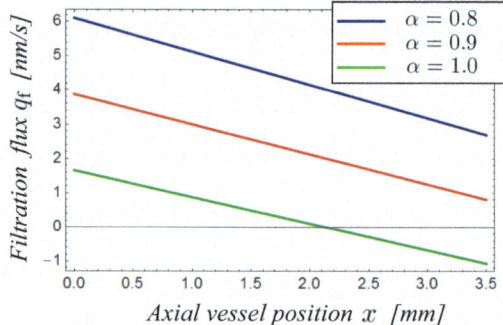

Fig. 2.47 Filtration flux across the capillary wall predicted by the glycocalyx-cleft model. The parameter α relates the Colloid Osmotic Pressure (COP) underneath the glycocalyx layer to the COP levels in the interstitium and the vascular lumen, respectively. Starling's filtration model (2.51) corresponds to $\alpha = 1.0$ ■

Example 3.1 (Deformation of Line, Volume, and Area Elements).

(a) The embedded fibers determine the structural tensor

$$\mathbf{A} = \mathbf{a}_0 \otimes \mathbf{a}_0 = \begin{bmatrix} a_{01}a_{01} & a_{01}a_{02} \\ a_{02}a_{01} & a_{02}a_{02} \end{bmatrix} = \begin{bmatrix} 0.9127 & 0.2823 \\ 0.2823 & 0.0873 \end{bmatrix} \tag{3.7}$$

that characterizes the material in the reference configuration Ω_0. The deformation of the continuum is described by the right Cauchy–Green strain

$$\mathbf{C} = \mathbf{F}^{\mathsf{T}}\mathbf{F} = \begin{bmatrix} 3 & 0 \\ 1 & 1.3 \end{bmatrix} \begin{bmatrix} 3 & 1 \\ 0 & 1.3 \end{bmatrix} = \begin{bmatrix} 9 & 3 \\ 3 & 2.69 \end{bmatrix}, \tag{3.8}$$

and yields the fourth invariant $I_4 = \lambda_\mathbf{a}^2 = \mathbf{A} : \mathbf{C} = A_{kl}C_{kl} = 10.142$, where the structural tensor (3.7) has been used. The stretch in the fibers is then $\lambda_\mathbf{a} = \sqrt{I_4} = 3.185$.

(b) Given (3.6), the area ratio, and thus the equivalence to the volume ratio in 3D, reads

$$J = \det\mathbf{F} = 3.9 \ .$$

The deformed body Ω occupies then the area $a = JA = 7.8 \text{ m}^2$.

(c) The inverse deformation gradient reads

$$\mathbf{F}^{-1} = \begin{bmatrix} 0.3333 & -0.2564 \\ 0 & 0.7692 \end{bmatrix},$$

and Nanson's formula then yields the normal vector, and thus the equivalence to the area vector in 3D,

$$\mathrm{d}\mathbf{s} = J\mathbf{F}^{-\mathsf{T}}\mathrm{d}\mathbf{S} = 3.9 \begin{bmatrix} 0.333333 & 0 \\ -0.25641 & 0.769231 \end{bmatrix} \begin{bmatrix} 0.7 \ \mathrm{d}x_1 \\ 3.2 \ \mathrm{d}x_2 \end{bmatrix}$$

$$= \begin{bmatrix} 0.91 \ \mathrm{d}x_1 \\ -0.7 \ \mathrm{d}x_1 + 9.6 \ \mathrm{d}x_2 \end{bmatrix} [\mathrm{m}] \ ,$$

where $|\mathrm{d}\mathbf{s}|[\mathrm{m}]$ denotes the length of the line element in Ω. ∎

Example 3.2 (Non-linear Deformation and Strain Measures).

(a) The motion $\chi(\mathbf{X}, t)$ determines the deformation gradient

$$\mathbf{F} = \mathrm{Grad}\mathbf{x} = \mathrm{Grad}\chi(\mathbf{X}, t) = \begin{bmatrix} \frac{\partial x_1}{\partial X_1} & \frac{\partial x_1}{\partial X_2} \\ \frac{\partial x_2}{\partial X_1} & \frac{\partial x_2}{\partial X_2} \end{bmatrix} = \begin{bmatrix} 3X_1 t & t \\ 0 & 1.3t \end{bmatrix}, \quad (3.14)$$

where (3.1) has been used. The determinant then reads $\det\mathbf{F} = 3.9X_1 t^2$, and given the constraint $\det\mathbf{F}(\mathbf{X}, t) > 0$ upon the deformation gradient, the motion $\chi(\mathbf{X}, t)$ is only valid for $X_1 > 0$ and $t \neq 0$.

(b) For $X_1 \neq 0$ and $t \neq 0$, the deformation gradient may be inverted. The inverse reads

$$\mathbf{F}^{-1} = \begin{bmatrix} \frac{0.333333}{X_1 t} & -\frac{0.25641}{X_1 t} \\ 0 & \frac{0.769231}{t} \end{bmatrix},$$

and, given the deformation gradient's material time derivative

$$\dot{\mathbf{F}} = \begin{bmatrix} 3X_1 & 1 \\ 0 & 1.3 \end{bmatrix} \; [\mathrm{s}^{-1}],$$

the velocity gradient yields

$$\mathbf{l} = \dot{\mathbf{F}}\mathbf{F}^{-1} = \begin{bmatrix} 3X_1 & 1 \\ 0 & 1.3 \end{bmatrix} \begin{bmatrix} \frac{0.333333}{X_1 t} & -\frac{0.25641}{X_1 t} \\ 0 & \frac{0.769231}{t} \end{bmatrix} = \begin{bmatrix} \frac{1}{t} & 0 \\ 0 & \frac{1}{t} \end{bmatrix} \; [\mathrm{s}^{-1}].$$

It is defined for the time $t \neq 0$.

(c) With the deformation gradient (3.14), the right and left Cauchy–Green strain tensors read

$$\mathbf{C} = \mathbf{F}^{\mathrm{T}}\mathbf{F} = \begin{bmatrix} 3X_1 t & 0 \\ t & 1.3t \end{bmatrix} \begin{bmatrix} 3X_1 t & t \\ 0 & 1.3t \end{bmatrix} = \begin{bmatrix} 9X_1^2 t^2 & 3X_1 t^2 \\ 3X_1 t^2 & 2.69t^2; \end{bmatrix} = \mathbf{C}^{\mathrm{T}};$$

$$\mathbf{b} = \mathbf{F}\mathbf{F}^{\mathrm{T}} = \begin{bmatrix} 3X_1 t & t \\ 0 & 1.3t \end{bmatrix} \begin{bmatrix} 3X_1 t & 0 \\ t & 1.3t \end{bmatrix} = \begin{bmatrix} t^2 + 9X_1^2 t^2 & 1.3t^2 \\ 1.3t^2 & 1.69t^2 \end{bmatrix} = \mathbf{b}^{\mathrm{T}},$$

where the definitions (3.10) and (3.11) have been used. Given the definition (3.12), the Green–Lagrange strain tensor reads

$$\mathbf{E} = \frac{1}{2}(\mathbf{C} - \mathbf{I}) = \begin{bmatrix} -0.5 + 4.5X_1^2 t^2 & 1.5X_1 t^2 \\ 1.5X_1 t^2 & -0.5 + 1.345t^2 \end{bmatrix} = \mathbf{E}^{\mathrm{T}}.$$

We may invert the left Cauchy–Green strain tensor

$$\mathbf{b}^{-1} = \begin{bmatrix} t^2 + 9X_1^2 t^2 & 1.3t^2 \\ 1.3t^2 & 1.69t^2 \end{bmatrix},$$

which then allows us to compute the Euler–Almansi strain tensor

$$\mathbf{e} = \frac{1}{2}(\mathbf{I} - \mathbf{b}^{-1}) = \begin{bmatrix} 0.5 - \frac{0.0556}{X_1^2 t^2} & \frac{0.0427}{X_1^2 t^2} \\ \frac{0.0427}{X_1^2 t^2} & 0.5 - \frac{0.2959}{t^2} - \frac{0.0329}{X_1^2 t^2} \end{bmatrix} = \mathbf{e}^{\mathrm{T}},$$

where the definition (3.13) has been used. ∎

Example 3.3 (Linear Versus Non-linear Strain Measures).

(a) The displacement vector reads

$$\mathbf{u} = \mathbf{x} - \mathbf{X} = \mathbf{RX} - \mathbf{X} = (\mathbf{R} - \mathbf{I})\mathbf{X} = \begin{bmatrix} \cos\alpha - 1 & -\sin\alpha \\ \sin\alpha & \cos\alpha - 1 \end{bmatrix} \begin{bmatrix} X_1 \\ X_2 \end{bmatrix}$$

and determines the engineering strain

$$\boldsymbol{\varepsilon} = \frac{1}{2}\left[\frac{\partial \mathbf{u}}{\partial \mathbf{X}} + \left(\frac{\partial \mathbf{u}}{\partial \mathbf{X}} \right)^{\mathrm{T}} \right] = \frac{1}{2}(\mathbf{R} + \mathbf{R}^{\mathrm{T}}) - \mathbf{I} = \begin{bmatrix} \cos\alpha - 1 & 0 \\ 0 & \cos\alpha - 1 \end{bmatrix},$$

where the definition (3.9) has been used. Rigid body rotation does not introduce any deformation in the body, and given the engineering strain is a linear strain measure, it yields the correct result $\boldsymbol{\varepsilon} = \mathbf{0}$ only at small rigid body rotations $\alpha \to 0$.

(b) The right Cauchy–Green strain (3.10) reads $\mathbf{C} = \mathbf{F}^{\mathrm{T}}\mathbf{F} = \mathbf{R}^{\mathrm{T}}\mathbf{R} = \mathbf{I}$, where the deformation gradient $\mathbf{F}$ has been substituted by the rigid body rotation $\mathbf{R}$. It allows us then to compute the Green–Lagrange strain (3.12), which is a geometrically exact strain measure and therefore yields the correct result of $\mathbf{E} = (\mathbf{C} - \mathbf{I})/2 = \mathbf{0}$. ∎

Example 3.4 (Independence of Strain Measures from Rigid Body Rotation).

(a) The use of the right $\mathbf{F} = \mathbf{RU}$ and left $\mathbf{F} = \bar{\mathbf{v}}\mathbf{Q}$ polar decompositions of the deformation gradient allows us to express the right and left Cauchy–Green strain tensors through

$$\mathbf{C} = \mathbf{F}^\mathrm{T}\mathbf{F} = (\mathbf{RU})^\mathrm{T}\mathbf{RU} = \mathbf{U}^\mathrm{T}\mathbf{U} = \mathbf{UU}^\mathrm{T} \; ; \quad \mathbf{b} = \mathbf{FF}^\mathrm{T} = \bar{\mathbf{v}}\mathbf{Q}(\bar{\mathbf{v}}\mathbf{Q})^\mathrm{T} = \overline{\mathbf{vv}}^\mathrm{T} = \bar{\mathbf{v}}^\mathrm{T}\bar{\mathbf{v}}.$$

The right $\mathbf{U}$ and left $\bar{\mathbf{v}}$ stretch tensors are independent of rigid body rotations upon Ω. The right and left Cauchy–Green strain tensors are then also independent from such rigid body rotation.

(b) The right $\mathbf{C} = \mathbf{U}^\mathrm{T}\mathbf{U}$ and left $\mathbf{b} = \overline{\mathbf{vv}}^\mathrm{T}$ Cauchy–Green strains determine the expressions

$$\mathbf{E} = (\mathbf{C} - \mathbf{I})/2 = (\mathbf{U}^\mathrm{T}\mathbf{U} - \mathbf{I})/2 \;\; \text{and} \;\; \mathbf{e} = (\mathbf{I} - \mathbf{b}^{-1})/2 = (\mathbf{I} - \bar{\mathbf{v}}^{-T}\bar{\mathbf{v}}^{-1})/2$$

of the Green–Lagrange and Euler–Almansi strains, respectively. The tensors $\mathbf{U}$ and $\bar{\mathbf{v}}$ are independent from rigid body rotations and so are $\mathbf{E}$ and $\mathbf{e}$. ∎

Example 3.5 (Simple and Pure Shear Deformation Kinematics).

(a) Given the principal stretches $\lambda_1 = \lambda$ and $\lambda_3 = 1$, the incompressibility condition $\lambda_1\lambda_2\lambda_3 = 1$ allows us to define the remaining principal stretch by $\lambda_2 = \lambda^{-1}$. With respect to the coordinate system $\{\hat{\mathbf{e}}_1, \hat{\mathbf{e}}_2, \hat{\mathbf{e}}_3\}$, pure shear kinematics are therefore determined by the motion

$$\hat{\chi}_{\text{ps}\,1} = \lambda X_1 \; ; \quad \hat{\chi}_{\text{ps}\,2} = X_2/\lambda \; ; \quad \hat{\chi}_{\text{ps}\,3} = X_3. \tag{3.15}$$

The deformation gradient then reads

$$\hat{\mathbf{F}}_{\text{ps}} = \text{Grad}\hat{\chi}_{\text{ps}} = \begin{bmatrix} \lambda & 0 & 0 \\ 0 & \lambda^{-1} & 0 \\ 0 & 0 & 1 \end{bmatrix}$$

and reflects the property $\det\hat{\mathbf{F}}_{\text{ps}} = 1$.

(b) Deducing from Fig. 3.7c, the motion

$$\chi_{\text{ss}\,1} = X_1 + \gamma X_2 \; ; \quad \chi_{\text{ss}\,2} = X_2 \; ; \quad \chi_{\text{ss}\,3} = X_3$$

expresses simple shear with respect to the coordinate system $\{\mathbf{e}_1, \mathbf{e}_2, \mathbf{e}_3\}$. The deformation gradient then reads

$$\mathbf{F}_{ss} = \text{Grad} \chi_{ss} = \begin{bmatrix} 1 & \gamma & 0 \\ 0 & 1 & 0 \\ 0 & 0 & 1 \end{bmatrix} \qquad (3.16)$$

with the property $\det \mathbf{F}_{ss} = 1$.

(c) The rotation tensor

$$\mathbf{R} = \begin{bmatrix} \widehat{\mathbf{e}}_1 \cdot \mathbf{e}_1 & \widehat{\mathbf{e}}_1 \cdot \mathbf{e}_2 & \widehat{\mathbf{e}}_1 \cdot \mathbf{e}_3 \\ \widehat{\mathbf{e}}_2 \cdot \mathbf{e}_1 & \widehat{\mathbf{e}}_2 \cdot \mathbf{e}_2 & \widehat{\mathbf{e}}_2 \cdot \mathbf{e}_3 \\ \widehat{\mathbf{e}}_3 \cdot \mathbf{e}_1 & \widehat{\mathbf{e}}_3 \cdot \mathbf{e}_2 & \widehat{\mathbf{e}}_3 \cdot \mathbf{e}_3 \end{bmatrix} = \begin{bmatrix} 1/\sqrt{2} & 1/\sqrt{2} & 0 \\ -1/\sqrt{2} & 1/\sqrt{2} & 0 \\ 0 & 0 & 1 \end{bmatrix}$$

allows us to rotate the coordinate system $\{\widehat{\mathbf{e}}_1, \widehat{\mathbf{e}}_2, \widehat{\mathbf{e}}_3\}$ into $\{\mathbf{e}_1, \mathbf{e}_2, \mathbf{e}_3\}$. The description of the pure shear motion $\chi_{ps}(\mathbf{X})$ within the rotated coordinate system $\{\mathbf{e}_1, \mathbf{e}_2, \mathbf{e}_3\}$ requires the rotation of the argument as well as the function itself, $\chi_{ps}(\mathbf{X}) = \mathbf{R}^T \widehat{\chi}_{ps}(\mathbf{RX})$. These operations may be interpreted as follows: The position $\mathbf{X}$ is first rotated to $\widehat{\mathbf{X}} = \mathbf{RX}$, where the function $\widehat{\chi}_{ps}(\widehat{\mathbf{X}})$ is evaluated within the principal stretch coordinates system $\{\widehat{\mathbf{e}}_1, \widehat{\mathbf{e}}_2, \widehat{\mathbf{e}}_3\}$, and the result is then rotated back to the coordinate system $\{\mathbf{e}_1, \mathbf{e}_2, \mathbf{e}_3\}$, $\chi_{ps} = \mathbf{R}^T \widehat{\chi}_{ps}$.

The components of $\widehat{\mathbf{X}} = \mathbf{RX}$ then read

$$\widehat{X}_1 = (X_1 + X_2)/\sqrt{2} \; ; \quad \widehat{X}_2 = (X_2 - X_1)/\sqrt{2} \; ; \quad \widehat{X}_3 = X_3 \; , \qquad (3.17)$$

and the kinematics (3.15) together with the rotation $\chi_{ps} = \mathbf{R}^T \widehat{\chi}_{ps}$ yield

$$\chi_{ps\,1} = \frac{X_1 - X_2 + (X_1 + X_2)\lambda^2}{2\lambda} \; ; \quad \chi_{ps\,2} = \frac{X_2 - X_1 + (X_1 + X_2)\lambda^2}{2\lambda} \; ; \quad \chi_{ps\,3} = X_3 \; ,$$

of pure shear with respect to $\{\mathbf{e}_1, \mathbf{e}_2, \mathbf{e}_3\}$. The deformation gradient then reads

$$\mathbf{F}_{ps} = \text{Grad} \chi_{ps} = \frac{1}{2\lambda} \begin{bmatrix} 1 + \lambda^2 & \lambda^2 - 1 & 0 \\ \lambda^2 - 1 & 1 + \lambda^2 & 0 \\ 0 & 0 & 1 \end{bmatrix} \qquad (3.18)$$

and has the property $\det \mathbf{F}_{ps} = 1$.

(d) With the deformation gradients (3.16) and (3.18), the right Cauchy–Green strains

$$\mathbf{C}_{ss} = \mathbf{F}_{ss}^T \mathbf{F}_{ss} = \begin{bmatrix} 1 & \gamma & 0 \\ \gamma & 1 + \gamma^2 & 0 \\ 0 & 0 & 1 \end{bmatrix} \; ; \quad \mathbf{C}_{ps} = \mathbf{F}_{ps}^T \mathbf{F}_{ps} = \frac{1}{2\lambda^2} \begin{bmatrix} 1 + \lambda^4 & \lambda^4 - 1 & 0 \\ \lambda^4 - 1 & 1 + \lambda^4 & 0 \\ 0 & 0 & 1 \end{bmatrix}$$

express simple shear and pure shear kinematics, respectively. The linear expansion of these expressions (for simple shear at $\gamma = 0$, and for pure shear at $\lambda = 1$) yields for both cases the right Cauchy–Green strain

$$\mathbf{C}_{\text{ss lin}} = \mathbf{C}_{\text{ps lin}} = \begin{bmatrix} 1 & \gamma & 0 \\ \gamma & 1 & 0 \\ 0 & 0 & 1 \end{bmatrix},$$

where $\lambda = 1 + \gamma/2$ has been used to substitute the pure shear parameter λ.

(e) The right polar decomposition theorem allows us to express the deformation gradient of pure shear through $\mathbf{F}_{\text{ps}} = \mathbf{R}_{\text{ps}}\mathbf{U}_{\text{ps}}$, where $\mathbf{U}_{\text{ps}}$ and $\mathbf{R}_{\text{ps}}$ denote the right stretch tensor and rigid body rotation, respectively. In addition, we use the eigenvalue representation $\mathbf{U}_{\text{ps}} = \widehat{\lambda}_{\text{ps}\,i}\widehat{\mathbf{N}}_{\text{ps}\,i} \otimes \widehat{\mathbf{N}}_{\text{ps}\,i}; i = 1, 2, 3$ of the right stretch tensor, where $\widehat{\lambda}_{\text{ps}\,i}^2$ and $\widehat{\mathbf{N}}_{\text{ps}\,i}$ denote the eigenvalues and eigenvectors of the right Cauchy–Green strain $\mathbf{C}_{\text{ps}} = \mathbf{F}_{\text{ps}}^{\text{T}}\mathbf{F}_{\text{ps}} = \mathbf{U}_{\text{ps}}^{\text{T}}\mathbf{U}_{\text{ps}}$, respectively. Here, the eigenvalues

$$\widehat{\lambda}_{\text{ps}\,1} = 1 \; ; \; \widehat{\lambda}_{\text{ps}\,2} = 1/\lambda^2 \; ; \; \widehat{\lambda}_{\text{ps}\,3} = \lambda^2$$

and eigenvectors

$$\widehat{\mathbf{N}}_{\text{ps}\,1} = \widehat{\mathbf{e}}_3 = \begin{bmatrix} 0 \\ 0 \\ 1 \end{bmatrix} \; ; \; \widehat{\mathbf{N}}_{\text{ps}\,2} = \widehat{\mathbf{e}}_1 = \begin{bmatrix} -1/\sqrt{2} \\ 1/\sqrt{2} \\ 0 \end{bmatrix} \; ; \; \widehat{\mathbf{N}}_{\text{ps}\,3} = \widehat{\mathbf{e}}_2 = \begin{bmatrix} 1/\sqrt{2} \\ 1/\sqrt{2} \\ 0 \end{bmatrix},$$

are the principal stretches and principal stretch directions that have been used to express $\widehat{\boldsymbol{\chi}}_{\text{ps}}$ in Task (a). The right stretch tensor and its inverse then read

$$\mathbf{U}_{\text{ps}} = \frac{1}{2}\begin{bmatrix} 1/\lambda + \lambda & \lambda^2 - 1/\lambda & 0 \\ \lambda - 1/\lambda & 1/\lambda + \lambda & 0 \\ 0 & 0 & 2 \end{bmatrix} \; ; \; \mathbf{U}_{\text{ps}}^{-1} = \frac{1}{2\lambda}\begin{bmatrix} 1 + \lambda^2 & \lambda^2 - 1 & 0 \\ \lambda^2 - 1 & 1 + \lambda^2 & 0 \\ 0 & 0 & 2\lambda \end{bmatrix},$$

and allow us to prove that pure shear kinematics is free of rigid body rotation. Given the deformation gradient (3.18), the property $\mathbf{R}_{\text{ps}} = \mathbf{F}_{\text{ps}}\mathbf{U}_{\text{ps}}^{-1} = \mathbf{I}$ holds, which in turn also explains the nomenclature "pure shear."

Given simple shear kinematics, $\mathbf{R}_{\text{ss}} = \mathbf{F}_{\text{ss}}\mathbf{U}_{\text{ss}}^{-1}$ expresses the rigid body rotation. The corresponding matrix operations lead to a very long expression, which is not shown here. We consider instead the particular case of $\gamma = 0.3$ that leads to the rotation

$$\mathbf{R}_{\text{ss}} = \begin{bmatrix} 0.988936 & 0.14834 & 0 \\ -0.14834 & 0.988936 & 0 \\ 0 & 0 & 1 \end{bmatrix},$$

where the deformation gradient (3.16) has been used. The rotation $\mathbf{R}_{\text{ss}}$ differs from the identity tensor, and simple shear kinematics therefore leads to superimposed rigid body rotation on top of the shear deformation. ∎

Example 3.6 (Symmetry of the Cauchy Stress Tensor).

(a) The sum of all forces in the $\mathbf{e}_1$ direction and the $\mathbf{e}_2$ direction read $(\sigma_{11} - \sigma_{11})\mathrm{d}x_1$ and $(\sigma_{22} - \sigma_{22})\mathrm{d}x_2$, respectively. Both expressions are zero, and linear equilibrium therefore holds.
(b) Angular equilibrium requires the sum of all moments to disappear. The moments taken around the left bottom corner of the material particle (shown by the dot in Fig. 3.9) read

$$-\sigma_{21}\mathrm{d}x_1\mathrm{d}x_2 + \sigma_{12}\mathrm{d}x_1\mathrm{d}x_2 + \frac{(\sigma_{11} - \sigma_{11})\mathrm{d}x_2\mathrm{d}x_1}{2} + \frac{(\sigma_{22} - \sigma_{22})\mathrm{d}x_2\mathrm{d}x_1}{2},$$

an expression that is zero for $\sigma_{21} = \sigma_{12}$. The Cauchy stress has therefore to be symmetric, $\boldsymbol{\sigma} = \boldsymbol{\sigma}^{\mathrm{T}}$. ∎

Example 3.7 (Cauchy Stress State in 2D).

(a) The stress components $\sigma_{11}, \sigma_{22}, \sigma_{12}$ with respect to the coordinate system $\{\mathbf{e}_1, \mathbf{e}_2\}$ are shown in Fig. 3.14a.
(b) Given the pairs $(-5, 3)$ [MPa] and $(10, 3)$ [MPa] of normal and shear stresses, Mohr's stress circle can be drawn, see Fig. 3.14b. It has the center

$$c = \frac{\sigma_{11} + \sigma_{22}}{2} = \frac{-5 + 10}{2} = 2.5 \text{ MPa},$$

and the radius

$$r = \sqrt{(\sigma_{22} - c)^2 + \sigma_{12}^2} = \sqrt{(10 - 2.5)^2 + 3^2} = 8.078 \text{ MPa}.$$

Thus,

$$\sigma_{\max} = \sigma_1 = c + r = 2.5 + 8.078 = 10.578 \text{ MPa},$$

$$\sigma_{\min} = \sigma_2 = c - r = 2.5 - 8.078 = -5.578 \text{ MPa},$$

are the extremal normal stresses, whilst

$$\tau_{\max} = +R = 8.078 \text{ MPa};$$

$$\tau_{\min} = -R = -8.078 \text{ MPa}$$

are the extremal shear stresses, respectively. As seen by Mohr's stress circle, at the extremal normal stresses $\sigma_{\max}, \sigma_{\min}$ the shear stress is zero, and therefore said stresses are the principal stresses, $\sigma_1 = \sigma_{\max}$; $\sigma_2 = \sigma_{\min}$.

Fig. 3.14 2D stress state. (a)
Stress components acting at
the faces of the material
particle with respect to the
coordinate system $\{e_1, e_2\}$.
(b) Mohr's stress circle that
corresponds to the stress state
shown in (a). (c) Rotation of
the base vector e_1 into the
first principal stress direction
$\widehat{n}_1$ in the stress space (left)
and physical space (right)

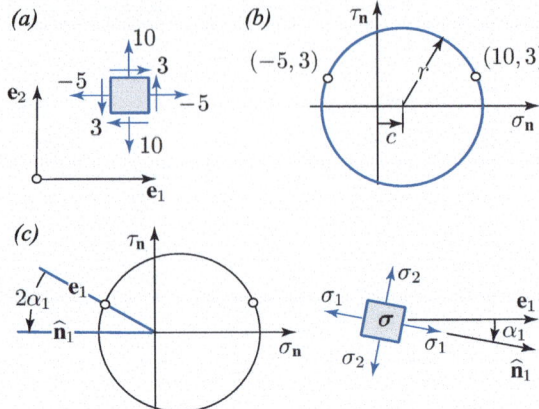

The coordinate base vector e_1 may be rotated into the principal stress
direction $\widehat{n}_1$ according to Fig. 3.14(c). Given the physical space, the coordinate
axes rotate by half the angle and in the opposite direction, as compared to the
rotation in the stress space. The condition

$$2\alpha_1 = \arctan \frac{\sigma_{12}}{c - \sigma_{11}} = \arctan \frac{3}{7.5} = 0.381$$

follows from Mohr's stress diagram, and the first principal stress direction $\widehat{n}_1$
appears rotated at $\alpha_1 = 0.190$ rad against the first coordinate base vector e_1, see
in Fig. 3.14c. In Mohr's diagram, the second principal stress direction $\widehat{n}_2$ appears
π [rad] rotated against $\widehat{n}_1$, and in the physical space $\widehat{n}_2$ is then perpendicular to
$\widehat{n}_1$.

(c) The principal stresses $\sigma_i; i = 1, 2$ are the solutions of the real and symmetric
eigenvalue problem $(\sigma - \sigma I)\widehat{n} = 0$. They are the roots of the characteristic
equation

$$\det[\sigma - \sigma I] = \det \begin{bmatrix} -5 - \sigma & 3 \\ 3 & 10 - \sigma \end{bmatrix} = \sigma^2 - 5\sigma - 59 = 0 \,,$$

and thus

$$\sigma_1 = 5/2 + \sqrt{(5/2)^2 + 59} = 10.578 \text{ MPa} \,;$$

$$\sigma_2 = 5/2 - \sqrt{(5/2)^2 + 59} = -5.578 \text{ MPa} \,.$$

The substitution of said roots in the eigenvalue problem $(\sigma - \sigma I)\widehat{n} = 0$ yields
the two linear systems

$$\begin{bmatrix} -15.578 & 3 \\ 3 & 0.578 \end{bmatrix}\begin{bmatrix} \hat{n}_{11} \\ \hat{n}_{12} \end{bmatrix} = \begin{bmatrix} 0 \\ 0 \end{bmatrix} \quad ; \quad \begin{bmatrix} 0.578 & 3 \\ 3 & 15.578 \end{bmatrix}\begin{bmatrix} \hat{n}_{21} \\ \hat{n}_{22} \end{bmatrix} = \begin{bmatrix} 0 \\ 0 \end{bmatrix}$$

of equations that determine the corresponding principal stress directions

$$\hat{\mathbf{n}}_1 = \begin{bmatrix} 0.982 \\ -0.189 \end{bmatrix} \quad ; \quad \hat{\mathbf{n}}_2 = \begin{bmatrix} 0.189 \\ 0.982 \end{bmatrix} . \tag{3.26}$$

(d) The rotation matrix

$$\mathbf{R} = [\hat{\mathbf{n}}_1 \ \hat{\mathbf{n}}_2]^{\mathrm{T}} = \begin{bmatrix} \hat{\mathbf{n}}_1 \cdot \mathbf{e}_1 & \hat{\mathbf{n}}_1 \cdot \mathbf{e}_2 \\ \hat{\mathbf{n}}_2 \cdot \mathbf{e}_1 & \hat{\mathbf{n}}_2 \cdot \mathbf{e}_2 \end{bmatrix} = \begin{bmatrix} \cos\alpha & \sin\alpha \\ -\sin\alpha & \cos\alpha \end{bmatrix} = \begin{bmatrix} 0.982 & -0.189 \\ 0.189 & 0.982 \end{bmatrix}$$

links the coordinate systems $\{\mathbf{e}_1, \mathbf{e}_2\}$ and $\{\hat{\mathbf{n}}_1, \hat{\mathbf{n}}_2\}$; it represents a rotation by the angle

$$\alpha = \arcsin(-0.189) = -0.190 \text{ rad} .$$

Figure 3.15 illustrates this rotation, and further details regarding the construction of $\mathbf{R}$ are given in the Sect. A.5.6.

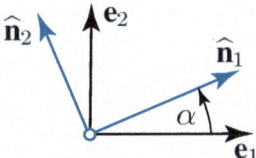

Fig. 3.15 Rotation of the coordinate system $\{\mathbf{e}_1, \mathbf{e}_2\}$ into the principal stress coordinate system $\{\hat{\mathbf{n}}_1, \hat{\mathbf{n}}_2\}$. The rotation matrix $\mathbf{R}(\alpha)$ specifies the transformation between the two Cartesian coordinate systems

(e) Whilst the stress state $\boldsymbol{\sigma}$ remains unchanged by a change of the coordinate system, the stress components σ_{ij} appear differently in different coordinate systems. Given the present example, the two matrices

$$\begin{bmatrix} -5 & 3 \\ 3 & 10 \end{bmatrix} \text{[MPa]} \quad \text{and} \quad \begin{bmatrix} -5.578 & 0 \\ 0 & 10.578 \end{bmatrix} \text{[MPa]}$$

represented the same stress, but with respect to the two coordinate systems $\{\mathbf{e}_1, \mathbf{e}_2\}$ and $\{\hat{\mathbf{n}}_1, \hat{\mathbf{n}}_2\}$. ∎

Example 3.8 (Octahedral Stress and von Mises Stress).

(a) Given the stress $\boldsymbol{\sigma}$, its first and second invariants read

$$I_1 = \mathrm{tr}\boldsymbol{\sigma} = \sigma_{11} + \sigma_{22} + \sigma_{33} = 10 + 25 - 14 = 21 \text{ MPa} ;$$

$$I_2 = \frac{1}{2}\left[(\mathrm{tr}\boldsymbol{\sigma})^2 - \mathrm{tr}\boldsymbol{\sigma}^2\right]$$

$$= \sigma_{11}\sigma_{22} + \sigma_{11}\sigma_{33} + \sigma_{22}\sigma_{33} - \sigma_{12}^2 - \sigma_{13}^2 - \sigma_{23}^2$$

$$= 10 \cdot 25 - 10 \cdot 14 - 25 \cdot 14 - 5^2 - 0^2 - 5^2 = -290 \text{ MPa}^2 ,$$

and allow us to compute the second invariant

$$J_2 = -I_1/3 + I_2 = -21/3 - 290 = -297 \text{ MPa}^2$$

of the deviatoric stress $\overline{\boldsymbol{\sigma}} = \mathrm{dev}\boldsymbol{\sigma}$.

(b) Given the invariants I_1 and J_2,

$$\sigma_{\mathrm{oct}} = I_1/3 = 7.0 \text{ MPa} ; \tau_{\mathrm{oct}} = \sqrt{-\frac{2}{3}J_2} = \sqrt{\frac{2}{3}297} = 14.07 \text{ MPa}$$

are the octahedral normal and shear stresses, whilst

$$\sigma_M = \sqrt{-3J_2} = \sqrt{3 \cdot 297} = 29.85 \text{ MPa}$$

expresses the material particle's von Mises stress. ∎

Example 3.9 (Octahedral and von Mises Stresses of Basic Stress States).

(a) The Cauchy stresses of uniaxial tension $\boldsymbol{\sigma}_{\mathrm{ut}}$, equi-biaxial tension $\boldsymbol{\sigma}_{\mathrm{bt}}$, and simple shear $\boldsymbol{\sigma}_{\mathrm{ss}}$ read

$$\boldsymbol{\sigma}_{\mathrm{ut}} = \begin{bmatrix} \sigma & 0 & 0 \\ 0 & 0 & 0 \\ 0 & 0 & 0 \end{bmatrix} ; \quad \boldsymbol{\sigma}_{\mathrm{bt}} = \begin{bmatrix} \sigma & 0 & 0 \\ 0 & \sigma & 0 \\ 0 & 0 & 0 \end{bmatrix} ; \quad \boldsymbol{\sigma}_{\mathrm{ss}} = \begin{bmatrix} 0 & \tau & 0 \\ \tau & 0 & 0 \\ 0 & 0 & 0 \end{bmatrix} , \tag{3.29}$$

where σ and τ are the respective normal and shear stress components that parameterize the stress states.

(b) With the von Mises stress definition $(3.28)_2$, the states (3.29) yield $\sigma_{\mathrm{M\,ut}} = \sigma$, $\sigma_{\mathrm{M\,bt}} = \sigma$, and $\sigma_{\mathrm{M\,ss}} = \sqrt{3}\tau$ for uniaxial tension, equi-biaxial tension, and simple shear, respectively. The corresponding octahedral stresses are $\tau_{\mathrm{oct\,ut}} = \sqrt{2}\sigma/3$, $\tau_{\mathrm{oct\,bt}} = \sqrt{2}\sigma/3$, and $\tau_{\mathrm{oct\,ss}} = \sqrt{6}\tau/3$. ∎

Example 3.10 (Stress Measures at Finite Deformations).

(a) The motion $\boldsymbol{\chi}(\mathbf{X})$ defines the deformation gradient $\mathbf{F}$, volume ration J, and inverse deformation gradient $\mathbf{F}^{-1}$ according to

$$\mathbf{F} = \mathrm{Grad}\chi\,(\mathbf{X}) = \begin{bmatrix} 3 & 1 \\ 0 & 1.3 \end{bmatrix} ; \ J = \det\mathbf{F}{=}3.9 ; \ \mathbf{F}^{-1}{=} \begin{bmatrix} 0.33333 & -0.25641 \\ 0 & 0.76923 \end{bmatrix}.$$

(b) The first Piola transform (3.31) then yields the expression

$$\mathbf{P} = J\boldsymbol{\sigma}\mathbf{F}^{-\mathrm{T}}$$

$$= 3.9 \begin{bmatrix} 1 & 5 \\ 5 & -10 \end{bmatrix} \begin{bmatrix} 0.33333 & 0 \\ -0.25641 & 0.76923 \end{bmatrix} = \begin{bmatrix} -3.7 & 15 \\ 16.5 & -30 \end{bmatrix} \text{ [MPa]}$$

for the first Piola–Kirchhoff stress. It is second-order two-point tensor.
The second Piola transform (3.32) yields the expression

$$\mathbf{S} = J\mathbf{F}^{-1}\boldsymbol{\sigma}\mathbf{F}^{-\mathrm{T}} = \mathbf{F}^{-1}\mathbf{P}$$

$$= \begin{bmatrix} 0.33333 & -0.25641 \\ 0 & 0.76923 \end{bmatrix} \begin{bmatrix} -3.7 & 15 \\ 16.5 & -30 \end{bmatrix} = \begin{bmatrix} -5.4641 & 12.6923 \\ 12.6923 & -23.0769 \end{bmatrix} \text{ [MPa]}$$

for the second Piola–Kirchhoff stress. It is a second-order one-point tensor that
is symmetric $\mathbf{S} = \mathbf{S}^{\mathrm{T}}$. ∎

Example 3.11 (The Physical Meaning of the Rate of Deformation Tensor).

(a) The simple shear motion (3.39) defines the deformation gradient $\mathbf{F}$, its inverse
$\mathbf{F}^{-1}$, and its material time derivative $\dot{\mathbf{F}}$ according to

$$\mathbf{F} = \mathrm{Grad}\chi = \begin{bmatrix} 1 & \gamma t \\ 0 & 1 \end{bmatrix} ; \ \mathbf{F}^{-1} = \begin{bmatrix} 1 & -\gamma t \\ 0 & 1 \end{bmatrix} ; \ \dot{\mathbf{F}} = \frac{\partial\chi}{\partial t} = \begin{bmatrix} 0 & \gamma \\ 0 & 0 \end{bmatrix}.$$

The velocity gradient (3.34) then reads

$$\mathbf{l} = \dot{\mathbf{F}}\mathbf{F}^{-1} = \begin{bmatrix} 0 & \gamma \\ 0 & 0 \end{bmatrix},$$

and the symmetric and skew-symmetric parts of $\mathbf{l}$ define the rate of deformation
tensor $\mathbf{d} = (\mathbf{l}+\mathbf{l}^{\mathrm{T}})/2$ and the spin tensor $\mathbf{w} = (\mathbf{l}-\mathbf{l}^{\mathrm{T}})/2$, respectively. They read

$$\mathbf{d} = \begin{bmatrix} 0 & \gamma/2 \\ \gamma/2 & 0 \end{bmatrix} \text{[s}^{-1}\text{]} ; \ \mathbf{w} = \begin{bmatrix} 0 & \gamma/2 \\ -\gamma/2 & 0 \end{bmatrix} \text{[s}^{-1}\text{]}$$

and are functions of the amount of shear γ. Given simple shear, the spin tensor
does not disappear, $\mathbf{w} \neq \mathbf{0}$, and the motion (3.39) is therefore not free of rigid
body rotation $\mathbf{R}$.
(b) With the linear transforms $\mathbf{m} = \mathbf{F}\mathbf{m}_0$ and $\mathbf{n} = \mathbf{F}\mathbf{n}_0$, the expression

$$\frac{d(\mathbf{m} \cdot \mathbf{n})}{dt} = \mathbf{m}_0 \cdot \dot{\mathbf{C}}\mathbf{n}_0 = 2\mathbf{m}_0 \cdot \dot{\mathbf{E}}\mathbf{n}_0 = 2\mathbf{m} \cdot \left(\mathbf{F}^{-\mathrm{T}}\dot{\mathbf{E}}\mathbf{F}^{-1}\right)\mathbf{n} = 2\mathbf{m} \cdot \mathbf{dn}$$

follows, where $\mathbf{m}_0$, $\mathbf{n}_0$ with $|\mathbf{m}_0| = |\mathbf{n}_0| = 1$ are convective vectors in their reference configurations, whilst $\mathbf{C} = \mathbf{F}^{\mathrm{T}}\mathbf{F}$ and $\mathbf{E} = (\mathbf{C} - \mathbf{I})/2$ denote the right Cauchy–Green and the Green–Lagrange strains, respectively. Given no stretch would appear along the $\mathbf{m}$ and $\mathbf{n}$ directions, $|\mathbf{m}| = |\mathbf{n}| = 1$, the rate of deformation $\mathbf{d}$ then determines how fast the angle between the two vectors $\mathbf{m}$ and $\mathbf{n}$ changes—it describes the rate of shearing.

Let us now consider the term $\mathbf{m} \cdot \mathbf{m}$. Its time derivative reads $d(\mathbf{m} \cdot \mathbf{m})/dt = 2\mathbf{m} \cdot \mathbf{dm} = 2\lambda_\mathbf{m}^2 \mathbf{m}/|\mathbf{m}| \cdot \mathbf{dm}/|\mathbf{m}|$, where the stretch $\lambda_\mathbf{m} = |\mathbf{m}|$ has been used. With $d(\mathbf{m} \cdot \mathbf{m})/dt = d\lambda_\mathbf{m}^2/dt = 2\lambda_\mathbf{m}d\lambda_\mathbf{m}/dt = 2\lambda_\mathbf{m}^2 d(\ln \lambda_\mathbf{m})/dt$ it leads to

$$\frac{\mathbf{m}}{|\mathbf{m}|} \cdot \mathbf{d}\frac{\mathbf{m}}{|\mathbf{m}|} = \frac{d(\ln \lambda_\mathbf{m})}{dt} \; ,$$

and the rate of deformation $\mathbf{d}$ therefore determines the change of logarithmic stretch along the direction $\mathbf{m}$.

(c) Given the pair ($\mathbf{m}_0 = [1\ 0]^\mathrm{T}$, $\mathbf{n}_0 = [0\ 1]^\mathrm{T}$) of convective vectors in Ω_0, the term

$$2\mathbf{m} \cdot \mathbf{dn} = 2\mathbf{Fm}_0 \cdot \mathbf{dFn}_0 = 2\begin{bmatrix} 1 & \gamma t \\ 0 & 1 \end{bmatrix}\begin{bmatrix} 1 \\ 0 \end{bmatrix} \cdot \begin{bmatrix} 0 & \gamma/2 \\ \gamma/2 & 0 \end{bmatrix}\begin{bmatrix} 1 & \gamma t \\ 0 & 1 \end{bmatrix}\begin{bmatrix} 0 \\ 1 \end{bmatrix} = \gamma$$

shows that the material shears at the shear rate γ, see Fig. 3.20a. From $\mathbf{m} \cdot \mathbf{dm} = \mathbf{Fm}_0 \cdot \mathbf{dFm}_0 = 0$ we conclude that no stretch appears along $\mathbf{m}$, whilst

$$\mathbf{n} \cdot \mathbf{dn} = 2\lambda_\mathbf{n}(d\lambda_\mathbf{n}/dt) = \mathbf{Fn}_0 \cdot \mathbf{dFn}_0 = \gamma^2 t$$

characterizes the development of stretch along $\mathbf{n}$. Using the initial condition $\lambda_\mathbf{n} = 1$ at $t = 0$, we may integrate this expression, which then yields the stretch $\lambda_\mathbf{n} = \sqrt{1 + \gamma^2 t^2/2}$ as a function of γ and t.

Given the pair ($\mathbf{m}_0 = [\sqrt{2}\ \sqrt{2}]^\mathrm{T}$, $\mathbf{n}_0 = [-\sqrt{2}\ \sqrt{2}]^\mathrm{T}$) of convective vectors in Ω_0, the expression $2\mathbf{m} \cdot \mathbf{dn} = 2\mathbf{Fm}_0 \cdot \mathbf{dFn}_0 = 4\gamma^2 t$ demonstrates that the material is exposed to a shear deformation at the shear rate of $4\gamma^2 t$, see Fig. 3.20b. In addition, the terms

$$\mathbf{m} \cdot \mathbf{dm} = 2\lambda_\mathbf{m}(d\lambda_\mathbf{m}/dt) = 2\gamma(\gamma t + 1) \; ; \quad \mathbf{n} \cdot \mathbf{dn} = 2\lambda_\mathbf{n}(d\lambda_\mathbf{n}/dt) = 2\gamma(\gamma t - 1)$$

indicate stretching along the $\mathbf{m}$ and $\mathbf{n}$ directions, respectively. Their integration gives the stretches $\lambda_\mathbf{m} = 1 + \gamma t$ and $\lambda_\mathbf{n} = 1 - \gamma t$, where the initial conditions $\lambda_\mathbf{m} = \lambda_\mathbf{n} = 1$ at $t = 0$ have been used.

In conclusion, simple shear deformation exposes the material to a combination of shearing and stretching—only at the limit $t = 0$, the stretching tends to zero and the material is then at a pure shear deformation. ∎

Example 3.12 (Coupling Between Material Parameters).

(a) The trigonometric relation $\tan(\pi/4 + \gamma) = (1+\varepsilon)/(1-\nu\varepsilon)$ follows directly from Fig. 3.23b, and the small-strain assumption allows us to approximate the left side by

$$\tan(\pi/4 + \gamma) = \frac{\sin(\pi/4 + \gamma)}{\cos(\pi/4 + \gamma)} \approx \frac{1+\gamma}{1-\gamma}.$$

It then leads to $(1+\gamma)(1-\nu\varepsilon) = (1-\gamma)(1+\varepsilon)$, and

$$2\gamma = (1+\nu)\varepsilon \tag{3.48}$$

determines the relation between ε, γ, and ν, where the second-order small term $\varepsilon\gamma$ has been neglected.

(b) Simple tension yields the relation $\sigma = E\varepsilon$ between the normal stress σ and the normal strain ε. The corresponding deformation is shown in Fig. 3.23c. We may also rotate the material particle by $\pi/4$, and derive the relation $\tau = 2G\gamma$ between the shear stress τ and the shear strain γ, see Fig. 3.23d. At simple tension, $\tau = \sigma/2$ determines the relation between the normal and shear stresses, a condition revealed by, for example, Mohr's stress circle. The substitution of these results in (3.48) then yields

$$E = 2G(1+\nu)$$

and describes the relation amongst Young's modulus E, shear modulus G, and Poisson ratio ν of the linear-elastic material. ∎

Example 3.13 (Hooke Material at Specific Load Cases).

(a) Given simple tension in the $\mathbf{e}_1$ direction, σ_{11} is the only non-vanishing stress component, and $\sigma_{22} = \sigma_{33} = \sigma_{12} = \sigma_{23} = \sigma_{13} = 0$ holds. Hooke's law (3.49) then reads

$$\varepsilon_{11} = \frac{\sigma_{11}}{E}; \ \varepsilon_{22} = -\nu\frac{\sigma_{11}}{E}; \ \varepsilon_{33} = -\nu\frac{\sigma_{11}}{E} \tag{3.52}$$

and expresses the normal strains of a material particle at simple tension.

(b) Given simple shear in the plane that is formed by the $\mathbf{e}_1$ and $\mathbf{e}_2$ directions, σ_{12} is the only non-vanishing stress component and $\sigma_{11} = \sigma_{22} = \sigma_{33} = \sigma_{23} = \sigma_{13} = 0$ holds. Hooke's law (3.49) together with relation (3.46) then yields

$$\varepsilon_{12} = \frac{\sigma_{12}}{2G} \tag{3.53}$$

and expresses the shear strain of a material particle at simple shear.

(c) Given plane stress conditions, the out-of-plane stress components are zero, whilst the out-of-plane strain can freely develop. Let us assume $\mathbf{e}_3$ is the out-of-plane direction, such that $\sigma_{33} = \sigma_{23} = \sigma_{13} = 0$ holds, and Hooke's law (3.49) reads

$$
\begin{bmatrix} \varepsilon_{11} \\ \varepsilon_{22} \\ \varepsilon_{33} \\ \varepsilon_{12} \end{bmatrix} = \frac{1}{E} \underbrace{\begin{bmatrix} 1 & -v & 0 \\ -v & 1 & 0 \\ -v & -v & 0 \\ 0 & 0 & 1+v \end{bmatrix}}_{\text{Compliance matrix}} \begin{bmatrix} \sigma_{11} \\ \sigma_{22} \\ \sigma_{12} \end{bmatrix} .
$$

Whilst the out-of-plane stress disappears, $\sigma_{33} = 0$, the Poison's effect results in an out-of-plane strain $\varepsilon_{33} \neq 0$. We may remove the line that corresponds to ε_{33} from the system, and invert it. The system

$$
\begin{bmatrix} \sigma_{11} \\ \sigma_{22} \\ \sigma_{12} \end{bmatrix} = \frac{E}{1-v^2} \underbrace{\begin{bmatrix} 1 & v & 0 \\ v & 1 & 0 \\ 0 & 0 & 1-v \end{bmatrix}}_{\text{Stiffness matrix}} \begin{bmatrix} \varepsilon_{11} \\ \varepsilon_{22} \\ \varepsilon_{12} \end{bmatrix}
$$

of equations then expresses the stress as function of the strain.

(d) Given plane strain, the out-of-plane strain components are zero, a constraint that in turn leads to out-of-plane stress components. Let us assume $\mathbf{e}_3$ is the out-of-plane direction, such that $\varepsilon_{33} = \varepsilon_{23} = \varepsilon_{13} = 0$ holds, and Hooke's law (3.49) reads

$$
\begin{bmatrix} \sigma_{11} \\ \sigma_{22} \\ \sigma_{33} \\ \sigma_{12} \end{bmatrix} = \frac{E}{(1+v)(1-2v)} \underbrace{\begin{bmatrix} 1-v & v & 0 \\ v & 1-v & 0 \\ v & v & 0 \\ 0 & 0 & 1-2v \end{bmatrix}}_{\text{Stiffness matrix}} \begin{bmatrix} \varepsilon_{11} \\ \varepsilon_{22} \\ \varepsilon_{12} \end{bmatrix} .
$$

We may remove the line that corresponds to the out-of-plane stress σ_{33} from the system, and invert it. The system

$$
\begin{bmatrix} \varepsilon_{11} \\ \varepsilon_{22} \\ \varepsilon_{12} \end{bmatrix} = \frac{1}{E} \underbrace{\begin{bmatrix} 1-v^2 & -v(1+v) & 0 \\ -v(1+v) & 1-v^2 & 0 \\ 0 & 0 & 1+v \end{bmatrix}}_{\text{Compliance matrix}} \begin{bmatrix} \sigma_{11} \\ \sigma_{22} \\ \sigma_{12} \end{bmatrix}
$$

of equations then expresses the strain as function of the stress. ∎

Example 3.14 (Linear Viscoelasticity: Kelvin–Voigt Element).

(a) The governing equation of the Kelvin–Voigt element may be multiplied with the integrating factor $g(t) = \exp[\int (1/\tau)dt] = \exp(t/\tau)$, resulting in

$$\frac{d\varepsilon}{dt} \exp(t/\tau) + \frac{\varepsilon}{\tau} \exp(t/\tau) = \frac{d}{dt}\left[\varepsilon \exp(t/\tau)\right] = \frac{\sigma}{\tau} \exp(t/\tau).$$

We may now integrate this expression, and

$$\varepsilon(t) = \frac{\sigma}{E} + C \exp(-t/\tau) \tag{3.70}$$

then determines the development of the strain of the Kelvin–Voigt element.

(b) Given a creep test, the stress increment $\Delta\sigma$ is applied at the infinitesimally short time period $0 < t < 0^+$. The dashpot then "locks" and the Kelvin–Voigt element does not develop any strain, see Fig. 3.25b. The initial condition $\varepsilon(t = 0^+) = 0$ holds, and the integration constant $C = -\Delta\sigma/E$ can be identified from (3.70). The expression

$$\varepsilon(t) = \Delta\sigma/E[1 - \exp(-t/\tau)] \tag{3.71}$$

then describes the evolution of strain of the creep test. Figure 3.28 illustrates this response of the Kelvin–Voigt element, where the normalized strain $\varepsilon(t)/\Delta\varepsilon$ and the normalized logarithmic strain $\log[1 - \varepsilon(t)/\Delta\varepsilon]$ are shown. The figure also illustrates the physical meaning of the retardation time τ.

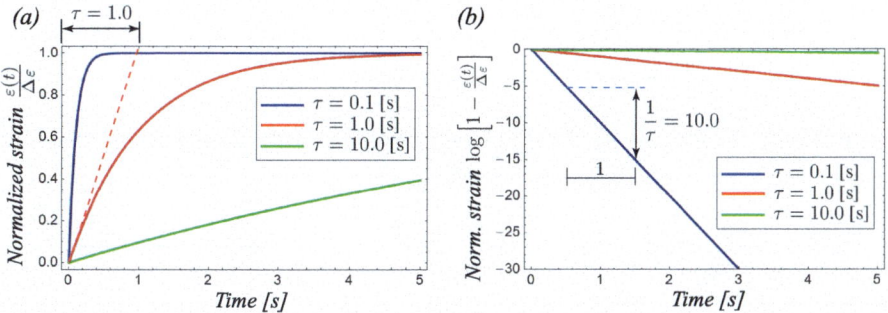

Fig. 3.28 Creep predicted by the Kelvin–Voigt rheology element. (**a**) Normalized strain $\varepsilon(t)/\Delta\varepsilon$ and (**b**) logarithmic normalized strain $\log[1 - \varepsilon(t)/\Delta\varepsilon]$ as a function of the time. The retardation time is denoted by τ, and its physical meaning illustrated in the diagrams

(c) The linearity of the Kelvin–Voigt element allows the superposition of strain responses. Given a discrete stress spectrum, the resulting strain therefore reads

$$\varepsilon(t) = \begin{cases} 0 & ; \quad t < t_0, \\ \Delta\varepsilon_0[1 - \exp(-(t-t_0)/\tau)] & ; \quad t_0^+ \le t < t_1, \\ \Delta\varepsilon_0[1 - \exp(-(t-t_0)/\tau)] + \Delta\varepsilon_1[1 - \exp(-(t-t_1)/\tau)] & ; \quad t_1^+ \le t < t_2, \\ \cdots & \end{cases}$$

a series that approximates the convolution integral

$$\varepsilon(t) = \frac{1}{E} \int_{\xi=-\infty}^{t} \{1 - \exp[-(t-\xi)/\tau]\}\dot{\sigma}\,d\xi \tag{3.72}$$

at the infinitesimally small "strain steps" $d\varepsilon = \dot{\sigma}\,dt/E$. The comparison of Eqs. (3.72) and (3.62) reveals that $\mathcal{J}(x) = [1 - \exp(-x/\tau)]/E$ expresses the creep function of the Kelvin–Voigt element.

(d) At the time interval $0 \le t \le 1$ s the stress rate $\dot{\sigma}(t) = k$ holds, and relation (3.72) then yields the strain

$$\varepsilon(t) = \frac{k}{E} \int_{\xi=0}^{t} \{1 - \exp[-(t-\xi)/\tau]\}d\xi = \frac{k}{E}\{\xi - \tau\exp[-(t-\xi)/\tau]\}_{\xi=0}^{t}$$

$$= \frac{k}{E}\{t - \tau[1 - \exp(-t/\tau)]\} \quad \text{for} \quad 0 \le t \le 1, \tag{3.73}$$

where the initial condition $\varepsilon(0) = 0$ has been used. At $t > 1$ the stress rate $\dot{\sigma}(t) = 0$ holds, and (3.72) then yields the strain

$$\varepsilon(t) = \frac{k}{E} \int_{\xi=0}^{1} \{1 - \exp[-(t-\xi)/\tau]\}d\xi = \frac{k}{E}\{\xi - \tau\exp[-(t-\xi)/\tau]\}_{\xi=0}^{1}$$

$$= \frac{k\tau}{E}\{1 - \tau\exp(-t/\tau)[\exp(-1/\tau) - 1]\} \quad \text{for} \quad 1 < t < \infty.$$

Figure 3.29 plots the normalized strain response of the Kelvin–Voigt rheology element for different retardation times τ.

Fig. 3.29 Development of the normalized strain $\varepsilon(t)$ according to the Kelvin–Voigt rheology element, where E and τ denote the elastic stiffness and the retardation time, respectively. At the time $0 < t < 1.0$ s, the constant stress rate k is prescribed, whilst the stress is constant for $t > 1.0$ s

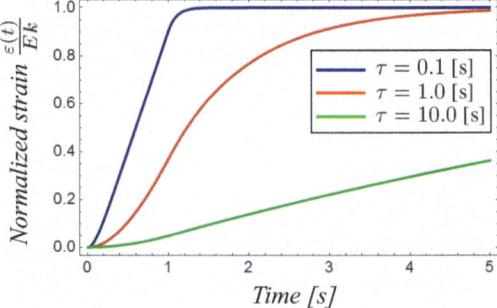

(e) At the limits $\tau \ll t$ and $\tau \gg t$ the relation (3.73) reads

$$\varepsilon(t) = \begin{cases} kt/E = \sigma/E & \text{for } \tau \ll t, \\ kt^2/\eta = \sigma/\eta t & \text{for } \tau \gg t, \end{cases}$$

where the series approximation $\exp(x) = 1 + x + x^2/2! + \cdots$ has been used in the derivation of this expression. Given a very slow process $\tau \ll t$, the Kelvin–Voigt element therefore responds like an elastic spring, whilst for a very fast process $\tau \gg t$, it responds like a viscous fluid. Such properties may also have been deduced from the underlying rheological model, see Fig. 3.25b or the plots in Fig. 3.29 ∎

Example 3.15 (Strain-Based Viscoelastic Generalization of the Incompressible neoHookean Material).

(a) At the thermodynamic limit, the viscous contribution to the free energy disappears, and $\Psi_{\text{iso}}(\mathbf{C}, \mathbf{C}_M = \mathbf{I}) = G(I_1 - 3)/2$ describes the material, where $I_1 = \text{tr}\mathbf{C}$ denotes the first invariant of $\mathbf{C}$. Coleman and Noll's relation of an incompressible material (3.131) allows us then to compute the second Piola–Kirchhoff stress

$$\mathbf{S}_E = 2\frac{\partial \Psi_{\text{iso}}(\mathbf{C})}{\partial \mathbf{C}} - \kappa \mathbf{C}^{-1}, \tag{3.80}$$

where $\mathbf{C} = \mathbf{F}^T\mathbf{F}$ denotes the right Cauchy–Green strain, and the pressure κ serves as Lagrange multiplier to enforce the incompressibility. With the free energy Ψ_{iso}, the relation $\partial \Psi_{\text{iso}}/\partial \mathbf{C} = G/2[\partial(I_1 - 3)/\partial I_1](\partial I_1/\partial \mathbf{C}) = G\mathbf{I}/2$ holds, and the second Piola–Kirchhoff stress contribution from the elastic body in Fig. 3.30 reads $\mathbf{S}_E = G\mathbf{I} - \kappa \mathbf{C}^{-1}$. The second Piola transform of incompressible materials allows us then to compute the Cauchy stress

$$\sigma_E = \mathbf{F}\mathbf{S}_E\mathbf{F}^T = G\mathbf{b} - \kappa\mathbf{I}, \tag{3.81}$$

where $\mathbf{b} = \mathbf{F}\mathbf{F}^T$ denotes the left Cauchy–Green strain.

Given simple tension of an incompressible material, the deformation gradient $\mathbf{F} = \text{Grad}\chi(\mathbf{X}) = \text{diag}[\lambda, \lambda^{-1/2}, \lambda^{-1/2}]$ results in the left Cauchy–Green strain $\mathbf{b} = \mathbf{F}\mathbf{F}^T = \text{diag}[\lambda^2, \lambda^{-1}, \lambda^{-1}]$. The Cauchy stress (3.81) then reads $\sigma = \text{diag}[G\lambda^2 - \kappa, G\lambda^{-1} - \kappa, G\lambda^{-1} - \kappa]$, and the Lagrange pressure κ may be identified from the stress $\sigma_{22} = \sigma_{33} = G\lambda^{-1} - \kappa = 0$ perpendicular to the tensile direction. The only non-trivial Cauchy stress component then reads $\sigma_{11} = G(\lambda^2 - \lambda^{-1})$.

(b) The governing equation (3.79) has the closed-form solution

$$\mathbf{E}_M = \int_{-\infty}^{t} \exp[-(t-\xi)/\tau]\dot{\mathbf{E}}d\xi \, , \qquad (3.82)$$

which allows us to compute the Green–Lagrange strain of the Maxwell spring. Given E_{11} by (3.79) and the definition (3.12),

$$\mathbf{E} = \text{diag}\left[kt, (2kt+1)^{-1/2} - 1, (2kt+1)^{-1/2} - 1\right] \qquad (3.83)$$

denotes the Green–Lagrange strain at $0 \le t \le 1$, whilst

$$\mathbf{E} = \text{diag}\left[k, (2k+1)^{-1/2} - 1, (2k+1)^{-1/2} - 1\right] \qquad (3.84)$$

is the strain at $t > 1$. The corresponding strain rates then read

$$\dot{\mathbf{E}} = k\,\text{diag}\left[1, -(2kt+1)^{-3/2}, -(2kt+1)^{-3/2}\right] \; ; \; \dot{\mathbf{E}} = \mathbf{0}\, , \qquad (3.85)$$

and (3.82) therefore yields the Green–Lagrange strain

$$\mathbf{E}_M = \int_{0}^{t} \exp[-(t-\xi)/\tau]\dot{\mathbf{E}}d\xi = \tau[1 - \exp(-t/\tau)]\dot{\mathbf{E}} \, ,$$

of the Maxwell body at $0 \le t \le 1$ s, and

$$\mathbf{E}_M = \int_{0}^{1} \exp[-(t-\xi)/\tau]\dot{\mathbf{E}}d\xi + \int_{1}^{t} \exp[-(t-\xi)/\tau]\dot{\mathbf{E}}d\xi$$
$$= k^{-1}\tau \exp(-t/\tau)[-1 + \exp(1/\tau)]\dot{\mathbf{E}}$$

at $t > 1$ s. Figure 3.31a illustrates the development of the Maxwell spring's normalized Green–Lagrange strain component $E_{M\,11}/k$ at different relaxation times τ.

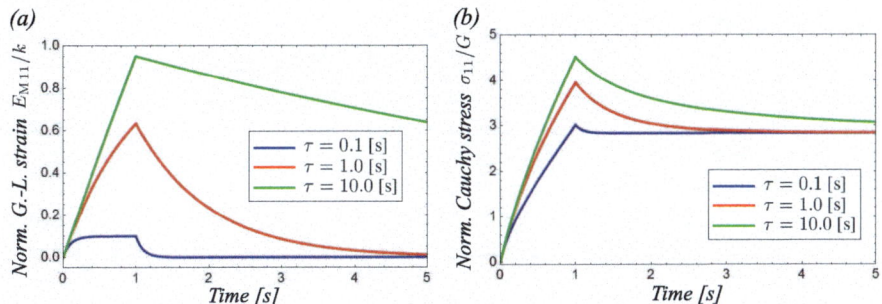

Fig. 3.31 Viscoelastic generalization of the incompressible neoHookean material at simple tension. (**a**) Development of the normalized Green–Lagrange strain component $E_{M\,11}/k$ for different relaxation times τ. (**b**) Development of the normalized Cauchy stress component σ_{11}/G for the parameter $k = 1$, $\beta = 0.6$, and at different relaxation times τ

(c) Coleman and Noll's relation of incompressible materials (3.131) determines the second Piola–Kirchhoff stress

$$\mathbf{S} = 2\frac{\partial \Psi_{\mathrm{iso}}(\mathbf{C}, \mathbf{C}_{\mathrm{M}})}{\partial \mathbf{C}} + 2\frac{\partial \Psi_{\mathrm{iso}}(\mathbf{C}, \mathbf{C}_{\mathrm{M}})}{\partial \mathbf{C}_{\mathrm{M}}} : \frac{\partial \mathbf{C}_{\mathrm{M}}}{\partial \mathbf{C}} - \kappa \mathbf{C}^{-1}$$

$$= G\left(\mathbf{I} + \beta \mathbf{I} : \frac{\partial \mathbf{C}_{\mathrm{M}}}{\partial \mathbf{C}}\right) - \kappa \mathbf{C}^{-1}, \tag{3.86}$$

where the Maxwell body's Green–Lagrange strain $\mathbf{E}_{\mathrm{M}}$ yields the right Cauchy–Green strain $\mathbf{C}_{\mathrm{M}} = 2\mathbf{E}_{\mathrm{M}} + \mathbf{I}$. Given the identity $\partial \mathbf{C}_{\mathrm{M}}/\partial \mathbf{C} = \partial \mathbf{E}_{\mathrm{M}}/\partial \mathbf{E}$ and relation (3.82), the factor $\partial \mathbf{C}_{\mathrm{M}}/\partial \mathbf{C}$ in (3.86) may be expressed by

$$\frac{\partial \mathbf{C}_{\mathrm{M}}}{\partial \mathbf{C}} = \int_{-\infty}^{t} \exp[-(t - \xi)/\tau]\frac{\partial \dot{\mathbf{E}}}{\partial \mathbf{E}}\mathrm{d}\xi, \tag{3.87}$$

such that

$$\mathbf{S} = G\left(\mathbf{I} + \beta \mathbf{I} : \int_{-\infty}^{t} \exp[-(t - \xi)/\tau]\frac{\partial \dot{\mathbf{E}}}{\partial \mathbf{E}}\mathrm{d}\xi\right) - \kappa \mathbf{C}^{-1} \tag{3.88}$$

expresses the second Piola–Kirchhoff stress. The Lagrange pressure κ may then be computed from the equilibrium perpendicular to the tensile direction, $S_{22} = S_{33} = 0$, and the expressions (3.83) to (3.85) provide the relation between the Green–Lagrange strain $\mathbf{E}$ and its rate $\dot{\mathbf{E}}$. At $0 \leq t \leq 1$, it reads

$$\dot{E}_{11} = \frac{E_{11}}{t}; \ \dot{E}_{22} = \frac{kE_{22}}{(1 + 2kt)(\sqrt{1 + 2kt} - 1)}; \ \dot{E}_{33} = \frac{kE_{33}}{(1 + 2kt)(\sqrt{1 + 2kt} - 1)},$$

and

$$\frac{\partial \dot{E}_{11}}{\partial E_{11}} = \frac{1}{t} \; ; \quad \frac{\partial \dot{E}_{22}}{\partial E_{22}} = \frac{\partial \dot{E}_{33}}{\partial E_{33}} = \frac{k}{(1+2kt)(\sqrt{1+2kt}-1)}$$

may therefore be used in the computation of (3.88). At $t > 1$, the strain is constant and $\partial \dot{E}_{11}/\partial E_{11} = \partial \dot{E}_{22}/\partial E_{22} = \partial \dot{E}_{33}/\partial E_{33} = 0$ determines the computation of (3.88).

Given this access, we can evaluate the integral (3.87), compute the second Piola–Kirchhoff stress (3.86), and identify the Lagrange pressure κ from the equation

$$S_{22} = S_{33} = 0 = G \left\{ 1 + \frac{k\beta\tau\alpha}{(1+2kt)(\sqrt{1+2kt}-1)} \right\} + \frac{\kappa}{1 - 2/\sqrt{1+2kt}} \, ,$$

where $\alpha = 1 - \exp(-t/\tau)$ at $0 \le t \le 1$, and $\alpha = \exp(-t/\tau)[\exp(1/\tau) - 1]$ at $t > 1$. Given κ, (3.86) and (3.87) may then be used to compute the stress S_{11} in the tensile direction.

The second Piola transform for incompressible materials defines the Cauchy stress

$$\sigma_{11} = S_{11}(1 + 2kt) = G(1 + 2kt)(1 + \beta\tau\alpha/t) - \kappa$$

in the tensile direction. Figure 3.31b shows the normalized Cauchy stress in tensile direction σ_{11}/G for a number of relaxation times τ. ∎

Example 3.16 (Stress-Based Viscoelastic Generalization of the Incompressible neoHookean Material).

(a) Coleman and Noll's relation (3.131) and the isochoric-volumetric split of the deformation discussed in Sect. 3.5.3.2 yield the elastic second Piola–Kirchhoff stress

$$\mathbf{S}_{\mathrm{E}} = \underbrace{2J^{-2/3}\mathrm{Dev}\left(\frac{\partial \Psi_{\mathrm{iso}}(\mathbf{C})}{\partial \mathbf{C}}\right)}_{\overline{\mathbf{S}}_{\mathrm{E}}} \underbrace{-p\mathbf{C}^{-1}}_{\mathbf{S}_{\mathrm{E\,vol}}}, \tag{3.93}$$

where $\mathbf{C}$ is the right Cauchy–Green strain, and $\mathrm{Dev}(\bullet) = (\bullet) - [\mathbf{C} : (\bullet)]\mathbf{C}^{-1}/3$ denotes the referential deviator operator. The negative hydrostatic pressure p is a Lagrange parameter that enforces the incompressibility.

The neoHookean potential (3.91) determines the relation $\partial \Psi_{\mathrm{iso\,E}}/\partial \mathbf{C} = G/2[\partial (I_1 - 3)/\partial I_1](\partial I_1/\partial \mathbf{C}) = G\mathbf{I}/2$, where the definition $I_1 = \mathrm{tr}\mathbf{C}$ of the strain invariant has been used. The elastic second Piola–Kirchhoff stress then reads

$$\mathbf{S}_{\mathrm{E}} = \overline{\mathbf{S}}_{\mathrm{E}} + \mathbf{S}_{\mathrm{E\,vol}} = G\mathrm{Dev}(\mathbf{I}) - p\mathbf{C}^{-1} \, , \tag{3.94}$$

and the second Piola transform (3.32) for incompressible materials determines the elastic Cauchy stress

$$\sigma_{\mathrm{E}} = \mathbf{F}\mathbf{S}_{\mathrm{E}}\mathbf{F}^{\mathrm{T}} = \underbrace{G\mathrm{dev}(\mathbf{b})}_{\overline{\sigma}_{\mathrm{E}}} \underbrace{-p\mathbf{I}}_{\sigma_{\mathrm{E\,vol}}}, \qquad (3.95)$$

where $\mathbf{b} = \mathbf{F}\mathbf{F}^{\mathrm{T}}$ is the (isochoric) left Cauchy–Green strain, and $\mathrm{dev}(\bullet) = (\bullet) - [\mathbf{I} : (\bullet)]\mathbf{I}/3$ denotes the spatial deviator operator.

Given simple tension of an incompressible material, the deformation gradient $\mathbf{F} = \mathrm{Grad}\boldsymbol{\chi}(\mathbf{X}) = \mathrm{diag}\left[\lambda, \lambda^{-1/2}, \lambda^{-1/2}\right]$ defines the left Cauchy–Green strain $\mathbf{b} = \mathbf{F}\mathbf{F}^{\mathrm{T}} = \mathrm{diag}\left[\lambda^2, \lambda^{-1}, \lambda^{-1}\right]$ with $I_1 = \lambda^2 + 2\lambda^{-1}$. The Cauchy stress (3.95) then reads

$$\sigma_{\mathrm{E}} = G/3\,\mathrm{diag}\left[2(\lambda^2 - \lambda^{-1}), (\lambda^{-1} - \lambda^2), (\lambda^{-1} - \lambda^2)\right] - p\,\mathrm{diag}\,[1, 1, 1]\,.$$

The hydrostatic pressure p may be identified from $\sigma_{\mathrm{E}22} = \sigma_{\mathrm{E}33} = G(\lambda^{-1} - \lambda^2)/3 - p = 0$, and the only non-trivial stress component then reads $\sigma_{\mathrm{E}11} = G(\lambda^2 - \lambda^{-1})$.

(b) Given simple tension at the prescribed stretch (3.92), the right Cauchy–Green strain $\mathbf{C} = \mathbf{F}^{\mathrm{T}}\mathbf{F}$ reads

$$\mathbf{C} = \mathrm{diag}\left[(1 + \zeta)^2, (1 + \zeta)^{-1}, (1 + \zeta)^{-1}\right]$$

with $\zeta = kt$ and $\zeta = k$ at the time intervals $0 \le t \le 1$ and $t > 1$, respectively. The elastic second Piola–Kirchhoff stress (3.94) then reads

$$\mathbf{S}_{\mathrm{E}} = G/3\,\mathrm{diag}\left[2[1 - (1 + \zeta)^{-3}], -\zeta[3 + \zeta(3 + \zeta)], -\zeta[3 + \zeta(3 + \zeta)]\right]$$
$$- p\,\mathrm{diag}\left[(1 + \zeta)^{-2}, (1 + \zeta), (1 + \zeta)\right]. \qquad (3.96)$$

The equilibrium perpendicular to the loading direction, $S_{\mathrm{E}22} = \overline{S}_{\mathrm{E}22} + S_{\mathrm{E\,vol}\,22} = 0$, determines the hydrostatic pressure

$$p = -\frac{G\zeta(3 + 3\zeta + \zeta^2)}{3(1 + \zeta)}, \qquad (3.97)$$

and (3.96) then yields explicit stress expressions.

Towards the derivation of the viscoelastic contribution (3.90), we use the material time derivative of the elastic isochoric second Piola–Kirchhoff stress. Given the corresponding term of (3.96), it reads

$$\dot{\overline{\mathbf{S}}}_{\mathrm{E}} = Gk\,\mathrm{diag}\left[2(1 + \zeta)^{-4}, -(1 + \zeta)^2, -(1 + \zeta)^2\right] \quad \text{and} \quad \dot{\overline{\mathbf{S}}}_{\mathrm{E}} = \mathbf{0}$$

at $0 \leq t \leq 1$ and $t > 1$, respectively. With these stress rates, the convolution integral (3.90) yields the over stress component in tensile direction of

$$\overline{S}_{M\,11}(t) = \frac{2Gk\beta\tau\alpha}{(1+kt)^4} \,,$$

where $\alpha = 1 - \exp(-t/\tau)$ at $0 \leq t \leq 1$, and $\alpha = \exp(-t/\tau)[\exp(1/\tau) - 1]$ at $t > 1$.

Figure 3.32a shows the development of the normalized over stress $\overline{S}_{M\,11}/G$ for a number of relaxation times τ. Given this problem, the governing equation (3.90) determines a decreasing over stress $S_{M\,11}/G$ even during the phase of the extension.

The application of the second Piola transform $\sigma = \mathbf{F}\mathbf{S}\mathbf{F}^{T} = \mathbf{F}(\overline{\mathbf{S}}_{E} + \overline{\mathbf{S}}_{M})\mathbf{F}^{T} - p\mathbf{I}$ for incompressible materials allows us to compute the Cauchy stress component

$$\sigma_{11} = \frac{Gk\{t(1+kt)[3+kt(3+kt)] + 6[1 - \tau\beta\exp(-t/\tau)]\}}{3(1+kt)^2}$$

at the time $0 \leq t \leq 1$, and

$$\sigma_{11} = \frac{Gk}{3(1+k)}\left\{3 + 3k + k^2 + \frac{6\tau\beta\exp(-t/\tau)[\exp(1/\tau) - 1](1+k)^3}{(1+kt)^4}\right\}$$

at the time $t > 1$. Figure 3.32b shows σ_{11}/G for a number of relaxation times τ.

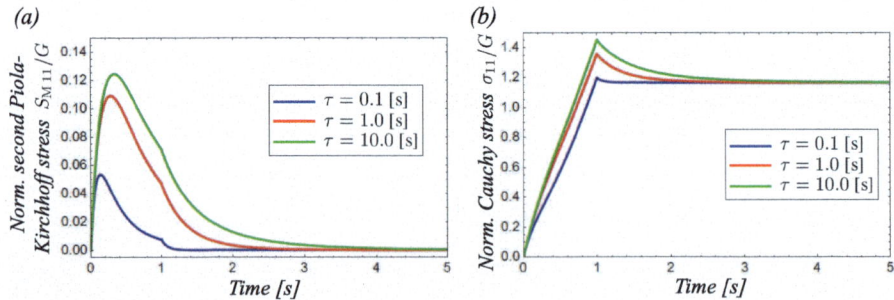

(a)

Norm. second Piola-Kirchhoff stress $\overline{S}_{M\,11}/G$

$\tau = 0.1$ [s]
$\tau = 1.0$ [s]
$\tau = 10.0$ [s]

Time [s]

(b)

Norm. Cauchy stress σ_{11}/G

$\tau = 0.1$ [s]
$\tau = 1.0$ [s]
$\tau = 10.0$ [s]

Time [s]

Fig. 3.32 Viscoelastic generalization of the incompressible neoHookean material at simple tension. (**a**) Development of the over stress in terms of the normalized second Piola–Kirchhoff component $S_{M\,11}/G$. (**b**) Development of the normalized Cauchy stress component σ_{11}/G. The computations used the parameter $k = 1\,\mathrm{s}^{-1}$, $\beta = 0.6$, and different relaxation times τ ∎

Example 3.17 (Continuity of the Incompressible Flow).

(a) The conservation of mass requires the inflow to match the outflow and thus

$$v_1 \Delta x_2 + v_2 \Delta x_1 = \left(v_1 + \frac{\partial v_1}{\partial x_1} \Delta x_1 \right) \Delta x_2 + \left(v_2 + \frac{\partial v_2}{\partial x_2} \Delta x_2 \right) \Delta x_1$$

holds. This relation may be divided by $\Delta x_1 \Delta x_2$ and then yields

$$\frac{\partial v_1}{\partial x_1} + \frac{\partial v_2}{\partial x_2} = 0 . \tag{3.105}$$

The comparison with the definition of the divergence in Sect. A.8.1.2 verifies that (3.105) represents the index notation of the flow continuity (3.104) in 2D Cartesian coordinates.

(b) The conservation of mass requires the inflow to match the outflow and thus

$$v_r r \Delta\theta + v_\theta \Delta r = \left(_r + \frac{\partial v_r}{\partial r} \Delta r \right) (r + \Delta r) \Delta\theta + \left(v_\theta + \frac{\partial v_\theta}{\partial \theta} \Delta\theta \right) \Delta r$$

holds. This relation may be divided by $\Delta r \Delta\theta$, resulting in

$$\frac{\partial v_r}{\partial r} + \frac{v_r}{r} + \frac{1}{r} \frac{\partial v_\theta}{\partial \theta} = \frac{1}{r} \left(\frac{\partial (r v_r)}{\partial r} + \frac{\partial v_\theta}{\partial \theta} \right) = 0 , \tag{3.106}$$

where the term of the order $\mathcal{O}(\Delta r)$ has been neglected. The comparison with the definition of the divergence in Sect. A.8.1.2 verifies that (3.106) represents the index notation of the flow continuity (3.104) in cylindrical coordinates. ∎

Example 3.18 (Equilibrium of the Material Particle in 2D).

(a) Figure 3.36a illustrates the Cauchy stress components, where the linear expansion $\sigma(\mathbf{x} + \Delta\mathbf{x}) = \sigma(\mathbf{x}) + \text{grad}\sigma(\mathbf{x}) : \Delta\mathbf{x}$ has been used.

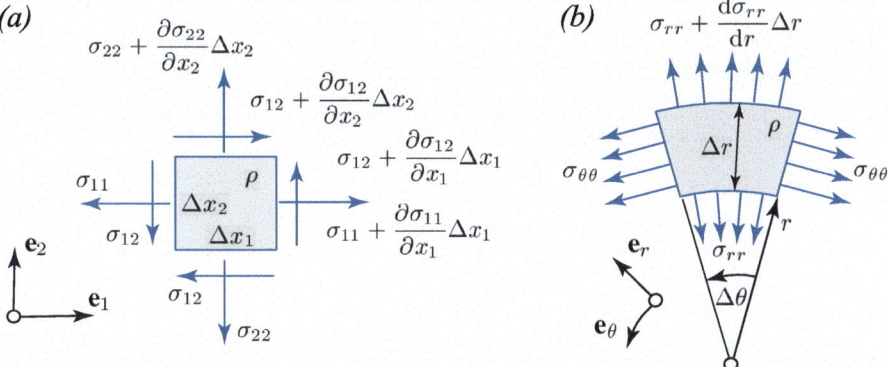

Fig. 3.36 Stresses acting at the material particle in (**a**) Cartesian 2D coordinates $\{\mathbf{e}_1, \mathbf{e}_2\}$ and (**b**) cylindrical coordinates $\{\mathbf{e}_r, \mathbf{e}_\theta\}$ at rotational symmetry

(b) The multiplication of the stress components with the respective edge lengths of
the material particle, and the introduction of the inertial force $\rho \Delta x_1 \Delta x_2 \dot{\mathbf{v}}$, yields
the equilibrium

$$\frac{\partial \sigma_{11}}{\partial x_1} \Delta x_1 \Delta x_2 + \frac{\partial \sigma_{12}}{\partial x_2} \Delta x_2 \Delta x_1 - \rho \Delta x_1 \Delta x_2 \dot{v}_1 = 0 \,,$$

along the $\mathbf{e}_1$ direction, as well as the equilibrium

$$\frac{\partial \sigma_{22}}{\partial x_2} \Delta x_2 \Delta x_1 + \frac{\partial \sigma_{12}}{\partial x_1} \Delta x_1 \Delta x_2 - \rho \Delta x_1 \Delta x_2 \dot{v}_2 = 0 \,,$$

along the $\mathbf{e}_2$ direction. These expressions may be divided by $\Delta x_1 \Delta x_2$, and the
partial differential equations

$$\frac{\partial \sigma_{11}}{\partial x_1} + \frac{\partial \sigma_{12}}{\partial x_2} - \rho \dot{v}_1 = 0 \;\; ; \;\; \frac{\partial \sigma_{22}}{\partial x_2} + \frac{\partial \sigma_{12}}{\partial x_1} - \rho \dot{v}_2 = 0 \qquad (3.109)$$

then determine the balance of linear momentum in 2D Cartesian coordinates.

The comparison with the definition of the divergence in Sect. A.8.1.2 verifies that (3.109) represents the index notation of Cauchy's momentum equation (3.107) in 2D Cartesian coordinates.

(c) At rotational symmetry, the derivatives along the circumferential direction θ,
together with the shear stress and strain in the $r - \theta$ plane disappear. The
conditions $\partial(\bullet)/\partial\theta = 0$ and $\sigma_{r\theta} = \varepsilon_{r\theta} = 0$ therefore hold. Figure 3.36b
shows the components of the rotational symmetric stress that act at the material
particle, where linear expansions of the stress components have been used.

(d) The multiplication of the stress components with the respective edge lengths of
the material particle, and the introduction of the inertial force $\rho r \Delta\theta \Delta r \dot{\mathbf{v}}$, yields
the equilibrium

$$-\sigma_{rr} r \Delta\theta + \left(\sigma_{rr} + \frac{\mathrm{d}\sigma_{rr}}{\mathrm{d}r} \Delta r \right)(r + \Delta R)\, \Delta\theta - 2\sigma_{\theta\theta} \frac{\Delta\theta}{2} \Delta r - \rho r \Delta\theta \Delta r \dot{v}_r = 0$$

along the $\mathbf{e}_r$ direction. It is the only non-trivial equilibrium relation of the
rotational symmetric problem. We may divide this equation by $r \Delta\theta \Delta r$, and
the differential equation

$$\frac{\mathrm{d}\sigma_{rr}}{\mathrm{d}r} - \frac{\sigma_{\theta\theta} - \sigma_{rr}}{r} - \rho \dot{v}_r = 0 \qquad (3.110)$$

then holds at the limit $\Delta\theta, \Delta r \rightarrow 0$. It expresses the balance of linear
momentum in cylindrical coordinates.

Given cylindrical coordinates, Cauchy's equation of motion (3.107) reads

$$\left[\begin{matrix} \frac{\partial \sigma_{rr}}{\partial r} + \frac{1}{r}\frac{\partial \sigma_{r\theta}}{\partial \theta} + \frac{\sigma_{rr} - \sigma_{\theta\theta}}{r} \\ \frac{\partial \sigma_{r\theta}}{\partial r} + \frac{1}{r}\frac{\partial \sigma_{\theta\theta}}{\partial \theta} + \frac{2\sigma_{r\theta}}{r} \end{matrix} \right] - \rho \left[\begin{matrix} \dot{v}_r \\ \dot{v}_\theta \end{matrix} \right] = \left[\begin{matrix} 0 \\ 0 \end{matrix} \right],$$

which, at rotational symmetry $\partial(\bullet)/\partial\theta = 0$ and $\sigma_{r\theta} = u_\theta = 0$, results in the only non-trivial differential equation (3.110). ∎

Example 3.19 (Inflated Thick-Walled Linear-Elastic Cylinder).

(a) Given plane stress $\sigma_{rz} = \sigma_{\theta z} = \sigma_{zz} = 0$, rotational symmetry $\partial(\bullet)/\partial\theta = 0$, and the definition of the divergence in cylindrical coordinates (A.19), Cauchy's momentum equation (3.107) then yields the only non-trivial ordinary differential equation

$$\frac{d\sigma_r}{dr} + \frac{\sigma_r - \sigma_\theta}{r} = 0. \tag{3.113}$$

The substitution of the stresses σ_r and σ_θ through the fundamental solution (3.111) then yields

$$-2a_0/r^3 + 2c_0/r + [2a_0/r^2 + c_0(1 + 2\log r) - c_0(3 + 2\log r)]/r = 0,$$

and proves that (3.111) satisfies Cauchy's momentum equation (3.107).

(b) Given Hooke's law at plane stress, we may substitute the strains $\varepsilon_r = (\sigma_r - \nu\sigma_\theta)/E$ and $\varepsilon_\theta = (\sigma_\theta - \nu\sigma_r)/E$ in the strain compatibility (3.112), which then yields the expression

$$\frac{\partial(r\sigma_\theta - r\nu\sigma_r)}{\partial r} - \sigma_r + \nu\sigma_\theta = 0.$$

The substitution of the stresses σ_r and σ_θ by (3.111), then yields $c = 0$, a condition that always holds for a linear-elastic material at plane stress and rotational symmetry.

Towards the identification of the remaining constants a_0 and b_0, we use the boundary conditions $\sigma_r(r_i) = -p_i$ at the inside, and $\sigma_r(r_o) = 0$ at the outside of the vessel. Given (3.111) and $c = 0$, these conditions yield $-p_i = b_0 + a_0/r_i^2$ and $0 = b_0 + a_0/r_o^2$, and results in $a_0 = p_i(r_i r_o)^2/(r_i^2 - r_o^2)$ and $2b_0 = p_i r_i^2/(r_o^2 - r_i^2)$. The stress in the vessel wall then reads

$$\sigma_r = \frac{r_i^2(r_o^2 - r^2)}{r^2(r_i^2 - r_o^2)} p_i \; ; \; \sigma_\theta = \frac{r_i^2(r^2 + r_o^2)}{r^2(r_o^2 - r_i^2)} p_i, \tag{3.114}$$

a state independent from the material parameters E and ν. Figure 3.38 plots the stresses (3.114) at diastolic p_{id} and systolic p_{is} blood pressures, respectively.

Fig. 3.38 Circumferential σ_θ and radial σ_r stress in the vessel wall at diastolic $p_{id} = 75$ mmHg and systolic $p_{is} = 120$ mmHg blood pressures

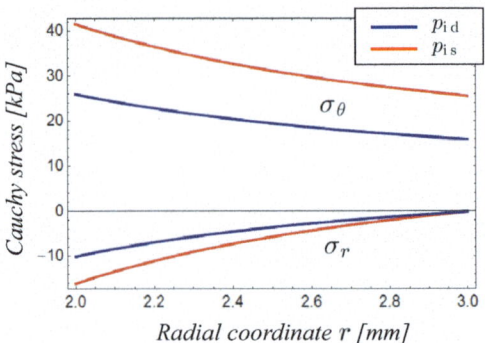

(c) At the outer surface, $r = r_o$, the stress state is uniaxial, $\sigma_{ro} = 0$ and $\sigma_{\theta o} = 2r_i^2 p_i/(r_o^2 - r_i^2) = 8p_i/5$. Hooke's law then yields the circumferential strain $\varepsilon_{\theta o} = \sigma_{\theta o}/E$, and $\Delta r_o = \varepsilon_{\theta o} r_o$ expresses the radial change at the vessel's outside. Between diastolic p_{id} and systolic p_{is} blood pressures, the radius then pulsates by $\Delta r = 0.046$ mm.

At the inner surface, $r = r_i$, the stress state is biaxial, $\sigma_{ri} = -p_i$ and $\sigma_{\theta i} = p_i(r_o^2 + r_i^2)/(r_o^2 - r_i^2) = 13p_i/5$. Hooke's law at plane stress

$$\begin{bmatrix} \varepsilon_{ri} \\ \varepsilon_{\theta i} \end{bmatrix} = \frac{p_i}{E} \begin{bmatrix} 1 & -\nu \\ -\nu & 1 \end{bmatrix} \begin{bmatrix} -1 \\ 13/5 \end{bmatrix}$$

then yields the circumferential strain $\varepsilon_{\theta i} = (\nu + 13/5)p_i/E$, and $\Delta r_i = \varepsilon_{\theta i} r_i$ expresses the radial change at the vessel's inside. The radial pulsatility is then $\Delta r = 0.059$ mm between diastolic p_{id} and systolic p_{is} inflations. ∎

Example 3.20 (Conservation of Energy in Material Description).

(a) With the internal energy U, the term $d\left(\int_{\Omega_{s0}} U dV\right)/dt$ expresses the change of the subdomain's system energy, where dV and Ω_{s0} denotes the referential volume element and the subdomain's reference configuration, respectively.

(b) With the heat flux $\mathbf{Q}_h$ and the heat source R_h, the subdomain's heat input reads

$$\int_{\Omega_{s0}} H_{input} dV = \int_{\Omega_{s0}} R_h dV - \int_{\partial\Omega_{s0}} \mathbf{Q}_h \cdot \mathbf{N} dS = \int_{\Omega_{s0}} (R_h - Div\mathbf{Q}_h) dV,$$

where $\mathbf{N}$ denotes the outward normal vector to $\partial\Omega_{s0}$, whilst dS is the referential area element. We used the divergence theorem to derive this expression, and $Div\mathbf{Q}_h$ denotes the divergence of the heat flux with respect to Ω_0. Given Cartesian coordinates, it reads $\partial Q_{hI}/\partial X_I$, where $\mathbf{X}$ denotes the referential position of the material particle.

(c) Given the definitions of the spatial volume element $dv = JdV$, we may pull-back the last integral in (3.119) to the subdomain's reference configuration Ω_{s0}, and the power input then reads

$$\int_{\partial\Omega_{s0}} p_{\text{input}} \mathrm{d}V = \int_{\Omega_{s0}} \mathbf{P} : \dot{\mathbf{F}} \mathrm{d}V\,,$$

where $\mathbf{P}$ and $\mathbf{F}$ denote the first Piola–Kirchhoff stress and the deformation gradient, respectively. To derive this expression, the first Piola transform (3.31), the definition of the velocity gradient (3.34), and the symmetry of the Cauchy stress have been used.

(d) With the results from Tasks (a) (b) (c), the conservation of energy of the subdomain Ω_{s0} reads

$$\int_{\Omega_{s0}} \left(\frac{\mathrm{d}U}{\mathrm{d}t} + \mathrm{Div}\mathbf{Q}_{\mathrm{h}} - R_{\mathrm{h}} - \mathbf{P} : \dot{\mathbf{F}} \right) \mathrm{d}V = 0\,,$$

an expression that holds for an arbitrary subdomain Ω_{s0}. Localization therefore results in the strong condition $\mathrm{d}U/\mathrm{d}t + \mathrm{Div}\mathbf{Q}_{\mathrm{h}} - R_{\mathrm{h}} - \mathbf{P} : \dot{\mathbf{F}} = 0$ that expresses the first law of thermodynamics at the material particle level within the body's reference configuration Ω_0. ∎

Example 3.21 (The Incompressible neoHookean Material).

(a) Coleman and Noll's relation of an incompressible material (3.133)

$$\boldsymbol{\sigma} = 2\mathbf{F} \frac{\partial \Psi_{\text{iso}}(\mathbf{C})}{\partial \mathbf{C}} \mathbf{F}^{\mathrm{T}} - \kappa \mathbf{I}$$

provides the relation between the Cauchy stress $\boldsymbol{\sigma}$ and the free energy Ψ_{iso} per unit (reference) volume. The Lagrange pressure κ enforces the incompressibility, and $\mathbf{F}$ and $\mathbf{C} = \mathbf{F}^{\mathrm{T}}\mathbf{F}$ denote the deformation gradient and the right Cauchy–Green strain, respectively.

Given the strain invariant $I_1 = \mathrm{tr}\mathbf{C}$ and the neoHookean free energy $\Psi_{\text{iso}} = G(I_1 - 3)/2$, the relation $\partial \Psi_{\text{iso}}/\partial \mathbf{C} = G/2[\partial(I_1 - 3)/\partial I_1](\partial I_1/\partial \mathbf{C}) = G\mathbf{I}/2$ holds, and

$$\boldsymbol{\sigma} = G\mathbf{b} - \kappa \mathbf{I} \tag{3.139}$$

expresses the Cauchy stress of the incompressible neoHookean material. Here, $\mathbf{b} = \mathbf{F}\mathbf{F}^{\mathrm{T}}$ denotes the left Cauchy–Green strain and represents the deformation kinematics.

Given simple tension, $\mathbf{F} = \mathrm{Grad}\chi_{\text{st}}(\mathbf{X}) = \mathrm{diag}[\lambda, \lambda^{-1/2}, \lambda^{-1/2}]$ describes the deformation gradient, and $\mathbf{b} = \mathbf{F}\mathbf{F}^{\mathrm{T}} = \mathrm{diag}[\lambda^2, \lambda^{-1}, \lambda^{-1}]$ expresses then the left Cauchy–Green strain. The Cauchy stress (3.139) therefore reads $\boldsymbol{\sigma} = \mathrm{diag}[G\lambda^2 - \kappa, G\lambda^{-1} - \kappa, G\lambda^{-1} - \kappa]$. The Lagrange pressure κ may be identified from the stress $\sigma_{22} = \sigma_{33} = G\lambda^{-1} - \kappa = 0$ perpendicular to the tensile direction, and $\sigma_{11} = G(\lambda^2 - \lambda^{-1})$ then yields the only non-trivial stress component, see Fig. 3.41a.

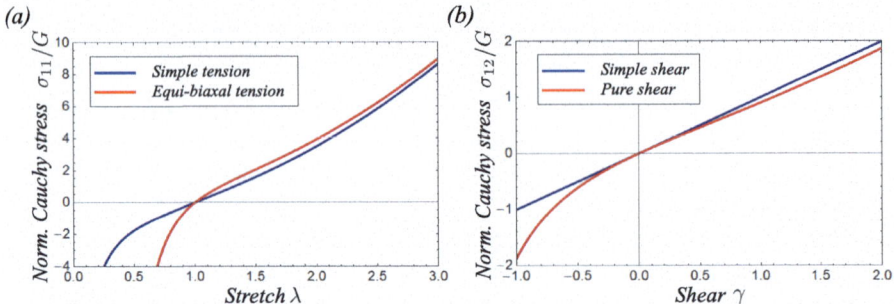

Fig. 3.41 Cauchy stress predictions of the incompressible neoHookean material upon a number of elementary deformation cases. (**a**) Simple tension and equi-biaxal tension. (**b**) Simple shear and pure shear. Given pure shear, the principal stretch $\lambda = 1 + \gamma/2$ has been substituted by the amount of shear γ

(b) Given equi-biaxial tension, $\mathbf{F} = \mathrm{Grad}\boldsymbol{\chi}_{et}(\mathbf{X}) = \mathrm{diag}[\lambda, \lambda, \lambda^{-2}]$ and $\mathbf{b} = \mathrm{diag}[\lambda^2, \lambda^2, \lambda^{-4}]$ denote the deformation gradient and the left Cauchy–Green strain, respectively. The Cauchy stress (3.139) then reads $\boldsymbol{\sigma} = \mathrm{diag}[G\lambda^2 - \kappa, G\lambda^2 - \kappa, G\lambda^{-4} - \kappa]$, and the identification of $\kappa = G\lambda^{-4}$ from $\sigma_{33} = 0$ then yields the two stress components $\sigma_{11} = \sigma_{22} = G(\lambda^2 - \lambda^{-4})$, see Fig. 3.41a

(c) Given simple shear,

$$\mathbf{F} = \begin{bmatrix} 1 & \gamma & 0 \\ 0 & 1 & 0 \\ 0 & 0 & 1 \end{bmatrix} \;\; ; \;\; \mathbf{b} = \begin{bmatrix} 1 + \gamma^2 & \gamma & 0 \\ \gamma & 1 & 0 \\ 0 & 0 & 1 \end{bmatrix}$$

expresses the deformation gradient $\mathbf{F} = \mathrm{Grad}\boldsymbol{\chi}_{ss}(\mathbf{X})$ and the left Cauchy–Green strain $\mathbf{b}$, respectively. The Cauchy stress (3.139) then reads

$$\boldsymbol{\sigma} = G \begin{bmatrix} \gamma^2 & \gamma & 0 \\ \gamma & 0 & 0 \\ 0 & 0 & 0 \end{bmatrix} ,$$

where the normal stress $\sigma_{33} = G - \kappa = 0$ has been used to substitute the Lagrange pressure $\kappa = G$, see Fig. 3.41b.

(d) Given pure shear,

$$\mathbf{F} = \frac{1}{2\lambda} \begin{bmatrix} 1 + \lambda^2 & \lambda^2 - 1 & 0 \\ \lambda^2 - 1 & 1 + \lambda^2 & 0 \\ 0 & 0 & 2\lambda \end{bmatrix} \;\; ; \;\; \mathbf{b} = \frac{1}{2\lambda^2} \begin{bmatrix} 1 + \lambda^4 & \lambda^4 - 1 & 0 \\ \lambda^4 - 1 & 1 + \lambda^4 & 0 \\ 0 & 0 & 2\lambda^2 \end{bmatrix}$$

expresses the deformation gradient $\mathbf{F} = \mathrm{Grad}\boldsymbol{\chi}_{ps}(\mathbf{X})$ and the left Cauchy–Green strain $\mathbf{b}$, respectively. The Cauchy stress (3.139) then reads

$$\sigma = \frac{G}{2\lambda^2} \begin{bmatrix} (\lambda^2 - 1)^2 & \lambda^4 - 1 & 0 \\ \lambda^4 - 1 & (\lambda^2 - 1)^2 & 0 \\ 0 & 0 & 0 \end{bmatrix} ,$$

where the normal stress $\sigma_{33} = G - \kappa = 0$ has been used to substitute the Lagrange pressure $\kappa = G$, see Fig. 3.41b. ∎

Example 3.22 (Equilibrium in Material Description).

(a) The linear momentum equilibrium of the subdomain Ω_s reads $\int_{\partial\Omega_s} \mathbf{t} ds + \int_{\Omega_s} \mathbf{b}_f dv = \mathbf{0}$, where $\mathbf{t}$ denotes the traction vector per unit spatial surface. We may use Cauchy's stress theorem $\mathbf{t} = \sigma \mathbf{n}$ to substitute the traction vector, which then yields

$$\int_{\partial\Omega_s} \sigma \mathbf{n} ds + \int_{\Omega_s} \mathbf{b}_f dv = \mathbf{0} . \tag{3.142}$$

With the divergence theorem (A.8.2), the two integrals may be combined, and $\int_{\Omega_s} (\text{div}\sigma + \mathbf{b}_f) dv = \mathbf{0}$ expresses the linear momentum equilibrium of the subdomain. Localization then yields

$$\text{div}\sigma + \mathbf{b}_f = \mathbf{0} ,$$

in accordance to Cauchy's static momentum equation (3.107).

(b) Given Nanson's formula (3.5) and the first Piola transform (3.31), the equilibrium relation (3.142) reads

$$\int_{\partial\Omega_{s0}} \mathbf{PN} dS + \int_{\Omega_{s0}} \mathbf{B}_f dV = \mathbf{0} , \tag{3.143}$$

where $dV = J^{-1} dv$ denotes the reference volume element, whilst $\mathbf{B}_f = J\mathbf{b}_f$ expresses the body force per unit reference volume. The divergence theorem (A.8.2) allows us to express the equilibrium of the subdomain through $\int_{\Omega_{s0}} (\text{Div}\mathbf{P} + \mathbf{B}_f) dV = \mathbf{0}$, and localization then yields

$$\text{Div}\mathbf{P} + \mathbf{B}_f = \mathbf{0}$$

in accordance to Cauchy's momentum equation (3.142) with respect to the reference configuration. Here, $\text{Div}\mathbf{P}$ denotes the divergence of the first Piola–Kirchhoff stress with respect to Ω_0. Given Cartesian coordinates, it reads $\partial P_{IJ}/\partial X_J$, where $\mathbf{X}$ denotes the referential position of the material particle. ∎

Example 3.23 (The Inflated Thin-Walled Linear-Elastic Circular Tube).

(a) The vessel may be sectioned longitudinally to free the circumferential stress σ_θ, as shown in Fig. 3.44a, and transversally to free the axial stress σ_z, as shown in Fig. 3.44b. The equilibrium along circumferential and axial directions then reads $2\sigma_\theta h = p_i d$ and $d\pi h\sigma_z = p_i d^2\pi/4$, respectively. In the derivation of these expressions, we assumed that blood pressure also translates in axial vessel wall stress, which applies to many practical applications. The biaxial stress state of the vessel wall therefore reads

$$\sigma_\theta = \frac{p_i d}{2h} \;;\; \sigma_z = \frac{p_i d}{4h} = \sigma_\theta/2.$$

The results $\sigma_{\theta\,d} = 62.5$ kPa and $\sigma_{\theta\,s} = 100.0$ kPa then determine the circumferential stresses in the wall at diastolic and systolic blood pressures, whilst $\sigma_{z\,d} = 31.3$ kPa and $\sigma_{z\,s} = 50.0$ kPa are the respective axial stresses.

Fig. 3.44 Sectioning of the inflated thin-walled tube towards the illustration of (**a**) circumferential stress σ_θ and (**b**) axial stress σ_z in the wall

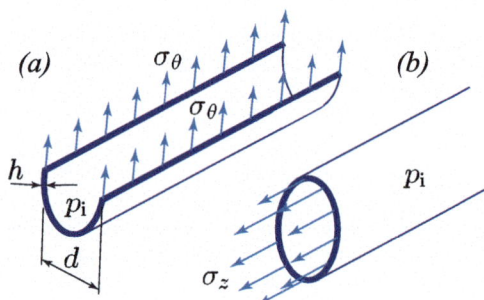

(b) The vessel wall of the inflated thin-walled tube represents a classical plane stress problem, where the circumferential and axial directions are the respective principal stress directions. Hooke's law (3.49) then reads

$$\begin{bmatrix} \varepsilon_\theta \\ \varepsilon_z \\ \varepsilon_r \end{bmatrix} = \frac{1}{E} \begin{bmatrix} 1 & -\nu \\ -\nu & 1 \\ -\nu & -\nu \end{bmatrix} \begin{bmatrix} \sigma_\theta \\ \sigma_z \end{bmatrix}, \tag{3.145}$$

where $\varepsilon_\theta, \varepsilon_z, \varepsilon_r$ denote the principal engineering strains in circumferential θ, axial z, and radial r directions, respectively.

Given relation (3.145), the vessel wall strain at diastolic and systolic blood pressures reads

$$\begin{bmatrix} \varepsilon_{\theta\,d} \\ \varepsilon_{z\,d} \\ \varepsilon_{r\,d} \end{bmatrix} = \frac{1}{625} \begin{bmatrix} 1 & -0.49 \\ -0.49 & 1 \\ -0.49 & -0.49 \end{bmatrix} \begin{bmatrix} 62.5 \\ 31.3 \end{bmatrix} = \begin{bmatrix} 0.0755 \\ 0.001 \\ -0.0735 \end{bmatrix},$$

$$\begin{bmatrix} \varepsilon_{\theta\,s} \\ \varepsilon_{z\,s} \\ \varepsilon_{r\,s} \end{bmatrix} = \frac{1}{625} \begin{bmatrix} 1 & -0.49 \\ -0.49 & 1 \\ -0.49 & -0.49 \end{bmatrix} \begin{bmatrix} 100.0 \\ 50.0 \end{bmatrix} = \begin{bmatrix} 0.1208 \\ 0.0016 \\ -0.1176 \end{bmatrix}.$$

The vessel's diameter pulsates then at $\Delta d = (\varepsilon_{\theta\,s} - \varepsilon_{\theta\,d})D = 0.223$ mm between the diastolic and systolic blood pressures. We also note that the axial strains are approximately two orders of magnitude smaller than the circumferential strains.

(c) Given the expression (2.32), the capacity of the elastic vessel segment is $C = 3D^3\pi L/(16HE) = 235.62$ mm^{3}Pa^{-1} = 3.1416 mm^3 mmHg^{-1}. ∎

Example 3.24 (Applications of the Principle of Virtual Work).

(a) We may introduce the arbitrary virtual displacement $\delta\mathbf{u}$ that moves the rigid body in space. Given "frozen" $\mathbf{P}_i$, the external virtual work reads $W_{\text{ext}} = \sum_{i=1}^{n} \mathbf{P}_i \cdot \delta\mathbf{u}$. A rigid body does not deform, and therefore no internal virtual work appears, $\delta W_{\text{int}} = 0$.

Given any arbitrary virtual displacement $\delta\mathbf{u}$, the PVW $\delta W_{\text{ext}} - \delta W_{\text{int}} = \delta W_{\text{ext}} = 0$ then yields $\sum_{i=1}^{n} \mathbf{P}_i = \mathbf{0}$ and resembles the classical balance of linear momentum.

(b) Let us introduce the force $F = ku$ that acts at the spring, where u denotes the elongation with respect to its load-free length. On top of the loaded configuration, we introduce the arbitrary virtual displacement δu. The force F is "frozen" upon this perturbation, and the corresponding external and internal virtual works then read $\delta W_{\text{ext}} = -G\delta u$ and $\delta W_{\text{int}} = -F\delta u = -ku\delta u$, respectively.

Given an arbitrary δu, the PVW $\delta W_{\text{ext}} - \delta W_{\text{int}} = -G\delta u + ku\delta u = 0$ then leads to $-G + ku = 0$ and resembles the classical balance of linear momentum.

(c) The shear stress $\tau = \eta\dot\gamma$ acts within the fluid layer, where $\dot\gamma = V/H$ denotes the shear rate. At "frozen" stresses τ and τ_0, the flow is perturbed by the virtual velocity $\delta v = \delta V y/H$, a field that satisfies the essential boundary condition $\delta v = 0$ at $y = 0$, also known as the no-slip condition.

The perturbation of the flow contributes to the external and internal virtual works per unit time, and thus to the external virtual power $\delta\dot W_{\text{ext}}$ as well as the internal virtual power $\delta\dot W_{\text{int}}$. The external virtual power reads $\delta\dot W_{\text{ext}} = \tau_0 A\delta V$, and the virtual shear rate $\delta\dot\gamma = \delta V/H$ allows us to compute the internal virtual power $\delta\dot W_{\text{int}} = \int_y \tau A\delta\dot\gamma \mathrm{d}y = \eta AV/H \int_y \delta\dot\gamma \mathrm{d}y = \eta AV\delta V/H$.

Given arbitrary δV, the PVW $\delta W_{\text{ext}} - \delta W_{\text{int}} = \tau_0 A\delta V - \eta AV\delta V/H = 0$ then leads to $\tau_0 - \eta V/H = \tau_0 - \tau = 0$ and resembles the classical balance of linear momentum. ∎

Example 3.25 (Strain Localization in a Rod at Tension).

(a) Let us consider the rod at State (III). The strain in the $n - 1$ elastically deformed sections reads $\varepsilon = \sigma/E$, whilst $\varepsilon^* = Y/E + (Y - \sigma)/H$ expresses the strain in the localized section, where σ denotes the stress. The averaged (smeared) strain then reads

$$\bar{\varepsilon} = u/L = \frac{1}{n}[(n - 1)\varepsilon + \varepsilon^*] = \frac{1}{n}\left[\frac{(n - 1)\sigma + Y}{E} + \frac{Y - \sigma}{H}\right].$$

This relation may be inverted, and

$$\sigma = \frac{\bar{\varepsilon}HEn - Y(H + E)}{H(n - 1) - E} \tag{3.153}$$

then expresses the stress as a function of the problem parameters. The dependence on the number of sections n is shown in Fig. 3.51a.

(b) The work $\int \sigma \, d\varepsilon$ per unit volume enters the mechanical system, energy that has been entirely dissipated upon State (IV). Given the average strain $\bar{\varepsilon}_1 = Y(H + E)/(HEn)$ at State (IV), the dissipation per unit volume reads $\mathcal{D} = \int_0^{\bar{\varepsilon}_1} \sigma \, d\varepsilon$. We may split the integral at $\bar{\varepsilon}_0 = Y/E$, the strain at the elastic limit Y, such that $\sigma = E\bar{\varepsilon}$ and (3.153) determine the stress at $0 \le \bar{\varepsilon} < \bar{\varepsilon}_0$ and $\bar{\varepsilon}_0 < \bar{\varepsilon} \le \bar{\varepsilon}_1$, respectively. The integration then yields the dissipation

$$\mathcal{D} = \frac{Y^2}{2E} + \int_{\bar{\varepsilon}_0}^{\bar{\varepsilon}_1} \sigma \, d\bar{\varepsilon} = \frac{Y^2}{2n}\left(\frac{1}{H} + \frac{1}{E}\right). \tag{3.154}$$

Given all dissipation appears in the section of the strain localization, we may also derive this expression through the dissipation of the localized section $\mathcal{D}_{loc} = Y^2(1/H + 1/E)/2$ weighted by the factor $1/n$.

As with the stress (3.153), the dissipation (3.154) depends on the number of sections n, and $\mathcal{D}$ even disappears for $n \to \infty$; the continuum solution of the problem. Figure 3.51a indicates this result—loading and unloading follow then the same path. This obviously non-physical result is a direct consequence of the non-polar continuum. The material volume, within which the localization develops, tends to zero and the dissipation $\mathcal{D}$ then disappears.

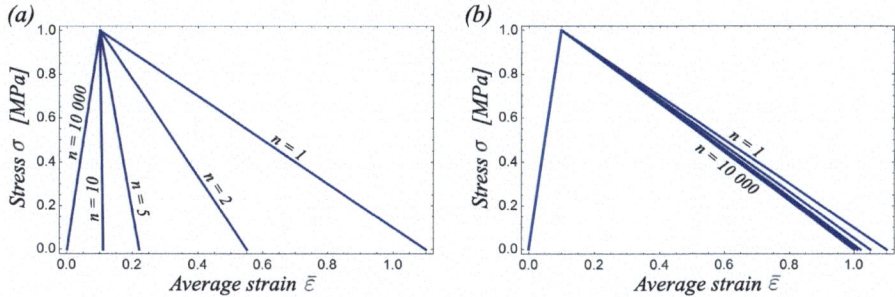

Fig. 3.51 Stress σ *versus* averaged strain $\bar{\varepsilon}$ response of a rod at simple tension that is discretized by n sections. (**a**) Non-regularized response. (**b**) Regularized response by using the "section size-dependent" softening modulus $H_{\text{reg}} = H/n$

(c) The substitution of H by $H_{\text{reg}} = H/n$ in Eqs. (3.153) and (3.154) regularizes the problem, and then yields

$$\bar{\varepsilon}_{\text{reg}} = \frac{1}{n} \left[\frac{(n-1)\sigma + Y}{E} + \frac{Y - \sigma}{H/n} \right] \;;\; \mathcal{D}_{\text{reg}} = \frac{Y^2}{2n} \left(\frac{1}{H/n} + \frac{1}{E} \right) .$$

It implicitly introduced a failure length-scale and therefore prevents from the development of the localization within an infinitesimal small volume of the material. The spurious dependence on n disappears, and the response converges towards $n \to \infty$, see Fig. 3.51b. The regularized dissipation of the continuum problem yields then the physically correct result of $\mathcal{D}_{\text{reg}\,n\to\infty} = Y^2/(2H)$. ∎

Example 4.1 (Transformation of the Quadrilateral Finite Element).

(a) Given the shape functions (4.5), the spatial interpolation reads

$$x = \left(\eta^- \xi^- x_1 + \eta^- \xi^+ x_2 + \eta^+ \xi^+ x_3 + \eta^+ \xi^- x_4 \right) /4 \;;$$
$$y = \left(\eta^- \xi^- y_1 + \eta^- \xi^+ y_2 + \eta^+ \xi^+ y_3 + \eta^+ \xi^- y_4 \right) /4 \,,$$

where the notation $\xi^+ = 1 + \xi$, $\xi^- = 1 - \xi$, $\eta^+ = 1 + \eta$, and $\eta^- = 1 - \eta$ has been used. The interpolation then reads

$$x = 1.5 - 0.25\eta + 1.75\xi \;;\; y = 1.3 + (1.05 - 0.15\xi)\eta + 0.1\xi$$

where the specific nodal coordinates $\mathbf{x}$ have been used.

(b) The spatial gradient, and thus the Jacobian of the transformation reads

$$\mathbf{J} = \frac{\partial \mathbf{x}}{\partial \boldsymbol{\xi}} = \begin{bmatrix} \partial x/\partial \xi & \partial x/\partial \eta \\ \partial y/\partial \xi & \partial y/\partial \eta \end{bmatrix}$$

$$
= \frac{1}{4} \begin{bmatrix} -\eta^- x_1 + \eta^- x_2 + \eta^+ x_3 - \eta^+ x_4 & -\xi^- x_1 - \xi^+ x_2 + \xi^+ x_3 + \xi^- x_4 \\ -\eta^- y_1 + \eta^- y_2 + \eta^+ y_3 - \eta^+ y_4 & -\xi^- y_1 - \xi^+ y_2 + \xi^+ y_3 + \xi^- y_4 \end{bmatrix} ,
$$

then results in

$$
\mathbf{J} = \begin{bmatrix} 1.75 & -0.25 \\ 0.1 - 0.15\eta & 1.05 - 0.15\xi \end{bmatrix} ,
$$

where the specific nodal coordinates $\mathbf{x}$ have been used.

(c) The determinant of the Jacobian transformation reads

$$
\det\mathbf{J} = 1.8625 - 0.0375\eta - 0.2625\xi ,
$$

a linear function with respect to the parent domain coordinates.

The area of the finite element in the physical space reads $A = (|\mathbf{a} \times \mathbf{b}| + |\mathbf{b} \times \mathbf{c}|)/2 = 7.45 \text{ cm}^2$, where $\mathbf{a} = [x_2 \ y_2 \ 0] - [x_1 \ y_1 \ 0]^T$, $\mathbf{b} = [x_3 \ y_3 \ 0] - [x_1 \ y_1 \ 0]^T$, and $\mathbf{c} = [x_4 \ y_4 \ 0] - [x_1 \ y_1 \ 0]^T$ represent vectors that are determined by the nodal coordinates. We used a 3D vector representation to allow for the use of the cross product.

The comparison of this result with the determinant of the Jacobian transformation in the center of the element reveals that $A = 4\det\mathbf{J}(\boldsymbol{\xi} = \mathbf{0}) = 4 \cdot 1.8625 = 7.45 \text{ cm}^2$ holds. ∎

Example 4.2 (Displacement and Strain Interpolation).

(a) The shape interpolation matrix of a quadrilateral element reads

$$
\mathbf{N} = \begin{bmatrix} N_1 & 0 & N_2 & 0 & N_3 & 0 & N_4 & 0 \\ 0 & N_1 & 0 & N_2 & 0 & N_3 & 0 & N_4 \end{bmatrix} .
$$

(b) With the definition of the linear strain, the strain interpolation matrix reads

$$
\mathbf{B} = \begin{bmatrix} \frac{\partial N_1}{\partial x} & 0 & \frac{\partial N_2}{\partial x} & 0 & \frac{\partial N_3}{\partial x} & 0 & \frac{\partial N_4}{\partial x} & 0 \\ 0 & \frac{\partial N_1}{\partial y} & 0 & \frac{\partial N_2}{\partial y} & 0 & \frac{\partial N_3}{\partial y} & 0 & \frac{\partial N_4}{\partial y} \\ \frac{1}{2}\frac{\partial N_1}{\partial y} & \frac{1}{2}\frac{\partial N_1}{\partial x} & \frac{1}{2}\frac{\partial N_2}{\partial y} & \frac{1}{2}\frac{\partial N_2}{\partial x} & \frac{1}{2}\frac{\partial N_3}{\partial y} & \frac{1}{2}\frac{\partial N_3}{\partial x} & \frac{1}{2}\frac{\partial N_4}{\partial y} & \frac{1}{2}\frac{\partial N_4}{\partial x} \end{bmatrix} ,
$$

and, given the Jacobian transformation $\mathbf{J} = \partial\mathbf{x}/\partial\boldsymbol{\xi}$, we may compute the spatial gradient

$$
\frac{\partial N_i}{\partial \mathbf{x}} = \frac{\partial N_i}{\partial \boldsymbol{\xi}} : \frac{\partial \boldsymbol{\xi}}{\partial \mathbf{x}} = \frac{\partial N_i}{\partial \boldsymbol{\xi}} : \mathbf{J}^{-1} ; \quad i = 1, \dots, 4 ,
$$

of the shape functions N_i.

(c) Given the parent domain position, the displacement interpolation matrix reads

$$
\mathbf{N} = \begin{bmatrix} 0.1925 & 0.0 & 0.3575 & 0.0 & 0.2925 & 0.0 & 0.1575 & 0.0 \\ 0.0 & 0.1925 & 0.0 & 0.3575 & 0.0 & 0.2925 & 0.0 & 0.1575 \end{bmatrix},
$$

and with the inverse Jacobian matrix

$$
\mathbf{J}^{-1} = \begin{bmatrix} 0.562238 & 0.13986 \\ -0.0643357 & 0.979021 \end{bmatrix},
$$

we can then compute the strain interpolation matrix

$$
\mathbf{B} = \begin{bmatrix} -0.1433 & 0.0 & 0.1755 & 0.0 & 0.1056 & 0.0 & -0.1378 & 0.0 \\ 0.0 & -0.2098 & 0.0 & -0.2797 & 0.0 & 0.3497 & 0.0 & 0.1399 \\ -0.1049 & -0.0717 & -0.1399 & 0.0878 & 0.1748 & 0.0528 & 0.0699 & -0.0689 \end{bmatrix}.
$$

∎

Example 4.3 (Heat Conduction Problem).

(a) Given the admissible variation of the temperature $\delta\theta$ with $\delta\theta = 0$ at $\partial\Omega_\theta$, the partial differential equation (4.12), multiplied with $\delta\theta$ and then integrated over Ω, yields

$$
\int_\Omega \delta\theta \left(\frac{\partial q_i}{\partial x_i} - r + \rho c\dot{\theta} \right) dv = 0. \tag{4.13}
$$

(b) The integration by parts $\int_\Omega (\partial q_i/\partial x_i)\delta\theta dv = \int_\Omega \partial(q_i\delta\theta)/\partial x_i dv - \int_\Omega q_i(\partial\delta\theta/\partial x_i)dv$ and the use of the divergence theorem $\int_\Omega \partial(q_i\delta\theta)/\partial x_i dv = \int_{\partial\Omega} q_i\delta\theta n_i ds$, allows us to express (4.13) through

$$
\int_{\partial\Omega} \delta\theta q_i n_i ds + \int_\Omega \delta\theta \left(\rho c\dot{\theta} - r \right) dv - \int_\Omega \frac{\partial\delta\theta}{\partial x_i} q_i dv = 0,
$$

where n_i denotes the components of the outward normal vector to the boundary $\partial\Omega$. Embedding the Dirichlet boundary condition $\delta\theta = 0$ at $\partial\Omega_\theta$ then results in the weak form

$$
\int_{\partial\Omega_q} \delta\theta q_i n_i ds + \int_\Omega \delta\theta \left(\rho c\dot{\theta} - r \right) dv + \int_\Omega \frac{\partial\delta\theta}{\partial x_i} k \frac{\partial\theta}{\partial x_i} dv = 0 \tag{4.14}
$$

of the transient heat conduction problem, where Fourier's law together with $\partial\Omega_q \cup \partial\Omega_\theta = \partial\Omega$ has been used. ∎

Example 4.4 (Linearization of a Spatial Variational Statement).

(a) The directional derivative of the relation $(4.18)_2$ reads

$$D_{\mathbf{u}} \int_{\Omega_0} \delta\mathbf{E} : \mathbf{S}\,dV = \int_{\Omega_0} D_{\mathbf{u}}(\mathbf{S} : \delta\mathbf{E})dV = \int_{\Omega_0} (D_{\mathbf{u}}\mathbf{E} : \mathbb{C} : \delta\mathbf{E} + \mathbf{S} : D_{\mathbf{u}}\delta\mathbf{E})\,dV \,,$$

where the chain rule has been used, and $\mathbb{C} = \partial\mathbf{S}/\partial\mathbf{E}$ denotes the material's stiffness in the reference configuration. Given the expressions $D_{\mathbf{u}}\mathbf{E} = \text{sym}(\mathbf{F}^{\mathrm{T}}\text{Grad}\Delta\mathbf{u})$, $\delta\mathbf{E} = \text{sym}(\mathbf{F}^{\mathrm{T}} \text{Grad}\delta\mathbf{u})$, and $D_{\mathbf{u}}\delta\mathbf{E} = \text{sym}(\text{Grad}^{\mathrm{T}}\Delta\mathbf{u}\,\text{Grad}\delta\mathbf{u})$ derived in Appendix D, the linearization then reads

$$D_{\mathbf{u}} \int_{\Omega_0} \delta\mathbf{E} : \mathbf{S}\,dV = \int_{\Omega_0} \Big[(\mathbf{F}^{\mathrm{T}} \text{Grad}\Delta\mathbf{u}) : \mathbb{C} : (\mathbf{F}^{\mathrm{T}} \text{Grad}\delta\mathbf{u})$$

$$+ \mathbf{S} : (\text{Grad}^{\mathrm{T}}\Delta\mathbf{u}\,\text{Grad}\delta\mathbf{u}) \Big] dV \,, \qquad (4.22)$$

where the symmetry of $\mathbb{C}$ and $\mathbf{S}$ has been used.

(b) With the second Piola transform $\mathbf{S} = J\mathbf{F}^{-1}\boldsymbol{\sigma}\mathbf{F}^{-\mathrm{T}}$ and the relation $C_{IJKL} = Jc_{ijkl}F_{iI}^{-1}F_{jJ}^{-1}F_{kK}^{-1}F_{lL}^{-1}$ between the stiffness in the reference and spatial configurations, the linearization (4.22) reads

$$D_{\mathbf{u}}\delta\Pi_{\text{int}} = \int_{\Omega} \big(\text{grad}_s\delta\mathbf{u} : c : \text{grad}_s\Delta\mathbf{u} + \text{grad}_s\delta\mathbf{u} : \text{grad}_s\Delta\mathbf{u}\boldsymbol{\sigma}\big)\,dv \,, \qquad (4.23)$$

where the kinematics relations $\text{grad}(\bullet) = \text{Grad}(\bullet)\mathbf{F}^{-1}$ and $dv = J\,dV$ have been used. ∎

Example 4.5 (The 1D Advection–Diffusion Finite Element).

(a) Given the shape functions N_1, N_2 together with the definitions $\mathbf{K}$, $\mathbf{D}$, and $\mathbf{f}$ according to (4.28), we get

$$\mathbf{K} = v \int_0^h \begin{bmatrix} N_1\frac{dN_1}{dx} & N_1\frac{dN_2}{dx} \\ N_2\frac{dN_1}{dx} & N_2\frac{dN_2}{dx} \end{bmatrix} dx = \frac{v}{2}\begin{bmatrix} -1 & 1 \\ -1 & 1 \end{bmatrix},$$

$$\mathbf{D} = v \int_0^h \begin{bmatrix} \frac{dN_1}{dx}\frac{dN_1}{dx} & \frac{dN_1}{dx}\frac{dN_2}{dx} \\ \frac{dN_2}{dx}\frac{dN_1}{dx} & \frac{dN_2}{dx}\frac{dN_2}{dx} \end{bmatrix} dx = \frac{v}{h}\begin{bmatrix} 1 & -1 \\ -1 & 1 \end{bmatrix},$$

and

$$\mathbf{f} = \xi \int_0^h \begin{bmatrix} N_1 \\ N_2 \end{bmatrix} dx = \frac{\alpha h}{2}\begin{bmatrix} 1 \\ 1 \end{bmatrix},$$

where the natural coordinate $\xi = 2x/h - 1$ has been used.

(b) Given the shape functions N_1, N_2 and S_1, S_2, the interpolations $c = N_i h_i$ and $\delta c = S_i \delta h_i$ express the concentration c and the test function δc. Consequently, the expressions

$$\mathbf{K} = v \int_0^h \begin{bmatrix} S_1 \frac{dN_1}{dx} & S_1 \frac{dN_2}{dx} \\ S_2 \frac{dN_1}{dx} & S_2 \frac{dN_2}{dx} \end{bmatrix} dx = \frac{v}{2} \begin{bmatrix} -1 & 1 \\ -1 & 1 \end{bmatrix} + \beta \frac{v}{2} \begin{bmatrix} 1 & -1 \\ -1 & 1 \end{bmatrix},$$

$$\mathbf{D} = v \int_0^h \begin{bmatrix} \frac{dS_1}{dx} \frac{dN_1}{dx} & \frac{dS_1}{dx} \frac{dN_2}{dx} \\ \frac{dS_2}{dx} \frac{dN_1}{dx} & \frac{dS_2}{dx} \frac{dN_2}{dx} \end{bmatrix} dx = \frac{v}{h} \begin{bmatrix} 1 & -1 \\ -1 & 1 \end{bmatrix},$$

and

$$\mathbf{f} = \xi \int_0^h \begin{bmatrix} S_1 \\ S_2 \end{bmatrix} dx = \frac{\alpha h}{2} \begin{bmatrix} 1 \\ 1 \end{bmatrix} + \beta \frac{\alpha h}{2} \begin{bmatrix} -1 \\ 1 \end{bmatrix}$$

define the Petrov–Galerkin 1D AD finite element, where the natural coordinate $\xi = 2x/h - 1$ has been used. ∎

Example 4.6 (Vessel Segment at Quasi-static Tension).

(a) The force P acts as an external force at node 2, which together with the stiffness matrix $\mathbf{K}$ of the linear truss element (4.35) then yields the system

$$\frac{EA}{l} \begin{bmatrix} 1 & -1 \\ -1 & 1 \end{bmatrix} \begin{bmatrix} u_1 \\ u_2 \end{bmatrix} = \begin{bmatrix} R \\ P \end{bmatrix}$$

of algebraic equations, where the (in general unknown) reaction force R at node 1 has been introduced. Given $\det\mathbf{K} = 0$, this system is singular and cannot be solved.

(b) The Dirichlet boundary condition $u_1 = 0$ at node 1 removes the first column and row from the system. The remaining equation $EAu_2/l = 342.3\,u_2 = P$ then yields the nodal displacement $u_2 = 0.876$ mm, where the cross-section $A = (d_o^2 - d_i^2)\pi/4 = 18.38\,\text{mm}^2$ has been used. With the linear interpolation, the displacement along the vessel's axial direction reads $u = 73.04\,x$ [mm], where x is given in meters. ∎

Example 4.7 (Nodal Forces and Stiffness of a Quadrilateral Finite Element).

(a) Given the shape functions (4.5) and the definition (4.7) together with the nodal coordinates, the strain interpolation matrix in the center $\xi = \eta = 0$ reads

$$\mathbf{B} = \begin{bmatrix} -0.1598 & 0 & 0.1031 & 0 & 0.1598 & 0 & -0.1031 & 0 \\ 0 & -0.1804 & 0 & -0.2062 & 0 & 0.1804 & 0 & 0.2062 \\ -0.09021 & -0.0799 & -0.1031 & 0.05155 & 0.09021 & 0.0799 & 0.1031 & -0.05155 \end{bmatrix},$$

a result that used the inverse Jacobian transformation

$$\mathbf{J}^{-1} = \begin{bmatrix} 0.525773 & -0.0515464 \\ 0.113402 & 0.773196 \end{bmatrix}$$

in the computation of the spatial gradients $\partial N_i / \partial \mathbf{x}$; $i = 1, \ldots, 4$.

(b) The integration order $2n - 1$ determines the Gauss–Legendre quadrature, where n denotes the number of integration points along a coordinate direction. The definitions (4.30) of $\mathbf{f}$ and $\mathbf{K}$ together with the quadrature rules (4.46) and (4.47) allow us to identify the number of integrations points needed for an exact integration. The traction $\bar{t}_i$ and the stiffness C_{ij} are constant, the strain interpolation B_{ij} includes linear terms, the spatial interpolation N_{ij} as well as the determinate of the Jacobian transformation det$\mathbf{J}$ include quadratic terms. The determination of $\mathbf{f}$ and $\mathbf{K}$ then represents the integration over fourth-order and respectively second-order polynomial expressions. The exact integration therefore needs a minimum of $n = 3$ and $n = 2 \times 2 = 4$ integration points, respectively.

(c) Tables 4.2 and 4.3 illustrate the algorithms that have been used to compute $\mathbf{f}$ and $\mathbf{K}$. With the provided data, the force vector reads

$$\mathbf{f} = [0.7275 \ 1.2125 \ 0.7275 \ 1.2125 \ 0.0 \ 0.0 \ 0.0 \ 0.0]^T \ [\text{N}],$$

whilst the stiffness matrix results in

$$\mathbf{K} = \begin{bmatrix} 584755. & 275405. & -357462. & 181732. \\ 275405. & 763267. & -72970. & 240848. \\ -357462. & -72970. & 547409. & -177679. \\ 181732. & 240848. & -177679. & 1197860 \\ -348424. & -259113. & -23801.2 & -208017. \\ -259113. & -349337. & 46685.5 & -908624. \\ 121131. & 56677.2 & -166146. & 203963. \\ -198025. & -654778. & 203963. & -530088. \end{bmatrix}$$

$$
\begin{bmatrix}
-348424. & -259113. & 121131. & -198025. \\
-259113. & -349337. & 56677.2 & -654778. \\
-23801.2 & 46685.5 & -166146. & 203963. \\
-208017. & -908624. & 203963. & -530088. \\
618703. & 277746. & -246478. & 189384. \\
277746. & 822726. & -65318.7 & 435235. \\
-246478. & -65318.7 & 291493. & -195322. \\
189384. & 435235. & -195322. & 749631.
\end{bmatrix}
\; [\mathrm{N\,mm^{-1}}] \, .
$$

Table 4.2 Algorithm to compute the nodal force vector **f** with Gauss–Legendre quadrature

(1) Parent domain location and weights for $n = 3$ integration points

$$
\mathbf{g} =
\begin{bmatrix}
-0.774597 & -1.0 & 0.555556 \\
0.0 & -1.0 & 0.888889 \\
0.774597 & -1.0 & 0.555556
\end{bmatrix}
$$

(2) Summation over the integration points

$f_i = 0; i = 1, \dots, 8$

Do $l = 1, \dots, 3$

    Compute Jacobian **J** and displacement interpolation matrix N_{ij} at $\xi = g_{l1}$ and $\eta = g_{l2}$

    Add integration point contribution to the nodal force

    $f_i \leftarrow f_i + N_{ai}\bar{t}_a h \, w \, \det\mathbf{J}$ with $w = g_{l3}$

End Do

Table 4.3 Algorithm to compute the element stiffness matrix **K** with Gauss–Legendre quadrature

(1) Parent domain location and weights for $n = 2 \times 2 = 4$ integration points

$$
\mathbf{g} =
\begin{bmatrix}
-0.57735 & -0.57735 & 1.0 \\
0.57735 & -0.57735 & 1.0 \\
0.57735 & 0.57735 & 1.0 \\
-0.57735 & 0.57735 & 1.0
\end{bmatrix}
$$

(2) Summation over the integration points

$k_{ij} = 0; i, j = 1, \dots, 8$

Do $l = 1, \dots, 4$

    Compute Jacobian **J** and shape function gradients $\mathrm{grad} N_i$ at $\xi = g_{l1}$ and $\eta = g_{l2}$

    Invert **J** and compute the strain interpolation matrix B_{ij} at $\xi = g_{l1}$ and $\eta = g_{l2}$

    Add integration point contribution to the element stiffness

    $K_{ij} \leftarrow K_{ij} + B_{ai}C_{ab}B_{bj} h \, w \, \det\mathbf{J}$ with $w = g_{l3}$

End Do

The set

$$\{2.827 \cdot 10^6, 1.001 \cdot 10^6, 7.506 \cdot 10^5, 6.358 \cdot 10^5,$$

$$3.61 \cdot 10^5, 3.993 \cdot 10^{-10}, -2.085 \cdot 10^{-10}, -1.197 \cdot 10^{-10}\},$$

then represents the eight eigenvalues of $\mathbf{K}$, out of which three are (approximately) zero. Given no boundary conditions have been prescribed, these three eigenvalues correspond to two rigid body translations and one rigid body rotation, respectively.

(d) The one-point Gauss–Legendre quadrature ($\xi = 0$, $\eta = 0$, $w = 4$) results in the element stiffness matrix

$$\mathbf{K} = \begin{bmatrix} 474508. & 267805. & -179604. & 193994. \\ 267805. & 570171. & -60708.4 & 552362. \\ -179604. & -60708.4 & 260479. & -197460. \\ 193994. & 552362. & -197460. & 695311. \\ -474508. & -267805. & 179604. & -193994. \\ -267805. & -570171. & 60708.4 & -552362. \\ 179604. & 60708.4 & -260479. & 197460. \\ -193994. & -552362. & 197460. & -695311. \end{bmatrix}$$

$$\begin{bmatrix} -474508. & -267805. & 179604. & -193994. \\ -267805. & -570171. & 60708.4 & -552362. \\ 179604. & 60708.4 & -260479. & 197460. \\ -193994. & -552362. & 197460. & -695311. \\ 474508. & 267805. & -179604. & 193994. \\ 267805. & 570171. & -60708.4 & 552362. \\ -179604. & -60708.4 & 260479. & -197460. \\ 193994. & 552362. & -197460. & 695311. \end{bmatrix} \quad [\mathrm{N\,mm^{-1}}] .$$

The set

$$\{2.731 \cdot 10^6, 8.015 \cdot 10^5, 4.681 \cdot 10^5, 4.472 \cdot 10^{-10},$$

$$-2.3 \cdot 10^{-10}, 9.153 \cdot 10^{-11}, -8.256 \cdot 10^{-11}, 3.7 \cdot 10^{-12}\}$$

represents the eigenvalues of $\mathbf{K}$, out of which five are (approximately) zero. They correspond to two rigid body translations, one rigid body rotation, and two hourglass modes of the finite element. The one-point integration therefore leads to an under-integrated finite element stiffness, and one expects spurious deformation modes from the application of such a finite element. ■

Example 4.8 (Hu-Washizu Variational Principles).

(a) The Hu-Washizu variational potential leads to the three variational statements

$$\delta_{\mathbf{u}} \Pi_{\text{HW}}(\mathbf{u}, p, \theta) = \int_{\Omega} \overline{\boldsymbol{\sigma}}(\mathbf{u}) : \text{grad}_s \delta\mathbf{u} \, dv + \int_{\Omega} p J(\mathbf{u}) \text{div}\delta\mathbf{u} \, dv - \delta_{\mathbf{u}} \Pi_{\text{ext}} = 0 ;$$

$$\delta_p \Pi_{\text{HW}}(\mathbf{u}, p, \theta) = \int_{\Omega} (J(\mathbf{u}) - \theta)\delta p \, dv - \delta_p \Pi_{\text{ext}} = 0 ;$$

$$\delta_\theta \Pi_{\text{HW}}(\mathbf{u}, p, \theta) = \int_{\Omega} (dU/d\theta - p)\delta\theta \, dv - \delta_\theta \Pi_{\text{ext}} = 0 ,$$

where $\delta_{\mathbf{u}} J(\mathbf{u}) = J(\mathbf{u})\text{div}\delta\mathbf{u}$ expressed the variation of the volume ratio.

(b) The variations of the augmented Hu-Washizu potential $\mathcal{L}(\mathbf{u}, p, \theta, \lambda)$ leads to the four variational statements

$$\delta_{\mathbf{u}}\mathcal{L}(\mathbf{u}, p, \theta, \lambda) = \delta_{\mathbf{u}} \Pi_{\text{HW}}(\mathbf{u}, p, \theta) ; \quad \delta_p\mathcal{L}(\mathbf{u}, p, \theta, \lambda) = \delta_p \Pi_{\text{HW}}(\mathbf{u}, p, \theta) ;$$

$$\delta_\lambda\mathcal{L}(\mathbf{u}, p, \theta, \lambda) = \int_{\Omega_0} h(\theta)\delta\lambda \, dV - \delta_\theta \Pi_{\text{ext}} = 0 ;$$

$$\delta_\theta\mathcal{L}(\mathbf{u}, p, \theta, \lambda) = \int_{\Omega_0} (dU/d\theta + \lambda dh/d\theta - p)\delta\theta \, dV - \delta_\lambda \Pi_{\text{ext}} = 0 . \qquad \blacksquare$$

Example 4.9 (SUPG-Stabilized 1D Advection–Diffusion Problem).

(a) Given the shape functions N_1, N_2, the artificial diffusivity $v^\star$ and

$$\mathcal{P}(\delta c) = vd\delta c/dx ; \quad R(c) = vdc/dx - vd^2c/dx^2 + \alpha$$

of the SUPG-stabilization, the expression (4.65) results in

$$\delta h_i \left[\int_{\Omega} N_i v B_j \, dv + \int_{\Omega} B_i v B_j \, dv + \int_{\Omega} B_i \frac{\beta h}{2} (v B_j + \alpha) \, dv \right] h_j$$

$$= -\delta h_i \int_{\Omega} N_i \alpha dv . \qquad (4.66)$$

The second derivative d^2c/dx^2 disappeared in this expression as a consequence of the linear shape functions. We rearrange the expressions towards

$$\int_{\Omega} \left[v \left(N_i + \frac{\beta h}{2} B_i \right) B_j + B_i v B_j \right] dv \, h_j = - \int_{\Omega} \alpha \left(N_i + \frac{\beta h}{2} B_i \right) dv ,$$

$$(4.67)$$

such that

$$\mathbf{K}_e = v \int_0^h \begin{bmatrix} S_1 B_1 & S_1 B_2 \\ S_2 B_1 & S_2 B_2 \end{bmatrix} \mathrm{d}x = \frac{v}{2} \left(\begin{bmatrix} -1 & 1 \\ -1 & 1 \end{bmatrix} + \beta \begin{bmatrix} 1 & -1 \\ -1 & 1 \end{bmatrix} \right)$$

and

$$\mathbf{D}_e = v \int_0^h \begin{bmatrix} B_1 B_1 & B_1 B_2 \\ B_2 B_1 & B_2 B_2 \end{bmatrix} \mathrm{d}x = \frac{v}{h} \begin{bmatrix} 1 & -1 \\ -1 & 1 \end{bmatrix}$$

denote the advection and the respective diffusion matrices, whilst

$$\mathbf{f}_e = \alpha \int_0^h \begin{bmatrix} S_1 \\ S_2 \end{bmatrix} \mathrm{d}x = \frac{\xi h}{2} \begin{bmatrix} 1 - \beta \\ 1 + \beta \end{bmatrix}$$

represents the right-hand-side vector. Here, the shape functions $S_i = N_i + \beta h B_i/2$; $i = 1, 2$ interpolates the test function δc, whilst $B_i = \mathrm{d}N_i/\mathrm{d}x$; $i = 1, 2$ interpolates the gradient of c. Given $\beta = 1$, the finite element expressions are identical to the 1D full upwind stabilization (4.61).

(b) Given the definition $S_i = N_i + \beta h B_i/2$ together with the property $\mathrm{d}B_i/\mathrm{d}x = 0$ of the linear finite element, we conclude that $\mathrm{d}S_i/\mathrm{d}x = B_i$ holds. The expression (4.67) therefore derives from the interpolations $\delta c = S_i \delta h_i$ and $c = N_i h_i$, and represents a consistent Petrov–Galerkin approach.

(c) Given the expressions $\mathbf{K}_e, \mathbf{D}_e$, and $\mathbf{f}_e$ at the finite element level, the global system $(\mathbf{K} + \mathbf{D})\mathbf{h} = \mathbf{f}$ may be assembled. Implementing the Dirichlet boundary conditions $h_1 = h_{11} = 0$, the system can be solved. Figure 4.16 illustrates the SUPG result at $\beta = 0.3$ in relation to the exact solution to this AD problem.

Fig. 4.16 SUPG-stabilized 1D Advection–Diffusion (AD) problem. The finite element solution has been stabilized with $\beta = 0.3$ ■

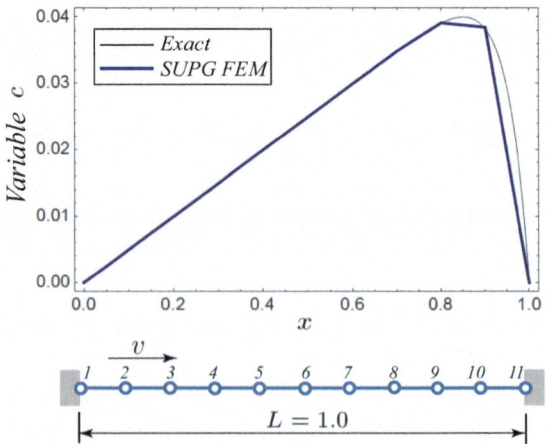

Example 4.10 (Solving a Linear System of Equations).

(a) The LU factorization of **K** reads

$$\mathbf{L} = \begin{bmatrix} 1 & 0 & 0 & 0 \\ 2 & 1 & 0 & 0 \\ 0 & -3 & 1 & 0 \\ 8 & 38 & \frac{-57}{5} & 1 \end{bmatrix} \; ; \; \mathbf{U} = \begin{bmatrix} 1 & 5 & 12 & 2 \\ 0 & -1 & 19 & -1 \\ 0 & 0 & -55 & 3 \\ 0 & 0 & 0 & \frac{281}{5} \end{bmatrix} \; ; \; \mathbf{I_P} = \begin{bmatrix} 0 & 0 & 1 & 0 \\ 0 & 1 & 0 & 0 \\ 0 & 0 & 0 & 1 \\ 1 & 0 & 0 & 0 \end{bmatrix},$$

where **L** and **U** are lower and upper triangular matrices, whilst $\mathbf{I_P}$ denotes the permutation matrix. Given the pivoted nodal forces $\bar{\mathbf{f}} = \mathbf{I_p}\mathbf{f} = [1\ 0\ 2\ 0]^T$, the intermediate vector $\mathbf{y} = [1\ -2\ -4\ \ 112/5]^T$ derives from the forward reduction of $\mathbf{Ly} = \bar{\mathbf{f}}$, which then allows us to compute the solution $\bar{\mathbf{h}} = [113/3091\ -598/3091\ 292/3091\ 112/281]^T$ from the back substitution of $\mathbf{y} = \mathbf{U}\bar{\mathbf{h}}$.

(b) Given the initialization $\mathbf{h}_0 = \mathbf{0}$, the iteration

$$\mathbf{Ah}_{n+1} = \mathbf{y}_n \text{ with } \mathbf{y}_n = \mathbf{Ah}_n + \mathbf{f} - \mathbf{Kh}_n \qquad (4.71)$$

may be used to iteratively solve the system. Here, **A** is a matrix that "lumps" the individual elements of the stiffness matrix **K** to the diagonal and reads $\mathbf{A} = \text{diag}[11, 19, 20, 11]$.

The iteration (4.71) then yields the convergence shown in Fig. 4.17, where the relative logarithmic error $\epsilon = \log(|\mathbf{h}_n|/|\mathbf{h}_{\text{exact}}|)$ quantifies the error of the iterative solution at the n-th iteration.

Fig. 4.17 Convergence of the iterative solution. Relative logarithmic error ϵ with respect to the number of iterations

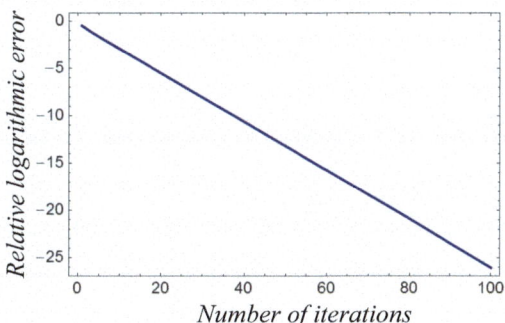

Example 4.11 (Stability of the Euler Integration).

(a) Given the differential equation (4.72), the time-marching iteration reads $y_{n+1} = y_n + \phi \Delta t$, where $\phi = -y_n t_n$ and $\phi = -y_{n+1} t_{n+1}$ describe the forward-Euler and backward-Euler integrations, respectively. After some algebraic manipulations, the iterations

$$y_{n+1} = y_n(1 - t_n \Delta t) \quad \text{and} \quad y_{n+1} = y_n/(1 + t_{n+1} \Delta t)$$

express the forward-Euler and backward-Euler integration.

(b) Figure 4.19 illustrates the numerical integrations, superimposed on the exact solution $y = \exp(-at^2/2)$. With the step size $\Delta t = 9/10$, the forward-Euler integration is unstable, whilst backward-Euler converges to the exact result $y(t \to \infty) = 0$. At the step size $\Delta t \leq 1$, also the forward-Euler integration is stable.

Fig. 4.19 Forward-Euler and backward-Euler integrations. Forward-Euler is unstable at the given step size ■

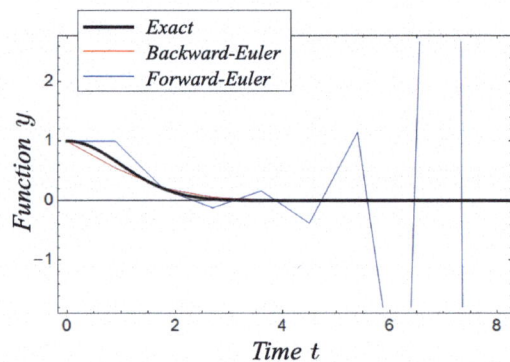

Example 4.12 (Spring Lever Structure).

(a) Given the kinematics relation $u = \sqrt{a^2 + b^2 - (a - v)^2} - b$ between the displacements u and v, together with the equilibrium in the horizontal direction $F_k = ku$, the governing equation

$$r = ku - \frac{F(b + u)}{a - v} = 0 \tag{4.86}$$

determines the system, where the trigonometric relation $F/F_k = (b+u)/(a-v)$ has been used.

(b) The prescribed displacement v allows us to express the displacement-controlled Newton–Raphson iteration by

$$F_{n+1} = F_n - r_n/K_n, \tag{4.87}$$

where $K_n = [\partial r / \partial F]_n = -(b+u)/(a-v)$ has been used. At a prescribed v, the stiffness K_n is a constant, and the Newton–Raphson iteration then converges within two steps. Figure 4.24 shows the structure's force *versus* displacement properties and Table 4.7 outlines the algorithm that has been used to compute said results.

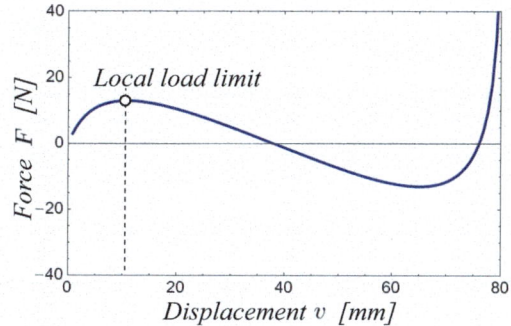

Fig. 4.24 Force *versus* displacement properties of the spring lever structure. A displacement-controlled Newton–Raphson algorithm is able to compute the displacement domain of $0 < v < 80$ mm. The force-controlled solution cannot solve the problem beyond $v \approx 10$ mm, where the structure shows a local load limit

Table 4.7 Algorithm to compute the force F at a prescribed displacement v of the spring lever structure and using the Newton–Raphson iteration

(1) Set displacement increment: $\Delta v = v_{\max}/n_l$
(2) Initialize displacement and force: $v = 0$, $F = 0$

(3) Loop over displacement increments
    Do While $v \leq v_{\max}$
        $v \leftarrow v + \Delta v$

(4) Loop over Newton–Raphson iteration
        $r = 1$
        Do While $|r| \geq 10^{-8}$
            $r = ku - F(b+u)/(a-v)$ with $u = \sqrt{a^2 + b^2 - (a-v)^2} - b$
            $\delta F = r/K$ with $K = -(u+b)/(a-v)$
            $F \leftarrow F - \delta F$
        End Do
    End Do

(c) Given the prescribed force F, the force-controlled Newton–Raphson iteration reads

$$v_{n+1} = v_n - r_n/K_n \; ; \quad K_n = \left[\frac{\partial r}{\partial v}\right]_n = \frac{k(a-v)-F}{u+b} - \frac{F(u+b)}{(a-v)^2} .$$

(4.88)

The stiffness K_n is a function of v, and the Newton–Raphson iteration requires a few iterations to converge.

At the displacement $v \approx 10$ mm a local limit point appears and $\partial F/\partial v = 0$ holds, see Fig. 4.24. The force can then no longer be increased and the force-controlled solution strategy cannot solve the problem beyond this point. ∎

Example 5.1 (Vessel Segment Characterization).

(a) Given the geometry of the stress-free vessel, the fluid properties and the linear-elastic description of the vessel wall, the relations (2.31), (2.32) (or (5.1)), and (2.33) yield

$$R = \frac{128\mu l}{\pi D^4} = 4.94368 \cdot 10^7 \text{ Pa s m}^{-3} ;$$

$$C = \frac{3D^3\pi l}{16HE} = 2.67406 \cdot 10^{-10} \text{ m Pa}^{-3} ;$$

$$L = \frac{4\rho l}{D^2\pi} = 1.47307 \cdot 10^7 \text{ kg m}^{-4} ,$$

expressions that determine the femoral artery's resistance R, capacity C, and inductance L, respectively.

(b) The right Cauchy–Green strain $\mathbf{C} = \text{diag}[\lambda_\theta^2, \lambda_z^2, \lambda_r^2]$ with the first invariant $I_1 = \text{tr}\mathbf{C} = \lambda_\theta^2 + \lambda_z^2 + \lambda_r^2$ determines the kinematics of vessel inflation, where θ, z, r denote the circumferential, axial, and radial vessel directions, respectively. Coleman and Noll's relation (5.5) allows us then to compute the circumferential second Piola–Kirchhoff stress

$$S_\theta = \frac{1}{\lambda_\theta}\frac{\partial\Psi}{\partial\lambda_\theta} - \frac{\lambda_r}{\lambda_\theta^2}\frac{\partial\Psi}{\partial\lambda_r} = 4c(I_1 - 3)\left[1 - (\lambda_r/\lambda_\theta)^2\right] ,$$

(5.13)

where the membrane condition $S_r = 0$ has been used to express the Lagrange pressure κ in (5.5). The Cauchy stress then reads $\sigma_\theta = \lambda_\theta^2 S_\theta$, and $p_i d = 2\sigma_\theta h$ expresses the static equilibrium along the circumferential vessel direction, where p_i denotes the inflation pressure. Given the stretches $\lambda_\theta = d/D$, $\lambda_z = 1$, and $\lambda_r = h/H = \lambda_\theta^{-1}$, the inflation pressure reads

$$p_i = 2S_\theta H/D ,$$

(5.14)

and the substitution of S_θ by (5.13) then results in

$$p_i = 114754.1 \frac{(\lambda_\theta^2 - 1)^3 (1 + \lambda_\theta^2)}{\lambda_\theta^6} \text{ [Pa]}. \tag{5.15}$$

We used the geometrical and constitutive data of the vessel in the derivation of this non-linear relation between p_i and λ_θ. At $p_i = 13333.3$ Pa (100.0 mmHg), it has one physically reasonable root, $\lambda_\theta = 1.24702$. This deformation corresponds to the vessel diameter $d = \lambda_\theta D = 7.607$ mm, and

$$R = \frac{128\mu l}{\pi d^4} = 2.04437 \cdot 10^7 \text{ Pa s m}^{-3} \text{ and}$$

$$L = \frac{4\rho L}{d^2 \pi} = 9.47279 \cdot 10^6 \text{ kg m}^{-4}$$

then determine the femoral artery's resistance and inductance at said pressure.

The calculation of the vessel's capacity $C = \Delta V / \Delta p$ at said pressure, and thus at $\lambda_\theta = 1.24702$, requires the volume increment $\Delta V = dl\pi \Delta d/2$. It represents the increase in vessel volume in response to the pressure increment Δp. We therefore linearize (5.14),

$$\Delta p = \frac{2 K_{\theta\theta} H}{D^2} \Delta d , \tag{5.16}$$

where $\Delta d = D\Delta\lambda_\theta$ is the diameter increment, and $K_{\theta\theta} = \partial S_\theta / \partial \lambda_\theta$ denotes the circumferential vessel stiffness coefficient. With the constitutive law (5.13), it reads $K_{\theta\theta} = 8c(1 - 2\lambda_\theta^2 + \lambda_\theta^8)/\lambda_\theta^7 = 7.9704 \cdot 10^5$ Pa, and

$$C = \frac{\Delta V}{\Delta p} = \frac{dl\pi \Delta d}{2\Delta p} = \frac{\lambda_\theta D^3 l\pi}{4 K_{\theta\theta} H} = 1.6735 \cdot 10^{-10} \text{ m Pa}^{-3}$$

then determines the capacity of the femoral artery at the pressure of 100.0 mmHg.

(c) With the substitution of the outflow q_{out} in (2.37) by (2.38), the three-element vessel model in Fig. 5.14 is governed by

$$p_{\text{in}} - p_{\text{out}} = R(q_{\text{in}} - C\dot{p}_{\text{in}}) + L(\dot{q}_{\text{in}} - C\ddot{p}_{\text{in}}) . \tag{5.17}$$

Towards the derivation of the model's impedance, we consider steady-state periodic conditions and introduce the flow $q_{\text{in}} = \mathbf{Q}\exp(i\omega t)$, where $\mathbf{Q}$ denotes its complex amplitude. The pressures $p_{\text{in}} = \mathbf{P}\exp(i\omega t)$ and $p_{\text{out}} = 0$ may be substituted in (5.17), and the impedance then reads

$$\mathbf{Z} = \frac{\mathbf{P}}{\mathbf{Q}} = \frac{R + i\omega L}{1 + i\omega RC - \omega^2 LC} . \tag{5.18}$$

Figure 5.15 shows the modulus $|\mathbf{Z}|$ and phase shift $\arg(\mathbf{Z})$ as predicted by the three-element lumped parameter model.

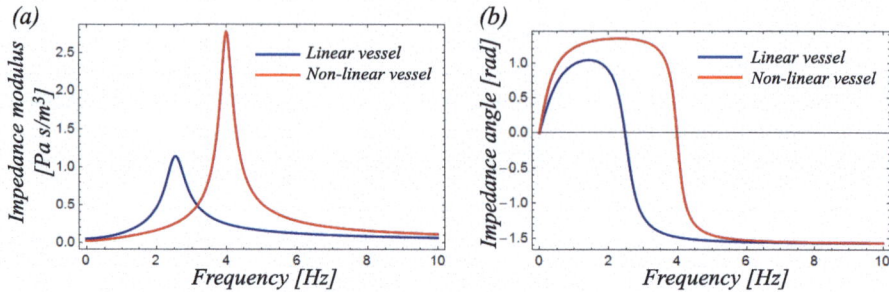

Fig. 5.15 Impedance of the three-element lumped parameter model. (**a**) Modulus $|\mathbf{Z}|$ and (**b**) phase shift $\arg(\mathbf{Z})$ as functions of the signal frequently. Blue and red curves correspond to the linear (Task (a)) and the non-linear (Task (b)) descriptions of the vessel wall ■

Example 5.2 (Residual Stresses of a Thick-Walled Artery).

(a) We consider the ring segment that is formed between the radii R_i and $R < R_o$. Given incompressibility, the expression $L(R^2 - R_i^2)k = l(r^2 - r_i^2)$ with $k = 2\pi/(2\pi - \alpha)$ holds and yields the kinematics relation

$$r = \sqrt{r_i^2 + (R^2 - R_i^2)/(k\lambda_z)}, \qquad (5.51)$$

where $\lambda_z = l/L$ denotes the axial stretch of the vessel segment.

The strain energy function (5.50) together with the relation (5.7) then determine the principal Cauchy stresses $\sigma_i = \bar{\sigma}_i - \kappa$, where $i = \theta, z, r$ denotes the circumferential, axial, and radial directions, respectively. Whilst the stretch determines the stress $\bar{\sigma}_i = 2c_1\lambda_i^2$, the Lagrange parameter κ contributes to the hydrostatic pressure and enforces the incompressibility.

(b) The only non-trivial equilibrium relation of an axisymmetric problem in cylindrical coordinates reads $r\,\mathrm{d}\sigma_r/\mathrm{d}r = \sigma_\theta - \sigma_r$. Its integration at the boundary conditions $\sigma_r(r_i) = \sigma_r(r_o) = 0$, yields the equilibrium expression

$$0 = \int_{r_i}^{r_o} \frac{\sigma_\theta - \sigma_r}{r} dr \qquad (5.52)$$

of the thick-walled tube problem.

(c) The axial stress σ_z results in the axial force $N = 2\pi \int_{r_i}^{r_o} (\overline{\sigma}_z - \kappa) r \, dr$, and with the relation $\kappa(r) = \overline{\sigma}_r(r) - \sigma_r(r) = \overline{\sigma}_r(r) - \int_{r_i}^{r} (\sigma_\theta - \sigma_r)/\xi \, d\xi$, it then reads

$$N = \pi \int_{r_i}^{r_o} (2\overline{\sigma}_z - \overline{\sigma}_r - \overline{\sigma}_\theta) r \, dr = 0. \qquad (5.53)$$

A detailed derivation of this relation is provided in Sect. 5.5.5.

(d) Given the fixed axial stretch $\lambda_z = 1$ and (5.51), the radial and circumferential stretches may be expressed as functions of the deformed inner radius r_i and the referential radius R, $\lambda_\theta = \lambda_\theta(r_i, R)$ and $\lambda_r = \lambda_z^{-1}\lambda_\theta^{-1} = \lambda_r(r_i, R)$. The stresses $\overline{\sigma}_i; i = \theta, z, r$ are then functions of r_i and R, and with the relation $r = \lambda_r R$ and $dr = \lambda_r dR$, the equilibrium expression (5.52) reads $0 = \int_{R_i}^{R_o} (\overline{\sigma}_\theta(r_i, R) - \overline{\sigma}_r(r_i, R))/R \, dR$, an expression that holds at $r_i = 7.26$ mm. In addition, (5.53) then reads $0 = \pi \int_{R_i}^{R_o} [2\overline{\sigma}_z(r_i, R) - \overline{\sigma}_r(r_i, R) - \overline{\sigma}_\theta(r_i, R)] R\lambda_r^2(r_i, R) dR$, and its numerical integration verifies $r_i = 7.26$ mm is a solution.

(e) The distribution of the stress differences across the wall is shown in Fig. 5.24.

(f) At the inside and the outside of the vessel, $\sigma_r = 0$ holds, resulting in the Lagrange parameter $\kappa = \overline{\sigma}_r$. It allows us to compute the respective circumferential stress $\sigma_\theta = \overline{\sigma}_\theta - \kappa = \overline{\sigma}_\theta - \overline{\sigma}_r$; $\sigma_\theta(R_i) = -3.307$ kPa and $\sigma_\theta(R_o) = 2.871$ kPa.

Fig. 5.24 Distribution of the stress differences $\overline{\sigma}_\theta - \overline{\sigma}_r = \sigma_\theta - \sigma_r$ and $\overline{\sigma}_\theta - \overline{\sigma}_z = \sigma_\theta - \sigma_z$ across the wall thickness of the load-free configuration Ω ∎

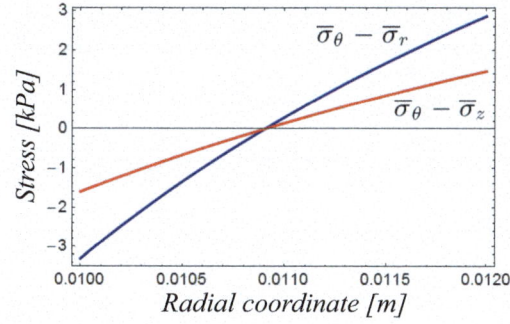

Example 5.3 (Discretization of the Convolution Integral).

(a) At the time t_{n+1} the convolution integral (5.57) may be split into

$$\overline{\mathbf{S}}_{\mathrm{M}}(t_{n+1}) = \sum_{i=1}^{N} \left\{ \beta_i \exp[-\Delta t/\tau_i] \int_{0}^{t_n} \exp[-(t_n - x)/\tau_i] \dot{\overline{\mathbf{S}}}_{\mathrm{E}} dx \right.$$

$$\left. + \beta_i \int_{t_n}^{t_{n+1}} \exp[-(t_{n+1} - x)/\tau_i] \dot{\overline{\mathbf{S}}}_{\mathrm{E}} dx \right\} ,$$

where $\Delta t = t_{n+1} - t_n$ denotes the time step. Given the over stress $\overline{\mathbf{S}}_{\mathrm{M}ni} = \beta_i \int_0^{t_n} \exp[-(t_n - x)/\tau_i] \dot{\overline{\mathbf{S}}}_{\mathrm{E}} dx$ at the time t_n, the over stress at the time t_{n+1} may be discretized by

$$\overline{\mathbf{S}}_{\mathrm{M}}(t_{n+1}) \approx \overline{\mathbf{S}}_{\mathrm{M}n+1} = \sum_{i=1}^{N} \left[\xi_i^2 \overline{\mathbf{S}}_{\mathrm{M}ni} + \beta_i \xi_i \left(\overline{\mathbf{S}}_{\mathrm{E}n+1} - \overline{\mathbf{S}}_{\mathrm{E}n} \right) \right] , \qquad (5.59)$$

where the abbreviation $\xi_i = \exp[-\Delta t/(2\tau_i)]$ has been introduced. The discretization is based on the second-order accurate mid-point integration rule

$$\int_{t_n}^{t_{n+1}} \exp\left(-\frac{t_{n+1} - x}{\tau_i} \right) \dot{\overline{\mathbf{S}}}_{\mathrm{E}} dx \approx \exp\left(\frac{-t_{n+1} + t_{n+1} - \Delta t/2}{\tau_i} \right) \frac{\overline{\mathbf{S}}_{\mathrm{E}n+1} - \overline{\mathbf{S}}_{\mathrm{E}n}}{\Delta t} \Delta t .$$

(b) Table 5.5 summarizes the different steps to solve the visco-hyperelastic thin-walled tube problem, and Fig. 5.27 presents the results that have been achieved with $n = 30$ and $n = 300$ time steps, respectively.

(c) Given the property $\partial(\bullet)_n/\partial \mathbf{C}_{n+1} = \mathbf{0}$ of all history terms $(\bullet)_n$, the derivative of the over stress (5.59) with respect to the right Cauchy–Green strain $\mathbf{C}$ yields

$$\overline{\mathbb{C}}_{\mathrm{M}n+1}^{\mathrm{algo}} = 2\partial \overline{\mathbf{S}}_{\mathrm{M}n+1}/\partial \mathbf{C}_{n+1} = \overline{\mathbb{C}}_{\mathrm{E}n+1} \sum_{i=1}^{N} \beta_i \xi_i ,$$

where $\overline{\mathbb{C}}_{\mathrm{E}n+1} = 2\partial \overline{\mathbf{S}}_{\mathrm{E}n+1}/\partial \mathbf{C}_{n+1}$ denotes the stiffness of the isochoric elastic stress. Note that $\overline{\mathbb{C}}_{\mathrm{M}n+1}^{\mathrm{algo}}$ depends on the algorithmic parameter $\xi_i = \exp[-\Delta t/(2\tau_i)]$—it is therefore called the algorithmic tangent.

Table 5.5 Algorithm to numerically solve the visco-hyperelastic thin-walled tube problem

(1) Specify geometry and constitutive parameters

$D = 9.0$ mm; $H = 0.8$ mm ; $c_3 = 50.0$ kPa

$\tau_1 = 0.2$ s^{-1}; $\tau_2 = 0.7$ s^{-1}; $\beta_1 = 0.3$; $\beta_2 = 0.2$

(2) Compute initial isochoric elastic second Piola–Kirchhoff stress

$\lambda_\theta = 1.4$; $\lambda_z = 1.2$; $\lambda_r = (\lambda_\theta \lambda_z)^{-2}$; $\alpha = 2c_3(\lambda_\theta^2 + \lambda_z^2 + \lambda_r^2 - 3)^2$

$\overline{S}_{E\theta n} = \alpha \left(2 - \frac{\lambda_r^2 + \lambda_z^2}{\lambda_\theta^2} \right)$; $\overline{S}_{Ezn} = \alpha \left(2 - \frac{\lambda_\theta^2 + \lambda_r^2}{\lambda_z^2} \right)$; $\overline{S}_{Ern} = \alpha \left(2 - \frac{\lambda_\theta^2 + \lambda_z^2}{\lambda_r^2} \right)$

(3) Set the time discretization and loop over the time increments

$t_e = 3.0$ s ; $n = 300$; $\Delta t = t_e/n$; $t = 0.0$

Do $i = 1, \ldots, n$

$t = t + \Delta t$

$\lambda_\theta = 1.4 + 0.1 \sin(2\pi t)$; $\lambda_z = 1.2$; $\lambda_r = (\lambda_\theta \lambda_z)^{-2}$; $\alpha = 2c_3(\lambda_\theta^2 + \lambda_z^2 + \lambda_r^2 - 3)^2$

$\overline{S}_{E\theta} = \alpha \left(2 - \frac{\lambda_r^2 + \lambda_z^2}{\lambda_\theta^2} \right)$; $\overline{S}_{Ez} = \alpha \left(2 - \frac{\lambda_\theta^2 + \lambda_r^2}{\lambda_z^2} \right)$; $\overline{S}_{Er} = \alpha \left(2 - \frac{\lambda_\theta^2 + \lambda_z^2}{\lambda_r^2} \right)$

(4) Integrate the convolution integral (5.57) over the time increment

$\xi_j = \exp(\frac{\Delta t}{2\tau_j})$; $\overline{S}_{Mkj} = \xi_j^2 \overline{S}_{Mkjn} + \beta_j \xi_j(\overline{S}_{Ek} - \overline{S}_{Ekn})$; $k = \theta, z, r$; $j = 1, 2$

(5) Express the hydrostatic pressure and compute the total second Piola–Kirchhoff stress

$p = \lambda_r^2 \left(\sum_{j=1,2} \overline{S}_{Mrj} + \overline{S}_{Er} \right)$

$S_k = \overline{S}_{Ek} + \sum_{j=1,2} \overline{S}_{Mkj} - p\lambda_k^{-2}$; $k = \theta, z$

(6) Compute the inflation pressure and the reduced axial force

$p_i = 2S_\theta/(D\lambda_z)$; $F = \lambda_z S_z DH\pi - p_i\pi/(4D^2\lambda_\theta^2)$

(7) Update the history variables

$\overline{S}_{Mkjn} \leftarrow \overline{S}_{Mkj}$; $k = \theta, z$; $j = 1, 2$

$\overline{S}_{Ekn} \leftarrow \overline{S}_{Ek}$; $k = \theta, z$

End Do

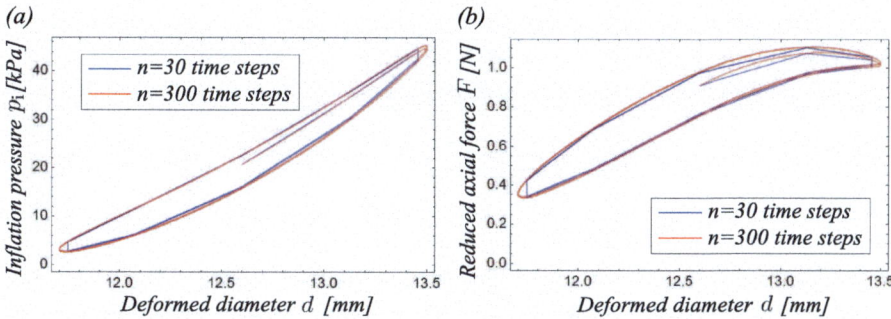

Fig. 5.27 Numerical results of the inflated visco-hyperelastic thin-walled tube at the prescribed circumferential and axial stretches of $\lambda_\theta = 1.4 + 0.1 \sin(2\pi t)$ and $\lambda_z = 1.2$, respectively. (**a**) Inflation pressure p_i *versus* the deformed diameter d of the tube. (**b**) Reduced axial force F *versus* the deformed diameter d of the tube. The curves cover the time interval $0 < t < 3$ s and results are presented for $n = 30$ and $n = 300$ time steps, respectively ∎

Example 5.4 (Progressive Engagement and Rupture of Collagen Fibrils).

(a) Let us consider the deformation of a single undulated fibril, where λ denotes the stretch of the tissue. It may be split into two parts, $\lambda = \lambda_s \lambda_e$, where λ_s defines the fibril's intermediate configuration, whilst the stretch λ_e elastically deforms the fibril relative to its intermediate configuration.

The first Piola–Kirchhoff stress of the vascular wall tissue may then be expressed by the integral

$$P = k \int_0^\lambda \Upsilon_s(x)\mathrm{d}x \,, \tag{5.61}$$

where $\Upsilon_s(\lambda) = \int_{-\infty}^\lambda \rho_s(x)\mathrm{d}x$ denotes the Cumulative Density Function (CDF). It determines the portion of collagen fibrils that are engaged at the stretch λ. The integration of (5.61) yields the piece-wise polynomial expressions

$$P_e(\lambda) = \begin{cases} 0 & ; \, 0 < \lambda \leq \lambda_1 \,, \\[2ex] \dfrac{k(\lambda - \lambda_1)^3}{3(\lambda_1 - \lambda_2)(\lambda_1 - \lambda_3)} & ; \, \lambda_1 \leq \lambda \leq \lambda_3 \,, \\[2ex] \dfrac{k[\lambda^3 - \lambda_1^2(\lambda_2 - \lambda_3) - \lambda_1(\lambda_2 - \lambda_3)(\lambda_3 - 3\lambda) - \lambda_2(\lambda_3^2 - 3\lambda_3\lambda + 3\lambda^2)]}{3(\lambda_1 - \lambda_2)(\lambda_2 - \lambda_3)} & ; \, \lambda_3 \leq \lambda \leq \lambda_2 \,, \\[2ex] \dfrac{k(3\lambda - \lambda_1 - \lambda_2 - \lambda_3)}{3} & ; \, \lambda_2 < \lambda < \infty \end{cases}$$

of the first Piola–Kirchhoff stress. The first Piola transform for incompressible materials allows us then to compute the Cauchy stress $\sigma_e = \lambda P_e$. Figure 5.29a shows the tissue's elastic properties.

(b) The ruptured fibrils do not contribute to the stress in the tissue, and we therefore subtract their contribution from the elastic stress (5.61). We may compute the PDF of the ruptured fibrils $\rho_f(x)$ by "stretching" the engagement PDF $\rho_s(\lambda)$ by the factor λ_f. The CDF $\Upsilon_f(\lambda) = \int_{\lambda_f \lambda_1}^\lambda \rho_f(x)\mathrm{d}x$ then determines the portion of ruptured fibrils, where λ_f denotes the failure stretch. We may alternatively have calculated the CDF by $\Upsilon_f(\lambda) = \int_{\lambda_1}^{\lambda/\lambda_f} \rho_s(x)\mathrm{d}x$, see Fig. 5.29b.

The first Piola–Kirchhoff stress to be retracted from the elastic stress then reads

$$P_f = k \left[\int_{\lambda_1}^{\lambda/\lambda_f} \Upsilon_s(x)\mathrm{d}x + \int_{\lambda/\lambda_f}^\lambda \Upsilon_s(\lambda/\lambda_f)\mathrm{d}x \right] \,, \tag{5.62}$$

and the lengthy piece-wise polynomial expressions from the integration of this relation are not explicitly shown. Given the stress $P_i = P_e - P_f$ of the irreversibly deformed (damaged) tissue, the first Piola transform then yields the Cauchy stress $\sigma_i = \lambda P_i$. Figure 5.29a illustrates the properties of the damaged tissue.

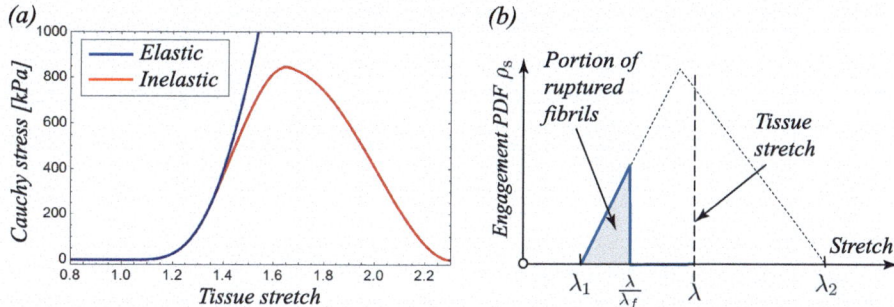

Fig. 5.29 (a) Cauchy stress *versus* stretch properties of the elastic (blue curve) and inelastic (red curve) properties of the medial tissue patch. (b) Probability Density Function (PDF) towards the computation of ruptured fibrils ∎

Example 5.5 (Simple Shear Testing in Vessel Tissue Characterization).

(a) The motion

$$x_1 = X_1 + \gamma X_2 \; ; \; x_2 = X_2 \; ; \; x_3 = X_3 \qquad (5.71)$$

determines simple shear kinematics, where $\mathbf{x} = \boldsymbol{\chi}(\mathbf{X}) = [x_1 \, x_2 \, x_3]^{\mathrm{T}}$ and $\mathbf{X} = [X_1 \, X_2 \, X_3]^{\mathrm{T}}$ denote the spatial and referential material particle positions, respectively. Figure 5.38 illustrates the kinematics, and the present experimental design yields $\gamma = u/h$.

Fig. 5.38 Simple shear kinematics relates the reference configuration Ω_0 and the spatial configuration Ω of the material particle

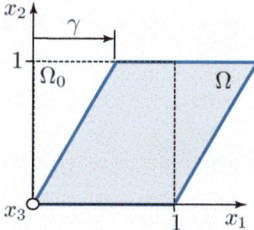

(b) The motion (5.71) defines the deformation gradient

$$\mathbf{F} = \mathrm{Grad}\,\boldsymbol{\chi}(\mathbf{X}) = \frac{\partial \boldsymbol{\chi}}{\partial \mathbf{X}} = \begin{bmatrix} 1 & \gamma & 0 \\ 0 & 1 & 0 \\ 0 & 0 & 1 \end{bmatrix},$$

which then leads to the right and left Cauchy–Green strains

$$\mathbf{C} = \mathbf{F}^T\mathbf{F} = \begin{bmatrix} 1 & \gamma & 0 \\ \gamma & 1+\gamma^2 & 0 \\ 0 & 0 & 1 \end{bmatrix} \; ; \; \mathbf{b} = \mathbf{F}\mathbf{F}^T = \begin{bmatrix} 1+\gamma^2 & \gamma & 0 \\ \gamma & 1 & 0 \\ 0 & 0 & 1 \end{bmatrix}, \quad (5.72)$$

respectively.

(c) The eigenvalue analysis of the symmetric second-order tensor $(5.72)_1$ allows us to express the right Cauchy–Green strain through $\mathbf{C} = \sum_{i=1}^{3} \lambda_{Ci} \mathbf{N}_i \otimes \mathbf{N}_i$, where

$$\lambda_{C1} = 1 \; ; \; \lambda_{C2} = 1 + \left(\gamma^2 - \gamma\sqrt{4+\gamma^2}\right)/2 \; ; \; \lambda_{C3} = 1 + \left(\gamma^2 + \gamma\sqrt{4+\gamma^2}\right)/2$$

are the eigenvalues, whilst

$$\mathbf{N}_1 = \begin{bmatrix} 0 \\ 0 \\ 1 \end{bmatrix} \; ; \; \mathbf{N}_2 = \begin{bmatrix} \frac{-\gamma-\sqrt{4+\gamma^2}}{2\sqrt{1+\lambda_{C2}}} \\ \frac{1}{\sqrt{1+\lambda_{C2}}} \\ 0 \end{bmatrix} \; ; \; \mathbf{N}_3 = \begin{bmatrix} \frac{-\gamma+\sqrt{4+\gamma^2}}{\sqrt{1+\lambda_{C3}}} \\ \frac{1}{\sqrt{1+\lambda_{C3}}} \\ 0 \end{bmatrix}$$

are the eigenvectors. The relation $\lambda_{Ci} = \lambda_i^2$ holds between the eigenvectors of $\mathbf{C}$ and the principal stretches λ_i. The eigenvectors depend on the deformation, and Fig. 5.39a illustrates the rotation of $\mathbf{N}_2$ and $\mathbf{N}_3$ with increasing levels of shear γ.

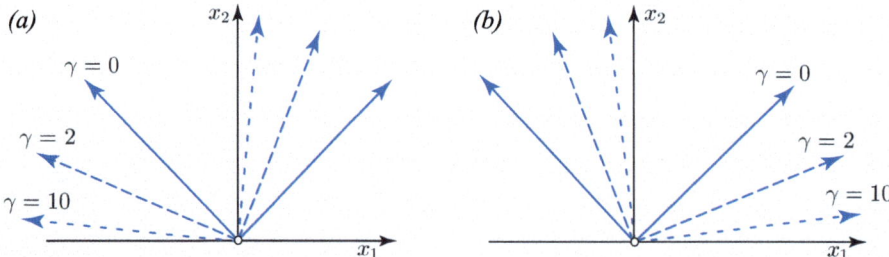

Fig. 5.39 Rotation of the principal strain directions for simple shear kinematics and at increasing levels of shear γ. (a) Principal strain directions $\mathbf{N}_2$ and $\mathbf{N}_3$ of the right Cauchy–Green strain $\mathbf{C}$. (b) Principal strain directions $\mathbf{n}_2$ and $\mathbf{n}_3$ of the left Cauchy–Green strain $\mathbf{b}$.

The eigenvalue analysis of the matrix $(5.72)_2$ expresses the left Cauchy–Green strain through $\mathbf{b} = \sum_{i=1}^{3} \lambda_{bi} \mathbf{n}_i \otimes \mathbf{n}_i$, where λ_{bi} and $\mathbf{n}_i$ denote its eigenvalues and eigenvectors, respectively. Whilst $\mathbf{b}$ and $\mathbf{C}$ have different eigenvectors, their eigenvalues are identical, $\lambda_{bi} = \lambda_{Ci}; i = 1, 2, 3$. The eigenvectors of $\mathbf{b}$ are

$$
\mathbf{n}_1 = \begin{bmatrix} 0 \\ 0 \\ 1 \end{bmatrix} \; ; \; \mathbf{n}_2 = \begin{bmatrix} \dfrac{\gamma - \sqrt{4+\gamma^2}}{2\sqrt{1+\lambda_{b2}}} \\[2ex] \dfrac{1}{\sqrt{1+\lambda_{b2}}} \\[2ex] 0 \end{bmatrix} \; ; \; \mathbf{n}_3 = \begin{bmatrix} \dfrac{\gamma + \sqrt{4+\gamma^2}}{2\sqrt{1+\lambda_{b3}}} \\[2ex] \dfrac{1}{\sqrt{1+\lambda_{b3}}} \\[2ex] 0 \end{bmatrix} \, ,
$$

and, as with $\mathbf{N}_i$, also $\mathbf{n}_i$ rotates with increasing deformation, see Fig. 5.39b.

(d) Coleman and Noll's relation for an incompressible solid (5.4) together with the strain energy (5.70) yields the Cauchy stress

$$
\boldsymbol{\sigma} = 2\alpha \mathbf{b} - \kappa \mathbf{I} \; ; \; \alpha = c_1 + 2c_2(I_1 - 3) = c_1 + 2c_2\gamma^2 , \qquad (5.73)
$$

where the first invariant $I_1 = \mathrm{tr}\,\mathbf{C} = 3 + \gamma^2$ follows from $(5.72)_1$. The Lagrange contribution $\kappa = 2\alpha$ to the hydrostatic pressure may be identified from the equilibrium condition $\sigma_{33} = 0$, and the experimental test then exposes the tissue to the shear stress $\sigma_{12} = 2\alpha\gamma = 2[c_1 + 2c_2(u/h)^2]u/h$ in the x_1–x_2 plane. The stress is distributed over the area wl, and the force

$$
P = 2wl \left[c_1 \left(\frac{u}{h} \right) + 2c_2 \left(\frac{u}{h} \right)^3 \right] \qquad (5.74)
$$

is therefore measured by the simple shear experiment shown in Fig. 5.37.

(e) The objective function

$$
\Phi(c_1, c_2) = \sum_{j=1}^{n} \left[P_j^{\mathrm{exp}} - P(u_j) \right]^2
$$

expresses the least-square error between model and experiment, where $n = 4$ denotes the number of experimental measurement points. With the experimental data, the objective function reads

$$
\begin{aligned}
\Phi(c_1, c_2) = {} & [3.0 - 283.333(c_1 + 0.222222c_2)]^2 \\
& + [5.0 - 566.667(c_1 + 0.888889c_2)]^2 \\
& + [12.0 - 850.0(c_1 + 2.0c_2)]^2 \\
& + [23.0 - 1133.33(c_1 + 3.55556c_2)]^2 ,
\end{aligned}
$$

and its minimum $\Phi \to$ MIN results in the parameters $c_1 = 6.352$ kPa and $c_2 = 3.904$ kPa. Figure 5.40 shows the force P as a function of the displacement u for these parameters.

Fig. 5.40 Force P *versus* displacement u of the simple shear experiment. The experimental data (circles) has been used for the least-square optimized model response (solid curve)

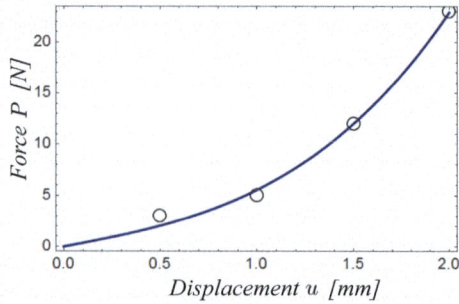

Example 5.6 (Pressure Inflation in Vessel Tissue Characterization).

(a) The deformation gradient $\mathbf{F} = \mathrm{diag}[\lambda_\theta, \lambda_z, \lambda_r]$ determines the kinematics of the inflation experiment, where $\lambda_\theta, \lambda_z$, and λ_r denote the stretches in the circumferential, axial, and radial directions, respectively. The right Cauchy–Green strain then reads $\mathbf{C} = \mathrm{diag}[\lambda_\theta^2, \lambda_z^2, \lambda_r^2]$ and $I_1 = \mathrm{tr}\mathbf{C} = \lambda_\theta^2 + \lambda_z^2 + \lambda_r^2$ is its first invariant.

Coleman and Noll's relation for incompressible solids (5.7) defines the principal Cauchy stresses $\sigma_i = \lambda_i \partial \Psi / \partial \lambda_i - \kappa$, where the indices $i = \theta, z, r$ denote axial, circumferential and radial vessel directions, respectively. Given the membrane assumption, the stress in radial direction is negligible as compared to the in-plane stresses, $\sigma_r \approx 0$. It allows us to substitute the Lagrange contribution to the hydrostatic pressure $\kappa = \lambda_r \partial \Psi / \partial \lambda_r$ and then leads to the circumferential and axial stresses

$$\left. \begin{aligned} \sigma_\theta &= \lambda_\theta \partial \Psi / \partial \lambda_\theta - \lambda_r \partial \Psi / \partial \lambda_r = \alpha(\lambda_\theta^2 - \lambda_r^2) \,; \\ \sigma_z &= \lambda_z \partial \Psi / \partial \lambda_z - \lambda_r \partial \Psi / \partial \lambda_r = \alpha(\lambda_z^2 - \lambda_r^2) \,, \end{aligned} \right\} \tag{5.76}$$

where the strain energy density (5.75) and the factor $\alpha = 4c(I_1 - 3)$ have been used.

(b) The equilibrium $p_i d = 2\sigma_\theta h$ along the circumferential direction and $F_z = \sigma_z \pi d h - d^2 \pi p_i / 4$ along the axial direction, together with the stress relations (5.76), determine the expressions

$$p_\mathrm{i} = \frac{2h\alpha}{d}(\lambda_\theta^2 - \lambda_r^2) \,; \quad F_z = \frac{dh\alpha\pi}{2}(2\lambda_z^2 - \lambda_r^2 - \lambda_\theta^2)$$

of the inflation pressure and the reduced axial force, respectively. Here, $\lambda_\theta = d/D$, $\lambda_z = l/L$ and $\lambda_r = DL(dl)^{-1}$ have been used to substitute the stretches.

(c) The least-square error between model and experiment is expressed by the objective function

$$\Phi = \sum_{j=1}^{n} \left\{ \alpha \left[p_{i j}^{\text{exp}} - p_i(d_j, l_j) \right]^2 + \left[F_{z j}^{\text{exp}} - F_z(d_j, l_j) \right]^2 \right\} ,$$

where $n = 2$ denotes the number of experimental measurement points. As the inflation pressure p_i and the reduced axial force F_z cover a similar range of data, we used the scaling parameter $\alpha = 1$ in this problem. Given the experimental data, the objective function reads

$$\Phi = (5.2 - 0.1016398c)^2 + (3.0 - 0.1466604c)^2$$

$$+ (21.0 - 0.5217602c)^2 + (15.0 - 0.3578883c)^2 ,$$

and its minimization $\Phi \rightarrow \text{MIN}$ determines the parameter $c = 40.0173$ kPa. Figure 5.42 shows the inflation pressure p_i and reduced axial force F_z as functions of the deformed vessel diameter d, and at the constant axial stretch $\lambda_z = l/L = 1.05$.

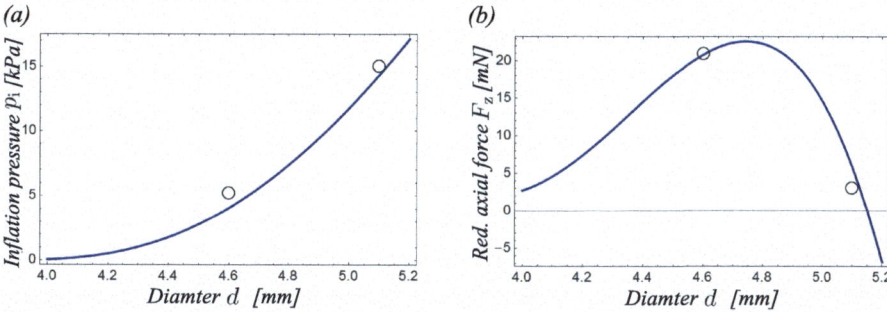

Fig. 5.42 The inflation pressure p_i and the reduced axial force F_z as functions of the deformed vessel diameter d. The vessel is inflated at the fixed deformed vessel length of $l = 73.5$ mm, and the circles show the experimental data that has been used in the least-square identification of the constitutive parameter c ∎

Example 6.1 (Walburn–Schneck Viscosity Model).

(a) We introduce the objective function

$$\Phi(c_1, c_2, c_3) = \sum_{i=1}^{n} (\eta_{\text{WS}}(\dot{\gamma}_i, \phi_i; c_1, c_2, c_3) - \eta_{\text{exp}_i})^2 ,$$

where $\eta_{\exp_i}$ denotes the viscosity measured at the i-th measurement point, whilst $\eta_{\mathrm{WS}}(\dot{\gamma}_i, \phi_i; c_1, c_2, c_3)$ is the viscosity predicted by the Walburn–Schneck model at the i-th measurement point. In addition to the model parameters, η_{WS} also depends on the shear rate $\dot{\gamma}_i$ and the hematocrit level ϕ_i. The minimization $\Phi(c_1, c_2, c_3) \rightarrow$ MIN then yields the set

$$c_1 = 2.971 \, \mathrm{mPa \, s} \; ; \quad c_2 = 4.729 \; ; \quad c_3 = 0.759$$

of best-fit parameters, and thus the parameters of the Walburn–Schneck viscosity model with the least error between the measured and predicted viscosity.

(b) Figure 6.11 illustrates the model's ability to represent the experimental data. Whilst the Walburn–Schneck model captures the viscosity at low shear rates, the inherent (and non-physical) property $\eta(\dot{\gamma} \rightarrow \infty) = 0$ of the Power Law model causes significant errors at high shear rates.

Fig. 6.11 Walburn–Schneck model prediction in relation to the experimentally measured viscosity. Model parameters have been identified by least-square optimization, and ϕ_i denotes hematocrit levels

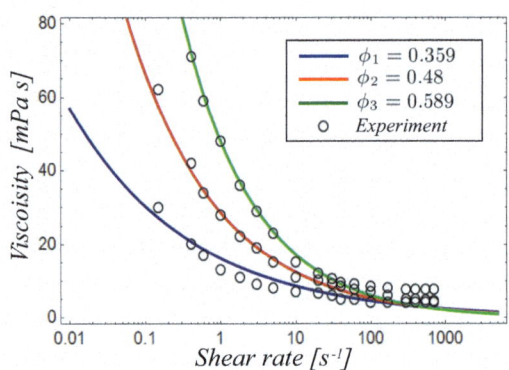

Example 6.2 (Collapse of a Constricted Vessel Segment).

(a) At steady-state the inflow rate $q_i = v_i d_i^2 \pi / 4$ matches the flow rate $q_c = v_c d_c^2 \pi / 4$ in the constriction. The blood then has the velocity $v_c = v_i \alpha$ in the constriction, where $\alpha = (d_i / d_c)^2$ denotes the constriction factor.

(b) Bernoulli's equation (6.19) reads $v_i^2 / 2 + p_i / \rho = v_c^2 / 2 + p_c / \rho$, and

$$p_c = p_i + \rho(v_i^2 - v_c^2)/2 = p_i + \rho v_i^2 (1 - \alpha^2)/2 , \qquad (6.21)$$

therefore expresses the pressure in the constriction.

(c) A vessel wall without bending stiffness buckles as soon as the pressure p_c in the constriction falls below the ambient pressure p_0—the vessel segment then collapses. Given $p_c = p_0$, the expression (6.21) determines the constriction factor

$$\alpha = \sqrt{1 - \frac{2(p_0 - p_i)}{\rho v_i^2}} = 17.189 \,,$$

a condition resulting in the diameter stenosis

$$S_{\text{diameter}} = \frac{d_i - d_c}{d_i} = (1 - \alpha^{-1/2}) = 0.7588$$

and the area stenosis

$$S_{\text{area}} = \frac{A_i - A_c}{A_i} = (1 - \alpha^{-1}) = 0.9418 \,,$$

where $A_k = d_k^2 \pi/4$; $k = i$, c denote the areas of the respective circular cross-sections. A constriction that occupies 76% of the diameter, or 94% of the cross-sectional area, therefore leads to the collapse of the vessel segment. ∎

Example 6.3 (Oscillating Plate on Top of a Fluid Layer).

(a) Given the assumptions of the velocity and pressure, the governing equation of 1D flows in Cartesian coordinates (6.22) results in the ordinary differential equation

$$\frac{d^2 V(y)}{dy^2} - \frac{\rho \omega V(y)}{\eta \tan(\omega t)} = 0 \,,$$

and integration then yields the velocity amplitude

$$V(y) = C_1(t) \exp(\alpha y) + C_2(t) \exp(-\alpha y) \ \text{ with } \ \alpha = \sqrt{\frac{\rho \omega}{\eta} \cot(\omega t)} \,.$$

The two no-slip boundary conditions $V(0) = 0$ and $V(H) = V_0$ allow us to identify the two integration constants

$$C_2(t) = -C_1(t) \,; \ \ C_1(t) = \frac{V_0}{\exp(\alpha H) - \exp(-\alpha H)} \,,$$

such that

$$v(y, t) = V_0 \frac{\exp(\alpha y) - \exp(-\alpha y)}{\exp(\alpha H) - \exp(-\alpha H)} \sin(\omega t)$$

expresses the velocity of the fluid layer.

(b) Given the provided parameters, the fluid profile $v(y, t)$ is plotted in Fig. 6.20.

Fig. 6.20 Fluid flow
underneath an oscillating
plate. Fluid velocity $v(y, t)$
versus fluid layer depth
coordinate y at different
times t

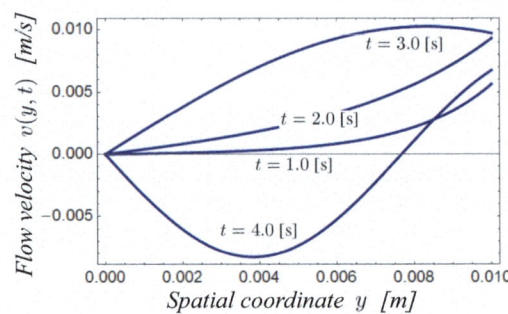

Example 6.4 (Steady-State Flow in a Vessel Segment).

(a) With the relation (6.35), the pressure gradient reads

$$-\frac{dp}{dz} = \frac{8\eta q}{\pi r_0^4} = 1990.2 \text{ Pa m}^{-1} \,, \tag{6.37}$$

and (6.34)$_2$ then yields the velocity $v_{max} = -(dp/dz)(r_0^2/4\eta) = 0.5486 \text{ m s}^{-1}$ in the center of the tube. At the radial coordinate $r[m]$, the Poiseuille flow velocity then reads

$$v(r) = v_{max}\left[1 - (r/r_0)^2\right] = 0.5486 - 1.2439 \cdot 10^5 r^2 \text{ [m s}^{-1}] \,, \tag{6.38}$$

where (6.34)$_1$ has been used in the computation of this result.

(b) Given the velocity profile (6.38) and r [m], the shear stress reads

$$\tau = \eta\frac{dv}{dr} = r\frac{dp}{dz} = -995.122r \text{ [kPa]} \,,$$

where (6.37) has been used.

(c) With the pressure gradient (6.37), $\Delta p = (dp/dz)l = 199.0 \text{ Pa}$ denotes the pressure drop over a 10.0 cm long segment of the iliac artery. ∎

Example 6.5 (Power Law Fluid at Steady-State Tube Flow).

(a) The flow is steady state $\partial(\bullet)/\partial t = 0$, fully developed $\partial(\bullet)/\partial z = 0$, and free of body forces $\mathbf{b}_f = \mathbf{0}$. The linear momentum (6.28) then determines the set

$$\frac{\partial \tau_{rz}}{\partial r} + \frac{\tau_{rz}}{r} - \frac{\partial p}{\partial z} = 0 \; ; \quad \frac{\partial p}{\partial r} = 0 \tag{6.39}$$

of partial differential equations that define the flow velocity v_z. The relation $(6.39)_2$ implies $p = p(z)$, and

$$\frac{1}{r}\frac{\mathrm{d}}{\mathrm{d}r}(r\tau_{rz}) = \frac{\mathrm{d}p}{\mathrm{d}z} \quad \text{with} \quad \tau_{rz} = \eta(\dot{\gamma})\frac{\mathrm{d}v_z}{\mathrm{d}r} \tag{6.40}$$

governs the problem, where the viscosity is a function of the scalar shear rate $\dot{\gamma}$. The product rule has been used in the derivation of this expression. Given the rate of deformation $\mathbf{d}$ of tube flow kinematics (6.27), the scalar shear rate reads $\dot{\gamma} = \sqrt{2\mathrm{tr}\mathbf{d}^2} = \sqrt{2\mathbf{d} : \mathbf{d}} = \mathrm{d}v_z/\mathrm{d}r$.

(b) The integration of the governing equation (6.40) yields

$$\frac{\mathrm{d}p}{\mathrm{d}z}\frac{r^2}{2} = r\tau_{rz} + C \, ,$$

where the symmetry conditions $\tau_{rz}(0) = 0$ result in the integration constant $C = 0$. With the Power law model (6.6) and the definition of the shear stress $\tau_{rz} = \eta(\dot{\gamma})(\mathrm{d}v_z/\mathrm{d}r)$, the differential equation

$$\left(\frac{\mathrm{d}p}{\mathrm{d}z}\frac{r}{2\eta_0\lambda^{n-1}}\right)^{1/n} = \frac{\mathrm{d}v_z}{\mathrm{d}r} \tag{6.41}$$

determines the problem. Integration and use of the no-slip boundary condition $v_z(r_0) = 0$, yields then the velocity profile

$$v_z = \alpha\left[1 - \left(\frac{r}{r_0}\right)^{\frac{n+1}{n}}\right] \quad \text{with} \quad \alpha = \frac{nr_0^{(n+1)/n}}{n+1}\sqrt[n]{\frac{-\mathrm{d}p/\mathrm{d}z}{2\eta_0\lambda^{n-1}}} \tag{6.41}$$

of the tube flow described by the Power Law rheology model.

(c) Figure 6.21 illustrates the normalized velocity profiles for a number of the power law constants n.

(d) Given the provided parameters, the integration over the velocity profile (6.41) yields the flow rate $q = 2\pi \int_0^{r_0} v_z r\,\mathrm{d}r = 4.06748$ ml s^{-1}.

Fig. 6.21 Velocity profiles
of the Power Law fluid that
flows through a circular tube.
Velocity and tube radius are
normalized and profiles for
different power law constants
n are shown. The Newtonian
fluid corresponds to $n = 1$
and determines the Poiseuille
flow profile ∎

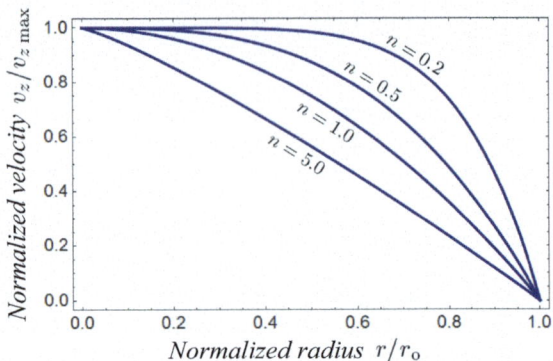

Example 6.6 (Stented Vessel).

(a) The flow is at steady state $\partial(\bullet)/\partial t = 0$, fully developed $\partial(\bullet)/\partial z = 0$, free of
body forces $\mathbf{b}_f = \mathbf{0}$, and governed by the linear momentum (6.28). It then leads
to

$$\frac{dp}{dz} = \frac{\eta}{r}\frac{d}{dr}\left(r\frac{dv}{dr}\right),$$

and twice integration results in the velocity

$$v = \frac{1}{4\eta}\frac{dp}{dz}r^2 - C_1 \ln r - C_2, \tag{6.43}$$

where C_1 and C_2 denote integration constants.
Given the native vessel, $C_1 = 0$, a condition that keeps the velocity bounded
at $r = 0$. The no-slip boundary condition $v(r_v) = 0$ identifies the second
integration constant $C_2 = r_v^2/(4\eta)(dp/dz)$, and the Poiseuille flow profile

$$v = v_{max}\left[1 - (r/r_v)^2\right] \quad \text{with} \quad v_{max} = \frac{r_v^2}{4\eta}\left(-\frac{dp}{dz}\right)$$

then describes the velocity, where $r_v = d_v/2$ denotes the vessel radius. The
velocity v_{max} in the center of the vessel is determined by the flow rate $q = 2\pi \int_0^{r_v} vr\,dr = \pi r_v^4/(8\eta)(-dp/dz)$, and reads $v_{max} = 2q/(\pi r_v^2)$.
The shear stress $\tau(r) = \eta dv/dr = -4\eta qr/(\pi r_v^4)$ then results in the WSS of
$-\tau(r_v) = 21.9$ Pa.

(b) The expression (6.43) also determines the flow through the ring formed between the stent hub and the vessel wall, see Fig. 6.23. In this case we may identify the two integration constants from the two no-slip conditions $v(r_v) = v(r_s) = 0$. The identification yields

$$C_1 = \frac{r_s^2 - r_v^2}{4\eta[\ln(r_s) - \ln(r_v)]} \left(\frac{dp}{dz}\right) \; ; \; C_2 = \frac{r_v^2 \ln(r_s) - r_s^2 \ln(r_v)}{4\eta[\ln(r_s) - \ln(r_v)]} \left(\frac{dp}{dz}\right) ,$$

and after some algebraic manipulations

$$v = \frac{1}{4\eta}\left[r^2 - \frac{r_s^2 - r_v^2}{\ln(r_s) - \ln(r_v)}\ln(r) - \frac{r_v^2 \ln(r_s) - r_s^2 \ln(r_v)}{\ln(r_s) - \ln(r_v)}\right]\left(\frac{dp}{dz}\right) \quad (6.44)$$

determines the velocity in the stented vessel. The integration over this velocity profile then yields the flow rate $q = 2\pi \int_{r_s}^{r_v} vr\,dr$, and allows us to express the pressure gradient

$$\frac{dp}{dz} = \frac{8q\eta[\ln(r_s) - \ln(r_v)]}{\pi(r_s^2 - r_v^2)[-r_s^2 + r_v^2 + r_s^2\ln(r_s) + r_v^2\ln(r_s) - r_s^2\ln(r_v) - r_v^2\ln(r_v)]}$$

$$= -35353.7 \; \text{Pa}\,\text{m}^{-1} . \quad (6.45)$$

Given the flow velocity (6.44) and the pressure gradient (6.45), the shear stress of the 1D ring-flow problem reads

$$\tau = \eta\frac{dv}{dr} = \left\{\frac{r}{2} - \frac{r_s^2 - r_v^2}{4r[\ln(r_s) - \ln(r_v)]}\right\}\left(\frac{dp}{dz}\right) = 0.0267072r^{-1} - 17676.8r \; [\text{Pa}] .$$

It is a function of r [m], an expression that and results in the WSS of $-\tau(r_v) = 32.41$ Pa of the stented vessel.

(c) Given the pressure gradient (6.45), the blood pressure drops by $\Delta p = (-dp/dz)l = 1979.81$ Pa over the length l of the stent. ∎

Example 6.7 (Pulsatile Blood Flow in a Vessel Segment).

(a) Given the vessel radius $r_o = d/2$ and the angular velocity ω, the Womersley number $Wo = r_o\sqrt{\omega/\nu} = 4.138$ determines the flow.

(b) The relation (6.46) allows us to compute the velocity profile at different time points, see Fig. 6.24. The graph shows the real part of the fluid velocity $Re[v(r, t)]$ *versus* the radial coordinate r. The velocity profiles deviate substantially from the parabolic profile of Poiseuille flow, and we note that reversal flow starts in the laminae near the wall.

Fig. 6.24 Blood velocity of pulsatile tube flow as described by a Newtonian fluid at the Womersley number $Wo = 4.138$. Velocity profiles are shown at a number of time points t ■

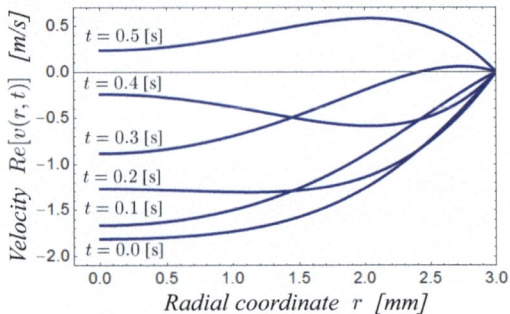

Example 7.1 (Biaxially Loaded Vessel Wall Patch).

(a) The right Cauchy–Green strain $\mathbf{C} = \text{diag}[\lambda_1^2, \lambda_2^2, \lambda_3^2]$ with the first invariant $I_1 = \lambda_1^2 + \lambda_2^2 + \lambda_3^2$ determines the principal strain state of the passive wall specimen. Coleman and Noll's procedure for incompressibly materials then determines the principal Cauchy stresses

$$\sigma_i = \lambda_i \partial \psi_{\text{ECM}} / \partial \lambda_i - \lambda_3 \partial \psi_{\text{ECM}} / \partial \lambda_3; \ i = 1, 2 , \quad \text{(no sumation)} \qquad (7.4)$$

where $\lambda_1 = \lambda_2 = \lambda$ and $\lambda_3 = \lambda^{-2}$ are the respective principal stretches.

The expression $\sigma = T/(ah) = 30.69$ kPa defines the Cauchy stress in the vessel wall, and the non-linear equilibrium relation $\sigma_1(\lambda) = \sigma$ (or alternatively $\sigma_2(\lambda) = \sigma$) determines the stretch λ. The only physically reasonable root is $\lambda = 1.1778$, and

$$\mathbf{F} = \begin{bmatrix} 1.1778 & 0 & 0 \\ 0 & 1.1778 & 0 \\ 0 & 0 & 0.7209 \end{bmatrix} ; \ \mathbf{P}_{\text{ECM}} = \begin{bmatrix} 26.06 & 0 & 0 \\ 0 & 26.06 & 0 \\ 0 & 0 & 0 \end{bmatrix} \text{[kPa]}$$

denotes the deformation gradient $\mathbf{F}$ and first Piola–Kirchhoff stress $\mathbf{P}_{\text{ECM}}$, respectively. At the spatial configuration Ω, the vessel wall specimen then stores the elastic energy $W = \psi(\lambda)a^2h = 0.00468$ J. Here, $A = a/\lambda = 1.9528$ cm and $H = h\lambda^2 = 2.3581$ mm denote the edge length and thickness of the stress-free specimen in the reference configuration Ω_0.

(b) Given the fixed configuration Ω, SMC-alignment remains unchanged upon activation. The SMC fiber direction vector $\mathbf{a}_0$ in the stress-free configuration Ω_0, is then identical to the fiber direction vector $\mathbf{a}$ in Ω. We may regard SMC-related stress as contributions from (active) stress fibers in the vessel wall. The related two-point first Piola–Kirchhoff stress then reads $\mathbf{P}_{\text{SMC}} = T_{\text{SMC}} \, \mathbf{a} \otimes \mathbf{a}_0$, and

$$\mathbf{P}_{active} = \mathbf{P}_{ECM} + \mathbf{P}_{SMC}$$

$$= \begin{bmatrix} 26.06 & 0 & 0 \\ 0 & 26.06 & 0 \\ 0 & 0 & 0 \end{bmatrix} + \begin{bmatrix} 20.31 & 11.73 & 0 \\ 11.73 & 6.77 & 0 \\ 0 & 0 & 0 \end{bmatrix} = \begin{bmatrix} 46.37 & 11.73 & 0 \\ 11.73 & 32.83 & 0 \\ 0 & 0 & 0 \end{bmatrix} \text{ [kPa]}$$

defines the stress state of the active vessel wall, where $T_{SMC}(\lambda = 1.1778) = 27.085$ kPa results from (7.3).   ∎

Example 7.2 (Ring Test to Characterize a Vessel Segment).

(a) The deformation gradient $\mathbf{F} = \text{diag}[\lambda_1, \lambda_2, \lambda_3]$ describes the kinematics, and Coleman and Noll's procedure $\sigma_{ECM\,i} = \lambda_i \partial\psi_{ECM}/\partial\lambda_i - \kappa; i = 1, 2, 3$ then yields the ECM-related principal Cauchy stresses

$$\sigma_{ECM\,i} = c_{ECM}(\lambda_i^2 - \lambda_3^2); i = 1, 2,$$

where the Lagrange parameter κ has been identified from $\sigma_3 = 0$.

At simple tension, the principal stretches $\lambda_1 = \lambda_\theta$ and $\lambda_2 = \lambda_3 = \lambda_\theta^{-1/2}$, determine the problem and $\sigma_{ECM\,\theta}^{st} = c_{ECM}(\lambda_\theta^2 - \lambda_\theta^{-1})$ expresses the Cauchy stress in the circumferential direction. At pure shear instead, $\lambda_1 = \lambda_\theta$, $\lambda_2 = 1$, $\lambda_3 = \lambda_\theta^{-1}$, and $\sigma_{ECM\,\theta}^{ps} = c_{ECM}(\lambda_\theta^2 - \lambda_\theta^{-2})$ is the circumferential Cauchy stress.

(b) Coleman and Noll's procedure results in the SMC-related principal Cauchy stresses $\sigma_{SMC\,i} = \lambda_i(\partial\Psi_{SMC}/\partial\lambda_i) - \kappa; i = r, \theta, z$, where κ denotes the Lagrange parameter. In addition, the kinematics relation $\lambda_{SMC} = \lambda_\theta \lambda_{pre}$ defines the expression

$$\frac{\partial\Psi_{SMC}}{\partial\lambda_i} = \frac{\partial\Psi_{SMC}}{\partial\lambda_{SMC}} \frac{\partial\lambda_{SMC}}{\partial\lambda_i} = 0 \; ; \; i = r, z, \tag{7.9}$$

a consequence of SMC fibers to be aligned in the circumferential direction. The equilibrium in radial direction $\sigma_{SMC\,r} = 0$ then implies $\kappa = 0$. In addition, the relation (7.9) and $\kappa = 0$ result in $\sigma_{SMC\,z} = 0$, and

$$\sigma_{SMC\,\theta} = \lambda_\theta \frac{\partial\Psi_{SMC}(\lambda_{SMC})}{\partial\lambda_{SMC}} \frac{\partial\lambda_{SMC}}{\partial\lambda_\theta} \tag{7.10}$$

then remains the only non-vanishing SMC stress contribution. It points along the SMC fiber direction and thus in the circumferential vessel direction. With the description (7.8) of SMC activation level, the stress (7.10) finally reads

$$\sigma_{SMC\,\theta} = c_{SMC}\lambda_d^{-1}\alpha\{(\lambda_{SMC} - 1)[\lambda_d + \lambda_{SMC}(\lambda_m - \lambda_{SMC})]$$
$$+ \lambda_{SMC}(\lambda_{SMC} - \lambda_m)\ln(\lambda_{SMC})\} \, .$$

SMC stress is identical for simple tension and pure shear kinematics, $\sigma_{\text{SMC}\theta}^{\text{st}} = \sigma_{\text{SMC}\theta}^{\text{ps}} = \sigma_{\text{SMC}\theta}$.

(c) The first Piola transform $\sigma_\theta = \lambda_\theta P_\theta$ for incompressible materials links the Cauchy stress σ_θ and first Piola–Kirchhoff stress P_θ, and with the result from Task (a), the force ratio

$$r_{\text{ECM}} = \frac{F_{\text{ECM}}^{\text{st}}}{F_{\text{ECM}}^{\text{ps}}} = \frac{P_{\text{ECM}\theta}^{\text{st}}}{P_{\text{ECM}\theta}^{\text{ps}}} = \frac{\sigma_{\text{ECM}\theta}^{\text{st}}}{\sigma_{\text{ECM}\theta}^{\text{ps}}} = \frac{\lambda_\theta(1 + \lambda_\theta + \lambda_\theta^2)}{(1 + \lambda_\theta)(1 + \lambda_\theta^2)}$$

expresses the relation between the ECM-related forces in simple tension and pure shear. Consequently, $r_{\text{ECM}} = 0.75$ at $\lambda_\theta = 1$ and approaches the limit $r_{\text{ECM}} = 1.0$ for $\lambda_\theta \to \infty$, see Fig. 7.12.

With the SMC-related stress σ_{active} independently from the particular deformation kinematics, $r_{\text{SMC}} = F_{\text{SMC}}^{\text{st}}/F_{\text{SMC}}^{\text{ps}} = 1$ holds.

Fig. 7.12 Force ratio r_{ECM} of the ring test experiment. It expresses the ratio of forces of simple tension and pure shear deformation kinematics of the ExtraCellular Matrix (ECM) ■

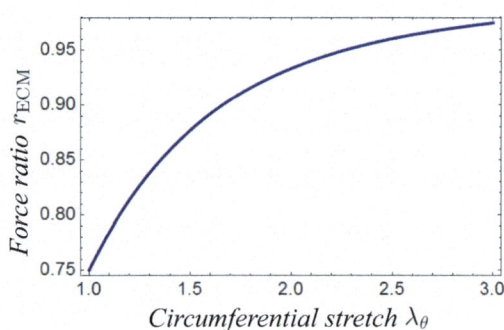

Example 7.3 (Calcium Determines the Functional Myosin States).

(a) Given the cytosolic free Ca^{2+} concentration of $\beta = 150$ nmol l^{-1}, the sigmoid function (7.12) takes the value sg(150) $= 0.0758582$. The phosphorylation rate $k_1(S)$ and the unphosphorylation rate $k_6(S)$ therefore yields $k_1 = k_6 = 0.0227575$ s^{-1}. The system (7.11) then reads

$$\begin{bmatrix} 0 \\ 0 \\ 0 \\ 0 \end{bmatrix} = \begin{bmatrix} -0.0227575 & 0.5 & 0 & 0.01 \\ 0.0227575 & -0.9 & 0.1 & 0 \\ 0 & 0.4 & -0.6 & 0.0227575 \\ 0 & 0 & 0.5 & -0.0327575 \end{bmatrix} \begin{bmatrix} \alpha_1 \\ \alpha_2 \\ \alpha_3 \\ \alpha_4 \end{bmatrix}, \qquad (7.14)$$

where the steady-state condition $d\alpha/dt = \mathbf{0}$ has been used. We attach the constraint relation $\alpha_1 + \alpha_2 + \alpha_3 + \alpha_4 = 1$ to the linear system (7.14), and its solution

$$\alpha_1 = 0.549207\ ;\ \alpha_2 = 0.0168519\ ;\ \alpha_3 = 0.0266816\ ;\ \alpha_4 = 0.40726$$

$$(7.15)$$

then determines the functional myosin states. Consequently, 54.9% of myosin is unphosphorylated, 1.7% is phosphorylated, 2.7% present phosphorylated cross-bridges attached to actin, and 40.7% present dephosphorylated cross-bridges attached to actin (latch bridges).

(b) With the time increment $\Delta t = t_{n+1} - t_n$, the forward-Euler discretization of (7.11) reads

$$\begin{bmatrix} \alpha_{1\,n+1} \\ \alpha_{2\,n+1} \\ \alpha_{3\,n+1} \\ \alpha_{4\,n+1} \end{bmatrix} = \begin{bmatrix} \alpha_{1\,n} \\ \alpha_{2\,n} \\ \alpha_{3\,n} \\ \alpha_{4\,n} \end{bmatrix} + \begin{bmatrix} -k_{1\,n} & k_2 & 0 & k_7 \\ k_{1\,n} & -k_2 - k_3 & k_4 & 0 \\ 0 & k_3 & -k_4 - k_5 & k_{6\,n} \\ 0 & 0 & k_5 & -k_{6\,n} - k_7 \end{bmatrix} \begin{bmatrix} \alpha_{1\,n} \\ \alpha_{2\,n} \\ \alpha_{3\,n} \\ \alpha_{4\,n} \end{bmatrix} \Delta t\ ,$$

$$(7.16)$$

where the notation $(\bullet)_n$ and $(\bullet)_{n+1}$ denote parameters at the time t_n and t_{n+1}, respectively. We use the expression

$$k_{1\,n} = k_{6\,n} = 0.3\mathrm{sg}(\beta_n) = \frac{0.3\exp\left(\frac{-400+\beta_n}{100}\right)}{1+\exp\left(\frac{-400+\beta_n}{100}\right)}\ [s^{-1}] \qquad (7.17)$$

to substitute the phosphorylation and unphosphorylation rates, respectively. Given the initial conditions

$$\alpha_{1\,0} = 0.549207\ ;\ \alpha_{2\,0} = 0.0168519\ ;\ \alpha_{3\,0} = 0.0266816\ ;\ \alpha_{4\,0} = 0.40726\ ,$$

$$(7.18)$$

the prescribed Ca^{2+} concentration (7.13), and $\beta_n = \beta(t_n)$, we iteratively solve the system with the algorithm shown in Table 7.3. The solution is practically converged at the time step $\Delta t = 0.1$ s, and Fig. 7.14 shows the evolution of the functional myosin states $\boldsymbol{\alpha}$.

Table 7.3 Algorithm to compute the functional myosin states at a prescribed $\beta = [Ca^{2+}]$ concentration

---

(1) Set time discretization

$\Delta t = 0.1$ s ; $t_{max} = 50$ s ; $n = t_{max}/\Delta t$

(2) Set initial conditions

$i = 0$; $\boldsymbol{\alpha} = [0.549207 \; 0.0168519 \; 0.0266816 \; 0.40726]^T$

(3) Solve equilibrium

Do While $i \leq n$

$t_i = i \Delta t$

$\beta = 630.0 \exp[(-t_i/9) + 150.0$ nmol $l^{-1}]$

$k_1 = k_6 = 0.3\{\exp[(\beta - 400)/100]\}/\{\exp[(\beta - 400)/100] + 1\}$ [s^{-1}]

$k_2 = k_5 = 0.5$ s^{-1} ; $k_3 = 0.4$ s^{-1} ; $k_4 = 0.1$ s^{-1} ; $k_7 = 0.01$ s^{-1}

$$\mathbf{K} = \begin{bmatrix} -k_1 & k_2 & 0 & k_7 \\ k_1 & -k_2 - k_3 & k_4 & 0 \\ 0 & k_3 & -k_4 - k_5 & k_6 \\ 0 & 0 & k_5 & -k_6 - k_7 \end{bmatrix}$$

$\boldsymbol{\alpha} \leftarrow \boldsymbol{\alpha} + \mathbf{K}\boldsymbol{\alpha}\,\Delta t$

$i \leftarrow i + 1$

End Do

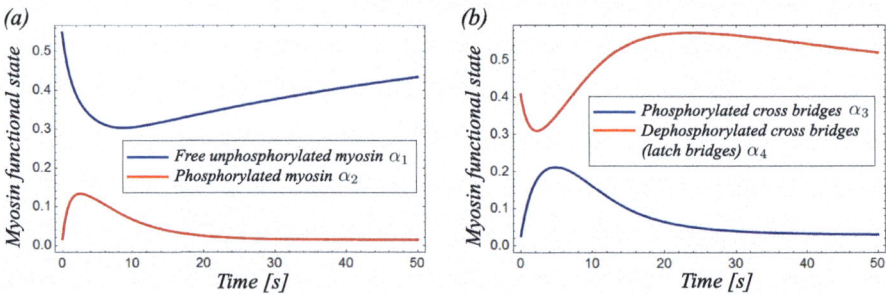

Fig. 7.14 Development of the four functional myosin states $\alpha_1, \alpha_2, \alpha_3, \alpha_4$ following the Ca^{2+} concentration prescribed by (7.13). (**a**) Non-cross-bridged myosin states α_1 and α_2 . (**b**) "Force generating" myosin states α_3 and α_4. The latch state α_4 relaxes much slower than α_3, the myosin state related to cycling cross-bridges ∎

Example 7.4 (Active Micro Vessel).

(a) Coleman and Noll's procedure $(7.24)_1$ and the use of the Piola transform, yield the circumferential Cauchy stress $\sigma_{\text{nMF}} = \lambda(\partial\Psi_{\text{nMF}1}/\partial\lambda)$. With the strain energy (7.26),

$$\sigma_{\text{nMF}} = k_a(\lambda^2 - \lambda)\exp[k_b(\lambda - 1)^2] \tag{7.37}$$

then expresses the circumferential Cauchy stress.

The equilibrium of the thin-walled tube $2\sigma_{\text{nMF}}h = p_i d$ allows us to express the inflation pressure through

$$p_i = 2k_a(\lambda^2 - \lambda)\exp[k_b(\lambda - 1)^2]h/(\lambda^2 D),$$

where $d = \lambda D$ and $h = \lambda_r H = H/\lambda$ denote the deformed diameter and the deformed wall thickness, respectively. At $\lambda = d/D = 1.5$, the vessel is at the pressure of $p_i = 7.56475$ kPa.

(b) The strain energy (7.27) allows us to compute the stress upon cross-bridge deformation. The Cauchy stress then reads

$$\sigma_{\text{MF}} = \lambda(\partial\Psi_{\text{MF}}/\partial\lambda) = (\alpha_3 k_{cb} + \alpha_4 k_{lb})(\lambda^2/\lambda_a - \lambda)\rho(\lambda_a) = 3.15283 \text{ kPa}. \tag{7.38}$$

Given the total deformation λ, expression (7.37) yields the stress $\sigma_{\text{nMF}} = 18.1301$ kPa from the non-myofibril-related structures, a stress that adds up to σ_{MF} and the pressure $p_i = 2(\sigma_{\text{nMF}} + \sigma_{\text{MF}})H/(\lambda^2 D) = 2.36478$ kPa then establishes in the vessel lumen.

(c) The force from the deformation of the cross-bridges and the force from the active sliding of filaments are to be in equilibrium, $P_{\text{MF}} = P_{\text{MFp}} = P_{\text{MFa}}$. The first Piola–Kirchhoff stress $P_{\text{MF}} = \sigma_{\text{MF}2}/\lambda$ follows from the Piola transform, whilst the potential (7.28) and the constitutive relation (7.24) allow us to compute the active first Piola–Kirchhoff stress P_{MFa}. With $\partial(\mathrm{d}X/\mathrm{d}t)/\partial X = \mathrm{d}(\partial X/\partial X)/\mathrm{d}t = 0$ and the product rule, it reads

$$P_{\text{MFa}} = \frac{\partial\Psi_a}{\partial\lambda_a} = \left[r_{lb}\alpha_3\dot{\lambda}_a + r_{cb}\alpha_4\left(\dot{\lambda}_a + \dot{\lambda}_c\right)\right]\rho(\lambda_a), \tag{7.39}$$

where the property $\Upsilon(X) = \int \rho(X)\mathrm{d}X$ among the CDF $\Upsilon(X)$ and the PDF $\rho(X)$ have been used. Given the present problem, the active stress reads

$$P_{\text{MFa}} = 199.809\dot{\lambda}_a + 1.35268/(1.0 - 42.8571\dot{\lambda}_a) - 0.0436349,$$

a function that is plotted in Fig. 7.19a. The equilibrium $P_{\mathrm{MFa}} = P_{\mathrm{MFp}}$ determines the rate of the active stretches

$$\dot{\lambda}_{\mathrm{a}1} = 0.00483 \quad \text{and} \quad \dot{\lambda}_{\mathrm{a}2} = 0.03187 \,,$$

out of which only $\dot{\lambda}_{\mathrm{a}1}$ appears to be physically reasonable, see Fig. 7.19a. The overlap between actin and myosin filaments therefore decreases at 0.48% per second, and the SMC relaxes accordingly.

(d) Table 7.5 shows the iterative algorithm to compute the vessel diameter d in response to the prescribed $\beta = [\mathrm{Ca}^{2+}]$ concentration (7.36). Given the "force generating" myosin states α_3 and α_4 through the kinetic relation (7.11), the wall stress contributions P_{nMF}, P_{MFp}, and P_{MFa} may be computed. The rate $\dot{\lambda}_{\mathrm{a}}$ of the actin–myosin stretch, as well as the total circumferential stretch λ are computed by a Newton–Raphson fixpoint iteration. It solves the internal $f_1 = P_{\mathrm{MFa}} - P_{\mathrm{MFp}} = 0$ and external $f_2 = p_{\mathrm{i}} - p_{\mathrm{o}}$ equilibrium, where p_{i} denotes the inflation pressure. Figure 7.16 shows the evolution of the calcium concentration β, the functional myosin states α_3, α_4, and the deformed vessel diameter $d = \lambda D$.

(e) Given the second law of thermodynamics (7.24)$_2$, the condition

$$S\dot{\theta} \leq -\frac{\partial \Psi}{\partial \lambda_{\mathrm{a}}}\dot{\lambda}_{\mathrm{a}} - \frac{\partial \Psi}{\partial \alpha_3}\dot{\alpha}_3 - \frac{\partial \Psi}{\partial \alpha_4}\dot{\alpha}_4 \qquad (7.40)$$

defines a thermodynamically admissible process, where S and θ denote the entropy per unit volume of undeformed tissue and the temperature, respectively. Figure 7.19b shows the domain of an admissible thermodynamical process shaded in grey.

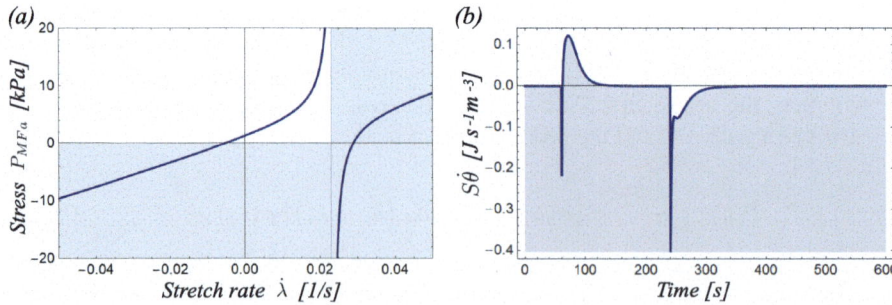

Fig. 7.19 (a) First Piola–Kirchhoff stress P_{MFa} that is generated by myofibrils, where the grey-shaded area indicates the domain of non-physical solutions. (b) Admissible thermodynamical process in response to a square wave of Ca^{2+} concentration. The product $S\dot{\theta}$ of entropy S and temperature rate $\dot{\theta}$ has to fall within the grey-shaded domain ∎

Table 7.5 Algorithm to compute the diameter d in response to a prescribed $\beta = [Ca^{2+}]$ concentration

(1) Set time discretization and problem parameters
$\Delta t = 0.1$ s ; $t_{max} = 600$ s ; $n = t_{max}/\Delta t$
$p_0 = 3$ kPa ; $k_a = 67$ kPa ; $k_b = 3$ kPa ; $k_{lb} = 10$ kPa ; $k_{cb} = 28$ kPa
$r_{lb} = 300$ kPa s ; $b = 30$ kPa ; $a = -1.0$ kPa ; $\dot{\lambda}_c = -0.7$; $\bar{\lambda}_a = 0.8$; $\sigma_a = 0.22$
$k_2 = k_5 = 0.5$ s^{-1} ; $k_3 = 0.4$ s^{-1} ; $k_4 = 0.1$ s^{-1} ; $k_7 = 0.01$ s^{-1}
(2) Set initial conditions
$i = 0$; $\boldsymbol{\alpha}_n = [0.549207\ 0.0168519\ 0.0266816\ 0.40726]^T$
$\lambda = 1.2762$; $\lambda_a = 1.0365$; $\dot{\lambda}_a = 0$
Do While $i \leq n$
$t = i\Delta t$
(3) Set Ca^{2+} concentration and compute functional myosin states
If$\{t > 60$, If$[t > 240, \beta = 150, \beta = 600], \beta = 150\}$ nmol l^{-1}
$k_1 = k_6 = 0.3\{\exp[(\beta - 400)/100]\}/\{\exp[(\beta - 400)/100] + 1\}$ [s^{-1}]
$\boldsymbol{\alpha} \leftarrow \boldsymbol{\alpha} + \mathbf{K}\boldsymbol{\alpha}\Delta t$ with $\mathbf{K}$ given by (7.11)
(4) Compute wall stress contributions
$\rho = \exp[-(\lambda_a - \bar{\lambda}_a)^2/(2\sigma_a^2)]/(\sqrt{2\pi}\sigma_a)$
$P_{nMF} = k_a(\lambda - 1)\exp[k_b(\lambda - 1)^2]$; $\sigma_{nMF} = \lambda P_{nMF}$
$P_{MFp} = (\alpha_3 k_{cb} + \alpha_4 k_{lb})(\lambda/\lambda_a - 1)\rho$; $\sigma_{MFp} = \lambda P_{MFp}$
$P_{MFa} = \left[\frac{\alpha_3 b(\dot{\lambda}_c - \dot{\lambda}_a)}{\dot{\lambda}_c - (b/a)\dot{\lambda}_a} + r_{lb}\alpha_4\dot{\lambda}_a\right]\rho$
(5) Solve internal and external equilibrium
$j = 0$; $f_1 = 1$; $f_2 = 1$
Do While $(f_1^2 + f_2^2 > 10^{-12})$
$f_1 = P_{MFa} - P_{MFp}$; $f_2 = 2(\sigma_{nMF} + \sigma_{MFp})HD^{-1}\lambda^{-2} - p_0$
$\begin{bmatrix} \Delta\dot{\lambda}_a \\ \Delta\lambda \end{bmatrix} = \begin{bmatrix} \partial f_1/\partial\dot{\lambda}_a & \partial f_1/\partial\lambda \\ \partial f_2/\partial\dot{\lambda}_a & \partial f_2/\partial\lambda \end{bmatrix}^{-1} \begin{bmatrix} f_1 \\ f_2 \end{bmatrix}$ $(\star)$
$\dot{\lambda}_a \leftarrow \dot{\lambda}_a - \Delta\dot{\lambda}_a$; $\lambda \leftarrow \lambda - \Delta\lambda$; $j \leftarrow j + 1$
If $j = 15$, solution not found: terminate
End Do
$d = \lambda D$
$\lambda_a \leftarrow \lambda_a + \dot{\lambda}_a\Delta t$; $i \leftarrow i + 1$
End Do

$(\star)$ singular for $\dot{\lambda}_a = 0$.

Example 7.5 (Evolution of Residual Stresses Through Vessel Growth).

(a) The ring of the radius $\widetilde{R}$ and the thickness $d\widetilde{R}$ in $\widetilde{\Omega}$ maps into the ring of the radius r and the thickness dr in Ω, and thus $2\pi r dr = 2\pi \widetilde{R} \det\mathbf{G}(\widetilde{R}) \det\mathbf{F}(R(\widetilde{R})) d\widetilde{R}$ holds. Given an incompressible elastic deformation, $\det\mathbf{F} = 1$, the integration of said expression yields

$$r^2 - r_{\mathrm{i}}^2 = 2 \int_{\widetilde{R}_{\mathrm{i}}}^{\widetilde{R}} x \det\mathbf{G}(x) dx ,$$

and

$$r = \sqrt{r_{\mathrm{i}}^2 + 2.4 \widetilde{R}^2 - 266.667 \widetilde{R}^3 - 1.44 \cdot 10^{-5}} \ [\mathrm{m}] \tag{7.47}$$

expresses the radius of the deformed vessel, where $\det\mathbf{G}(x) = G_\theta$ and R [m] have been used.

(b) We may express the principal Cauchy stresses in circumferential θ and radial r directions of the incompressible vessel by $\sigma_i = \overline{\sigma}_i - \kappa; i = \theta, r$, where $\overline{\sigma}_i = \lambda_i \partial\Psi/\partial\lambda_i = \lambda_i^2 c$ and κ denotes the Lagrange pressure.

The only non-trivial equilibrium relation of an axisymmetric problem in cylindrical coordinates reads $r d\sigma_r/dr = \sigma_\theta - \sigma_r = \overline{\sigma}_\theta - \overline{\sigma}_r$. Its integration between the boundaries $\sigma_r(r_{\mathrm{i}}) = -p_{\mathrm{i}}$ and $\sigma_r(r_{\mathrm{o}}) = 0$ then results in

$$p_{\mathrm{i}} = \int_{r_{\mathrm{i}}}^{r_{\mathrm{o}}} \frac{\overline{\sigma}_\theta - \overline{\sigma}_r}{r} dr = \int_{\widetilde{R}_{\mathrm{i}}}^{\widetilde{R}_{\mathrm{o}}} \frac{\overline{\sigma}_\theta - \overline{\sigma}_r}{\widetilde{R}} d\widetilde{R} , \tag{7.48}$$

where the kinematics relation $r = \lambda_{\mathrm{tot}\,r} \widetilde{R}$ has been used.

With the multiplicative kinematics relation $\lambda_{\mathrm{tot}\,\theta} = \lambda_\theta G_\theta$, the elastic incompressibility $\lambda_\theta \lambda_r = \det\mathbf{F} = 1$, the definition of the circumferential total stretch $\lambda_{\mathrm{tot}\,\theta} = r/\widetilde{R}$, and expression (7.47), we may express the integrand of (7.48)$_2$ as a function of $\widetilde{R}$ and r_{i}. Given r_{i}, (7.48)$_2$ can be integrated (at least numerically), and Fig. 7.22a shows the results for a number of radii r_{i}.

(c) Equation (7.48) may be solved with a quasi-Newton–Raphson fixpoint iteration

$$r_{\mathrm{i}} \leftarrow 1.3 \widetilde{R}_{\mathrm{i}}$$

Do Until $|p_{\mathrm{i}}| < \varepsilon$

$$p_{\mathrm{i}} = \int_{\widetilde{R}_{\mathrm{i}}}^{\widetilde{R}_{\mathrm{o}}} \frac{\overline{\sigma}_\theta - \overline{\sigma}_r}{\widetilde{R}} d\widetilde{R}$$

$$r_{\mathrm{i}} \leftarrow r_{\mathrm{i}} - k p_{\mathrm{i}}$$

End Do ,

where $k = \Delta r_{\mathrm{i}}/\Delta p_{\mathrm{i}} = 0.00002817 \; \mathrm{m \; kPa^{-1}}$ denotes a stiffness estimate. The fixpoint iteration yields the radius $r_{\mathrm{i}} = 2.9776$ mm, and defines the configuration of the pressure-free vessel segment. Figure 7.22b shows the distribution of stretches in the vessel wall at the pressure-free configuration.

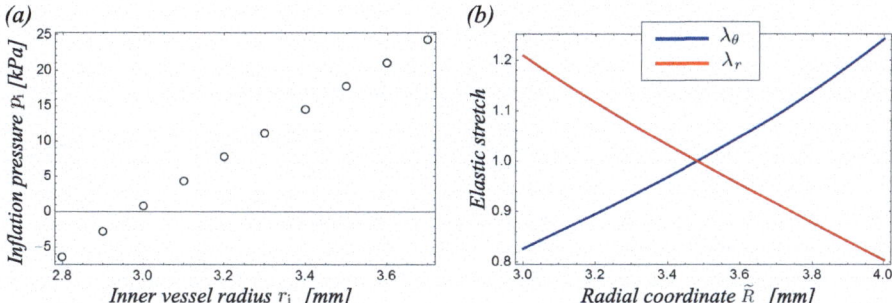

Fig. 7.22 (**a**) Inflation pressure p_{i} as a function of the inner radius r_{i} of a thick-walled vessel segment. (**b**) Circumferential stretch λ_θ and radial stretch λ_r across the wall of the vessel segment at its pressure-free configuration ∎

Example 7.6 (Growth that Maintains a Homeostatic Wall Stress Level).

(a) With the growth-related principal stretches G_θ, G_z, G_r, the growth tensor reads

$$\mathbf{G} = G_\theta (\mathbf{E}_\theta \otimes \widetilde{\mathbf{E}}_\theta) + G_z (\mathbf{E}_z \otimes \widetilde{\mathbf{E}}_z) + G_r (\mathbf{E}_r \otimes \widetilde{\mathbf{E}}_r),$$

where, $\mathbf{E}_\theta, \mathbf{E}_z, \mathbf{E}_r$ and $\widetilde{\mathbf{E}}_\theta, \widetilde{\mathbf{E}}_z, \widetilde{\mathbf{E}}_r$ denote the circumferential, axial, and radial vessel directions in the reference configuration Ω_0 and the initial reference configuration $\widetilde{\Omega}_0$, respectively. Given constant-density volumetric growth, the mass balance (7.50) allows us to derive

$$G_\theta^{-1}\dot{G}_\theta + G_z^{-1}\dot{G}_z + G_r^{-1}\dot{G}_r = \varsigma_{\mathrm{v}}, \tag{7.51}$$

an expression that relates G_θ, G_z, G_r and the rate of volume change ς_{v}.

(b) The equilibrium $2\sigma_\theta h = p_{\mathrm{i}} d$ of the inflated thin-walled cylindrical tube results in the spatial wall thickness $h = kd$, where $k = p_{\mathrm{i}}/(2\sigma_\theta)$ is a constant, and $d = \lambda_\theta D$. With the incompressibility of the non-growth-related deformation $\lambda_\theta \lambda_z \lambda_r = 1$, the thickness of the vessel in Ω_0 reads $H = \lambda_\theta \lambda_z h$. The radial growth may then be expressed by

$$G_r = H/\widetilde{H} = \lambda_\theta^2 \lambda_z k D/\widetilde{H}, \tag{7.52}$$

where $\widetilde{H}$ is the wall thickness in the initial reference configuration $\widetilde{\Omega}_0$.

At isotropic growth, $G_\theta = G_z = G_r = G$ holds and (7.51) results in

$$\varsigma_v = 3G^{-1}\dot{G} = 6\dot{\lambda}_\theta/\lambda_\theta ,$$

where (7.52) and $\lambda_z = l/L = \text{const}$ has been used. It determines the rate of volume change ς_v that is needed in response to $\dot{\lambda}_\theta$. ∎

Example 7.7 (The Thin-Walled Vessel at Tissue Turnover).

(a) The description of the mass density (7.58) may be integrated, and

$$\rho_0(\tau) = \rho_0(0)\exp(-\tau/\varsigma) + \varsigma_M^+[\varsigma - \varsigma\exp(-\tau/\varsigma)] \tag{7.65}$$

then expresses the density per unit reference volume at the time τ.

(b) Given the free energy density (7.59), Coleman and Noll's procedure for incompressible materials (3.131) followed by the second Piola transform yields the i-th principal Cauchy stress

$$\sigma_i(\tau) = \frac{\rho_0(0)\exp(-\tau/\varsigma)}{\rho_0(\tau)}\lambda_i(\tau)\frac{\partial\Psi(\mathbf{F}(\tau,0))}{\partial\lambda_i(\tau)}$$

$$+ \int_{\xi=0}^{\tau} \frac{\varsigma_M^+\exp[(\xi-\tau)/\varsigma]}{\rho_0(\xi)}\lambda_i(\tau,\xi)\frac{\partial\Psi(\mathbf{F}(\tau,\xi))}{\partial\lambda_i(\tau,\xi)}\mathrm{d}\xi - \kappa , \tag{7.66}$$

where κ is a Lagrange contribution to the hydrostatic stress that enforces the elastic incompressibility.

The deformation gradient of the inflated incompressible thin-walled tube reads

$$\mathbf{F}(\tau,\xi) = \mathrm{diag}\left[\frac{\lambda_{\mathrm{tot}}(\tau)\lambda_{\mathrm{pre}}}{\lambda_{\mathrm{tot}}(\xi)} , 1 , \frac{\lambda_{\mathrm{tot}}(\xi)}{\lambda_{\mathrm{tot}}(\tau)\lambda_{\mathrm{pre}}}\right] , \tag{7.67}$$

where multiplicative kinematics of the turnover-based growth description have been used, see Fig. 7.24. Given the constant λ_{tot}, the principal stretches read

$$\lambda_1 = \lambda_{\mathrm{pre}} , \quad \lambda_2 = 1 , \quad \lambda_3 = \lambda_{\mathrm{pre}}^{-1} , \tag{7.68}$$

such that (7.66) finally reads

$$\sigma_1(\tau) = c\left\{\frac{\rho_0(0)\exp\left(\frac{-\tau}{\varsigma}\right)}{\rho_0(\tau)}\left(\lambda_{\mathrm{pre}}^2 - \lambda_{\mathrm{pre}}^{-2}\right)\right.$$

$$+ \int_{\xi=0}^{\tau} \frac{\varsigma_{\mathrm{M}}^{+} \exp\left(\frac{\xi-\tau}{\zeta}\right)}{\rho_0(\xi)} \left[\lambda_{\mathrm{pre}}^2 - \lambda_{\mathrm{pre}}^{-2}\right] \mathrm{d}\xi \Bigg\} , \tag{7.69}$$

where the equilibrium in radial direction $\sigma_3 = 0$ has been used to identify κ.

(c) Splitting the time domain $0 < t < \tau$ in n increments of the size $\Delta\tau = \tau/n$ and using backward-Euler discretization of the integral in (7.69), allows us to express the Cauchy stress by

$$\sigma(\tau) \approx c \left\{ \frac{\rho_0(0) \exp\left(\frac{-\tau}{\zeta}\right)}{\rho_0(\tau)} \left(\lambda_{\mathrm{pre}}^2 - \lambda_{\mathrm{pre}}^{-2}\right) \right.$$

$$\left. + \sum_{l=1}^{n} \frac{\varsigma_{\mathrm{M}}^{+} \exp\left(\frac{(l-n)\Delta\tau}{\zeta}\right)}{\rho_0(l\Delta\tau)} \left[\lambda_{\mathrm{pre}}^2 - \lambda_{\mathrm{pre}}^{-2}\right] \Delta\tau \right\} . \tag{7.70}$$

(d) The equilibrium of the thin-walled tube determines the inflation pressure

$$p_{\mathrm{i}} = 2\frac{\sigma h}{d} = 2\frac{\sigma H}{D\lambda_{\mathrm{pre}}^2} , \tag{7.71}$$

where $\lambda_{\mathrm{tot}} = \lambda_{\mathrm{pre}}$ has been used. Given the mass density (7.65) and the stress approximation (7.70), the inflation pressure p_{i} can be computed for different time discretizations, see Fig. 7.26. The convolution integral in (7.69) can also be numerically integrated, which then yields the result labeled by "exact solution" in Fig. 7.26.

The inflation pressure predicted by the thin-walled tube made of neoHookean material yields $p_{\mathrm{i}} = 2cH(1 - \lambda_{\mathrm{pre}}^{-4})/D = 11.1924$ kPa.

Fig. 7.26 Development of the pressure p_{i} in a cylindrical tube, which wall is described by continuous tissue turnover. The convergence of the numerical solution towards the exact solution and as a function of the number of time increments n, is shown ■

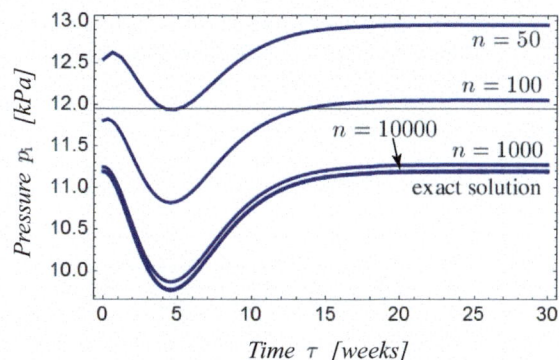

Example 7.8 (Vessel Wall Growth Through Tissue Turnover).

(a) Given the free energy density (7.59), Coleman and Noll's procedure for incompressible materials (3.131) followed by the second Piola transform yields the Cauchy stress

$$\boldsymbol{\sigma}(\tau) = \underbrace{\frac{\rho_0(0)S(\tau)}{\rho_0(\tau)}\boldsymbol{\sigma}(\mathbf{F}_{\mathrm{pre}}) + \int_{\xi=0}^{\tau} \frac{\varsigma_{\mathrm{M}}^{+}S(\tau-\xi)}{\rho_0(\xi)}\boldsymbol{\sigma}(\mathbf{F}(\tau,\xi))\mathrm{d}\xi}_{\bar{\boldsymbol{\sigma}}_i} - \kappa\mathbf{I}, \qquad (7.72)$$

where κ denotes a Lagrange contribution to the hydrostatic stress to enforce the elastic incompressibility.

(b) The kinematics of the problem determine the principal stretches

$$\lambda_\theta(\tau,\xi) = \lambda_{\theta\,\mathrm{tot}}(\tau)\lambda_{\theta\,\mathrm{pre}}/\lambda_{\theta\,\mathrm{tot}}(\xi)\,, \quad \lambda_z = 1\,, \quad \lambda_r(\tau,\xi) = \lambda_\theta^{-1}\,, \qquad (7.73)$$

where multiplicative kinematics as shown in Fig. 7.27 have been used. The relation (7.72), together with the Cauchy stress $\sigma_i = c\lambda_i^2 - \kappa$ of the neoHookean model then leads to the principal Cauchy stresses

$$\sigma_\beta(\tau) = c\left\{\frac{\rho_0(0)}{\rho_0(\tau)}\exp\left(\frac{-\tau}{\zeta}\right)\lambda_{\beta\,\mathrm{pre}}^2 \right.$$

$$\left. + \int_{\xi=0}^{\tau} \frac{\varsigma_{\mathrm{M}}^{+}}{\rho_0(\xi)}\exp\left(\frac{\xi-\tau}{\zeta}\right)\left(\frac{\lambda_{\beta\,\mathrm{tot}}(\tau)}{\lambda_{\beta\,\mathrm{tot}}(\xi)}\lambda_{\beta\,\mathrm{pre}}\right)^2\mathrm{d}\xi\right\} - \kappa\,,$$

where $\beta = \theta, z, r$ denotes the circumferential, axial, and radial vessel directions, respectively. The referential mass density

$$\rho_0(\tau) = \rho_0(0)\exp(-\tau/\zeta) + \varsigma_{\mathrm{M}}^{+}[\zeta - \zeta\exp(-\tau/\zeta)]\,, \qquad (7.74)$$

follows from the integration of the governing equation (7.58).

The discretization of the time interval $0 < t < \tau$ by n equidistant increments $\Delta\tau$ results in the approximation

$$\sigma_{\beta\alpha} = c\left\{\frac{\rho_{0\,1}}{\rho_{0\,\alpha}}\exp\left(\frac{-\alpha\Delta\tau}{\zeta}\right)\lambda_{\beta\,\mathrm{pre}}^2 \right.$$

$$\left. + \sum_{k=1}^{\alpha} \frac{\varsigma_{\mathrm{M}}^{+}}{\rho_{0\,k}}\exp\left[\frac{(k-\alpha)\Delta\tau}{\zeta}\right]\left(\frac{\lambda_{\beta\,\mathrm{tot}\,\alpha}}{\lambda_{\beta\,\mathrm{tot}\,k}}\lambda_{\beta\,\mathrm{pre}}\right)^2\Delta\tau\right\} - \kappa \qquad (7.75)$$

of the principal Cauchy stresses $\sigma_{\beta\alpha}$; $\beta = \theta, z, r$ at the time $\tau_\alpha = \alpha\Delta\tau$. Here, the mass density $\rho_{0\alpha}$ at the time τ_α is given by (7.74).

(c) The equilibrium of the thick-walled tube problem reads

$$p_{\rm i} = \int\limits_{r_{\rm i}}^{r_{\rm o}} \frac{\sigma_\theta - \sigma_r}{r}\,{\rm d}r = \int\limits_{R_{\rm i}}^{R_{\rm o}} \frac{\sigma_\theta - \sigma_r}{R}\,{\rm d}R, \qquad (7.76)$$

where the kinematics relation $r = \lambda_{{\rm tot}\,r} R$ has been used. The tube may be discretized by m rings of the thickness $\Delta R = (R_{\rm o} - R_{\rm i})/m$, which then approximates the integral (7.76) by

$$p_{\rm i} \approx \sum_{j=1}^{m} \frac{\sigma_{\theta\,j} - \sigma_{r\,j}}{R_j}\,\Delta R.$$

We may use (7.75) to substitute the principal stresses, and

$$p_{\rm i} \approx \sum_{j=1}^{m} \frac{c\Delta R}{R_j} \left\{ \frac{\rho_{0\,1\,j}}{\rho_{0\,i\,j}} \exp\left(\frac{-\alpha\Delta\tau}{\zeta}\right) (\lambda_{\theta\,\rm pre}^2 - \lambda_{\theta\,\rm pre}^{-2}) \right.$$
$$\left. + \sum_{k=1}^{\alpha} \frac{s_{\rm M}^+}{\rho_{0\,k\,j}} \exp\left[\frac{(k - \alpha)\Delta\tau}{\zeta}\right] \left(\frac{\lambda_{\theta\,{\rm tot}\,\alpha\,j}^4 \lambda_{\theta\,\rm pre}^4 - \lambda_{\theta\,{\rm tot}\,k\,j}^4}{\lambda_{\theta\,{\rm tot}\,k\,j}^2 \lambda_{\theta\,{\rm tot}\,\alpha\,j}^2 \lambda_{\theta\,\rm pre}^2}\right) \Delta\tau \right\}, \qquad (7.77)$$

represents the equilibrium of the thick-walled vessel problem. In the derivation of this expression, we used the elastic incompressibility $\lambda_\theta\lambda_r = 1$ to substitute the radial stretch λ_r, whilst the indices k and j refer to the discretization in time and space, respectively.

The incompressibility links the circumferential stretch of the individual layers to the tube's radii. At the time τ_α, the circumferential stretch in the j-th layer then reads $\lambda_{\theta\,\alpha\,j} = R_j^{-1}\sqrt{r_{{\rm i}\alpha}^2 + R_j^2 - R_{\rm i}^2}$, where $r_{{\rm i}\alpha}$ denotes the inner radius of the spatial configuration at the time τ_α.

(d) With the pressure $p_{\rm i}$ and time τ_α, a fixpoint iteration may be used in the solution of (7.77) towards the computation of $r_{{\rm i}\alpha}$. Table 7.7 reports the applied algorithm, and Fig. 7.28 shows some results.

Table 7.7 Algorithm to solve an inflated thick-walled vessel at continuous tissue turnover at the time $\tau_\alpha = \alpha \Delta \tau$

---

(1) Specify parameters

$R_i = 3.0$ mm; $R_o = 4.0$ mm; $m = 30$

$c = 120$ kPa; $\zeta = 6.0$ weeks^{-1}; $\overline{p}_i = 13.33$ kPa

$\rho_0 = 0.6$ kg dm^{-3}; $\varsigma_M^+ = 0.1$ kg dm^{-3} week^{-1}

---

(2) Set spatial discretization

$\Delta R = (R_o - R_i)/m$

$R_j = R_i + (j - 0.5)\Delta R$ for $j = 1, \ldots, m$

---

(3) Available history information

$\rho_{0\,l\,j}$; $\lambda_{\theta\,l\,j}$ for $l = 1, \ldots, \alpha - 1, \ j = 1, \ldots, m$

---

(4) Compute density

$\rho_{0\alpha\,j} = \rho_0 \exp\left[\frac{-\alpha\Delta\tau}{\zeta}\right] + \varsigma_M^+ \zeta[1 - \exp(-\alpha\Delta\tau/\zeta)]$ for $j = 1, \ldots, m$

---

(5) Solve equilibrium

$r_i = 1.2R_i$; $r_{in} = R_i$; $p_{in} = 0$; $k = 1$

Do While $k < k_{max}$

    $p_i = 0$

    Do $j = 1, \ldots, m$

$$\lambda_{\theta\alpha\,j} = \frac{\sqrt{r_i^2 + R_j^2 - R_i^2}}{R}$$

$$s = \sum_{l=1}^{\alpha} \frac{\varsigma_M^+}{\rho_{0\,l\,j}} \exp\left[(l - \alpha)\frac{\Delta\tau}{\zeta}\right] \left(\frac{\lambda_{\theta\,\mathrm{tot}\,\alpha\,j}^4 \lambda_{\theta\,\mathrm{pre}}^4 - \lambda_{\theta\,\mathrm{tot}\,l\,j}^4}{\lambda_{\theta\,\mathrm{tot}\,\alpha\,j}^2 \lambda_{\theta\,\mathrm{tot}\,l\,j}^2 \lambda_{\theta\,\mathrm{pre}}^2}\right) \Delta\tau$$

$$\sigma_\theta - \sigma_r = c\left[\frac{\rho_{0\,1\,j}}{\rho_{0\alpha\,j}}\left(\lambda_{\theta\,\mathrm{pre}}^2 - \lambda_{\theta\,\mathrm{pre}}^{-2}\right) + s\right]$$

$$p_i \leftarrow p_i + \frac{\sigma_\theta - \sigma_r}{R_j}\Delta R$$

    End Do

    $r_i \leftarrow r_i - \beta(p_i - \overline{p}_i)$ with $\beta = \frac{r_i - r_{in}}{p_i - p_{in}}$

    $r_{in} = r_i$; $p_{in} = p_i$

    If $|p_i - \overline{p}_i| < 10^{-6}$, $k = k_{max} + 1$

    If $k = k_{max}$, solution not found: terminate

    $k \leftarrow k + 1$

End Do

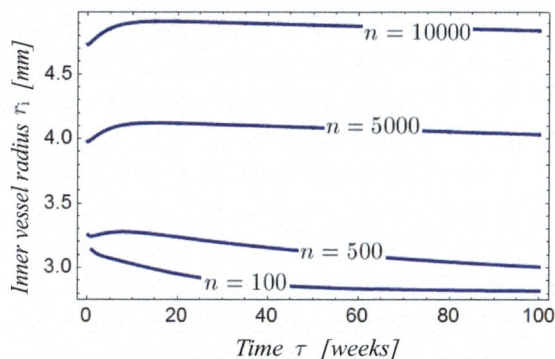

Fig. 7.28 Development of the inner radius r_i over time of the inflated thick-walled vessel at continuous tissue turnover. Solutions for different time discretizations n are shown. Even at $n = 10,000$, the solution is not converged ∎

Example A.1 (Population Mean of the Vessel Wall Strength).

(a) Given the Quantile-Quantile plot (QQ-plot) shown in Fig. A.4a, the strength data falls around the diagonal and therefore supports the assumption of a normal distributed sample.

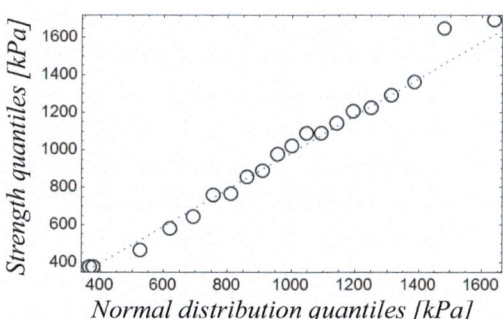

Fig. A.4 Quantile-Quantile plots (QQ-plots) illustrating normal distribution of the vessel wall strength

(b) The sample has the mean

$$\bar{x} = \frac{\sum_{i=1}^{n} x_i}{n} = 1003.28 \text{ kPa},$$

where $n = 19$ denotes its size. Its standard deviation s and the standard error e_s read

$$s = \sqrt{\frac{\sum_{i=1}^{n} (x_i - \bar{x})^2}{n-1}} = 362.65 \text{ kPa}; \quad e_s = \frac{s}{\sqrt{n}} = 83.2 \text{ kPa}.$$

(c) The confidence levels $c_{90} = 90\%$ and $c_{95} = 95\%$ correspond to the probabilities of $p_{90}^\star = (1 + c_{90}/100)/2 = 0.95$ and $p_{95}^\star = (1 + c_{95}/100)/2 = 0.975$, respectively. The solution of $p^\star = \Upsilon(t^\star)$ yields $t_{90}^\star = 1.73406$ and $t_{95}^\star =$

2.10092, where $\Upsilon(x)$ denotes the CDF of the student t-distribution with $\nu = n - 1 = 18$ degrees of freedom. Given the sample data computed by Task (b),

$$90\%: \ \bar{x} - t^*_{90}e_s \leq \mu \leq \bar{x} + t^*_{90}e_s \ ; \ 859.0\,\text{kPa} \leq \mu \leq 1147.6\,\text{kPa} \,,$$

$$95\%: \ \bar{x} - t^*_{95}e_s \leq \mu \leq \bar{x} + t^*_{95}e_s \ ; \ 828.5\,\text{kPa} \leq \mu \leq 1178.1\,\text{kPa}$$

express the population mean and the respective CIs.   ∎

Example A.2 (Trace and Determinant of a Matrix). The trace of $\mathbf{A}$ is the sum of the diagonal matrix components, and reads

$$\text{tr}\mathbf{A} = 1 + 5 + 9 = 15 \,.$$

Given the alternating symbol e_{ijk} (A.8),

$$\det\mathbf{A} = -3 \cdot 5 \cdot (-7) + (-2) \cdot (-6) \cdot (-7)$$
$$+ 3 \cdot 4 \cdot 8 - 1 \cdot (-6) \cdot 8 - (-2) \cdot 4 \cdot 9 + 1 \cdot 5 \cdot 9 = 282$$

expresses the determinant of $\mathbf{A}$.   ∎

Example A.3 (Linear Vector Transform). The operations

$$\mathbf{b} = \mathbf{A}\mathbf{a} = \begin{bmatrix} 7 & 2 & -1 \\ 4 & 3 & 4 \\ 5 & 6 & 9 \end{bmatrix} \begin{bmatrix} 1 \\ 2 \\ 3 \end{bmatrix} = \begin{bmatrix} 7 + 4 - 3 \\ 4 + 6 + 12 \\ 5 + 12 + 27 \end{bmatrix} = \begin{bmatrix} 8 \\ 22 \\ 44 \end{bmatrix}$$

constitute the linear vector transform.   ∎

Example A.4 (Eigenvalue Problem). The necessary condition

$$\det(\mathbf{A} - \lambda\mathbf{I}) = \det \begin{bmatrix} 0.8 - \lambda & 0.3 \\ 0.2 & 0.7 - \lambda \end{bmatrix} = 0$$

determines the characteristic equation

$$\lambda^2 - \frac{3}{2}\lambda + \frac{1}{2} = (\lambda - 1)(\lambda - 1/2) = 0 \,,$$

and the two roots $\lambda = 1$ and $\lambda = 1/2$ are therefore the eigenvalues of $\mathbf{A}$.

Given these eigenvalues, the eigenvalue problem $\mathbf{A}\mathbf{x} = \lambda\mathbf{x}$ yields the two linear systems of equations:

$$\begin{bmatrix} 0.8 & 0.3 \\ 0.2 & 0.7 \end{bmatrix} \begin{bmatrix} x_{1\,1} \\ x_{1\,2} \end{bmatrix} = \begin{bmatrix} x_{1\,1} \\ x_{1\,2} \end{bmatrix} \Rightarrow \mathbf{x}_1 = \begin{bmatrix} 0.6 \\ 0.4 \end{bmatrix} \,,$$

and

$$\begin{bmatrix} 0.8 & 0.3 \\ 0.2 & 0.7 \end{bmatrix} \begin{bmatrix} x_{2\,1} \\ x_{2\,2} \end{bmatrix} = \frac{1}{2} \begin{bmatrix} x_{2\,1} \\ x_{2\,2} \end{bmatrix} \Rightarrow \mathbf{x}_2 = \begin{bmatrix} 0.5 \\ -0.5 \end{bmatrix} ,$$

where $\mathbf{x}_1$ and $\mathbf{x}_2$ are the eigenvectors of $\mathbf{A}$. ■

Example A.5 (Norm of a Vector and Vector Products). Given the definition in Sect. A.6.1, the norms $|\mathbf{a}| = \sqrt{1+4+9} = 3.742$ and $|\mathbf{b}| = \sqrt{16+25+36} = 8.775$ represent the magnitudes of $\mathbf{a}$ and $\mathbf{b}$, respectively.

The dot vector product yields

$$\mathbf{a} \cdot \mathbf{b} = \begin{bmatrix} 1 \\ 2 \\ 3 \end{bmatrix} \cdot \begin{bmatrix} 4 \\ 5 \\ -6 \end{bmatrix} = 1 \cdot 4 + 2 \cdot 5 - 3 \cdot 6 = -4 ,$$

the cross vector product yields

$$\mathbf{a} \times \mathbf{b} = \det \begin{bmatrix} \mathbf{e}_1 & \mathbf{e}_2 & \mathbf{e}_3 \\ 1 & 2 & 3 \\ 4 & 5 & -6 \end{bmatrix} = \begin{bmatrix} -3 \cdot 5 + 2 \cdot (-6) \\ 3 \cdot 4 - 1 \cdot (-6) \\ -2 \cdot 4 + 1 \cdot 5 \end{bmatrix} = \begin{bmatrix} -27 \\ 18 \\ -3 \end{bmatrix} ,$$

the dyadic vector product yields

$$\mathbf{a} \otimes \mathbf{b} = \begin{bmatrix} 1 \\ 2 \\ 3 \end{bmatrix} \otimes \begin{bmatrix} 4 \\ 5 \\ -6 \end{bmatrix} = \begin{bmatrix} 1 \cdot 4 & 1 \cdot 5 & 1 \cdot (-6) \\ 2 \cdot 4 & 2 \cdot 5 & 2 \cdot (-6) \\ 3 \cdot 4 & 3 \cdot 5 & 3 \cdot (-6) \end{bmatrix} = \begin{bmatrix} 4 & 5 & -6 \\ 8 & 10 & -12 \\ 12 & 15 & -18 \end{bmatrix} ,$$

and the triple scalar product yields

$$(\mathbf{a} \times \mathbf{b}) \cdot \mathbf{c} = \det \begin{bmatrix} 1 & 2 & 3 \\ 4 & 5 & -6 \\ -7 & 8 & 9 \end{bmatrix} = -3 \cdot 5 \cdot (-7) + 2 \cdot (-6) \cdot (-7)$$

$$+ 3 \cdot 4 \cdot 8 - 1 \cdot (-6) \cdot 8 - 2 \cdot 4 \cdot 9 + 1 \cdot 5 \cdot 9 = 306 ,$$

where the definitions in Sect. A.6.1 have been used. ■

Example A.6 (Coordinate Transformation). Both systems are Cartesian coordinate systems, and their respective base vectors satisfy the normality condition $|\mathbf{e}_i| = |\tilde{\mathbf{e}}_i| = 1$ together with the orthogonality condition $\mathbf{e}_i \cdot \mathbf{e}_j = \tilde{\mathbf{e}}_i \cdot \tilde{\mathbf{e}}_j = \delta_{ij}$, where δ_{ij} denotes the Kronecker delta.

Both systems are related by the rotation matrix

$$\mathbf{R} = \mathbf{e}_i \cdot \widetilde{\mathbf{e}}_j = \begin{bmatrix} \cos\theta_{11} & \cos\theta_{12} & \cos\theta_{13} \\ \cos\theta_{21} & \cos\theta_{22} & \cos\theta_{23} \\ \cos\theta_{31} & \cos\theta_{32} & \cos\theta_{33} \end{bmatrix} = \begin{bmatrix} -1/\sqrt{2} & 1/\sqrt{2} & 0 \\ 1/\sqrt{2} & 1/\sqrt{2} & 0 \\ 0 & 0 & -1 \end{bmatrix},$$

which, given this particular case, is symmetric.

The coefficients of $\mathbf{R}$ represent the direction cosines, as it is illustrated in Fig. A.10.

Fig. A.10 The coefficients $R_{ij} = \mathbf{e}_i \cdot \widetilde{\mathbf{e}}_j$ of the rotation matrix $\mathbf{R}$ represent the direction cosines $\cos\theta_{ij}$, and thus the cosines of the angles between the two base vectors $\mathbf{e}_i$ and $\widetilde{\mathbf{e}}_i$, respectively ■

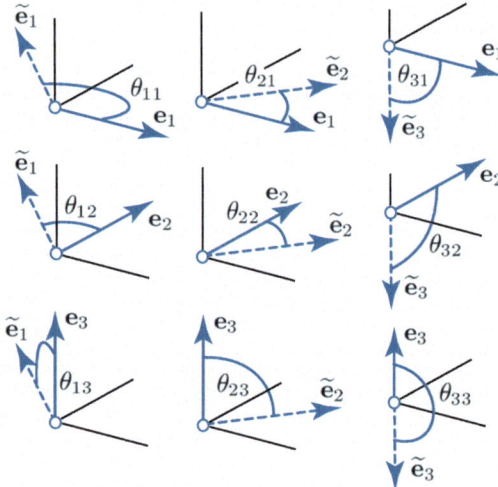

Example A.7 (Vector Components in Different Coordinate Systems). Given the Cartesian system $\{\mathbf{e}_1, \mathbf{e}_2\}$, the components of the base vectors $\widetilde{\mathbf{e}}_1 = [\sqrt{3}/2\ 1/2]^\mathrm{T}$ and $\widetilde{\mathbf{e}}_2 = [-1/2\ \sqrt{3}/2]^\mathrm{T}$ can be deduced from Fig. A.9, and the rotation matrix

$$\mathbf{R} = \begin{bmatrix} \mathbf{e}_1 \cdot \widetilde{\mathbf{e}}_1 & \mathbf{e}_1 \cdot \widetilde{\mathbf{e}}_2 \\ \mathbf{e}_2 \cdot \widetilde{\mathbf{e}}_1 & \mathbf{e}_2 \cdot \widetilde{\mathbf{e}}_2 \end{bmatrix} = \begin{bmatrix} \sqrt{3}/2 & -1/2 \\ 1/2 & \sqrt{3}/2 \end{bmatrix}$$

with $\det\mathbf{R} = 1$ then specifies the mapping between $\{\mathbf{e}_1, \mathbf{e}_2\}$ and $\{\widetilde{\mathbf{e}}_1, \widetilde{\mathbf{e}}_2\}$, respectively.

The coordinate transformation changes the vector components according to

$$\begin{bmatrix} \widetilde{a}_1 \\ \widetilde{a}_2 \end{bmatrix} = \mathbf{R}^\mathrm{T} \begin{bmatrix} a_1 \\ a_2 \end{bmatrix} = \begin{bmatrix} \sqrt{3}/2 & 1/2 \\ -1/2 & \sqrt{3}/2 \end{bmatrix} \begin{bmatrix} 1 \\ 1 \end{bmatrix} = \frac{1}{2} \begin{bmatrix} \sqrt{3}+1 \\ \sqrt{3}-1 \end{bmatrix},$$

where the use of $\mathbf{R}^\mathrm{T}$ is noticed.   ■

Example A.8 (Objectivity of Vectors). Given the transformation $\widetilde{\mathbf{e}}_i = \mathbf{R}\mathbf{e}_i$ of the base vectors, as well as $\widetilde{a}_i = \mathbf{R}^T(\mathbf{a} \cdot \mathbf{e}_i)$ of the vector components, the relation $\mathbf{a} = \widetilde{a}_i\widetilde{\mathbf{e}}_i = (\mathbf{R}^T\mathbf{a} \cdot \mathbf{e}_i)\mathbf{R}\mathbf{e}_i$ expresses the vector $\mathbf{a}$ within the rotated coordinate system $\{\widetilde{\mathbf{e}}_1, \widetilde{\mathbf{e}}_2, \widetilde{\mathbf{e}}_3\}$.

The index notation $\widetilde{a}_i\widetilde{e}_{ij} = R_{mk}a_m e_{ik}R_{jn}e_{in}$ of this expression, together with the relations $e_{ij} = \delta_{ij}$ and $R_{ki}R_{kj} = \delta_{ij}$, yields then $\mathbf{a} = \widetilde{a}_i\widetilde{\mathbf{e}}_i = a_i\mathbf{e}_i$. Aside from the invariance of the vector $\mathbf{a}$, this relation also implies that the vector product $\mathbf{a} \cdot \mathbf{b} = a_i\mathbf{e}_i \cdot b_j\mathbf{e}_j = \widetilde{a}_i\widetilde{\mathbf{e}}_i \cdot \widetilde{b}_j\widetilde{\mathbf{e}}_j$ is invariant with respect to coordinate transformation. ∎

Example A.9 (Objectivity of Tensors). The first invariant $(A.17)_1$ of $\mathbf{A} = \mathbf{a} \otimes \mathbf{a}$ yields $I_1 = \mathrm{tr}(\mathbf{a} \otimes \mathbf{a}) = \mathbf{I} : (\mathbf{a} \otimes \mathbf{a}) = \mathbf{a} \cdot \mathbf{a}$. The coordinate transformation $\mathbf{R}$ rotates the vector $\mathbf{a}$ into $\widetilde{\mathbf{a}} = \mathbf{R}\mathbf{a}$, and the first invariant then reads $I_1 = \widetilde{\mathbf{a}} \cdot \widetilde{\mathbf{a}} = \mathbf{a}\mathbf{R}^T\mathbf{R}\mathbf{a}$. Given the property $\mathbf{R}^T\mathbf{R} = \mathbf{I}$ of the rotation tensor, the condition $\widetilde{\mathbf{a}} \cdot \widetilde{\mathbf{a}} = \mathbf{a} \cdot \mathbf{a}$ holds, which in turn implies that I_1 remains unchanged upon coordinate transformation.

The second invariant $(A.17)_2$ of $\mathbf{A} = \mathbf{a} \otimes \mathbf{a}$ reads $I_2 = 1/2[I_1^2 - \mathrm{tr}(\mathbf{a} \otimes \mathbf{a})^2]$, where the term $\mathrm{tr}(\mathbf{a} \otimes \mathbf{a})^2$ may be expressed by $\mathrm{tr}(\mathbf{a} \otimes \mathbf{a} \cdot \mathbf{a} \otimes \mathbf{a}) = |\mathbf{a}|^2\mathrm{tr}(\mathbf{a} \otimes \mathbf{a}) = |\mathbf{a}|^2 I_1$. The norm $|\mathbf{a}|$ is invariant and therefore also $I_2 = 1/2(I_1^2 - |\mathbf{a}|^2 I_1)$ remains unchanged upon coordinate transformation.

The third invariant $(A.17)_3$ of $\mathbf{A} = \mathbf{a} \otimes \mathbf{a}$ reads $I_3 = \det(\widetilde{\mathbf{a}} \otimes \widetilde{\mathbf{a}}) = \det(\mathbf{R}\mathbf{a} \otimes \mathbf{a}\mathbf{R}^T)$. Given the property $\det[\mathbf{R}(\mathbf{a} \otimes \mathbf{a})\mathbf{R}^T] = \det\mathbf{R}\det(\mathbf{a} \otimes \mathbf{a})\det\mathbf{R}^T$ of the determinate and the condition $\det\mathbf{R} = \det\mathbf{R}^T = 1$ of the rotation $\mathbf{R}$, the third invariant $I_3 = \det(\mathbf{a} \otimes \mathbf{a}) = \det(\widetilde{\mathbf{a}} \otimes \widetilde{\mathbf{a}})$ remains unchanged upon coordinate transformation. ∎

References

1. P.A. Aarts, S.A. van den Broek, G.W. Prins, G.D. Kuiken, J.J. Sixma, R.M. Heethaar, Blood platelets are concentrated near the wall and red blood cells, in the center in flowing blood. Arteriosclerosis **8**, 819–824 (1988)
2. H. Abè, K. Hayashi, M. Sato (eds.), *Data Book on Mechanical Properties of Living Cells, Tissues, and Organs* (Springer, New York, 1996)
3. M. Absinta, S.K. Ha, G. Nair, P. Sati, N.J. Luciano, M. Palisoc, A. Louveau, K.A. Zaghloul, S. Pittaluga, J. Kipnis, D.S. Reich, Human and nonhuman primate meninges harbor lymphatic vessels that can be visualized noninvasively by MRI. Elife **6**, e29738 (2017)
4. M. Adham, J.P. Gournier, J.P. Favre, E. De La Roche, C. Ducerf, J. Baulieux, X. Barral, M. Pouyet, Mechanical characteristics of fresh and frozen human descending thoracic aorta. J. Surg. Res. **64**, 32–34 (1996)
5. R. Adolph, D.A. Vorp, D.L. Steed, M.W. Webster, M.V. Kameneva, S.C. Watkins, Cellular content and permeability of intraluminal thrombus in Abdominal Aortic Aneurysm. J. Vasc. Surg. **25**, 916–926 (1997)
6. F. Ahmed, M. Mehrabadi, Z. Liu, G.A. Barabino, C.K. Aidun, Internal viscosity-dependent margination of red blood cells in microfluidic channels. J. Biomed. Eng. **140**, 2605–2607 (2018)
7. B. Alberts, D. Bray, J. Lewis, M. Raff, K. Roberts, J.D. Watson, *Molecular Biology of the Cell* (Garland Publishing, New York, 1994)
8. P.W. Alford, J.D. Humphrey, L.A. Taber, Growth and remodeling in a thick-walled artery model: effects of spatial variations in wall constituents. Biomech. Model. Mechanobio. **7**, 245–262 (2008)
9. M. AlGhatrif, E.G. Lakatta, The conundrum of arterial stiffness, elevated blood pressure, and aging. Curr. Hypertens. Rep. **17**, 12 (2015)
10. D. Ambrosi, G.A. Ateshian, E.M. Arruda, S.C. Cowin, J. Dumais, A. Goriely, G.A. Holzapfel, J.D. Humphrey, R. Kemkemer, E. Kuhl, J.E. Olberding, L.A. Taber, K. Garikipati, Perspectives on biological growth and remodeling. J. Mech. Phys. Solids **59**, 863–883 (2011)
11. M. Bäck, T.C. Gasser, J.-B. Michel, G. Caligiuri, Review. biomechanical factors in the biology of aortic wall and aortic valve diseases. Cardiovasc. Res. **99**, 232–241 (2013)
12. L. Beir ao Da Veiga, F. Brezzi, A. Cangiani, L.D. Marini, G. Manzini, A. Russo, Basic principles of virtual element methods. Math. Models Methods Appl. Sci. **23**, 199–214 (2013)
13. P. Aparicio, M. Thompson, P.N. Watton, A novel chemo-mechano-biological model of arterial tissue growth and remodelling. J. Biomech. **49**, 2321–233 (2016)
14. R.L. Armentano, J. Levenson, J.G. Barra, E.I. Fischer, G.J. Breitbart, R.H. Pichel, A. Simon, Assessment of elastin and collagen contribution to aortic elasticity in conscious dogs. Am. J. Physiol. **260**, H1870–H1877 (1991)
15. ASME, *Assessing Credibility of Computational Modeling through Verification and Validation: Application to Medical Devices* (The American Society of Mechanical Engineers (ASME), New York, 2018)

T. C. Gasser, *Vascular Biomechanics*, https://doi.org/10.1007/978-3-030-70966-2

16. H. Astrand, J. Stålhand, J. Karlsson, M. Karlsson, B. Sonesson Band T. Länne. In vivo estimation of the contribution of elastin and collagen to the mechanical properties in the human abdominal aorta: effect of age and sex. J. Appl. Physiol. **110**, 176–187 (2011)

17. M. Auer, T.C. Gasser, Reconstruction and finite element mesh generation of Abdominal Aortic Aneurysms from computerized tomography angiography data with minimal user interaction. IEEE T. Med. Imaging **29**, 1022–1028 (2010)

18. A. Ayyalasomayajula, J.P. Vande Geest, B.R. Simon, Porohyperelastic finite element modeling of abdominal aortic aneurysms. J. Biomech. **132**, 371–379 (2010)

19. C.C.F.M.J. Baaten, H.T. Cate, P.E.J. van der Meijden, J.W.M. Heemskerk, Platelet populations and priming in hematological diseases. Blood Rev. **31**, 389–399 (2017)

20. S. Baek, R. Gleason, K.R. Rajagopal, J.D. Humphrey, Theory of small on large: Potential utility in computations of fluid-solid interactions in arteries. Comput. Meth. Appl. Mech. Eng. **196**, 3070–3078 (2009)

21. S. Baek, K.R. Rajagopal, J.D. Humphrey, A theoretical model of enlarging intracranial fusiform aneurysms. ASME. J. Biomech. Eng. **128**, 142–149 (2006)

22. N. Baeyens, M.A. Schwartz, Biomechanics of vascular mechanosensation and remodeling. Mol. Biol. Cell **27**, 7–11 (2016)

23. A.J. Bailey, R.G. Paul, L. Knott, Mechanism of maturation and ageing of collagen. Mech. Ageing Dev. **106**, 1–56 (1998)

24. D. Balzani, J. Schröder, D. Gross, Simulation of discontinuous damage incorporating residual stresses in circumferentially overstretched atherosclerotic arteries. Acta Biomater. **2**, 609–618 (2006)

25. A.J. Bank, D.R. Kaiser, S. Rajala, A. Cheng, In vivo human brachial artery elastic mechanics: effects of smooth muscle relaxation. Circulation **100**, 41–47 (1999)

26. G.I. Barenblatt, The mathematical theory of equilibrium of cracks in brittle fracture. Adv. Appl. Mech. **7**, 55–129 (1962)

27. S.R.H. Barrett, M.P.F. Sutcliffe, S. Howarth, Z.-Y. Li, J.H. Gillard, Experimental measurement of the mechanical properties of carotid artherothrombotic plaque fibrous cap. J. Biomech. **42**, 1650–1655 (2009)

28. C.A. Basciano, C. Kleinstreuer, Invariant-based anisotropic constitutive models of the healthy and aneurysmal abdominal aortic wall. ASME. J. Biomech. Eng. **131**, 021009 (2009)

29. O. Baskurt, B. Neu, H. Meiselman, *Red Blood Cell Aggregation* (CRC Press, Boca Raton, 2011)

30. K.-J. Bathe, *Finite Element Procedures in Engineering Analysis* (Prentice Hall, Englewood Cliffs, 1982)

31. Z. P. Bažant, Concrete fracture models: Testing and practice. Eng. Fract. Mech. **69**, 165–205 (2002)

32. Z.P. Bažant, G. Pijaudier-Cabot, Nonlocal continuum damage, localization instability and convergence. J. Appl. Mech. **55**, 287–293 (1988)

33. J.V. Beck, K.A. Woodbury, Inverse problems and parameter estimation: integration of measurements and analysis. Meas. Sci. Tech. **9**, 839–847 (1998)

34. T. Belytschko, W.K. Liu, B. Moran, *Nonlinear Finite Elements for Continua and Structures* (Wiley, Chichester, 2000)

35. T. Belytschko, J.S.-J. Ong, W.K. Liu, J.M. Kennedy, Hourglass control in linear and nonlinear problems. Comput. Meth. Appl. Mech. Eng. **43**, 251–276 (1984)

36. N. Berg, L. Fuchs, L. Prahl Wittberg, Blood flow simulations of the renal arteries—effect of segmentation and stenosis removal. Flow Turbulence Combust **102**, 27–41 (2019)

37. D.H. Bergel, The static elastic properties of the arterial wall. J. Physiol. **156**, 445–457 (1961)

38. G. Bergers, S. Song, The role of pericytes in blood-vessel formation and maintenance. Neuro Oncol. **7**, 452–464 (2005)

39. C.L. Berry, S.E. Greenwald, Effect of hypertension on the static mechanical properties and chemical composition of the rat aorta. Cardiovasc. Res. **10**, 437–451 (1976)

40. G. Bhave, E.G. Neilson, Body fluid dynamics: Back to the future. J. Am. Soc. Nephrol. **22**, 2166–2181 (2011)

41. A.P. Bhole, B.P. Flynn, M. Liles, N. Saeidi, C.A. Dimarzio, J.W. Ruberti, Mechanical strain enhances survivability of collagen micronetworks in the presence of collagenase: implications for loadbearing matrix growth and stability. Philos. T. R. Soc. A **367**, 3339–3362 (2009)

42. J. Biasetti, F. Hussain, T.C. Gasser, Blood flow and coherent vortices in the normal and aneurysmatic aortas. A fluid dynamical approach to Intra-Luminal Thrombus formation. J. R. Soc. Interface **8**, 1449–1461 (2011)

43. J. Biasetti, P.G. Spazzini, T.C. Gasser, An integrated fluido-chemical model towards modeling the formation of intra-luminal thrombus in abdominal aortic aneurysms. Front. Physiol. **3**, 266 (2011)

44. J. Biehler, M.W. Gee, W.A. Wall, Towards efficient uncertainty quantification in complex and large scale biomechanical problems based on a Bayesian multi fidelity scheme. Biomech. Model. Mechanobio. **14**, 489–513 (2014)

45. J. Biehler, W.A. Wall, The impact of personalized probabilistic wall thickness models on peak wall stress in abdominal aortic aneurysms. Int. J. Numer. Meth. Bioeng. **34**, e2922 (2018).

46. D. Bigoni, T. Hueckel, Uniqueness and localization—I. Associative and non-associative elastoplasticity. Int. J. Solids Struct. **28**, 197–213 (1991)

47. C. Bingham, An antipodally symmetric distribution on the sphere. Ann. Stat. **2**, 1201–1225 (1974)

48. M.A. Biot, General theory of three-dimensional consolidation. J. Appl. Phys. **12**, 155–164 (1941)

49. E. Biros, G. Gäbel, C.S. Moran, C. Schreurs, J.H. Lindeman, P.J. Walker, M. Nataatmadja, M. West, L.M. Holdt, I. Hinterseher, C. Pilarsky, J. Golledge. Shear deformable shell elements for large strains and rotations. Oncotarget **30**, 12984–12996 (2015)

50. J.E. Bishop, G. Lindahl, Regulation of cardiovascular collagen synthesis by mechanical load. Cardiovasc. Res. **42**, 27–44 (1999)

51. D. Bluestein, E. Rambod, M. Gharib, Vortex shedding as a mechanism for free emboli formation in mechanical heart valves. ASME. J. Biomech. Eng. **122**, 125–134 (2000)

52. M.-L. Bochaton-Piallat, C.J.M. de Vries, G.J. van Eys, Vascular smooth muscle cells, in *The ESC Textbook of Vascular Biology*, ed. by R. Krams, M. Bäck, chapter 7 (Oxford University Press, Oxford, 2017), pp. 91–103

53. J. Bonet, R.D. Wood, *Nonlinear Continuum Mechanics for Finite Element Analysis* (Cambridge University Press, Cambridge, 1997)

54. G.V.R. Born, P.D. Richardson, Mechanical properties of human atherosclerotic lesions, in *Pathology of Human Atherosclerotic Plaques*, ed. by S. Glagov, W.P. Newman, S.A. Schaffer (Springer, New York, 1990), pp. 413–423

55. R.M. Bowen, Theory of mixtures, in *Continuum Physics*, ed. by A.C. Eringen, vol. III (Academic Press, New York, 1976)

56. S. Bozzi, U. Morbiducci, D. Gallo, R. Ponzini, G. Rizzo, C. Bignardi, G. Passoni, Uncertainty propagation of phase contrast-MRI derived inlet boundary conditions in computational hemodynamics models of thoracic aorta. Comput. Meth. Biomech. Biomed. Eng. **20**, 1104–1111 (2017)

57. F.A. Braeu, A. Seitz, R.C. Aydin, C.J. Cyron, Homogenized constrained mixture models for anisotropic volumetric growth and remodeling. Biomech. Model. Mechanobio. **16**, 9889–906 (2016)

58. F.A. Braeu, R.C. Aydin, C.J. Cyron, Anisotropic stiffness and tensional homeostasis induce a natural anisotropy of volumetric growth and remodeling in soft biological tissues. Biomech. Model. Mechanobio. **18**(2), 327–345 (2019)

59. A.N. Brooks, T.J.R. Hughes, Streamline upwind/Petrov-Galerkin formulations for convection dominated flows with particular emphasis on the incompressible Navier–Stokes equations. Comput. Meth. Appl. Mech. Eng. **32**, 199–259 (1982)

60. D.E. Brooks, J.W. Goodwin, G.V. Seaman, Interactions among erythrocytes under shear. J. Appl. Physiol. **28**, 172–177 (1970)

61. L. Bruder, J. Pelisek, H.-H. Eckstein, M.W. Gee, Biomechanical rupture risk assessment of abdominal aortic aneurysms using clinical data: a patient-specific, probabilistic framework and comparative case-control study. PLOS ONE **15**, e0242097 (2020)

62. R. Burattini, K.B. Campbell, Comparative analysis of aortic impedance and wave reflection in ferrets and dogs. Am. J. Physiol. **282**, H244–H255 (2002)

63. F.M. Burnet, The concept of immunological surveillance. Prog. Exp. Tumor. Res. **13**, 1–27 (1970)

64. B. Calvo, E. Peña, P. Martins, T. Mascarenhas, M. Doblaré, R.M. Natal Jorge, A. Ferreira, On modelling damage process in vaginal tissue. J. Biomech. **42**, 642–651 (2009)

65. B. Calvo, E. Peña, M.A. Martínez, M. Doblaré, An uncoupled directional damage model for fibred biological soft tissues. formulation and computational aspects. Int. J. Numer. Meth. Bioeng. **69**, 2036–2057 (2007)

66. W.R. Cannon, Organization for physiological homeostasis. Physiol. Rev. **9**, 399–431 (1929)

67. I. Cantat, C. Misbah, Lift force and dynamical unbinding of adhering vesicles under shear flow. Phys. Rev. Lett. **83**, 880–883 (1999)

68. L. Cardoso, S. Weinbaum, Changing views of the biomechanics of vulnerable plaque rupture: a review. Ann. Biomed. Eng. **42**, 415–431 (2014)

69. T.E. Carew, R.N. Vaishnav, D.J. Patel, Compressibility of the arterial wall. Circ. Res. **23**, 61–68 (1968)

70. D. Carey, Control of growth and differentiation of vascular cells by extracellular matrix proteins. Annu. Rev. Physiol. **53**, 161–177 (1991)

71. C.G. Caro, D.J. Doorly, M. Tarnawaski, K.T. Scott, Q. Long, C.L. Dumoulin, Non-planar curvature and branching of arteries and non-planar type flow. Proc. R. Soc. Lond. A **452**, 185–197 (1996)

72. B. Carrey, *How We Learn and Why It Happens* (Random House, New York, 2015)

73. N. Casson, *A Flow Equation for Pigment Oil-Suspensions of the Printing Ink Type* (Pergamon Press, Oxford, 1959)

74. Y. Castier, R.P. Brandes, G. Leseche, A. Tedgui, S. Lehoux, p47phox-dependent NADPH oxidase regulates flow-induced vascular remodeling. Circ. Res. **97**, 533–540 (2005)

75. A.W. Caulk, J.D. Humphrey, S.-I. Murtada, Fundamental roles of axial stretch in isometric and isobaric evaluations of vascular contractility. ASME. J. Biomech. Eng. **141**, 031008 (2018)

76. E.L. Chaikof, R.L. Dalman, M.K. Eskandari, B.M. Jackson, W.A. Lee, M.A. Mansour, et al., The society for vascular surgery practice guidelines on the care of patients with an abdominal aortic aneurysm. J. Vasc. Surg. **87**, 2–77 (2018)

77. D. Chapelle, J.-F. Gerbeau, J. Sainte-Marie, I.E. Vignon-Clementel, A poroelastic model valid in large strains with applications to perfusion in cardiac modeling. Comput. Mech. **46**, 91–101 (2010)

78. Y.S. Chatzizis, A.U. Coskun, M. Jonas, E.R. Edelman, C.L. Feldman, P.H. Stone, Role of endothelial shear stress in the natural history of coronary atherosclerosis and vascular remodeling: molecular, cellular, and vascular behavior. J. Am. Coll. Cardiol. **49**(25), 2379–2393 (2007)

79. X.-L. Chen, S.E. Varner, A.S. Rao, J.Y. Grey, S. Thomas, C.K. Cook, M.A. Wasserman, R.M. Medford, A.K. Jaiswal, C. Kunsch, Laminar flow induction of antioxidant response element-mediated genes in endothelial cells. a novel anti-inflammatory mechanism. J. Bio. Chem. **278**, 703–711 (2003)

80. G.C. Cheng, H.M. Loree, R.D. Kamm, M.C. Fishbein, R.T. Lee, Distribution of circumferential stress in ruptured and stable atherosclerotic lesions: a structural analysis with histopathological correlation. Circulation **87**, 1179–1187 (1993)

81. G.P. Cherepanov, The propagation of cracks in a continuous medium. J. Appl. Math. Mech. **31**, 503–512 (1967)

82. E.M. Cherry, J.K. Eaton, Shear thinning effects on blood flow in straight and curved tubes. Phys. Fluids **25**, 073104 (2013). https://doi.org/10.1063/1.4816369

83. S. Chien, Shear dependence of effective cell volume as a determinant of blood viscosity. Science **168**, 977–979 (1970)

84. C. Chnafa, *Using image-based large-eddy simulations to investigate the intracardiac flow and its turbulent nature.* Ph.D. thesis, Universite de Montpellier, Montpellier, France, 2014

85. H.S. Choi, R.P. Vito, Two-dimensional stress-strain relationship for canine pericardium. ASME. J. Biomech. Eng. **112**, 153–159 (1990)
86. E. Choke, G. Cockerill, W.R. Wilson, S. Sayed, J. Dawson, I. Loftus, M.M. Thompson, A review of biological factors implicated in Abdominal Aortic Aneurysm rupture. Eur. J. Vasc. Endovasc. Surg. **30**, 227–244 (2005)
87. A.J. Chorin, The numerical solution of the Navier–Stokes equations for an incompressible fluid. Bull. Am. Math. Soc. (N.S.) **73**, 928–931 (1967)
88. C.J. Choung, Y.C. Fung, Residual stress in arteries, in *Frontiers in Biomechanics*, ed. by G.W. Schmid-Schoenbein, S.L. Woo, B.W. Zweifach (Springer, New York, 1986), pp. 117–129
89. C.J. Chuong, Y.C. Fung, Three-dimensional stress distribution in arteries. J. Biomed. Eng. **105**, 268–274 (1983)
90. M. Cilla, E. Pe na, M.A. Martínez, 3d computational parametric analysis of eccentric atheroma plaque: influence of axial and circumferential residual stresses. Biomech. Model. Mechanobio. **11**, 1001–1013 (2012)
91. J.M. Clark, S. Glagov, Transmural organization of the arterial media: the lamellar unit revisited. Arteriosclerosis **5**, 19–34 (1985)
92. E.G. Cleary, The microfibrillar component of the elastic fibers. morphology and biochemistry, in *Connective Tissue Disease. Molecular Pathology of the Extracellular Matrix*, ed. by J. Uitto, A.J. Perejda, vol. 12 (Dekker, New York, 1987), pp. 55–81
93. R.W. Clough, The finite element method in plane stress analysis, in *Proceedings of the Second ASCE Conference on Electronic Computation* (American Society of Civil Engineers, Reston, 1960)
94. A.J. Cocciolone, J.Z. Hawes, M.C. Staiculescu, E.O. Johnson, M. Murshed, J.E. Wagenseil, Elastin, arterial mechanics, and cardiovascular disease. Am. J. Physiol. Heart Circ. Physiol. **315**, H189–H205 (2018)
95. G.R. Cokelet, J.R. Brown, S.L. Codd, J.D. Seymour, Magnetic resonance microscopy determined velocity and hematocrit distributions in a Couette viscometer. Biorheology **42**, 385–399 (2005)
96. G. Colantuoni, J.D. Hellums, J.L. Moake, C.P. Alfrey Jr., Response of human platelets to shear stress at short exposure times. Trans. Am. Soc. Artif. Intern. Organs **23**, 626–630 (1977)
97. B.D. Coleman, W. Noll, The thermodynamics of elastic materials with heat conduction and viscosity. Arch. Rat. Mech. Anal. **13**, 167–178 (1963)
98. E. Comellas, T.C. Gasser, F.J. Bellomo, S. Oller, A homeostatic-driven turnover remodelling constitutive model for healing in soft tissues. J. R. Soc. Interface **13**, 20151081 (2016)
99. F. Condemi, S. Campisi, M. Viallon, P. Croisille, S. Avril, Relationship between ascending thoracic aortic aneurysms hemodynamics and biomechanical properties. IEEE T. Biomed. Eng. **67**(4), 949–956 (2019)
100. R. Courant, K. Friedrichs, H. Lewy, über die partiellen differenzengleichungen der mathematischen physik. MATANNA **100**, 32–74 (1928)
101. S.C. Cowin, Tissue growth and remodeling. Ann. Rev. Biomed. Eng. **6**, 77–107 (2004)
102. S.C. Cowin, L. Cardoso, Mixture theory-based poroelasticity as a model of interstitial tissue growth. Mech. Mat. **44**, 47–57 (2012)
103. R.H. Cox, Effects of norepinephrine on mechanics of arteries in vitro. Am. J. Physiol. **23**, 420–425 (1976)
104. R.H. Cox, Mechanics of canine iliac artery smooth muscle in vitro. Am. J. Physiol. **230**, 462–470 (1976)
105. R.H. Cox, Regional variation of series elasticity in canine arterial smooth muscles. Am. J. Physiol. **234**, H542–H551 (1978)
106. M.A. Crisfield, *Non-linear Finite Element Analysis of Solids and Structures, Essentials*, vol. 1 (Wiley, Chichester, 1991)
107. C.J. Cyron, R.C. Aydin, J.D. Humphrey, A homogenized constrained mixture (and mechanical analog) model for growth and remodeling of soft tissue. Biomech. Model. Mechanobio. **15**, 1389–1403 (2016)

108. P. Dale, J. Sherratt, P. Maini, A mathematical model for collagen fibre formation during foetal and adult dermal wound healing. Proc. R. Soc. Lond. B **263**, 653–660 (1996)

109. R.L. Dalman, A. Wanhainen, K. Mani, B. Modarai, Top 10 candidate aortic disease trials. J. Intern. Med. https://doi.org/10.1111/joim.13042

110. R.C. Darling, C.R. Messina, D.C. Brewster, L.W. Ottinger, Autopsy study of unoperated Abdominal Aortic Aneurysms. Circulation **56**(II suppl), 161–164 (1977)

111. P.F. Davies, Flow-mediated endothelial mechanotransduction. Physiol. Rev. **75**, 519–560 (1995)

112. M.J. Davis, Aortic aneurysm formation: lessons from human studies and experimental models. Circulation **98**, 193–195 (1998)

113. T. Davis, *Direct Methods for Sparse Linear Systems* (SIAM, Philadelphia, 2006)

114. S. de Putter, B.J.B.M. Wolters, M.C.M. Rutten, M. Breeuwer, F.A. Gerritsen, F.N. van de Vosse, Patient-specific initial wall stress in Abdominal Aortic Aneurysms with a backward incremental method. J. Biomech. **40**, 1081–1090 (2007)

115. W.R. Dean, Fluid motion in a curved channel. Proc. R. Soc. Lond. A **121**, 402–420 (1928)

116. W.F. Decraemer, M.A. Maes, V.J. Vanhuyse, An elastic stress-strain relation for soft biological tissues based on a structural model. J. Biomech. **13**, 463–468 (1980)

117. A. Delfino, N. Stergiopulos, J.E. Moore Jr., J.-J. Meister, Residual strain effects on the stress field in a thick wall finite element model of the human carotid bifurcation. J. Biomech. **30**, 777–786 (1997)

118. H. Demiray, Large deformation analysis of some soft biological tissues. ASME. J. Biomech. Eng. **103**, 73–78 (1981)

119. S.J. Denardo, R. Nandyala, G.L. Freeman, G.L. Pierce, W.W. Nichols, Pulse wave analysis of the aortic pressure waveform in severe left ventricular systolic dysfunction. Circ. Heart Fail. **3**, 149–156 (2010)

120. B.G. Derubertis, S.M. Trocciola, E.J. Ryer, F.M. Pieracci, J.F. McKinsey, P.L. Faries, K.C. Kent, Vascular endothelium responds to fluid shear stress gradients. J. Vasc. Surg. **46**, 630–635 (2007)

121. P. Di Achille, G. Tellides, C.A. Figueroa, J.D. Humphrey, A haemodynamic predictor of intraluminal thrombus formation in abdominal aortic aneurysms. Proc. R. Soc. Lond. A **470**, 20140163 (2014)

122. E.S. DiMartino, A. Bohra, J.P. Vande Geest, N. Gupta, M.S. Makaroun, D.A. Vorp, Biomechanical properties of ruptured versus electively repaired Abdominal Aortic Aneurysm wall tissue. J. Vasc. Surg. **43**, 570–576 (2006)

123. E.S. DiMartino, D.A. Vorp, Effect of variation in intraluminal thrombus constitutive properties on abdominal aortic wall stress. Ann. Biomed. Eng. **31**, 804–809 (2003)

124. P.F. Dillon, M.O. Aksoy, S.P. Driska, R.A. Murphy, Myosin phosphorylation and the crossbridge cycle in arterial smooth muscle. Science **211**, 495–497 (1981)

125. J. Ding, S. Niu, Z. Chen, T. Zhang, B.P. Griffith, Z.J. Wu, Shear-induced hemolysis: species differences. Artif. Organs **30**, 419–429 (2015)

126. K.P. Dingemans, P. Teeling, J.H. Lagendijk, A.E. Becker, Extracellular matrix of the human aortic media: an ultrastructural histochemical and immunohistochemical study of the adult aortic media. Anat. Rec. **258**, 1–14 (2000)

127. J.B. Dixon, A.A. Gashev, D.C. Zawieja, J.E. Moore, G.L. Cote, Image correlation algorithm for measuring lymphocyte velocity and diameter changes in contracting microlymphatics. Ann. Biomed. Eng. **35**, 387–396 (2007)

128. J.B. Dixon, S.T. Greiner, A.A. Gashev, G.L. Cote, J.E. Moore, Lymph flow, shear stress, and lymphocyte velocity in rat mesenteric prenodal lymphatics. Microcirculation **13**, 597–610 (2006)

129. P.B. Dobrin, T.R. Canfield, Elastase, collagenase, and the biaxial elastic properties of dog carotid artery. Am. J. Physiol. **247**, H124–H131 (1984)

130. P.B. Dobrin, A.A. Rovick, Influence of vascular smooth muscle on contractile mechanics and elasticity of arteries. Am. J. Physiol. **217**, 1644–1651 (1969)

131. J.T. Dodge Jr., B.G. Brown, E.L. Bolson, H.T. Dodge, Lumen diameter of normal human coronary arteries. influence of age, sex, anatomic variation, and left ventricular hypertrophy or dilation. Circulation **86**, 232–246 (1992)

132. D. Doorly, S. Sherwin, Geometry and flow, in *Cardiovascular Mathematics*, ed. by L. Formaggia, A. Quarteroni, A. Veneziani. MS&A, vol. 1 (Springer, Milan, 2009), pp. 177–209

133. N.J.B. Driessen, M.A.J. Cox, C.V.C. Bouten, F.P.T. Baaijens, Remodeling of the angular collagen fiber distribution in cardiovascular. Biomech. Model. Mechanobio. **7**, 93–103 (2008)

134. N.J.B. Driessen, W. Wilson, C.V.C. Bouten, F.P.T. Baaijens, A computational model for collagen fibre remodelling in the arterial wall. J. Theor. Biol. **226**, 53–64 (2004)

135. D.S. Dugdale, Yielding of steel sheets containing slits. J. Mech. Phys. Solids **8**, 100–104 (1960)

136. A. Duprey, K. Khanafer, M. Schlicht, S. Avril, D.M. Williams, R. Berguer, In vitro characterization of physiological and maximum elastic modulus of ascending thoracic aortic aneurysm using uniaxial tensile testing. Eur. J. Vasc. Endovasc. Surg. **39**, 700–707 (2010)

137. M. Eastwood, R. Porter, U. Khan G. McGrouther, R. Brown, Mechanistic micro-structural theory by dermal fibroblasts and the relationship to cell morphology. J. Cell. Physiol. **166**, 33–42 (1996)

138. M. Ellmerer, L. Schaupp, G.A. Brunner, G. Sendlhofer, A. Wutte, P. Wach, T.R. Pieber, Measurement of interstitial albumin in human skeletal muscle and adipose tissue by open-flow microperfusion. Am. J. Physiol. Endocrinol. Metab. **278**, E352–E356 (2000)

139. J.C. Ellsmere, R.A. Khanna, J.M. Lee, Mechanical loading of bovine pericardium accelerates enzymatic degradation. Biomaterials **20**, 1143–1150 (1999)

140. H. Elman, D. Silvester, A. Wathen, *Finite Elements and Fast Iterative Solvers*, vol. 2 (Oxford University Press, New York, 2014)

141. J.L. Emery, J.H. Omens, A.D. McCulloch, Biaxial mechanics of the passively overstretched left ventricle. Am. J. Physiol. **272**, H2299–H2305 (1997)

142. J.L. Emery, J.H. Omens, A.D. McCulloch, Strain softening in rat left ventricular myocardium. ASME. J. Biomech. Eng. **119**, 6–12 (1997)

143. P. Erhart, C. Grond-Ginsbach, M. Hakimi, F. Lasitschka, S. Dihlmann, D. Böckler, and A. Hyhlik-Dürr, Finite element analysis of abdominal aortic aneurysms: predicted rupture risk correlates with aortic wall histology in individual patients. J. Endovas. Ther. **21**, 556–564 (2014)

144. P. Erhart, J. Roy, J-P. de Vries, M. Lindquist Liljeqvist, C. Grond-Ginsbach, A. Hyhlik-Dürr, D. Böckler, Prediction of rupture sites in abdominal aortic aneurysms after finite element analysis. J. Endovas. Ther. **23**, 121–124 (2016)

145. B.J. Erickson, P. Korfiatis, Z. Akkus, T.L. Kline, Machine learning for medical imaging. RadioGraphics 37(2), (2017). https://doi.org/10.1148/rg.2017160130

146. R. Fåhræus, The suspension stability of the blood. Physiol. Rev. **9**, 241–274 (1929)

147. R. Fåhræus, T. Lindqvist, The viscosity of the blood in narrow capillary tubes. Am. J. Physiol. **96**, 562–568 (1931)

148. A. Fasano, A. Sequeira, *Hemomath—The Mathematics of Blood* (Springer, Berlin, 2017)

149. G. Faury, Function-structure relationship of elastic arteries in evolution: from microfibrils to elastin and elastic fibres. Pathol. Biol. **49**, 310–325 (2001)

150. R.E. Feaver, B.D. Gelfand, B.R. Blackman, Human haemodynamic frequency harmonics regulate the inflammatory phenotype of vascular endothelial cells. Nat. Commun. **4**, 1525 (2013)

151. S. Federico, A. Grillo, G. Giaquinta, W. Herzog, Convex fung-type potentials for biological tissues. Meccanica **43**, 279–288 (2008)

152. D.A. Fedosov, M. Dao, G.E. Karniadakis, S. Suresh, Computational biorheology of human blood flow in health and disease. Ann. Biomed. Eng. **42**, 368–387 (2013)

153. J. Feher, *Quantitative Human Physiology. An Introduction*, 2nd edn. (Elsevier, Amsterdam, 2017)

154. A. Ferrara, A. Pandolfi, Numerical modeling of fracture in human arteries. Comput. Meth. Biomech. Biomed. Eng. **11**, 553–567 (2008)

155. J. Ferruzzi, P. mboxDi Achille, G. Tellides, J.D. Humphrey, Combining in vivo and in vitro biomechanical data reveals key roles of perivascular tethering in central artery function. PLOS ONE **13**, e0201379 (2018)

156. J. Ferruzzi, D.A. Vorp, J.D. Humphrey, On constitutive descriptors of the biaxial mechanical behaviour of human abdominal aorta and aneurysms. J. R. Soc. Interface **8**, 435–450 (2011)

157. G. Fessel, J.G. Snedeker, Equivalent stiffness after glycosaminoglycan depletion in tendon— an ultra-structural finite element model and corresponding experiments. J. Theor. Biol. **268**, 77–83 (2011)

158. A. Fick, Über diffusion. Ann. Phys. **94**, 59–86 (1855)

159. M.F. Fillinger, S.P. Marra, M.L. Raghavan, F.E. Kennedy, Prediction of rupture risk in Abdominal Aortic Aneurysm during observation: wall stress versus diameter. J. Vasc. Surg. **37**, 724–732 (2003)

160. H.M. Finlay, L. McCullough, P.B. Canham, Three-dimensional collagen organization of human brain arteries at different transmural pressures. J. Vasc. Res. **32**, 301–312 (1995)

161. H.M. Finlay, P. Whittaker, P.B. Canham, Collagen organization in the branching region of human brain arteries. Stroke **29**, 1595–1601 (1998)

162. J.T. Flaherty, J.E. Pierce, V.J. Ferrans, D.J. Patel, W.K. Tucker, D.L. Fry, Endothelial nuclear patterns in the canine arterial tree with particular reference to hemodynamic events. Circulation **30**, 23–33 (1972)

163. I. Fleming, B.R. Kwak, M.J. Meens, The endothelial cell, in *The ESC Textbook of Vascular Biology*, ed. by R. Krams, M. Bäck, chapter 6 (Oxford University Press, Oxford, 2017), pp. 73–89

164. C.M. Fleeter, G. Geraci, D.E. Schiavazzi, A.M. Kahnd, A.L. Marsden, Multilevel and multifidelity uncertainty quantification for cardiovascular hemodynamics. Comput. Methods Appl. Mech. Eng., **365** (2020)

165. D. Flormann, K. Schirra, T. Podgorski, C. Wagner, On the rheology of red blood cell suspensions with different amounts of dextran: separating the effect of aggregation and increase in viscosity of the suspending phase. Rheol. Acta **55**, 477–483 (2016)

166. P.J. Flory, Thermodynamic relations for highly elastic materials. Trans. Faraday Soc. **57**, 829–838 (1961)

167. P.-W. Fok, N.M. Mirzaei, Modeling the Glagov's compensatory enlargement of human coronary atherosclerotic plaque, in *Biomechanics of Living Organs e Hyperelastic Constitutive Laws for Finite Element Modeling*, ed. by Y. Payan, J. Ohayon (Academic Press, London, 2017)

168. M. Folkesson, A. Silveira, P. Eriksson, J. Swedenborg, Protease activity in the multi-layered intra-luminal thrombus of abdominal aortic aneurysms. Atherosclerosis **218**, 294–299 (2011)

169. J.S. Forrester, The pathogenesis of atherosclerosis and plaque instability, in *Atherosclerosis and Oxidant Stress: A New Perspective*, ed. by J.L. Holtzman (Springer, New York, 2007)

170. C. Forsell, T.C. Gasser, Numerical simulation of the failure of ventricular tissue due to deep penetration: the impact of constitutive properties. J. Biomech. **44**, 45–51 (2011)

171. C. Forsell, T.C. Gasser, J. Swedenborg, J. Roy, The quasi-static failure properties of the abdominal aortic aneurysm wall estimated by a mixed experimental-numerical approach. Ann. Biomed. Eng. (2012). https://doi.org/10.1007/s10439-012-0712-3

172. C. Forsell, H. M. Björck, P. Eriksson, A. Franco-Cereceda, T.C. Gasser, Biomechanical properties of the thoracic aneurysmal wall; differences between bicuspid aortic valve (BAV) and tricuspid aortic valve (TAV) patients. Ann. Thorac Surg. **98**, 65–71 (2014)

173. O. Frank, Die grundform des arteriellen pulses. Z. Bio. **37**, 483–526 (1899)

174. K.H. Fraser, T. Zhang, M.E. Taskin, B.P. Griffith, Z.J. Wu, A quantitative comparison of mechanical blood damage parameters in rotary ventricular assist devices: Shear stress, exposure time and hemolysis index. J. Biomed. Eng. **134**, 081002 (2012)

175. P. Fratzl (ed.), *Collagen—Structure and Mechanics* (Springer, New York, 2008)

176. M.H. Friedman, D.L. Fry, Arterial permeability dynamics and vascular disease. Arteriosclerosis **104**, 189–194 (1993)

177. H. Fu, Y. Jiang, D. Yang, F. Scheiflinger, W.P. Wong, T.A. Springer, Flow-induced elongation of von Willebrand factor precedes tension-dependent activation. Nat. Commun. **8**, 324 (2017)

178. G. Fuchs, N. Berg, L.M. Broman, L. Prahl Wittberg, Modeling sensitivity and uncertainties in platelet activation models applied on centrifugal pumps for extracorporeal life support. Sci. Reports **9**, 8809 (2019)

179. G. Fuchs, N. Berg, L. Prahl Wittberg, Pulsatile aortic blood flow—a critical assessment of boundary conditions. J. Engr. Mech. **4**, 011002 (2021)

180. T. Fujiwara, Y. Uehara, The cytoarchitecture of the medial layer in rat thoracic aorta: a scanning electron-microscopic study. Cell Tissue Res. **270**, 165–172 (1992)

181. Y. Fukumoto, T. Hiro, T. Fujii, G. Hashimoto, T. Fujimura, J. Yamada, T. Okamura, M. Matsuzaki, Localized elevation of shear stress is related to coronary plaque rupture: a 3-dimensional intravascular ultrasound study with in-vivo color mapping of shear stress distribution. J. Am. Coll. Cardiol. **51**, 645–650 (2008)

182. Y.C. Fung, *Biomechanics: Mechanical Properties of Living Tissue* (Springer, New York, 1981)

183. Y.C. Fung, What are the residual stresses doing in our blood vessels? Ann. Biomed. Eng. **19**, 237–249 (1991)

184. Y.C. Fung, K. Fronek, P. Patitucci, Pseudoelasticity of arteries and the choice of its mathematical expression. Am. J. Physiol. **237**, H620–H631 (1979)

185. Y.C. Fung, P. Tong, *Classical and Computational Solid Mechanic* (World Scientific Publishing, Singapore, 2001)

186. M.T. Gallagher, R.A.J. Wain, S. Dari, J.P. Whitty, D.J. Smith, Non-identifiability of parameters for a class of shear-thinning rheological models, with implications for haematological fluid dynamics. J. Biomech. **85**, 230–238 (2019)

187. D. Gallo, G. De Santis, F. Negri, D. Tresoldi, R. Ponzini, D. Massai, M.A. Deriu, P. Segers, B .Verhegghe, G. Rizzo, U. Morbiducci, On the use of in vivo measured flow rates as boundary conditions for image-based hemodynamic models of the human aorta: implications for indicators of abnormal flow. Ann. Biomed. Eng. **40**, 729–741 (2012)

188. C.M. Garcìa-Herrera, J.M. Atienza, F.J. Rojo, E. Claes, G.V. Guinea, D.J. Celentano, C. Garcìa-Montero, R.L. Burgos, Mechanical behaviour and rupture of normal and pathological human ascending aortic wall. Med. Biol. Eng. Comput. **50**, 559–566 (2012)

189. A.A. Gashev, D.C. Zawieja, Hydrodynamic regulation of lymphatic transport and the impact of aging. Pathophysiology **17**, 277–287 (2010)

190. O.Y. Gasheva, D.C. Zawieja, A.A. Gashev, Contraction-initiated no-dependent lymphatic relaxation: a self-regulatory mechanism in rat thoracic duct. J. Physiol. **575**, 821–832 (2006)

191. T.C. Gasser, An irreversible constitutive model for fibrous soft biological tissue: a 3d microfiber approach with demonstrative application to Abdominal Aortic Aneurysms. Acta Biomater. **7**, 2457–2466 (2011)

192. T.C. Gasser, The biomechanical rupture risk assessment of abdominal aortic aneurysms—method and clinical relevance, in *Biomedical Technology*, ed. by P. Wriggers, T. Lenarz, chapter x, pages x–x (Springer, New York, 2016)

193. T.C. Gasser, M. Auer, F. Labruto, J. Swedenborg, J. Roy, Biomechanical rupture risk assessment of Abdominal Aortic Aneurysms. Model complexity versus predictability of finite element simulations. Eur. J. Vasc. Endovasc. Surg. **40**, 176–185 (2010)

194. T.C. Gasser, C. Forsell, The numerical implementation of invariant-based viscoelastic formulations at finite strains. an anisotropic model for the passive myocardium. Comput. Methods Appl. Mech. Eng. **200**, 3637–3645 (2011)

195. T.C. Gasser, S. Gallinetti, X. Xing, C. Forsell, J. Swedenborg, J. Roy, Spatial orientation of collagen fibers in the Abdominal Aortic Aneurysm wall and its relation to wall mechanics. Acta Biomater. **8**, 3091–3103 (2012)

196. T.C. Gasser, G. Görgülü, M. Folkesson, J. Swedenborg, Failure properties of intra-luminal thrombus in Abdominal Aortic Aneurysm under static and pulsating mechanical loads. J. Vasc. Surg. **48**, 179–188 (2008)

197. T.C. Gasser, G.A. Holzapfel, A rate-independent elastoplastic constitutive model for (biological) fiber-reinforced composites at finite strains: Continuum basis, algorithmic formulation and finite element implementation. Comput. Mech. **29**, 340–360 (2002)

198. T.C. Gasser, G.A. Holzapfel, Modeling 3D crack propagation in unreinfoced concrete using PUFEM. Comput. Methods Appl. Mech. Eng. **194**, 2859–2896 (2005)

199. T.C Gasser, G.A. Holzapfel, 3D crack propagation in unreinforced concrete. A new smoothing algorithm for tracking 3D crack surfaces. Comput. Methods Appl. Mech. Eng. **195**, 5198–5219 (2006)

200. T.C. Gasser, G.A. Holzapfel, Modeling dissection propagation in soft biological tissues. Eur. J. Mech. A/Solids **25**, 617–633 (2006)

201. T.C. Gasser, G.A. Holzapfel, Modeling dissection failure during balloon angioplasty. Ann. Biomed. Eng. **35**, 711–723 (2007)

202. T.C. Gasser, R.W. Ogden, G.A. Holzapfel, Hyperelastic modelling of arterial layers with distributed collagen fibre orientations. J. R. Soc. Interface **3**, 15–35 (2006)

203. T.C. Gasser, Biomechanical rupture risk assessment: a consistent and objective decision-making tool for abdominal aortic aneurysm patients. AORTA **4**, 42–60 (2016)

204. T.C. Gasser, U. Hedin, J. Roy, in *Biomechanics of Coronary Atherosclerotic Plaque: From Model to Patient*, chapter 7: The Interaction of Biochemical, Biomechanical, and Clinical Factors of Coronary Disease. Biomechanics of Living Organs (Elsevier, Amsterdam, 2020), pp. 171–186

205. T.C. Gasser, G. Martufi, M. Auer, M. Folkesson, J. Swedenborg, Micro-mechanical characterization of intra-luminal thrombus tissue from abdominal aortic aneurysms. Ann. Biomed. Eng. **38**, 371–379 (2010)

206. T.C. Gasser, A. Nchimi, J. Swedenborg, J. Roy, N. Sakalihasan, D. Böckler, A. Hyhlik-Dürr, A novel strategy to translate the biomechanical rupture risk of abdominal aortic aneurysms to their equivalent diameter risk: method and retrospective validation. Eur. J. Vasc. Endovasc. Surg. **47**, 288–295 (2014)

207. R.T. Gaul, D.R. Nolan, T. Ristori, C.V.C. Bouten, S. Loerakker, C. Lally, Pressure-induced collagen degradation in arterial tissue as a potential mechanism for degenerative arterial disease progression. J. Mech. Behav. Biomed. Mater. **109**, 103771 (2020)

208. A.A. Gavrilov, K.A. Finnikov, Y.S.Ignatenko, O.B. Bocharov, R. May, Drag and lift forces acting on a sphere in shear flow of power-law fluid. Fluid. J. Engin. Thermophys. **27**, 474–488 (2018)

209. M.W. Gee, C. Reeps, H.-H. Eckstein, W.A. Wall , Prestressing in finite deformation abdominal aortic aneurysm simulation. J. Biomech. **42**, 1732–1739 (2009)

210. A. Ghavamian, S.J. Mousavi, S. Avril, Computational study of growth and remodeling in ascending thoracic aortic aneurysms considering variations of smooth muscle cell basal tone. Front. Med. Biol. Eng. **8**, 1230 (2020)

211. M. Giersiepen, L.J. Wurzinger, R. Opitz, H. Reul, Estimation of shear stress-related blood damage in heart valve prosthesis. In vitro comparison of 25 aortic valves. Int. J. Artificial Organs **13**, 300–306 (1990)

212. F.J.H. Gijsen, J.J. Wentzel, A. Thury, F. Mastik, J.A. Schaar, J.C.H. Schuurbiers, C.J. Slager, W.J. van der Giessen, P.J. de Feyter, A.F.W. van der Steen, P.W. Serruys, Strain distribution over plaques in human coronary arteries relates to shear stress. Am. J. Physiol. Heart Circ. Physiol. **295**, H1608–H1614 (2008)

213. S. Glagov, E. Weisenberg, C.K. Zarins, R. Stankunavicius, G.J. Kolettis, Compensatory enlargement of human atherosclerotic coronary arteries. N. Engl. J. Med. **316**, 1371–1375 (1987)

214. R.L. Gleason, J.D. Humphrey, A mixture model of arterial growth and remodeling in hypertension: altered muscle tone and tissue turnover. J. Vasc. Res. **41**, 352–363 (2004)

215. R.L. Gleason, E. Wilson, J.D. Humphrey, Biaxial biomechanical adaptations of mouse carotid arteries cultured at altered axial extension. J. Biomech. **40**, 766–776 (2007)

216. M. Göktepe, O.J. Abilez, E.A. Kuhl, A generic approach towards finite growth with examples of athlete's heart, cardiac dilation, and cardiac wall thickening. J. Mech. Phys. Solids **58**, 1661–1680 (2010)

217. H.L Goldsmith, J.C Marlow, Flow behavior of erythrocytes. II. particle motions in concentrated suspensions of ghost cells. J. Colloid Interface Sci. **71**, 383–407 (1979)
218. D. Gomez, A. Al Haj Zen, L.F. Borges, M. Philippe, P.S. Gutierrez, G. Jondeau, et al, Syndromic and non-syndromic aneurysms of the human ascending aorta share activation of the smad2 pathway. J. Pathol. **218**, 131–142 (2009)
219. P. Gondret, L. Petit, Dynamic viscosity of macroscopic suspensions of bimodal sized solid spheres. J. Rheol. **41**, 1261–1274 (1997)
220. A. Goriely, *The Mathematics and Mechanics of Biological Growth*. Interdisciplinary Applied Mathematics, vol. 45 (Springer, New York, 2017)
221. L. Goubergrits, K. Affeld, Numerical estimation of blooddamage in artificial organs. Artif. Organs **28**, 499–507 (2004)
222. S. Govindjee, J.C. Simo, Mullins' effect and the strain amplitude dependence of the storage modulus. Int. J. Solids Struct. **29**, 1737–1751 (1992)
223. B.S. Gow, *Handbook of Physiology. Section 2: The Cardiovascular System*, chapter Vascular Smooth Muscle, vol. 2 (American Physiological Society, Rockville, 1980), pp. 353–408
224. S.E. Greenwald, Ageing of the conduit arteries. J. Pathol. **211**, 157–172 (2007)
225. R.M. Greenhalgh, J.T. Powell, Endovascular repair of abdominal aortic aneurysm. N. Engl. J. Med. **358**, 494–501 (2008)
226. S.E. Greenwald, J.E. Moore, Jr., A. Rachev, T.P.C. Kane, J.-J. Meister, Experimental investigation of the distribution of residual strains in the artery wall. ASME. J. Biomech. Eng. **119**, 438–444 (1997)
227. H.C. Groen, F.J. Gijsen, A. van der Lugt, M.S. Ferguson, T.S. Hatsukami, A.F. van der Steen, C. Yuan, J.J. Wentzel, Plaque rupture in the carotid artery is localized at the high shear stress region: a case report. Stroke **38**, 2379–2381 (2007)
228. K. Grote, I. Flach, M. Luchtefeld, E. Akin, S.M. Holland, H. Drexler, B. Schieffer, Mechanical stretch enhances mRNA expression and proenzyme release of matrix metalloproteinase-2 (MMP-2) via NAD(P)H oxidase-derived reactive oxygen species. Circ. Res. **92**, e80–e86 (2003)
229. A. Grytsan, *Abdominal Aortic Aneurysm Inception and Evolution—A Computational Study*. Ph.D. thesis, KTH Royal Institute of Technology, Stockholm, Sweden, 2016
230. A. Grytsan, T.S.E. Eriksson, P.N. Watton, T.C. Gasser, Growth description for vessel wall adaptation: a thick-walled mixture model of abdominal aortic aneurysm evolution. Materials **10**, 994 (2017)
231. N. Gundiah, A.R. Babu, L.A. Pruitt, Effects of elastase and collagenase on the nonlinearity and anisotropy of porcine aorta. Physiol. Meas. **34**, 1657–73 (2013)
232. X. Guo, G.S. Kassab, Variation of mechanical properties along the length of the aorta. Am. J. Physiol. Heart Circ. Physiol. **285**, H2614–H2622 (2003)
233. A. Gustafsson, M. Tognini, F. Bengtsson, T.C. Gasser, H. Isaksson, L. Grassi, Subject-specific FE models of the human femur predict fracture path and bone strength under single-leg-stance loading. J. Mech. Behav. Biomed. Mater. **113**, 104118 (2021)
234. L. Gustafsson, L. Appelgren, H.E. Myrvold, Effects of increased plasma viscosity and red blood cell aggregation on blood viscosity in vivo. Am. J. Physiol. **241**, H513–18 (1981)
235. A.C. Guyton, J.E. Hall, *Human Physiology and Mechanisms of Disease*, 6th edn. (W. B. Saunders, Philadelphia, 1996)
236. R.J. Guzman, K. Abe, C.K. Zarins, Flow-induced arterial enlargement is inhibited by suppression of nitric oxide synthase activity in vivo. Surgery **122**, 273–279 (1997)
237. D.G. Guzzardi, A.J. Barker, P. Van Ooij, S.C. Malaisrie, J.J. Puthumana, D.D. Belke, H.E.M. Mewhort, D.A. Svystonyuk, S. Kang, S. Verma, et al., Valve-related hemodynamics mediate human bicuspid aortopathy: insights from wall shear stress mapping. J. Am. Coll. Cardiol. **66**(8), 892–900 (2015)
238. C. Hahn, M.A. Schwartz, Mechanotransduction in vascular physiology and atherogenesis. Nat. Rev. Mol. Cell Bio. **10**, 53–62 (2008)
239. C.M. Hai, H.R. Kim, An expanded latch-bridge model of protein kinase c-mediated smooth muscle contraction. J. Appl. Physiol. **98**, 1356–1365 (2005)

240. C.M. Hai, R.A. Murphy, Cross-bridge phosphorylation and regulation of latch state in smooth muscle. Am. J. Physiol. **254**, C99–C106 (1988)

241. C.N. Hall, C. Reynell, B. Gesslein, N.B. Hamilton, A. Mishra, B.A. Sutherland, F.M. O'Farrell, A.M. Buchan, M. Lauritzen, D. Attwell, Capillary pericytes regulate cerebral blood flow in health and disease. NAT **508**, 55–60 (2014)

242. A. Hamedzadeh, T.C. Gasser, S. Federico, On the constitutive modelling of recruitment and damage of collagen fibres in soft biological tissues. Eur. J. Mech. A/Solids **72**, 483–496 (2018)

243. S.S. Hans, O. Jareunpoon, M. Balasubramaniam, G.B. Zelenock, Size and location of thrombus in intact and ruptured Abdominal Aortic Aneurysms. J. Vasc. Surg. **41**, 584–588 (2005)

244. I. Hariton, G. de Botton, T.C. Gasser, G.A. Holzapfel, Stress-driven collagen fiber remodeling in arterial walls. Biomech. Model. Mechanobio. **6**, 163–75 (2007)

245. A.K. Harris, D. Stopak, P. Wild, Fibroblast traction as a mechanism for collagen morphogenesis. Nature **290**, 249–251 (1981)

246. R. Haverkamp, M.W. Williams, J.E. Scott, Stretching single molecules of connective tissue glycans to characterize their shape-maintaining elasticity. Biomacromols **6**, 1816–1818 (2005)

247. M. Heikkinen, J. Salenius, R. Zeitlin, J. Saarinen, V. Suominen, R. Metsanoja, O. Auvinen, The fate of AAA patients referred electively to vascular surgical unit. Scand. J. Surg. **91**, 354–352 (2002)

248. V.M. Heiland, C. Forsell, J. Roy, U. Hedin, T.C. Gasser, Identification of carotid plaque tissue properties using an experimental-numerical approach. J. Mech. Behav. Biomed. Mater. **27**, 226–238 (2013)

249. J.D. Hellums, 1993 Whitaker lecture: Biorheology in thrombosis research. Ann. Biomed. Eng. **22**, 445–455 (1994)

250. E.S. Helton, S. Palladino, E.E. Ubogu, A novel method for measuring hydraulic conductivity at the human blood-nerve barrier in vitro. Microvasc. Res. **109**, 1–6 (2017)

251. G. Heuser, R.A. Opitz, A Couette viscometer for short timeshearing of blood. Biorheology **17**, 17–24 (1980)

252. A.V. Hill, The heat of shortening and dynamics constants of muscles. Proc. R. Soc. Lond. A **126**, 136–195 (1938)

253. R.A. Hill, L. Tong, P. Yuan, S. Murikinati, S. Gupta, J. Grutzendler, Regional blood flow in the normal and ischemic brain is controlled by arteriolar smooth muscle cell contractility and not by capillary pericytes. Neuron **87**, 95–110 (2015)

254. A. Hillerborg, M. Modeer, P.E. Petersson, Analysis of crack formation and crack growth in concrete by means of fracture mechanics and finite elements. Cement Concr. Res. **6**, 773–782 (1976)

255. B.P. Ho, L.G. Leal, Inertial migration of rigid spheres in two-dimensional unidirectional flows. J. Fluid Mech. **65**, 365–400 (1974)

256. A.L. Hodgkin, A.F. Huxley, A quantitative description of membrane current and its application to conduction and excitation in nerve. J. Physiol. **117**, 500–544 (1952)

257. J. Hoffman, *Methods in Computational Science*. Computational Science and Engineering, vol 24, SIAM, 2021. ISBN 1611976723, 9781611976724

258. J. Hokanson, S. Yazdani, A constitutive model of the artery with damage. Mech. Res. Commun. **24**, 151–159 (1997)

259. J.P. Holt, Flow through collapsible tubes and through in situ veins. IEEE T. Biomed. Eng. **16**, 274–283 (1969)

260. G.A. Holzapfel, T.C. Gasser, A viscoelastic model for fiber-reinforced composites at finite strains: continuum basis, computational aspects and applications. Comput. Methods Appl. Mech. Eng. **190**, 4379–4403 (2001)

261. G.A. Holzapfel, T.C. Gasser, R.W. Ogden, A new constitutive framework for arterial wall mechanics and a comparative study of material models. J. Elast. **61**, 1–48 (2000)

262. G.A. Holzapfel, T.C. Gasser, M. Stadler, A structural model for the viscoelastic behavior of arterial walls: continuum formulation and finite element analysis. Eur. J. Mech. A/Solids **21**, 441–463 (2002)

263. C.O. Horgan, G. Saccomandi, A description of arterial wall mechanics using limiting chain extensibility constitutive models. Biomech. Model. Mechanobio. **1**, 251–266 (2003)

264. L. Horny, T. Adamek, R. Zitny, Age-related changes in longitudinal prestress in human abdominal aorta. Arch. Appl. Mech. **83**, 875–888 (2013)

265. D.R. Hose, P.V. Lawford, W. Huberts, L.R. Hellevik, S.W. Omholt, F.N. van de Vosse, Cardiovascular models for personalised medicine: where now and where next? Med. Eng. Phys. **72**, 38–48 (2019)

266. T.J.R. Hughes, *The Finite Element Method: Linear Static and Dynamic Finite Element Analysis* (Prentice-Hall, Englewood Cliffs, 1987)

267. T.J.R. Hughes, *The Finite Element Method: Linear Static and Dynamic Finite Element Analysis* (Dover, New York, 2000)

268. T.J.R. Hughes, A.N. Brooks, A multi-dimensional upwind scheme with no crosswind diffusion, in *Finite Element Methods for Convection Dominated Flows, AMD*, ed. by T.J.R. Hughes, vol. 34 (ASME 1979), pp. 19–35

269. T.J.R. Hughes, J.A. Cottrell, Y. Bazilevs, Isogeometric analysis: CAD, finite elements, NURBS, exact geometry and mesh refinement. Comput. Methods Appl. Mech. Eng. **194**, 4135–4195 (2005)

270. T.J.R. Hughes, L.P. Franca, G.M. Hulbert, A new finite element formulation for computational fluid dynamics: VIII. The galerkin/least-squares method for advective-diffusive equations. Comput. Methods Appl. Mech. Eng. **73**, 173–189 (1989)

271. J.D. Humphrey, *Cardiovascular Solid Mechanics. Cells, Tissues, and Organs* (Springer, New York, 2002)

272. J.D. Humphre. Constrained mixture models of soft tissue growth and remodeling - twenty years after. J Elast (2021). https://doi.org/10.1007/s10659-020-09809-1

273. J.D. Humphrey, K.R. Rajagopal, A constrained mixture model for growth and remodeling of soft tissues. Math. Model. Methods Appl. Sci. **12**, 407–430 (2002)

274. J.D. Humphrey, E. Wilson, A potential role of smooth muscle tone in early hypertension: a theoretical study. J. Biomech. **36**, 1595–1601 (2003)

275. J.D. Humphrey, E.R. Dufresne, M.A. Schwartz, Mechanotransduction and extracellular matrix homeostasis. Nat. Rev. Mol. Cell Bio. **15**, 802–812 (2014)

276. S.J. Hund, M.V. Kameneva, J.F. Antaki, A quasi-mechanistic mathematical representation for blood viscosity. Fluids **2**(1), 10 (2017)

277. C. Hurschler, B. Loitz-Ramage, R. Vanderby Jr., A structurally based stress-stretch relationship for tendon and ligament. ASME. J. Biomech. Eng. **119**, 392–399 (1997)

278. F. Hussain, Coherent structures and turbulence. J. Fluid Mech. **173**, 303–356 (1986)

279. A. Hyhlik-Dürr, T. Krieger, P. Geisbüsch, D. Kotelis, T. Able, D. Böckler, Reproducibility of deriving parameters of AAA rupture risk from patient-specific 3d finite element models. J. Endovas. Ther. **18**, 289–298 (2011)

280. K. Imoto, T. Hiro, T. Fujii, A. Murashige, Y. Fukumoto, G. Hashimoto, T. Okamura, J. Yamada, K. Mori, M. Matsuzaki, Longitudinal structural determinants of atherosclerotic plaque vulnerability: a computational analysis of stress distribution using vessel models and three-dimensional intravascular ultrasound imaging. J. Am. Coll. Cardiol. **46**, 1507–1515 (2005)

281. I. Ionescu, J.E. Guilkey, M. Berzins, R.M. Kirby, J.A. Weiss, Simulation of soft tissue failure using the material point method. ASME. J. Biomech. Eng. **128**, 917–94 (2006)

282. P.A. Jackson, B.R. Duling, Myogenic response and wall mechanics of arterioles. Am. J. Physiol. Heart Circ. Physiol. **257**, H1147–H1155 (1989)

283. Z.S. Jackson, A.I. Gotlieb, B.L. Langille, Wall tissue remodeling regulates longitudinal tension in arteries. Circ. Res. **90**, 918–925 (2002)

284. M. Jadidi, M. Habibnezhad, E. Anttila, K. Maleckis, A. Desyatova, J. MacTaggart, A. Kamenskiy, Mechanical and structural changes in human thoracic aortas with age. Acta Biomater. **103**, 172–188 (2020)

285. S. Jamalian, M. Jafarnejad, S.D. Zawieja, C.D. Bertram, A.A. Gashev, D.C. Zawieja, M.J. Davis, J.E. Moore, Demonstration and analysis of the suction effect for pumping lymph from tissue beds at subatmospheric pressure. Sci. Rep. **7**, 12080 (2017)

286. J. Jeong, F. Hussain, On the identification of a vortex. J. Fluid Mech. **285**, 69–94 (1995)

287. M. Jirásek, M. Bauer, Numerical aspects of the crack band approach. Comput. Struct. **110**, 60–78

288. M.J. Joyner, D.P. Casey, Regulation of increased blood flow (hyperemia) to muscles during exercise: a hierarchy of competing physiological needs. Physiol. Rev. **95**, 549–601 (2015)

289. C.M. Judd, G.H. McClelland, C.S. Ryan, *Data Analysis: A Model Comparison Approach To Regression, ANOVA, and Beyond*, 3rd edn. (Taylor & Francis, London, 2017)

290. L.M. Kachanov, *Introduction to Continuum Damage Mechanics* (Martinus Nijhoff Publishers, Dordrecht, 1986)

291. M. Kaliske, H. Rothert, Formulation and implementation of three-dimensional viscoelasticity at small and finite strains. Comput. Mech. **19**, 228–239 (1997)

292. A. Kamenskiy, A. Seas, G. Bowen, P. Deegan, A. Desyatova, N. Bohlim, W. Poulson, J. MacTaggart, In situ longitudinal pre-stretch in the human femoropopliteal artery. Acta Biomater. **32**, 231–237 (2016)

293. A.V Kamenskiy, Y.A. Dzenis, S.A.J. Kazmi, M.A. Pemberton, I.I.I. Pipinos, N.Y. Phillips, K. Herber, T. Woodford, C.S. Lomneth R.E. Bowen, J.N. MacTaggart, Biaxial mechanical properties of the human thoracic and abdominal aorta, common carotid, subclavian, renal and common iliac arteries. Biomech. Model. Mechanobio. **13**, 1341–1359 (2014)

294. A.V. Kamenskiy, M. Jadidi, S. A. Razian, M. Habibnezhad, E. Anttila, Mechanical, structural, and physiologic differences in human elastic and muscular arteries of different ages: comparison of the descending thoracic aorta to the superficial femoral artery. Acta Biomater. **119**, 268–283 (2021)

295. A. Kamiya, R. Bukhari, T. Togawa, Adaptive regulation of wall shear stress optimizing vascular tree function. Bull. Math. Biol. **46**, 127–137 (1984)

296. A. Kamiya, T. Togawa, Adaptive regulation of wall shear stress to flow change in the canine carotid artery. Am. J. Physiol. **239**, H14–H21 (1980)

297. A. Kaplan, M. Haenlein, Siri, Siri, in my hand: Who's the fairest in the land? On the interpretations, illustrations, and implications of artificial intelligence. Bus. Horiz. **62**, 15–25 (2019)

298. E. Karlöf, Seime T, Dias N, Lengquist M, Witasp A, Almqvist H, Kronqvist M, Gåadin JR, Odeberg J, Maegdefessel L, Stenvinkel P, Matic LP, Hedin U. Correlation of computed tomography with carotid plaque transcriptomes associates calcification with lesion-stabilization. Atherosclerosis **288**, 175–185 (2019). https://doi.org/10.1016/j.atherosclerosis.2019.05.005 Epub 2019 May 11

299. M. Kazi, J. Thyberg, P. Religa, J. Roy, P. Eriksson, U. Hedin, J. Swedenborg, Influence of intraluminal thrombus on structural and cellular composition of Abdominal Aortic Aneurysm wall. J. Vasc. Surg. **38**, 1283–1292 (2003)

300. A. Kelly-Arnold, N. Maldonado, D. Laudier, E. Aikawa, L. Cardoso, S Weinbaum, Revised microcalcification hypothesis for fibrous cap rupture in human coronary arteries. Proc. Natl. Acad. Sci. U.S.A. **110**, 10741–10746 (2013)

301. P.P.N. Kemper, S. Mahmoudi, I. Apostolakis, E. Konofagou, Feasibility of bi-linear mechanical characterization of the abdominal aorta in a hypertensive mouse model. Ultrasound Med. Biol. (2021)

302. J.T. Keyes, D.R. Lockwood, U. Utzinger, L.G. Montilla, R.S. Witte, J.P. Vande Geest, Comparisons of planar and tubular biaxial tensile testing protocols of the same porcine coronary arteries. Ann. Biomed. Eng. **41**, 1579–1791 (2013). https://doi.org/10.1007/s10439-012-0679-0

303. A. Kheradvar, C. Rickers, D. Morisawa, M. Kim, G.-R. Hong, G. Pedrizzetti, Diagnostic and prognostic significance of cardiovascular vortex formation. J. Cardiol. **74**, 403–411 (2019)

304. A.W. Khira, K.H. Parker, Wave intensity in the ascending aorta: effects of arterial occlusion. J. Biomech. **38**, 647–655 (2005)

305. S. Khosla, D.R. Morris, J.V. Moxon, P.J. Walker, T.C. Gasser, J. Golledge, Meta-analysis of peak wall stress in ruptured, symptomatic and intact abdominal aortic aneurysms. Br. J. Surg. **101**, 1350–1357 (2014)

306. P.J. Kilner, G.Z. Yang, R.H. Mohiaddin, D.N. Firmin, D.B. Longmore, Helical and retrograde secondary flow patterns in the aortic arch studied by three-directional magnetic resonance velocity mapping. Circulation **88**, 2235–2247 (1993)

307. M.H. Kim, N.R. Harris, J.M. Tarbell, Regulation of capillary hydraulic conductivity in response to acute change in shear. Am. J. Physiol. **289**, H2126–H2136 (2005)

308. E. Knörzer, W. Folkhard, W. Geercken, C. Boschert, M. H. Koch, B. Hilbert, H. Krahl, E. Mosler, H. Nemetschek-Gansler, T. Nemetschek, New aspects of the etiology of tendon rupture. An analysis of time-resolved dynamic-mechanical measurements using synchrotron radiation. Arch. Orthop. Trauma. Surg. **105**, 113–120 (1986)

309. A. Kolmogorov, The local structure of turbulence in incompressible viscous fluid for very large Reynolds' numbers. Doklady Akademiia Nauk SSSR **30**, 301–305 (1941)

310. R. Krams, M. Bäck (eds.), *The ESC Textbook of Vascular Biology* (Oxford University Press, Oxford, 2017)

311. J. Krejza, M. Arkuszewski, S.E. Kasner, J. Weigele, A. Ustymowicz, R.W. Hurst, B.L. Cucchiara, S.R. Messe, Carotid artery diameter in men and women and the relation to body and neck size. Stroke **37**, 1103–1105 (2006)

312. I.M. Krieger, T.J. Dougherty, A mechanism for non-Newtonian flow in suspensions of rigid spheres. Trans. Soc. Rheo. **3**, 137–152 (1959)

313. E. Kröner, Allgemeine Kontinuumstheorie der Versetzungen und Eigenspannungen. Arch. Rat. Mech. Anal. **4**, 273–334 (1960)

314. M. Kroon, G.A. Holzapfel, A theoretical model for fibroblast-controlled growth of saccular cerebral aneurysms. J. Theor. Biol. **257**, 73–83 (2009)

315. D.N. Ku, D.P. Giddens, C.K. Zarins, S. Glagov, Pulsatile flow and atherosclerosis in the human carotid bifurcation. Positive correlation between plaque location and low oscillating shear stress. Arteriosclerosis **5**, 293–302 (1985)

316. E. Kuhl, R. Maas, G. Himpel, A. Menzel, Computational modeling of arterial wall growth. Biomech. Model. Mechanobio. **6**, 321–331 (2007)

317. H. Kuivaniemi, E.J. Ryer, J.R. Elmore, G. Tromp, Understanding the pathogenesis of abdominal aortic aneurysms. Expert Rev. Cardiovasc. Ther. **13**, 975–987 (2015)

318. P.K. Kundu, I.M. Cohen, *Fluid mechanics*, 4th edn. (Academic Press, London, 2008)

319. M.H. Kural, M. Cai, D. Tang, T. Gwyther, J. Zheng, K.L. Billiar, Planar biaxial characterization of diseased human coronary and carotid arteries for computational modeling. J. Biomech. **45**, 790–798 (2012)

320. D. Lacasse, A. Garon, D. Pelletier, Mechanical hemolysis in blood flow: user-independent predictions with the solution of a partial differential equation. Comput. Meth. Biomech. Biomed. Eng. **10**, 1–12 (2007)

321. E.M. Landis, Micro-injection studies of capillary permeability. II. The relation between capillary pressure and the rate at which fluid passes through the walls of single capillaries. Am. J. Physiol. **82**, 217–238 (1927)

322. E.M. Landis, Micro-injection studies of capillary blood pressure in human skin. Heart **30**, 209–228 (1930)

323. G.J. Langewouters, K.H. Wesseling, W.J.A. Goedhard, The static elastic properties of 45 human thoracic and 20 abdominal aortas in vitro and the parameters of a new model. J. Biomech. **17**, 425–435 (1984)

324. B.L. Langille, M.P. Bendeck, F.W. Keeley, Adaptations of carotid arteries of young and mature rabbits to reduced carotid blood flow. Am. J. Physiol. **256**, H931–H939 (1989)

325. Y. Lanir, Constitutive equations for fibrous connective tissues. J. Biomech. **16**, 1–12 (1983)

326. Y. Lanir, Mechanistic micro-structural theory of soft tissues growth and remodeling: tissues with unidirectional fibers. Biomech. Model. Mechanobio. **14**, 245–266 (2015)

327. T. Länne, B. Sonesson, D. Bergqvist, H. Bengtsson, D. Gustafsson, Diameter and compliance in the male human abdominal aorta: influence of age and aortic aneurysm. Eur. J. Vasc. Surg. **6**, 178–184 (1992)

328. E. Larsson, *Abdominal Aortic Aneurysm—Gender Aspects on Risk Factors and Treatment*. Ph.D. thesis, Karolinska Institute, Stockholm, Sweden, 2011

329. M. Latorre, J.D. Humphrey, A mechanobiologically equilibrated constrained mixture model for growth and remodeling of soft tissues. Z. Angew. Math. Mech. **98**(12), 2048–2071 (2018)

330. M. Latorre, J.D. Humphrey, Fast, rate-independent, finite element implementation of a 3d constrained mixture model of soft tissue growth and remodeling. Comput. Meth. Appl. Mech. Eng. **368**, 113156 (2020)

331. J.D. Laubrie, J.S. Mousavi, S. Avril, A new finite-element shell model for arterial growth and remodeling after stent implantation. Int. J. Numer. Meth. Biomed. Eng. **36**(1), e3282 (2020)

332. S. Laurent, J. Cockcroft, L. Van Bortel, P. Boutouyrie, C. Giannattasio, D. Hayoz, B. Pannier, C. Vlachopoulos, I. Wilkinson, H. Struijker-Boudier, Expert consensus document on arterial stiffness: methodological issues and clinical applications. Eur. Heart J. **27**, 2588–2605 (2006)

333. M.G. Lawlor, M.R. O'Donnell, B.M. O'Connell, M.T. Walsh, Experimental determination of circumferential properties of fresh carotid artery plaques. J. Biomech. **44**, 1709–1715 (2011)

334. J.R. Leach, M.R. Kaazempur Mofrad, D. Saloner, *Computational Models of Vascular Mechanics, in Computational Modeling in Biomechanics* (Springer, Berlin, 2010)

335. C.L. Lendon, M.J. Davies, G.V.R. Born, P.D. Richardson, Atherosclerotic plaque caps are locally weakened when macrophages density is increased. Atherosclerosis **87**, 87–90 (1991)

336. A.S. Les, J.J. Yeung, G.M. Schultz, R.J. Herfkens, R.L. Dalman, C.A. Taylor, Supraceliac and infrarenal aortic flow in patients with abdominal aortic aneurysms: mean flows, waveforms, and allometric scaling relationships. Cardiovasc. Eng. Technol. (2010). https://doi.org/10.1007/s13239-010-0004-8

337. D.Y. Leung, S. Glagov, M.B. Mathews, Cyclic stretching stimulates synthesis of matrix components by arterial smooth muscle cells in vitro. Science **191**, 475–477 (1976)

338. J.R. Levick, C.C. Michel, Microvascular fluid exchange and the revised starling principle. Cardiovasc. Res. **87**, 198–210 (2010)

339. G. Li, M. Wang, A.W. Caulk, N.A. Cilfone, S. Gujja, L. Qin, P.-Y. Chen, Z. Chen, S. Yousef, Y. Jiao, C. He, B. Jiang, A. Korneva, M.R. Bersi, G. Wang, X. Liu, S. Mehta, A. Geirsson, J.R Gulcher, T.W. Chittenden, M. Simons, J.D. Humphrey, G. Tellides, Chronic mTOR activation induces a degradative smooth muscle cell phenotype. J. Clin. Invest. **130**, 1233–1251 (2020)

340. Z-Y. Li, S.P. Howarth, T. Tang, J.H. Gillard, How critical is fibrous cap thickness to carotid plaque stability? A flow-plaque interaction model. Stroke **37**, 1195–1199 (2006)

341. Z.-Y. Li, S.P.S. Howarth, T. Tang, M.J. Graves, J. U-King-Im, R.A. Trivedi, P.J.K., J.H. Gillard, Structural analysis and magnetic resonance imaging predict plaque vulnerability: a study comparing symptomatic and asymptomatic individuals. J. Vasc. Surg. **45**, 768–775 (2007)

342. J. Liao, I. Vesely, Skewness angle of interfibrillar proteoglycans increases with applied load on mitral valve chordae tendineae. J. Biomech. **40**, 390–398 (2007)

343. M.L. Liljeqvist, Geometric, biomechanical and molecular analyses of abdominal aortic aneurysms, PhD thesis, Karolinska Institutet, ISBN: 978-91-7831-918-3 (2020)

344. M. Liu, L. Liang, W. Sun, A generic physics-informed neural network-based constitutive model for soft biological tissues. Comput. Methods Appl. Mech. Eng. **372**, 113402 (2020)

345. S. Loerakker, C. Obbink-Huizer, F.P.T. Baaijens, A physically motivated constitutive model for cell-mediated compaction and collagen remodeling in soft tissues. Biomech. Model. Mechanobio. **13**, 985–1001 (2014)

346. S. Loerakker, T. Ristoria, F.P.T. Baaijens, A computational analysis of cell-mediated compaction and collagen remodeling in tissue-engineered heart valves. J. Mech. Behav. Biomed. Mater. **58**, 173–187 (2016)

347. S.V. Lopez-Quintero, R. Amaya, M. Pahakis, J.M. Tarbell, The endothelial glycocalyx mediates shear-induced changes in hydraulic conductivity. Am. J. Physiol. **296**, H1451–H1456 (2009)

348. H.M. Loree, R.D. Kamm, R.G. Stringfellow, R.T. Lee, Effects of fibrous cap thickness on peak circumferential stress in model atherosclerotic vessels. Circ. Res. **71**, 850–858 (1992)
349. J. Lu, X. Zhou, M.L. Raghavan, Inverse elastostatic stress analysis in pre-deformed biological structures: Demonstration using Abdominal Aortic Aneurysms. J. Biomech. **40**, 693–696 (2007)
350. W. Lyne, Unsteady viscous flow in a curved pipe. J. Fluid Mech. **45**, 13–31 (1971)
351. I.M. Machyshyn, P.H.M. Bovendeerd, A.A.F. van de Ven, P.M.J. Rongen, F.N. van de Vosse, A model for arterial adaptation combining microstructural collagen remodeling and 3d tissue growth. Biomech. Model. Mechanobio. **9**, 671–687 (2010)
352. A. Maier, M. Essler, M.W. Gee, H.H. Eckstein, W.A. Wall, C. Reeps, Correlation of biomechanics to tissue reaction in aortic aneurysms assessed by finite elements and [18F]-fluorodeoxyglucose-PET/CT. Int. J. Numer. Meth. Biomed. Eng. **28**, 456–471 (2012)
353. A. Maier, M.W. Gee, C. Reeps, J. Pongratz, H.H. Eckstein, W.A. Wall, A comparison of diameter, wall stress, and rupture potential index for abdominal aortic aneurysm rupture risk prediction. Ann. Biomed. Eng. **38**, 3124–3134 (2010)
354. L.E. Malvern, *Introduction to the Mechanics of a Continuous Medium* (Prentice-Hall, Englewood Cliffs, 1969)
355. K.N. Margaris, R.A. Black, Modelling the lymphatic system: challenges and opportunities. J. R. Soc. Interface **9**, 601–612 (2012)
356. G. Marini, A. Maier, C. Reeps, H.-H. Eckstein, W.A. Wall, M.W. Gee, A continuum description of the damage process in the arterial wall of abdominal aortic aneurysms. Int. J. Numer. Methods Bioeng. **28**, 87–99 (2012)
357. V. Marque, P. Kieffer, B. Gayraud, I. Lartaud-Idjouadiene, F. Ramirez, J. Atkinson, Aortic wall mechanics and composition in a transgenic mouse model of Marfan syndrome. Arterioscl. Thromb. Vasc. Biol. **21**, 1184–1189 (2001)
358. J.N. Marsh, S. Takiuchi, S.J. Lin, G.M. Lanza, S.A. Wickline, Ultrasonic delineation of aortic microstructure: the relative contribution of elastin and collagen to aortic elasticity. J. Acoust. Soc. Am. **115**, 2032–2040 (2004)
359. J.M. Marshall, H.C. Tandon, Direct observations of muscle arterioles and venules following contraction of skeletal muscle fibres in the rat. J. Physiol. **350**, 447–459 (1984)
360. G. Martufi, T.C. Gasser, A constitutive model for vascular tissue that integrates fibril, fiber and continuum levels. J. Biomech. **44**, 2544–2550 (2011)
361. G. Martufi, T.C. Gasser, Turnover of fibrillar collagen in soft biological tissue with application to the expansion of abdominal aortic aneurysms. J. R. Soc. Interface **9**, 3366–3377 (2012)
362. G. Martufi, M. Lindquist Liljeqvist, N. Sakalihasan, G. Panuccio, R. Hultgren, J. Roy, T.C. Gasser, Local diameter, wall stress and thrombus thickness influence the local growth of abdominal aortic aneurysms. J. Endovas. Ther. **23**, 957–966 (2016)
363. T. Matsumoto, K. Hayashi, Stress and strain distribution in hypertensive and normotensive rat aorta considering residual strain. J. Biomech. **118**, 62–73 (1996)
364. M.I. Mäyränpää, J.A. Trosien, V. Fontaine, M. Folkesson, M. Kazi, P. Eriksson, J. Swedenborg, U. Hedin, Mast cells associate with neovessels in the media and adventitia of abdominal aortic aneurysms. J. Vasc. Surg. **50**, 388–396 (2009)
365. D.A. McDonald, *Blood Flow in Arteries*, 6th edn. (Edward Arnold, London, 2011)
366. A. Melcher, D.E. Donald, Maintained ability of carotid baroreflex to regulate arterial pressure during exercise. Am. J. Physiol. Heart Circ. Physiol. **241**, H838–H849 (1981)
367. A.V. Melnik, H.B. Da Rocha, A. Goriely, On the modeling of fiber dispersion in fiber-reinforced elastic materials. Int. J. Non-Linear Mech. **75**, 92–106 (2015)
368. A. Menzel, Modelling of anisotropic growth in biological tissues. A new approach and computational aspects. Biomech. Model. Mechanobio. **3**, 147–171 (2005)
369. A. Menzel, E. Kuhl, Frontiers in growth and remodeling. Mech. Res. Commun. **42**, 1–14 (2012)
370. C.C. Michel, Fluid movements through the capillary wall, in *Handbook of Physiology; Section 2: The Cardiovascular System*, ed. by E.M. Renkin, C.C. Michel. The Microcirculation, vol. 4 (American Physiological Society, Rockville, 1984), pp. 375–409

371. C.C. Michel, M.E. Phillips, Steady-state fluid filtration at different capillary pressures in perfused frog mesenteric capillaries. J. Physiol. **388**, 421–435 (1987)

372. J.-B. Michel, G. Jondeau, D.M. Milewicz, From genetics to response to injury: vascular smooth muscle cells in aneurysms and dissections of the ascending aorta. Cardiovasc. Res. **114**(4), 578–589 (2018)

373. J.-B. Michel, J.L. Martin-Ventura, J. Egido, N. Sakalihasan, V. Treska, J. Lindholt, E. Allaire, U. Thorsteinsdottir, G. Cockerill, J. Swedenborg, Novel aspects of the pathogenesis of aneurysms of the abdominal aorta in humans. Cardiovasc. Res. **90**, 18–27 (2011)

374. C. Miller, T.C. Gasser, A microstructurally motivated constitutive description of collagenous soft biological tissue towards the description of their non-linear and time-dependent properties. J. Mech. Phys. Solids, **154**, 104500 (2021).

375. D.M. Milewicz, C.S. Kwartler, C.L. Papke, E.S. Regalado, J. Cao, A.J. Reid, Genetic variants promoting smooth muscle cell proliferation can result in diffuse and diverse vascular diseases: evidence for a hyperplastic vasculomyopathy. Genet. Med. **12**, 196–203 (2010)

376. M. Mirramezani, S.C. Shadden, A distributed lumped parameter model of blood flow. Ann. Biomed. Eng. **48**, 2870–2886 (2020)

377. T. Mitchell, *Machine Learning* (McGraw Hill, New York, 2017)

378. H. Miyazaki, K. Hayashi, Tensile tests of collagen fibers obtained from the rabbit patellar tendon. Biomed. Microdevices **2**, 151–157 (1999)

379. A.I. Moens, *Die Pulskurve* (E.J. Brill, Leiden, 1878)

380. D. Mohan, J.W. Melvin, Failure properties of passive human aortic tissue I—uniaxial tension tests. J. Biomech. **15**, 887–902 (1982)

381. J.A.T. Monteiro, E.S. da Silva, M.L. Raghavan, P. Puech-Leão, M. de Lourdes Higuchi, J.P. Otoch, Histologic, histochemical, and biomechanical properties of fragments isolated from the anterior wall of abdominal aortic aneurysms. J. Vasc. Surg. **59**, 1393–1401 (2013)

382. H. Motulsky, A. Christopoulos, *Fitting Models to Biological Data Using Linear and Nonlinear Regression* (Oxford University Press, New York, 2004)

383. S.J. Mousavi, S. Farzaneh, S. Avril, Patient-specific predictions of aneurysm growth and remodeling in the ascending thoracic aorta using the homogenized constrained mixture model. Biomech. Model. Mechanobio. **18**(6), 1895–1913 (2019)

384. R.A. Murphy, C.M. Rembold, The latch-bridge hypothesis of smooth muscle contraction. Canad. J. Physiol. Pharm. **83**, 857–864 (2005)

385. S.-I. Murtada, J. Ferruzzi, H. Yanagisawa, J.D. Humphrey, Reduced biaxial contractility in the descending thoracic aorta of fibulin-5 deficient mice. ASME. J. Biomech. Eng. **138**, 051008 (2016)

386. S.-I. Murtada, J.D. Humphrey, Regional heterogeneity in the regulation of vasoconstriction in arteries and its role in vascular mechanics, in *Molecular, Cellular, and Tissue Engineering of the Vascular System*, ed. by B.M. Fu, N.T. Wright (Springer, Switzerland, 2018), pp. 105–128

387. S.-I. Murtada, S. Lewin, A. Arner, J.D. Humphrey, Adaptation of active tone in the mouse descending thoracic aorta under acute changes in loading. Biomech. Model. Mechanobio. **15**, 579–592 (2016)

388. Y. Nabeshima, E.S. Grood, A. Sakurai, J.H. Herman, Uniaxial tension inhibits tendon collagen degradation by collagenase in vitro. J. Orthop. Res. **14**, 123–130 (1996)

389. M.R. Najjari, C. Cox, M.W. Plesniak, Formation and interaction of multiple secondary flow vortical structures in a curved pipe: transient and oscillatory flows. J. Fluid Mech. **876**, 481–526 (2019)

390. T. Nakahara, M.R. Dweck, N. Narula, D. Pisapia, J. Narula, H.W. Strauss, Coronary artery calcification: From mechanism to molecular imaging. JACC Cardiovasc. Imaging **10**, 582–593 (2017)

391. A. Nchimi, J.-P. Cheramy-Bien, T.C. Gasser, G. Namur, P. Gomez, A. Albert, L. Seidel, J.O. Defraigne, N. Labropoulos, N. Sakalihasan, Multifactorial relationship between 18F-fluoro-deoxy-glucose positron emission tomography signaling and biomechanical properties in unruptured aortic aneurysms. Circ. Cardiovasc. Imaging **7**, 82–91 (2014)

392. I. Newton, *Philosophiæ Naturalis Principia Mathematica* (1687)

393. S.C. Nicholls, J.B. Gardner, M.H. Meissner, H.K. Johansen, Rupture in small abdominal aortic aneurysms. J. Vasc. Surg. **28**, 884–888 (1998)
394. W.W. Nichols, M.F. O'Rourke, C. Vlachopoulos, *McDonald's Blood Flow in Arteries. Theoretical, Experimental and Clinical Principles*, 6th edn. (Arnold, London, 2011)
395. R. Nissen, G.J. Cardinale, S. Udenfriend, Increased turnover of arterial collagen in hypertensive rats. Proc. Natl. Acad. Sci. U.S.A. **75**, 451–453 (1978)
396. I.T. Nizamutdinova, D. Maejima, T. Nagai, C.J. Meininger, A.A. Gashev, Histamine as an endothelium-derived relaxing factor in aged mesenteric lymphatic vessels. Lymphat. Res. Biol. **15**, 136–145 (2017)
397. C. Noble, N. Smulders, N.H. Green, R. Lewis, M.J. Carré, S.E. Franklin, S. MacNeil, Z.A. Taylor, Creating a model of diseased artery damage and failure from healthy porcine aorta. J. Mech. Behav. Biomed. Mater. **60**, 378–393 (2016)
398. M. Noris, M. Morigi, R. Donadelli, S. Aiello, M. Foppolo, M. Todeschini, S. Orisio, G. Remuzzi, A. Remuzzi, Nitric oxide synthesis by cultured endothelial cells is modulated by flow conditions. Circ. Res. **76**, 536–543 (1995)
399. A. Öchsner, *Continuum Damageand Fracture Mechanics* (Springer, Singapore, 2016)
400. M.K. O'Connell, S. Murthy, S. Phan, C. Xu, J. Buchanan, R. Spilker, R.L. Dalman, C.K. Zarins, W. Denk, C.A. Taylor, The three-dimensional micro- and nanostructure of the aortic medial lamellar unit measured using 3d confocal and electron microscopy imaging. Matrix Biol. **27**, 171–181 (2008)
401. R.W. Ogden, *Non-Linear Elastic Deformations* (Dover, New York, 1997)
402. R.W. Ogden, Elements of the theory of finite elasticity, in *Nonlinear Elasticity. Theory and Applications*, ed. by Y.B. Fu, R.W. Ogden (Cambridge University Press, Cambridge, 2001), pp. 1–57
403. J. Ohayon, O. Dubreuil, P. Tracqui, S. Le Floc'h, G. Rioufol, L. Chalabreysse, F. Thivolet, R.I. Pettigrew, G. Finet, Influence of residual stress/strain on the biomechanical stability of vulnerable coronary plaques: potential impact for evaluating the risk of plaque rupture. Am. J. Physiol. Heart Circ. Physiol. **293**, H1987–H1996 (2007)
404. J. Ohayon, G. Finet, A.M. Gharib, D.A. Herzka, P. Tracqui, J. Heroux, G. Rioufol, M.S. Kotys, A. Elagha, R.I. Pettigrew, Necrotic core thickness and positive arterial remodeling index: emergent biomechanical factors for evaluating the risk of plaque rupture. Am. J. Physiol. Heart Circ. Physiol. **295**, H717–H727 (2008)
405. J. Ohayon, G. Finet, R.I. Pettigrew (eds.), *Biomechanics of Coronary Atherosclerotic Plaque: From Model to Patient*. Biomechanics of Living Organs (Elsevier, 2020)
406. J. Ohayon, A.M. Gharib, A. Garcia, J. Heroux, S.K. Yazdani, M. Malvè, P. Tracqui, M.A. Martinez, M. Doblare, G. Finet, R.I. Pettigrew, Is arterial wall-strain stiffening an additional process responsible for atherosclerosis in coronary bifurcations? Am. J. Physiol. Heart Circ. Physiol. **301**, H1097–H1106 (2011)
407. J. Ohayon, N. Mesnier, A. Broisat, J. Toczek, L. Riou, P. Tracqui, Elucidating atherosclerotic vulnerable plaque rupture by modeling cross substitution of ApoE$^{-/-}$ mouse and human plaque components stiffness. Biomech. Model. Mechanobio. **11**, 801–813 (2011)
408. J. Ohayon, Y. Payan (eds.), *Biomechanics of Living Organs: Hyperelastic Constitutive Laws for Finite Element Modeling* (Elsevier, Amsterdam, 2017)
409. S. Oka, T. Azuma, Physical theory of tension in thick walled blood vessels in equilibrium. Biorheology **7**, 109–118 (1970)
410. H.S. Oktay, T. Kang, J.D. Humphrey, G.G. Bishop, Changes in the mechanical behavior of arteries following balloon angioplasty, in *ASME 1991 Biomechanics Symposium, AMD-Vol. 120* (American Society of Mechanical Engineers, New York, 1991)
411. S.A. O'Leary, D. Healy, E.G. Kavanagh, M.T. Walsh, T.M. McGloughlin, B.J. Doyle, The biaxial biomechanical behavior of abdominal aortic aneurysm tissue. Ann. Biomed. Eng. **42**, 2440–2450 (2014)
412. S.A. O'Leary, J.J. Mulvihill, H.E. Barrett, E.G. Kavanagh, M.T. Walsh, T.M. McGloughlin, B.J. Doyle, Determining the influence of calcification on the failure properties of abdominal aortic aneurysm (AAA) tissue. J. Mech. Behav. Biomed. Mater. **42**, 154–167 (2015)

413. M. Ortiz, A. Pandolfi, Finite-deformation irreversible cohesive elements for three-dimensional crack-propagation analysis. Int. J. Numer. Meth. Eng. **44**, 1267–1282 (1999)

414. S. Oyre, E.M. Pedersen, S. Ringgaard, P. Boesiger, W.P. Paaske, In vivo wall shear stress measured by magnetic resonance velocity mapping in the normal human abdominal aorta. Eur. J. Vasc. Endovasc. Surg. **13**, 263–271 (1997)

415. A. Paknahad, M. Goudarzi, N.W. Kucko, S.C.G. Leeuwenburgh, L.J. Sluys, Calcium phosphate cement reinforced with poly (vinyl alcohol) fibers: an experimental and numerical failure analysis. Acta Biomater. **119**, 458–471 (2021)

416. R. Pal, Rheology of concentrated suspensions of deformable elastic particles such as human erythrocytes. J. Biomech. **36**, 981–989 (2003)

417. J.R. Pappenheimer, A. Soto-Rivera, Effective osmotic pressure of the plasma proteins and other quantities associated with the capillary circulation in the hind-limbs of cats and dogs. Am. J. Physiol. **152**, 471–491 (1948)

418. Pardiso Reference Manual (2018). https://www.pardiso-project.org/

419. A. Passos, J.M. Sherwood, E. Kaliviotis, R. Agrawal, C. Pavesio, S. Balabani, The effect of deformability on the microscale flow behavior of red blood cell suspensions. Phys. Fluids **31**, 1–11 (2019)

420. R.C. Pasternak, M.H. Criqui, E.J. Benjamin, F.G.R. Fowkes, E.M. Isselbacher, P.A. McCullough, P.A. Wolf, Z.-J. Zheng, Atherosclerotic vascular disease conference: writing group I: epidemiology. Circulation **109**, 2605–2612 (2004)

421. D.J. Patel, J.S. Janicki, T.E. Carew, Static anisotropic elastic properties of the aorta in living dogs. Circ. Res. **25**, 765–779 (1969)

422. S.W. Patterson, E.H. Starling, On the mechanical factors which determine the output of the ventricles. J. Physiol. **48**, 357–379 (1914)

423. O.M. Pedersen, A. Aslaksen, H. Vik-Mo, Ultrasound measurement of the luminal diameter of the abdominal aorta and iliac arteries in patients without vascular disease. J. Vasc. Surg. **17**, 596–601 (1993)

424. T.J. Pedley, *The Fluid Mechanics of Large Blood Vessels*. Cambridge Monographs on Mechanics and Applied Mathematics (Cambridge University Press, New York, 1980)

425. R.H.J. Peerlings, R. De Borst, W.A.M. Brekelmans, J.H.P. De Vree, Gradient enhanced damage for quasi-brittle materials. Int. J. Numer. Methods Eng. **39**(19), 3391–3403 (1996)

426. C. Petit, S.J. Mousavi, S. Avril, Review of the essential roles of SMCs in ATAA biomechanics, in *Advances in Biomechanics and Tissue Regeneration*, ed. by M.H. Doweidar (Elsevier, Amsterdam, 2019), pp. 95–114

427. R.J. Phillips, R.C. Armstrong, R.A. Brown, A.L. Graham, J.R. Abbott, A constitutive equation for concentrated suspensions that accounts for shearinduced particle migration. Phys. Fluids **4**, 30–40 (1991)

428. J.E. Pichamuthu, J.A. Phillippi, D.A. Cleary, D.W. Chew, J. Hempel, D.A. Vorp, T.G. Gleason, Differential tensile strength and collagen composition in ascending aortic aneurysms by aortic valve phenotype. Ann. Thorac Surg. **96**, 2147–2154 (2013)

429. P.J. Carreau, D. DeKee, R.P. Chhabra, *Rheology of Polymeric Systems: Principles and Applications* (Hanser, Munich, 1997)

430. D.S. Pleouras, A.I. Sakellarios, P. Tsompou, V. Kigka, S. Kyriakidis, S. Rocchiccioli, D. Neglia, J. Knuuti, G. Pelosi, L.K. Michalis, D.I. Fotiadis, Simulation of atherosclerotic plaque growth using computational biomechanics and patient-specific data. Sci. Rep. **10**, 101–106 (2020)

431. S. Polzer, J. Bursa, T.C. Gasser, R. Staffa, R. Vlachovsky, Numerical implementation to predict residual strains from the homogeneous stress hypothesis with application to abdominal aortic aneurysms. Ann. Biomed. Eng. **41**, 1516–1527 (2013)

432. S. Polzer, T.C. Gasser, Biomechanical rupture risk assessment of abdominal aortic aneurysms based on a novel probabilistic rupture risk index. J. R. Soc. Interface **12**, 20150852 (2015)

433. S. Polzer, T.C. Gasser, B. Markert, J. Bursa, P. Skacel, Impact of poroelasticity of the intraluminal thrombus on the wall stress of abdominal aortic aneurysms. Biomed. Eng. Online **11**, 62 (2012). https://doi.org/10.1186/1475--925X--11--62

434. S. Polzer, T.C. Gasser, K. Novak, V. Man, M. Tichy, P. Skacel, J. Bursa, Structure-based constitutive model can accurately predict planar biaxial properties of aortic wall tissue. Acta Biomater. **14**, 133–145 (2015)

435. S. Polzer, T.C. Gasser, R. Vlachovský, L. Kubíček, L. Lambert, V. Man, K. Novaák, M. Slažanský, J. Burša, R. Staffa, Biomechanical indices are more sensitive than diameter in predicting rupture of asymptomatic abdominal aortic aneurysms. J. Vasc. Surg. **71**, 617–626 (2020)

436. S. Polzer, T.C. Gasserb, J. Bursa, R. Staffa, R. Vlachovsky, V. Mana, P. Skacela, Importance of material model in wall stress prediction in abdominal aortic aneurysms. Med. Eng. Phys. **35**, 1282–1289 (2013)

437. S. Polzer, V. Man, R. Vlachovský, L. Kubíček, J. Kracík, R. Staffa, T. Novotny, J. Bursa, M.L. Raghavan, Failure properties of abdominal aortic aneurysm tissue are orientation dependent. J. Mech. Behav. Biomed. Mater. **114**, 104181 (2020). https://doi.org/10.1016/j.jmbbm.2020.104181:104181

438. S.B. Pope, *Turbulent flows* (Cambridge University Press, Cambridge, 2012)

439. J. Qiu, Y. Zheng, J. Hu, D. Liao, H. Gregersen, X. Deng, Y. Fan, G. Wang, Biomechanical regulation of vascular smooth muscle cell functions: from in vitro to in vivo understanding. J. R. Soc. Interface **11**, 20130852 (2014). https://doi.org/10.1098/rsif.2013.085

440. D. Quemada, Rheology of concentrated disperse systems and minimum energy dissipation principle. Rheol. Acta **16**, 82–94 (1977)

441. K.P. Quinn, B.A. Winkelstein, Altered collagen fiber kinematics define the onset of localized ligament damage during loading. J. Appl. Physiol. **105**, 1881–1888 (2008)

442. R.R. Ross, P. Bornstein, The elastic fiber: I. The separation and partial characterization of its macromolecular components. J. Cell Biol. **40**, 366–381 (1969)

443. A. Rachev, A model of arterial adaptation to alterations in blood flow. J. Elast. **61**, 83–111 (2000)

444. A. Rachev, S.E. Greenwald, Residual strains in conduit arteries. J. Biomech. **36**, 661–670 (2003)

445. A. Rachev, S.E. Greenwald, T. Shazly, Are geometrical and structural variations along the length of the aorta governed by a principle of 'optimal mechanical operation'? ASME. J. Biomech. Eng. **135**, 81006 (2013). https://doi.org/10.1115/SBC2013-14427

446. A. Rachev, K. Hayashi, Theoretical study of the effects of vascular smooth muscle contraction on strain and stress distributions in arteries. Ann. Biomed. Eng. **27**, 459–468 (1999)

447. A. Rachev, N. Stergiopulos, J.-J. Meister, Theoretical study of dynamics of arterial wall remodeling in response to changes in blood pressure. J. Biomech. **29**, 635–642 (1996)

448. M.L. Raghavan, J. Kratzberg, E. Castro de Tolosa, M. Hanaoka, P. Walker, E. da Silva, Regional distribution of wall thickness and failure properties of human Abdominal Aortic Aneurysm. J. Biomech. **39**, 3010–3016 (2006)

449. M.L. Raghavan, D.A. Vorp, Toward a biomechanical tool to evaluate rupture potential of Abdominal Aortic Aneurysm: identification of a finite strain constitutive model and evaluation of its applicability. J. Biomech. **33**, 475–482 (2000)

450. M.L. Raghavan, M.W. Webster, D.A. Vorp, Ex vivo biomechanical behavior of Abdominal Aortic Aneurysm: assesment using a new mathematical model. Ann. Biomed. Eng. **24**, 573–582 (1996)

451. M.L. Raghavan, M.M. Hanaoka, J.A. Kratzberg, M. de Lourdes Higuchi, E.S. da Silva, Biomechanical failure properties and microstructural content of ruptured and unruptured abdominal aortic aneurysms. J. Biomech. **44**, 2501–2507 (2011)

452. J.M. Ramstack, L. Zuckerman, L.F. Mockros, Activation of platelets. J. Biomech. **12**, 113–125 (1979)

453. R. Rapadamnaba , M. Ribatet, B. Mohammadi, Global sensitivity analysis for assessing the parameters importance and setting a stopping criterion in a biomedical inverse problem. Int. J. Numer. Methods Biomed. Eng., **37** (2021)

454. M.S. Razavi, J.B. Dixon, R.L. Gleason, Characterization of rat tail lymphatic contractility and biomechanics: incorporating nitric oxide-mediated vasoregulation. J. R. Soc. Interface **17**, 20200598 (2020)

455. D.A. Reasor, M. Mehrabadi, D.N. Ku, C.K. Aidun, Determination of critical parameters in platelet margination. Ann. Biomed. Eng. **41**, 238–249 (2012)

456. A. Redaelli, S. Vesentini, M. Soncini, P. Vena, S. Mantero, F.M. Montevecchi, Possible role of decorin glycosaminoglycans in fibril to fibril force transfer in relative mature tendons—a computational study from molecular to microstructural level. J. Biomech. **36**, 1555–1569 (2003)

457. J.N. Reddy (ed.), *An Introduction to the Finite Element Method*, 2nd edn. (McGraw Hill, Boston, 1993).

458. C. Reeps, A. Maier, J. Pelisek, F. Hartl, V. Grabher-Maier, W.A. Wall, M. Essler, H.-H. Eckstein, M.W. Gee, Measuring and modeling patient-specific distributions of material properties in abdominal aortic wall. Biomech. Model. Mechanobio. **12**, 717–733 (2013)

459. RESCAN Collaborators, M.J. Bown, M.J. Sweeting, L.C. Brownand J.T. Powell, S.G. Thompson, Surveillance intervals for small abdominal aortic aneurysms. J. Am. Med. Assoc. **309**, 806–813 (2013)

460. J.R. Rice, A path independent integral and the approximate analysis of strain concentration by notches and cracks. J. Appl. Mech. **35**, 379–386 (1968)

461. P.D. Richardson, Biomechanics of plaque rupture: progress, problems, and new frontiers. Ann. Biomed. Eng. **30**, 524–536 (2002)

462. S. Rigozzi, R. Mueller, J.G. Snedeker, Local strain measurement reveals a varied regional dependence of tensile tendon mechanics on glycosaminoglycan content. J. Biomech. **42**, 1547–1552 (2009)

463. S. Rigozzi, R. Mueller, J.G. Snedeker, Collagen fibril morphology and mechanical properties of the achilles tendon in two inbred mouse strains. J. Anat. **216**, 724–731 (2010)

464. B. Rippe, B. Haraldsson, Transport of macromolecules across microvascular walls: the two-pore theory. Physiol. Rev. **74**, 163–219 (1994)

465. F. Riveros, S. Chandra, E.A. Finol, T.C. Gasser, J.F. Rodriguez, A pull-back algorithm to determine the unloaded vascular geometry in anisotropic hyperelastic AAA passive mechanics. Ann. Biomed. Eng. **41**, 694–708 (2013)

466. M.L. Rizzini, D. Gallo, G. De Nisco, F. D'Ascenzo, C. Chiastra, P.P. Bocchino, F. Piroli, G.M. De Ferrari, U. Morbiducci, Does the inflow velocity profile influence physiologically relevant flow patterns in computational hemodynamic models of left anterior descending coronary artery? Med. Eng. Phys. **82**, 58–69 (2020)

467. R.J. Rizzo, W.J. McCarthy, S.N. Dixit, M.P. Lilly, V.P. Shively, W.R. Flinn, J.S.T. Yao, Collagen types and matrix protein content in human abdominal aortic aneurysms. J. Vasc. Surg. **10**, 365–373 (1989)

468. M.R. Roach, A.C. Burton, The reason for the shape of the distensibility curve of arteries. Canad. J. Biochem. Physiol. **35**, 681–690 (1957)

469. A.M. Robertson, A. Sequeira, R.G. Owens, Rheological models for blood, in *Cardiovascular Mathematics*, ed. by L. Formaggia, A. Quarteroni, A.Veneziani (Springer, Milano, 2009)

470. P.S. Robinson, T.F. Huang, E. Kazam, R.V. Iozzo, D.E. Birk, L.J. Soslowsky, Influence of decorin and biglycan on mechanical properties of multiple tendons in knockout mice. ASME. J. Biomech. Eng. **127**, 181–185 (2005)

471. S. Roccabianca, C. Bellini, J.D. Humphrey, Computational modelling suggests good, bad and ugly roles of glycosaminoglycans in arterial wall mechanics and mechanobiology. J. R. Soc. Interface **11**, 20140397 (2014)

472. S. Roccabianca, C.A. Figueroa, G. Tellides, J.D. Humphrey, Quantification of regional differences in aortic stiffness in the aging human. J. Mech. Behav. Biomed. Mater. **29**, 618–634 (2014)

473. E.K. Rodriguez, A. Hoger, A.D. McCulloch, Stress-dependent finite growth in soft elastic tissues. J. Biomech. **27**, 455–467 (1994)

474. J.F. Rodríguez, G. Martufi, M. Doblaré, E. Finol, The effect of material model formulation in the stress analysis of abdominal aortic aneurysms. Ann. Biomed. Eng. **37**, 2218–2221 (2009)
475. J.F. Rodríguez, C. Ruiz, M. Doblaré, G.A. Holzapfel, Mechanical stresses in abdominal aortic aneurysms: influence of diameter, asymmetry, and material anisotropy. ASME. J. Biomech. Eng. **130**, 021023 (2008)
476. A. Romo, P. Badel, A. Duprey, J.P. Favre, S. Avril, In vitro analysis of localized aneurysm rupture. J. Biomech. **189**, 607–616 (2014)
477. L.B. Rowell, G.L. Brengelmann, J.R. Blackmon, R.A. Bruce, J.A. Murray, Disparities between aortic and peripheral pulse pressures induced by upright exercise and vasomotor changes in man. Circulation **37**, 954–964 (1968)
478. C.S. Roy, The elastic properties of the arterial wall. J. Physiol. **3**, 125–159 (1881)
479. Z.M. Ruggeri, Von willebrand factor, platelets and endothelial cell interactions. Thromb. Haemost. **1**, 1335–1342 (2003)
480. Z.M. Ruggeri, G.L. Mendolicchio, Adhesion mechanisms in platelet function. Circ. Res. **100**, 1673–1685 (2007)
481. D.E. Rumelhart, G.E. Hinton, R.J. Williams, Learning representations by back-propagating errors. Nature **323**, 533–536 (1986)
482. M.S. Sacks, Biaxial mechanical evaluation of planar biological materials. J. Elast. **61**, 199–246 (2000)
483. P. Sáez, A. García, E. Pe a, T.C. Gasser, M.A. Martínez, Microstructural quantification of collagen fiber orientations and its integration in constitutive modeling of the porcine carotid artery. Acta Biomater. **33**, 1183–1193 (2016)
484. H. Sakai, F. Ikomi, T. Ohhashi, Effects of endothelin on spontaneous contractions in lymph vessels. Am. J. Physiol. **277**, H459–H466 (1999)
485. N. Sakalihasan, H. Van Damme, P. Gomez, P. Rigo, CM. Lapiere, B. Nusgens, R. Limet, Positron emission tomography (pet) evaluation of abdominal aortic aneurysm (AAA). Eur. J. Vasc. Endovasc. Surg. **23**, 431–436 (2002)
486. N. Sakalihasan, J.-B. Michel, A. Katsargyris, H. Kuivaniemi, J.-O. Defraigne, A. Nchimi, J.T. Powell, K. Yoshimura, R. Hultgren, Abdominal aortic aneurysms. Nat. Rev. Dis. Primers **4**, 34 (2018)
487. K.S. Sakariassen, P.A. Holme, U. Orvin, R.M. Barstad, N.O. Solum, F.R. Brosstad, Shear-induced platelet activation and platelet microparticle formation in native human blood. Thromb. Res. **92**, S33–S41 (1998)
488. A. Saltelli, M. Ratto, T. Andres, F. Campolongo, J. Cariboni, D. Gatelli, M. Saisana, S. Tarantola, *Global Sensitivity Analysis. The Primer*. John Wiley & Sons Ltd, (2008).
489. N.V. Salunke, L.D.T. Topoleski, Biomechanics of atherosclerotic plaque. Crit. Rev. Biomed. Eng. **25**, 243–285 (1997)
490. Z.J. Samila, S.A. Carter, The effect of age on the unfolding of elastin lamellae and collagen fibers with stretch in human carotid arteries. Canad. J. Physiol. Pharm. **59**, 1050–1057 (1981)
491. R.G. Sargent, Verification and validation of simulation models, in *Proceedings of the 2011 Winter Simulation Conference*, ed. by S. Jain, R.R. Creasey, J. Himmelspach, K.P. White, M. Fu (Institute of Electrical and Electronic Engineers, Piscataway, 2011), pp. 183–198
492. N. Sasaki, S. Odajima, Elongation mechanism of collagen fibrils and force-strain relations of tendon at each level of the structural hierarchy. J. Biomech. **29**, 1131–1136 (1996)
493. J. Scallan, V.H. Huxley, R.J. Korthuis, *Capillary Fluid Exchange: Regulation, Functions, and Pathology* (Morgan & Claypool Life Sciences, University of Missouri-Columbia, 2010)
494. H. Schmid, P.N. Watton, M.M. Maurerand J. Wimmer, P. Winkler, Y.K. Wang, O. Roehrle, M. Itskov, Impact of transmural heterogeneities on arterial adaptation. Biomech. Model. Mechanobio. **9**, 295–315 (2010)
495. H. Schmid-Schönbein, E. Volger, H.J. Klose, Microrheology and light transmission of blood. Pflügers Arch. – Eur. J. Physiol. **333**, 140–155 (1972)
496. A.J. Schriefl, G. Zeindlinger, D.M. Pierce, P. Regitnig, G.A. Holzapfel, Determination of the layer-specific distributed collagen fiber orientations in human thoracic and abdominal aortas and common iliac arteries. J. R. Soc. Interface **7**, 1275–1286 (2012)

497. J. Schröder, P. Neff, Invariant formulation of hyperelastic transverse isotropy based on polyconvex free energy functions. Int. J. Solids Struct. **40**(2), 401–445

498. R. Schubert, M.J. Mulvany, The myogenic response: established facts and attractive hypotheses. Clin. Sci. **96**, 313–326 (1999)

499. J.E. Scott, Elasticity in extracellular matrix 'shape modules' of tendon, cartilage, etc. A sliding proteoglycan-filament model. J. Physiol. **553**(2), 335–343 (2003)

500. J.E. Scott, Cartilage is held together by elastic glycan strings. Physiological and pathological implications. Biorheology **45**, 209–217 (2008)

501. G. Segrè, A. Silberberg, Behavior of macroscopic rigid spheres in Poiseuille flow. J. Fluid Mech. **14**, 136–157 (1962)

502. K.W. Seo, S.J. Lee, Y.H. Kim, J.U. Bae, S.Y. Park, S.S. Bae, C.D. Kim, Mechanical stretch increases MMP-2 production in vascular smooth muscle cells via activation of PDGFR-beta/AKT signaling pathway. PLOS ONE **8**, e70437 (2013)

503. B. Sharifimajd, J. Stålhand, A continuum model for excitation-contraction of smooth muscle under finite deformations. J. Theor. Biol. **355**, 1–9 (2014)

504. C.Y. Shim, I.J. Cho, W.-I. Yang, M.-K. Kang, S. Park, J.-W. Ha, Y. Jang, N. Chung, Central aortic stiffness and its association with ascending aorta dilation in subjects with a bicuspid aortic valve. J. Am. Soc. Echoradiogr. **24**, 847–852 (2011)

505. Y.-T. Shiu, J.A. Weiss, J.B. Hoying, M.N. Iwamoto, I.S. Joung, C.T. Quam, The role of mechanical stresses in angiogenesis. Crit. Rev. Biomed. Eng. **33**, 431–510 (2005)

506. J.C. Simo, A J_2-flow theory exhibiting a corner-like effect and suitable for large-scale computation. Comput. Methods Appl. Mech. Eng. **62**, 169–194 (1987)

507. J.C. Simo, T.J.R. Hughes, *Computational Inelasticity* (Springer, New York, 1998)

508. T.P. Singh, J.V. Moxon, V. Iyer, T.C. Gasser, J. Jenkins, J. Golledge, Comparison of peak wall stress and peak wall rupture index in ruptured and asymptomatic intact abdominal aortic aneurysms. Br. J. Surg. (2020). https://doi.org/10.1002/bjs.11995

509. T.P. Singh, J. Moxon, T.C. Gasser, J. Golledge, Systematic review and meta-analysis of peak wall stress and peak wall rupture index in ruptured and asymptomatic intact abdominal aortic aneurysms. J. Am. Heart Ass. **10**, e019772 (2021)

510. R. Skalak, Growth as a finite displacement field, in *Proceedings of the IUTAM Symposium on Finite Elasticity, 1981*, ed. by D.E. Carlson, R.T. Shield (The Hague, Martinus Nijhoff Publishers, 1981)

511. R. Skalak, S. Zargaryan, R.K. Jain, P.A. Netti, A. Hoger, Compatibility and the genesis of residual stress by volumetric growth. J. Math. Biol. **34**, 889–914 (1996)

512. F.T. Smith, Fluid flow into a curved pipe. Proc. R. Soc. Lond. A **351**, 71–87 (1968)

513. I.M. Sobol, Global sensitivity indices for nonlinear mathematical models and their Monte Carlo estimates. Math. Comp. Simul. **55**, 271–280 (2001)

514. D.P. Sokolis, Passive mechanical properties and structure of the aorta: segmental analysis. Acta Physiol. **190**, 277–289 (2007)

515. D.P. Sokolis, E.M. Kefaloyannis, M. Kouloukoussa, E. Marinos, H. Boudoulas, P.E. Karayannacos, A structural basis for the aortic stress-strain relation in uniaxial tension. J. Biomech. **39**, 1651–1662 (2006)

516. B. Sonesson, F. Hansen, H. Stale, T. Länne, Compliance and diameter in the huma abdominal aorta—the influence of age and sex. Eur. J. Vasc. Surg. **7**, 690–697 (1993)

517. M. Sonka, J.M. Fitzpatrick, *Handbook of Medical Imaging. Volume 2. Medical Image Processing and Analysis* (Spie Press, Bellingham, 2000)

518. N.T. Soskel, L. B.Sandberg, Pulmonary emphysema. from animal models to human diseases, in *Connective Tissue Disease. Molecular Pathology of the Extracellular Matrix*, ed. by J. Uitto, A.J. Perejda, vol. 12 (Dekker, New York, 1987), pp. 423–453

519. A.J.M. Spencer, Theory of invariants, in *Continuum Physics*, ed. by A.C. Eringen, vol. I (Academic Press, New York, 1971)

520. A.J.M. Spencer, Constitutive theory for strongly anisotropic solids, in *Continuum Theory of the Mechanics of Fibre-Reinforced Composites*, ed. by A.J.M. Spencer (Springer, Wien, 1984), pp. 1–32. CISM Courses and Lectures No. 282, International Centre for Mechanical Sciences

521. A.F. Stalder, A. Frydrychowicz, M.F. Russe, J.G. Korvink, J. Hennig, K. Li, M.Markl, Assessment of flow instabilities in the healthy aorta using flow-sensitive MRI. J. Musculoskel. Neuron. Interact. **33**, 839–846 (2011)

522. J. Stålhand, A. Klarbring, Aorta in vivo parameter identification using an axial force constraint. Biomech. Model. Mechanobio. **3**, 191–199 (2005)

523. J. Stålhand, A. Klarbring, M. Karlsson, Towards in vivo material identification and stress estimation. Biomech. Model. Mechanobio. **2**, 169–186 (2004)

524. E.H. Starling, The production and absorption of lymph, in *Textbook of Physiology*, ed. by E.A. Schäfer, vol. 1 (Pentland, London, 1898), pp. 285–311

525. H.C. Stary, *Atlas of Atherosclerosis: Progression and Regression*, 2nd edn. (The Parthenon Publishing Group Limited, Boca Raton, London, New York, Washington, 2003)

526. A.J. Staverman, The theory of measurement of osmotic pressure. Recueil Trav. Chim. **70**, 344–352 (1951)

527. N. Stergiopulos, J.-J. Meister, N Westerhof, Evaluation of methods for estimation of total arterial compliance. Am. J. Physiol. **268**, H1540–H1548 (1995)

528. R. Stojanovic, On the stress relation in non-linear thermoelasticity. Int. J. Non-Linear Mech. **4**, 217–233 (1969)

529. G. Stoll, M. Bendszus, Minflammation and atherosclerosis: novel insights into plaque formation and destabilization. Stroke **37**, 1923–1932 (2006)

530. P.H. Stone, S. Saito, S. Takahashi, Y. Makita, S. Nakamura, T. Kawasaki, A. Takahashi, T. Katsuki, S. Nakamura, A. Namiki, A. Hirohata, T. Matsumura, S. Yamazaki, H. Yokoi, S. Tanaka, S. Otsuji, F. Yoshimachi, J. Honye, D. Harwood, M. Reitman, A.U. Coskun, M.I. Papafaklis, C.L. Feldman, Prediction of progression of coronary artery disease and clinical outcomes using vascular profiling of endothelial shear stress and arterial plaque characteristics: the prediction study. Circulation **126**, 172–181 (2012)

531. P. Suetens, *Fundamentals of Medical Imaging*, 3rd edn. (Cambridge, 2017)

532. B. Sundström (eds.), *Handbook of Solid Mechanics* (KTH Solid Mechanics, KTH Royal Institute of Technology, Stockholm, 2010)

533. J. Swedenborg, P. Eriksson, The intraluminal thrombus as a source of proteolytic activity. Ann. N.Y. Acad. Sci. **1085**, 133–138 (2006)

534. L.A. Taber, D.W. Eggers, Theoretical study of stress-modulated growth in the aorta. J. Theor. Biol. **180**, 343–357 (1996)

535. L.A. Taber, J.D. Humphrey, Stress-modulated growth, residual stress, and vascular heterogeneity. ASME. J. Biomech. Eng. **123**, 528–535 (2001)

536. K. Takamizawa, K. Hayashi, Strain energy density function and uniform strain hypothesis for arterial mechanics. J. Biomech. **20**, 7–17 (1987)

537. N. Takeishi, M.E. Rosti, Y. Imai, S. Wada, L. Brandt, Haemorheology in dilute, semi-dilute and dense suspensions of red blood cells. J. Fluid Mech. **872**, 818–848 (2019)

538. A.M. Tamburro, A. DeStradis, L. D'Alessio, Fractal aspects of elastin supramolecular structure. J. Biomol. Struct. Dyn. **12**, 1161–1172 (1995)

539. E. Tanaka, H. Yamada, An inelastic constitutive model of blood vessels. Acta Mech. **82**, 21–30 (1990)

540. T.T. Tanaka, Y.C. Fung, Elastic and inelastic properties of the canine aorta and their variation along the aortic tree. J. Biomech. **7**, 357–370 (1974)

541. D. Tang, C. Yang, S. Kobayashi, D.N. Ku, Effect of a lipid pool on stress/strain distributions in stenotic arteries: 3-d fluid-structure interactions (FSI) models. ASME. J. Biomech. Eng. **126**, 336–370 (2004)

542. R. Taniguchi, K. Hoshina, A. Hosaka, T. Miyahara, H. Okamoto, K. Shigematsu, T. Miyata, T. Watanabe, Strain analysis of wall motion in abdominal aortic aneurysms. Ann. Vasc. Dis. **7**, 393–398 (2014)

543. A.E. Taylor, D.N. Granger, Exchange of macromolecules across the circulation, in *Handbook of Physiology; Section 2: The Cardiovascular System*, ed. by E.M Renkin and C.C. Michel, vol. 4: The Microcirculation (American Physiological Society, 1984), pp. 467–520

544. Z. Terzian, T.C. Gasser, F. Blackwell, F. Hyafil, L. Louedec, C. Deschildre, W. Ghodbane, R. Dorent, A. Nicoletti, M. Morvan, M. Nejjari, L. Feldman, G. Pavon-Djavid, J.-B. Michel, Peri-strut micro-hemorrhages: a possible cause of in-stent neoatherosclerosis? Cardiovasc. Pathol. **26**, 30–38 (2016). https://doi.org/10.1016/j.carpath.2016.08.007

545. A. Teutelink, E. Cancrinus, D. van de Heuvel, F. Moll, J.P. de Vries, Preliminary intraobserver and interobserver variability in wall stress and rupture risk assessment of abdominal aortic aneurysms using a semiautomatic finite element model. J. Vasc. Surg. **55**, 326–330 (2012)

546. T.E. Tezduyar, Stabilized finite element formulations for incompressible flow computations'—this research was sponsored by NASA-Johnson space center (under grant nag 9–449), NSF (under grant MSM-8796352), U.S. Army (under contract DAAL03-89-C-0038), and the University of Paris VI. vol. 28 of Advances in Applied Mechanics (Elsevier, Amsterdam, 1991), pp. 1–44

547. The UK Small Aneurysm Trial Participants, Mortality results for randomised controlled trial of early elective surgery or ultrasonographic surveillance for small abdominal aortic aneurysms. Lancet **352**, 1649–1655 (1998)

548. M. Thiriet (ed.), *Anatomy and Physiology of the Circulatory and Ventilatory Systems* (Springer, New York, 2014)

549. R. Thoma (ed.), *Untersuchungen über die Histogenese und Histomechanik des Gefässystems* (Enke Verlag, Stuttgart, 1893)

550. M.J. Thubrikar, M. Labrosse, F. Robicsek, J. Al-Soudi, B. Fowler, Mechanical properties of abdominal aortic aneurysm wall. J. Med. Eng. Technol. **25**, 133–142 (2001)

551. J.R. Thunes, S. Pal, R.N. Fortunato, J.A. Phillippi, G. Gleason, D.A. Vorp, S. Maiti, A structural finite element model for lamellar unit of aortic media indicates heterogeneous stress field after collagen recruitment. J. Biomech. **49**, 1562–1569 (2016)

552. J.S. Torday, Homeostasis as the mechanism of evolution. Biology (Basel) **4**, 573–590 (2015)

553. O. Trabelsi, V. Dumas, E. Breysse, N. Larochet, S. Avril, In vitro histomechanical effects of enzymatic degradation in carotid arteries during inflation tests with pulsatile loading. J. Mech. Behav. Biomed. Mater. **103**, 103550 (2020)

554. F. Tronc, M. Wassef, B. Esposito, D. Henrion, S. Glagov, A. Tedgui, Role of no in flow-induced remodeling of the rabbit common carotid artery. Arterioscl. Thromb. and Vasc. Biol. **16**, 1256–1262 (1996)

555. M. Truijers, J.A. Pol, L.J. Schultzekool, S.M. van Sterkenburg, M.F. Fillinger, J.D. Blankensteijn, Wall stress analysis in small asymptomatic, symptomatic and ruptured Abdominal Aortic Aneurysms. Eur. J. Vasc. Endovasc. Surg. **33**, 401–407 (2007)

556. J.L. Tuttle, R.D. Nachreiner, A.S. Bhuller, K.W. Condict, B.A. Connors, B.P. Herring, M.C. Dalsing, J.L. Unthank, Shear level influences resistance artery remodeling: wall dimensions, cell density, and eNOS expression. Am. J. Physiol. Heart Circ. Physiol. **281**, H1380–H1389 (2001)

557. E.M. Urbina, R.V. Williams, B.S. Alpert, R.T. Collins, S.R. Daniels, L. Hayman, M. Jacobson, L. Mahoney, M. Mietus-Snyder, A. Rocchini, J. Steinberger, B. McCrindle, Noninvasive assessment of subclinical atherosclerosis in children and adolescents: recommendations for standard assessment for clinical research: a scientific statement from the American Heart Association and on behalf of the American Heart Association Atherosclerosis, Hypertension, and Obesity in Youth Committee of the council on cardiovascular disease in the young. Hypertension **54**, 919–950 (2009)

558. R.N. Vaishnav, J. Vossoughi, Residual stress and strain in aortic segments. J. Biomech. **20**, 235–239 (1987)

559. R.N. Vaishnav, J.T. Young, J.S. Janicki, D.J. Patel, Nonlinear anisotropic elastic properties of the canine aorta. Biophys. J. **12**, 1008–1027 (1972)

560. A.G. van der Giessen, *Coronary Atherosclerosis and Wall Shear Stress—Towards Application of CT Angiography*. Ph.D. thesis, Eindhoven Technical University, 2010. ISBN: 978-90-90-25382-4

561. P.E.J. van der Meijden, J.W.M. Heemskerk, Platelet biology and functions: new concepts and clinical perspectives. Nat Rev Cardiol. **16**, 166–179 (2019). https://doi.org/10.1038/s41569-018-0110-0

562. J.W. van Teeffelen, S.S. Segal, Rapid dilation of arterioles with single contraction of hamster skeletal muscle. Am. J. Physiol. Heart Circ. Physiol. **290**, H119–H127 (2006)

563. S. van Wyk, L. Prahl Wittberg, K.V. Bulusu, L. Fuchs, M.W. Plesniak, Non-newtonian perspectives on pulsatile blood-analog flows in a 180-degree curved artery model. Phys. Fluids **27**, 071901 (2015)

564. S. van Wyk, L. Prahl Wittberg, L. Fuchs, Wall shear stress variations and unsteadiness of pulsatile blood-like flows in 90-degree bifurcations. Comp. Biol. Med. **43**, 1025–1036 (2013)

565. S. van Wyk, L. Prahl Wittberg, L. Fuchs, Atherosclerotic indicators for blood-like fluids in 90-degree arterial-like bifurcations. Compos. Biol. Med. **50**, 56–69 (2014)

566. J.P. Vande Geest, E.D. Dillavou, E.S. DiMartino, M. Oberdier, A. Bohra, M.S. Makaroun, D.A. Vorp, Gender-related differences in the tensile strength of Abdominal Aortic Aneurysm. Ann. N.Y. Acad. Sci. **1085**, 400–402 (2006)

567. J.P. Vande Geest, E.S. DiMartino, A. Bohra, M.S. Makaroun, D.A. Vorp, A biomechanics-based rupture potential index for Abdominal Aortic Aneurysm risk assessment: demonstrative application. Ann. N.Y. Acad. Sci. **1085**, 11–21 (2006)

568. J.P. Vande Geest, M.S. Sacks, D.A. Vorp, The effects of aneurysm on the biaxial mechanical behavior of human abdominal aorta. J. Biomech. **39**, 1324–1334 (2006)

569. J.P. Vande Geest, M.S. Sacks, D.A. Vorp, A planar biaxial constitutive relation for the luminal layer of intra-luminal thrombus in Abdominal Aortic Aneurysms. J. Biomech. **39**, 2347–2354 (2006)

570. J.P. Vande Geest, D.H.J. Wang, S.R. Wisniewski, M.S. Makaroun, D.A. Vorp, Towards a noninvasive method for determination of patient-specific wall strength distribution in Abdominal Aortic Aneurysms. Ann. Biomed. Eng. **34**, 1098–1106 (2006)

571. J. von Kries, On the relationships between pressure and velocity, which exist in connection with wave motion in elastic tubing, in *Festschrift of the 56th Convention of German Scientists and Physicians (in German)*, Tübingen, Germany, 1883 (Akademische Verlagsbuchhandlung), pp. 67–88

572. J. Vappou, J. Luo, E.E. Konofagou, Pulse wave imaging for noninvasive and quantitative measurement of arterial stiffness in vivo. Am. J. Hypertens. **23**, 393–398 (2010)

573. S.S. Varghese, S.H. Frankel, P.F. Fischer, Direct numerical simulation of stenotic flows. Part 1. Steady flow. J. Fluid Mech. **582**, 253–280 (2007)

574. S.S. Varghese, S.H. Frankel, P.F. Fischer, Direct numerical simulation of stenotic flows. Part 2. Pulsatile flow. J. Fluid Mech. **582**, 281–318 (2007)

575. A.K. Venkatasubramaniam, M.J. Fagan, T. Mehta, K.J. Mylankal, B. Ray, G. Kuhan, I.C. Chetter, P.T. McCollum, A comparative study of aortic wall stress using finite element analysis for ruptured and non-ruptured Abdominal Aortic Aneurysms. Eur. J. Vasc. Surg. **28**, 168–176 (2004)

576. S. Vesentini, A. Redaelli, F.M. Montevecchi, Estimation of the binding force of the collagen molecule-decorin core protein complex in collagen fibril. J. Biomech. **38**, 433–443 (2005)

577. D.C. Viano, A.I. King, J.W. Melvin, K. Weber, Injury biomechanics research: an essential element in the prevention of trauma. J. Biomech. **22**, 403–417 (1989)

578. R. Virmani, A.P. Burke, A. Farb, F.D. Kolodgie, Pathology of the vulnerable plaque. Am. J. Cardiol. **47**, C13–C18 (2006)

579. J.F. Vollmar, E. Paes, P. Pauschinger, E. Henze, A. Friesch, Aortic aneurysms as late sequelae of above-knee amputation. Lancet **2**, 834–835 (1989)

580. K.Y. Volokh, D.A. Vorp, A model of growth and rupture of abdominal aortic aneurysm. J. Biomech. **41**, 1015–1021 (2008)

581. D.A. Vorp, P.C. Lee, D.H. Wang, M.S. Makaroun, E.M. Nemoto, S. Ogawa, M.W. Webster, Association of intraluminal thrombus in Abdominal Aortic Aneurysm with local hypoxia and wall weakening. J. Vasc. Surg. **34**, 291–299 (2001)

582. D.A. Vorp, M.L. Raghavan, S.C. Muluk, M.S. Makaroun, D.L. Steed, R. Shapiro, M.W. Webster, Wall strength and stiffness of aneurysmal and nonaneurysmal abdominal aorta. Ann. N.Y. Acad. Sci. **800**, 274–276 (1996)

583. D.A. Vorp, B.J. Schiro, M.P. Ehrlich, T.S. Juvonen, M.A. Ergin, B.P. Griffith, Effect of aneurysm on the tensile strength and biomechanical behavior of the ascending thoracic aorta. Ann. Thorac Surg. **800**, 1210–1214 (2003)

584. J. Vossoughi, Longitudinal residual strains in arteries, in *Proceedings of the 11th Southern Biomedical Engineering Conference* (Memphis, 1992), pp. 17–19

585. B. Vrhovski, A.S. Weiss, Biochemistry of tropoelastin. Eur. J. Biochem. **258**, 1–18 (1998)

586. J.E. Wagenseil, R.P. Mecham, Vascular extracellular matrix and arterial mechanics. Physiol. Rev. **89**, 957–989 (2009)

587. C. Wagner, P. Steffen, S. Svetina, Aggregation of red blood cells: from Rouleaux to Clot formation. C. R. Phys. **14**(6), 459–469 (2013)

588. F.J. Walburn, D.J. Schneck, A constitutive equation for whole human blood. Biorheology **13**, 201–210 (1976)

589. S.L. Waters, J. Alastruey, D.A. Beard, P.H.M. Bovendeerd, P.F. Davies, G. Jayaraman, O.E. Jensen, J. Lee, K.H. Parker, A.S. Popel, T.W. Secomb, M. Siebes, S.J. Sherwin, R.J. Shipley, N.P. Smith, F.N. van de Vosse, Theoretical models for coronary vascular biomechanics: progress & challenges. Prog. Biophys. Mol. Biol. **104**, 49–76 (2011)

590. P.N. Watton, N.A. Hill, Evolving mechanical properties of a model of Abdominal Aortic Aneurysm. Biomech. Model. Mechanobio. **8**, 25–42 (2009)

591. P.N. Watton, N.A. Hill, M. Heil, A mathematical model for the growth of the Abdominal Aortic Aneurysm. Biomech. Model. Mechanobio. **3**, 98–113 (2004)

592. R.C Webb, Smooth muscle contraction and relaxation. Adv. Physiol. Educ. **27**, 201–206 (2003)

593. H.W. Weizsäcker, H. Lambert, K. Pascale, Analysis of the passive mechanical properties of rat carotid arteries. J. Biomech. **16**, 703–715 (1983)

594. G.N. Wells, R. de Borst, L.J. Sluys, A consistent geometrically non-linear approach for delamination. Int. J. Numer. Methods Eng. **54**, 1333–1355 (2002)

595. Z. Werb, M.J. Banda, J.H. McKerrow, R.A. Sandhaus, Elastases and elastin degradation. J. Invest. Dermatol. **79**, 154–159 (1982)

596. N. Westerhof, J.-W. Lankhaar, B.E. Westerhof, The arterial windkessel. Med. Biol. Eng. Comput. **47**, 131–141 (2009)

597. J.S. Wilson, S. Baek, J.D. Humphrey, Importance of initial aortic properties on the evolving regional anisotropy, stiffness and wall thickness of human abdominal aortic aneurysms. J. R. Soc. Interface **9**, 2047 (2012)

598. H. Wolinsky, Effects of hypertension and its reversal on the thoracic aorta of male and female rats. Circ. Res. **28**, 622–637 (1971)

599. H. Wolinsky, S. Glagov, Structural basis for the static mechanical properties of the aortic media. Circ. Res. **14**, 400–413 (1964)

600. J.R. Womersley, Method for the calculation of velocity, rate of flow and viscous drag in arteries when the pressure gradient is known. J. Physiol. **127**, 553–563 (1955)

601. P. Wriggers, *Nonlinear Finite Element Methods* (Springer, Berlin, 2008)

602. P. Wu, S. Groß-Hardt, F. Boehning, P.-L. Hsu, An energy-dissipation-based power-law formulation for estimating hemolysis. Biomech. Model. Mechanobio. **19**, 591–602 (2019)

603. L.J. Wurzinger, R. Opitz, P. Blasberg, H. Schmid-Schonbein, Platelet and coagulation parameters following millisecond exposure to laminar shear stress. Thromb. Haemost. 54, 381–386 (1985)

604. F.L. Wuyts, V.J. Vanhuyse, G.J. Langewouters, W.F. Decraemer, E.R. Raman, S. Buyle, Elastic properties of human aortas in relation to age and atherosclerosis: A structural model. Phys. Med. Biol. **40**, 1577–1597 (1995)

605. J. Xiong, S.M. Wang, W. Zhou, J.G. Wu, Measurement and analysis of ultimate mechanical properties, stress-strain curve fit, and elastic modulus formula of human abdominal aortic aneurysm and nonaneurysmal abdominal aorta. J. Vasc. Surg. **48**, 189–195 (2008)

606. C. Xu, D.L. Pham, J.L. Prince, Image segmentation using deformable models, in *Handbook of Medical Imaging. Volume 2. Medical Image Processing and Analysis*, ed. by M. Sonka, J.M. Fitzpatrick (Spie Press, Bellingham, 2000), pp. 129–174

607. J. Yang, J.W. Clark Jr., R.M. Bryan, C. Robertsson, The myogenic response in isolated rat cerebrovascular arteries: smooth muscle cell model. Med. Eng. Phys. **25**, 691–709 (2003)

608. O.H. Yeoh, Some forms of strain energy functions for rubber. Rubber Chem. Technol. **66**, 754–771 (1993)

609. T. Young, The Croonian lecture: on the functions of the heart and arteries. Philos. T. R. Soc. B **99**, 1–31 (1809)

610. H. Yu, S. Engel, G. Janiga, D. Thévenin, A review of hemolysis prediction models forcomputational fluid dynamics. Artif. Organs **41**, 603–621 (2017)

611. P.D. Yurchenco, J.J. O'Rear, Basal lamina assembly. Curr. Opin. Cell Biol. **6**, 674–681 (1994)

612. C.K. Zarins, D.P. Giddens, B.K. Bharadvaj, V.S. Sottiurai, R.F. Mabon, S. Glagov, Carotid bifurcation atherosclerosis. quantitative correlation of plaque localization with flow velocity profiles and wall shear stress. Circ. Res. **53**, 502–514 (1983)

613. S. Zeinali-Davarani, S. Baek, Medical image-based simulation of abdominal aortic aneurysm growth. Mech. Res. Commun. **42**, 107–117 (2012)

614. S. Zeinali-Davarani, M.-J. Chow, R. Turcotte, Y. Zhang, Characterization of biaxial mechanical behavior of porcine aorta under gradual elastin degradation. Ann. Biomed. Eng. **41**, 1528–1538 (2013)

615. J.-N. Zhang, A.L. Bergeron, Q. Yu, C. Sun, L. McBride, P.F. Bray, J.F. Dong, Duration of exposure to high fluid shear stress is critical in shear-induced platelet activation-aggregation. Thromb. Haemost. **90**, 672–678 (2003)

616. Y. Zhou, W.P. Dirksen, J.L. Zweier, M. Periasamy, Endothelin-1-induced responses in isolated mouse vessels: the expression and function of receptor types. Am. J. Physiol. Heart Circ. Physiol. **287**, H573–H578 (2004)

617. O.C. Zienkiewicz, R.L. Taylor, *The Finite Element Method. Fluid Dynamics*, vol. 3, 5th edn. (Butterworth Heinemann, Oxford, 2000)

618. O.C. Zienkiewicz, R.L. Taylor, *The Finite Element Method. Solid Mechanics*, vol. 2, 5th edn. (Butterworth Heinemann, Oxford, 2000)

619. O.C. Zienkiewicz, R.L. Taylor, *The Finite Element Method. The Basis*, vol. 1, 5th edn. (Butterworth Heinemann, Oxford, 2000)

620. M.A. Zulliger, P. Fridez, K. Hayashi, N. Stergiopulos, A strain energy function for arteries accounting for wall composition and structure. J. Biomech. **37**, 989–1000 (2004)

621. M.A. Zulliger, N.T. Kwak, T. Tsapikouni, N. Stergiopulos, Effects of longitudinal stretch on VSM tone and distensibility of muscular conduit arteries. Am. J. Physiol. Heart Circ. Physiol. **283**, H2599–H2605 (2002)

622. M.A. Zulliger, A. Rachev, N. Stergiopulos, A constitutive formulation of arterial mechanics including vascular smooth muscle tone. Am. J. Physiol. Heart Circ. Physiol. **287**, H1335–H1343 (2004)

623. M.A. Zulliger, N. Stergiopulos, Structural strain energy function applied to the ageing of the human aorta. J. Biomech. **40**, 3061–3069 (2007)

624. A.L. Zydney, J.D. Oliver, C.K. Colton, A constitutive equation for the viscosity of stored red cell suspensions: Effect of hematocrit, shear rate, and suspending phase. J. Rheol. **35**, 1639–1680 (1991)

Index

The manufacturer's authorised representative in the EU is Springer
Nature Customer Service Centre GmbH, Europaplatz 3, 69115 Heidelberg,
Germany. If you have any concerns regarding our products, please
contact ProductSafety@springernature.com

Printed and bound by CPI Group (UK) Ltd, Croydon, CR0 4YY
29/04/2026
02099528-0005